Adelaide

WATER of a CITY

Adelaide

WATER of a CITY

This book was produced by
Barbara Hardy Centre for
Sustainable Urban Environments

 University of
South Australia | **Barbara Hardy Centre**
for Sustainable Urban Environments

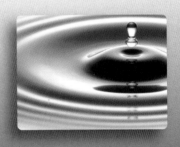

Editor in chief
Christopher B. Daniels

Photography by
John Hodgson

Foreword by
Barbara Hardy

Wakefield
Press

Wakefield Press
1 The Parade West
Kent Town
South Australia 5067
www.wakefieldpress.com.au

First published 2010

Graphic design and format by David Radzikiewicz, A7 Designs
Figures and illustrations by Carolyn Herbert, University of South Australia
Copyedited by Kathy Sharrad, Wakefield Press
Printed in China at Everbest Printing Co. Ltd

National Library of Australia Cataloguing-in-Publication entry

Title:	Adelaide: water of a city/edited by Christopher B. Daniels; photography by John Hodgson.
ISBN:	978 1 86254 861 9 (hbk.).
Notes:	Includes index.
	Bibliography.
Subjects:	Water conservation – South Australia.
	Water-supply – South Australia – Management.
	Water quality management – South Australia.
	Water resources development – South Australia.
Other Authors/	Daniels, Christopher B.
Contributors:	Hodgson, John.
Dewey Number:	333.9116099423

This book was produced by
Barbara Hardy Centre for
Sustainable Urban Environments

 University of
South Australia | **Barbara Hardy Centre**
for Sustainable Urban Environments

Thanks to our sponsors for helping to make this book possible.

Sponsors

GOLD

 University of
South Australia | School of
**Natural and
Built Environments**

Stormwater
INDUSTRY ASSOCIATION

A7Designs
WEBSITE & GRAPHIC DESIGN
www.a7designs.com.au

BURT'S BEES
Earth Friendly Natural Personal Care Products

SILVER

*Silver sponsorships were contributed by the
following three government agencies:*

**Government
of South Australia**

SA Water

Adelaide and Mount Lofty Ranges
Natural Resources Management Board

Department for Environment and Heritage

URS
engineering and environmental professionals

Flinders
UNIVERSITY
ADELAIDE • AUSTRALIA

CITY OF
Playford

AWA
australian water association

Table of contents

CHAPTER 1

CHAPTER 2

CHAPTER 3

CHAPTER 4

CHAPTER 5

Foreword

Think of the word 'water' – that 'everyday' commodity – and it is hard to realise that every person and living species on Planet Earth depends for their very existence on the presence and accessibility of this precious resource. Water is essential, and it needs to be pure, uncontaminated and easily accessible from a relatively assured source. Water comes to us as rain from clouds in near space, and accumulates in its fresh form as rivers, lakes, swamps and wetlands. In its saline form, of course, water is the sea covering the major proportion of the surface of our planet.

Where we now live was determined (in large part) by the availability of water. The location of Adelaide town, and later the city, was decided by Colonel Light because of the freshwater potential of what was to be called the River Torrens. As time progressed after settlement, freshwater availability was very high on the list of priorities for locating Adelaide's reservoirs in the nearby hills. The long-term use of and attitude towards water during probably the last 40,000 years by the Kaurna people, who preceded we white settlers on the Adelaide Plains, reflected the scarcity and value of this precious resource. The Kaurna did not have 'mains water supply' but they coped and flourished by careful, diligent and non-polluting use of the water available such as in the River Torrens.

Adelaide is said to be the driest city in the driest state in the driest continent of Planet Earth, and the current long-standing drought is proof of this daunting statement. Drought is a fearsome word, and the people of Adelaide are suffering under severe shortages of water such that restrictions have had to be set in place in order to ensure at least some water for everyone. Farmers and vegetable growers are particularly hard hit, and those of us who are vegetarians are doing our bit to help in the knowledge that the direct consumption of plants, vegetables and fruit eliminates the need for water to pass through the body of an animal in order to produce flesh for carnivores to eat! Yet all the food products that we consume have water 'embedded' within them resulting from the process of their production. However, we also use water for a myriad of other activities, not just for consumption.

Most people, 'love' water, not only for consumption, but also to play in – sail boats, go kayaking, swimming, to create gardens, support wildlife, as part of artworks (fountains), and as part of the attractive nature of our communities. There is also groundwater as well, lurking beneath our feet and which we often extract for use in our gardens. Here in Adelaide, South Australia, we do all these things, and use water in all these ways, even while acknowledging that water is a scarce, very precious and vital commodity. Hence, the recycling of water is becoming a very important method of making available water go a little further. The use of greywater is also being investigated and implemented by many people, organisations and industries.

This great collection of articles from many people on a huge range of aspects of the 'watery' world we live in is, in itself, a precious resource. Scanning this wonderfully large and comprehensive book comprising 28 chapters, 10 case studies and 128 boxes confirms that our rainfall is not what it used to be. Whilst we are used to floods, the regular floods of 1917, 1931 and 1956 do not appear to have happened for several decades. The River Murray waters are being called upon increasingly, and are also in trouble, and there are very many other aspects of water and its Adelaide story that make fascinating reading.

This book contains a magnificent collection of knowledge, researched and described by a large number of appropriate and eminent authors. It would indeed be hard to better the planning, the preparation, the implementation and the fulfilment of the story of Adelaide's water. My fervent recommendation is, 'read it!'

Barbara Hardy
Seacliff, South Australia

Preface and acknowledgements

This is the second book in the series concerning the environmental nature of Adelaide. The first book traced the changes to the ecology of the city and its landscape as it developed from 1836. Adelaide is a medium-sized city of about 1.2 million people, situated in the central southern part of Australia. *Adelaide: Water of a City* has water as its theme, and since water is so central to ecology and landscape, it has strong parallels with the first book. This book is authored by a team of water specialists from Adelaide, bringing together a vast number of 'voices'. It represents the collective knowledge, wisdom and views of individuals involved with water, but from a wide range of occupations, ages, and view points. Authors are academics, researchers, policy-makers, business, industry and government workers, and members of our society who have contributed to the water debate in Adelaide. The chapters represent the scholarly research of the authors and in many cases they have been given permission to express opinions that do not necessarily represent their employers' official views. With so many contributors, it was possible that the cacophony and clamor could obscure the simple melody – that of sustainability. That everyone worked so hard to 'sing as choir' is testament to their passion, zeal and positive involvement in this work. To all the contributors to this book, I am extremely grateful.

Water is a huge topic. It required a wonderful Editorial Board to undertake the identification and sorting of subject matter and presentation, the selection and recruitment of authors and referees, the sourcing of funding, and to 'open doors' to make this publication happen. This book could not have been achieved without the wisdom, support and dedication of its Editorial Board, comprising Jerome Argue (URS Australia), Simon Beecham (UniSA), Richard Clark (Stormwater Industry Australia Inc.), John Howard (SA Water), David Jones (University of Adelaide), Richard Marks (Interwater Pty Ltd), Jennifer McKay (UniSA), Philip Roetman (UniSA), and Keith Smith (Adelaide and Mount Lofty Ranges National Resource Management (AMLRNRM) Board). They have extensive knowledge of the subject matter of this book and how to best present it; this book simply would not exist without their guidance and hard work.

As with the first in this series, the method employed to generate the book was to commission papers on set topics, in this case using the collective networks of the Editorial Board. Chapters represent detailed analyses of specific subjects, while boxes represent short vignettes on interesting exemplars related to the chapter. We also commissioned case studies, which involved focused analyses of issues relating to a chapter.

All chapters and vignette boxes were subject to independent, academic, blind peer referee by at least two referees, a member of the Editorial Board, and at least one external referee. I thank these referees for vastly improving the quality of the work: Matthew Abraham, Glyn Ashman, David Bevan, Karla Billington, Monique Blason, Trevor Christensen, Philip Clarke, David Cole, Brian Daniels, Peter Day, Brian Finlayson, Jacqueline Frizenschaf, Victor Gostin, Robert Gregory, John Hatch, James Hayter, Jo Henshall, Geoff Henstock, Andrew Humpage, Kylie Hyde, Patrick James, Andrew Johnson, Tim Johnson, David Jones, Janet Kelly, Shaun Kennedy, Simon Langsford, Grant Lewis, Nicole Lewis, Steve Meredith, David Moyle, Rebecca Neumann, John Radcliffe, Julian Robertson, John Sabine, Claus Schonfeldt, Stuart Simpson, Bob Such, Keith Walker, Kate Walsh, Adrian Werner, Kelly Westell, Ian Wheaton, and Lou Wilson.

As with the first book, *Adelaide: Nature of a City*, this text is beautifully illustrated and produced. The design team at A7 Designs of David Radzikiewicz and Barbara Mussared, with help from Eliza Tieman, produced another graphic masterpiece. Carolyn Herbert (re)created all the figures, tables and graphs in the book, often interpreting and creating beautiful pictures from scribbles on napkins or bits of paper. Carolyn created the index, Julia Roetman edited and 'endnoted' the references while Philip Roetman managed the huge database of material, and I owe them all a huge debt of gratitude for their hard work.

Stephanie Johnston and the team at Wakefield Press, particularly Kathy Sharrad, all supported and encouraged this project from its inception. David Bruce prepared the map, and Sandra Orgeig and Carolyn Herbert proofread and edited parts of the manuscript. The magnificent photography of John Hodgson makes this book visually spectacular. John's painstaking efforts to make each photograph so stunning, while also completing such a wide and vast collection of photographs, is awe-inspiring. I also thank our financial sponsors (listed separately), but particularly Professor Peter Hoj, Vice Chancellor and President of the University of South Australia (UniSA), and Professor Patrick James, Head of the School of Natural and Built Environments at UniSA. Without the active support of UniSA this project would have been impossible.

This book is an argument and strategy for change. It focuses on changes to the manner in which the inherited water systems should be changed through the adoption of better plans and better planning systems. Moreover, it is strongly related to sustainability, which is now emerging as possibly the greatest challenge that has ever faced humanity. This book demonstrates that for such a complex system as the watercycle system of a city, there is a huge amount of activity currently being undertaken. We have come a long way in our understanding and realisation that we are part of the environment, and we must act accordingly.

Finally, to my family, Sandy, Sam and Alex, and to my parents and friends and colleagues who have supported and encouraged my obsession with this project over the past three years – thankyou.

Christopher B. Daniels

Barbara Hardy Centre for Sustainable Urban Environments
School of Natural and Built Environments
University of South Australia

Editor in chief
December 2009

Editor in chief

Christopher B. Daniels, BSc (Hons), PhD

School of Natural and
Built Environments
City East Campus
University of South Australia
Adelaide SA 5000
Phone: (08) 8302 2317
Email: chris.daniels@unisa.edu.au

Chris Daniels is Professor of Urban Ecology, Director of Research in the School of Natural and Built Environments, and Head of Geospatial and Environmental Management at the UniSA. He is also the Director of the Barbara Hardy Centre for Sustainable Urban Environments. Chris was educated at the University of Adelaide and the University of New England. He has held academic positions at the University of California and Flinders University before moving to the University of Adelaide, and then to UniSA. He is married with two children and lives in Belair.

Chris has been a prolific scientist and author, having edited four books and contributing to more than 150 scientific publications. He has been published on issues regarding the natural and built environments, and has an abiding interest in reptiles, particularly lizards. Chris is also an award-winning science communicator with three Premier's Science Excellence Awards in Science Communication and Education. He has regular sessions on 891 ABC Radio and discusses a range of science and environmental topics.

Chris and Catherine Tait edited *Adelaide: Nature of a City, the Ecology of a Dynamic City*, which was published in 2005 and received the Whitley Award from the Royal Zoological Society of NSW, and a commendation from the Planning Institute of Australia (South Australia).

Editorial Board

Jerome J. Argue, BE (Civil), MBA

URS Australia
Level 4, 70 Light Square
Adelaide SA 5000
Phone: (08) 8366 1000
Email: jerome_argue@urscorp.com

Jerome Argue has had more than 20 years' experience in the investigation and analysis of flooding, stormwater and drainage systems including hydrological and hydraulic modelling, flood inundation mapping, stormwater masterplanning, and system design. Over the past 14 years he has developed a specialty in water sensitive design and its application, particularly in an urban context. This has involved designing and documenting a range of treatment and reuse systems for residential, commercial and institutional facilities. In addition, he has been involved in masterplanning the watercycle management for large-scale developments, involving subdivisions of up to 3500 lots, through sensitive environments bordering wetlands and river systems.

Simon Beecham, BSc (Hons), PhD, GradCert HEd

School of Natural and
Built Environments
Mawson Lakes Campus
University of South Australia
South Australia 5095
Phone: (08) 8302 5141
Email: simon.beecham@unisa.edu.au

Professor Simon Beecham is Managing Director of the SA Water Centre for Water Management and Reuse at UniSA. His research interests include integrated water management, water sensitive urban design, and climate change impacts on urban water systems. Simon is author of the Syfon computer program, which was used to design the roofwater harvesting system at the Sydney Olympic Stadium, as well as the airport terminal roof drainage systems at Hong Kong (Chek Lap Kok), Kuala Lumpur, Sydney and Adelaide. He is also author of the SWITCH and Switch2 water sensitive urban design software packages, which are recommended in Engineers Australia's *Australian Runoff Quality (ARQ) Manual*.

Richard D.S. Clark, MA, DIC (Hydrology)

Richard Clark and Associates
21 Ann Street
Stepney SA 5069
Phone: (08) 8362 9265
Email: rdsclark@optusnet.com.au

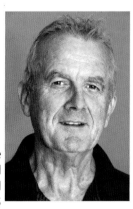

Richard Clark has 45 years experience in investigations, planning and design of natural and engineered water systems. For the past 15 years he has worked as a water systems modeller, using his renowned WaterCress software package on the innovative schemes at New Haven Village, Mawson Lakes, Parafield Airport, Grange Golf Club, Morphettville Racecourse, and many other locations.

John R. Howard, BSc (Hons), PhD

SA Water Corporation
250 Victoria Square
Adelaide SA 5001
Phone: (08) 7424 1066
Email: john.howard@sawater.com.au

John Howard has 24 years experience in water quality and water resources management, with roles in the United Kingdom and South Africa. Before joining SA Water, John was Water Quality and Environmental Manager for a South African regional authority responsible for providing water to over 4 million people. Joining SA Water as Principal Water Quality Scientist in 1998, John was subsequently appointed General Manager of the Australian Water Quality Centre, and in August 2005 was appointed Head of Water, Quality and Environment, which provides leadership in scientific and technical advisory services to SA Water. John has played a key role in the establishment of Water Quality Research Australia, of which he is presently a Director. He has recently been appointed a Director of the Australian Water Association.

David S. Jones, BA, MLArch, MUP (Prelim), PhD

Landscape Architecture and
Urban Design
School of Architecture
University of Adelaide
Adelaide SA 5005
Phone: (08) 8303 4589
Email: david.jones@adelaide.edu.au

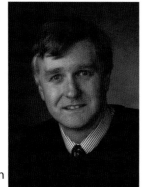

Dr David Jones is Foundation Director of the Landscape Architecture Program at the University of Adelaide and past Head of the School of Architecture, Landscape Architecture and Urban Design. A leading researcher in cultural landscape practice and theory, he has been Project Leader for the Adelaide Botanic Garden Conservation Study, the Mount Lofty Botanic Garden Conservation Study, the Government House Grounds, Adelaide, Landscape Conservation Study, and the Adelaide Parklands Cultural Landscape Assessment Study for the respective government agencies.

Richard Marks, BE (Civil), GDip ChemEng, CPEng

InterWater Pty Ltd
PO Box 199
Stepney SA 5069
Phone: (08) 8301 1494
Email: watermarks@ozemail.com.au

Richard Marks is a civil and chemical engineer with over 40 years experience in water and wastewater engineering. Richard and Dr June Marks, a sociologist, established InterWater Pty Ltd in 2000 to focus on projects involving water sensitive designs integrating natural systems and recycling, and including community engagement strategies.

Since coming to Adelaide in 1996, Richard has had significant engineering roles with KBR on the Virginia Pipeline Scheme, Salisbury stormwater harvesting schemes, Olympic Dam expansion watercycle, Point Boston Peninsula water self-sufficiency scheme, and numerous other advanced water-sensitive design projects. From 1993 to 1995, Richard was Federal President of the Australian Water Association and from 1996 to 1998 was Chair of Asia Pacific Group of the International Water Association. He served on the Anglicare-SA Council from 1996 to 2006 and has been Chair of the Anglican Housing Association since 2001.

Jennifer M. McKay, BA (Hons), PhD, LLB, GDLP

School of Commerce
City West Campus
University of South Australia
Adelaide SA 5000
Phone: (08) 8302 0887
Email: jennifer.mckay@unisa.edu.au

Jennifer McKay is a Professor of Business Law at UniSA. She is also the Director of the Centre for Comparative Water Policies and Laws. Jennifer has served on the SA Water Resource Council and the Natural Resources Management Council, and has been invited to make law reform suggestions to commissions in three states and the Commonwealth. She has refereed many publications on national and international sustainable development law, has edited books, been on the editorial board of national and international journals, and currently has a Fulbright Senior Fellowship to the University of California, Berkeley.

Philip E.J. Roetman, BAppSc

School of Natural and
Built Environments
City East Campus
University of South Australia
Adelaide SA 5000
Phone: (08) 8302 2359
Email: philip.roetman@unisa.edu.au

Philip Roetman has a Bachelor of Applied Science from UniSA. He conducts research within the Barbara Hardy Centre for Sustainable Urban Environments, and is particularly interested in people's attitudes towards the natural environment. Philip is an integral part of the multidisciplinary team behind the Barbara Hardy Centre's Citizen Science Research and Education Program.

Keith E. Smith, BAppSc (Hons), MSc

Adelaide and Mount Lofty Ranges
Natural Resources Management Board
205 Greenhill Road
Eastwood SA 5063
Phone: (08) 8273 9115
Email: keith.smith@adelaide.nrm.sa.gov.au

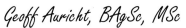

Keith Smith is an ecologist with the Adelaide and Mount Lofty Ranges Natural Resources Management Board. He has worked in the field of urban environmental management for more than 20 years, with local government as the Senior Wetlands Officer with the City of Salisbury, and more recently with the NRM Board. His primary interests are in the areas of urban ecosystem management, urban vegetation dynamics, and stormwater management. He is also Adjunct Lecturer at the School of Natural and Built Environments, UniSA.

Photographs

John Hodgson BA (Hons), MURP, MEnvLaw, LFPIA, EFIAP, FAPS

Sir Samuel Way Building
Victoria Square
Adelaide SA 5000
Phone: (08) 8204 0273
Email: jhodgson4@bigpond.com

In addition to his work as a part-time Commissioner in the Environment, Resources and Development Court, John is a distinguished amateur photographer who has achieved considerable success at both national and international levels over more than 30 years. He is a former President of the Australian Photographic Society and has been an invited judge and presenter at photographic exhibitions and conventions in South Africa and China, as well as in most states of Australia. His photographic interests are broad, encompassing colour and monochrome prints, audio-visuals, and digital electronic images.

Chapter authors

Nina Allen, BE (C&Env) (Hons)

Department of Health
11–15 Hindmarsh Square
Adelaide SA 5000
Phone: (08) 8226 7162
Email: nina.allen@health.sa.gov.au

Nina Allen has a degree in Civil and Environmental Engineering (Honours) from the University of Adelaide. She currently works at SA Health, in the Applied Environmental Health branch as Manager of the Wastewater Section. She is responsible for the review and development of wastewater management policy, legislation, guidelines, assessment and approval of community wastewater management systems under the *Public and Environmental Health Act 1987*.

Geoff Auricht, BAgSc, MSc

South Australian Research and Development Institute
Pastures Group, Waite Campus
Adelaide SA 5001
Phone: (08) 8303 9498
Email: auricht.geoff@saugov.sa.gov.au

Geoff Auricht is Principal Scientist of the Pasture Group at the South Australian Research and Development Institute (SARDI). He is the breeder of the current suite of SARDI lucernes and has been instrumental in the development and release of a wide variety of other pasture cultivars. Geoff currently manages a team of 25 people managing breeding, agronomic research, and commercialisation.

Edward W. Banks, BEnvSc (Hons)

PhD student, Flinders University
Adelaide SA 5001
Phone: 0409 097 970
Email: edward.banks@flinders.edu.au

Edward Banks has worked as a hydrogeologist for the Department of Water, Land and Biodiversity Conservation (DWLBC) from 2005 until 2007, researching groundwater recharge processes and the interactions between surfacewater and groundwater in the Mount Lofty Ranges. This research has been undertaken for the development of the water allocation plans in South Australia. He has now taken leave from the department to undertake a PhD, which is investigating surfacewater/groundwater interactions in fractured rock catchments.

Steve Barnett, MSc

Department of Water, Land and Biodiversity Conservation
Adelaide SA 5001
Phone: (08) 8463 6950
Email: barnett.steve@saugov.sa.gov.au

Steve Barnett is Principal Hydrogeologist at the DWLBC. He has been involved with hydrogeological investigations and management of groundwater resources throughout South Australia over the last 30 years, with particular emphasis on the Murray-Darling Basin.

Peng Bi, MBBS, MMedSc, PhD

School of Population Health and Clinical Practice
North Terrace Campus
University of Adelaide
Adelaide SA 5005
Phone: (08) 8303 3583
Email: peng.bi@adelaide.edu.au

Dr Bi is a Senior Lecturer in the School of Population Health and Clinical Practice at the University of Adelaide. He has intensive research experience in the study of climate change and its impact on population health including waterborne diseases, with excellent track records in publication and attracting category 1 research grants.

Robert (Bob) P. Bourman, BA (Hons), MA, DipEd, DipT, PhD

School of Natural and Built Environments
Mawson Lakes Campus
University of South Australia
Adelaide SA 5095
Phone: (08) 8302 5177
Email: bbourman@uow.edu.au

From January 2009, Professor Bourman will be a Visiting Professorial Fellow in the School of Earth and Environmental Sciences (Faculty of Science) at the University of Wollongong. Professor Bourman is a geomorphologist whose research interests include coastal evolution, the genesis of 'laterite' and landscape evolution, glacial sedimentation during the Late Palaeozoic, palaeomagnetic and thermoluminescence dating of Pleistocene weathering and sediments, mathematical modelling of the River Murray mouth, and the role of neotectonics in explaining apparent global sea level variations.

Simon Bryars, BSc (Hons), PhD

Department for Environment and Heritage
GPO Box 1047
Adelaide SA 5001
Phone: (08) 8222 9424
Email: bryars.simon@saugov.sa.gov.au

Dr Simon Bryars has 15 years research experience in the marine environment, including recent investigations into the impacts of land-based discharges on Adelaide's marine ecosystems. Dr Bryars has worked at the University of Adelaide, the Department of Primary Industries and Resources South Australia (PIRSA), and SARDI (Aquatic Sciences). He is currently working as the Marine Ecologist for Threatened Species within the South Australian Department for Environment and Heritage (DEH).

Michael Burch, BSc (Hons)

Australian Water Quality Centre
South Australian Water Corporation
PO Box 1751
Adelaide SA 5001
Phone: (08) 7424 1012
Email: mike.burch@sawater.com.au

Mike Burch is Research Leader for Biology Research at the Australian Water Quality Centre for SA Water, and an Affiliate Associate Professor in the School of Earth and Environmental Sciences at the University of Adelaide. In the last 20 years he has been involved in a wide range of research projects and management activities relating to water quality in reservoirs and the River Murray in South Australia. He has particular expertise both locally and internationally in control and management of toxic algal blooms.

Nelson F. Cano

University of Sydney
Sydney NSW 2006
Phone: (02) 9351 6457
Email: n.cano@usyd.edu.au

Nelson Cano is the Water, Sediment and Chemical Laboratory Manager at the University of Sydney.

David Cunliffe, PhD

SA Department of Health
Citi Centre Building
11 Hindmarsh Square
Adelaide SA 5000
Phone: (08) 8226 7153
Email: cunliffe.david@saugov.sa.gov.au

David Cunliffe is the Principal Water Quality Adviser at SA Health, with over 25 years experience in dealing with the public health aspects of water quality. He contributed to the development of a range of national and international guidelines dealing with drinking water quality, recycled water reuse, desalination and recreational water quality. He is also a member of the World Health Organization's Drinking Water Quality Committee, the National Health and Medical Research Council's Water Quality Advisory Committee, and the Joint Steering Committee for the Australian Guidelines for Water Recycling.

Elizabeth Curran, BSc (Hons)

20 Grandview Avenue
Urrbrae SA 5064
Phone: (08) 8388 7839; 0438 916 499
Email: walt1718@mac.com

Beth's career spanned nearly 30 years with the Bureau of Meteorology, first as a weather forecaster but more recently in the field of climate. She retired as Manager of Climate Services South Australia in 2007. Beth has been a reviewer for the climate science assessments of the Intergovernmental Panel on Climate Change and spent several years as a Science Adviser on climate change matters to the Australian government.

Loc G. Do, BDS, MSc (Dent), PhD

Australian Research Centre for Population Oral Health
School of Dentistry
University of Adelaide
Adelaide SA 5005
Phone: (08) 8303 3964
Email: loc.do@adelaide.edu.au

Loc Do is a researcher with a dental background. His main research interests are epidemiology of oral conditions and prevention of dental cavities using fluorides including water fluoridation.

Neill J. Dorrington, BSc (Hons)

University of Sydney (student)
Honours Room, Madsen F09
University of Sydney
Sydney NSW 2000
Phone: (02) 4343 1301; 0431 985 855
Email: ndor6400@usyd.edu.au; neilldorrington@hotmail.com

Neill Dorrington completed a Bachelor of Science majoring in geosciences/physics and is now an honours student at the University of Sydney. His honours project involves studying palaeomagnetic secular variation signatures in lake sediments, and using these variations as a means of dating the recent past.

Joe Flynn

Water Industry Alliance
PO Box 1751
Adelaide SA 5000
Phone: (08) 7424 1892
Email: ceo@waterindustry.com.au

Joe Flynn has an international infrastructure background at the Managing Director and GM level, having led water and electricity utilities, and run companies for some of Australia's largest infrastructure service businesses including Leighton and Tenix. He has chaired government/industry economic development initiatives and is currently the CEO of the Water Industry Alliance, almost 200 companies who collectively have accumulated more than $2 bn in exports of South Australian water-related services and technologies.

Richard J.B. Gale

Prince Alfred College
Kent Town SA 5071
Phone: (08) 8334 1200

Richard J.B. Gale is a year 10 student at Prince Alfred College who undertook a sediment analysis of the River Torrens for which he received an Oliphant Science Award Prize.

Stephen J. Gale, MA, PhD

School of Geosciences
University of Sydney
Sydney NSW 2006
Phone: (02) 9351 3308
Email: s.gale@usyd.edu.au

S.J. Gale teaches in the School of Geosciences at the University of Sydney. His has undertaken research within the general area of environmental change. Of most relevance here is his work on the behaviour of hydrological systems, particularly in response to human activity.

James Gehling, BSc (Hons), MSc, PhD

South Australian Museum
North Terrace
Adelaide SA 5000
Phone: (08) 8207 7441
Email: gehling.jim@saugov.sa.gov.au

Dr James Gehling lectured in Earth and Environmental Sciences at UniSA and its former institutions for 26 years, specialising in marine and earth sciences. After two years on a research fellowship, working on the Ediacara fossils and palaeo-environments of the Avalon Peninsula in Newfoundland, he joined the South Australian Museum first as a Honorary and then as a Senior Research Scientist in Palaeontology. His research projects centre on the ancient life and environments of the Ediacaran and Cambrian periods in South Australia, Newfoundland and Namibia.

Peter Gell, MEnvSc, DipEd, PhD

Centre for Environmental Management
University of Ballarat
University Drive
Mount Helen, Ballarat, VIC 3353
Phone: (03) 532 76155
Email: p.gell@ballarat.edu.au

Peter Gell is a Professor of Environmental Science and Director of the Centre for Environmental Management at the University of Ballarat. He is also an Adjunct Associate Professor of Geographical and Environmental Studies at the University of Adelaide. He is a palaeoecologist who examines century-scale records of wetland change to understand the influence of natural climate change relative to the impact of industrialised society.

Nabil Z. Gerges, BSc, PhD

Australian Groundwater Technologies and
NZG Groundwater, Consultant
1 Shamrock Road
Hallett Cove SA 5158
Phone: (08) 8352 4262; (08) 8381 6268
Email: gergesnz@gmail.com; ngerges@agwt.com.au

Nabil Gerges is the recognised authority on geology and hydrogeology of St Vincent Basin, which includes the Adelaide metropolitan, Northern Adelaide Plains and Willunga basins. He is also recognised as an expert in basin analysis 'sedimentary basin' and fractured rock aquifers, and is one of the leading authorities in Australia on aquifer storage and recovery techniques.

Christine Goodwin, AdvDipT, DipDesign, GradDip Arts Admin, MArch (research)

Fifth Creek Studio
PO Box 515
Montacute SA 5134
Phone: (08) 8390 2292
Email: fifthcreek@optusnet.com.au

Christine Goodwin has a background in education, the arts and landscape design, and is a Director of Fifth Creek Studio, specialising in living architecture: landscape architecture, green roofs and walls, and urban design. She is currently co-authoring a book on living architecture, with a focus on the integration of green roofs and living walls into sustainable green urban infrastructure.

Catherine Gray, BEnvSc (Hons)

Department for Transport, Energy and Infrastructure
33–37 Warwick Street
Walkerville SA 5081
Phone: (08) 8343 2986
Email: catherine.gray@saugov.sa.gov.au

Catherine Gray has worked with Chris Daniels at BioCity: Centre for Urban Habitats (where, under her maiden name (Tait), she was an editor and author on the book *Adelaide: Nature of a City*. At the University of Adelaide she was a Lecturer and Course Coordinator for Urban Biodiversity Management. She works as the Eastern Region Environmental Officer at the Department for Transport, Energy and Infrastructure.

Pamela Gurner-Hall

TAFE SA
Urrbrae Campus
Netherby SA 5062
Phone: 0419 213 237
Email: serendipity@picknowl.com.au

Pamela Gurner-Hall is a Lecturer in garden and landscape design, horticulture, permaculture and landscape architecture at TAFE SA. She has a special interest in the interface between sustainable housing and garden/environment development. Pamela has 34 years of horticulture industry experience as a grower, retailer, marketer and designer, and has been an international speaker in South America, Japan, and Africa.

Nicholas Harvey, BA (Hons), BEd, MPlan, PhD

Faculty of Humanities and Social Sciences
North Terrace Campus
University of Adelaide
Adelaide SA 5005
Phone: (08) 8303 4163
Email: nick.harvey@adelaide.edu.au

Professor Nick Harvey is the Executive Dean of the Faculty of Humanities and Social Sciences at the University of Adelaide. Nick is one of the five Australian lead authors on the Australia and New Zealand Chapter of the IPCC, Working Group II report on climate change impacts and vulnerability. He has written over 150 scientific papers and books including his latest book *Global Change and Integrated Coastal Management*, and has coauthored a book on coastal management in Australia.

Graeme Hopkins, BSc (Arch), BArch, FAILA, ARAIA

Fifth Creek Studio
PO Box 515
Montacute SA 5134
Phone: (08) 8390 2292
Email: fifthcreek@optusnet.com.au

Graeme Hopkins, a Registered Architect and Registered Landscape Architect, is a Director of Fifth Creek Studio, specialising in living architecture: landscape architecture, green roofs and walls, and urban design. He is coauthoring a book on green roofs and walls that builds on his recent Churchill Fellowship overseas studies. He is also Principal Urban Designer for Planning SA, and Adjunct Associate Professor at the University of Adelaide.

Sara Hughes, MSc

University of California, Santa Barbara
Bren School of Environmental Science and Management
4526 Bren Hall
Santa Barbara, CA 93106 USA
Phone: +1 805 893 8437
Email: shughes@bren.ucsb.edu

Sara Hughes is a PhD student at the University of California, Santa Barbara. Her research focuses on the institutions surrounding urban water supply and management in California and Australia. She holds an MSc from Michigan State University.

Graeme Hugo, BA (Hons), MA, PhD, FASSA

Department of Geographical and Environmental Studies
North Terrace Campus
University of Adelaide
Adelaide SA 5005
Phone: (08) 8303 5646
Email: graeme.hugo@adelaide.edu.au

Graeme Hugo is University Professorial Research Fellow, Professor of the Department of Geographical and Environmental Studies and Director of the National Centre for Social Applications of Geographic Information Systems at the University of Adelaide. His research interests are in population issues in Australia and South-East Asia, especially migration. In 2002 he secured an ARC Federation Fellowship over five years for his research project, 'The new paradigm of international migration to and from Australia: dimensions, causes and implications'.

Andrew Humpage, BSc, GradDip (Med. Lab. Sci.), MAppSc, PhD

Australian Water Quality Centre
SA Water House
Victoria Square
Adelaide SA 5000
Phone: (08) 7424 2064
Email: andrew.humpage@sawater.com.au

Andrew Humpage is Senior Biochemist at the Australian Water Quality Centre. His main research interest is the toxicology of chemicals of concern in water and wastewaters, and application of this knowledge to the development of toxicity-based bioassays. He is also an Affiliate Senior Lecturer in the Discipline of Pharmacology at the University of Adelaide.

Greg Ingleton, BAppSc (Hons)

Environmental Management Group
SA Water Corporation
PO Box 1751
Adelaide SA 5001
Phone: 0438 457 543
Email: greg.ingleton@sawater.com.au

Greg Ingleton was employed as a graduate at SA Water in 2003. Within the company, he has worked as a Catchment Management Officer and as a Source Water Quality Specialist, where he employed new methods of managing risk reduction in source water quality. He is currently working as the Receiving Environment Assessment Specialist, looking at techniques for quantifying and reducing the environmental impacts of SA Water's wastewater discharges.

Cara Jenkin, BA (Journalism)

The Advertiser
31 Waymouth Street
Adelaide SA 5000
Phone: (08) 8206 2397
Email: jenkinc@theadvertiser.com.au

Cara Jenkins has worked in print and television media in regional and urban South Australia since 2001. She has worked as the *Advertiser*'s Environment Reporter since 2006, and has a particular focus on water issues.

Greg R. Johnston, BSc (Hons), PhD

School of Biological Science
Flinders University
GPO Box 2100
Adelaide SA 5000
Phone: (08) 8201 5677
Email: greg.johnston@flinders.edu.au

Dr Greg Johnston is a Senior Lecturer in Biodiversity and Conservation at Flinders University and a Conservation Biologist with the Royal Zoological Society of SA. He has worked in the mining industry, academia and zoos in Australia, New Guinea and the Middle East. His major research projects are a field study of the evolutionary ecology of brood reduction in birds, and colour variation in reptiles. He is also Adjunct Senior Lecturer at the School of Natural and Built Environments, UniSA.

Jon Kellett, BA, MCD, PhD, MPIA

School of Natural and Built Environments
City East Campus
University of South Australia
Adelaide SA 5000
Phone: (08) 8302 1701
Email: jon.kellett@unisa.edu.au

Dr Jon Kellett has taught urban planning and environmental management for over 25 years at Sheffield Hallam University in the United Kingdom and UniSA. A focus of his research over the last few years is how cities respond to climate change, particularly in respect to their use of natural resources such as energy and water. He is particularly interested in how urban behaviour may change to accommodate the imperatives of climate change.

Tim Kelly, BAppSc

SA Water Corporation
77 Grenfell Street
Adelaide SA 5000
Phone: (08) 7424 1278
Email: tim.kelly@sawater.com.au

Tim Kelly is SA Water's Principal Climate Change Advisor responsible for coordinating efforts to reduce its emissions, prepare strategies to adapt to climate change and support research, in collaboration with SA Water business units, the South Australian government and other stakeholders. Tim has 16 years experience in the water industry, and has led SA Water's voluntary participation in the Greenhouse Challenge Plus program since 2003. In 2006, Tim worked with Professor Stephen Schneider, Adelaide's Thinker in Residence on Climate Change as one of two research assistants exploring climate change: risks and opportunities for Adelaide.

Grant Lewis, BE (Civil) MEngSc (Public Health Engineering)

SA Water
Level 16, 77 Grenfell Street
Adelaide SA 5000
Phone: (08) 8204 1922
Email: grant.lewis@sawater.com.au

Grant Lewis is a Senior Process Engineer at SA Water. He has worked in construction, operations and design areas within the company. Grant has worked in the area of wastewater design since 1979 and has been involved in projects such as the Heathfield WWTP (wastewater treatment plant) upgrade, Whyalla water reclamation plant, and the Port Pirie WWTP upgrade.

Andrew Lothian, DipTech (Town Planning), MSc (Env. Res.), PhD, CEnvP, FEIANZ
Scenic Solutions
39B Marion Street
Unley SA 5061
Phone: 0439 872 226
Email: alothian@aapt.net.au

Andrew Lothian worked in DEH for many years and was involved in environmental policy at the state and national levels covering such issues as climate change, ecological sustainability, state of environment reporting, environmental valuation, environmental industry, waste and land contamination. Since 2002 he has worked as a consultant specialising in the measurement and mapping of landscape quality and assessing the visual impact of developments on scenic quality. He is a Visiting Lecturer at several universities and is a part-time Commissioner on the Environment, Resources and Development Court.

Andrew J. Love, BSc (Hons), MSc, PhD
School of Chemistry, Physics and Earth Sciences
Flinders University
GPO Box 2100
Adelaide SA 5001
Phone: (08) 8201 2198
Email: andy.love@flinders.edu.au

Dr Andrew Love is the inaugural Senior Industry Research Fellow (Hydrogeology) at Flinders University School of Chemistry, Physics and Earth Sciences. Dr Love is currently the Chief Investigator on two National Water Commission projects; fractured rock aquifers in the Mount Lofty Ranges and the hydrogeology of the Great Artesian Basin. His research interests and expertise include flow through fractured rock, surfacewater/groundwater interactions, and the hydrogeology of sedimentary basins.

June Marks, BA, PhD
InterWater Pty Ltd
PO Box 199
Stepney SA 5069
Phone: (08) 8363 4430
Email: junesmarks@gmail.com

June Marks is a Director of InterWater. She conducted research on Australian attitudes to water recycling through an ARC Linkage Project with Flinders University, supported by United Water International. Her findings on potable and non-potable reuse have informed the Australian Water Recycling Guidelines.

Timothy McBeath BA, MA
Tim McBeath Greywater/Stormwater Consulting
Main South Road
Croydon SA 5008
Phone: 0421 739 701
Email: tim.mcbeath@yahoo.com.au

Tim McBeath completed a Bachelor of Arts at Flinders University in 2005 majoring in Biology and Philosophy. He completed a Masters in Urban and Regional Planning at UniSA in 2007, which included a 10,000 word sustainable urban design project on the viability of retrofitted greywater, as part of the waterproofing strategy for Adelaide. He is currently working as a private Greywater/Stormwater Consultant.

Clare Peddie, BHSc, GradDipScJourn
The Advertiser
31 Waymouth Street
Adelaide SA 5000
Phone: (08) 8206 2204
Email: peddiec@theadvertiser.com.au

Clare Peddie is a Science Journalist at the *Advertiser*. She is passionate about the environment and hopes to make a difference by sharing the latest in science with everyday people. A degree in health sciences from the University of Adelaide and graduate diploma in science journalism, combined with seven years experience as a Science Communicator in research organisations such as the CSIRO and Cooperative Research Centres, have provided Clare with valuable knowledge, experience and personal networks.

Sheryn Pitman, BA, DipEd, MES
Botanic Gardens of Adelaide
North Terrace
Adelaide, South Australia 5000
Phone: (08) 8222 9490
Email: pitman.sheryn@saugov.sa.gov.au

Sheryn Pitman manages the Sustainable Landscapes Project, a collaborative partnership between the Botanic Gardens of Adelaide, the Land Management Corporation, SA Water, the Innovations and Economic Opportunities Group and the Adelaide and Mount Lofty Ranges NRM Board.

Sheryn has a multidisciplinary background in natural resources management, education, communication and project management. She has worked extensively in the environmental industry in areas such as native vegetation rehabilitation, biodiversity conservation, and coastal management.

Darren Ray, BSc (Hons)

SA Regional Office of the Commonwealth Bureau of Meteorology
25 College Road
Kent Town SA 5067
Phone: (08) 8366 2664
Email: d.ray@bom.gov.au

Darren's career at the Bureau of Meteorology began in regional weather forecasting, before he followed a long-term interest in drivers of climate and climate change in Australia into the South Australian Climate Section, which he now manages.

Matthew W. Rofe, BEd, BA (Hons), PhD

School of Natural and Built Environments
City East Campus
University of South Australia
Adelaide SA 5000
Phone: (08) 8302 2358
Email: matthew.rofe@unisa.edu.au

Matthew Rofe is the Program Director of Postgraduate Studies in Urban and Regional Planning at UniSA. The central theme of his research agenda involves unravelling the complexity of human landscapes and the often conflicting meanings that are attached to place. Presently Matthew is researching globalisation and the emergence of international educating cities (focusing upon Adelaide), urban revitalisation projects, and entrepreneurial governance (focusing on Port Adelaide).

Martin Shanahan, BA, LLB, PhD

Centre for Regulation and Market Analysis
City West Campus
University of South Australia
Adelaide SA 5000
Phone: (08) 8302 0745
Email: martin.shanahan@unisa.edu.au

Martin Shanahan is Associate Professor in Economics at UniSA. He researches in a variety of fields including economic history, water markets, market regulation, and wealth and income distribution. He is co-editor of the *Australian Economic History Review*.

Craig T. Simmons, BE (Hons), BSc, PhD

School of Chemistry, Physics and Earth Sciences
Flinders University
GPO Box 2100
Adelaide SA 5001
Phone: (08) 8201 2348
Email: craig.simmons@flinders.edu.au

Craig Simmons is Professor of Hydrogeology at Flinders University. He is one of Australia's foremost groundwater academics. He currently serves as an Associate Editor on the editorial boards of prestigious international scientific journals including the *Journal of Hydrology* and *Ground Water*, and has also served previously as an Associate Editor for the *Hydrogeology Journal* (Managing Editor 2005–2008), *Ground Water* and *Water Resources Research*.

A. John Spencer, MDSc, MPH, PhD

ARCPOH, School of Dentistry
University of Adelaide
Adelaide SA 5005
Phone: (08) 8303 5438
Email: john.spencer@adelaide.edu.au

John Spencer is the Professor of Social and Preventive Dentistry and Director of the Australian Research Centre for Population Oral Health at the University of Adelaide. He is a leading contributor of information on oral health, dental practice, dental services and policy in Australia. He is also Editor of *Community Dentistry and Oral Epidemiology*, International Advisor to *Community Dental Health*, and a member of the Editorial Advisory Board of the *Australian Dental Journal*.

Gertrude Szili, PhD (Planning) candidate

School of Natural and Built Environments
City East Campus
University of South Australia
Adelaide SA 5000
Phone: (08) 8302 2356
Email: gertrude.szili@unisa.edu.au

Gertrude Szili has an educational background in the disciplines of environmental studies, human geography and anthropology. Her current doctoral research involves the investigation of the various ways environmental discourses are mobilised within contemporary urban landscapes. Other research interests include urban governance, environmental ethics, place making and city marketing, qualitative research methodologies, media analysis, ecotourism and biodiversity conservation and restoration.

Sandra Taylor, BA (Hons), MSc, PhD

School of Natural and Built Environments
City East Campus
University of South Australia
Adelaide SA 5000
Phone: (08) 8302 2359, 0428 116 749
Email: sandra.taylor@unisa.edu.au

Dr Sandra Taylor lectured in environmental studies at the University of Adelaide from 1975 to 2007. Since her retirement she has been appointed to Adjunct Senior Lectureships at the University of Adelaide and UniSA, where she teaches postgraduate courses in urban ecology. She is also currently a member of the Adelaide and Mount Lofty Ranges NRM Board.

Kelly Westell, BA (Journalism)

SA Water Corporation
77 Grenfell Street
Adelaide SA 5000
Phone: (08) 8204 1436
Email: kelly.westell@sawater.com.au

Kelly Westell has worked in various roles as a journalist, media adviser and communications manager prior to her current position as SA Water's Manager of Stakeholder Relations.

Hilary P.M. Winchester, MA, PhD

City West Campus
University of South Australia
Adelaide SA 5000
Phone: (08) 8302 0209
Email: hilary.winchester@unisa.edu.au

Professor Hilary Winchester is Pro Vice Chancellor: Strategy and Planning at UniSA. She has a PhD in geography from the University of Oxford and was a member of the Torrens Taskforce 2006–2007.

Christopher Wright, BA, BAI, MSc, MEngSc

Bureau of Meteorology
Box 421
Kent Town SA 5071
Phone: (08) 8278 8818
Email: c.wright@bom.gov.au

Chris Wright is the Engineering Hydrologist for the Bureau of Meteorology, in charge of the Flood Warning Program for South Australia since 1988. He has taken a close interest in climate and water data.

Ying Zhang MMed Sc, MBBS, PhD

Discipline of Public Health, QUMPRC
University of Adelaide
Adelaide SA 5005
Phone: (08) 8302 2817
Email: ying.zhang@adelaide.edu.au

Dr Zhang is a Postdoctoral Research Fellow in epidemiology. Her PhD thesis uses Australian and Chinese data to elucidate the relationship between climate variations and selected infectious diseases.

John Zwar, DipCLM, HCert (Hort.)

TAFE SA
Urrbrae Campus
505 Fullarton Road
Netherby SA 5062
Phone: (08) 8372 6876
Email: john.zwar@tafesa.edu.au

John Zwar is a Lecturer in horticulture and environmental management at TAFE SA. Throughout his career he has worked as a horticulturalist, a Curator of the National Botanic Gardens of Papua New Guinea, and a Senior Environmental Scientist at Roxby Downs. He also proposed the establishment of the Australian Arid Lands Botanic Garden in Port Augusta and is still involved with this exciting development which opened in 1996.

Case study authors
(if not in the chapter author list)

Martin Ely, BArch, AILA, MPIA

School of Architecture, Landscape Architecture and Urban Design
North Terrace Campus
University of Adelaide
Adelaide SA 5005
Phone: 0407 809 984
Email: martin.ely@adelaide.edu.au

Martin is an Urban Designer and Registered Landscape Architect, with extensive experience in urban and streetscape design.

Scott Heyes, BDesSt, BL Arch (Hons), MLarch, PhD

Faculty of Architecture, Building and Planning,
University of Melbourne
Parkville VIC 3010
Phone: (03) 8344 4982
Email: sheyes@unimelb.edu.au

Scott Heyes is a Landscape Architect and Cultural Geographer at the University of Melbourne.

Ganesh Keremane, BSc, MSc, PhD

School of Commerce
City West Campus
University of South Australia
South Australia 5001
Phone: (08) 8302 0148
Email: ganesh.keremane@unisa.edu.au

Ganesh is working as a Research Assistant at the Centre for Comparative Water Policies and Laws on the social, cultural, institutional and policy issues related to water management in Australia.

Thorsten Mosisch, BSc, MSc, PhD

SA Water
242–260 Victoria Square
PO Box 1751
Adelaide SA 5000
Phone: (08) 7424 1984
Email: thorsten.mosisch@sawater.com.au

Thorsten Mosisch is Source Water Quality and Special Projects Manager with SA Water as part of the Water Quality and Integrated Management Group.

Colin Pitman, BAgr, BE

City of Salisbury
12 James Street
Salisbury SA 5108
Phone: (08) 8406 8215
Email: cpitman@salisbury.sa.gov.au

Colin Pitman works at the City of Salisbury where he has led a team who have been innovative in energy management, environmental design, waste management and the traditional civil engineering works of local government.

Richard Thomson, BE (Chem) MSc, FAICD

Rinnai Australia
Mobile: 0411 146 436
Email: pamelath@optusnet.com.au

Richard Thomson is Company Director and Consulting Engineer at Rinnai Australia.

Box authors
(if not in the chapter or case study lists)

John Argue

39 Anglesey Avenue
St Georges, SA 5064
Phone: (08) 8379 6272
Email: john.argue@unisa.edu.au

John has retired from his role as Associate Professor of Water Engineering at UniSA, but continues a research association with the Urban Water Resources Centre at UniSA.

Karla Billington

SA Water Corporation
SA Water House
Victoria Square
Adelaide SA 5000
Phone: (08) 7424 2468
Email: karla.billington@sawater.com.au

Working at SA Water Corporation as Catchments Manager, Karla focuses on source water protection for drinking water supplies.

Amy Blaylock

KESAB Environmental Solutions/Adelaide Mount Lofty Ranges Natural Resources Management Board
214 Grange Road
Flinders Park SA 5025
Phone: (08) 8234 7255
Email: amy_b@kesab.asn.au

Amy has worked for the past seven years as a Waterwatch Education Officer and Regional Coordinator in the Patawalonga, Torrens and Port catchments.

Jennifer Bonham

Geographical and Environmental Studies
North Terrace Campus
University of Adelaide
Adelaide SA 5005
Phone: (08) 8303 4655
Email: jennifer.bonham@adelaide.edu.au

Jennifer is Lecturer in Geographical and Environmental Studies at the University of Adelaide.

Barry Brook

School of Earth and Environmental Sciences
North Terrace Campus
University of Adelaide
Adelaide SA 5005
Phone: (08) 8303 3745
Email: barry.brook@adelaide.edu.au

Barry is the Foundation Sir Hubert Wilkins Professor of Climate Change, working on climate change, computational and statistical modelling, and the synergies between human impacts on earth systems.

Justin Brookes

School of Earth and Environmental Sciences
North Terrace Campus
University of Adelaide
Adelaide SA 5005
Phone: (08) 8303 3747
Email: justin.brookes@adelaide.edu.au

Justin is a Senior Research Fellow in the School of Earth and Environmental Sciences at the University of Adelaide.

Melissa Brown

Gemtree Vineyards
Kangarilla Road
McLaren Flat SA 5171
Phone: (08) 8383 0409
Email: melissa@theterraces.com.au

Melissa is a viticulturist for 112 ha of Gemtree Vineyards, which is being farmed biodynamically, and oversees the development and maintenance of the Gemtree Wetlands.

Jake Bugden

Sustainable Focus
11 Gething Crescent
Bowden SA 5007
Phone: (08) 8340 8666
Email: jake@sustainablefocus.com.au

Jake is the Director of Sustainable Focus, a company that helps organisations move towards sustainability.

Don Cameron

School of Natural and Built Environments
Mawson Lakes Campus
University of South Australia
Mawson Lakes SA 5095
Phone: (08) 8302 3128
Email: donald.cameron@unisa.edu.au

Don is a Senior Lecturer in geotechnical engineering at UniSA.

Ian Clark

School of Natural and Built Environments
Mawson Lakes Campus
University of South Australia
Mawson Lakes SA 5095
Phone: (08) 8302 5245
Email: ian.clark@unisa.edu.au

Ian is Associate Head, Teaching and Learning in the School of Natural and Built Environments at UniSA.

Peri Coleman

Delta Environmental Consulting
12 Beach Road
St Kilda SA 5110
Phone: (08) 8280 5910
Email: peri@deltaenvironmental.com.au

Peri owns, and is Senior Consultant for, Delta Environmental Consulting.

Jock Conlon

Urban Biodiversity Unit
Department of Environment and Heritage
Wittunga House
328 Shepherds Hill Road
Blackwood SA 5051
Phone: (08) 8278 0608
Email: conlon.johnathon@saugov.sa.gov.au

Jock is the Central Project Officer at the Urban Biodiversity Unit of the DEH.

Sean D. Connell

School of Earth and Environmental Sciences
North Terrace Campus
University of Adelaide
Adelaide SA 5005
Phone: (08) 8303 6125
Email: sean.connell@adelaide.edu.au

Sean is Associate Professor in Marine Biology in the Discipline of Ecology and Evolution at the University of Adelaide.

Asta Cox

Phone: 0412 892 273
Email: asta.cox@nt.gov.au

Asta completed an honours project in water ecology, specialising in paleoliminology, at the University of Adelaide.

Andrew Crompton

City of Burnside
Conservation and Land Management
401 Greenhill Road
Tusmore SA 5065
Phone: (08) 8366 4267
Email: acrompton@burnside.sa.gov.au

Andrew is Group Team Leader in Conservation and Land Management at the City of Burnside.

Peter Cullen

(deceased)
c/o Wentworth Group of Concerned Scientists
Suite 3, 3b Maquarie Street
Sydney NSW 2000
Phone: (02) 9251 3811
Email: information@wentworthgroup.org

Peter was a Professor at the University of Canberra and worked in the field of natural resources management for over 35 years.

David Deer

City of Mitcham
5 Winston Court
Melrose Park SA 5039
Phone: (08) 8374 7793
Email: ddeer@mitchamcouncil.sa.gov.au

David has worked in local government for over 30 years in the area of horticulture, including open space development and management, and streetscapes maintenance and development.

Peter Devitt

Department of Surgery
Royal Adelaide Hospital
North Terrace
Adelaide SA 5000
Phone: (08) 8222 5517
Email: peter.devitt@adelaide.edu.au

Peter is a Consultant Gastrointestinal Surgeon and author of surgical texts.

Keith Downard

Adelaide and Mount Lofty Ranges
Natural Resources Management Board
205 Greenhill Road
Eastwood SA 5063
Phone: (08) 8273 9116
Email: keith.downard@adelaide.nrm.sa.gov.au

Keith works in stormwater management, especially catchment-scale planning and implementation of stormwater reuse schemes, for the Adelaide and Mount Lofty Ranges NRM Board.

Paul Downton

Ecopolis Architects
109 Grote Street
Adelaide SA 5000
Phone: (08) 8410 9218
Email: paul@ecopolis.com.au

Paul is an architect, writer and independent researcher specialising in ecological architecture and urban ecology.

Zoe Drechsler

Adelaide City Council
25 Pirie Street
Adelaide SA 5000
Phone: (08) 8203 7203
Email: z.drechsler@adelaidecitycouncil.com

Zoe works in the Sustainability Department, which is responsible for the delivery of key environmental strategy and policy for the City of Adelaide as well as strategic environmental projects.

Tim Fisher, BA

Land and Water Australia
Canberra ACT 2601

Tim is the Director (non-executive) at Land and Water Australia.

Pauline Frost

Playford Greening and Landcare Group Inc.
PO Box 438
Elizabeth SA 5112
Phone: 0400 084 828

Pauline is the Chair of the Playford Greening and Landcare Group.

Mike Gemmell

South Australian Museum
North Terrace
Adelaide SA 5000
Phone: (08) 8207 7404
Email: gemmell.mike@saugov.sa.gov.au

Mike is the Information Officer at the South Australian Museum.

Sarah Gilmour

City of Mitcham
131 Belair Road
Torrens Park SA 5062
Phone: (08) 8372 8139
Email: sgilmour@mitchamcouncil.sa.gov.au

Sarah is employed as a Strategic Planning Officer, Sustainability, to assist the City of Mitcham in undertaking long-term planning that is environmentally, socially and economically responsible.

John Goldfinch

Fmg Engineering
42 Fullarton Road
Norwood SA 5067
Phone: (08) 8363 0222
Email: johng@fmgengineering.com.au

John works in the investigation of building and engineering problems, including report preparation for repair strategy.

Jennifer Harrison

Institute for Environmental Research, Australian Nuclear Science and Technology Organisation
PMB 1
Menai NSW 2234
Phone: (02) 9717 3919
Email: jennifer.harrison@ansto.gov.au

Jennifer is a Radiochemist who specialises in the measurement of environmental radioactivity in soil, sediments and water.

David Harvey

Tauwitchere Pastoral Co
Tauwitchere
Narrung SA 5259
Phone: (08) 8575 4255
Email: tauwitch@lm.net

David manages and works a 2300 ha in-conversion biodynamic farm, which was irrigated but is now dryland, milking 200 cows, 350 beef breeders and cropping.

Carolyn Herbert

School of Natural and Built Environments
City East Campus
University of South Australia
Adelaide SA 5000
Phone: (08) 8302 2359
Email: carolyn.herbert@unisa.edu.au

Carolyn is a Research Assistant at UniSA.

Kate Hutson

School of Earth and Environmental Sciences
North Terrace Campus
University of Adelaide
Adelaide SA 5005
Phone: (08) 8303 5282
Email: kate.hutson@adelaide.edu.au

Kate is a Postdoctoral Research Fellow in Marine Parasitology at the University of Adelaide.

Chris Kaufmann

Water Proofing Northern Adelaide Regional Subsidiary
Salisbury Downs SA 5108
Phone: 0401 984 771
Email: ckaufmann@salisbury.sa.gov.au

Chris is the Executive Officer of the Water Proofing Northern Adelaide Regional Subsidiary.

Hugh Kneebone

Adelaide and Mount Lofty Ranges
Natural Resources Management Board
205 Greenhill Road
Eastwood SA 5063
Phone: (08) 8273 1933, 0427 184 744
Email: hugh.kneebone@adelaide.nrm.sa.gov.au

As Manager of the NRM Education team, Hugh oversees the delivery of environmental education for sustainability through a whole-of-school approach to learning, understanding and action.

Kay Lawrence

South Australian School of Art
City West Campus
University of South Australia
Adelaide SA 5001
Phone: (08) 8302 0333
Email: kay.lawrence@unisa.edu.au

Kay is a Professor of Visual Arts and Head of the South Australian School of Art.

David Lawry

TREENET
Waite Campus
University of Adelaide
Adelaide SA 5064
Phone: (08) 8303 7078
Email: david@treenet.com.au

David is the Founder and Director of TREENET, which is based at the University of Adelaide's Waite Campus.

Nigel Long

South Australian Chamber of Mines and Energy
290 Glen Osmond Road
Fullarton SA 5063
Phone: (08) 8202 9999
Email: nlong@sacome.org.au

Nigel is Director, Environment and Sustainability, at the South Australian Chamber of Mines and Energy.

Sir Charles Madden

WindWater P/L
79 Birksgate Drive
Urrbrae SA 5064
Phone 0433 565 039:
Email: charlie.madden@internode.on.net

Charles is the Managing Director of WindWater P/L.

Adrian Marshall

Facilities Management Unit
City West Campus
University of South Australia
Adelaide SA 5000
Phone: (08) 8302 2131
Email: adrian.marshall@unisa.edu.au

Adrian is the Environmental Manager for UniSA; there he is working on an environmental management system with a major focus on energy efficiency and reducing the university's greenhouse gas emissions.

James McColl

CSIRO Land and Water
Private Bag 2
Glen Osmond SA 5064
Phone: (08) 8303 8453
Email: jim.mccoll@csiro.au

James works at the CSIRO, in water policy research and communication, with an emphasis on the management of the Murray-Darling Basin.

Christel Mex

Adelaide and Mount Lofty Ranges Natural Resources Management Board
205 Greenhill Road
Eastwood SA 5069
Phone: (08) 8273 9138
Email: christelmex@adelaide.nrm.sa.gov.au

Christel is Director, Communications and Engagement at the Adelaide and Mount Lofty Ranges NRM Board.

John Murphy

SA Water Corporation
Victoria Square
Adelaide SA 5000
Phone: (08) 7424 1951
Email: john.murphy@sawater.com.au

John is the Manager of Portfolio Reporting at SA Water, providing a corporate-wide system to consistently capture and report on key data associated with its capital works program.

Ashley Natt

School of Earth and Environmental Sciences
North Terrace Campus
University of Adelaide
Adelaide SA 5005
Phone: 0422 691 878
Email: ashley.natt@adelaide.edu.au

Ashley is currently doing a PhD using palaeolimnological techniques to investigate late Holocene El Niño variability and human impacts in internationally significant Galapagos archipelago wetlands.

Trevor Nottle

Department of Earth and Environmental Sciences/
Department of Architecture
University of Adelaide
Adelaide SA 5005
Phone: (08) 8339 4210
Email: trevor.nottle@adelaide.edu.au

Trevor is currently writing up a PhD thesis on the impact of climate change on home gardens on the Adelaide Plains.

Ira Pant

Tarac Technologies Pty Ltd
PO Box 78
Nuriootpa SA 5355
Phone: (08) 8562 1522
Email: irap@tarac.com.au

Ira is Project Manager, Environment and Sustainability, with major responsibility for the strategic environmental initiatives within Tarac Technologies.

David Pezzaniti

School of Natural and Built Environments
Mawson Lakes Campus
University of South Australia
Adelaide SA 5095
Phone: (08) 8302 3652
Email: david.pezzaniti@unisa.edu.au

David leads the Urban Water Resource Group at the SA Water Centre for Water Management and Reuse, located at UniSA.

Jian G. Qin

School of Biological Sciences
GPO Box 2100
Flinders University
Adelaide SA 5001
Phone: (08) 8201 3045
Email: jian.qin@flinders.edu.au

Jian is a Senior Lecturer in Biology at Flinders University, doing research in aquaculture, aquatic ecology and fish biology.

Belinda Rawnsley

South Australian Research and Development Institute
Waite Research Precinct
Hartley Grove
Urrbrae SA 5064
Phone: (08) 8303 9409
Email: rawnsley.belinda@saugov.sa.gov.au

Belinda is a Senior Plant Pathologist at SARDI, specialising in grapevine diseases.

Christopher Reynolds

Phone: (08) 83449372
Email: chris.reynolds@flinders.edu.au

Before retirement, Chris taught environmental law at Flinders Law School from 1995–2007, and at UniSA in 2008.

Carmel Schmidt

South Australian Research and Development Institute
Sustainable Systems, Waite Research Institute
Hartley Grove
Urrbrae SA 5064
Phone: (08) 8303 9676, 0422 008 584
Email: schmidt.carmel@saugov.sa.gov.au

Carmel works at SARDI, identifying and examining the economic opportunities for a range of natural resources including wastewater, agricultural carbon and native vegetation.

Peter (Pedro) Schultz BSc (Hons)

Adelaide and Mount Lofty Ranges
Natural Resources Management Board
205 Greenhill Road
Eastwood SA 5063
Phone: (08) 8273 9102
Email: peter.schultz@adelaide.nrm.sa.gov.au

Peter is a Senior Freshwater Biologist with the Adelaide and Mount Lofty Ranges NRM Board, managing projects on monitoring water quality, biodiversity, data analysis, aquatic weeds, environmental flows, and low-flow bypasses.

Michael Sierp

Fisheries Division
Primary Industries and Resources South Australia
Level 14, 25 Grenfell Street
Adelaide SA 5001
Phone: (08) 8226 2889
Email: sierp.michael@saugov.sa.gov.au

Michael is Manager of Marine Biosecurity Policy at PIRSA.

Martha Sinclair

Department of Epidemiology and Preventive Medicine
School of Public Health and Preventive Medicine
Monash University
c/o The Alfred, Commercial Road
Melbourne VIC 3004
Phone: (03) 9903 0571
Email: martha.sinclair@med.monash.edu.au

Martha is from the Department of Epidemiology and Preventive Medicine at Monash University, and undertakes public health risk assessments for water supplies.

James Smith

fauNature
4 Teringie Drive
Teringie SA 5072
Phone: 0406 400 933
Email: james.smith@faunature.com.au

James runs fauNature, which brings people and nature together, by helping to attract the wildlife we would like to enjoy as part of our daily lives.

Kate Smith

Urban Biodiversity Unit
Department for Environment and Heritage
328 Shepherds Hill Road
Blackwood SA 5051
Phone: (08) 8278 0606
Email: smith.kate@saugov.sa.gov.au

Kate is involved in implementing on-ground restoration work, providing technical advice and community education programs for the Urban Biodiversity Unit.

Pamela Smith

Department of Archaeology
Flinders University
GPO Box 2110
Adelaide SA 5001
Phone: (08) 8278 8172
Email: smithric@tpg.com.au

Pamela Smith is a Senior Research Associate in the Department of Archaeology at Flinders University.

John Tibby

Geographical and Environmental Studies
North Terrace Campus
University of Adelaide
Adelaide SA 5000
Phone: (08) 8303 5146
Email: john.tibby@adelaide.edu.au

John is a Senior Lecturer in the Geographical and Environmental Studies Unit at the University of Adelaide.

Steve Walker

KESAB/Adelaide and Mount Lofty Ranges Natural Resources
Management Board
214 Grange Road
Flinders Park SA 5025
Phone: (08) 8234 7255
Email: swalker@kesab.asn.au

Steve is a NRM Education Officer, supporting education for sustainability via the Australian Sustainable Schools Initiative SA, Waterwatch SA and Weed Warriors SA.

Craig Williams

School of Pharmacy and Medical Sciences
City East Campus
University of South Australia
Adelaide SA 5000
Phone: (08) 8302 1906
Email: craig.williams@unisa.edu.au

Craig is a Lecturer in Biological Science and the Head of Mosquito Research at UniSA.

Zhifang Wu

School of Commerce
City West Campus
University of South Australia
Adelaide SA 5001
Phone: (08) 8302 0884
Email: zhifang.wu@unisa.edu.au

Zhifang is currently undertaking a PhD, which consists of an evaluation of a natural resources management policy in South Australia.

Mike Young

School of Earth and Environmental Sciences
North Terrace Campus
University of Adelaide
Adelaide SA 5005
Phone: (08) 8303 5279
Email: mike.young@adelaide.edu.au

Mike holds a Research Chair in Water Economics and Management and specialises in the design and development of market-based approaches to water management.

Atun Zawadzki

Australian Nuclear Science and Technology Organisation (ANSTO)
New Illawarra Road
Lucas Heights NSW 2234
Phone: (02) 9717 3480
Email: atun.zawadzki@ansto.gov.au

Atun is Manager of the ANSTO Low Level Radiochemistry Laboratory, specialising in the measurement of radioactivity in environmental samples.

Introductory map

David Bruce

School of Natural and Built Environments
City East Campus
University of South Australia
Adelaide SA 5000
Phone: (08) 8302 1856
Email: david.bruce@unisa.edu.au

David Bruce is Associate Head of Academic Programs in the Division of Information Technology, Engineering and the Environment at UniSA.

Cartoonists

Michael Atchison

(deceased)
c/o the *Advertiser*
31 Waymouth Street
Adelaide SA 5000

Michael was the Political Cartoonist for the *Advertiser* for 40 years.

Ross Bateup

Ross Bateup Pty Ltd Urban Design, Architecture
Chesser Street
Adelaide SA 5000
Phone: (08) 8339 2360, 0400 284 479
Email: bateupcartoons@bigpond.com

Ross Bateup is a Political Cartoonist for the Adelaide *Advertiser* and other Australian newspapers, and is also a freelance Political Cartoonist syndicating cartoons and caricatures through the *New York Times* and CWS Syndicates worldwide to international papers.

Abbreviations used throughout the book

Abbreviation	Word
ha	ha
hPa	hectopascal
g	gram
GL	gigalitre
ka	Thousand years ago
kg	kilogram
kL	kilolitre
km	kilometre
kt	kilotonne
L	litre
m	metre
Ma	Million years
mb	millibar
mg	milligram
mL	millilitre
ML	megalitre
t	tonne

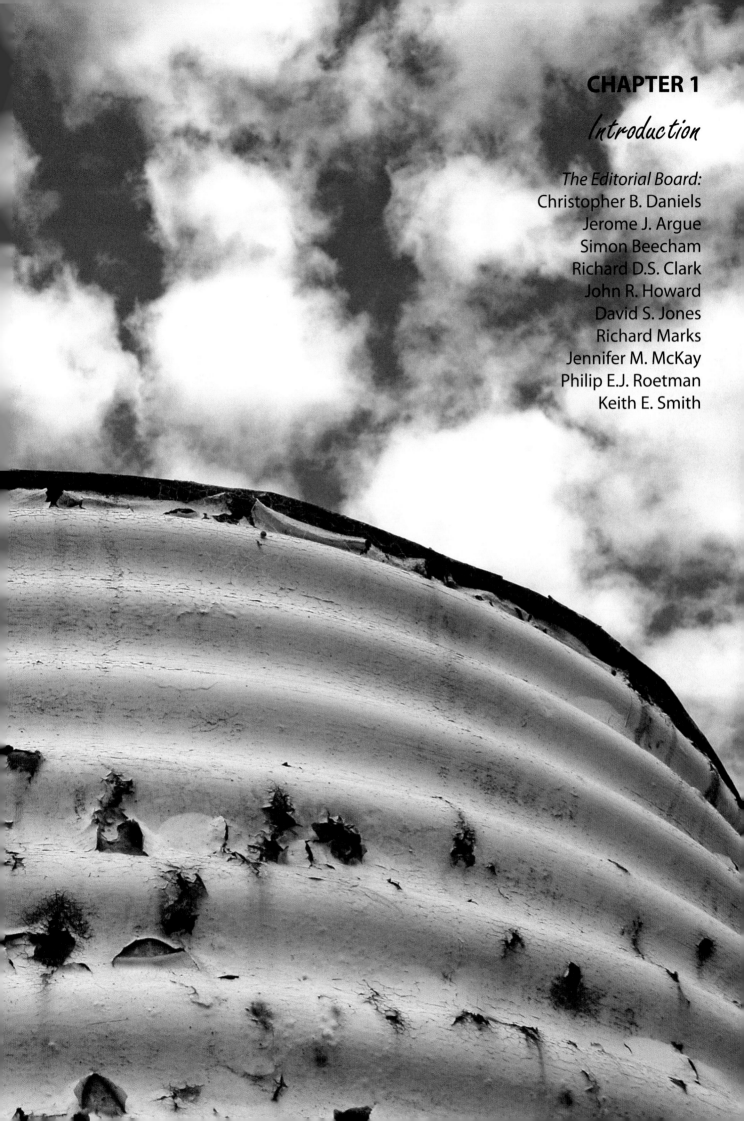

CHAPTER 1

Introduction

The Editorial Board:
Christopher B. Daniels
Jerome J. Argue
Simon Beecham
Richard D.S. Clark
John R. Howard
David S. Jones
Richard Marks
Jennifer M. McKay
Philip E.J. Roetman
Keith E. Smith

The Editorial Board

Christopher B. Daniels, Jerome J. Argue, Simon Beecham, Richard D.S. Clark, John R. Howard, David S. Jones, Richard Marks, Jennifer M. McKay, Philip E.J. Roetman and Keith E. Smith

It is raining outside today. It is late November and we are getting one of those infrequent but heavy showers that define spring in Adelaide. Unfortunately, the rain today is not enough. As with the last few years, our winter rains were less than average and we face a long, hot summer and the prospect of a longer, drier autumn. While Adelaide is in a dry year, and has been for the last few years, we are by no means alone as the drought extends over the eastern states of Australia. And that is our problem.

Adelaide, and in fact the state of South Australia, relies heavily on the Murray River system for its water. On average, 52% of all the water used in the state comes from the Murray; Adelaide receives on average 42% of its water from the Murray but our intake can be as high as 90% in drought years (see tables 1.1 and 1.2). The devastating drought in eastern Australia has massively decreased flows in the Murray-Darling System and revealed how reliant Adelaide has become on this source of water, and therefore how precariously we are perched, located as we are at the mouth of a drying system. Until recently, water has been an available and apparently easily renewable resource, but now, suddenly, the water shortage is the primary environmental issue in this era of significant changes in climate. Most climate change scenarios predict that annual temperatures for the Adelaide region will increase by 0.4–1.2°C by 2030 (with an increase in the number of hot days over 35°C and the annual number of hot spells over 35°C), with a reduction in annual rainfall of 1–10%, coupled with increases in extreme rainfall events. Potential impacts to riparian systems (medium impact), surfacewater resources (medium–high) and groundwater (medium–high) indicate that riparian systems have a limited capacity to adapt to these changes and are very vulnerable.[1,2]

Despite the drought, however, floods will still be with us, even if occurring less frequently. Indeed, even in the midst of this general drought, we have experienced several flood events of unusual intensity. Even if things magically returned to 'normal' for a short time, the unpredictable recurrence of droughts and floods in southern Australia, and changes in the frequency and intensity of these events, should warn us that we need to continually revise our water plans to ensure they keep up with the changing climate, with the expectation of future change.

Yet, we should not be in this situation. Enough rain falls to suit the needs of the approximately 1.2 million people living in the Adelaide and Mount Lofty Ranges region. Further, as Clark and Argue demonstrate in chapter 21, our catchments could potentially support a population of 2 million people in Adelaide, as targeted by the state government, without requiring water from the River Murray. However, Planning SA's (renamed the Department of Planning and Local Government in October 2008) recent review for growth to 2036 – *Directions for creating a new plan for Greater Adelaide* (2008)[3] with its prediction of up to 537,000 new residents (potentially requiring 247,000 additional dwellings) – does not address water sustainability. Water sustainability is over-ridden in this document, which is driven by conventional theories for the planning of growth corridors to accommodate traditional forms of dwelling and their spatial locations. This plan compromises the inherent productive landscape, its natural resource qualities, water catchment and harvesting, and the scenic landscape imperatives of the peri-urban regions surrounding the Adelaide metropolitan area.

So how did we get ourselves into this mess, and how will we get out of it? This book describes the nature of the Adelaide Mount Lofty Catchment and its climate – to better understand the environment we inhabit. It takes multiple perspectives in viewing the range of issues of water management past and present. At the broadest level, this is a book about plans and planning for a 'brave new world', in which we, as individuals, communities and nations, will have embraced sustainability.

Table 1.1 Water use in an average year

| | Supplies | | Uses | |
Source	Amount supplied (GL)		Rural consumption (GL)	Adelaide consumption (GL)
River Murray	88		8	80
Adelaide Hills catchments	137		16	121
Rural groundwater	53		53	0
Metro groundwater	9		0	9
Metro rainwater tanks	1		0	1
Stormwater, recycled water	22		17	5
Total	**310**		**94**	**216**

Table 1.1 Sources of water and amount consumed (in GL) by the city of Adelaide compared to rural consumption during a year with average rainfall. Source: adapted from Government of South Australia, 'Water proofing Adelaide: a thirst for change 2005–2025', 2005.

Table 1.2 Water use in a dry year

Supplies		Uses	
Source	Amount supplied (GL)	Rural consumption (GL)	Adelaide consumption (GL)
River Murray	179	8	171
Adelaide Hills catchments	43	13	30
Rural groundwater	72	72	0
Metro groundwater	9	0	9
Metro rainwater tanks	1	0	1
Stormwater, recycled water	22	17	5
Total	**326**	**110**	**216**

Table 1.2 Sources of water and amount consumed (in GL) by the city of Adelaide compared to rural consumption during a year with below average rainfall. Source: adapted from Government of South Australia, 'Water proofing Adelaide', 2005.

At the local level, this book concentrates on water plans and planning for Adelaide, and therefore only addresses this one aspect of sustainability, albeit an important and vital component. The specific challenge of the Editorial Board was to reach a consensus definition of 'sustainability', as far as water within the Adelaide area is concerned.

The most general definition of sustainability is a system that:

Meets the needs of the present without compromising the ability of future generations to meet their own needs (Brundtland 1987).[4]

After much discussion, the board determined that the best identification of a sustainable system (that applied to urban water), can be gained by first reviewing the desired outcomes for that system. These outcomes could then be analysed to determine whether they suggest a preferred management model. Listing the desired outcomes is relatively simple and non-controversial, using the general principle of 'water sustainability'.

Under future scenarios of most likely population growth, climate change, technological advances and social adaptations, the urban water system should continue to provide, or be readily adapted to provide, all of the following outcomes into the foreseeable future:

1. A reliable, affordable, equitable and healthy supply of water that supports the social and physical environment.
2. Efficient management of excess water and wastewater to avoid disease, inconvenience and harm to people, biota, and the built environment.
3. Minimisation of damages from floods (or sea storms) up to a risk level acceptable to the community, whilst also building into the system the ability to survive more infrequent but 'catastrophic' storm events.
4. Avoidance of damage to ecosystems caused by the excessive diversion of water from them, or to them, by the amount or quality of the water.

5. Minimum contributions to greenhouse gases involved with the construction and operation of water supply and management systems.
6. Water and wastewater systems that provide a maximum net benefit and consider full lifecycle costs.
7. Management of water and wastewater systems that encourage innovative responses to local conditions.
8. A community that embraces and contributes to water management in practice and in decision-making.
9. Sufficient access to water for recreation, amenity and aesthetic satisfaction.
10. Learning from our historical mistakes and policies to provide a more secure and sustainable urban habitat.

The first issue the board faced was to determine water usage. Calculating a city's water usage is very difficult because water usage and the fractional extractions from different sources can vary greatly from year to year. Adelaide's usage varies between wet, dry or average years. Moreover, the extent to which agriculture and industry are included, and where the boundaries for the city lie, influences the numbers. Throughout this book, different authors will use different numbers to support their various arguments, as they present specific discussions on specific situations. We have adopted the numbers presented in the 'Water Proofing Adelaide' (2005)[5] document (tables 1.1 and 1.2). We also use the SA Water figures[6] for water usage by different components of our community, and home usage from Planning SA's 'Directions for creating a new plan for Greater Adelaide' (2008)[3], which notes that present residential water use comprises gardens (40%), bath (20%), laundry (16%), kitchen (11%), toilet (11%), and other domestic use (2%), and that average water consumption tends to remain the same as housing density increases from 10 to 14 dwellings per hectare.

Secondly, what is Adelaide? In terms of its water and for the purposes of this study, Adelaide is defined as the region delineated by the urban growth boundary (although 'Directions for creating a new plan' allows for significant

BOX 1

The watercycle

Earth is often called the 'blue planet'. Indeed, viewed from space, earth is primarily blue – of water in oceans and lakes which cover around 71% of its surface. Water is also evident as white clouds in the atmosphere and as the solid ice and snow that cover the poles and mountaintops of the planet. Earth is a wet planet and the volume of water is virtually constant at an estimated 1.4 billion km^3. However, this water is not distributed evenly, nor is its distribution static. The movement of water around the planet is driven mostly by solar energy, known as the 'watercycle' (or the hydrologic cycle).

Water is stored and moves through the oceans, the atmosphere, over land and underground. Around 97% of the water on earth is held within the saline seas and oceans The heating of these waterbodies by the sun causes evaporation, where water is vaporised and moves into the atmosphere.

Additionally, around 10% of atmospheric water vapour is gained through evapotranspiration over land – a combination of the evaporation of water from the soil and the transpiration of water from plants In cooler climates, a comparatively small volume of atmospheric water vapour is also gained directly from ice and snow through subtion (a change of state that excludes the liquid phase). Water held in the atmosphere accounts for around 0.001% of the water on earth. As this water vapour cools, it condenses and forms clouds. Water vapour and clouds are transported through the atmosphere by wind.

Water returns to the earth's surface as precipitation when clouds cool or become saturated. There are numerous forms of precipitation, including snowfall and rainfall. In cold locations, with regular snowfalls, icecaps and glaciers can form, which can store around 1.8% of the water on earth (about 69% of the global freshwater resource). Around Adelaide, snowfall occurs only occasionally in the upper parts of the Mount Lofty Ranges. More commonly, cloud is transported from the west and rainfall occurs as it lifts to pass the ranges ('orographic lift'). This water has often been transported from either the Southern or Indian Ocean.

When it rains, the water either infiltrates into the ground or flows as surface runoff. Infiltration replenishes groundwater aquifers. Globally, these bodies hold 1–2% of the planet's total volume of water. Much of this water is saline, but the freshwater fraction accounts for around 30% of the global freshwater resource. Water can percolate deep underground and be stored for many thousands of years. However, shallow groundwater flows interact with surfacewaters, and discharge into streams and rivers, lakes and oceans, and as natural springs.

Surface runoff flows into waterways such as streams and rivers, and wetlands such as lakes, marshes and swamps. This surface water accounts for only 0.02% of the water on earth, and less than 1% of the freshwater. In places with cool climates, streamflow is often fed by melting ice and snow during the warmer months of the year. Adelaide's reservoirs are replenished by native streamflows through the Mount Lofty Ranges (with additional waters pumped from the River Murray). Most of the streamflow on the Adelaide

Plains, including urban stormwater, is discharged into Gulf St Vincent.

The discharge of water into the gulf continues the cycling of water around the planet; however, it is not indicative of how water cycled through the Adelaide region prior to European settlement when the River Torrens drained into a marsh area before being deposited into the coastal dunes known as the Reedbeds. Water from the Reedbeds would have evaporated, seeped through the dunes to the gulf or into subterranean aquifers, or, at times, floodwaters would be discharged to the gulf via waterways to the north and south. We cannot re-establish the Reedbeds, but, with new aquifer management and water sensitive urban design practices we can mimic the groundwater recharge and reuse the freshwater rather than let it flow out to sea. For, although we live on the blue planet, readily available supplies of freshwater are limited, and we are remiss to waste them.

Philip Roetman

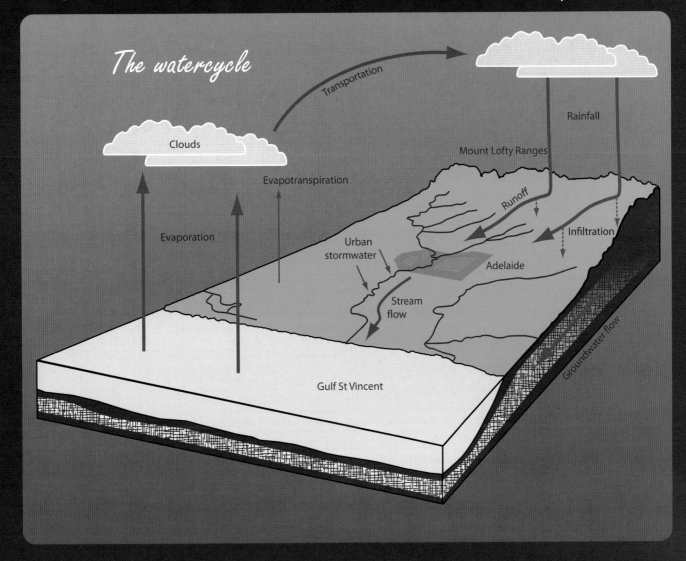

The watercycle

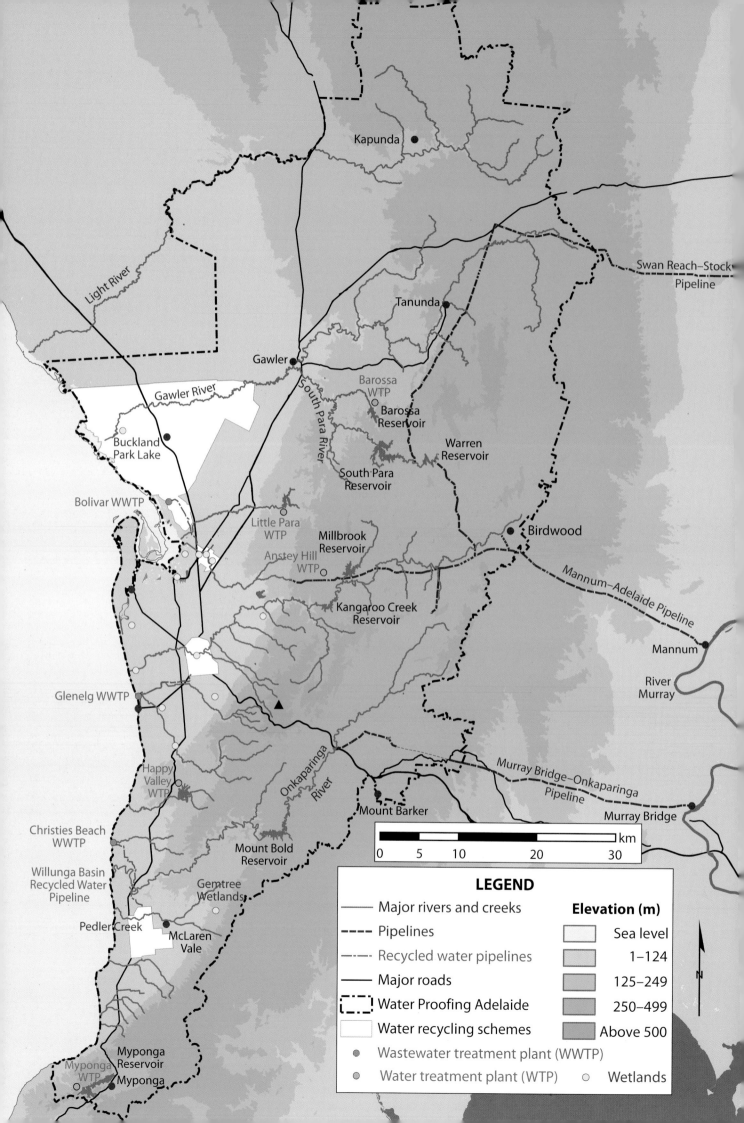

Kapunda

Swan Reach–Stock
Pipeline

Light River

Tanunda

Gawler

Barossa
WTP

Barossa
Reservoir

Gawler River

South Para River

Warren
Reservoir

Buckland
Park Lake

South Para
Reservoir

Bolivar WWTP

Little Para
WTP

Millbrook
Reservoir

Birdwood

Anstey Hill
WTP

Mannum–Adelaide Pipeline

Kangaroo Creek
Reservoir

Mannum

Glenelg WWTP

River
Murray

Happy
Valley
WTP

Onkaparinga River

Murray Bridge–Onkaparinga
Pipeline

Mount Barker

Christies Beach
WWTP

Murray Bridge

Willunga Basin
Recycled Water
Pipeline

Mount Bold
Reservoir

km

0   5   10   20   30

Gemtree
Wetlands

Pedler Creek

McLaren
Vale

LEGEND

Major rivers and creeks

Pipelines

Recycled water pipelines

Major roads

Water Proofing Adelaide

Water recycling schemes

Wastewater treatment plant (WWTP)

Water treatment plant (WTP)

Elevation (m)

Sea level

1–124

125–249

250–499

Above 500

Wetlands

Myponga
WTP

Myponga
Reservoir

Myponga

N

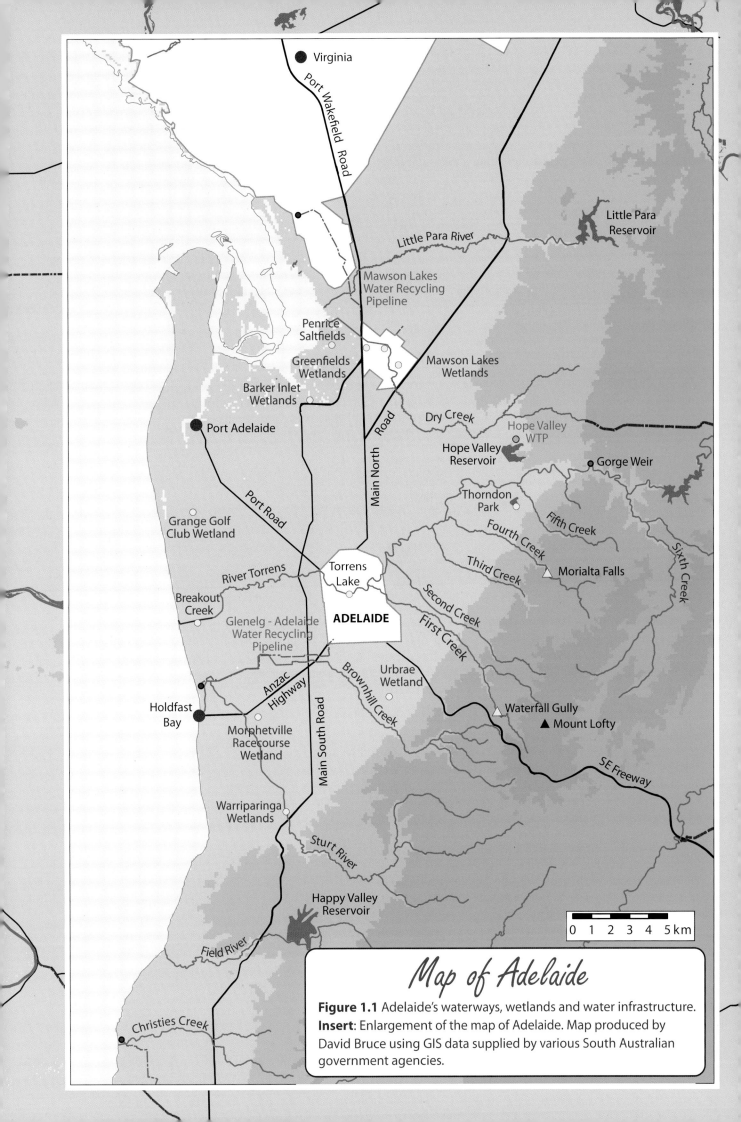

Virginia

Port Wakefield Road

Little Para River

Little Para Reservoir

Mawson Lakes Water Recycling Pipeline

Penrice Saltfields

Greenfields Wetlands

Mawson Lakes Wetlands

Barker Inlet Wetlands

Main North Road

Dry Creek

Hope Valley WTP

Port Adelaide

Hope Valley Reservoir

Gorge Weir

Port Road

Thorndon Park

Grange Golf Club Wetland

Fifth Creek

Fourth Creek

Third Creek

Morialta Falls

River Torrens

Torrens Lake

Second Creek

Sixth Creek

Breakout Creek

First Creek

ADELAIDE

Glenelg - Adelaide Water Recycling Pipeline

Anzac Highway

Brownhill Creek

Urbrae Wetland

Holdfast Bay

Waterfall Gully

Mount Lofty

Morphetville Racecourse Wetland

Main South Road

Warriparinga Wetlands

SE Freeway

Sturt River

Happy Valley Reservoir

Field River

0 1 2 3 4 5 km

Christies Creek

Map of Adelaide

Figure 1.1 Adelaide's waterways, wetlands and water infrastructure. **Insert**: Enlargement of the map of Adelaide. Map produced by David Bruce using GIS data supplied by various South Australian government agencies.

growth outside this boundary). Metropolitan/urban Adelaide therefore extends from the Gawler River in the north to Sellicks Beach in the south, bordered to the east by the Hills Face Zone of the Mount Lofty Ranges and to the west by Gulf St Vincent (Figure 1.1). As most of its catchments lie on or just outside this boundary, and those catchments also serve significant agricultural areas in the Mount Lofty Ranges, the Mount Lofty Ranges Catchment regions are included in the organisational discussions where appropriate. Figure 1.1 provides all the major water-related structures in the region covered by this book.

Using the criteria described on p. 36, we concluded, as a 'rule of thumb', that the Adelaide metropolitan area consumes just over 200 GL of water a year. The adjacent Mount Lofty Ranges agricultural area consumes another 100 GL (tables 1.1 and 1.2). In an average year, Adelaide uses about 80 GL of Murray River water, increasing to approximately 171 GL in drought years. If the dire situation arises where we can no long rely on water from the Murray, we must either find this water from elsewhere or cut our usage accordingly. We must also manage our water according to the principles of sustainable water systems outlined above. This is a tough ask but not an insurmountable one.

The aim of this book is to examine in detail the watercycle of Adelaide from a wide variety of perspectives. The definition of 'use' includes the sharing of the water with the supporting environments. In coming to grips with sustainability, it will become apparent through this book that it is the design of the city itself that is the main focus. In a nutshell, our core water challenge becomes our ability to progressively change our city to become more water sensitive, particularly as the city grows and the climate changes in response to past unsustainable practices.

The first chapters (in Part One) describe the geo- and biophysical environment of Adelaide. Part Two identifies the relationships between the form of the city and the movement of water into, through, and out of it, examining the development of water supply, sewerage and drainage systems. The manner and extent to which the present systems fail to meet many of the criteria listed on p. 37 (particularly the damage to ecosystems) will become apparent, as will the changes that will be required for systems to meet all the listed criteria in a comprehensive and integrated manner. To understand the complex relationships between the history, engineering, social, aesthetic, structural, economic, environmental, commercial, and political requirements for water, the Editorial Board asked the authors to take a personal approach. Therefore, in Part Two, the flow of information is not linear in that one chapter does not logically lead to the next; rather, when reading this section, imagine yourself in the centre of a circle with the chapter authors arranged around you. You listen to one, then move to the next. Each chapter is independent, each tells you its view. No chapter is worth more than another, and each is equally valid.

Please note that many places are mentioned in this book that have Aboriginal place names as well as names ascribed since European settlement. In some instances both names are included, but, in order to remain concise, in most cases authors have used only the post-settlement name. The Editorial Board recognises the need to celebrate and respect the knowledge, beliefs and heritage of both Indigenous and non-Indigenous inhabitants of our state. Our undertaking here is to include a representative view of the water systems in the area around the City of Adelaide; this book provides a detailed discussion of Kaurna place names and heritage values related to water in chapter 6.

Our aim is to provide the broad community with the information required to understand the issues around water in Adelaide. This book is not a policy document, but will inform policy makers by providing the information required to manage our water sustainably, now and into the future.

References

1. Adelaide and Mount Lofty Ranges Natural Resources Management Board, 'Climate change in the Adelaide and Mount Lofty Ranges region', Adelaide, 2007 (online).
2. D. Bardsley, 'There's a change on the way: an initial integrated assessment of projected climate change impacts and adaptation options for natural resource management in the Adelaide and Mount Lofty Ranges region', Department of Water, Land and Biodiversity Conservation report, Adelaide, 2006 (online).
3. Planning SA, 'Directions for creating a new plan for Greater Adelaide: better planning, better future', Adelaide, 2008 (online).
4. G.H. Brundtland, *Our common future*, Oxford University Press, Oxford, 1897.
5. Government of South Australia, 'Water proofing Adelaide: a thirst for change 2005–2025', 2005 (online).
6. SA Water, 'About us', homepage.

Box references

Box 2: The four colours of freshwater

1. V. Finlay, *Colour*, Sceptre, London, 2002.
2. M. Falkenmark and J. Rockström, *Balancing water for humans and nature: the new approach in ecohydrology*, Earthscan Publications Ltd, London, 2004.
3. J.C. Radcliffe, 'Water recycling in Australia', review report, Australian Academy of Technological Sciences and Engineering, Victoria, 2004.
4. A.C. Hurlimann and J.M. McKay, 'What attributes of recycled water make it fit for residential purposes?: the Mawson Lakes experience', *Desalination*, 187, 2006, pp. 167–177.

BOX 2

The four colours of freshwater

There are a number of popular terms to describe the quantity and quality of water. In terms of quantity, 1 kL is 1000 L or 1 m^3 of water. One kilolitre weighs 1 t. The average household uses 250 ML/year. One megalitre is 1000 kL or 1 million L, or the volume (in rough terms) of most 50-m swimming pools. A gigalitre is 1000 ML or 1 billion L and represents the volume of 1 km^2 by 1 m deep. Hope Valley Reservoir holds 2.8 GL and Happy Valley Reservoir holds 11 GL.

Commonsense tells us that freshwater is not colourless in its raw form. Many of us have examined water in a clear glass. It can be full of sediment, like water from the Yellow River, or it can be clear. Some days water is clearer than other days in Adelaide. For water professionals, four colours of water have specific quality meanings, and often also have cultural meanings. Water is described in the media and by water professionals as green, blue, grey and black.

Greenwater is water in the soil from rainfall or groundwater, used directly to feed the ecology of national parks, domestic gardens and all biomass. Greenwater is used indirectly for economic production in rain-fed agriculture, and for plants and wetlands.

Cultural interpretation: positive connotations as green is the colour of nature.[1,2] Replaced the term 'soil water' in 2000. The term has been used to name towns in the United States and Canada.

Bluewater

Bluewater is the water in rivers, streams, lakes, aquifers, and in modern channels, used for irrigation of agricultural crops and pasture for food and biofuel production. Bluewater directly supports municipal domestic supplies (which are stored in dams such and Kangaroo Creek and the Murray River channel). Bluewater is treated to a high standard for municipal use (then it is called 'potable water').

Raw, untreated water is used for most irrigation, and indirectly supports ecosystem functions. These days farmers are given an allocation of bluewater; because of the drought, most using River Murray water have been given less than 30% of the amount they have had in the

past. In other regions where groundwater has been used, the reduction in allocations has not been so severe.

Cultural interpretation: in Adelaide, bluewater is likely to have a positive connotation as blue has positive associations in most cultures.[1] However, bluewater is also used to describe the outputs of copper corrosion in pipes which can leave blue stains and lead to a metallic taste in water.

Greywater

Greywater is water from showers, bathrooms and washing machines in homes, and water from factories. Since 2006 this water has been used by some residents of Adelaide to water their gardens.

Cultural interpretation: positive – there seems to be a great willingness of residents in Adelaide to bucket this water on to their gardens.

Blackwater

Blackwater is sewerage water which is always treated to a high standard. There are many wastewater treatment plants in Adelaide, the biggest being Bolivar in the north. The treated sewage is mixed with stormwater and other waters and used for agriculture in the Virginia Triangle and McLaren Vale. South Australia has the highest reuse of blackwater in Australia.[3] The treated

sewage can be 'shandied', where it also mixed with stormwater and other waters and filtered through wetlands. Treated wastewater is mixed with stormwater and reused at Mawson Lakes for flushing toilets and outside watering.

Cultural interpretation: likely to be negative; much research locally[4] has shown that Adelaide residents have an aversion to drinking this water even if treated. However, many cities around the world now drink treated wastewater or shandied water.

Jennifer McKay

BOX
3

Environmentally sustainable development (ESD)

Definitions

The over-allocation of water has been recognised and the concept of environmentally sustainable development (ESD) has been introduced into water management laws since 1992, and into to Natural Resources Management (NRM) laws generally since late 1990s in each state and at federal level. The definition of ESD is 'using, conserving and enhancing the community's resources so that ecological processes, on which life depends, are maintained, and the total quality of life, now and in the future, can be increased'.[1] Many states had introduced water law reforms with the objective of ESD, and these were given impetus by the first Council of Australian Governments (CoAG)[2] reforms in 1994. The National Water Initiative (NWI) followed these in 2004. The first set of reforms required:

- corporatisation of water utilities;
- separation of the functions of economic and/or environmental regulators from the supply of water;
- full cost recovery;
- ESD; and
- plans to address over-allocation of river and groundwater systems.

The NWI reforms also aim to increase productivity and efficiency of Australia's water use, to service rural and urban communities, and to ensure the health of river and groundwater systems. In particular, the NWI has 80 key actions all aimed at ESD. Under broad headings these are:

- A common lexicon for water words.
- Water access entitlements and planning framework.
- Risk assignment.
- Indigenous access.
- Water markets and trading.
- Intra- and inter-state trade compatible registers and institutional and regulatory arrangements.
- Best practice in water pricing and institutional arrangements.
- Integrated management of water for environmental and other public benefits.
- Water resource accounting.
- Urban water use reform through demand management and water-sensitive design.
- Community partnerships and adjustments.
- Knowledge and capacity building.

The reforms bring to the fore the limitations of power over water in the Constitution of Australia, and the extra territorial limitations in state constitutions. These limitations have engendered novel methods of federal/state relations and power sharing. The quest to implement ESD in water management and use has generated a mosaic of innovative laws, regulations, and institutions, mainly at the state level, because of Section 100. (In the ACT the National Capital Authority manages water). Each state has recently created regional NRM bodies (there are 56) with different legal structures and reporting relationships, and using different definitions of ESD, surfacewater, groundwater and just about everything else. These NRM boards are often charged with drafting water allocation plans (as occurs in South Australia). While they all aim to achieve ESD in water, the NRM definitional differences are a real impediment to inter-state learning and national consistency. The lack of consistency may lead to a water regulation forum, and shopping by water-users to select the least intrusive water regulation systems.

Jennifer McKay

Sustainable development timeline

BOX 4

1950s
• Conceptualisation of the world as a singular and interdependent entity, brought about by the first photographs of earth taken from space.

1962
• *Silent Spring* by Rachel Carson was released, drawing attention to the effects of pesticides and pollution on the natural environment. This book is often cited as a catalyst in the development of the environmental movement.

1968
• *Population Bomb* by Paul Ehrlich was released and drew attention to the dangers of human overpopulation.
• the Club of Rome organisation was established to investigate the dynamic relationship between human societies and the natural environment.

1972
• United Nations' first Conference on the Human Environment in Stockholm.
• *Limits to Growth* report published by the Club of Rome, highlighting the potential negative impacts of human population growth and technological advancement.

1983
• World Commission on Environment and Development (WCED) was formed to address the social and economic consequences of the continuing degradation of the natural environment.

1987
• WCED published *Our Common Future*, which launched the idea of sustainable development, characterised as development that 'meets the needs of the present without compromising the ability of future generations to meet their own needs'. The report is commonly known as the 'Brundtland Report' after the chair of the WCED.

1988
• Intergovernmental Panel on Climate Change (IPCC) was established to assess how human societies would be affected by anthropocentric climate change.

1989
• Australian prime minister Bob Hawke released 'Our Country our Future', which delivered a range of measures for dealing with environmental problems.

1992
• UN Earth Summit was held in Rio De Janeiro. The idea of sustainable development was firmly established as a global policy objective.
• National Strategy for Ecologically Sustainable Development was constructed.
• Intergovernmental agreement on the National Greenhouse Strategy.

1997
• The Kyoto Protocol was established as an international agreement that set targets to reduce greenhouse gas emmissions.

2002
• UN Johannesburg summit attempted to further clarify the concept of sustainable development.

2007
• Australia ratified the Kyoto Protocol.
• The UN climate change conference in Bali continued negotiations in dealing with climate change.

Jennifer McKay

BOX 5

Adelaide's water supply timeline

1836 • European settlement of South Australia.

1839 • River Torrens is the principal source of water for Adelaide, used for watering stock, bathing, disposing of rubbish and sewerage.

• Dysentery epidemic grips Adelaide – five children die in one day.

1848 • 36 water-carters operate in the city of Adelaide, earning an average of £3 per week.

1850 • Adelaide's population reaches 11,000; residents pay £8484 annually for impure water supplied by water-carters.

1856 • The Waterworks and Drainage Commission appointed.

1857 • Work commences on Thorndon Park Reservoir.

1860 • Adelaide receives its first supply of reticulated water when the valve house at Kent Town is turned on.

1862 • Thorndon Park Reservoir commissioned.

1869 • Work commences on Hope Valley Reservoir.

1872 • Hope Valley Reservoir commissioned.

1878 • establishment of Hydraulic Engineer's Department.

1881 • Deep Drainage System commissioned – Adelaide is the first city in Australia to enjoy the benefits of a water-borne sewerage system.

1883 • North Adelaide is connected to the sewerage system.

1888 • Hindmarsh, Thebarton and St Peters are connected to the sewerage system and the work of connecting Kensington and Norwood is well underway.

1892 • Work commences on Happy Valley Reservoir.

1897 • Happy Valley Reservoir commissioned.

1899 • Work commences on the Barossa Reservoir.

1902 • Barossa Reservoir commissioned.

1903 • Work commences on sewers for Glenelg.

1907 • Glenelg treatment plant commissioned.

1910 • Work commences on sewers for Port Adelaide.

1914 • Work commences on Millbrook Reservoir.

• Work commences on Warren Reservoir.

1914 to 1915 • Severe drought hits South Australia – the Murray River is reduced to a series of waterholes. River Murray Commission decides to regulate the river flow through construction of a series of locks, weirs and barrages.

1918 • Millbrook Reservoir commissioned.

1919 • Spanish influenza epidemic sweeps Australia. In the first half of 1919 normal life comes to a standstill. Desperate governments try to stem the spread of the disease and 12,000 Australian lives are lost.

1920 • Work commences on the construction of locks and weirs along the River Murray.

1929 • Engineering and Water Supply Department (E&WS) established.

• Sewerage Act empowers E&WS to construct and operate sewerage systems.

• A committee forms in South Australia to discuss bacteriological examinations of various water supplies.

1930 • Construction on locks and weirs along the Murray River completed.

• Adelaide's population reaches 315,000.

1931 • River Murray Commission recommends that barrages be built on the channels leading from Lake Alexandrina to the Murray Mouth at the Coorong.

• Government appoints an Advisory Committee on Water Supplies Examination to guide departmental actions on improving the protection of catchments and quality of water supplied to the public. The committee operates for two decades.

1932 • Waterworks Act introduced.

Work commences on Mount Bold Reservoir.

1933 • Glenelg sewerage treatment works commissioned.

• State Water Laboratory established by the E&WS.

1935 • Port Adelaide treatment plant commissioned.

• Work commences on the construction of barrages.

1938 • Mount Bold Reservoir commissioned.

• Most of Adelaide's suburbs are sewered.

1940 • Barrages on the River Murray completed.

1947 • 1000 miles of sewer have been laid in the metropolitan area – double the sewerage of 20 years previously.

1949 • Work commences on South Para Reservoir.

• Work begins on the construction of the Mannum–Adelaide Pipeline.

1950s • Chlorination stations installed at all metropolitan reservoirs.

1955 • Mannum–Adelaide Pipeline commissioned.

1956 • Work begins on the duplication of the Glenelg sewerage treatment works.

• Salisbury sewerage treatment works – constructed by the E&WS Department during World War II – is acquired from the Commonwealth.

1958 • South Para Reservoir commissioned.

1961 • Work commences on Bolivar treatment works. Caters for the sewerage and trade wastes of a population of 600,000. Can be expanded if necessary.

1962 • Doubling the size of the Glenelg sewerage works completed.
• Myponga Reservoir commissioned.

1965 • Partial operation of Bolivar treatment works begins with the diversion of sewage from Salisbury and Elizabeth. Additional stage of the scheme also underway.
• Adelaide is nearly 100% sewered; no other Australian city has more than 75% of its population served by sewerage.

1966 • Bolivar treatment works officially opened.
• Work commences on Kangaroo Creek Reservoir.

1968 • Work begins on the construction of the Murray Bridge–Onkaparinga Pipeline.
• The government announces its decision to fluoridate South Australia's water supplies.

1971 • Kangaroo Creek Reservoir commissioned. Fluoridation of metropolitan Adelaide water

1969 • supplies begins.

1973 • Murray Bridge–Onkaparinga Pipeline commissioned.

1974 • Work commences on the Hope Valley water treatment plant (WTP).
• Work commences on Little Para Reservoir.

1976 • Innovative Water Resources Act enacted. The Act enables water resources to be conserved, developed and managed for the benefit of the people of South Australia.

1977 • Hope Valley WTP commissioned.
• Work on Little Para Reservoir completed.

1978 • Six point River Murray Salinity Control Program is adopted. The main elements completed by 1984.

1979 • Little Para Reservoir commissioned.

1980 • Anstey Hill WTP commissioned.

1982 • Barossa WTP commissioned.

1984 • Little Para WTP commissioned.

1989 • Stage 1 Happy Valley WTP completed.

1991 • Happy Valley WTP commissioned.

1993 • Myponga WTP commissioned.

1995 • E&WS becomes SA Water.

1994 • First Council of Australian Governments (CoAG) reforms. Market will allocate water to improve efficiency: consumption based on 2-part tariffs (urban 1998, rural 2001); full cost recovery; separate identification and funding of community service obligations; trading in rural water entitlements;

allocation of water for the environment; broader social values embraced.
• State Water Laboratory and the Australian Centre for Water Quality Research are combined. Renamed the Australian Water Quality Centre to recognise the significant expertise in water quality and treatment research at the Bolivar facilities.

1997 • *South Australia Water Resources Act* s 61 Catchment Management Boards Section 61 provides that a board must create a catchment management plan to manage the water resources of a region and ecosystems dependent in close cooperation with the community, local government and state and federal government the State Water Plan 2000 addresses statewide resource issues and all regional plans must be consistent with it.

1999 • Virginia Pipeline Scheme commissioned. The first and largest reclaimed water scheme of its type in Australia.

2000 • CoAG governments has also imposed a regional model for the delivery of NAP and NHT funding of environmental activities. The principle driver for regional delivery was to 'harness the capacity of those closest to the problem on the ground, building on local knowledge, experience and expertise, and enabling flexible and responsive solutions to local NRM challenges'.
• There are 56 NRM regions in Australia and these have been agreed by state and Commonwealth.
• *South Australia Natural Resource Management Act 2004* creates 11 NRM regional bodies with wider remit than Catchment Management Boards.

2003 • Permanent water conservation measures introduced in Adelaide.

2005 • Mawson Lakes Recycled Water Scheme launched.
• Water Proofing Adelaide strategy published.

---

[1] The role of regional bodies as per the Commonwealth is expressed as 'undertaking regional natural resource management planning, prioritising regional level investments, co-coordinating actions at the landscape scale, getting community ownership in decision making and reporting on progress.'

Thorsten Mosisch

CHAPTER 2

The variable climate

Elizabeth Curran
Christopher Wright
Darren Ray

Introduction

The climate of a location is described by its temperature and rainfall characteristics, as well as other variables such as wind, humidity and evaporation – all of which have a significant impact on water resources. The story of climate, though, is incomplete without considering the causes or major influences that determine *local* climate. Of prime importance are the atmospheric and oceanic general circulations, and the regional factors that impact on these. In the southern Australian region these impacts include sea surface temperature patterns in the Indian, Southern and Pacific oceans, the Southern Annular Mode (SAM), which modulates the intensity and latitudinal extent of the westerly wind regime in mid and high latitudes, and the El Niño-Southern Oscillation (ENSO), which impacts on extreme climatic events in the Australian region. Other local factors such as proximity to the coast, topography, aspect and exposure, and even vegetation type and other ground characteristics, are also important in determining smaller-scale variations in climate and weather such as rainfall distribution, local winds and frost frequency.

People often ask how weather and climate are linked. Probably the simplest answer is to say that weather is what we experience day to day, and climate is the synthesis of the weather over a long period of time. In other words, climate describes the average or most common conditions, the extremes, and the frequency of specific weather events.

This chapter shows how the general atmospheric circulation drives the seasonal responses of Adelaide's climate and describes its key features, including its natural variability and extremes. It includes a description of the weather patterns that have a significant bearing on the seasonal characteristics. Recent climate trends and potential future climate changes due to human activities, especially those activities resulting in increasing atmospheric greenhouse gas emissions, are also discussed.

Note: Unless otherwise stated, the mean statistics used in the text are from the Bureau of Meteorology's official observing site at Kent Town, just to the east of the Adelaide city centre, where observations commenced in 1977. Extreme values may be from the Kent Town record or, if prior to 1977, from the West Terrace record, the official Adelaide city observing site until 1977, just west of the city centre. Rainfall records were commenced in the vicinity of this site in 1839 and standardised temperature records commenced in 1888. Observing sites are selected to meet standard criteria to ensure they are representative of the wider locale.

Chapter photograph: Adelaide, from Montefiore Hill.

Impact of the general circulation and weather patterns

The earth's weather and climate patterns are driven by solar heating of the earth's surface. Most of the sun's energy is intercepted in the equatorial and tropical regions of the globe, leading to a surplus of energy in the tropics and a deficit at the poles. The general circulation of both the atmosphere and ocean act to transfer excess energy from lower to higher latitudes.

Near the equator, heating of the lower atmosphere causes large-scale vertical motion and convective cloud formation. As the air rises it expands, and commences its journey towards the poles, transporting both heat and mass away from the tropics. This upper outflow results in a region of low pressure at mean sea level known as the Inter Tropical Convergence Zone (ITCZ). The south-easterly trade winds which prevail across northern Australia flow into this zone. As the upper air moves southwards in the Southern Hemisphere, it accelerates relative to the earth's surface, and turns westerly in the fast-moving subtropical jet stream. Of significance to Adelaide's climate, air in this region sinks, returning to the lower atmosphere as the gently subsiding air of the subtropical ridge, a belt of high atmospheric pressure girdling the globe. The ridge occurs at much the same latitudes as southern South Australia, around 30–40°S.

General circulation of the atmosphere

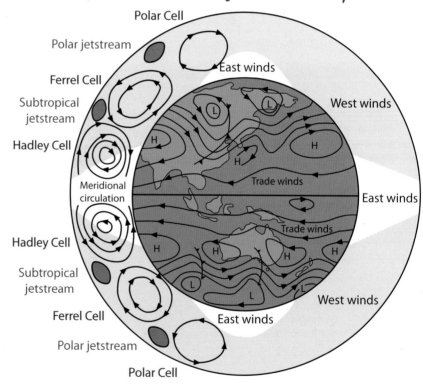

Figure 2.1 Main features of the general circulation of the atmosphere, showing the jet streams and cell structure of the meridional circulation transporting heat from the tropics to the poles. The Hadley Cell, closest to the equator, and the Ferrel Cell, in mid latitudes, circulate in the opposite sense so that both produce subsidence in the region of the subtropical ridge. Source: Bureau of Meteorology, 'The controls on Australian climate, chapter 1 introduction', 2009, copyright Commonwealth of Australia, reproduced by permission.

Mediterranean climates are characterised by dry, warm to hot summers and temperate winters, with a distinct rainfall maximum during the winter months. These are the general features of Adelaide's climate. Kent Town, with an average annual rainfall of 545 mm, in most years receives less than 70 mm in summer (December to February) while average winter (June to August) rainfall is around 220 mm. Average rainfalls in the transitional seasons, spring and autumn, are much less than winter but still play a vital role in providing soil moisture for plant growth and resilience. Early European immigrants were quick to recognise the similarity of the new settlement's climate with those of countries like Italy. It is no mere coincidence that this is so, although our geography suggests, at least in terms of the temperature regime, lying on the southern coastline of Australia, there may be a closer similarity with coastal locations on the southern shores of the Mediterranean and along the northern coastline of Africa. Other regions with Mediterranean climates include south-west Western Australia and California.

Like the Mediterranean, Adelaide lies in the mid latitudes, where – in both the Northern and Southern hemispheres – the distinct dry summers and wet winters are due to the impact of contrasting features of the general circulation.

It rarely drizzles in Adelaide. Mostly, rains tend to be moderate to heavy but relatively brief. Like other locations with a Mediterranean climate, we have about 90 rain days a year. (London and Paris, which are not Mediterranean climates, have similar yearly rainfalls but have 250–300 rain days a year). In winter it is not uncommon for a front and accompanying moist stream to produce rainfall events of 30–40 mm over two or three days. These events are usually separated by extended fine periods, often lasting a week or longer. While less reliable, good rainfall events may also occur in autumn and spring. In fact, history shows it is the period from April to June when daily rainfalls are most likely to exceed 25 mm. Average rainfalls in autumn and spring are very similar but in recent times autumn totals have declined slightly. The reason for this, whether it be climate variability, climate change or a combination of the two, is not clear. Infrequent, short-lived thunderstorms are responsible for most of the limited rainfall in summer.

Adelaide's Mediterranean climate makes managing water for gardens a challenge, especially in drought years. While the annual rainfall (an average of 550 mm/year ranging to in excess of 1100 mm/year in the hills and down to near 400 mm/year at Semaphore) is reasonable, its distribution mainly in the cooler months and its considerable variability from year to year requires careful

In summer the mean position of the subtropical high-pressure belt lies in the mid latitudes (near 35–40 degrees latitude north and south), and the descending arm of the Hadley Cell brings stable and generally fine conditions. Clear skies and average daily temperatures in the mid 20°Cs, with maximum temperatures often above 30°C, promote evapotranspiration, further reducing available water reserves in the soil. In winter, as the subtropical ridge retreats equator-ward, these regions come under the influence of the westerly wind belt with its travelling weather systems. Reaching landfall after an extended journey over water, these frontal systems and accompanying moist, unstable westerly streams bring the welcome winter rains. Forced lifting of the westerlies over coastal ranges serves to enhance the rainfall, an important ingredient in defining a Mediterranean climate.

planning to minimise water stress in summer. When late-winter and spring rains fail, this severely reduces soil moisture availability in the summer. High evaporation rates caused by hot summer temperatures and exacerbated by windy conditions add to the often harsh plant environment. Gardens well suited to Adelaide not only have deep-rooted and drought-tolerant plants, many of which may be of Mediterranean origin, but they are also designed to reduce soil moisture loss through evaporation.

Elizabeth Curran and Darren Ray

Mean sea level pressure (hPa)

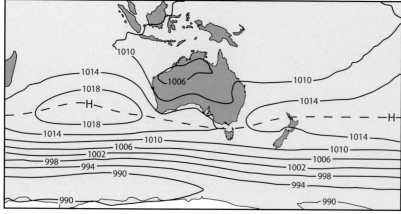

January: 1950 to 2007

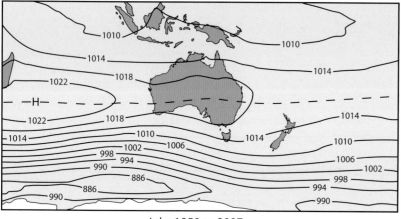

July: 1950 to 2007

Figure 2.2 Average mean sea level pressure for January and July showing the position of the subtropical ridge and associated high pressure systems (H). Source: NOAA/ESRK Physical Sciences Division, Boulder Colorado (online).

Summer

During the summer months (December to February), maximum solar heating occurs in the tropics south of the equator, and the mean position of the subtropical ridge lies south of Adelaide. The relatively cool ocean waters of the Great Australian Bight and the Western Pacific are preferred locations for the large high-pressure cells making up the subtropical ridge. Adelaide's day-to-day weather during this season is largely determined by whichever high is more dominant (see Figure 2.2). Most commonly, this is the high centred in the Bight. Fresh sea breezes and overnight down-slope winds can occur under this regime area, but otherwise winds are generally light. The southerly onshore stream keeps relative humidity in the mid range and temperatures in the mid 20°Cs. Days are generally fine. Sea breezes along the coast keep coastal suburbs 2 to 3°C cooler than the northern and eastern suburbs, but these steady breezes also increase potential evaporation in exposed localities.

On the flip side, when the Western Pacific high-pressure centre anchors in the Tasman Sea and extends a ridge of subsiding air over Queensland and New South Wales, a hot, dry northerly stream prevails over Adelaide – a potential weather pattern for heatwaves. Once established, daily maximum temperatures in the mid to high 30°Cs are common, with each successive day a little warmer. In extreme events, several days with temperatures exceeding 40°C may occur. The sequence is broken by a front or trough moving in from the west.

Brief thunderstorms and showers may occur with summertime fronts, but most often there is little cloud or precipitation. Rather, the frontal passage brings a sharp wind shift, the much-awaited 'cool change', well known colloquially to Adelaide residents, and a return to a southerly stream as the high pressure rebuilds in the Bight.

Significant rainfalls in summer are generally limited to infrequent, short-lived severe thunderstorm events. These require an infeed of tropical moisture. The most suitable conditions exist when a persistent trough extends south from Queensland, allowing air with high water vapour content (evident as high humidities) to feed into southern South Australia over several days.

Winter

In winter (June to August) the belt of maximum global heating lies in the Northern Hemisphere, shifting the southern Hadley Cell northwards. The mean position of the subtropical ridge now lies to the north of Adelaide, and the prevailing airstream is from the west. Travelling frontal systems in the stream are responsible for the generally reliable rainfall in these months. The Mount Lofty Ranges,

This large-scale overturning of the atmosphere, from the tropics to the subtropics, is known as the Hadley Cell and is one of the most robust features of the atmospheric circulation. Two other meridional (north/south) circulations assist in the transport of mass and heat between the equator and the poles. Figure 2.1 is a simple representation of these circulations. The Ferrel Cell, a smaller cell in mid latitudes, contributes to the subsidence in the subtropical ridge and to the strength of the surface westerly winds in the mid latitudes, another important feature affecting Adelaide's climate. The third cell, the Polar Cell, is important in transferring mass from the poles to the mid to high latitudes, i.e., in determining the relative strengths of the pressure cells and zonal wind systems in this region. It contributes to the intensity of the SAM.

The latitudinal position of the subtropical ridge and mid-latitude westerly belt will vary, depending on the time of the year and day-to-day perturbations of the atmosphere. Adelaide's distinct seasons – its hot, dry summers and cool, wet winters – are largely a product of the mean seasonal position of these major features.

running roughly perpendicular to the stream, force the air to rise, enhancing the precipitation processes and effectively wringing extra rainfall from these fronts and moist airstreams as they move over Adelaide. It is common for rainfall in the Adelaide Hills to be twice the rainfall that occurs over the city, but this is not a hard-and-fast rule, and in some storm situations the rain along the coast could be greater than at Mount Lofty.

Many of the more significant winter rainfall events occur when moist tropical air is drawn into a weather system such as a north-west cloud band or a cut-off depression (cut-off low). The air originates over warm oceans and holds much more water vapour than air originating south of Australia.

In a north-west cloud band the moisture is sourced to the north and north-west of Western Australia. As the air moves southwards it rises gently. As the pressure decreases the air cools to saturation and extensive bands of middle-level cloud can form. Not all cloud bands produce rain, but if the moisture and cooling are sufficient, good, widespread falls will result.

The frequency of north-west cloud bands and cut-off lows varies greatly from year to year. Sea surface temperature patterns in the northern Indian Ocean and in the mid Pacific appear to be important in their respective development.

Cut-off low-pressure systems that entrain moisture from the Coral and/or Tasman Sea have the potential to produce flooding. Divorced from the westerly belt, these slow-moving systems can produce rain events lasting 12 to 24 hours. A widespread rainband can occur, as well as convective clouds including thunderstorms with heavy downpours. While the systems are slow moving, the associated winds can be fresh to strong and recirculate the moist air over Adelaide as well as entraining moist air from Bass Strait. Successive rainfall bursts can result, and the cumulative effect of these rainfall bursts can lead to major flooding.

Widespread droughts brought on by failed winter and spring rains across southern and eastern Australia are often attributed to El Niño events, associated with warm sea surface temperature anomalies in the eastern equatorial Pacific and cool anomalies to the north-east of Australia. These ocean anomalies are associated with extremes in the ENSO, a zonal (east/west) circulation across the Pacific Ocean in low latitudes. One of the most severe El Niño-related droughts to affect South Australia was in 1982. However, not all El Niño events are associated with drought in the Adelaide region, and some of the most significant local droughts have occurred in neutral or even La Niña years when sea surface temperature anomalies in the Pacific are opposite to those observed in El Niño years (see Box 9: Significant droughts). In these situations it is likely that the Indian Ocean sea surface temperature anomalies can be a deciding factor, along with the seasonal position of the subtropical ridge. For more information on atmospheric and oceanic patterns associated with El Niño and La Niña events and Indian Ocean sea surface anomalies, the reader is referred to the climate pages of the Bureau of Meteorology website.

Across southern South Australia, dry winters typically occur when the subtropical ridge is stronger than usual, with the westerly wind belt then favouring more southerly latitudes. Hence the subtropical ridge, and associated blocking highs in the Australian region, play a major role in steering frontal systems further south of South Australia than usual. Changes in the subtropical ridge occur over short time scales, through natural variability, but a broadening and strengthening of the subtropical ridge has been observed in recent decades, most strongly in autumn and early winter, with a resultant drying trend seen in rainfall in southern Australia. This effect is further complicated by changes in the frequency of rain-bearing cut-off lows, which may actually increase under a strengthened subtropical ridge, though these systems need good connections with tropical moisture to produce substantial rain.

Autumn and spring

During autumn and spring, Adelaide's mean daily temperatures change rapidly as the belt of maximum global heating follows the sun across the equator. Autumn brings balmy days and light winds under the influence of the subtropical ridge. History shows that by late autumn Adelaide has normally received its first good rains for the year. These may be the signal that the winter westerlies are settling in over southern South Australia, but occasionally the early rains prove to be 'false starts'. In these years the subtropical ridge may persist near Adelaide latitudes well into winter, delaying and sometimes even preventing the full onset of the winter westerlies and associated travelling weather systems.

In spring, the Australian landmass warms rapidly as the zone of maximum global heating again moves into the southern hemisphere. This produces a strong temperature gradient between the land and the cool Southern Ocean, setting up conditions for the 'continental sea-breeze' effect. This is the season of maximum windiness in Adelaide as fresh and gusty south-westerlies, most prevalent during the afternoon, move in from the Southern Ocean. As this occurs, the mean position of the belt of maximum westerly winds circling the globe retracts to the south, and the average monthly rainfall and raindays per month decrease.

Selected climate statistics for Adelaide

Temperature

In the warmer part of the year, November to March, maximum temperatures occasionally exceed 40°C, while in winter overnight minimums can dip to near zero, especially in sheltered areas of the Mount Lofty Ranges. Average maximum temperatures are typically higher and winter minimums colder in drought years when there is less water vapour in the lower atmosphere.

In summer and early autumn, mean maximum temperatures are in the high 20°Cs. On average, the warmest month is February, although the highest daily temperature in the full instrumental record (dating back to 1887) is 46.1°C on 12 January 1939 (see Table 2.1). Both January and February are warmer than December, with more than 10% of days having a maximum temperature exceeding 37°C.

BOX 7

The nature of flooding

Paradoxical though it may seem in a city like Adelaide, which is prone to water shortages, flooding constitutes a natural hazard and has done so since the earliest days of settlement. Risks are heightened where inappropriate development has occurred on floodplains. In many instances it is not a question of *if* properties will be flooded but *when*. Planning is of paramount importance in avoiding problems with inundation by floodwaters.

Floods are essentially a consequence of rainfall events but the costs to society largely result from human activity and economic development. Wise management of flood-plains is imperative in reducing the costs of damage by flooding. In many instances, humans increase the flood risk by actions such as channelisation and construction of flood-control embankments. Furthermore, urbanisation invariably increases the volume, speed and height of runoff from storm events, exposing properties on flood-plains to even greater damage from inundation. Flood-proofing activities may actually produce an unwarranted sense of security on floodplains of considerable risk. There are several major types of flood events that affect the Adelaide area:

1. Flooding from perched river channels

The lower reaches of many streams flowing across the gently sloping alluvial plains diminish in channel capacity, flood out, and construct natural levees along the channels. In this fashion, the stream becomes perched above the surrounding floodplain so that if the levee is breached or overtopped, dramatic flooding may occur. The Gawler River, Little Para River, Dry Creek and the Torrens River all become perched streams in their lower reaches. Ironically, places close to the river built on the high levees are safer from flooding than lower-lying areas sometimes kilometres from the streams. The Gawler River has a catchment area of 1070 km^2 with most of its lower 30 km channel, between Gawler and the sea, being perched above its floodplain. Severe flooding occurred in 1992 when three 1 in 100 year floods occurred. Similar flooding has occurred in the lower Torrens Valley, with the 1931 flood being especially severe.

2. Localised stormwater flooding

Localised stormwater flooding may occur at any location throughout the year. These types of events invariably lead to ponding and flooding where the drainage system is blocked or is inadequate to deal with the volume of water, which needs to be evacuated. The impacts of localised stormwater flooding have been magnified by the increased amounts of impermeable surfaces such as roadways, paths and houses. A localised flood in the Waterfall Gully area occurred during 7–8 November 2005 following a storm event in the upper catchment area. Not only were guard rails and walking tracks damaged but houses below the falls were flooded and Wilson Bog failed, resulting in the formation of a debris avalanche that flowed down the valley and filled the weir at the base of the First Waterfall with some 10,000 t of peat, sand, pebbles and boulders. Without the trapping effect of the weir, extremely serious damage would have ensued, and possibly loss of life.

3. Coastal flooding – storm surge effects

Storm surges occur when there is a coincidence of a high astronomical tide (e.g., a spring tide when the earth, sun and moon are aligned and the gravitational pulls of the sun and moon are maximised), low barometric pressure, which causes the level of the sea to rise, and strong, persistent onshore winds pile up water at the shoreline. These weather-generated enhancements of predicted tide levels can lead to direct flooding of low-lying coastal areas such as tidal creeks, and indirectly intensify coastal erosion on sandy beaches. The effectiveness of waves is increased during storm surges as there is less frictional retardation at the wave base, and major storm damage can occur.

Port Adelaide has suffered storm surge flooding on many occasions. In 1865 many dwellings were flooded with seawater after embankments were breached. Four-metre-high tides, including the storm surge component, have a recurrence interval of 100 years or in other terms, there is a one in a hundred chance of a 4 m tide occurring in any year.[1] Tides near the 4 m level occurred in 1960, 1972, June 1981 and July 1981 at Port Adelaide.[1] The area is known to be subsiding due to compaction of sediments and tectonic dislocation so the risk from coastal flooding is likely to increase in the future even if the predicted sea level rise associated with global warming does not occur. It is relatively easy to identify areas that are prone to tidal flooding using biogeomorphic units across tide-dominated coastal sedimentary environments. Unfortunately much development has already occurred in flood-prone areas so that protective embankments may need to be improved in the future.

Combined storm surge and runoff from land

Often the conditions that give rise to storm surges will also cause rainfall that exacerbates the flooding problem, as the floodwaters cannot easily escape to the sea because of the elevated tidal levels. The Patawalonga Basin has been affected by such flooding, which is sometimes heightened by management issues related to the operation of the gates and lock system.

Concrete channelisation of the lower Sturt River has accelerated the delivery of floodwaters to the basin, thereby increasing the flood risk. Additional lock gates have been installed and floodwaters have been diverted directly out to sea via the Barcoo Outlet in attempts to overcome the flooding problem. However, increasing the time floodwaters take to rise, and reducing the height of the flood peak by retaining more water on the land, would also assist. A flood-control dam has also been built in Sturt Gorge to smooth out the flow of floodwaters down to the Patawalonga Basin.

Port Noarlunga is sometimes threatened with flooding as a consequence of a combination of high tides, storm surges and flood runoff from the Onkaparinga River. The impacts of flooding have sometimes been exacerbated by management strategies, whereby water retention is maximised in reservoirs such as Mount Bold. When reservoirs are near to maximum capacity, storm events upstream result in reservoir managers 'pulling the plug' so that individual flood events are accentuated over and above what would have occurred if the river had been allowed to run freely. In more recent times, management of reservoir levels seems to have been more sensitive, utilising the reservoir as a flood-control measure. However, the water shortages of the past few years may once again encourage the maximisation of water holding within the reservoirs so that if the reservoir is full and a storm occurs, the response may well be to once again pull the plug, with minimum warning. Mount Bold last overflowed in November 2005 when little warning was given of the potential flood risk.

Robert Bourman

Table 2.1 Temperature statistics for Adelaide

	Mean monthly maximum (°C)	Extreme daily maximum (°C) date		Mean monthly minimum (°C)	Extreme daily minimum (°C) date	
		Lowest	Highest		Lowest	Highest
Summer	26.9–29.3	15.4 3 Dec 1955	46.1 12 Jan 1939	15.5–17.1	6.8 3 Dec 1955	33.9 29 Jan 2009
Autumn	19.0–26.2			10. 2–15.1		
Winter	15.3–16.6	8.3 29 Jun 1922	29.1 31 Aug 1911	7.4–8.2	-0.4 8 Jun 1982	18.4 30 Aug 1993
Spring	18.9–24.8			9.6–13.8		

Table 2.1 Temperature statistics for Adelaide. Mean monthly temperatures are based on records from the Bureau of Meteorology, Kent Town, from 1977 to 2009. The range of temperature shows the variation between the mean monthly temperature of the coolest and warmest month in the season. For example, in summer, December is the coolest and February the warmest month, with a mean maximum temperatures of 26.90°C and 29.30°C respectively. Extreme daily temperatures shown in the table have been derived from the daily temperature record of both Kent Town and West Terrace (comprising the period 1887 to March 2009). Source: Bureau of Meteorology, 'Climate change statistics for Australian locations, monthly climate statistics', copyright Commonwealth of Australia, reproduced by permission.

Mean minimum temperatures in summer are in the mid teens, although the monthly averages mask the overall range. The lowest overnight minimum of 6.8°C occurred in December 1955 (when snow was observed on the Mount Lofty Ranges), and the warmest of 33.9°C was in January 2009.

In winter months daily temperatures are in the mid teens, and overnight temperatures average below 10°C. The coldest month is July with a mean monthly maximum of 15.3°C and a mean minimum of 7.4°C. In the city temperatures rarely fall low enough for frost to form (zero degrees at ground level). At Kent Town frost occurrence is much less than one day per month during winter, but frosts are more common in exposed low-lying areas on the plains and in the Mount Lofty Ranges.

Rainfall

Adelaide's mean annual rainfall is around 550 mm in the city, but there is a gradual increase across the plains from around 400 mm on Lefevre Peninsula to over 600 mm in the foothills. This gradient is accentuated on the western slopes of the Mount Lofty Ranges where, at higher elevations such as Uraidla and Mount Lofty, the mean annual rainfall exceeds 1100 mm, and then decreases sharply, demonstrating a classical orographic rain shadow to the east. Moving north across the city, annual rainfall totals decrease from near 630 mm at Happy Valley in the south to around 430 mm at Elizabeth, and then 500 mm at Gawler on the lower slopes of the Ranges (see Figure 2.3). Consequently, the average rainfall across local river catchments varies markedly. In the Ranges the highest annual rainfall in a calendar year is 1853 mm, recorded at Aldgate in 1917. However, in its wettest 12 month period (May 1923 to April 1924) Mount Lofty Summit recorded 2130 mm.

In any year the increase in rainfall across the plain is evident, but from year to year the annual total at any one locality will vary greatly. Records dating back to 1839 show

Adelaide's official annual rainfall totals have varied from a low of 257.8 mm in 1967 to a high of 883.2 mm in 1992. As a guide to rainfall reliability, comparison of annual totals from Kent Town shows approximately one third are below 500 mm, one third lie above 600 mm, and the remainder lie between 500 mm and 600 mm, with the mid point (or median) near 550 mm.

Rainfall totals

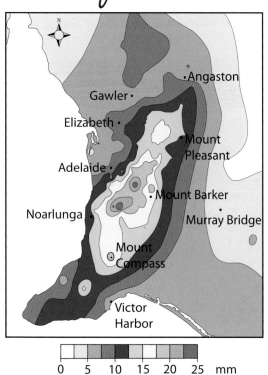

Figure 2.3 Map of rainfall totals in the Adelaide region for the 24 hours to 9 am on 17 July 2007.
Source: Bureau of Meteorology.

Table 2.2 Rainfall statistics for Adelaide

Month (no. of years of data) Kent Town (West Terrace)	Median monthly rainfall (mm) Kent Town (West Terrace)	Mean number of raindays Kent Town (West Terrace)	Highest total (mm) and year Kent Town (West Terrace)		Lowest total (mm) and year Kent Town (West Terrace)	
January (30)	21.7	4.6	42.2	1996	0.4	1992
(140)	*(13.3)*	*(4.4)*	*(84.0*	*1941)*	*(0.0*	*1957)*
February (31)	10.4	3.4	63.0	2003	0.0	2007
(140)	*(10.2)*	*(3.9)*	*(154.7*	*1925)*	*(0.0*	*1965)*
March (31)	20.2	5.5	106.0	1983	0.0	1994
(140)	*(17.0)*	*(5.5)*	*(116.6*	*1878)*	*(0.0*	*1870)*
April (31)	32.6	7.9	105.6	1998	0.8	1981
(140)	*(38.3)*	*(9.4)*	*(154.7*	*1971)*	*(0.0*	*1945)*
May (31)	56.6	12.0	128.0	1987	8.2	2005
(140)	*(61.5)*	*(13.3)*	*(196.9*	*1875)*	*(2.6*	*1934)*
June (31)	82.2	15.0	174.6	1981	12.4	2006
(140)	*(65.0)*	*(14.8)*	*(217.9*	*1916)*	*(6.0*	*1958)*
July (31)	68.8	16.4	159.8	1986	22.2	1997
(140)	*(65.0)*	*(16.2)*	*(138.5*	*1890)*	*(10.1*	*1899)*
August (31)	69.2	16.0	129.0	1992	11.4	2006
(140)	*(58.6)*	*(15.8)*	*(157.7*	*1852)*	*(8.4*	*1944)*
September (31)	58.0	13.4	151.4	1992	16.0	1987
(140)	*(47.4)*	*(13.2)*	*(148.1*	*1923)*	*(7*	*1951)*
October (31)	43.0	10.5	105.0	1980	1.0	2006
(140)	*(42.2)*	*(10.9)*	*(133.2*	*1949)*	*(1.0*	*1969)*
November (30)	30.8	8.1	107.0	1992	1.0	1996
(140)	*(24.0)*	*(7.8)*	*(113.2*	*1839)*	*(1.4*	*1967)*
December (30)	23.8	6.9	72.8	1993	5.8	1991
(140)	*(20.6)*	*(6.2)*	*(101.1*	*1861)*	*(0.0*	*1904)*
Annual (29)	560.8	119.7	883.2	1992	287.6	2006
(140)	*(526.2)*	*(121.4)*	*(786.4*	*1851)*	*(257.8*	*1967)*

Table 2.2 Rainfall statistics for Adelaide city using records from Kent Town (1977–2007) and West Terrace (1839–1978). The median is the mid-point of all observations at the station; i.e., in 50% of years the rainfall is greater than the median. Kent Town is around 3 km closer to the Mount Lofty Ranges and so has a higher annual median than West Terrace. The longer record at West Terrace gives a more representative range of extreme rainfalls in the Adelaide region than does the limited 30 year record at Kent Town. Source: Bureau of Meteorology.

Between November and March rainfall is unreliable, with February showing the greatest year-to-year variability. In the 168 years of combined record for West Terrace and Kent Town, February's monthly rainfall has ranged from zero in several years to a total of 154.7 mm in 1925. Most of this 1925 total was recorded in just one severe thunderstorm lasting several hours (see Box 8: Flooding in the Adelaide region).

The transition from dry to wet conditions typically occurs in April/May, when in most years the westerly stream and associated travelling fronts move into Adelaide latitudes. Adelaide's most reliable rainfalls occur under the westerly regime. From May to September the average Kent Town rainfall exceeds 50 mm in each month.

Table 2.2 summarises the rainfall statistics for the city using records from Kent Town and West Terrace. Due to its proximity to the hills, Kent Town has a higher annual median rainfall than West Terrace. However, the records show that in the 30 year period, 1978–2007, the median monthly rainfalls and mean number of rain days in April and May were less than at West Terrace. This is consistent with a belated start to the winter rains in recent years, and less prominence of the westerly stream in these months.

This year-to-year variability highlights the importance of using more than just a few years to assess the average rainfall at any location, and why it is important to use standardised data when comparing climate at different locations or over time. The standard climate period, used

Wind roses

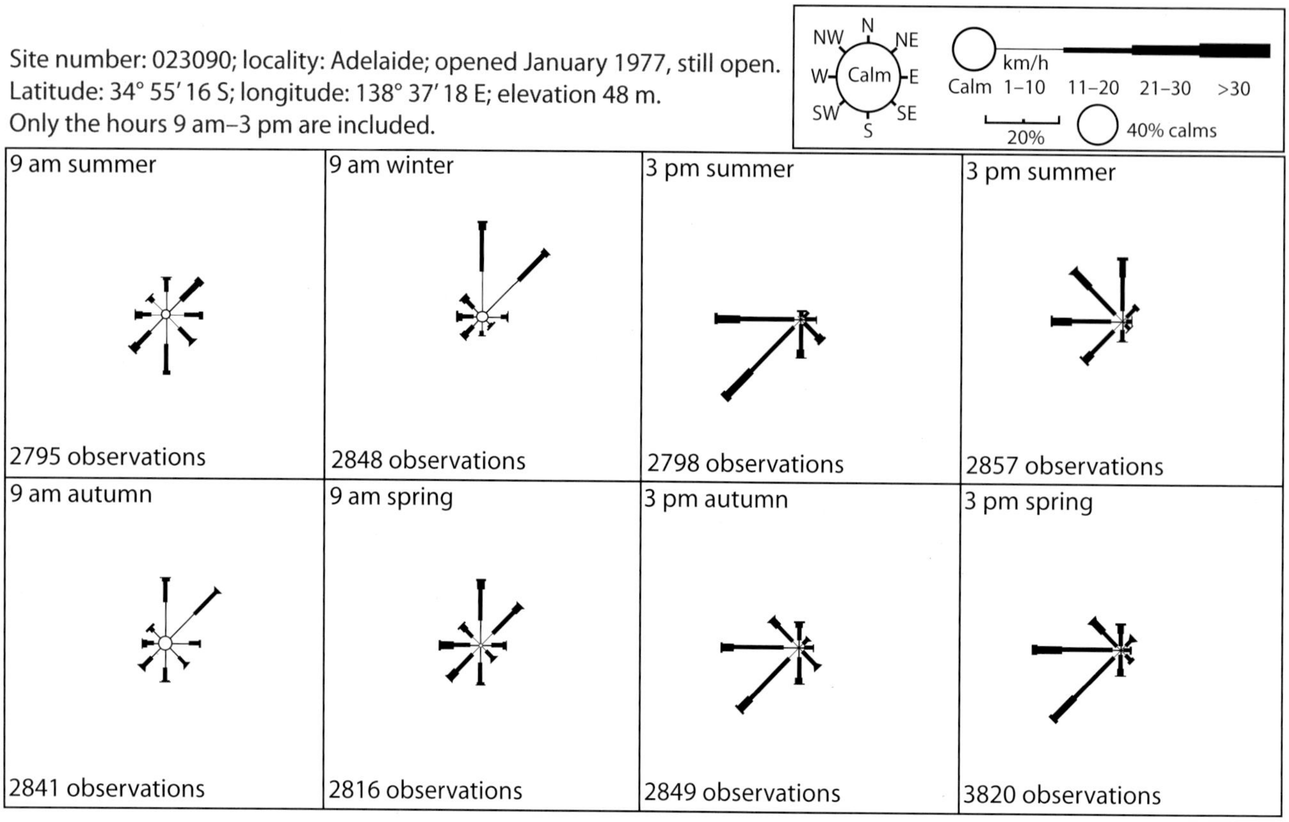

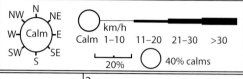

Site number: 023090; locality: Adelaide; opened January 1977, still open.
Latitude: 34° 55' 16 S; longitude: 138° 37' 18 E; elevation 48 m.
Only the hours 9 am–3 pm are included.

9 am summer	9 am winter	3 pm summer	3 pm summer
2795 observations	2848 observations	2798 observations	2857 observations
9 am autumn	9 am spring	3 pm autumn	3 pm spring
2841 observations	2816 observations	2849 observations	3820 observations

Figure 2.4 Wind roses from Kent Town (Adelaide) showing the prevailing north-east to south-westerly winds in winter, and south to south-easterlies in summer with the afternoon south-westerly sea breeze. The data was collected between 1977 and 2008. Source: Australian Bureau of Meteorology.

by the World Meteorological Organization, is 30 years, with 1961–1990 the current baseline period used for global comparisons. Longer records are required to assess and attribute climate trends.

Wind

During the day when heating from the sun mixes the lower layers of the atmosphere, the surface wind is generally representative of the prevailing wind pattern. Overnight cooling at ground level usually separates the surface layer from the prevailing air-stream, and localised wind regimes become important. In winter, prevailing north to south-west winds dominate in an overall westerly air-stream, though overnight a low level north-easterly drift is common over the plains as the westerly stream is blocked by the ranges. During summer southerly winds prevail. The afternoon south-westerly sea breeze, driven by the heating of the land surface, is superimposed on this stream, often increasing the wind strength, especially along the coast. Overnight gully winds may occur along the foothills. These refreshing breezes result from local cooling after dusk of air in contact with the higher ground and sloping sides of the gullies. As this cooler air is denser than the surrounding 'free' air, it rolls down the slopes and along the gully floors until it spills out on to the plains and its momentum is dispersed. Occasionally these light easterly winds are strengthened and extend much further westwards across the suburbs as they are reinforced by strong and gusty down-slope winds resulting

from the surge of a cool, south-easterly stream ahead of a new high-moving into the Bight. Wind roses for January and July are shown in Figure 2.4.

Water loss to the atmosphere

Water loss to the atmosphere occurs through evaporation from water surfaces and through transpiration from plants. Various methods are used to estimate such water loss. Here, our discussion is confined to 'pan evaporation' and 'evapotranspiration'. Every day significant amounts of water vapour are lost to the atmosphere through evaporation. It occurs from open water surfaces such as rivers, lakes and reservoirs and through less obvious sources such as bare soil and other wet surfaces. Pan or 'potential evaporation' is a measure per unit area of the amount of evaporation that would occur if an unlimited water source was available. Measured using a metal pan evaporimeter, it gives a good indication of the evaporative loss from a large water source such as a lake or reservoir. Pan evaporation rate is dependent on temperature of the surrounding air and water, solar irradiation or sunshine, relative humidity and wind strength. In fact, wind strength is of critical importance, and is the major factor in the differing evaporation rates between a sheltered spot such as a garden, and an exposed field site or water expanse.

At Kent Town, an open site but somewhat sheltered from extreme winds, mean monthly pan evaporation exceeds

Floods in the Adelaide region

BOX 8

Floods are natural events but the relatively low rainfall in Adelaide and the frequent dry periods mean they do not rank highly against other risks such as bushfires. People who live in Adelaide are often unaware that floods might be a risk to themselves, their animals and their property. Floods are most common in the winter months, July through October, but can occur in any season. In fact, the most extreme short-duration rainfall rates – capable of causing flash flooding – are most likely to result from severe summer storms feeding off tropical moisture. The most extreme rainfall on record was in February 1925 when 150 mm of rain fell in three hours.

Cut-off lows have the capacity to produce devastating flooding of waterways in the Adelaide Hills and across the plains. Divorced from the westerly belt, these slow-moving systems can produce rain events lasting 12–24 hours – far longer than travelling fronts. A mixture of stratiform and convective clouds, including thunderstorms, can result in extensive rain interspersed with heavy downpours. The winds associated with the system can be strong and recirculate moist air over Adelaide, giving successive rainfall bursts. When these winds are westerly the rainfall intensities over the Adelaide Hills will be enhanced. The cumulative effect of all these factors can lead to exceptional rainfall and subsequent flooding.

One such event occurred in November 2005. Falls of 100 to 120 mm in six hours on already saturated ground led to severe flash flooding of a number of creeks across Adelaide, with the worst being Brownhill Creek and First Creek/Waterfall Gully. Hundreds of homes were affected.

Flooding of the River Torrens was a regular event in the early days of settlement, when bridges and houses were periodically washed away. Even as development progressed, builders of housing and infrastructure rarely realised the risk floods posed. As a result, a number of

creeks and watercourses were treated as stormwater drains, not as carriers of floods. In particular, many privately owned creeks run through backyards. There are many examples of houses built near and over creeks, and these have a varying standard of capacity for flood flows. Flood waters can be blocked or diverted and will readily break out from the channel, affecting adjacent buildings and facilities. While the creeks function well under normal flows, the result could be disastrous during major flood events.

Remedies to reduce the potential for floods are limited where residents have built over creeks and/or where the natural flow of the creeks has been ignored. The cost to the government to buy back high-risk properties and land is greater than what the community will pay. In 2007, a trial program commenced to better educate the community. All residents in high-risk locations were visited, the flood risk explained, and practical approaches were demonstrated to minimise damage when a flood occurs.

As flood hydrology became better understood, planning and development became more flood-conscious. During the 1980s, the Torrens was recognised as a significant flood risk, especially where its banks were narrow and steep and provided little room for the increased flow of floods. To counter this, the Kangaroo Creek Dam was raised to provide more flood storage, while downstream the banks were laid back and much of the introduced vegetation that caused blockages was cleared out. At the same time, construction of the O-Bahn along the river assisted in providing an open channel suitable to contain floods. The resulting Linear Park has resulted in a reduction in the 1 in 100 year flood risk.

Chris Wright

Evaporation and rainfall

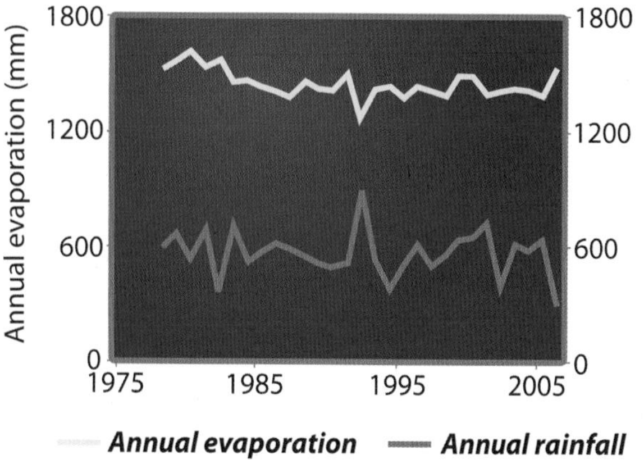

Figure 2.5 A comparison of the annual rainfall and evaporation at Kent Town. In general during wet years, such as 1992, evaporation is reduced, while in dry years, e.g., 1982 and 2006, it is increased. Evaporation is also sensitive to the strength and persistence of the wind run. Variations in wind run from year to year have not been analysed. Courtesy of Bruce Books, Australian Bureau of Meteorology.

mean monthly rainfall in all months except during late autumn and winter. At Adelaide Airport, a more exposed site, mean monthly evaporation exceeds mean monthly rainfall in all months. Pan evaporation rates at these two sites are shown in Table 2.3.

Figure 2.5 shows how the ravaging effects of drought on water storages and the landscape is exacerbated by the inverse relationship between potential evaporation and rainfall. Clear skies and lack of rain in a drought year maximises incoming solar irradiation and reduces relative humidity, thereby enhancing evaporative losses in an already dry environment. In wet years, increased cloud cover and higher relative humidity reduce daily evaporation.

Pan evaporation measurements are difficult to automate and, as a result, evapotranspiration calculated from other meteorological variables is becoming a popular tool to estimate daily water losses. Evapotranspiration represents the combined water loss through evaporation from the soil (and water surfaces) and transpiration from plants. It is dependent on vegetation type but can be calculated using temperature, estimated water vapour pressure gradient between the plant and the atmosphere, solar irradiation, and wind strength. Daily values for a standard vegetative cover (such as lucerne or mown grass) can be readily computed and forecast using numerical weather prediction model output. The calculated value is then used to compute water deficits for different crops and groundcovers. Daily evapotranspiration measurements are increasingly being used by irrigators to determine watering schedules.

Snow

Snow is very occasionally observed on elevated parts of the Mount Lofty Ranges. The weather systems responsible are known as cold outbreaks and bring very cold air from high latitudes over the state. The rapidly moving cold stream becomes unstable as it travels over relatively warmer water, producing moderately deep convective cloud. Water vapour sublimes directly to ice crystals in these clouds, resulting in snowfalls in higher elevations. The snow is short lived on the ground, as temperatures greater than 0°C cannot sustain it. Winter is the most likely time for snow but it has been observed on the higher parts of the Ranges in autumn (April 1916) and even in early summer (December 1955). The rare sight attracts a crowd of sightseers.

Thunderstorms

The Adelaide region typically experiences about 15 days per year when thunder is heard. Most of these storms have lifecycles of up to an hour and produce brief heavy downpours. Occasionally, more organised severe thunderstorms lasting one to three hours may produce flash floods, large hail, tornadoes, and/or severe wind gusts. These storms are most likely to occur during late spring and summer, with the majority occurring in the afternoon and evening.

Climate variability, climate trends and climate change

Adelaide's climate has a high natural variability, characteristic of much of Australia, and evident on all manner of timescales; inter-annual, decadal and periods of centuries and longer. Some of the early graphs and tables, based on the instrumental record, give an insight to the considerable month to month and year to year variability in temperature and rainfall patterns. Extreme events such as droughts, flash floods, heatwaves and bushfires, although infrequent, are also natural features of this climate. These events can have a huge impact on the community and economy, and put excessive stress on local infrastructure.

Superimposed on the natural climate variability, there is now clear evidence of recent global warming at rates outside the experience of modern society. In 2007, the Intergovernmental

Table 2.3 Daily evaporation (mm/day)

	Jan	Feb	Mar	Apr	May	Jun	Jul	Aug	Sep	Oct	Nov	Dec	Annual
Kent Town	7.1	6.7	4.9	3.1	1.8	1.4	1.5	2.1	3.0	4.3	5.7	6.5	4.0
Adelaide Airport	8.9	8.4	6.5	4.3	2.7	1.9	2.0	2.8	3.9	5.5	7.2	8.2	5.2

Table 2.3 Daily evaporation (mm/day) at Kent Town and Adelaide Airport. The observing sites are approximately 10 km apart. Due to its exposure and greater wind run, evaporation rates are greater at the airport than at Kent Town in all months. Source: Bureau of Meteorology.

Panel on Climate Change (IPCC, established in 1988 by the World Meteorological Organization and the United Nations Environment Programme to provide a scientific assessment of climate change) concluded that globally, a warming trend was 'unequivocal', and most of the observed increase in temperature since the mid-20th century is very likely due to increases in atmospheric concentrations of greenhouse gases attributable to human activities. Paleo-climate information supports the finding that the global warmth observed in the last half century is unusual in at least the previous 1300 years.

In the relatively short historical climate record across southern Australia, it is possible to detect recent trends in temperature and various other climate parameters. While the warming is consistent with global trends attributed to human activities, the changes in other parameters may be due to external factors such as human activities or they may be part of the natural climate variability, or a combination of external and internal factors. For some climate parameters, such as severe storms, there is incomplete data to draw global conclusions.

Much research is being undertaken internationally to better understand the observed trends and enable attribution of the cause of these changes. To underpin this research a high-quality network of Australian long-term climate reporting stations, with records going back to at least the early 20th century, has been identified by the Bureau of Meteorology. The data from this network has been tested to ensure it is free of contamination from effects such as urban heating and poor observational practices. Apart from enabling the detection of climate trends and change, the data are used as input for the modelling of future climate and for general climate studies.

An Australian database of observed trends in selected extreme events, and a metadata database for Australia documenting the availability and location of natural systems data holdings, are accessible on the Bureau of Meteorology website. These data holdings, mostly held by organisations other than the bureau, include information on ecosystems and individual species, snow, ice, tree-rings, caves, crops, pollen, sediment, coral, and ice cores.

Recent trends in local climate parameters

Using the high quality climate data sets now available, the climate record can be analysed to detect recent trends in our region. Where trends are detected, more complex studies are required to determine the cause of the trend, i.e., whether it is part of the natural cycle of variability or whether human activities are responsible. Studies undertaken to determine the cause of a trend are known as 'attribution' studies. Below is a summary of recent major findings based primarily on detection studies:

1. An increase of 0.9°C is observed in mean temperature for Australia from 1910 to 2006, with most of the warming (0.16°C per decade) occurring since 1950 (Figure 2.6). There has been an increase in the number of hot days (>35°C) and hot nights (>20°C) and a decrease in the number of cold days (<15°C) and cold nights (<5°C).

2. Annual rainfall in Adelaide shows great interannual variability, with the wettest year being 1992 and the driest 1967 (Figure 2.7). On a seasonal basis a recent decline in autumn rainfall is evident across south-eastern South Australia including Adelaide. Current research has linked the cause of the decline to changes in mean sea level pressure across southern Australia. The change in the rainfall regime bears some similarities with the more extended and pronounced drying in south-west Western Australia since the 1970s, shown to be due to changes in the large-scale winter pressure and wind patterns, as well as changes in moisture inflow over the region. Any decline in rainfall translates into a greater percentage reduction in infiltration and streamflow, and this amplification increases as the drying conditions are extended in time. This can have serious implications for water storage. The results of extensive research and model simulations of the south-west Western Australian rainfall decline have led researchers to attribute a combination of natural variability, ozone depletion in the stratosphere, and increasing greenhouse gas concentrations as the likely cause of this regional change.

3. Droughts, measured in terms of rainfall deficiencies, are peppered throughout the instrumental record (since 1839), but, of significance, the average temperature in recent drought years has been warmer than in droughts of the early 20th century. Warmer temperatures increase heat stress and potential evaporation, impacting on open water storages and soil moisture, reinforcing the severity of the rainfall deficiencies. The timing of rainfall, for example a 'false autumn break' early in the season, can also play a critical role in determining water stress for crops.

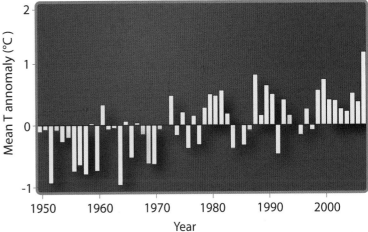

Annual mean temperature

Figure 2.6 Annual mean temperature anomaly for the Adelaide Mount Lofty Ranges region. The temperature shows a clear warming trend since 1950, and an exceptionally warm year in 2007.
Source: Australian Bureau of Meteorology.

4. Due to their intermittent nature, changes in the frequency of extreme events are generally more difficult to assess than changes in mean temperature and rainfall. The likelihood of an individual extreme event is influenced by various climatic factors. This means it is almost impossible to attribute a particular event to a single cause. The occurrence and its intensity is more than likely due to a combination of factors, which may include external factors such as increasing greenhouse gases.

Future climate projections

While there is confidence that increasing atmospheric greenhouse gas concentrations will result in further warming in our region over the next century, for other climate variables, such as rainfall and extreme events the link to global warming is far more complex. It is not feasible to simply extrapolate recent trends. To a large extent, changes in these elements will be driven by changes in the local features of the global circulation.

Credible climate projections for the 21st century and beyond can only be made using global numerical models, which can replicate the current climate features including the atmospheric and ocean circulations. These models use a range of plausible greenhouse gas emission pathways to produce a suite of future climate projections. Using expert analysis of the output from these model runs, the IPCC, in its 4th Assessment (2007), drew the following conclusions:

Most models project global warming over the next two decades to be around 0.2°C per decade. Continued greenhouse gas emissions at or above the current rates would cause further warming and induce many changes in the global climate system during the 21st century that would very likely be larger than those observed over the last 100 years.

1. Sea level pressure is projected to increase in the subtropics and mid-latitudes (Adelaide latitudes), and decrease over high latitudes due to a poleward expansion of the Hadley Circulation and a poleward shift in storm tracks;
2. No consistent discernible change in the ENSO frequency or intensity is evident in the projections;
3. Globally averaged mean atmospheric water vapour, evaporation and precipitation are projected to increase (i.e., an intensification of the hydrological cycle), but precipitation is expected to decrease in the subtropics and mid latitudes. Even where mean precipitation decreases, precipitation intensity (mm/unit time) in specific events, such as severe storms, is projected to increase, but the duration between such events will also increase.

To enable the impacts of climate change at the regional level to be assessed, a set of climate projections prepared by the international Coupled Model Intercomparison Project is available through the IPCC process. The projections have been developed using 23 different international models, representing state-of-the-art climate modelling. Simulations are available for both the past (20th century) and the future (21st century). Output includes monthly temperature and precipitation data, solar radiation, wind speed and relative humidity. Good representation of the current regional climate gives researchers confidence in the model's skill in simulating the future climate.

A key limitation in developing projections beyond 2030 is that actual future global greenhouse gas emissions are unknown. The emission pathways used in the models are based on those in the IPCC Special Report on Emission Scenarios (2000). Since its publication, emissions have been tracking at the high end of the emission scenarios. This infers that unless significant reductions are made to current global emissions trajectories, the most likely future climates are those derived using the higher emission rates in the report. Added to this, and little realised by many, is that global warming over the next 20 years or so will largely be driven by past emissions, due both to the length of time that greenhouse gases reside in the atmosphere and to the thermal inertia in the ocean/atmosphere systems. This means that an overall warming trend and consequent climate impacts over the next few decades are inevitable.

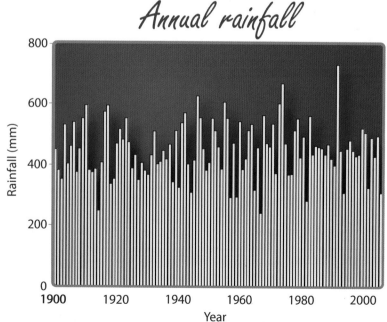

Annual rainfall

Figure 2.7 Annual rainfall in the Adelaide region, using a composite of stations. The rainfall record shows a high amount of interannual and multidecadal variation. Source: Australian Bureau of Meteorology.

Using output from the 13 models that most closely replicate climate features in the Australian region, the Centre for Australian Weather and Climate Research (CAWCR) has prepared local regional projections

Adelaide experiences a highly variable climate. While there have been some devastating droughts of short duration, such as the 10-month El Niño-induced drought of 1982–1983, the more significant droughts are those associated with dry periods lasting several years. Accumulated rainfall totals fall well below average during these generally widespread events, often extending across most of south-eastern Australia.

Extended dry periods severly impacting on Adelaide include:

1864–1869: very low rainfalls occurred in Adelaide during 1865 and 1869 as part of a more extensive drought in South Australia. In all six years of this period Adelaide had very low annual totals, averaging just 434 mm per year, compared with an annual average of 547 mm in the previous 25 years. George Kingston, who had kept meticulous daily records from 1839, assured the colonists that this short-term downturn in rainfall was no more than natural variability. His confidence was justified with the annual total in each of the next three years exceeding 565 mm.

1895–1902: almost all of Australia was affected by this dry period, culminating in the Federation drought of 1902. The period 1895–1898 was one of the most severe droughts in South Australia's history, with devastating crop, stock and pasture losses. Adelaide experienced serious rainfall deficiencies (in the lowest 10% of records) in 1896 and 1897, and below average rainfall in all but one year during the eight-year period.

1914: during the widespread 1911–1916 drought, 1914 saw one of the lowest annual rainfalls on record for Adelaide, and one of the worst years experienced across the state's agricultural areas. Sections of the River Murray were reduced to a series of waterholes.

1940–1950: during this period south-eastern Australia recorded its driest decade in more than 100 years. In six of the seven years, Adelaide recorded below average rainfall, with 1940 and 1950 the worst years. Interestingly, no single year recorded serious deficiencies.

1959–1967: in this nine-year period there were only two years when Adelaide's annual rainfall exceeded the

average. The period was a major drought event across inland Australia and in south-eastern Australia from 1964. In 1967 Adelaide recorded its lowest annual rainfall, and 1959 was the third lowest recorded at West Terrace.

1982–1983: this short drought affected all of south-eastern Australia and is regarded as one of the region's most severe. Adelaide had good autumn rains prior to the event, and good years on either side. The

extremely dry weather through winter and spring in 1982 and the summer of 1982–1983 turned the surrounding country to a tinder box, culminating in the devastating Ash Wednesday bushfires in February 1983; 26 people lost their lives and nearly 200 homes were destroyed in the state. The total cost of the fires was estimated to be in excess of $350 million.

2006: saw one of the lowest annual rainfalls on record for Adelaide, with winter and spring rains failing after promising autumn falls. The impact of this one very dry year was exacerbated by an extended seven-year period of serious to severe rainfall deficiencies over much of agricultural South Australia and the Murray-Darling Basin. Water restrictions were introduced in Adelaide for the first time since 1967. Rainfall improved to some extent in 2007 and 2008 but deficiencies over much of south-eastern Australia remained. Adelaide recorded below average rainfall in four years (2002, 2006, 2007, 2008).

Elizabeth Curran

for 2030, 2050 and 2070. The projections give a range of possible values to allow for differences between each model output. These are the best estimates of future climate, based on our current understanding, and should be used in the knowledge that they will be refined over time.

Based on the CAWCR projections for 2030, the best estimate for annual warming in the Adelaide region, relative to 1990, is 0.9°C, with slightly less warming in winter. The number of days with temperatures over 35°C is projected to increase from 17 currently to 23, and there is expected to be a small increase (~2%) in annual potential evaporation. The best estimates suggest continued increases in mean temperature and potential evaporation throughout the century.

The projected change in annual rainfall in 2030 is an overall decline of 4%, with the most pronounced declines, of 8% and 6% in the wetter seasons of spring and winter. The range in the model output projections for annual rainfall spans 'zero change' and varies from -11% to +2%. The range is determined by ranking all model results from the lowest to the highest value – the lowest 10% and the highest 10% (the extreme values) are then excluded. Similarly, the projections for seasonal rainfall vary from a decline to a small increase. The uncertainty in sign (an increase or decrease) is most pronounced in summer and autumn. The projections for 2030 also show a small decline (<1%) in relative humidity and very small increase of +0.2% in solar radiation.

These projections for Adelaide are based on gridded output from the global climate models. They do not take into account local topographical effects due to Adelaide's proximity to the coast and Mount Lofty Ranges, which may have a significant impact on changes, especially in rainfall.

Bibliography

J.M. Arblaster and G.A. Meehl, 'Contributions of external forcings to Southern Annular Mode trends', *Journal of Climate*, 19, 2006, pp. 2896–2905.

H.J. Bolle, *Mediterranean climate: variability and trends (regional climate studies)*, Springer, Germany, 2003.

Bureau of Meterology, homepage, 2008 (online).

Bureau of Meteorology, 'The greenhouse effect and climate change', 2003 (online).

Bureau of Meteorology (Canberra), 'Climate of South Australia', 1998.

Bureau of Meteorology (Melbourne), 'Drought in Australia', 1989.

Commonwealth Scientific and Industrial Research Organisation and the Bureau of Meteorology, 'Climate Change in Australia', technical report, 2007.

W. Drosdowsky, 'The latitude of the subtropical ridge over eastern Australia: the L index revisited', *International Journal of Climatology*, 25, 2005, pp. 1291–1299.

L.J. Dwyer, 'Results of rainfall observations made in South Australia and the Northern Territory 1839–1950', Bureau of Meteorology (Melbourne), 1950.

A.D. Evans, 'South Australian rainfall variability and climate extremes', thesis, Flinders University of South Australia, Adelaide, 2007.

W. Grace and E. Curran, 'A binormal model of frequency distributions of daily maximum temperature', *Australian Meteorological Magazine*, 42 (4), 1993, pp. 151–161.

Intergovernmental Panel on Climate Change, *Climate change 2007: the physical science basis. Summary for policymakers. Contribution of Working Group I to the Fourth Assessment Report of the IPCC*, Cambridge University Press, Cambridge, 2007, p. 996 (online).

N. Nicholls, 'Developments in climatology in Australia: 1946–1996', *Australian Meteorological Magazine*, 46, 1997, pp. 127–135.

W.L. Physick and R.A.D. Bryron-Scott, 'Observations of the sea breeze in the vicinity of a gulf', *Weather*, 32 (10), 1977, pp. 373–381.

A. Sturman and N. Tapper, *The weather and climate of Australia and New Zealand* (2nd edn), Oxford University Press, Melbourne, 2006.

R. Suppiah, B. Preston, P.H. Whetton, K.L. McInnes, R.N. Jones, I. Macadam, J. Bathols and D. Kirono, 'Climate change under enhanced greenhouse conditions in South Australia. An updated report on: assessment of climate change, impacts and risk management strategies relevant to South Australia', CSIRO Marine and Atmospheric Research, Canberra, 2006.

R.J. Szkup and B.H. Brooks, 'Heatwave trends in Adelaide', *Bulletin of the Australian Meteorological and Oceanographic Society*, 19, 2006, p. 73.

B. Timbal and B. Murphy, 'Observed climate change in the south-east of Australia and its relation to large scale modes of variability', *Bureau of Meteorology Research Centre*, research letter, 6, 2007, pp. 6–11.

Box references

Box 7: The nature of flooding
1. R.P. Bourman and N. Harvey, 'chapter 4: landforms' in Nance, C. and Speight, D.L. (eds) *A land transformed: environmental change in South Australia*, Longman Cheshire Pty Ltd, Adelaide, 1986, pp. 78–125.

CHAPTER 3

Catchments and waterways

Robert P. Bourman
Nicholas Harvey
Simon Bryars

Introduction

The major part of the city of Adelaide occupies the alluvial lowlands between Gulf St Vincent and the Mount Lofty Ranges. The landscape rises to the east and south along the fault lines underlying Adelaide (Figure 3.1). The vast majority of the streams traversing the plains originate in the ranges, as illustrated in Figure 3.1. There is a marked asymmetry of the River Torrens across the plains, where it is dominated by left-bank tributaries. Relatively high precipitation in the ranges feeds the streams of the lower country. The original streams rarely reached the coast but petered out on the plains or formed wetlands, dammed back by the dune barriers flanking the shoreline.

Catchments of the Adelaide region

There are four major catchments occurring in the metropolitan region: the Northern Adelaide and Barossa; the Torrens; the Patawalonga; and the Onkaparinga. A drainage basin or catchment may be defined as:

An area of land that collects water, which drains to the lowest point in the area which could be either a lake, a dam, or the sea. Rain falling on the land will make its way to this lowest point, via creeks, rivers and stormwater systems. As well as rivers, creeks, lakes and dams, a catchment also includes groundwater, stormwater, wastewater and water-related infrastructure.[1]

The Patawalonga Catchment comprises 230 km² and encompasses a large part of metropolitan Adelaide. It includes Brownhill Creek, Keswick Creek, the Sturt River and the Airport Drain, all of which flow into the Patawalonga Basin at Glenelg. The comparatively larger catchment area of Northern Adelaide and Barossa covers almost 2000 km² and includes the Gawler River, which forms the northern boundary of the Adelaide metropolitan region, the South Para and North Para rivers, Little Para River, and Dry Creek. The Onkaparinga Catchment comprises approximately 920 km² and includes the Onkaparinga and Field rivers, Christie Creek, Pedler Creek, and the Washpool Lagoon, 'one of the last remaining coastal lagoons in metropolitan Adelaide'.[2] The catchment of the River Torrens comprises 620 km² with the upper part arising at Mount Pleasant, beyond the Adelaide metropolitan area. In the lower part of the catchment, the Torrens has five major tributaries flowing from the Adelaide Hills: First, Second, Third, Fourth and Fifth creeks. It is significant to note that the River Torrens was one of the major reasons for Colonel Light's selection of the location of Adelaide.[3]

Channel and floodplain modifications and management

Human impact varies between the different catchments. Before discussing the four major catchments within the region, it is worth noting impacts on the Gawler River at its northern boundary. This river forms an administrative

Chapter photograph: The outlet of Christie Creek.

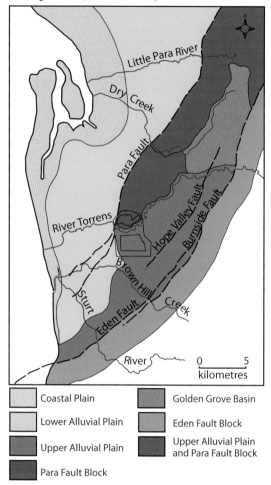

Major faults of Adelaide

Coastal Plain

Lower Alluvial Plain

Upper Alluvial Plain

Para Fault Block

Golden Grove Basin

Eden Fault Block

Upper Alluvial Plain and Para Fault Block

Figure 3.1 Major faults and geomorphic zones of the Adelaide area. Source: modified from M.J. Sheard and G.M. Bowman, 'Soils, stratigraphy and engineering geology of nearsurface material of the Adelaide Plains', Mines and Energy South Australia, Journal 2, Primary Industries and Resources South Australia, 1996.

boundary between two councils, but unilateral flood protection works by either council may have major flood impacts on the adjoining council. This highlights the need to manage flood issues on a catchment basis rather than by administrative boundaries. The two northern catchments within the Adelaide planning region are quite different, with the Little Para being predominantly rural and the Dry Creek Catchment significantly altered by industrial and urban development. The sedimentation in the Dry Creek Catchment has led to management problems such as the accumulation of 1.5 kt of sediment at the Walkleys Road culvert in 1993, and another 0.7 kt in 1994.[4] Elsewhere, there have been attempts to improve the quality of the catchment, such as through the construction of the Salisbury Greenfields Wetlands in the 1980s.

To the south, the Torrens Catchment has been heavily modified. While the catchment has a higher conservation value in its upper reaches, this becomes patchy downstream and many lower sections no longer have any conservation value because of extensive modification. The Torrens used to discharge into a low-lying marsh area called the

The biodiversity of Buckland Park

BOX 10

Buckland Park Lake is located approximately 30 km north of Adelaide on the floodplain of the Gawler River. It is thought to have been created by the damming of several of the stream channels of the Gawler River in around 1920.[1]

Although the hydrology of the mouth of the Gawler River has been significantly altered, Buckland Park Lake is one of the few remaining semi-natural places on the Northern Adelaide Plains where rivers meet the sea. The draft Estuaries Policy and Action Plan for South Australia[2] listed Buckland Park Lake as one of only four recognised estuaries north of Adelaide in the Gulf St Vincent (others included the nearby Salt Creek to the north and the Port River Barker Inlet system to the south).

The area surrounding Buckland Park Lake consists of low, flat alluvial plains with widely spaced stream channels feeding a number of shallow terminal lakes. There are complex and diverse ecological characteristics present, including coastal dune systems, estuarine and intertidal environments, river floodplains, seasonal river channel flows, open freshwater bodies, and mudflats.

Few other places near Adelaide support as diverse a range of habitats within such a small area. Saltmarsh and samphire communities surround Buckland Park Lake, forming a link between coastal and terrestrial vegetation communities. Lignum shrublands exist throughout the terrestrial areas, and river red gum woodlands are found along the Gawler River channel. On the low sand ridges, coastal dune vegetation dominates the landscape, and mangrove forests of conservation significance (*Avicennia marina* ssp. *marina*) are present at the shoreline.

At the landscape scale, the area has natural links with protected areas including the Port Gawler Conservation Park and the dolphin sanctuary to the west. Neighbouring Thompson Creek, Bolivar treatment works land and Middle Beach provide variety between freshwater and brackish water, and contain the rare gahnia filum sedgeland, a known habitat for the endangered yellowish sedge skipper butterfly (*Hesperilla flavescens flavia*).[3] Extensive salt evaporation pans occur to the south of the lake, and to the north and east agricultural uses have resulted in cleared and fragmented native vegetation. A few small remnants near Buckland Park Lake indicate that the surrounding areas would have once supported extensive mallee box (*Eucalyptus porosa*) low woodland.

One of the most notable features of the Buckland Park Lake area is its importance for waterbirds. During a 1989–1990 study over 60 species of waterbirds were recorded as using the lake and surrounding habitat, including many species rarely encountered in the Adelaide region.[1] The study recorded nine species breeding at the lake and cited a further 12 species reported to have bred there in previous years, making it the single most important breeding habitat for a range of waterfowl within the Adelaide region.[1] It is thought that the habitat is highly suitable for waterbirds due to the large annual flood of freshwater that reaches the system, then remains amongst the lignum, bulrush (*Typha domingensis*) and samphire.

Within the lake area there is over 120 ha of lignum shrubland, excluding the areas of river red gum woodland along the Gawler River, representing 96% of the currently mapped remnant lignum shrubland in the Southern Mount Lofty region 4. The importance this places on the role of Buckland Park Lake in biodiversity conservation cannot be underestimated. These shrublands, combined with the flooding regime, contribute significantly to the protection of waterbird habitat and breeding cycles.

Buckland Park Lake must be considered an important biodiversity asset at both a site level and at the landscape scale. It is a valuable link between the freshwater riparian system on the Gawler River and the intertidal saltmarsh and mangrove communities at the coast. The area contains vegetation communities of conservation significance. Preserving the integrity of this habitat contributes to the protection of many native bird species, some of which travel between the lake and areas such as the Port River/Barker Inlet wetlands and the Coorong and Lower Lakes.

Acknowledgement: this information has been compiled from the report M. Durant, 'Buckland Park: summary of native vegetation and threats to ecological assets', Greening Australia, Adelaide, 2007.

Kate Smith

'Reedbeds', landward of the West Beach dunes. In times of high flow, the water drained both to the north through the Port River estuary and also to the south through the Patawalonga Outlet. However, in 1937 an artificial sea outlet was constructed for the Torrens at West Beach, diverting freshwater flow from entering the Patawalonga and Port River estuaries. Given the lack of flow from the Torrens to the Port River, that area was modified due to pressure for urban growth. This area was originally a mangrove swamp with tidal drainage, but the mangrove and samphire vegetation was cleared in the mid 1970s to make way for the construction of the West Lakes waterfront residential development. Although extensive soft sediment (mud, sand and seagrass detritus) was removed, this did not entirely prevent building problems. At times, there have also been water quality problems in the area, with issues such as marine weed outbreaks related to stormwater inflows and the capacity of the marine flushing system.

The section of the Torrens closest to the Adelaide CBD has been modified by the construction of a weir and creation of the artificial Torrens Lake. While this has very little conservation value, it does have a high amenity value. Upstream from the lake, impacts include the O-Bahn transport system along the Torrens Linear Park and significant modification to all five major tributaries flowing from the hills – First, Second, Third, Fourth and Fifth creeks. Warburton[5] described in some detail the state of the five creeks in the mid 1970s.

First Creek begins in Cleland Conservation Park and works its way through Hazelwood Park and Marryatville, but is constrained by urban development and in places has an engineered channel. It travels underground in a few sections as it approaches the CBD before disappearing under Prince Alfred College and re-emerging in the Botanic Gardens before entering the River Torrens near Frome Road. Second Creek has suffered even greater modification. It starts in the hills to the north of Greenhill Road and travels through Stonyfell, Burnside and Leabrook before largely disappearing underground until it enters the Torrens at St Peters.

Third Creek has three main tributaries in the hills near Norton Summit, and travels towards Magill and Tranmere where there are sections of concrete channel. It travels underground as it enters the suburb of Firle before moving through a narrow drainage reserve or concrete channel until it joins the Torrens in Felixstow. Fourth Creek, which has the largest catchment of the creeks, rises in the hills above Morialta Conservation Park and traverses the park's waterfalls, then passing through Rostrevor. Fourth Creek still has a number of river gums and is discernible as a creek course although constrained in many places by urban development. Fifth Creek starts in the hills near Cherryville before traversing Black Hill Conservation Park beside Montacute Road and then turning north-west through Foxfield Reserve, Athelstone and Paradise. Although the creek is undergrounded at Gorge Road, and has some narrow creek reserve sections, like Fourth Creek it benefits from a number of open spaces with river gums. This is not true in the final section of the Athelstone Recreation Reserve, where native vegetation was non existent in the 1970s.[5]

South of the Torrens is the Patawalonga Catchment. In its upper reaches, in the vicinity of the Sturt River Catchment, are some sections of higher conservation value, such as the Sturt Gorge Conservation Park, but in the lower reaches the river is highly impacted to the extent that west of Marion Road it now resembles an open concrete drain. Closer to the hills, construction of the Warraparinga Wetlands has provided an area of amenity value linked to Indigenous heritage values in the area. A flood-control dam was constructed in the Sturt Gorge, and the lower section across the plain was set in a concrete channel to ameliorate flooding problems.

In the lower reaches of the Patawalonga Catchment there has been significant human impact, particularly with the creation and subsequent modifications to the Airport Drain, which merges with the Sturt River, Brownhill Creek and the much smaller Patawalonga Creek to flow into the Patawalonga Lake, formerly a tidal estuary, created by the construction of lock gates at Glenelg. The Patawalonga Basin is important for the ponding of stormwater. Enlarged in 1972 to store 450 ML, subsequent sediment accumulation reduced the storage volume to 300 ML.[6] The basin thus acts as a storage sink for sediment and pollutants, with severe impacts on water quality. Some of the sediment is flushed out of the system under conditions of low tide or during flood events; the sea becomes stained with terrestrial materials, which drift northward along the coast and sometimes require the closure of beaches.

Significant redevelopment projects at Glenelg, near the mouth of the Patawalonga, necessitated flood mitigation works for Patawalonga Lake. The original Patawalonga Basin was formed by construction of a weir and lock system in the 1960s, with the number of flood gates increased from five to eight in the early 1970s following flood issues from stormwater runoff. Later, a weir was built at the northern end of the lake with a stormwater outlet system cut through to the sea to allow the regular flood events to bypass the Patawalonga Outlet. The Barcoo Outlet, comprising underground culverts, was completed in 2001. In the first rains following its opening it was reported that a 'fine black waste' was deposited at West Beach[7] and warnings were made about undertaking 'recreational activities' at both West and Henley beaches.[8] In addition to pollution, the issue of flooding came to the fore in February 2003 when the gates of the Barcoo Outlet did not open during heavy rains, causing flooding in Glenelg North, and partially responsible for flooding in West Beach.[9] In June 2003, severe flooding in Glenelg North with 'an estimated $20 million damage' bill occurred when the Patawalonga weir gate system reportedly failed.[10]

Many of the human impacts have happened in a piecemeal manner, and it was not until the 1990s that the state government adopted a more integrated approach with the establishment of four Catchment Water Management boards in metropolitan Adelaide, beginning with the creation of the board for the Patawalonga and the Torrens catchments in 1995. Two years later, a board was created for the Northern Adelaide and Barossa Catchment, and in 1998 for the Onkaparinga Catchment. The boards were dissolved when the federal government introduced the Natural Resource Management (NRM) system, and separate legislation was introduced for the creation of NRM boards within South Australia. Most of the streams and rivers in the area now fall within the Adelaide and Mount Lofty Ranges NRM region.

Characteristics of Adelaide streams

In many other parts of the world, channel capacities increase downstream with additional water contributed by inputs from tributaries; many Adelaide streams display a diminution in channel capacity downstream. This phenomenon is reflected in Adelaide streams over thousands of years, as fossil channels display downstream diminution in channel size. Thus, there is a tendency for channel capacities to be inadequate to contain the streamflow, with the response that natural levees in the lower stream sections are overtopped and the excess channel capacity of a flow is accommodated by flood-outs.

There are few perennial streams in the Adelaide area. Waterfall Gully, fed from water stored in high-level swamps such as Wilson Bog, is perennial for at least part of its course. However, the vast majority of streams are intermittent or ephemeral. Originally, many streams did not have continuous channels but constituted the 'Chain of Ponds', linked during flood events. Following European settlement, grazing and trampling by stock and deliberate drainage works led to the erosion of continuous channels and the drainage of the ponds, with important impacts on stream hydrology.

A common feature of many Adelaide streams is that valley dimensions are largest near the junction of the hills and the plain. Streams are incised into alluvial deposits, flanked by high-level red-coloured river terraces and alluvial fans, with lower-level grey-coloured terraces, and floodplains occupying valleys cut into the red alluvium. In the lower stream reaches there are further decreases in valley size; there are no terraces and the sediments that originally formed the low-level terraces (grey alluvium) are washed out over the older red-coloured alluvium, forming natural levee banks flanking the stream. In this way, the stream channel can become perched above the general level of the plain so that if a levee is breached, widespread flooding can occur at lower levels even in areas remote from the actual channel. In some instances, downstream channel capacity has been reduced by accelerated sedimentation following vegetation clearance and accelerated erosion on the valley slopes. It is uncommon on the Adelaide Plains for eroded sediments to be transported directly out to sea, but they are stored within the valley as sediment slugs and may be moved further down-valley by successive flood events.

Under natural conditions, many of the streams of the Adelaide region did not reach the sea; they either simply ran out of water as they flowed across the plain, losing water by evaporation or seepage into the underlying aquifers (losing or influent streams), or, due to low gradients, ended in wetlands sometimes dammed back by coastal sand dunes (Figure 3.2). Occasionally, some freshwater was delivered to

the marine environment through streams such as the Gawler River, Smith Creek, and the Port River/Barker Inlet estuary system, which originally comprised inflows from the Torrens River and several other smaller creeks. The Port River estuary has been constrained by the northward development of LeFevre Peninsula over the last 6000–7000 years.[11] Most coastal sediment here was derived from offshore sources during the rise in sea level following the last glacial maximum, and there is a significant biogenic contribution to Adelaide's coastal sediments. However, very little sediment was delivered to the coast by Adelaide streams.

To the south, the original extensive wetland of the Reedbeds in the western suburbs was formed by the blocking effect of coastal dunes. Most freshwater adjacent to Holdfast Bay, including the River Torrens, was initially caught in the Reedbeds.[10, 11, 12, 13] This water would have either recharged the below-ground aquifer, seeped into the nearshore marine environment through the dunes, evaporated, or trickled into the Port River to the north or into the Patawalonga Creek to the south. Nevertheless, any freshwater flowing into the Port River would have been well mixed with seawater by the time it reached the open coast. Further south, the Sturt River flowed into the Patawalonga Creek, which opened directly into Holdfast Bay at Glenelg,[12] and was originally the only break in the sand dunes between Outer Harbor and Seacliff.[13] However, coastal inflows from the Patawalonga Creek would have been minimal.

The southernmost part of the Adelaide region historically would have received some direct freshwater runoff from several small creeks (Field River, Christies Creek, Pedler Creek, Maslin Creek, Willunga Creek, and Aldinga Creek), with the Onkaparinga River being the major source of freshwater inflow[14] into a significant estuary.

Evolution of the major relief features of the Adelaide area

The main relief features of the Adelaide region have resulted from faulting of resistant, ancient rocks of the Adelaide Geosyncline that caused uplift of the Mount Lofty Ranges and downfaulting of the Adelaide Plains and Gulf St Vincent. For example, the Croydon bore penetrates some 600 m below sea level (bsl) before intersecting resistant Precambrian bedrock, which extends up to 700 m above sea level (asl) in the Mount Lofty Ranges – suggesting differential offset in excess of 1000 m. Views from Mount Lofty over the Adelaide Plains give a dramatic impression of the extent of tectonic dislocation responsible for the geomorphic evolution of this landscape, which in turn has established the framework for waterflow across the city.

Figure 3.1 shows the major faults affecting the topography of the Adelaide area. The arcuate faults trend broadly north–south, forming relative lowlands such as the Willunga, Noarlunga and Adelaide-Golden Grove embayments and the Adelaide Plains Sub-basin. The basic tectonic structure of the Mount Lofty Ranges existed prior to the Tertiary, as fault scarps and valleys provided avenues for Tertiary marine incursions and influenced the deposition of terrestrial sediments. Faulting continued throughout the Tertiary

Figure 3.2 'Map of the country between Adelaide and the sea coast', 1882. Image courtesy of the Bodleian Library, University of Oxford, reference I3:26 (3).

and beyond, as revealed by the differential dislocation of sediments, especially marine limestones, with the oldest rocks displaying the greatest offsets. The earth movements continue to today, demonstrated by ongoing minor tremors and more severe earthquakes such as that of March 1954, which seriously damaged buildings near its epicentre at Darlington.

Originally conjoined to Antarctica as part of the super continent Gondwana, Australia had moved sufficiently north by the Eocene (55 Ma ago) to allow a seaway to develop along the southern margin of Australia. Basalt erupted on Kangaroo Island and in the Polda Basin of Eyre Peninsula about 165 Ma ago, possibly a stress precursor of the impending continental separation.

Geological background of the Adelaide area

The geological history of the Adelaide area is significant in explaining and understanding the evolution of Adelaide's drainage network and the ongoing geological changes impacting on it, which, in turn, can influence management of the system. In addition, the geological setting influences such things as the siting of dams, locations of aquifers, and sources of building materials.

The Adelaide Geosyncline, which contains rocks of Precambrian and Cambrian ages, is now occupied by the upland areas of Kangaroo Island and the Mount Lofty and Flinders ranges.[15] Extensive basaltic extrusions accompanied initial rifting and subsidence. A sedimentary sequence exceeding 24 km in thickness was deposited in a gradually subsiding geosynclinal depression. The succession consists largely of shallow water, marine and glacigenic sediments, and was derived from higher surrounding areas.

During the Delamerian Orogeny of Late Cambrian and Early Ordovician age (480–500 Ma ago), the sediments of the Geosyncline were subjected to compressive tectonic forces that resulted in folding, faulting, intrusion of granitic rocks (e.g. Encounter Bay granites), and variable metamorphism. These forces culminated in the formation of a vast fold mountain range, the Delamerides,[16] which probably extended into the then juxtaposed Antarctic continent.

For the next 220 Ma, the geological record is sparse; the area was probably subjected to prolonged erosion, during which the Delamerides were eroded to expose granites emplaced at depths of 10 km.[17]

The next major geological event was that of extensive glaciation during the Permian, some 280 Ma ago, excellent evidence of which is preserved at Hallett Cove and on the Fleurieu Peninsula. The major direction of ice movement across southern South Australia was from the south-east towards the north-west, and striations on granite at Port Elliot demonstrate that the granite was exposed prior to or during the Permian ice advance.

During the Triassic and Jurassic, the entire landmass of the state stood asl so that only fluviatile and lacustrine sediments were deposited, and basalts of Jurassic age[18]

on Kangaroo Island and in the Polda Trough[19] were possibly extruded in response to stresses related to the separation of Australia from Antarctica about 165 Ma ago. As most of the southern part of the state remained asl throughout the Mesozoic, the land surface was subjected to prolonged weathering and erosion.[20] Throughout the Tertiary, reactivated faults resulted in the further uplift of the denuded Delamerides to initiate the formation of the Mount Lofty–Flinders Ranges, a process which continues today.

The succession of Tertiary sediments in the St Vincent Basin began with the deposition of terrestrial freshwater sediments (North Maslin Sand) in the Middle Eocene, followed by the marginal marine South Maslin Sand and later marine sediments including the Port Willunga Formation of the Late Eocene to Oligocene age. A marine transgression occurred during the Pliocene, resulting in deposition of the Late Pliocene Hallett Cove sandstone, which occurs at levels up to 30 m asl and underlies the Adelaide CBD.

The Tertiary was a period of considerable geological change as differential and episodic tectonism led to the broad-scale development of uplands and plains. Weathered land surfaces were uplifted, initiating dissection and the stripping of previously deposited sediments by streams, which cut gorges and deposited widespread channel and lake sediments throughout the Mount Lofty Ranges.

During the Pleistocene, extensions and retreats of glaciers, mainly in the Northern Hemisphere, were accompanied by worldwide glacio-eustatic sea level fluctuations. The earliest Pleistocene marine deposit in the Adelaide area is the Burnham Limestone, which crops out at Marino, Port Willunga and Sellicks Beach. During the majority of the Pleistocene Period, shorelines stood below modern sea level. However, excursions of up to 6 m above the present shoreline occurred during Late Pleistocene interglacials. The marine Glanville Formation indicates the level of the sea during the Last Interglacial some 125 ka ago, when sea level was ~2 m higher and the climate was warmer and wetter than today. During the Last Glacial Maximum of about 20,000 years ago, sea level stood as much as 120 m below present sea level, under which conditions there would have been a land link to Kangaroo Island. Gulf St Vincent was also dry land, and engrafted streams from the Adelaide area would have continued across the exposed seabed and may have linked with the River Murray.

The geomorphic units of the immediate Adelaide area

The immediate Adelaide area has been divided into major geomorphic zones[21] (Figure 3.1).

The coastal zone consists of the St Kilda Formation, which includes both the white sand dunes that back the coastline as well as the estuarine and lagoonal mud facies landward of the dunes. Now much modified by human intervention, the dunes were originally up to 15 m high and 400 m wide and began to form ~7000 years ago when sea level rose and stabilised near its present level. The formation of the sand dunes and the northward drift of sand along the coast formed LeFevre Peninsula and the estuaries of the

Port River and the Patawalonga, as well as the original Reedbeds Wetlands, which existed near West Beach up until the 1930s.

Inland of the estuarine zone is an older line of distinctive red- to yellow-coloured sand dunes, the Fulham sand, on which many of the city's golf courses are built. These older dunes do not display the same linearity of the modern coastal dunes. They broadly parallel the coastline but also seem to have migrated inland from the coast. The age of the Fulham sand has been determined as ~75, 000 years,[22] at which time sea level was possibly 80 m lower than today. Thus the Fulham sand was probably blown from the exposed sea floor of the gulf, and parts of it would have been reincorporated into the modern dunes as sea level rose to near its present position. Other smaller and older relics of dune deposits occur along the Para Fault near Kilburn and Gepps Cross.

The Lower Alluvial Plain occurs to the west of the Para Fault Zone and is underlain by Late Pleistocene alluvial sediments, dominantly those of the Pooraka Formation and younger deposits that include the Waldeila Formation and post-European settlement sediments. These alluvial sediments variously grade to and overlie the marine Glanville Formation of Last Interglacial age (~125 ka). Originally marking a shoreline ~2 m higher than present, the Glanville Formation in the Adelaide area has undergone substantial tectonic depression, especially in the Port Adelaide area.

The Upper Alluvial Plain is also underlain by alluvial sediments of Pleistocene age, west of the Eden–Burnside Fault and occupying part of the fault angle depression between the Burnside and Para faults. Numerous streams draining from the uplifted ranges have deposited alluvium and debris avalanche deposits at the base of the scarp, where the stream channels broaden and become less hydraulically efficient. The Pleistocene sediments of both the Lower and Upper alluvial plains are underlain by older Tertiary deposits such as the Eocene North and South Maslin Sand, and the marine fossiliferous Pliocene Hallett Cove Sandstone.

The Eocene sand of the Golden Grove Basin is preserved in the northern extension of a fault angle depression between the Burnside and Para faults. The Para Fault Block and the Eden Fault Block represent the uplifted, weathered and eroded ancient rocks of the Adelaide Geosyncline. At Anstey Hill, on the eastern margin of the Eden Escarpment, a small bench cemented with iron oxides occurs about halfway up the scarp at an elevation of about 210 m. It is underlain by sands of Eocene age and is a small remnant (~5000 m²) of a formerly far more extensive land surface. Major Edmunds, a cartographer and Boer War veteran who saw its potential as a gun site, named it the 'Gun Emplacement' as it commands splendid views of the city and the coastline. However, it was never used for this purpose.

A series of alluvial fans fronts the major fault-generated escarpments, particularly the Eden-Burnside Fault, the Para Fault and the Willunga Fault, and river terraces flank the major streams flowing across the plains. Stream incision following uplift of the ranges has produced the waterfalls and gorges within the ranges, providing sites for water storages.

River terraces of the Adelaide area

The streams draining the Mount Lofty Ranges have deposited a series of alluvial sediments at the bases of the escarpments as alluvial fans, also forming river terraces and natural levees flanking the modern streams. River terraces are former floodplains now situated above usual flood levels, and have been stranded due to downcutting by the streams following tectonic uplift, a fall in sea level, a climatic change that leads to channel incision, or combinations of these factors. Direct and indirect impacts of human activity may also accelerate stream incision and lead to terrace formation. River terraces record past hydrological events and are very useful in explaining landscape evolution.

In the Adelaide region the river terraces are most commonly formed on alluvial fill and have evolved through alternating aggradation and erosion. For example, initially large valleys were excavated into pre-existing rocks and sediments, following which they were variably infilled with red/brown-coloured Pooraka Formation sediments. Subsequent erosion of the Pooraka Formation developed smaller valleys, which in turn were partially infilled with younger sediments of the grey/black-coloured Waldeila Formation (see Table 3.1). In general, the older terraces occupy the broader, deeper palaeovalleys, and successively younger alluvial fills are progressively smaller and infilled to lower levels, an indication that fluvial activity in the past was considerably greater than at present.

High red river terraces

The Pooraka Formation is one of the most widespread alluvial units across the Adelaide Plains, comprising significant portions of the alluvial fan deposits and river terraces. The Pooraka Formation is a distinctive red- to yellow-coloured alluvial deposit often with a pronounced red-brown soil underlain by a light-coloured calcareous horizon. Commonly, the highest river terraces of the Adelaide region are underlain by sediments of the Pooraka Formation, which Twidale[23] had also named the Klemzig Sand along the valley of the River Torrens. It forms the major terrace formations along the Torrens (Diagram 3.1); Adelaide Oval is sited on a major terrace (Parklands Terrace) underlain by the Pooraka Formation. The deposition of the Pooraka Formation represents a major pluvial event, and at this time the Adelaide Plains would have presented a picture of aggrading, well-vegetated swamps grazed by now-extinct megafauna such as the Diprotodon. Tate noted the occurrence of bones of the giant marsupial in the sediments, describing them in 1879 as the 'mammaliferous drift'.[24] The Pooraka Formation has been demonstrated to be of Last Interglacial age or 125 ka old,[25] and formed when sea level was ~2 m higher than present and when the climate was considerably warmer and wetter than today.

Intermediate terraces

Terraces at intermediate elevations between the highest and lowest terraces occur in many streams having been eroded into the Pooraka Formation sediments. For example, in the lower Torrens Valley the Walkerville, Parklands and Hackney/Felixstow terraces are all formed on the Pooraka Formation

BOX 11

Tulya Wodli
Riparian Restoration Project

Location

This project encompasses a length of approximately 500 m of the River Torrens with its banks in Tulya Wodli (Park 27), within Bonython Park. The site is 100 m at its widest and covers an area of approximately 4.5 ha.

Project background and objectives

Almost all native vegetation was cleared from the site during construction of the Torrens Weir in the 1860s. Since that time, woody weeds and other planted exotics flourished, leaving only small patches of native sedges and rushes along the water's edge.

This project aims to improve habitat diversity by replacing weeds with local native trees, shrubs and sedges. Areas of the site are fairly inaccessible to the public, which will help provide good refuge for fauna once fully restored.

All vegetation works are designed to require minimal ongoing maintenance, provide an attractive destination for residents and visitors, and contribute to the Adelaide City Council's commitment of planting 100,000 indigenous plants within the parklands through its partnership with the SA Urban Forests Million Trees Program.

On-ground works

Stage one of the project started in 2004 with the removal of woody weeds. The riparian zone was smoothed via earthworks and laid with biodegradable erosion matting prior to the beginning of planting.

Stage two, in 2005/2006, included planting the area with local indigenous seedlings and continuing weed control. These plantings included a range of local native trees, shrubs, groundcovers and grasses. Sedges were also used with erosion matting to help stabilise areas.

The final stage was conducted during the 2006/2007 financial year with more plantings and weed control. Consideration was also given to promoting the educational value of the site for schools and community groups.

Original flora

Originally the site was part of a *Eucalyptus camaldulensis* (river red gum) and *Eucalyptus leucoxylon* (South Australian blue gum) riparian association, which would have transitioned into a *Eucalyptus porosa* (mallee box) woodland on the upper banks of the river.

Fauna

Native waterbirds, aquatic reptiles, frogs and various common birds have been observed at the site. When completed, the project is expected to attract additional wildlife to the area.

Key issues
Bank erosion

Due to its proximity to the Torrens Weir, the site is frequently inundated by high-velocity flow after heavy rainfall events.

Woody weeds

There were dense infestations of woody weeds as well as a number of deciduous exotic trees scattered across the site.

Indigenous consultation

Consultation has taken place with the Kaurna Heritage Board for the preparation of community land management plans in the parklands.

Monitoring and evaluation

Staff have established a series of photograph points, which will be used to monitor plant survival. A frog census will be conducted at regular intervals while other fauna will be recorded opportunistically. The site is also being considered as a bird monitoring site as part of a wider program to assess the habitat benefits of revegetation/restoration.

Project partners

This project has been developed with support from the SA Urban Forests Million Trees Program, Adelaide City Council, Adelaide and Mount Lofty Ranges Natural Resources Management Board, and Youth Conservation Corps.

Jock Conlon

Table 3.1 Sediment characteristics

Sediments	Characteristics
Post-European Settlement Aggradation (PESA)	• Presence of European artefacts
Waldeila Formation	• Grey black alluvium of Mid Holocene age (4–6 ka)
Un-named minor alluvial fill unit	• Light grey in colour (40–50 ka)
Pooraka or Christies Beach Formation	• Red-brown of last interglacial age (125 ka)
Taringa Formation (Keswick Clay)	• Columnar, green-grey clay containing angular clasts and calcium carbonate in the upper part of the sequence. In places it may be a mud flow deposit
Ochre Cove Formation (Hindmarsh Clay)	• Red and orange mottled sandy sediment Middle Pleistocene. Contains Bruhnes/Matuyama geomagnetic reversal of 780 ka (Pillans and Bourman, 1995)
Kurrajong Formation (Hindmarsh Clay)	• Lithified, resistant breccia that merges with the base of the Ochre Cove Formation (Middle Pleistocene)
Seaford Formation (Hindmarsh Clay)	• Green grey clayey sand with weak orange mottles. Base interfingers with Burnham Limestone and is Early Pleistocene in age
Burnham Limestone	• Fossils present suggest an Earliest Pleistocene age

Table 3.1 Alluvial/colluvial sediments, basic characteristics and ages. Source: R.P. Bourman, 'River terraces of the Fleurieu Peninsula, South Australia', *Australian Geographical Journal*, 105, 2006, pp. 1–24.

but are at different elevations.[23] Erosion of the unpaired terraces occurred during stream rejuvenation, possibly due to tectonic uplift, but most likely a result of climatic change that favoured erosion of the alluvial deposits. Sporadic occurrences of a light-grey coloured alluvium occur in valleys eroded into the Pooraka Formation. This younger unit has been dated at ~40–50 ka and represents a subpluvial event about this time.

Low grey-black terraces

The next alluvial unit is grey-black in colour and forms a lower set of paired river terraces within valleys carved out of the Pooraka Formation. In the lower reaches of many streams, the grey-black alluvium occurs in natural levees overlapping the older red alluvium. Named the Waldeila Formation in the Noarlunga and Willunga basins by Ward[26] and the Walkerville Sand[23] in the Torrens Valley, these sediments have been dated as of mid-Holocene age (4–6 ka), and reflect a warmer, wetter episode, a minor subpluvial event. There is widespread occurrence of terraces on the mid-Holocene Waldeila Formation, both at the coast and well inland.

In some cases, such as the lower Onkaparinga River, extensive terraces formed on this alluvial material are closely associated with a former (slightly higher) shoreline (1–2 m asl). When sea level stabilised ~7 ka years ago, the lower reaches of the Onkaparinga Valley comprised a broad, shallow, sheltered estuary, as revealed by the presence of estuarine shells dated at ~4.6 ka and incorporated within the Waldeila Formation.[27]

Relationship between the grey and red terraces

When the elevations of the grey and red terraces are traced downstream, three main sets of relationships are revealed. Near uplands, the red terrace is always highest and the low, grey alluvium is set within a valley cut into the red alluvium. Where there is a large stream, such as the Onkaparinga River, and the high country is reasonably close to the coast, the elevational distance between the two sets of terraces is maintained to the coastline. Characteristically, the red terrace grades to the last interglacial shoreline and the grey terrace to the mid-Holocene shore. These terraces have probably been formed by the influences of changing global sea level. However, this simple pattern may have been interrupted by faulting in some localities.

In other situations, particularly where there is a long, low-gradient flow to the coast over a plain (e.g., Gawler River), or the stream has only a small drainage basin (e.g., Sellicks Creek), the terraces converge. Initially occupying a valley cut in the red alluvium, the grey alluvium progressively fills in more and more of the valley, which also decreases in size downstream, eventually spilling out of the valley and forming natural levees that can blanket the red alluvium. In this way, a perched stream can form, increasing the flood risk to areas away from the stream as the water flowing through the channel is well above the level of the surrounding floodplain. In some instances, the location of a cross-over point may be influenced by faulting. For example, the abrupt western margin of the River Torrens terraces is coincident with the Para Fault, where downfaulting to the west would have lowered the level of the red alluvium and initiated its burial by grey alluvium. Notwithstanding this, the formation of cross-over terraces can occur simply by climatic change and variations in sediment availability and the stream's flow regime.

Converging terraces can also occur high up in the ranges, grading to a local base level well asl. These terraces almost certainly reflect the influences of climatically induced changes in bedload/discharge relationships that caused erosion or sedimentation with warmer, wetter pluvial events encouraging deposition in valleys, and colder, drier conditions favouring valley incision.

Terraces and sediments predating the Pooraka Formation

The alluvial sequences of the Adelaide area and their associated terraces are dominated by the Pooraka (~125 ka) and Waldeila Formations (4–6 ka). Rocks and sediments that underlie the Pooraka Formation are of Pleistocene, Tertiary, Permian or older ages. Table 3.1 lists the major Pleistocene fluvial sediments and their possible ages as recorded in the Noarlunga and Willunga embayments. Although occasional relics of sediments and terraces occur older than those associated with the Pooraka Formation, the vast majority of older, terrace-forming Pleistocene sediments have been removed by erosion.

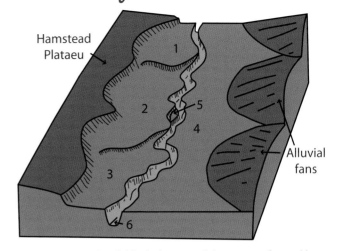

Terraces of the River Torrens

Diagram 3.1 Sketch block diagram of the terraces formed by lateral corrosion during slow incision: 1) Klemzig Terrace; 2) Walkerville Terrace; 3) Parklands Terrace; 4) Hackney/Felixstowe Terrace; 5) depositional terrace remnant of Walkerville Sand; 6) present channel of the River Torrens. Source: C.R. Twidale, *Geomorphology*, Thomas Nelson Aust Ltd, Melbourne, 1968, p. 406.

Human impact on rivers and streams

Human impact has affected streams in a number of ways, such as vegetation clearance in the catchment causing localised erosion, gullying and accelerated sedimentation in downstream sections. In areas of intensive urban development, such as the Torrens, Brownhill and Patawalonga catchments, many streams have been channelised, redirected or even piped underground. Dams and reservoirs have altered flow regimes, and, together with artificial levee banks, such as the construction of the Torrens sea outlet, have altered flood patterns. While many of the earlier modifications were piecemeal, more recent human impacts have been planned or controlled either through catchment or Natural Resources Management authorities.

Gully evolution and management

Pre-European episodes of erosion and sedimentation were triggered by factors such as climate change, tectonic dislocation and global movements in sea level. However, a new and dramatic episode of landscape change was precipitated by the impact of European settlers. The most obvious manifestations of these changes are erosion in stream channels and accelerated sedimentation in downstream sections of river valleys, marked by the burial of European artefacts such as fence posts, wire, metallic objects, bullock wagon chains, cattle bones, horse bones, pottery, bottles, other household items, bridges, building materials, and even building structures.

Spectacular examples of human-induced gullying occur in the Sellicks Creek drainage basin where prominent gullies up to 10 m deep have formed since the arrival of Europeans.[28] The erosion gullies are a consequence of widespread landscape modification in an area of easily erodible

alluvial and colluvial sediments that were associated with swamps and wetlands. The incision of the gullies has drained the swamps, lowered groundwater levels, and dramatically altered the local hydrology. The upper part of the Sellicks Creek drainage basin lies within the Mount Lofty Ranges and extends up to 300 m asl. A break in slope between the ranges and the piedmont alluvial fan zone occurs at ~120 m asl. Base-level controls have been extremely important in limiting the impacts of gullying in Sellicks Creek. Downward erosion of colluvium in the upper part of the basin is limited by outcrops of resistant bedrock and quartz veins. In the alluvial lower drainage basin, resistant silicified Quaternary sediments also limit stream incision, as do artificial base levels where bridges and drains have been constructed across the creek.

Gully development does not seem to be simply related to vegetation clearance within the catchment, but requires some direct or indirect form of channel disturbance. Moreover, not all gullies within a drainage basin need to form at the same time. In Sellicks Creek, several different mechanisms for gully development have occurred at different times. In the upper part of the basin, well-vegetated and stable swamps and wetlands were disturbed with burial by sediment washed down from the valley sides following land clearance. The loss of stabilising channel vegetation, together with slightly steepened slopes, lead to channel incision some time in the late 1850s. Further downstream at Main South Road, erosion was precipitated by changed channel and streamflow characteristics, where pipes constricted flow under the roadway. Channel disturbance by stock trampling and grazing also fostered erosion here. Further downstream at Justs Road, a bridge with a severely restricted sluiceway accelerated spectacular erosion

BOX 12   *Challenges to environmental flows*

Impact of small dams

The impact of farm dams on stream baseflow is severe and increasing. The current dam permit system limits the volume farm dams in a sub-catchment can capture to 50% of the median annual rainfall runoff. This means that streams get flow but only after all upstream dams are filled, resulting in significant change to the period of zero flow in streams in summer and autumn. This places stress on native fish, macroinvertebrates, in-stream vegetation and riparian vegetation. For example, the gauging station records at Mount Pleasant on the upper River Torrens show the average period of summer zero flow in the first decade of records, 1974–1983, was 74.2 days (10.6 weeks), while for the most recent decade, 1998–2007, this has almost doubled to 137.3 days (19.6 weeks)

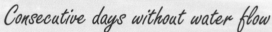

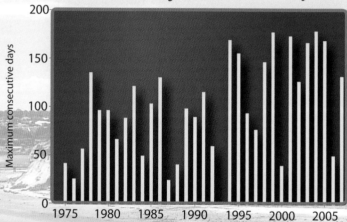

Figure 1. Consecutive days with zero flow at Mount Pleasant flow gauging station, from summer to autumn 1974–2007.

Impact of large storages

A second major impact occurs downstream of the major water harvesting storages. Our early South Australian governments in the 19th and early 20th centuries, with great wisdom invested in small diversion weirs on the two main rivers that capture and divert water to much larger off-stream storages. The first on the River Torrens, the Gorge Weir, completed in 1860, diverted water to Thorndon Park and (later) Hope Valley reservoirs. Clarendon Weir on the Onkaparinga River, completed in 1896, diverts water to Happy Valley Reservoir. In 1918 the Gumeracha Weir began diverting River Torrens water to Millbrook Reservoir. The diversion pipes or aquifers have quite large capacities, hence downstream of these three diversion weirs all the water from small and medium flows is harvested and none flows downstream. Only rare truncated flows from major rain events still pass downstream when the stream flow exceeds the capacity of the diversion pipes. The hydrograph for the flows downstream of the Clarendon Weir for the years 1997–2007 show this clearly (Figure 2) when compared with the hydrograph for Scott Creek, a naturally flowing stream only 8 km distant (Figure 3). The main Onkaparinga River has had to survive on groundwater, and has only had natural flows in spring in five of these 10 years.

The loss of small flows in summer threatens biodiversity as refuge pools dry. The loss of medium to large flushing flows is altering the river sediment composition, geomorphology, and in-stream and riparian vegetation. Important breeding cues for native fish are missed, favouring introduced generalist species like European carp, redfin and trout.

The flow changes wrought by the three large on-stream storages – Warren Reservoir (1917) on the South Para River, Mount Bold Reservoir (1937) on the Onkaparinga River, and the Kangaroo Creek Reservoir (1967) on the River Torrens – are different. As with the diversion weirs, they restrict rain passing downstream. However, they are used to deliver water to downstream diversion weirs in summer. This creates a reserve pattern to the natural river flow. Instead of the natural climatic pattern of low summer flows and high winter flows – which the native fauna and flora have evolved with – we now have zero or low winter flows and unseasonably high summer flows in the reaches downstream of these reservoirs. Downstream of the South Para Reservoir there is zero flow unless the storage is full, but this is a rare occurrence.

A similar situation exists in the reaches downstream of the terminus of the Murray Bridge–Onkaparinga Pipeline at Hahndorf and the dissipaters at Mount Pleasant and Angus Creek that feed Mannum–Adelaide Pipeline water into the River Torrens. Again there are high and very variable summer flows in these reaches of the rivers used as aquifers to transmit and treat River Murray water to off-stream storages.

Environmental flow solutions for small storages

The change caused by farm dams in base stream flows threatens both biodiversity and river ecosystem function. Environmental flows could alleviate the impact but for most of the streams in the Mount Lofty Ranges there are no stored water sources to supply environmental flow releases. The only practical method to alleviate the impact of this critical lack of flow is to allow some flows to bypass on-stream and off-stream farm dams. This can be achieved by either a bypass option or returning a set percentage of all flows downstream. Simple bypass mechanisms exist that pass all flows below a set threshold around a dam. Investigations are needed to find the most cost-effective way to allow flow to start in the small headwater streams and tributaries of the main rivers as soon as the rain begins to fall. Then streams will begin to flow when rain falls and the dams will fill during larger rainfall events, reversing the current situation where dams must fill first. The only impact on farms will be a delay of a few weeks in the time taken to fill.

For larger storages

A trial of environmental flow releases downstream of these weirs is planned to commence soon in the Western Mount Lofty Ranges. At each stream, a variety of base flows, freshets and flushing flows is planned. An associated monitoring program is underway monitoring the predicted changes to stream morphology, vegetation, productivity and biodiversity.

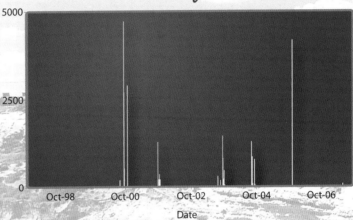

Flow downstream of Clarendon Weir

Figure 2. Gauged flows downstream of the Clarendon diversion weir from 1996 to 2006. ML/d = megalitres per day.

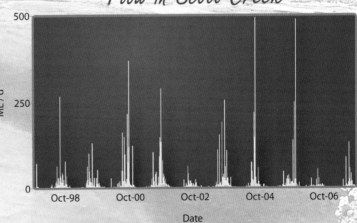

Flow in Scott Creek

Figure 3. Gauged flows in Scott Creek from 1996 to 2006. ML/d = megalitres per day.

Source (figures 1–3): Adelaide and Mount Lofty Ranges NRM Board.

Peter Schultz

downstream. The bridge was built in the 1920s and erosion began shortly after that. The lowest section of about 1 km was rapidly eroded into a box-like canyon up to 8 m deep in 1911, when a local landowner diverted ponded water from his land onto the roadway. Artificial wetlands have recently been constructed in this section of the creek.

The following generalised model of gully development has been reported by Bourman and Harvey[29] and Bourman and James.[28]

1. Gully development is related to landscape modifications such as vegetation clearance, destruction of groundcover, cultivation and engineering works, thereby increasing the susceptibility of the landscape to erode.

2. A lag time ensues between the initial disturbance and an erosional response until the optimum conditions for gullying are achieved, e.g., very wet years, storm events, time for plant roots to decay.

3. Sudden and dramatic gully initiation is possibly related to the occurrence of intense, drought-breaking rains.[30]

4. Rapid gully development can occur with some gullies forming in response to individual storm events or forming over short-time periods.

5. Slowing down of erosion as adjustments to the new conditions are established. Erosion proceeds but at a much reduced rate.

6. Under a new regime, some gullies gradually stabilise and some may actually start to heal, especially where stock are excluded and the channel floors become vegetated. Successive aerial photographs (1949 to present) of Sellicks Creek reveal only minor changes in the gully system, and that a new quasi-equilibrium condition is being approached.

The main cause of gullying at Sellicks Creek appears to be related to channel vegetation disturbance and agricultural activities, though various other factors such as road, culvert and bridge structures, which impact on streamflow, are locally significant. Disturbance of the valley floor vegetation is the dominant cause of gully development in the Sellicks Creek Catchment. Without disturbance of the channel vegetation catchment, clearance on its own will probably not initiate erosion. Consequently, an effective mechanism for stabilising gullies appears to be by reducing stream velocity by plantings to increase channel roughness. The alluvial sediments forming alluvial fans at the foot of the Mount Lofty Ranges are very susceptible to erosion and great care needs to be taken in managing waterflow across them. Within the last decade, a new gully was initiated north of the Sellicks Creek drainage basin by inappropriate diversion of water from Main South Road, resulting in the formation of a long V-shaped gully up to 6 m deep that prevented access to a large area of a property.

Post-European Settlement Aggradation (PESA)

Prior to recent landscaping, a spectacular example of Post-European Settlement Aggradation (PESA) occurred along Dry Creek, Mawson Lakes, about a kilometre downstream of Main North Road. Accelerated erosion had cut a vertical channel some 6 m deep through alluvial deposits, including 2–3 m of PESA sediments, overlaying Pooraka Formation deposits. Dry Creek appears to be relatively insignificant but it actually drains an area of 109 km^2 or some 40% of Adelaide's suburban area. Originally, Dry Creek dissipated into the alluvial deposits of the plains, rarely reaching the sea; today the creek is discharged via artificial drains through samphire flats and mangroves into Barker Inlet.

Sedimentation in this section of Dry Creek followed land clearance for agricultural purposes in the few decades after European settlement, which began in 1836. Large amounts of charcoal occur in the PESA sediments, and a tree trunk, dated at about 400 years before present, could suggest that the charcoal was a product of Aboriginal burning practices. However, the presence of a glass bottle, a European artefact of the 1850s, also at the base of the PESA deposits, indicates that European agricultural activities were responsible for the accelerated erosion and deposition. Following deposition of the PESA sediments, largely as levees flanking the stream channel, Dry Creek incised into its own floodplain, cutting a deep and steep-sided channel. Various factors may have influenced this erosion phase – the initial deposition of sediments would have buried and killed riparian vegetation, which is so important in maintaining channel stability, thereby initiating erosion. Alternatively, as suggested by Schumm,[31] following a reduction in the delivery of sediments to the stream channel, as erosion slows down on the valley sides of the catchment, with enhanced runoff from the hillsides, the stream is likely to cut back down through its own recently deposited sediments. In many other instances across the Adelaide Plains, PESA deposits are perched at the top of the present stream channels after having completely infilled, choked and overtopped pre-existing channels.

In Dry Creek there have been several events of PESA, marked by different vintages of European artefacts, the most recent episode involving plastic materials. The Dry Creek drainage basin is now intensively urbanised, which has increased discharges, reduced immediate stream loads, exacerbated erosion downstream, and caused accelerated erosion even further downstream. Artificial structures across streams invariably have unwanted impacts. Concrete drains under roads, for example, often lead to erosion immediately downstream, and that eroded material is dumped further downstream.

Changes in waterflow to the coast

Today, the channelisation of streams, the impermeable nature of much of the urban development, the discharge of stormwater into the sea, and the cutting of an artificial outlet for the River Torrens from Breakout Creek directly to the sea at West Beach, have vastly increased waterflow into the sea. The increased freshwater flows containing pollutants and nutrients, along with discharge of treated sewage, have had detrimental impacts on the stabilising seagrass meadows

The flood of 1931

BOX 13

It is most appropriate to describe the impact and scope of the 'great' River Torrens flood of 1931 in the words of a reporter for the *Chronicle* at the time, as published on 10 September 1931:

Hundreds of houses flooded

Sweeping down from the hills, stormwaters, fed by torrential rain in the Mount Lofty Ranges ... devastated wide areas in the eastern, southern and western suburbs causing damage estimated at over £30,000. Hundreds of people were rendered homeless through the flooding of their houses, many square miles of country between Adelaide and the sea were inundated, scores of market gardeners irretrievably ruined, and miles of road placed under water or blocked by landslides.

Overflowing its banks, First Creek, swept in two streams through Marryatville, Norwood and Hackney, flooding 30 or 40 houses a foot deep in King William and Rundle Streets, Kent Town ...

Transformed by the swollen waters of Third, Fourth, Fifth and Sixth Creeks, and by innumerable freshets which grew into steams overnight, the Torrens swept down chains wide. It broke its banks in many places, cutting off road and tram traffic to Henley and flooding 150 houses.

In most places the stormwater passed away quickly; but, owing to the immense quantity flowing down the Torrens, which was 60 or 70 yards [54–64 m] wide at the back of the Zoo, the low lying country near Henley Beach will be inundated for many days. There the flood was worse than that of 1917, for a count last night showed that at least 150 houses were flooded ...

David Jones

Floods at George Street, Norwood, 1931 (SLSA: B 5995). Image courtesy of the State Library of South Australia.

in Adelaide's coastal waters. Loss of seagrasses increase the impacts of waves, which erode sand from the surf zone and beaches. In addition, urbanisation has created impermeable surfaces such that many stormwater drains discharge directly into the sea across sandy beaches.

Coastal inflows can be grouped into three broad geographic zones: Port Gawler to Outer Harbor; Holdfast Bay (defined here as Outer Harbor to Seacliff); and Seacliff to Sellicks Beach. Much of the information here is sourced from recent work completed by the Task IS1 researchers from the Adelaide Coastal Waters Study.[14, 32, 33, 34]

Coastal inflows from the Gawler River have been modified significantly, with the total volume greatly reduced by the construction of South Para Reservoir.[14] There are now only occasional inflows during major rainfall events or when the reservoir overflows. Due to diversion of the Torrens River directly to Holdfast Bay (see opposite page), the construction of evaporation ponds, effluent treatment ponds and artificial wetlands, the Port River/Barker Inlet estuary now receives very little freshwater. For 70 years, 1935–2005, the Port Adelaide wastewater treatment plant (WWTP) discharged treated effluent into the Port River near Ethelton. The Port Adelaide WWTP no longer discharges to the Port River but is instead redirected to Bolivar WWTP.

One of the most ecologically significant changes to coastal freshwater inflows in the Port Gawler/Outer Harbour zone was the opening of the Bolivar WWTP outfall, north of St Kilda in 1967. This outfall currently delivers 35.3 GL/yr of treated effluent directly to the marine environment (average volume for period 1995–2005).[34] Reuse of effluent for irrigation from Bolivar did reduce flows to the sea for a while, but redirection of effluent from the Port Adelaide WWTP has maintained flows from Bolivar at recent volumes.

Direct freshwater inflows to Holdfast Bay have increased dramatically since European settlement, especially since the 1930s. Significant 'events' in the Holdfast Bay zone include:

- Opening of three treated effluent outfalls from Glenelg WWTP in 1943, 1958 and 1973. The Glenelg WWTP now delivers 17.3 GL/yr of treated effluent to the nearshore marine environment at West Beach (average volume for period 1995–2005).[34]
- Operation of the Glenelg WWTP sludge outfalls several kilometres offshore from Glenelg, 1961–1993.
- Operation of the Port Adelaide WWTP sludge outfall ~4.5 km offshore from Semaphore, 1978–1993.
- Diversion of the Torrens River away from the Reedbeds and through the sand dunes directly to the sea to form Breakout Creek during the 1930s.
- Construction of concrete stormwater drains and an efficient stormwater system from Adelaide into Holdfast Bay (since the 1950s).
- Construction of a weir at Patawalonga Creek outlet in 1890, which allowed some regulation of freshwater inflows to the sea.

Collectively, these events have resulted in the most dramatic changes to freshwater inflows along the entire Adelaide coast, with an average of 114.9 GL/yr entering the marine environment as stormwater, and an additional 62.0 GL/yr from WWTPs during 1995–2005[34] – in an area where historical inflows were minimal.

Direct flows from the Onkaparinga River have been dramatically reduced and flow regimes altered since the construction of Mount Bold Reservoir.[14] The smaller creeks flow to the sea intermittently with relatively small annual volumes,[14] but when they do it is often with slugs of highly turbid water during major rainfall events. The opening of the Christies Beach WWTP in the 1970s[33] marked a significant new source of freshwater to the Seacliff/Sellicks Beach zone; it delivered 9.4 GL/yr for the period 1995–2005.[34]

It can be seen that the nature of freshwater flows to Adelaide's coastal waters has changed dramatically since European settlement. For the period 1995–2005, the total freshwater inflow per annum to the marine environment between Port Gawler and Sellicks Beach averaged 177 GL, containing 2357 t of nitrogen and 8428 t of suspended solids.[34] Much of this freshwater discharge is from sources that did not exist prior to European settlement. For example, the 83.4 GL/yr of freshwater discharged directly to the sea from the Bolivar, Glenelg and Christies Beach WWTPs, and the Torrens River is completely unnatural (average for 1995–2005).[34] However, in some cases, such as the Gawler and Onkaparinga rivers, it is likely that the volume of natural catchment flows has decreased.[32]

Coastal impact of stream and river flow

Stormwater and wastewater discharges contain various pollutants including nutrients (viz. nitrogen and phosphorus), suspended solids, colour-dissolved organic matter (CDOM), and toxicants such as heavy metals. Stormwater is often highly turbid (suspended sediments and/or CDOM), while treated wastewater is clearer but with much higher nutrient concentrations. Digested sludge, which is no longer discharged, mostly suspended solids rather than freshwater, high in nutrients and also highly turbid. Toxicants such as pesticides are detected intermittently and/or at generally low levels in stormwater.[14, 35]

Elevated levels of nutrients have been regularly detected in Holdfast Bay since at least the 1970s; these are caused by WWTP and industrial outfalls (and to a much lesser extent by stormwater).[35] While dissolved nutrients can be difficult to detect in marine waters, a survey of nitrogen-stable isotopes in seagrass tissue indicates that seagrasses along the entire coast from Port Gawler to Port Noarlunga are receiving nitrogen from the three local WWTPs and other industrial sources.[35] Highly turbid coastal waters in Holdfast Bay, caused by pulsed stormwater events, have been observed since the 1930s.[34] Indeed, photosynthetically active radiation levels have recently been measured at zero on the seabed off Henley Beach during some stormwater events[36] (e.g., Figure 3.3). Significantly, any discharges into Adelaide's nearshore coastal waters are trapped nearshore by prevailing winds and moved up and down the coast by tides.[37] Elevated levels of heavy metals in sediments of the Port River near Ethelton are due to decades of discharges from the Port Adelaide WWTP.

Figure 3.3 Stormwater from the Torrens River discharging into the sea, October 2005. Photographer: S. Bryars.

Alteration to freshwater inflows to Adelaide's coastal waters, in terms of both the quantity and quality of water, have had an impact on the local marine habitats. There are three main subtidal benthic habitats off Adelaide's coast, at depths of <20 m: seagrass meadows (dominated by *Amphibolis* and *Posidonia*); reefs; and sand.[38, 39] Seagrass habitat is dominant in the north from Port Gawler to Seacliff, with reefs and sand dominant in the south from Seacliff to Sellicks Beach.[39] Sand is also now common in some parts of the north due to several localised zones of seagrass loss. North of Outer Harbor, intertidal habitats consist of extensive tidal mud/sandflats (vegetated with seagrasses or unvegetated), tidal creeks, mangrove forests (*Avicennia marina*), and saltmarshes.[38] South of Outer Harbor, the main intertidal habitats are sandy beaches and rocky shores.[38] There are also several natural estuaries along Adelaide's coast including the Gawler River, the Port River/Barker Inlet estuary, the Patawalonga Creek, and the Onkaparinga River.[38] The pelagic habitat (or overlying water column) may also be classified as a habitat in its own right.

Freshwater discharges can potentially be detrimental to marine habitats in a number of ways: freshwater per se (i.e., causing a decrease in salinity); nutrients (viz. nitrogen and phosphorus); sediments (causing a decrease in water clarity and/or sedimentation); CDOM (causing a decrease in water clarity); and toxicants (e.g., pesticides, heavy metals). With changes to freshwater inflows, sediment and nutrient levels in Adelaide's marine waters have increased dramatically. Recent experimental work has clearly shown the detrimental effects of elevated nutrient levels on

Amphibolis and *Posidonia*, even at very low concentrations.[40] It is believed that the nutrients stimulate epiphyte growth on the seagrasses, eventually causing the seagrasses to die. It is also apparent that *Amphibolis* is more sensitive than *Posidonia* to elevated nutrients,[39, 41] which explains some of the patterns of selective seagrass loss. It is thought that increased nutrient levels have mainly affected local seagrasses, rather than factors such as reduced water clarity, reduced salinity or exposure to toxicants.[41] Experimental work has also shown a link between the degradation of temperate reefs and the addition of nutrients and/or sediments.[42, 43, 44] Indeed, the degradation of Adelaide's reefs is thought to have been caused by increased nutrients and increased sedimentation.[45]

Besides seagrasses and reefs, some other marine habitats can also be affected by increased nutrients. For example, increased nutrient levels can stimulate growth of the macroalga *Ulva*, which can then degrade intertidal mangroves (*Avicennia marina*) by smothering both juvenile plants and the aerial roots of adult plants. Increased nutrient levels can also cause phytoplankton blooms in the water column, including those comprised of toxic dinoflagellates.

The loss of seagrasses (estimated 9000 ha) off Adelaide since the 1930s is well documented[35, 46, 47, 46] with much of this linked to freshwater discharges from WWTP outfalls and stormwater outlets.[47] As with seagrass meadows, reefs off Adelaide have suffered from the effects of freshwater discharges, but the decline of reefs is not as well documented. Nonetheless, it is evident that over the past few decades many of Adelaide's reefs have become degraded, most notably with a significant reduction in the cover of large canopy-forming macroalgae.[45] The sections over the page explore specific cases of habitat degradation in more detail.

- The discharge of treated effluent and digested sludge from WWTPs is linked with habitat degradation in every location where it has occurred along Adelaide's coast.
- Bolivar WWTP effluent outfall caused complete loss of 900 ha of intertidal and subtidal seagrasses,[46] selective loss of *Amphibolis* further offshore,[39] and loss of intertidal mangroves.[46]
- Port Adelaide WWTP effluent outfall linked to toxic algal blooms in the Port River.[46]
- Port Adelaide WWTP sludge outfall caused complete loss of 365 ha of seagrasses and selective loss of 1100 ha of *Amphibolis*.[39, 46]
- Glenelg WWTP effluent outfalls linked to complete loss of 2800 ha of seagrasses in nearshore zone from Semaphore to Seacliff.[46]
- Glenelg WWTP sludge outfalls caused small area of complete seagrass loss,[48] but a much larger area of selective loss of *Amphibolis*.[39] Reefs in the area that are degraded (e.g., Broken Bottom, Seacliff Reef)[45] could also have been affected.
- Christies Beach WWTP effluent outfall; seagrass is not a dominant habitat in this area but there may have been some small losses.[35] Degradation of nearby Horseshoe and Port Noarlunga reefs has occurred (e.g., loss of canopy-forming macroalgae), and this may be linked to the effluent discharge;[45] impacts on the benthic infauna of sand habitat that have been documented.

Due to vegetation clearance, the proliferation of impervious surfaces, and a network of stormwater drains, freshwater flows now generally occur as major pulses of turbid water. Nonetheless, while the Torrens River and the Patawalonga Outlet have been linked to the nearshore seagrass loss in Holdfast Bay, they are unlikely to have been the major cause of nearshore loss.[41] In contrast, the Torrens River and other stormwater sources are thought to have had a major impact on coastal reefs through increased sedimentation.[45] While reefs further south, such as Port Noarlunga Reef, would historically have been exposed to some freshwater flows from the Onkaparinga River, the nature and quality of the flows would have changed dramatically with the construction of Mount Bold Reservoir, clearance of the catchment, and the input of turbid freshwater from the River Murray into the Onkaparinga River system. There is also speculation that turbid discharges from Christies Creek are having a detrimental impact on Horseshoe and Port Noarlunga reefs.

Apart from impacts on marine habitats, stormwater causes other issues for Adelaide's coastal waters, such as increased turbidity and rubbish (an amenity issue), and increased levels of faecal bacteria (a public health issue).

Changes to the nature and extent of freshwater inflows can also affect estuarine fish that rely on the marine/freshwater interface for part of their lifecycle (e.g., congolli) – an issue that is not explored further in this chapter but which is probably significant. In this respect, it is worth noting that in estuarine locations, where freshwater inflows were a natural occurrence, then some level of freshwater input is still required. In contrast, in locations where freshwater inflows are an unnatural occurrence, inflows should be ceased or have nutrient and sediment levels within them drastically reduced.

While freshwater inflows have caused or been linked to the majority of degradation of Adelaide's seagrasses and reefs, some degradation is due to other processes. For example, dredge spoil dumping off Outer Harbour ('spoil ground') has smothered many hectares of seagrasses; increased sedimentation in Largs Bay since construction of the Outer Harbour break walls has caused the loss of many hectares of seagrasses;[46] increased erosion and expansion of blowouts due to erosional processes continues to cause seagrass loss in southern Holdfast Bay[47] (although the initial and ongoing link with freshwater discharges and seagrass degradation in this process is unclear); and reef degradation in the Port Noarlunga area has been linked to increased sedimentation caused by sand-dredging operations off Port Stanvac.[45]

The significant degradation that has occurred in Adelaide's marine habitats as a result of freshwater inflows has many ecological and management consequences. For example, the loss of seagrass and macroalgal cover will mean a loss of habitat for many species (including commercially and recreationally important fish species), a loss of primary production (i.e., photosynthetic services), and, in the case of seagrasses, loss of an important sediment binder. Indeed, the loss of seagrasses in the nearshore of Holdfast Bay has exacerbated the longshore drift of sand in the area, which is a major management problem in itself. A further problem with seagrass loss (viz. *Amphibolis* and *Posidonia*) is that recovery is a very slow process, if it occurs at all. For example, while some recovery of *Posidonia* is occurring at the disused sludge outfall sites, recovery to original coverage at the Port Adelaide WWTP sludge outfall will take many decades.[48, 49] (On the positive side, it also means that if discharges are ceased then seagrass recovery may be possible.) It is apparent that natural recovery of *Amphibolis* is not occurring at either of the sludge outfalls,[39] and attempts at artificial rehabilitation of *Amphibolis* are currently underway in Holdfast Bay.[48]

Unnatural freshwater inflows containing elevated nutrients and sediments have had a devastating impact on Adelaide's marine habitats, and yet we continue to discharge many gigalitres of freshwater into the marine environment that could otherwise be reused on land. If we are to have any chance of natural recovery or rehabilitation of our seagrasses and reefs, all unnatural freshwater flows (and the pollutants they carry) need to be reduced dramatically; a move that would appear obvious given the major damage they have caused, and that freshwater is such a precious commodity for Adelaide.

Why conserve the Field River?

Urban ecosystems are usually defined as areas 'in which people live at high densities (equal to or greater than 186 people/km²), or where the built infrastructure covers a large proportion of the land surface'. Currently, slightly less than half the world's population live in dense urban areas, particularly cities. In industrialised countries this percentage is closer to 80%. The situation in Australia is particularly significant, as nearly 85% of Australians live in towns with 1000 or more inhabitants, while two thirds of the population live in the eight capital cities. Two of the principal concerns for promoting biological diversity within urban environments are the maximisation of biodiversity within the city and the management of problem species (such as introduced and native weeds). In addition, as the world's population increasingly includes people with an urban life-style, it is largely in this urban environment that they will gather their views on nature.

These issues are particularly marked in Adelaide. Adelaide is a unique city, best described as a garden city with relatively few parks and even fewer natural remnants (Table 1). Historically, the Adelaide region has supported a diverse range of natural habitats. Before 1836, the Adelaide Plains supported approximately 1130 species of native plants and around 285 native species of bird (including migratory and nomadic species). The Adelaide area has now lost many of its native plant, bird, mammal and reptile species over the last 166 years, because the area was not managed to protect biological diversity. Now, less than 3% of the natural vegetation remains on the Adelaide Plains. The larger patches of remnant vegetation are situated within the national, conservation and recreation parks located throughout the Adelaide area and a few degraded regions such as the Field River (Table 1).

Presently, the major threat to local biodiversity is the limited space. In locations such as the Field River, where the space *is* available, uncontrolled recreational activities, inappropriate management and rubbish dumping are also threatening the continued survival of this ecosystem. It is now vital that the importance of appropriate land management strategies be highlighted in order to combat and mitigate the harmful and long-lasting effects of these particular practices. Development will continue, but, as stressed by the National Conservation Strategy of Australia, development needs to be within the sustainable capacity of the natural environment. The four objectives of the national strategy are: preserving genetic diversity; ensuring the sustainable use of species and ecosystems; preserving essential ecological processes and life-support systems; and sustaining and enhancing environmental qualities, thus improving quality of life. Conservation leading to the sustainability of the few remaining open spaces such as the Field River is now becoming a priority for councils and communities, with management of the remaining species and habitat types the main priority. Maintenance and development of the Field River could be a world first – the retention of a natural system inside a city – to highlight the importance of and need to conserve biodiversity.

Park type	Park name	Park area (ha)
National park	Belair	816
	Onkaparinga River	1490
Conservation park	Cleland	993
	Horsnell Gully	245
	The Knoll	1.7
	Torrens Island	79
	Morialta	540
	Port Gawler	433
	Montacute	201
	Black Hill	701
	Fort Glanville	5
	Ferguson	8
	Hallett Cove	50.2
	Eurilla	7.5
	Angove Scrub	5.19
	Kenneth Stirling	253
	Mount George	66.7
	Mark Oliphant	181
	Mylor	45.7
	Scott Creek	706
	Marino	30
	Moana Sands	21.4
	Aldinga Scrub	265
Recreation park*	Onkaparinga estuary	290
	Para Wirra	1426
	Brownhill Creek	51
	Greenhill	27
	Shepherds Hill	88
	Sturt Gorge	178
	Cobbler Creek	288
	Anstey Hill	307
	O'Halloran Hill	289

Source: Turner, 'Conserving Adelaide's biodiversity resources', SA Urban Biodiversity Program, Department for Environment and Heritage, 2001.

Table 1. Summary of the size of the national, conservation and major recreation parks in the Adelaide metropolitan area.

Chris Daniels

Conclusion

There has been a marked change in attitude towards waterflow across the plains from the hills over the past decade. Prior to that time, Adelaide's water supply seemed assured from the reservoirs in the hills, and through pumping from the River Murray. Rainwater tanks were deemed to be unnecessary or even a health risk. Drainage schemes were developed to evacuate stormwater from the suburbs to the sea as rapidly as possible to ameliorate the flood risk, a risk exacerbated by the very development of paved roads and houses. Water in the suburbs was seen as a nuisance and pollutants associated with the runoff made storage of the water an undesirable activity.

Under pre-European conditions, flooding occurred but the majority of floodwaters were probably retained on the land, forming swamps and wetlands and recharging aquifers. Some freshwater did reach the sea and ecosystems had become established in balance with variable salinities experienced in different parts of the estuaries and near-shore waters.

The long-term geological view reveals that past stream discharges were considerably greater than at present, and that there has been an general diminution in waterflow since 125 ka when wet and warm conditions favoured the deposition of vast alluvial deposits (the Pooraka Formation) across the Adelaide region. Sea level at this time was some 2 m higher than present. The general drying trend was punctuated by minor pluvial events about 40 ka and 4–6 ka ago. Colder episodes were associated with drier and windier conditions. These severe conditions culminated about 20 ka ago when sea level was up to 150 m lower than now. We are currently experiencing mild interglacial conditions that have favoured human activities.

Today there is a move towards returning the hydrological regime of the Adelaide region back to that of pre-European times. Even though extensive areas of the city of Adelaide are covered with hard surfaces, water is being retained on the land by the use of increasing numbers of rainwater tanks, and there have been some highly successful wetland developments with associated aquifer recharge schemes developed, such as those at Parafield Airport. We are concerned about discharging sediment and pollution-laden waters into the gulf, where past events have severely impacted on the nearshore ecology. Care needs to be taken in planning to avoid flood-prone areas and to maintain any surviving wetland environments, which provide an amenity for humans, a habitat for wildlife, and for the preservation of biodiversity.

Acknowledgements

We would like to thank Kris James for her research and editing assistance.

References

1. Onkaparinga Waterwatch Network, 'What is a catchment?' (online).
2. QED Pty Ltd, Green Environmental Consultants, Epawe and G. Carpenter, 'Washpool Lagoon and Environs Management Plan', report no. 07–104, City of Onkaparinga and Planning SA, endorsed April 2008 (online).
3. AMLRNRM Board, 'History of the Torrens Catchment', 2008 (online).
4. PPK Environment and Infrastructure Pty Ltd, Dry Creek and Little Para Drainage Authorities and MFP Corporation, 'Dry Creek and Little Para catchments: Integrated Catchment Water Management Plan', 1997.
5. J.W. Warburton, *Five metropolitan creeks of the River Torrens, South Australia: an environmental and historical study*, Civic Trust of South Australia and Department of Adult Education, University of Adelaide, 1977.
6. B.C. Tonkin and Associates, 'Patawalonga Catchment Water Management Plan', Patawalonga Catchment Water Management Board, Adelaide, 1996.
7. S. Fewster, 'Coastal pollution outlet for anger', *Advertiser*, 28 March 2002, p. 19.
8. D. Gordon and B. Hurrell, 'It was built to save the beaches, but today it is threatening them: our pipeline to pollution', *Advertiser*, 27 March 2002, p. 1.
9. E. Miller, 'Gate failure remains a mystery', *Guardian Messenger*, 26 March 2003, p. 11.
10. C. Hockley and M. Clemow, 'Floods of tears', *Advertiser*, 28 June 2003.
11. G.M. Bowman and N. Harvey, 'Geomorphic evolution of a Holocene beach-ridge complex, LeFevre Peninsula, South Australia', *Journal of Coastal Research*, 2, 1986, pp. 345–362.
12. J.B. Cleland, 'The flora between Outer Harbour and Sellick's Beach, South Australia' in Fenner, C. and Cleland, J.B. (eds) *The geography and botany of the Adelaide coast*, Field Naturalists' Section of the Royal Society of South Australia, Adelaide, publication no. 3, 1935.
13. S.A. Lewis, 'Gulf St Vincent water pollution studies 1972–1975', report of the Committee on the Effects of Land-based Discharges from metropolitan Adelaide Upon the Marine Environment of Gulf St Vincent, EWS 75/14, E&WS, Adelaide, 1975.
14. J. Wilkinson, J. Hutson, E. Bestland and H. Fallowfield, 'Audit of contemporary and historical quality and quantity data of stormwater discharging into the marine environment, and field work programme', ACWS technical report no. 3 prepared for the Adelaide Coastal Waters Study Steering Committee, Department of Environmental Health, Flinders University of South Australia, Adelaide, July 2005.
15. W.V. Preiss, 'The Adelaide Geosyncline, late Proterozoic stratigraphy, sedimentation, paleoontology and tectonics: geological survey of South Australia', *Bulletin*, 1987, p. 53.
16. B. Daily, J.B. Firman, B.G. Forbes and J.M. Lindsay, 'Geology' in Twidale, C.R., Tyler, M.J. and Webb, B.P. (eds) *Natural history of the Adelaide region*, Royal Society of South Australia, 1976, pp. 5–42.
17. A.R. Milnes, W. Compston and B. Daily, 'Pre- to syn-tectonic emplacement of early Palaeozoic granites in south-eastern Australia', *Journal Geological Society of Australia*, 24, 1977, pp. 87–106.
18. B. Daily, C.R. Twidale and A.R. Milnes, 'The age of the lateritized summit surface on Kangaroo Island and adjacent regions of South Australia', *Journal Geological Society of Australia*, 21, 1974, pp. 387–392.
19. A.R. Milnes, B.J. Cooper and J.A. Cooper, 'The Jurassic wisanger basalt of Kangaroo Island, South Australia', *Transactions Royal Society of South Australia*, 106, 1982, pp. 1–13.
20. A.R. Milnes, R.P. Bourman and K.H. Northcote, 'Field relationships of ferricretes and weathered zones in southern South Australia: a contribution to "laterite" studies in Australia', *Australian Journal of Soil Research*, 23, 1985, pp. 441–465.
21. M.J. Sheard and G.M. Bowman, 'Soils, stratigraphy and engineering geology of nearsurface materials of the Adelaide Plains', report book 94/9, Mines and Energy South Australia, Adelaide, 1996.
22. R.P. Bourman, A.P. Belperio, C.V. Murray-Wallace and J.H. Cann, 'A last Interglacial embayment fill at Normanville, South Australia, and its neotectonic implications', *Transactions Royal Society of South Australia*, 123 (1), 1999, pp. 1–15.
23. C.R. Twidale, *Geomorphology*, Thomas Nelson (Aust) Ltd, Melbourne, 1968.
24. R. Tate, 'On the discovery of marine deposits of Pliocene Age in Australia', *Transactions of the Royal Society of South Australia*, 13, 1890, p. 172.
25. R.P. Bourman, P. Martinaitis, J.R. Prescott and A.P. Belperio, 'The age of the Pooraka Formation and its implications, with some preliminary results from luminescence dating', *Transactions Royal Society of South Australia*, 121 (3), 1997, pp. 83–94.

26. W.T. Ward, 'Geology, geomorphology and soils of the south-western part of county Adelaide, South Australia', CSIRO soil publication no. 23, 1966.

27. R.P. Bourman, 'Some aspects of the geomorphology of the Onkaparinga River', *Taminga*, 9 (2), 1972, pp. 117–143.

28. R.P. Bourman and K.J. James, 'Gully evolution and management: a case study of the Sellicks Creek drainage basin', *South Australian Geographical Journal*, 94, 1995, pp. 81–105.

29. R.P. Bourman and N. Harvey, 'chapter 4: landforms' in Nance, C. and Speight, D.L. (eds) *A land transformed: environmental change in South Australia*, Longman Cheshire Pty Ltd, Adelaide, 1986, pp. 78–125.

30. D. Dragovich, 'Gullying in the Mount Lofty Ranges', *Australian Journal of Science*, 29, 1966, pp. 80–81.

31. S.A. Schumm, *The fluvial system*, John Wiley & Sons, New York, 1977.

32. J. Wilkinson, 'Reconstruction of historical stormwater flows in the Adelaide metropolitan area', ACWS technical report no. 10 prepared for the Adelaide Coastal Waters Study Steering Committee, Department of Environmental Health, Flinders University of South Australia, Adelaide, September 2005.

33. J. Wilkinson, M. Pearce, N. Cromar and H. Fallowfield, 'Audit of the quality and quantity of treated wastewater discharging from wastewater treatment plants (WWTPs) into the marine environment', ACWS technical report no. 1 prepared for the Adelaide Coastal Waters Study Steering Committee, Department of Environmental Health, Flinders University of South Australia, Adelaide, November 2003.

34. J. Wilkinson, N. White, L. Smythe, H. Fallowfield, J. Hutson, E. Bestland, C. Simmons and S. Lamontagne, 'Volumes of inputs, their concentrations and loads received by Adelaide metropolitan coastal waters', ACWS technical report no. 12 prepared for the Adelaide Coastal Waters Study Steering Committee, Flinders Centre for Coastal and Catchment Environments, Flinders University of South Australia, Adelaide, September 2005.

35. S. Bryars, D. Miller, G. Collings, M. Fernandes, G. Mount and R. Wear, 'Field surveys 2003–2005: assessment of the quality of Adelaide's coastal waters, sediments and seagrasses', ACWS technical report no. 14, publication no. RD01/0208–15, South Australian Research and Development Institute (Aquatic Sciences), Adelaide, 2006.

36. G. Collings, D. Miller, E. O'Loughlin, A. Cheshire and S. Bryars, 'Turbidity and reduced light responses of the meadow forming seagrasses Amphibolis and Posidonia, from the Adelaide metropolitan coastline', ACWS technical report no. 12 prepared for the Adelaide Coastal Waters Study Steering Committee, publication no. RD01/0208–17, South Australian Research and Development Institute (Aquatic Sciences), Adelaide, 2006.

37. C. Pattiaratchi, J. Newgard and B. Hollings, 'Physical oceanographic studies of Adelaide coastal waters using high resolution modelling, in-situ observations and satellite techniques: sub task 2 final technical report', ACWS technical report no. 20 prepared for the Adelaide Coastal Waters Study Steering Committee, School of Environmental Systems Engineering, University of Western Australia, 2006.

38. S. Bryars, 'An inventory of important coastal fisheries habitats in South Australia', Fish Habitat Program, Primary Industries and Resources South Australia, Adelaide, 2003.

39. S. Bryars and K. Rowling, 'Benthic habitats of eastern Gulf St Vincent: major changes in seagrass distribution and composition since European settlement of Adelaide' in Bryars, S. (ed.) *Restoration of coastal seagrass ecosystems: Amphibolis antarctica in Gulf St Vincent, South Australia*, publication no. F2008/000078–1, South Australian Research and Development Institute (Aquatic Sciences), Adelaide, 2008, p. 90.

40. G. Collings, S. Bryars, S. Nayar, D. Miller, J. Lill and E. O'Loughlin, 'Elevated nutrient responses of the meadow forming seagrasses, Amphibolis and Posidonia, from the Adelaide metropolitan coastline', ACWS technical report no. 11 prepared for the Adelaide Coastal Waters Study Steering Committee, publication no. RD01/0208–16, South Australian Research and Development Institute (Aquatic Sciences), Adelaide, 2006.

41. S. Bryars, G. Collings, S. Nayar, G. Westphalen, D. Miller, E. O'Loughlin, M. Fernandes, G. Mount, J. Tanner, R. Wear, Y. Eglington and A. Cheshire, 'Assessment of the effects of inputs to the Adelaide coastal waters on the meadow forming seagrasses, Amphibolis and Posidonia', Task EP1 final technical report, publication no. RD01/0208–19, South Australian Research and Development Institute (Aquatic Sciences), Adelaide, 2006.

42. S.K. Gorgula and S.D. Connell, 'Expansive covers of turf-forming algae on human-dominated coast: the relative effects of increasing nutrient and sediment loads', *Marine Biology*, 145, 2004, pp. 613–619.

43. B.D. Russell and S.D. Connell, 'A novel interaction between nutrients and grazers alters relative dominance of marine habitats', *Marine Ecology Progress Series*, 289, 2005, pp. 5–11.

44. B.K. Eriksson, A. Rubach and H. Hillebrand, 'Biotic habitat complexity controls species diversity and nutrient effects on net biomass production', *Ecology*, 87, 2006, pp. 246–254.

45. D.J. Turner, T.N. Kildea and G. Westphalen, 'Examining the health of subtidal reef environments in South Australia, part 2: status of selected South Australian reefs based on the results of the 2005 surveys', publication no. RD03/0252–6, South Australian Research and Development Institute (Aquatic Sciences), Adelaide, 2007.

46. K.S. Edyvane, 'Coastal and marine wetlands in Gulf St Vincent, South Australia: understanding their loss and degradation', *Wetlands Ecology and Management*, 7, 1999, pp. 83–104.

47. G. Westphalen, G. Collings, R. Wear, M. Fernandes, S. Bryars and A. Cheshire, 'A review of seagrass loss on the Adelaide metropolitan coastline', ACWS technical report no. 2 prepared for the Adelaide Coastal Waters Study Steering Committee, publication no. RD04/0073, South Australian Research and Development Institute (Aquatic Sciences), Adelaide, 2004.

48. S. Bryars and V. Neverauskas, 'Marked contrasts in seagrass loss and recovery at two nearby sewage sludge outfalls' in Bryars, S. (ed.) *Restoration of coastal seagrass ecosystems*, p. 90.

49. S. Bryars and V. Neverauskas, 'Natural recolonisation of seagrasses at a disused sewage sludge outfall', *Aquatic Botany*, 80, 2004, pp. 283–289.

Box references

Box 10: The biodiversity of Buckland Park

1. D.C. Paton, L.P. Pedler and W.D. Williams, 'The ecology and management of Buckland Park Lake', World Wide Fund for Nature, 1991.

2. Department for Environment and Heritage, 'Estuaries of South Australia: our vision for the future, draft policy and action plan for public consultation', Coast and Marine Conservation Branch, Adelaide, 2005.

3. P. Coleman and F. Coleman, 'Local recovery plan for the yellowish sedge-skipper and thatching grass', report for the South Australian Urban Biodiversity Program, Delta Environmental Consulting, 2000.

4. D.M. Armstrong, S.J. Croft and J.N. Foulkes, 'A biological survey of the south Mount Lofty Ranges, South Australia 2000–2001', Department of Environment and Heritage, Adelaide, 2003.

CHAPTER 4

Aquifers and groundwater

Steve Barnett
Edward W. Banks
Andrew J. Love
Craig T. Simmons
Nabil Z. Gerges

Introduction

We have become accustomed to the belief that the most common sources of useful freshwater on our planet are from rivers, streams, lakes and dams, with a less obvious and minor source being from beneath the ground. The underground (subsurface) component is extremely important. At any one time, less than 3% of the available fluid freshwater on our planet is stored above the ground (surfacewater), and more than 97% is stored below the ground. Hydrogeology encompasses the inter-relationships of geologic materials and processes with water in subsurface environments. Surfacewater, by virtue of its visible occurrence, tends to be more easily understood. Groundwater, on the other hand, because of its hidden nature, is shrouded by mystery and superstition. Although human observation and use of groundwater predates biblical times, an appreciation for the occurrence and movement of it is more recent. The overwhelming increase in our knowledge of the groundwater or subsurface hydrology discipline has only been achieved in the 20th century.

Aquifers are subsurface geologic units, such as sands or gravels, that are capable of both storing and transmitting significant quantities of water to a bore, spring or watercourse such as a river. For many, they have often been thought of as underground rivers, but rarely is this the case. Usually, groundwater is stored in and moves through a geological material, such as sand in the pores between the grains of sand. Groundwater storage and movement is therefore much more like water moving through a sponge than a pipeline. Groundwater moves very slowly through these aquifers; several metres per year at most. Often the water in these aquifers can be very old, possibly hundreds or even hundreds of thousands of years. The water we extract out of the ground today is often a very old vintage indeed! As pressure continues to mount on our surfacewater supplies, increasing demand will be placed on our groundwater resources. Our water management approaches will need to consider groundwater and surfacewater as connected and integral parts of the hydrologic cycle. Water debates in a nation facing continued drought and the serious prospects of climate change will necessarily involve groundwater now more than ever before. This chapter is dedicated to a discussion of groundwater in the Adelaide city and its surrounds, and raises awareness on groundwater as a vital resource that needs to be understood, scientifically and technically, as a basis for making critical decisions regarding its management as a water resource. Groundwater is now featuring much more in water resources management debates. Most recently, this has included a complete moratorium on the installation of groundwater wells for domestic water supply.

In addition to the water supply from the River Murray and catchments in the Adelaide Hills, groundwater, recycled wastewater and stormwater are also used to supply Adelaide. It is important to note that 20% of the total water supply for Adelaide and its surrounds is groundwater (compared to mains water at 68%). Further information about Adelaide's water supply and its uses can be found at the Water Proofing Adelaide website.

Groundwater is used in a variety of commercial, industrial, residential and public activities. Whether it is for watering the garden, parks and sportsgrounds, or for industrial and agricultural use, groundwater continues to play an important and increasingly prominent role – and is a precious resource requiring protection.

In order to understand Adelaide's hydrogeology, we must understand not only the aquifers beneath the metropolitan area, but also the role of the Mount Lofty Ranges (MLR) in controlling recharge to these aquifers. Thus, this chapter is divided into two sections: the first deals with the Adelaidean metropolitan aquifers; and the second deals with the hydrogeology of the MLR. The vital connection between these two groundwater resources is explored and elucidated.

Hydrogeology of the Adelaide metropolitan area

The aquifer systems beneath the greater urban area of Adelaide are experiencing increasing demand pressure as drought and increasing costs for reticulated water make groundwater a more economically attractive option for water supply. This demand is not new; up until 1970, SA Water extracted up to 10,000 ML/yr from the aquifers beneath Adelaide to supplement the mains water supply during drought years. There are two main types of aquifer beneath the metropolitan Adelaide area: the shallow Quaternary aquifers; and the deep, confined Tertiary aquifers. Each have differing yield characteristics, and, as a result, different users. A comprehensive review and assessment of the geology and hydrogeology of the Adelaide metropolitan area has been completed by Gerges,[1] which provides a critical foundation for this analysis.

Shallow Quaternary aquifers

These aquifers were formed in the Quaternary Period, the geologic time period from the end of the Pliocene Era, approximately 1.8 Ma ago, to the present. The main lithology of the Quaternary sediments beneath Adelaide is mottled clay and silt, known as the Hindmarsh clay, which contains up to six thin, interbedded sand and gravel layers. These layers form thin aquifers, which are designated Q1 to Q6 in order of increasing depth. The shallowest Q1 aquifer lies at depths between 3 and 10 m below ground level, with an average thickness of 2 m; the deepest aquifer occurs up to 50 m below ground.

Domestic water-users take water from 'backyard bores', tapping into these aquifers, which provide relatively low yields of up to 3 L/sec. Since 1990, about 2600 backyard bores have been drilled in the metropolitan area; of these, an estimated 2000 are thought to be operational. Figure 4.1 shows the distribution of these bores and the location of selected observation wells in the Q1 aquifer. Water levels from these wells are presented in Figure 4.2, together with the cumulative monthly deviation rainfall graph (shown in white). This graph shows the difference between observed rainfall and the long-term average on a monthly basis. An upward trend indicates greater than average rainfall – average rainfall is represented by a horizontal trend – while a downward trend indicates below average rainfall.

Soil profiles and soil types in the Adelaide region

BOX 15

The north to north-easterly trending faults that separate the coastal strip from the Mount Lofty Ranges largely dictate the soils and their layering (i.e., soil profile) within the sediments that form the Adelaide soils. The coastal strip is sandy (St Kilda Formation), while red-brown, saline, silty clays of low to moderate expansive behaviour extend throughout the low-lying coastal plains to the north-west (alluvial soils and the soils of the Pooraka Formation). The hills contain shallow soils over rock with weathered rock in the transition zone between the soil and sound rock. Adelaide clays, which exhibit significant expansive soil behaviour, are found near the surface in the intermediate areas of greater Adelaide. The Pooraka Formation often underlies much of this area at depth.

The Adelaide clays vary from red-brown earths of moderate to high expansive behaviour to highly expansive black earths and very expansive Pleistocene clays (Keswick and Hindmarsh clays). During summer and autumn, shrinkage cracking is obvious at the surface of these more expansive soils. Fortunately, outcrops of Pleistocene clay are relatively rare although their names suggest that they were uncovered in Keswick and in Hindmarsh, suburbs near the CBD. Indeed, these soil deposits are troublesome for building construction in the city. Fortunately the clays are usually overlain by other soil layers. Well outside the city area, the Pleistocene clays are often found below the design depth of movement for the Adelaide region (4 m) and so may have little impact on small building construction.

A distinctive feature of Adelaide compared with other cities is that most soil profiles are deep, with an underlying deep water table (>10 m). Also, soil profiles are alkaline, as, long ago, lime from shell remains and limestone was swept up by the prevailing westerly winds and deposited on higher ground. Subsequently, limestone has accumulated in the seasonal zone of soil moisture change within soil profiles. The accumulated lime can be seen in excavations or trenches as permeable marl layers, or clay soils with calcium carbonate nodules. The greater the amount of lime, the less expansive the soil will be, but lime-rich soils can be collapsible under the influence of loading and wetting, making effective site drainage imperative for such soil profiles.

In this short description of Adelaide soils and soil profiles, it is evident that each building site requires deep site investigation for evaluation of the combination of soils that may be present. A further feature of the Adelaide region that complicates site investigation is that 'gilgai' landforms exist, which can result in changes in soil profiles over relatively short distances (a few metres). The moon-like surface of gilgai country consists of a recurring pattern of mounds and small depressions. Generally gilgais occur over extremely expansive soils. The soil profile tends to be less expansive below the depressions than below the mounds. Gilgais may not always be obvious if the land has been reworked; the changes in soil profile are best seen in excavations for large basements or for pipelines.

Note: 'Bay of Biscay soils' is a romantic non-technical term applied by past generations to infer expansive soils causing ground mounds in Adelaide. The expression should be discarded.

Don Cameron

Backyard bores

Figure 4.1 Shallow backyard bores in Adelaide.

The graph shows a disturbing trend of below-average rainfall since 1988, apart from wet years in 1992, 2002 and 2005. The water levels in Figure 4.2 show little change, with some showing a close relationship to rainfall patterns. All wells show a drop in level due to the 2006 dry winter, but when compared to earlier trends it was not significant. Because of the large number of these bores and the relatively low volumes of extraction, management of the shallow Quaternary aquifers (through licensing and metering) would not be cost-effective, given the limited resources available.

Confined Tertiary aquifers

These deep aquifers are the largest and most important groundwater resource in the Adelaide metropolitan area. The Tertiary geological time interval occurred approximately 65 million to 1.8 million years ago (when these aquifers were formed). They comprise layers of sand and limestone separated by thin, clay-confining layers. The area can be divided into two main provinces – the Golden Grove Embayment and the Adelaide Plains Sub-basin – as presented in Figure 4.3, with the Para Fault forming the boundary. The geological cross-section in Diagram 4.1 shows significant differences in the thickness and extent of the various Tertiary aquifers in these

provinces. As with the Quaternary aquifers, the Tertiary aquifers are numbered with increasing depth below the groundsurface, the shallowest being the T1 aquifer.

Golden Grove Embayment

The Tertiary sediments within the Golden Grove Embayment form a complex sedimentary sequence, which thickens from about 20 m in the Tea Tree Gully area to over 300 m in the south-west, near the coast at Brighton. Most extractions come from the T1 and T2 aquifers, which consist mainly of sand units receiving natural recharge laterally across the Eden-Burnside Fault (Diagram 4.1), from the fractured rock aquifers of the MLR. Towards the coast, the deeper T3 and T4 aquifers are intersected below a depth of 150 m and contain brackish groundwater over 5000 mg/L.

Adelaide Plains Sub-basin

The sedimentary units in this province are much thicker and more consistent than in the Golden Grove Embayment, as shown in Diagram 4.1. The Tertiary aquifers are up to 400 m thick, with most extractions coming from the T1 and T2 aquifers, which consist of interbedded limestones, sandstones and fossiliferous sands. The deeper T3 and T4 aquifers are less productive and contain saline groundwater.

Groundwater use

To the north-west of the Adelaide city centre, large industrial and commercial users (such as Penrice Soda Products, Coopers Brewery, Coca Cola Amatil, and various golf courses north of Grange), extract approximately 8000 ML/yr from about one hundred wells. Groundwater movement is towards these major pumping centres. Most extraction takes place in summer, and groundwater levels are lowest at the end of this season. Because of the significant potential demand for industrial use, groundwater resources were prescribed under the *Natural Resources Management Act 2004* in 2007. In future, all extractions for industrial and commercial purposes will be licensed and metered, with a sustainable limit placed on the volumes that can be extracted annually.

Figure 4.4 displays results from long-term groundwater level monitoring, with two different trends caused by different types of extraction. Observation well YAT 42 shows major seasonal fluctuations, with declines in groundwater levels during summer caused by extractions for the irrigation of golf courses, and recovery in winter when pumping ceases. ADE 2 shows the impacts of industrial pumping in the Thebarton area, where the recovery of water levels during winter is hampered by all-year-round extractions. Consequently, seasonal fluctuations are smaller than those caused by summer irrigation. Both hydrographs show a larger drawdown due to increased pumping during the dry winter and spring of 2006, with a strong recovery during 2007.

Drought pumping

On several occasions up until the 1970s, SA Water extracted up to 10,000 ML/yr from a number of bores in the western suburbs of Adelaide to supplement the MLR reservoir water

BOX 16

Why do Adelaide houses crack in summer?

Urban infrastructure such as houses and other buildings, as well as roads and drains, can start to crack over summer. The damage is likely to become more severe through autumn as the dryness of the ground reaches greater depths. A common form of damage is caused by relative movement under sections of the footing system of a house, which in most cases has not been adequately designed to meet the challenges the ground and climate can create. Heavy clay soils shrink (settle) and swell (lift or heave) in response to evaporation and rainfall. These are the expansive soils of Adelaide (typically grey-brown or red-brown plastic soils), which underlie large parts of the greater metropolitan area. Interestingly, outcrops of black earths are regarded as highly expansive soils, but it is the soils usually found below these near-surface soils that cause most of the movement.

The response of the ground to a seasonal impact, such as the first rains of winter, can be quite slow. Movements occur relatively deep within Adelaide's soil profiles; the design depth of soil movement for houses is four metres, which is about twice the seasonal depth of movement. Since expansive soils have very low permeability to water, rainfall takes considerable time to have much effect on the ground movement. Vertical shrinkage cracks in the ground may help to get the water down initially.

Construction of houses on previously non-urban sites creates an environmental change of the soil moisture state beneath each allotment. With houses supported on raft slabs, the central portions of the floor are not directly influenced by seasonal wetting and drying of the bare ground outside the slab. In the early years after construction, edge wetting may lead to edge heave of concrete slabs, while in the later years soil moisture may eventually be drawn beneath the slab to cause centre heave, in combination with seasonal drying about the slab edges. Centre heave can be more damaging if it is exacerbated by other edge drying influences such as dense or significant vegetation feeding deeply on the soil water reserves.

If your house sits over expansive soils in greater Adelaide, the likelihood of building damage is affected by:

- **The year it was built**
  Very old houses were built without any recognition of the potential threat of expansive soil profiles. These houses tend to have solid masonry walls, which can not readily flex with the ground movement. Footings were designed simply to spread building loads to safe levels. Less old houses may have incorporated some recognition of expansive clay soils and have footings, which may resist some bending due to ground movement. Houses built after the mid 1980s should have been appropriately designed according to engineering principles, which take account of the flexibility of the structure and the protection provided by the stiffened footing system. The risk of damage for these houses is much diminished but not eliminated entirely.
- **The month it was built**
  There tends to be less expansive soil movement if the house was constructed in winter/spring, although the majority of houses are built outside this period.
- **The type of construction**
  Solid walls are less flexible and so require greater protection. Render is particularly sensitive to ground movements. Vertical-control joints in walls (articulation joints) can help provide more flexibility.
- **How much maintenance of the site has been carried out**
  Excessive garden watering, dense plantings of trees and shrubs, poor site drainage, blocked drains and leaky plumbing all add to the problem of coping with expansive clay soils. Engineering design considers the unavoidable environmental soil moisture change, including an estimate of the impact of trees.
- **The variability of the soil profile on the allotment**
  Site investigation for housing is limited, and parts of Adelaide are notorious for sudden changes in soil profiles over short distances ('gilgai' landform structures in highly expansive clay soil profiles).

Finally, in some isolated parts of Greater Adelaide, cracking may suddenly appear in summer if there has been a water leak near or under the footings, or if an area adjacent to the house has been flooded, e.g. a hose has been left on inadvertently. Sudden settlements (collapse settlement) can occur if clay silts that are rich in lime (calcium carbonate) have become excessively wet. These soils seem hard enough when they are dry, but because they are not well compacted, flooding can cause softening of the lime and loading can then make the soil more compact. Excess site water has to be well managed within these areas to avoid collapse settlement. In summary, care should be taken to manage house sites as far as practical in order to avoid building damage. Otherwise, any damage has to be accepted by the owner; the danger with this strategy is that aesthetic damage may one day develop into more serious damage. With older building stock, some modest but even watering around the site is necessary to offset the lack of stiffness in the footings and the readiness of some walls to crack.

Don Cameron

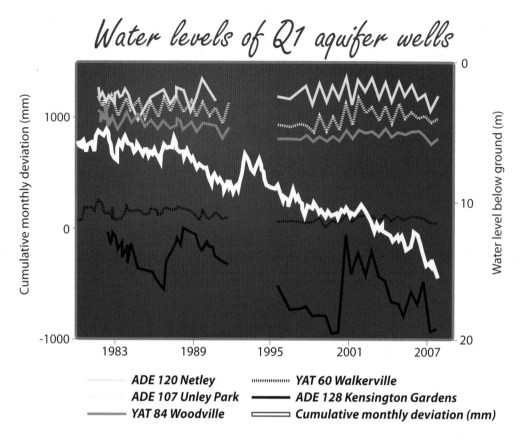

Water levels of Q1 aquifer wells

Figure 4.2 Water levels from five wells in the Q1 aquifer, as well as the cumulative monthly deviation shown by the white line. The cumulative monthly deviation shows the difference between observed rainfall and the long-term average. This line shows a trend of below-average rainfall since 1988, with some wetter years in the early 1990s, 2002 and 2005. The water levels of all wells show little change, except for ADE 128 Kensington Gardens, which drops during 1987, 2000 and 2007.

supply during drought years. This water was directly injected into the water distribution network. Unfortunately, the supply bores have since been decommissioned or used for other purposes. Figure 4.5 shows that recent estimates of extraction are only just approaching these previous volumes.

Proposals for using these confined aquifers again, as an emergency drought supply, are being examined, using wells around of 100–250 m deep, each yielding around 1 ML/day. The potential capacity of the T1 and T2 aquifers to provide emergency drought relief water supplies will be constrained by the resource management conditions for pressure drawdown limits, and management of groundwater salinity impacts.

Groundwater provinces

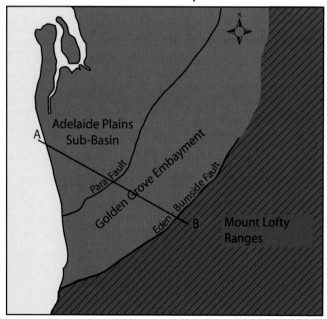

Figure 4.3 Groundwater provinces in the Adelaide area.

Hydrogeology of the Mount Lofty Ranges

Groundwater is a significant and valuable resource in the MLR. It provides water for agricultural industries, local communities, natural ecosystems, as well as sustaining metropolitan Adelaide's mains water supply. Understanding the extent of the groundwater system and how it interacts with the surfacewater systems is imperative, as these resources come under increasing pressure for development. The development of appropriate management strategies is crucial to ensure that these resources will continue to provide quality water into the future.

How groundwater is stored and moves within the MLR plays an important role in the hydrology and hydrogeology of not only the mountain ranges themselves, but also, just as importantly, the hydrogeology of groundwater stored beneath the northern and Central Adelaide Plains. Groundwater acts as both an aqueduct (water channel) and a water resource. Groundwater in the MLR acts as an aqueduct by recharging the Tertiary aquifer systems beneath the Adelaide Plains, across the Eden-Burnside Fault. This provides water for mainly recreational and industrial use in Adelaide, but in the past has been used as a water supply during drought. The important role of groundwater as a resource for horticulture and viticulture in the MLR is well recognised. However, less well recognised is the role that groundwater has in sustaining a number of important ecosystems fed from the underground water reservoirs.

This is particularly important in the lower reaches of the catchments where numerous permanent pools are located within stream channels.

A topographic divide separates the MLR into the Eastern and Western Mount Lofty Ranges (EMLR and WMLR), where watersheds in the east drain towards the River Murray, and in the west transverse across fractured rock catchments into the Adelaide Plains, with the ultimate terminus into Gulf St Vincent. The WMLR can be divided into three main regions: the Central Hills; McLaren Vale; and Fleurieu Peninsula. The importance of the western margin of the Central Hills region as a recharge zone to the sedimentary aquifers of the northern and Central Adelaide Plains areas has a significant impact on water supplies for metropolitan Adelaide.

Central Hills, Western Mount Lofty Ranges

The Central Hills of the WMLR includes three major surfacewater catchments, which are characterised by the surface features of the Gawler, Torrens and Onkaparinga rivers. The catchments from north to south are: the Gawler River (South Para) Catchment; the River Torrens Catchment; and the Onkaparinga River Catchment. The surface topography of these catchments broadly coincides with the major groundwater catchments, and the topographic highs correspond to the groundwater flow divides of the aquifers beneath.

Aquifer types, yields and salinity

The major aquifer types in the Central Hills of the WMLR are fractured rock aquifers, with minor unconsolidated sedimentary aquifers (Diagram 4.2). Fractured rock aquifers consist of two principle components – the fractures and the matrix blocks. Fractures can include cracks, joints and faults, which can vary in scale from centimetres to kilometres. Groundwater flows through these fractures, whereas the pore space within matrix blocks act as storage reservoirs. Unweathered fractured rock aquifers are best classified as a 'fractured formation', where the matrix porosity is low but the storage within the matrix is larger than in the fracture network. The unaltered fractured formations transition into a weathered zone of higher porosity and permeability as it approaches the groundsurface. The three major fractured rock provinces within the Central Hills of the WMLR are the Adelaidean Metasediments, the Barossa Complex and the Kanmantoo Group. In general, the Kanmantoo Group dominates the eastern section of the Central Hills region, and the western section is characterised by Adelaidean Metasediments. The Barossa Complex is exposed across a large area of the River Torrens Catchment, with minor occurrences in the Onkaparinga River Catchment.

Cross-section of aquifers

A
5 km
B
Eden–Burnside Fault
Para Fault
0
Quaternary
100
T1 aquifer
200
T2 aquifer
300
T3 and T4 aquifers
400
Basement
500

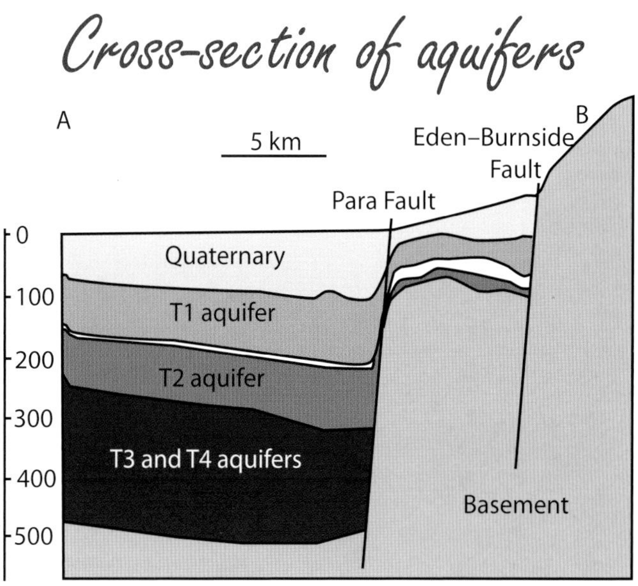

Diagram 4.1 Geological cross-section across the Adelaide area.

Adelaidean Metasediments

The majority of the hydrogeology of the WMLR consists of sedimentary and metasedimentary rocks of the Neoproterozoic Adelaidean sequence, which forms part of the large geological entity referred to as the Adelaide Geosyncline, encompassing the present-day MLR, Flinders Ranges and Olary Province.[2] The Geosyncline was formed in the Precambrian as a result of rifting and extensive subsidence. The depositional sequence can be in excess of 20 km thick. The sedimentary sequence subsequently has been folded and uplifted as well as being subjected to predominately low-grade metamorphism during the Delamerian Orogeny. Deformation has resulted in north–south trending, south-plunging folds, and complex faulting. The predominant rock units in the Adelaidean sequence are sandstone, siltstone, shale, slate and quartzite. These are the most productive aquifers in the MLR, as a result of the open and permeable fracture formation. The aquifers also contain the best-quality groundwater suitable for domestic and irrigation purposes.

Well hydrographs – T1 aquifer

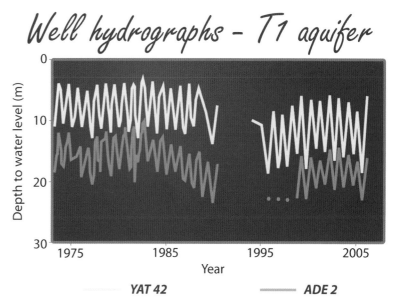

Figure 4.4 Hydrographs from the long-term monitoring of two observation wells in the T1 aquifer – YAT 42 and ADE 02.

Historical drought pumping

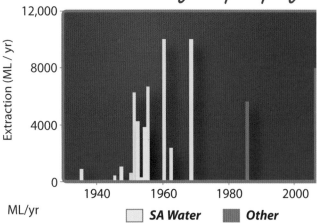

Figure 4.5 Historical drought pumping from bores in the western suburbs of Adelaide.

Barossa Complex

The Barossa Complex forms a series of basement inliers within the WMLR and are the oldest rocks in the MLR. The rocks consist of gneisses, schists and pegmatites, formed at high temperatures and pressures and subjected to extensive ductile deformation, caused by the accumulation of strain in fault zones.[3] The Barossa Complex is generally considered to be a poor-quality aquifer with low well yields and high-salinity water as a result of the tight and impermeable rock fracture formation. Over time, the fractures and joints have been filled with clayey, weathered material, which can reduce rainfall infiltration to the aquifer and also dissolve into the water, raising salinity levels.

Aquifer types

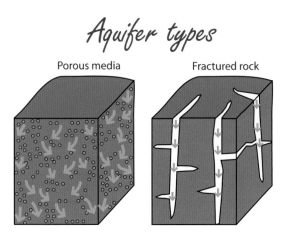

Diagram 4.2 The difference between a sand aquifer (porous media) and a fractured rock aquifer. In the former, water moves between interconnected gaps between the sand grains; in the latter, water must find its way through irregularly spaced and often unconnected fractures and cracks in the rocks. The arrows indicate groundwater flow-pathways.

Kanmantoo Group

The Kanmantoo Group is a sequence of clastic felspathic sandstones, schists and gneiss, deposited during the Cambrian period, about 500 Ma ago. Similar to the Barossa Complex, the Kanmantoo Group is considered a poor aquifer with typically low well yields and high salinity. (Although, areas in this aquifer have been found with high yields and low salinity.)

Sedimentary aquifers of the Central Hills, Western Mount Lofty Ranges catchments

Minor, unconsolidated sedimentary aquifers are associated with alluvium and colluvial deposits in the WMLR catchments, often found adjacent to the drainage pattern. Groundwater in these aquifer systems is stored in sediments of gravel, sand, silt and clays, often with high organic content, and the groundwater moves through gaps between the clastic-sized grains. Salinity and well yields vary significantly in these systems.

Movement of water through the rocks

Groundwater recharge occurs throughout the WMLR when a percentage of rainfall percolates through the soil zone and reaches the watertable of the aquifer. It is important to note that most of the rainfall is lost via evaporation from the soil, transpiration by plants, or diverted via overland flow into streams, with very little recharging the aquifer. Estimating groundwater recharge is an essential component of the water balance of the WMLR. However, to determine recharge to a fractured rock aquifer is particularly difficult because groundwater is stored in, and moves through, fractures and joints in what is essentially impervious rock. A number of new techniques – involving characterising the fractured rock aquifer and groundwater dating – were conducted between 2005 and 2007 at a number of sites in the WMLR and EMLR by the Department of Water, Land and Biodiversity Conservation (DWLBC) [4, 5, 6] to estimate annual recharge rates. Sites were selected to be representative of a range of the geology and climate zones typical of the MLR.

Groundwater will travel in the subsurface until it reaches its discharge zone from the aquifer system. This can occur via a number of different flowpaths of varying depths and residence time of the groundwater. The shallowest flowpath is referred to as 'interflow'; this is the flow that discharges into a local creek or watercourse. The other types of subsurface flow are referred to as local, intermediate, and regional flow systems. The propensity for a particular type of system to occur depends upon a number of factors including rainfall, geology and topography. In essence, the higher the rainfall and steeper the topography, the greater the propensity for local flow systems. Conversely, with lower rainfall and lower topographic gradient, the likelihood of regional systems increases. Intermediate flow systems, as the name suggests, fall between these other two.

What is salt damp?

Salt damp is a progressive form of deterioration that can occur in porous building materials unless preventative measures are taken. It is characterised by the appearance of white to yellowish-white efflorescent salt deposits on the surfaces of materials such as brick, concrete, stone, mortar and plaster. In South Australia, the term salt damp can be regarded as being synonymous with the more internationally recognised term 'rising damp'.

The difference between the two is that for salt damp, water-soluble salts are carried in solution; the saline solution is drawn upwards from the soil into footings and walls, picking up salts along the way that may also be derived from the building materials themselves. The mechanism causing this upwards progression of salinity is a combination of capillary and osmotic suctions. As the saline solution rises above ground level a stage is reached where the building components (footings and walls) are then exposed to the evaporative effects of temperature, wind and humidity. This causes water to leave the saline solution, and salt crystals then concentrate in the near-surface pore spaces of building materials. Because the salts that cause salt damp are hygroscopic (moisture attracting) they can exist in both their anhydrous and hydrated forms, depending upon variations in atmospheric temperature and humidity.

The resulting crystalline phase change from anhydrous to hydrated condition brings with it a volume increase in the crystals of salt concentrated just beneath the material's surface. The salt crystals can increase in size by more than 300% for some salts such as sodium sulphate (Thenardite–Mirabilite transformation = 317% volume change).

The significant crystalline size change exerts a bursting pressure on the small pore spaces at the surface of the porous material in which the salt crystals are left behind after the water component of the saline solution evaporates. This pressure causes the surface of the material to progressively spall away (exfoliate) and drop to the ground below.

It is important to understand that where porous building materials are completely embedded below ground level, in conditions where humidity remains approximately constant and evaporation cannot occur, salt damp deterioration will not occur. It is only where the surfaces of building materials are exposed to atmospheric evaporation of the water component of the saline solution that salt damp attack takes place.

Salt damp attack is therefore a slow and progressive form of building deterioration that occurs first at the surface of building materials and progressively works its way inwards rather than being analogous to other forms of building deterioration, such as termite attack of timber where the termites will eat out the core of the timber member and work their way outwards leaving behind only a paint film in some cases. Salt damp attack is completely the reverse. It can be prevented by the appropriate placement of damp-proofing membranes to form physical barriers.

Why is salt damp so prevalent in Adelaide?

Adelaide has saline soils as a consequence of there once being a shallow saltwater lake covering the plains area extending to the base of the Mount Lofty Ranges. It also has a semi-arid climate with cool, wet winters and hot, dry summers – conditions conducive to salt damp.

John Goldfinch

Interaction between groundwater and surfacewater

Interactions between groundwater and surfacewater form one component of the hydrological cycle, and are largely controlled by the effects of topography, geology and climate. In areas of steep topographic relief and dry temperate climate, such as the MLR, groundwater discharge into streams is a significant component to the catchment water balance, particularly during the summer months and drought conditions. Many of the surfacewater features are perennial, dependent on groundwater, and support a diverse range of flora and fauna. Conversely, many of the ephemeral creeks throughout the WMLR may be a source of groundwater recharge during times of high flow. A reduction in surface runoff to streams – hence a potential reduction in groundwater recharge – has been attributed to the construction of farm dams. Very little is known about the importance of the interactions between groundwater and surfacewater in a fractured rock environment, the dominant aquifer type throughout the WMLR.

Groundwater resources in the Central Hills, Western Mount Lofty Ranges

Due to the nature of fractured rock aquifers, the quantity and quality of the groundwater resources beneath the three main surfacewater catchments of the Central Hills vary considerably. The physical characteristics of the underlying geology influence the well yield and water quality; the climate and rainfall determine the amount of recharge to the aquifer systems; and the topography influences the movement of groundwater through the aquifer. Rainfall across the Central Hills is winter-dominated, falling between April and October. The amount tends to decrease from the western boundary towards the east, with the highest annual rainfall occurring around Mount Lofty. The pronounced topographic relief results in dominant local flow systems compared to the flatter relief of the plains, which tend to have dominant intermediate and regional flow systems.

The Gawler (South Para) River Catchment is located approximately 60 km north-east of Adelaide, extending from Mount Pleasant in the east to Gawler in the west. The catchment covers an area of about 337 km² and is characterised by steep topography with elevations ranging from 630 mAHD (metres above the Australian Height Datum ~ mean sea level) in the north-east to 50 mAHD on the western boundary. The Barossa Complex, Kanmantoo Group and Adelaidean Metasediments underlie the Gawler River Catchment in the Central Hills. As mentioned earlier, higher yields and lower-salinity groundwater are found in the Adelaidean Metasediments, which also receives the higher recharge in this catchment. Estimated recharge rates range from 56 to 73 mm/yr.[7]

The River Torrens Catchment covers an area of approximately 501 km². The catchment slopes in a westerly direction and ranges in elevation from about 700 mAHD at its most southerly margin, dropping sharply to 140 mAHD on the western side of the Eden-Burnside Fault, from where the River Torrens traverses the Adelaide Plains and discharges into the sea at West Beach. The catchment contains two important reservoirs at Kangaroo Creek and Millbrook.

Rainfall varies considerably between 500 to 1000 mm/yr, with higher rainfall in the southern part of the catchment in areas of higher elevation. The Kanmantoo Group underlies the eastern section of this catchment, with the Adelaidean Metasediments underlying the rest of the catchment except for a small section in the north-west, the Barossa Complex. Estimated annual recharge for this catchment ranges from 24 to 87 mm/yr.[7]

The Onkaparinga River Catchment covers an area of approximately 555 km² and adjoins the southern boundary of the River Torrens Catchment. The catchment ranges in elevation from 500 mAHD near Mount Torrens to 10 mAHD at Old Noarlunga where the Onkaparinga River discharges into the Onkaparinga estuary. Mount Bold Reservoir in the south-west captures the majority of surface runoff from the catchment, with only minor flows making it to the Onkaparinga estuary. The average annual rainfall in this catchment ranges from 800 to 900 mm/yr except for in the north-west near Lobethal where the annual average is up to 1200 mm/yr. The major land uses in the region are dairying, irrigated pasture for grazing, and some viticulture. Estimated annual recharge rates are higher in the areas of the catchment underlain by the Adelaidean Metasediments, and areas of higher rainfall, ranging from 32 to 111 mm/yr.[7] There are only small areas underlain by the lower-yielding and poorer-quality aquifers of the Kanmantoo Group and Barossa Complex in the east and north-west of the catchment.

Some concluding remarks

Effective management of water is a major challenge facing Adelaide, and Australia, today. Various options and strategies for waterproofing Adelaide in times of drought and increasing water demand must be considered as part of a 'whole of watercycle' approach. This necessarily includes groundwater resources. Current groundwater use is estimated at approximately 20% of the total water supply for Adelaide. It is used for a variety of commercial, industrial, residential and public activities. Whether it is for watering the garden, parks and sportsgrounds, or industrial and agricultural use, groundwater continues to play an important and increasingly prominent role, and is a precious resource.

In this chapter we have seen how groundwater levels are responding to drought conditions, and we have explored the vital connection between groundwater resources in the MLR and the Adelaide metropolitan aquifers – with the MLR serving as the vital control on groundwater recharge to the aquifers underlying the metropolitan area. Groundwater will continue to be an important and necessary part of water debates in this state. In a time of drought, groundwater will also continue to face increasing pressure as an alternative water resource. Groundwater must be managed carefully and be seen as a critical, and connected, part of the watercycle. Water management plans and policies must recognise that groundwater, like surfacewater, is limited and requires attention and protection – from both water quantity and quality points of view. We have already seen increasing political and scientific debate and discussion, as well as media attention in recent times, about our precious

Saltwater intrusion

BOX 18

Saltwater intrusion is the contamination of formerly freshwater aquifers by seawater near the coastline. It happens when freshwater is withdrawn at a faster rate than it can be replenished, which causes a lowering of the watertable and a decrease in hydrostatic pressure. The watertable is the level to which water will rise in a well. When it is below the level of the land its shape is similar to the topographic shape of the land surface. Generally, its elevation decreases towards the coastline, which means that under natural conditions groundwater in an aquifer moves seaward and prevents saltwater from intruding into coastal aquifers. The boundary between freshwater and saltwater is maintained near the coast, and freshwater can be drawn from wells on the land.

Saltwater can contaminate a freshwater aquifer through several pathways, including lateral intrusion from the ocean; by upward intrusion from deeper, more saline zones of a groundwater system; and by downward intrusion from coastal waters.

Most of the Australian population and almost two thirds of the world's population (4 billion people) live within 400 km of the ocean shoreline and just over half of the world's population live within 200 km. Many of these coastal regions rely on groundwater as their main source of freshwater for drinking, industry and agricultural purposes. As the population continues to grow, freshwater supplies are constantly depleted, causing issues such as saltwater intrusion to increase the importance of groundwater monitoring and management.

The key to controlling saltwater intrusion is to maintain the proper balance between water being pumped from the aquifer and the amount of water recharging it. Constant monitoring of the saltwater interface is necessary to maintain this balance.

Control of saltwater intrusion

The increased use of groundwater has caused the saltwater interface to move inland and closer to the ground surface in many parts of the world. In the past the only solution to the problem was to drill new bores further inland, which only complicated the issue. Today, the strategies for management of saltwater intrusion to ensure a sustainable source of groundwater for the future can be grouped into three general categories: engineering techniques; regulatory (legislative) approaches; and scientific monitoring and assessment.

Often, several actions are taken simultaneously or as part of a more comprehensive strategy for managing both the groundwater and surface water supplies of a coastal community.

A common approach for managing saltwater intrusion has been to reduce the rate of pumping from coastal wells or to move the locations of withdrawals further inland. Reductions in coastal withdrawals allow groundwater levels to recover from their lowered (or stressed) levels and fresh groundwater to displace the intruded saltwater. An alternative to reducing groundwater withdrawals is to artificially recharge freshwater into an aquifer to increase groundwater levels and hydraulically control the movement of the intruding saltwater. Artificial recharge can be accomplished through injection wells or by infiltration of freshwater at the land surface. In both cases, the recharged water creates hydraulic barriers to saltwater intrusion.

Other, more innovative approaches include aquifer storage and recovery (ASR) systems, desalination systems, and the blending of waters of different qualities. ASR is a process by which water is recharged through wells into a suitable aquifer, stored for a period of time, and then extracted from the same wells when needed. Typically, water is stored during wet seasons and extracted during dry seasons. ASR systems are implemented in northern Adelaide.

Desalination refers to water treatment processes that produce freshwater by removing dissolved salts from brackish or saline waters. Desalination systems are being implemented and/or proposed in the drier parts of Australia as new sources of freshwater become more difficult to develop. It is possible that desalination will also be applied to brackish surface water and saline groundwater, thus reducing the demand for groundwater and balancing recharge with discharge.

Scientific monitoring and assessment provides basic characterisation of the groundwater resources of an area, an understanding of the different pathways by which saltwater may intrude an aquifer, and a basis for management of water supplies. Water quality monitoring networks are particularly important in serving as early warning systems of saltwater movement towards freshwater supply wells.

Ian Clark

groundwater resources. There has been much discussion on the potential threat of aquifer water levels declining in the drought, and, more recently, a complete moratorium on the installation of groundwater wells for domestic water supply. The debates surrounding water, and groundwater in particular, will undoubtedly continue.

Groundwater is often called the 'forgotten resource', but must be firmly placed on the radar in all discussions, debates and policies pertaining to water resources management issues in this state, and Australia more generally. Some of the current challenges facing groundwater management in Australia are also applicable to the Adelaide region and were outlined in a recent position paper by Evans and associates.[8] These include:

- How to best identify sustainable levels of extraction and how to return over-allocated systems back to sustainable levels.
- How to licence and meter groundwater usage (for non-domestic users) more systematically, and to ensure compliance to stop unauthorised use of groundwater; consideration of pricing strategies that will assist with ongoing groundwater resource assessment, monitoring and management.
- To further consider, where feasible, opportunities for surfacewater to be stored in aquifers as an alternative to dams and reservoirs, in order to minimise evaporative losses. (Adelaide is a national leader in aquifer storage and recovery schemes, a number of which are already being operated and trialled throughout the Adelaide region.)
- Exploring ways of better managing groundwater resources in Adelaide and its surrounds (including organisational arrangements within government and policy) to ensure they are not at risk of becoming over-exploited.

References

1. N. Gerges, 'The geology and hydrogeology of the Adelaide metropolitan area', thesis, Flinders University of South Australia, Adelaide, 1999.
2. W.V. Preiss, 'The Adelaide Geosyncline, late Proterozoic stratigraphy, sedimentation, paleoontology and tectonics', *Geological Survey of South Australia, Bulletin*, 53, 1987.
3. J.F. Drexel, W.V. Preiss and A.J. Parker, 'The geology of South Australia, vol. 1, the Precambrian, South Australia', *Geological Survey of South Australia, Bulletin*, 54, 1993.
4. E.W. Banks, T. Wilson, G. Green and A.J. Love, 'Groundwater recharge investigations in the Eastern Mount Lofty Ranges', *South Australia*, Department of Water, Land and Biodiversity Conservation report 2007/20, Adelaide, 2006.
5. E.W. Banks, D. Zulfic and A.J. Love, 'Groundwater recharge investigation in the Tookayerta Creek Catchment, South Australia', Department of Water, Land and Biodiversity Conservation report 2007/14, Adelaide, 2006.
6. G. Green, E.W. Banks, T. Wilson and A.J. Love, 'Groundwater recharge and flow investigations in the Western Mount Lofty Ranges, South Australia', Department of Water, Land and Biodiversity Conservation report 2007/29, Adelaide, 2007.
7. G. Green and D. Zulfic, 'Summary of groundwater recharge estimates for the catchments of the Western Mount Lofty Ranges prescribed water resources area', Department of Water, Land and Biodiversity Conservation technical note 2008/16, Adelaide, 2008.
8. R. Evans, W.R. Evans, P. Jolly, S. Barnett, T. Hatton, N. Merrick and C.T. Simmons, 'National groundwater reform', position paper submitted to state and federal governments, Sydney, 2006.

CHAPTER 5
Biodiversity of the waterways

Christopher B. Daniels
Greg R. Johnston
Catherine Gray

Introduction

The catchments and waterways within Adelaide are home to a great many native and introduced plants and animals. Significantly, these streams, wetlands and catchments provide valuable habitat for aquatic and riparian species, and thus promote the survival of many vulnerable communities of plants and animals. Some of the major 'rivers' and streams within the Adelaide region include, from north to south:

- Light River;
- Gawler River (including the South Para River);
- Little Para River;
- Dry Creek;
- River Torrens (including the six suburban creeks, named First to Sixth creeks);
- Brownhill Creek;
- Sturt River;
- Field River;
- Christie Creek;
- Onkaparinga River; and
- Pedlar Creek.

None of Adelaide's waterways are large or even significant by world standards, and most have ephemeral flow. However, they provide corridors of more or less natural habitats between the Mount Lofty Ranges and the coast. In many parts of Adelaide, riparian corridors represent the only remnants of natural habitat on the Adelaide Plains.[1]

Adelaide's streams, unlike those in many other cities, cover a remarkably short distance between their headwaters and their mouth. This small distance (often as little as 20 to 30 km), has resulted in significant flooding at both headwaters and mouths during periods of unusually wet weather. The geology and geography underpinning our waterways (described in chapters 3 and 4), together with high winter rainfall (chapter 2) has had powerful effects on the types of plants and animals that inhabit these regions, and the structure of their ecological communities. As a result, our riverine and estuary ecological communities differ markedly from those found elsewhere in Australia. For example, our fish are small in size and capable of tolerating a wide range of water temperatures and streamflow rates; while some of our frogs are able to exist away from water for extended periods. The alteration of parts of most of Adelaide's waterways from natural systems to highly modified, concreted drains has greatly exacerbated the peculiar flow pattern, with devastating effects on the riparian wildlife.[1]

The unusual features of riparian wildlife in Adelaide also result, at least in part, from the unusual biogeographic position of Adelaide and the Mount Lofty Ranges. Adelaide sits on the boundary between the dry plains and the wetter ranges, which represent an isolated refugium of relatively cool, wet Bassian habitats inhabited by species adapted to these conditions. The major part of the Bassian biogeographic province occurs in south-eastern and south-western Australia. Each of these areas, and the

Mount Lofty Ranges, contain endemic species which have evolved in isolation in these places. Thus many of the animals and plants inhabiting the waterways of the Mount Lofty Ranges are most closely related to species as far away as Victoria and Western Australia rather than to species in the surrounding drier parts (termed the Eyrean region) of South Australia. The Eyrean biogeographic province is characterised by desert-adapted wildlife. In travelling from the north-west to the south-east across the Adelaide Plains, say from Two Wells to Beaumont, one is essentially moving from an area dominated by Eyrean fauna and flora to one dominated by Bassian fauna and flora. This transition from one biogeographic province to another has a profound influence on species inhabiting the waterways and wetlands.

This chapter will describe the environmental nature of the creek, wetland and estuarine systems as ecological communities. We will focus on the three 'zones': riparian, wetland and estuary. We will briefly describe the plant assemblages and types of vertebrates and invertebrates found in the catchment systems, and then describe the threats to biodiversity in our waterways, concluding with solutions to ensure the future of biodiversity for our aquatic freshwater ecosystems.

The riparian zone

A riparian zone is the area of land directly adjoining or surrounding, and influenced by, a body of water (river, creek, lake, wetland, etc), such as the river or streambank.[2] As such, riparian zones can vary in width from less than a metre on a steep bank to many metres on a wide floodplain. Riparian zones provide a unique habitat that is the interface with surrounding terrestrial and aquatic habitats. They are typically more biologically productive than surrounding habitats because of increased soil, water and nutrient availability, and higher humidity.

Riparian zones influence the water quality and ecology of the waterbodies they adjoin. Plants provide a refuge habitat for fish and other aquatic life when waters are flowing fast and deeply during floods. Plants also provide shade and thereby reduce water temperatures, important for the breeding cycles of many aquatic species and for minimising algal growth. Additionally, plants supply organic material (leaf litter) and nutrients that feed invertebrates (which in turn become food for other animals), shelter from predators, places to breed and hunt for food (for both terrestrial and aquatic species), and corridors for movement and migration from one patch of habitat to another (particularly important in urban areas such as Adelaide). Riparian vegetation, such as reeds and rushes, also help reduce streambank erosion by slowing the flow (or velocity) of the water, and stabilising the banks and trapping sediment. The plants in the riparian zone are also able to act as a filter system for the runoff water from a catchment by trapping pollutants, such as sediment, herbicides, pesticides and nutrients, before they enter the associated waterbody.[2]

The nature of the riparian zone is largely determined by the amount, frequency and seasonality of water flow. Eyrean creeks generally have small, intermittent flows of water,

Aquatic invertebrates

BOX 19

To most people, a creek or a pond looks like a nice peaceful and tranquil place. Insects like dragonflies, damselflies, mayflies and caddis flies (none of which are actually true flies), fly above the water and seem to perform no obvious purpose except to provide an occasional feed for the odd bird, lizard or spider. Now, if we look below the water surface it is a completely different matter. It is a veritable warzone down there with hunters like water tigers and mud eyes, some of the scariest creatures on the planet. These are the larvae of insects such as dragonflies and water beetles, and most are carnivores with a voracious appetite. They are predatory on other invertebrates and some small vertebrates but are still in the middle of the food chain.

On the surface of the water there is a film made up of floating dead matter and the creatures that feed on it (plankton). Plankton include springtails (collembola), water fleas (daphnia) and cyclops (copepods). Among the larger invertebrates that inhabit this area are the water boatmen, back swimmers, water striders, pond skaters and fishing spiders. There are also the adults, nymphs and larvae of the various aquatic insects that have to come to the surface to breathe, including water beetles, whirligig beetles, water scorpions and mosquito wrigglers. The larvae of some species have gills and do not need to surface to replenish their air supply.

Down on the creek bed or pond floor, in amongst the pebbles and detritus, there are myriad creatures that make this a dangerous place to live. A high percentage of the creatures that live here are predators or scavengers. The predators among the insect population include water beetle larvae, predacious diving beetle larvae, whirligig beetle larvae, and scavenger water beetle larvae (commonly called water tigers due to their predatory nature and appetite). The adults of these water beetles are also predators and tend to be bottom feeders.

The larvae of dragonflies, damselflies, mayflies, stoneflies and caddisflies also all live and hunt as detritus or bottom dwellers. Most are free living (but the caddis fly larvae builds a shelter of silk, sand grains and pieces of twig that protects it from some predators). The larvae may undergo 10–15 instar stages which may take two to three years. In its final instar, the larvae crawls from the water and ascends some suitable material where it undergoes full metamorphosis to emerge as a flying adult. The shed skins of these flies are frequently seen attached to water plant stems, on stones and under bridges.

Water scorpions, pond skaters, water boatmen and water bugs are all 'true bugs', feeding by impaling their prey with a stabbing mouth part or proboscis. The immature stages are all miniature versions of the adult but without wings. All stages are predatory and they may hunt anywhere in the water column. The adults will occasionally fly to another pond if conditions demand it. All stages must come to the surface to breath. Water scorpions have a pair of long filaments on the end of the abdomen which, when brought together, can be used as a breathing tube. Water boatmen spend most of their time swimming at the pond surface and only submerge when danger presents itself. At the surface, they capture a bubble of air in a depression on their abdomen, covered with fine hairs, they then take their air supply with them.

Of the 'true flies', the species commonly found associated with a water source are crane flies, mosquitoes, midges and sand flies. The 'rat-tailed maggot' may be found in water with a very high organic content. The maggot itself has a very long breathing tube that looks like a tail, hence the name. The adult stage is a drone or hover fly. Mosquitoe larvae are the most easily recognised of the fly larvae. They are commonly called wrigglers and may be seen on the surface of the pond, breathing through their tail. The larvae feeds on decaying organic matter on the floor of the pond. The larvae of midges are commonly called bloodworms. They are a red-coloured 'worm' up to about 15 mm long and live in the mud and feed on decaying vegetable matter. The adult midge looks similar to a mosquito but does not bite.

Leeches are commonly encountered. These are usually dark green in colour with one or two pale-cream lines along the body. They will swim in the water and are readily attracted to birds, fish and humans. Leeches will bite with a painless bite then suck blood from the host. They may be transported from pond to pond on the legs of waterbirds.

Also living in the mud on the bottom of the pond are myriad microscopic creatures including nematodes, hydra, turbellaria and protozoa, all of which require the use of a microscope for identification.

All of these creatures are affected pollutants to some extent, some more than others. If a pond has a good population of all of these creatures then it is a very healthy pond.

Mike Gemmell

which may occur at any time of the year. On the other hand, waterways in the wetter Bassian province have larger, more reliable and consistent waterflow during winter. Thus waterways in the Eyrean province have a depauperate specialist flora and fauna in their riparian zones, whereas waterways in Bassian regions have better-developed specialist riparian ecological communities.

Vegetation of the riparian zone

Natural riparian vegetation has an understorey that includes native grasses, herbs and shrubs, with an overstorey of larger trees. Creek-lines flowing across the plains were once well vegetated with river red gums (*Eucalyptus camaldulensis*), stands of which may still be seen along the River Torrens, First and Second creeks, and along the Sturt River. Apart from the River Torrens, Fourth Creek is the only stream on the plains that has small colonies of river bottlebrush (*Callistemon sieberi*) growing along its banks. A small tree, the blackwood (*Acacia melanoxylon*), once grew along this stream at Rostrevor, and a large shrub, the sticky boobialla (*Myoporum viscosum*), still lines the Morialta Creek at the grounds of Stradbrooke School grounds.[1, 3]

In the southern areas of Adelaide, some of the major riparian plant species include grey box (*Eucalyptus microcarpa*), South Australian blue gum (*Eucalyptus leucoxylon*), sweet bursaria or Christmas Bush (*Bursaria spinosa*), native lilac (*Hardenbergia violacea*), golden wattle (*Acacia pycnantha*), kangaroo thorn (*Acacia paradoxa*), kangaroo grass (*Themeda triandra*), and other native grasses, drooping sheoak (*Allocasuarina verticillata*), and river red gum.[3, 4, 5] The remnant vegetation in the Patawalonga Catchment (the largest of the southern catchments) is mostly located in the eastern region of the catchment. There are 14 vegetation associations historically growing in the catchment, including grey box woodland across the suburban foothills, as well as a mix of blue gum woodlands and messmate stringybark woodlands (*Eucalyptus obliqua*) around Belair National Park and Brownhill Creek Recreation Park.[3] Other major vegetation associations historically located within the southern catchments include brown stringybark forest (*Eucalyptus baxteri*), grey box woodland, candle bark (*Eucalyptus rubida*) or mountain gum forest (*Eucalyptus dalrympleana*), rough-barked manna gum (*Eucalyptus viminalis* spp. *cygnetensis*), river red gum open woodland, mallee box open woodland (*Eucalyptus porosa*), and blue gum open woodland.[1, 3, 5, 6]

To the north there were many different natural communities and habitats located within the catchment area. This diversity reflects the mixture of species from the Eyrean and Bassian biogeographic provinces, and range from areas of stringy bark open forests in wetter (Bassian) areas to open savanna woodlands of mallee box in the drier, more Eyrean areas.[7] The riparian vegetation along Dry Creek and Little Para River was generally dominated by river red gums and South Australian blue gums. There was a slightly different riparian community along the Gawler River, with river red gum and river box (*Eucalyptus largiflorens*) open forest towards the coastal areas, and pink gum (*Eucalyptus fasciculosa*), blue gum and drooping sheoaks communities occurring towards the foothills. The catchment also had areas of tea-tree swamp (*Leptospermum* sp.), samphire swamp (*Halosarcia* sp.), coastal shrublands (with coast daisy bush, *Olearia axillaris*, and umbrella bush, *Acacia ligulata*), and *Gahnia filum* swamp.[3, 7]

Changes to riparian vegetation

There are now few riparian areas that have not been altered in some way. Vegetation clearance, grazing, agriculture and urbanisation have all been major causes of change. Grazing and trampling, predominantly by domestic stock, impede the natural regeneration of riparian vegetation. In addition, in response to river channel modification, increased streambank erosion also changes the riparian vegetation structure. Accidental or purposeful release and introduction of exotic riparian species, such as weeping willows, has changed the structure and function of riparian zones around Adelaide. In particular, in all areas of Adelaide the riparian zone is now dominated by introduced deciduous trees, including poplars, willows, ash, blackberry, hawthorn, and many others. The high occurrence of deciduous exotic trees along rivers and creeks causes extensive shading, which can make the aquatic ecosystem unsuitable for local native flora and fauna. In addition, the leaf litter loads from the exotic trees can be very high for short periods in late autumn/ early winter when the leaves decay in the water. The rotting vegetation causes the oxygen levels to decrease greatly; eutrophication and changes in water chemistry are also associated with the release of acidic tannins from autumn leaves. These factors cause a reduction in species' richness.[8] The combination of deoxygenation and eutrophication can also cause proliferation of problematic algae, which in turn cause fish kills.

In the water itself, introduced aquatic weeds such as water plantain (*Alisma lanceolatum*) and Cape pond lily (*Aponogeton distachyus*) can spread over the water surface of a stream or wetland, changing the physical conditions of the water (by reducing light penetration) thereby making it unsuitable for many animal and plant species. Six introduced aquatic species are now well established: *Alisma lanceolatum* (water plantain); *Sagittaria graminea* (grass-leaved arrowhead), *Aponogeton distachyus* (Cape pond lily), *Cabomba caroliniana* (cabomba), *Callitriche stagnalis* (common starwort), and *Rorippa nasturtium-aquaticum* (watercress). (Five of these species were recorded as present before 1900.[9]) These introduced plants have caused four local aquatic species – *Myriophyllum muelleri* (slender milfoil), *Ranunculus inundatus* (river buttercup), *Ranunculus amphitricha* (small river buttercup) and *Ottelia ovalifolia* (swamp lily) – to become locally extinct.[9] Other impacts on riparian vegetation include tree removal or branch lopping, road construction, weed infestations (and subsequent weed control), and clearing associated with recreation practices.[5]

Wetlands and estuaries

Wetlands are generally shallow areas of slow-moving or standing water (i.e., swamps and marshes), found along or adjacent to rivers, creeks and floodplains. Wetlands are not always comprised of freshwater; they can be brackish or saline and found along coasts as saltmarshes and mudflats.[10] They are either temporarily or permanently flooded with water and usually have abundant vegetation

Mosquitoes are a natural part of Australian landscapes. They lay eggs in or near water, which hatch into larvae ('wrigglers') that graze on algae, vegetable matter, or even each other. Female mosquitoes feed on blood for the development of new eggs. South Australia is home to around 50 different mosquito species, each with its own habitat requirements and behaviour. There are nearly 20 species in the Adelaide region.

The arrival of Europeans and the construction of cities created a change in Australian mosquito populations. Urban and suburban developments change the way water flows and accumulates, creating artificial aquatic habitats in tanks, drains, gutters, buckets and pot plant bases. Mosquitoes, ever the opportunists, have found ways of adapting to these artificial environments. There are specialist urban mosquito varieties that have become common in cities such as Adelaide. Some of these types can be found throughout the world, such as *Culex quinquefasciatus* (also known as 'quinques'). This species bites indoors in the dead of night, and is the cause of numerous sleepless nights, whining in people's ears.

Native Australian mosquitoes have also been very successful in adapting to human habitation. The most successful types have been able to migrate from their woodland habitats, where they breed in small, water-filled tree holes and root buttresses, and colonise rainwater tanks. *Aedes notoscriptus* is common in tanks, and with the increase in domestic water storage for use on gardens, mosquitoes like this will probably become more common.

In cities like Adelaide there are still large amounts of natural habitat for native mosquitoes. This includes the pools that form in drying creek beds, groundpools in grassy depressions, water collecting in plants, and intertidal saltmarshes. Mosquitoes continue to breed in these places, which are often very close to suburban developments. While most of the time there are not enough mosquitoes to cause problems, the right weather conditions can lead to plagues of biting mosquitoes. In low-lying coastal areas close to saltmarsh and mangrove forests, the northern and southern saltmarsh mosquitoes (*Aedes vigilax* and *Aedes camptorhynchus*) can attack in their thousands. Controlling these outbreaks sometimes involves large-scale insecticide application to ecologically important aquatic habitats – something to be avoided.

It is worth remembering that it is humans who are encroaching on native mosquito habitats, and in some cases we are creating new breeding grounds for these pests. Living with mosquitoes is something we will have to get used to.

Craig Williams

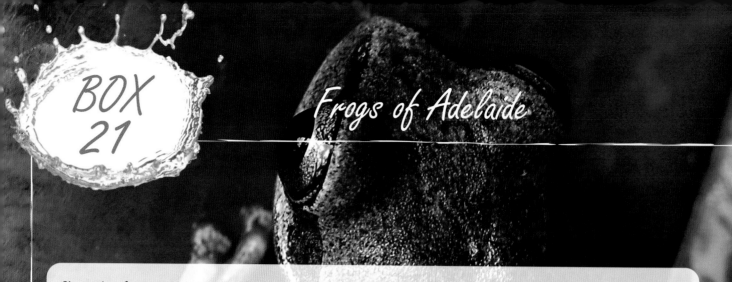

BOX 21

Frogs of Adelaide

Six native frog species are endemic to the Adelaide region. Despite significant habitat alteration none of the native amphibian species have become extinct, nor have there been any exotic introductions. Frogs are widespread and relatively common through the Adelaide region, although their presence is not always evident during the hot, dry periods of the year. Some species have specific habitat requirements, such as Bibron's toadlet, while the spotted grass frog and the common froglet are widespread and found in a range of habitats. Bibron's toadlet is threatened because of limited distribution and habitat destruction. Frogs are dependant on temporary and/or permanent freshwater; habitat destruction and water quality are the major threats to their survival.

Tree frogs – Hylidae

Brown tree frog (Ewing's frog, southern brown tree frog or whistling tree frog) *Litoria ewingi*

Identification: female 45 mm, male 40 mm. Very pale to dark brown with a broad, darker-brown stripe down the centre of the back. A narrow brown/black stripe extends from the tip of the snout, through the eye to the shoulder. A pale stripe is commonly present beneath the eye. The back of the thighs are yellow-orange, often with black spots. Discs present on fingers and toes. Fingers lack webbing, while toes are half webbed.

Habitat and range: various habitats including in vegetation, on ground or under rocks near permanent water in creeks, ponds, wetlands or garden pools. Prefers undulating or hilly country. May also be found clinging to windows or even in bathrooms. Widespread.

Mating call: loud high-pitched call described as 'creek-creek-creek' or 'weep-eep-eep'. Male calls from the water's edge or in terrestrial or aquatic vegetation.

Eggs and breeding site: principally March–October, although can be anytime of the year. Eggs are laid in clumps attached to submerged vegetation. Tadpoles hatch after 4–5 days. Initially tadpoles are cream in colour, but darken as they mature.

The southern bell frog (golden frog, warty swamp frog or growling grass frog) *Litoria raniformis*

Identification: female 104 mm, male 65 mm. Colour varies from light green to brown with green, yellowish or brown markings. A broad, pale-green mid-dorsal strip and a large tympanum (ear drum) are its most consistent features. The thighs may be bright blue, the skin above often warty. Belly is granular and white. Very small discs

present on fingers and toes. Fingers lack webbing, while toes are almost fully webbed.

Habitat and range: this species was released into the Mount Lofty Ranges in the 1960s but has not been recorded in the region since the mid 1980s. Found in permanent waterbodies, including rivers, lakes and wetlands, with abundant vegetation (especially bulrushes). Vulnerable in South Australia but populations still exist in the Murray Valley and the south-east.

Mating call: Loud, drawn out 'cra-ack, cra-ack' followed by short grunts. Male calls while floating in water .

Eggs and breeding site: Breeding takes place August–April; female lays 2000–4000 eggs, deposited as floating rafts which subsequently sink. Tadpoles are pinkish-grey with yellow fins, and can grow up to 100 mm in length, taking 2–12 months to develop.

Ground frogs – Leptodactylidae (Myobatrachidae)

Common froglet (Brown froglet or clicking froglet) *Crinia signifera*

Identification: female 30 mm, male 25 mm. Colour and patterning is highly variable. Back and sides cream, sandy, grey, brown, brick-red or black. Stripes, spots or blotchy markings may also be present. Skin on back smooth, raised ridges or warty. Belly is granular with dark-grey and white marbling. Throat to breast may be grey or light brown, and can include a pale stripe or series of dots. Indistinct tympanum. Fingers and toes unwebbed.

Habitat and range: found at the edge of creeks, ponds, rivers, wetlands, drains and areas of seepage; beneath rocks, vegetation and debris, including garden ponds. Common in Adelaide and south-eastern South Australia.

Mating call: variable call that may be a single, slow 'crick' or a rapid, high-pitched 'crick- crick- crick'. Male calls from ground or leaf litter by water's edge or from submerged vegetation.

Eggs and breeding site: Year-round breeder, excluding summer. Lays 100–300 small eggs in clumps attached to vegetation in shallow water. Breeds in permanent or temporary water. Tadpoles take approximately 4–8 weeks to complete development.

Eastern banjo frog (pobblebonk, four-bob frog or bullfrog) *Limnodynastes dumerili*

Identification: female 83 mm, male 70 mm. A large frog with a broad snout, rounded head and 'toad-like' appearance, with rough and warty skin over its body. Can be pale grey, olive green, purple or black, but typically dark brown in colour. Golden ridge runs from the nostril to the shoulder. Often irregular darker markings on back. Sides may be coloured purple, bronze, orange or black. White ventral surface, sometimes mottled grey. Limbs are short and stout. A prominent gland can be found on the tibia of the hind limb. Toes limited webbing. Females have flanges (flaps of skin) on their fingers.

Habitat and range: common and widespread in southern South Australia near rivers and wetlands. It lives in small holes beneath damp wood or rocks where it is often encountered by gardeners. During dry summer months this species digs a burrow and aestivates. Following good rains it emerges to feed and breed.

Mating call: a deep, melodic 'bonk' or 'plonk'; when numerous frogs are chorusing, each typically at a slightly different pitch, the effect is like an out-of-tune banjo or 'pobblebonk', its onomatopoeic name. Male calls from vegetation in water.

Eggs and breeding site: may breed throughout the year but typically in spring or summer. Location is static or slow-moving water, where up to 4000 eggs are laid in foam nests, the bubbles for which are created by the female's flanges. The nests are secured to vegetation or stones at the edge of pools or streams. The tadpole stage takes from 3–15 months.

Spotted grass frog (spotted marsh frog or marbled frog) *Limnodynastes tasmaniensis*

Identification: female 47 mm, male 42 mm. Dark-green or brown blotches on a background of grey, brown or any shade of green with similar coloured bars on its limbs. A cream-yellow through to rusty-red line is frequently present running from the tip of the snout down the middle of the back. A broad, dark band runs from the snout through the nostril, below the eye to the forelimb. Skin is smooth or with small warts above, smooth and white below. The eye has a golden iris. Males have a dark yellow-green throat while breeding. Females have large flanges (flaps of skin) on the first two fingers.

Habitat and range: common through the south and east of South Australia. It lives in or near water in creeks, wetlands, marshes, dams or reliable pools, sometimes garden ponds. During the summer it will take refuge beneath stones or logs in the bed of dry creeks or pools.

Mating call: three races exist in South Australia. The Adelaide race has a rapid, staccato burst of soft 'uk-uk-uk-uk', with 3–6 notes per call. Males call while floating, lying partially submerged among vegetation or sitting in shallow water.

Eggs and breeding site: may breed throughout the year but typically in August to March. Eggs are laid in a circular white foam nest, which floats on the water surface or among vegetation in marshes, swamps, ponds and temporary waters. Nests of 90–1300 eggs are created by the female's flanges. The tadpole takes 3–5 months to develop, up to a year in colder environments.

Painted frog *Neobatrachus pictus*
Identification: female 55 mm, male 60 mm. Moderately sized, squat frog with rounded head and body. Background colour is a dull olive-green to yellow with extensive dark-green or brown blotches over the dorsal aspect. Thin yellow or cream stripe often runs down back. Skin of back is covered in small warts or bumps which in males develop into spikes during the mating season. Belly is white and smooth. The eye is prominent with a vertical pupil. Fingers lack webbing, toes are webbed.

Habitat and range: open woodland, mallee, farmland and grassland through the south-east of South Australia. During summer this species digs a burrow and aestivates.

Mating call: A rapid, melodic 'woodpecker-like' drumming or trill. Male usually calls while floating in water, usually after heavy rain.

Eggs and breeding site: may breed any time of the year after heavy rains in temporary pools. Eggs are laid in chains of up to 1000, hidden among submerged vegetation. The tadpole typically takes two months to develop to a maximum of 75 mm. Once metamorphosis has taken place, the young frogs bury themselves in moist soil nearby, emerging to disperse following heavy rains.

Bibron's toadlet (brown toadlet) *Pseudophryne bibroni*
Identification: female 32 mm, male 30 mm. Small toad with short legs. Grey or dark brown to black on back with varying darker warts and orange or reddish blotches, particularly on head. Pale-yellow or orange area on upper arm. Belly with bold dark-grey and white marbling. Often a yellow spot on rear end. Tympanum hidden. Skin of belly smooth, while back, sides and limbs are granular with warts. Toes are unwebbed.

Habitat and range: grassland, forest or woodland habitats on damp hill slopes or near creeks, under cover such as leaf litter, rocks or logs. Numbers are declining.

Mating call: drawn-out 'ank-ank' or 'ark'. Male calls from burrows or on ground from cover.

Eggs and breeding site: breeding from February to August. Up to 200 eggs are laid in a damp cavity or depression under leaf litter, rocks or logs. Eggs remain in situ, developing to an advanced stage until heavy rains flood the nesting site or wash them into ponds, ditches, soaks or creeks. Tadpoles are dark brown, with a mottled tail and grow to 30 mm in length.

James Smith and Steve Walker

and animal species.[11] Natural wetlands improve water quality, help alleviate the impact of floods, and stabilise the banks of rivers and streams.[10] Even when wetlands are temporarily dry, they provide habitat for some animals, e.g., lizards that utilise cracks in the dry surface.[11] Other plants and animals survive through dry periods by remaining inactive as seeds, spores or eggs until they are stimulated to grow when the wetland fills again with water. A square metre of mud in a typical wetland will hold thousands of aquatic plant seeds.[11]

Perhaps the most famous of Adelaide's wetlands was a large area of freshwater swamps known as the Reedbeds, which collected water during winter from most creeks including the River Torrens that flowed from the Mount Lofty Ranges across the Adelaide Plains. As the wetlands filled, they drained either northward into the Port River/Barker Inlet estuary, or southward through the Patawalonga. They generally dried during summer, although some areas contained more permanent swamps. The proclamation of South Australia took place adjacent to the Reedbeds; the old gum tree at St Leonards is a river red gum (*Eucalyptus camaldulensis*), which was the dominant large tree around the swamps. Beneath the river red gums were areas of lignum (*Muehlenbeckia florulenta*), common reed (*Phragmites australis*), and bulrush (*Typha domingensis*). These reedbeds probably merged with a sedgeland containing rushes (*Juncus* spp.), sedges (*Cyperus* spp.), and club rushes (*Isolepis* spp.).[1, 3] The swamps were home to vast numbers of birds including many waterfowl and the now very rare orange-bellied parrot (*Neophema chrysogaster*), which no longer occurs in the area since the swamps were destroyed. The cumulative destruction of the parrots' winter feeding areas on mainland Australia, and of its summer breeding habitat in Tasmania, has reduced the numbers to a world population of 180.

The waterways where the swamps drained into the Port River and Patawalonga Creek were brackish and lined with salt-tolerant South Australian swamp paperbark (*Melaleuca halmaturorum*) low, open forest. Beyond the paperbarks, where the highest tides reach, was a wide expanse of samphire (*Halosarcia* sp. and *Sarcocornia* sp.) and saltbush (*Atriplex paludosa*) low shrubland.[1, 3] These saline, swampy backwaters are largely flat but minor variations in topography influence the frequency of inundation and soil salinity, and correlate with the plant species that occur at any given location. For example, the samphire, *Halosarcia arbuscula*, dominates where the tides flood regularly and the soil contains 16–18% soluble salt. Grey samphire (*Halosarcia halocnemoides*) occurs where the soil is flooded once or twice a year and contains 3–5% soluble salt, whereas the marsh saltbush (*Atriplex paludosa*) occurs in areas rarely flooded and contains 1.5–3% soluble salt.[12] The samphire is the first of the truly estuarine communities, and contains relatively few species, each of which is relatively abundant. This reflects the harsh environmental conditions in estuaries, where few species can survive (but those that can, flourish). Found within all these wetland habitats are quite a few threatened species – two saltmarsh plants, the bead samphire (*Halosarcia flabelliformis*) and the cushion centrolepis (*Centrolepis cephaloformis*), which can only be found in the wetlands.[1, 3]

The two major estuaries (Barker Inlet and Onkaparinga) are structurally composed of low-lying barrier dunes backed by samphire flats, intertidal mangrove forest, intertidal mud and sandflats, tidal channels, and subtidal seagrass meadows. At Barker Inlet, a dense mangrove forest occurs where the tide floods daily, in the littoral zone. Aerial surveys show that about 90% of the Barker Inlet Aquatic Reserve area, including 1000 ha of mangrove forest drained by a network of tidal creeks, is exposed by low spring tides. The cool, temperate coasts of southern Australia are colonised by only one species of mangrove, *Avicennia marina* var. *resinifera*, in contrast to tropical latitudes where many species may be present. These mangroves colonise almost the entire shoreline of Barker Inlet at the upper intertidal level, including the eastern side of Torrens and Garden islands, which separate Barker Inlet from the Port River. These trees range from stunted shrubs at the dry, landward margin of the inlet to a dense forest, 5 m high, where good tidal circulation or freshwater runoff is available. Some of the largest trees grow along Swan Alley Creek. Pneumatophores and algal mats capture and stabilise sediments within and around the edges of the mangrove forest. Pneumatophores are the aerial roots of the mangrove, which poke above the wet, sometimes anoxic, mud, collecting oxygen for the root system.[13] Mangrove sediments in the Barker Inlet/Port River estuary are normally covered with a slippery algal mat community (which includes diatoms, blue-green algae and algae).

Seaward from the mangroves are extensive areas of more or less bare sandflats. These flats may be several hundred metres wide. Near the low-tide level, there are zones, first of eel grass beds (*Heterozostera*), and then subtidal seagrasses (*Posidonia*) between 0.5 m and 12 m below low-tide level. The seagrasses support an extensive epiflora, which grow on them, including macroalgae (*Ulva*, *Enteromorpha* and *Cladophora*) and red algae.[1] Generally, there are few hard substrates on the intertidal sandflats. The surface of the sediments supports abundant microscopic life, especially diatoms and blue-green algae. Where hard substrates do occur, such as rocks or mussel beds (e.g., *Austromytilus*), sea grapes (*Hormosira banksii*) may form dense communities.[1]

Riparian, aquatic and estuarine animals

Riparian vegetation provides habitat for in-stream animals including fish, invertebrates, reptiles, amphibians, birds and mammals. The frequency and duration of inundation is a major factor affecting the nature of riparian communities. Most waterways in the Adelaide area flow only during winter, drying gradually during spring, becoming a series of waterholes, and drying out completely during summer. The only exceptions to this are some waterholes that may persist through the summer in the deep gullies of the Mount Lofty Ranges. This pattern of flow determines the types of animals that inhabit the riparian zone.

Invertebrates

A diverse array of invertebrates live in seasonal creeks (Diagram 5.1).[1] The speed of waterflow has a large impact on which species can occur at different places along creeks. In areas were flow is seasonally swift, caddis fly larvae may occur in vast numbers, living in protective casings among rocky riffles. Mayfly larvae are also common in swift-flowing

Native fish extinct in Adelaide

The southern purple-spotted gudgeon and the river blackfish are locally extinct in Adelaide but still found in other parts of Australia, although they remain sparsely distributed. Local exctinction of a species can lead to a number of problems. For example, losing a population from one area may mean the loss of significant genetic diversity because there may be genetic differentiation between isolated populations of the same species (particularly with widely distributed species). It is also possible that subspecies exist that are geographically separate to one another.

The southern purple-spotted gudgeon (pictured) was previously found in the streams of the Mount Lofty Ranges and in the Murray-Darling Basin. It is a medium-sized fish (up to 13 cm), heavily predated upon by the introduced eastern *Gambusia* (introduced around the late 1920s) and the redfin perch (introduced around the turn of the 20th century). It seems that the double impact of introduced predators and habitat modification led to the extinction of the gudgeon, approximately 20–50 years after the arrival of the *Gambusia* and perch. The gudgeon prefers to live in slow-flowing water among aquatic plants, where hard substrates are used for spawning. It is being considered for nationally endangered status and has protection under the *Fisheries Act 1982*.

The river blackfish was previously found in streams of the Mount Lofty Ranges (particularly the Torrens and Onkaparinga rivers) and in the Murray-Darling Basin. It is a large fish (up to 35 cm in this region), which possibly experienced less predation from introduced fish except perhaps on eggs or young, but a heavier human-predation load. The blackfish spawns in hollow logs and has a lifecycle completed entirely in freshwater. Integrity of habitat is crucial to its survival; it is particularly susceptible to increased sediment loads. The river blackfish is an elusive, nocturnal species found in permanently flowing, clear, cooler mountain streams to medium-sized rivers with high dissolved oxygen content. It is only found where abundant cover such as snags, reeds, boulders and hollow logs exist. The species has protection under the *Fisheries Act 1982*.

Catherine Gray

streams. In calmer waters, more species are able to survive and complex communities of many different types of animals may develop seasonally. Many microorganisms, including plants, animals and protozoa, may form biofilms on most inundated surfaces. These support communities of grazers including freshwater snails and the peculiar planarian (a simple flat worm that can split in two and regrow matching halves). In relatively still, often hypoxic water, the red-coloured chironomid larvae (chironomids include the common midge) may be present in huge numbers, while water striders (a type of bug – Hemiptera) skate across the surface. These animals support predators, such as damselfly and dragonfly larvae, dytiscid water beetles, backswimmers and water scorpions, omnivores such as yabbies (*Cherax destructor*), and other shrimp-like decapod crustaceans (such as *Macrobranchium australiense*). All these invertebrates, in turn, may be consumed by higher-order vertebrate predators such as water rats (*Hydromys chrysogaster*) or kingfishers. Freshwater plankton may also develop in calmer parts of waterways and include microsopic copepods, mites, rotifers and ostracods. Different planktonic species require different conditions. The planktonic communities reflect the health and condition of waterways and can be used as bioindicators to monitor waterway and wetland health. This complex mix of species does not usually include mosquito larvae because they are easy prey for so many other animals. Most of Adelaide's mosquitoes breed in brackish or saline water or in very small waterbodies such as the trays beneath pot plants, birdbaths and the water collected after rain in undisturbed buckets and other containers around the house.

Reptiles

Several widespread and conspicuous reptiles occur in the riparian habitats.[1, 4, 14] The water skink (*Eulamprus quoyii*) is a lizard that occurs only in the riparian zone and can be seen (or heard) rustling about in reedbeds on the banks of many waterways around Adelaide, notably Torrens Lake. Being ectothermic ('cold-blooded'), water skinks must spend a great deal of time warming themselves in the sun or even underwater. During the summer, these skinks converge on the remaining ponds and may occur in very high numbers. They have a complex social life and can often be seen chasing each other and engaging in fights, or basking on top of each other peacefully.

The Adelaide riparian system hosts three species of poisonous snakes: the relatively common red-bellied black snake (*Psuedechis porphyriacus*), the now uncommon tiger snake (*Notechis scutatus*), and the abundant brown snake (*Psuedonaja textilis*).[4, 14] The red-bellied black and tiger snakes feed on frogs and other riparian animals, while the brown snake uses waterways as corridors for dispersal and is found in these locations because there are often abundant rats and mice, which form the major portion of their diet. All three species are large, front-fanged, cobra-like snakes capable of inflicting a lethal (to humans) bite.

Two common species of tortoises (now generally called turtles) are found around Adelaide. The long-necked tortoise (*Chelodina longicollis*) is native to the River Torrens and the region, but the short-necked tortoise (*Emydura macquarii*) was probably introduced into Adelaide's streams as released

pets. On occasion, released turtles from elsewhere in Australia turn up in Adelaide's waterbodies. Some ponds and enclosed waterways now contain so many tortoises that the populations are at risk of starvation (e.g., the pond in the Wittunga Gardens at Blackwood).

Mammals

While water rats (*Hydromys chrysogaster*) occur in all permanent waterbodies, from the saltwater mangroves to freshwater swamps, other native mammals have proven less adaptable. Water rats are a special group of rodents with behaviours more closely resembling otters than other rats. Water rats can still be seen hunting for yabbies and fish in the Torrens Lake at dusk or dawn. The native swamp (*Rattus lutreolus*) and bush (*Rattus fuscipes*) rats are restricted to the remnant bushland in the Mount Lofty Ranges. Platypus once occurred in the few permanent fresh waterholes in the Mount Lofty Ranges; they now appear to be locally extinct with the last recorded sightings in the waterholes of the Onkaparinga Gorge. In urban areas, both species of introduced rats, the Norwegian (*Rattus norvegicus*) and ship rat (*Rattus rattus*), and the house mouse (*Mus musculus*), are abundant and therefore attract large populations of introduced predators such as cats and foxes. Foxes in particular traverse the riparian linear parks and take a heavy toll on birds and reptiles.

Fish

The fish assemblage inhabiting the waterways of urban Adelaide has been severely disrupted in recent history. Several species no longer occur here and many new species have been introduced. In the Mount Lofty Ranges and Adelaide, where the native fish species have evolved in highly variable waterways, they tend to be small in size (less than 35 cm long) and generally tolerant of large changes in temperature.[15, 16] Eleven species of native freshwater fish were originally found within the Adelaide region, including two lamprey species.[17] Although several of these species are listed as threatened at the state or regional level, none are yet considered nationally threatened.[18] Currently there are nine native species still located in Adelaide's rivers and streams but only three are considered common:[15, 16] flathead gudgeon (*Phylipnodon grandiceps*); Swan River goby (*Pseudogobius olorum*); and common galaxias (*Galaxias maculatus*). The river blackfish has not been recorded in rivers on the Adelaide Plains since 1902,[9, 17] while the native southern purple-spotted gudgeon was not recorded in Adelaide after the mid 1800s and officially recorded as locally extinct in the early 1950s.[9] Lampreys once bred in the Torrens River every year but they no longer do so (or can). These unusual 'jawless fish' travel up freshwater creeks to breed but spend much of their lives in marine habitats. The regulation of many rivers and creeks by dams has precluded these breeding migrations, and resulted in local extinction of species with this type of life-history. Living alongside these native species are three translocated natives (from outside the South Australian Gulf Division), and seven exotic freshwater fish.[17]

The first introduced species was the brown trout (*Salmo trutta*), in 1871.[9] Two other introduced species (goldfish, *Carassius auratus* and tench, *Tinca tinca*) arrived by 1876, and two more (the redfin perch, *Perca fluviatilis,* and rainbow

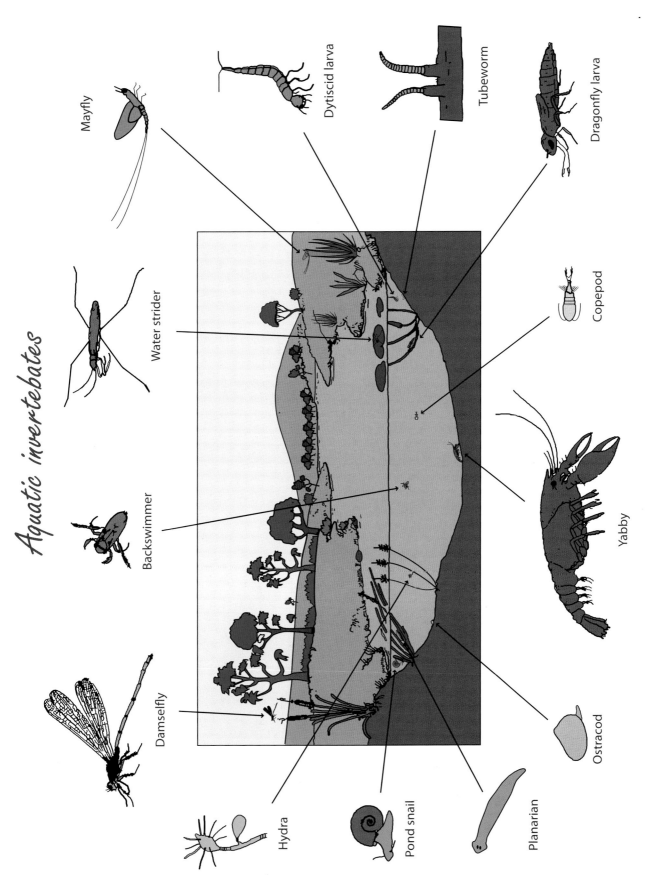

Aquatic invertebates

Mayfly

Dytiscid larva

Tubeworm

Dragonfly larva

Water strider

Backswimmer

Copepod

Damselfly

Yabby

Hydra

Pond snail

Planarian

Ostracod

Diagram 5.1 Invertebrates that are found in and around creeks and waterbodies in the Adelaide area.

BOX
23

Evolution of waterbirds

Waterbirds evolved to exploit different food resources in the world's rivers, streams, coastal and estuarine environments, and the seas and oceans. Birds can be generalists in these environments or occupy one of a large number of specialised niches. To a great extent, the lifestyles and natural history of a bird species is determined by its physiology and behaviour, which in turn has been shaped by its diet. Evolutionary forces have often caused species in different families and/or orders to evolve similar strategies and adaptations to the same problems, a process known as 'convergent evolution'. Also, the evolutionary development of particular adaptations can be observed in related species of birds. In this latter case it is allometry (the relationship between body size and some morphological characteristic) that reveals evolutionary patterns.

Morphology of feet
Anseriformes (ducks) have webbed feet, which are more or less flattened. They are good swimmers and their plumage is excellent for repelling water, due to special oils they produce. Birds with webbed feet can paddle through the water and walk on mud. As a duck pushes its feet back, the web spreads out to provide more surface area to thrust the water. Then, as the duck draws its foot forward and the toes come together the web folds up so there is less resistance to the water.

Gruiformes (moorhens and coots) are very diverse. Most species live in damp habitats near lakes, swamps and rivers. The lobed feet of coots make them good swimmers and enable them to walk on marshes. The moorhen's long-toed, unlobed feet enable it to walk on floating vegetation as well as climb. Jacanas' toes are even longer, as they don't only walk on floating vegetation but also make their nests on foating plants.

Waders and stilts (Charadriiformes) tend to have long legs and unwebbed feet, although avocets have partially webbed feet that permit them to swim easily in deeper water. While most seabirds (Procellariiformes) have webbed feet for swimming, the pelicans (Pelecaniformes) have totipalmate feet. Totipalmate feet have four toes, which are connected to form a large paddle useful for swimming or propelling pelicans under water as they feed by pursuit diving. Some boobies have colourful feet, which they display during mating. Raptors (birds of prey) have powerful hooked toes (talons), while penguins (Sphenisciformes) adopt an upright posture because their short feet are set far back on their body. They walk with difficulty and often move by pushing themselves along on their breast with wings and feet.

Bill morphology
Bill morphology is correlated to foraging behaviour. All ducks share the same bill structure, characterised by a special internal shape of the bill and a modified tongue that sucks in water at the tip of the bill and expels it from the sides and the rear. Along the sides of the bill is an array of lamellae (filter plates) designed to trap small particles, which are then licked off and swallowed. The nail at the tip of the bill is used for biting off vegetation. Ducks are primarily herbivorous and feed on leaves, stems, flowers, roots, etc. Aside from the differences in body size and neck length, waterfowl species have developed different bill sizes and shapes. Dabbling ducks have a flat bill, deeper than they are broad at the base. The edges of the bill are soft because these ducks find their food by touch. The nail on the tip of the bill serves to hook or move food. The interior lamellae act like sieves, retaining seeds, bugs and other food. Shovelers have extra-wide bills with well-developed lamellae to skim crustaceans and other invertebrates from the water

surface. The bills of geese are adapted to uproot grasses to eat roots, rhizomes and bulbs (snow geese) or to shear off short grasses (Cape Barren geese).

Marine predators such as penguins, cormorants or albatrosses have bills with curved projections that direct fish towards the oesophagus. Some Rallidae species, such as coots or moorhens, have a frontal shield on the upper mandible, which often becomes duller in colour after the mating season. Gulls do not have specialised bills and are often omnivorous. Many members of the Pelecaniformes (pelicans and cormorants) have developed a gular pouch. Pelicans are large cormorants (note the similarity in beak shape). The giant bill of the pelican is a result of allometry, and is essentially an overgrown version of the beak of a coromorant, large because pelicans are large birds.

Waders have bills of different lengths and curvatures to reach different invertebrates when they probe into the mud or soft soil (or by upturning stones like the turnstone, or skimming the water surface like the avocet). In this way, different species feed in the same habitat without competition for food. Prey is detected by means of the sensitive nerve endings at the end of the bill. Plovers feed on small invertebrates they find on the surface. Redshanks and waders, with their similar medium-length bills, probe the top 4 cm of substratum and feed on worms, bivalves and crustaceans. Curlews, godwits and other long-billed birds can reach lugworms and other deep-burrowing animals. The avocet swings its bill back and forth underwater, looking for aquatic insects, crustaceans and molluscs, and also feeds on plants and seeds it finds on the water surface.

Chris Daniels and Greg Johnston

trout, *Oncorhynchus mykiss*) were present by 1895.[9] The crimson-spotted rainbowfish (*Melanotaenia fluviatilis*), an Australian native fish that was not found locally, has been recently introduced. The introductions of trout for amateur angling have wrought many changes to waterways around Adelaide; the maintenance of trout stocks in the seasonal creeks requires annual introductions. With the arrival of the brown and rainbow trout, native populations of mountain galaxias have declined through direct predation. European carp (*Cyprinus carpio*), now known as carp or common carp, were introduced in the 1960s and persist in a few waterholes and wetlands that remain over summer, increasing the turbidity of the water by stirring up mud while foraging. The eastern *Gambusia*, also known as the plague minnow and, misleadingly, the mosquitofish (*Gambusia holbrooki*), was introduced in the mid 1950s.[9] (*Gambusia* do not eat significantly more mosquito larvae ('wrigglers') than other fish, but do eat the eggs and larvae of fish and frogs that *do* eat wrigglers.) The presence of the eastern *Gambusia* has therefore been attributed to declines in native frog populations, while mosquito numbers have been on the increase since the 1950s.

These introduced species require access to some types of aquatic vegetation but are not dependent on objects such as snags, hollow logs and other solid substrates. The high tolerances of these introduced species to varying environmental conditions, including temperature and habitat, has led to them often becoming major threats to the remaining native species through direct competition and predation. Exotic fish may also spread disease and parasite infections through native fish populations, which are often more susceptible to the symptoms of introduced diseases. For example, the introduced redfin perch carries a pathogenic virus that is lethal to at least four native Australian species. Of the introduced species, the tench and possibly the native translocated species are not as harmful to native freshwater fish as are the carp, brown trout, rainbow trout, goldfish, redfin perch, and *Gambusia*.

Amphibians

Amphibians are characteristically ectothermal vertebrates (no internal temperature regulation) with soft, glandular skin, four fingers and five toes, and a lifecycle predominantly involving an aquatic larval stage that metamorphoses to become an air-breathing, semi-terrestrial adult. The eggs tend to have no shell and are usually laid in water.[19] All amphibians found in Australia are members of the order Anura (the frogs), except the American axolotl (*Ambystoma mexicanum*), which has so far only been recorded as an aquarium pet.[19]

There are only seven different species of amphibian found in the Adelaide metropolitan area including two from the genus *Limnodynastes*, the marbled or spotted grass frog (*Limnodynastes tasmaniensis*) and the bullfrog or the eastern banjo frog (*Limnodynastes dumerilli*). The smallest species of frog are the common froglet (*Crinia signifera*) and Bibron's toadlet (*Pseudophryne bibroni*). The brown tree frog (*Litoria ewingi*) lives within dense vegetation in suburban gardens and prefers well-watered areas.[1, 20] Until the mid 1950s, the few records in Adelaide of the southern bell frog (*Litoria raniformis*) came from an introduced population in the northern suburb

of Salisbury. The seventh frog species in Adelaide is the water-holding or painted frog (*Neobatrachus pictus*), a burrower that digs itself below the surface of the soil to avoid dry summer conditions and occasionally mistaken for the introduced cane toad (*Bufo marinus*), which does not occur locally.

Adelaide's frog species are dependent on surfacewater for reproduction, and many species spend part of their lives close to water. Riparian lands, particularly those containing a range of vegetation types and structures (shrubs, ground layers and trees), are a suitable habitat for many frog species because of their humid climate, moist soils and dense vegetation. Abundant vegetation, fallen timber and flood debris also provide vital crevices and hollows for resting, hunting and hiding from predators. Also, there is often a high density of insect prey in these zones.[21] Some frogs reside permanently in riparian areas, however many species can survive considerable distances from isolated water (although these species are usually found in highest population densities within riparian areas).[21, 22] Generally, any disturbance to Adelaide's riparian zones and loss of riparian vegetation will have a significant impact on many frog species due to their high dependence on these areas.[21]

Despite the water requirement for many Australian frog species during breeding periods, one of the more remarkable characteristics of most Australian amphibians is their ability to survive independently of permanent waterbodies. The main trend among Australian species is to breed in temporary streams, creeks or pools.[23] Therefore, when judging whether an environment is suitable for frogs, the type of soil (whether it is permeable to water or not), and the presence of refuges (to get out of extreme climatic situations and to avoid predation), need to be considered. For burrowing frogs, these two considerations are linked.[23] Frogs have a great ability to find any crevice where the temperature is cooler.

Recently, frog numbers have been declining in both urban and non-urban areas.[22, 24] No clear pattern has yet emerged as to the definitive causes. Some human activities that result in negative impacts to frogs include:[25]

- use of insecticides, particularly via aerial spraying;
- draining wetland areas for urban development;
- converting ephemeral ponds to dams for use by domestic stock (destroying shelter for frogs); and
- release of the freshwater mosquito fish, the eastern gambusia (*Gambusia holbrooki*), and trout, which have been observed feeding on frog eggs and tadpoles.

Some other general reasons for declines in frog populations may include[25] altered air and water quality, pollution, spread of disease-causing organisms, pesticides that are hormonally active, climate change, large outbreaks of fungal infection (*Rhizopus microsporus*) among frog spawn, and possibly the advent and enlargement of a hole in the ozone layer leading to increased ultraviolet radiation.

Within the last few decades it has been observed that frog eggs and tadpoles are often sensitive to the presence of a wide range of environmental pollutants. These pollutants are able to produce abnormalities in the soft and skeletal tissues, as well as causing the death or altered behaviour of tadpoles. However, it is doubtful that a particular pollutant can be directly related to a particular abnormality.[23] Altering tadpole behaviour may make them more susceptible to predation. Some of the chemical substances polluting South Australia's (including Adelaide's) waterways include insecticides, heavy metals, herbicides and fungicides.[23] The abnormalities in eggs and tadpoles are often caused because many of these pollutants are able to pass through the jelly coating of frog eggs directly to the developing embryo. Consequently, to successfully complete their lifecycle, frogs can only tolerate low levels of pollutants.[22]

Birds

Of all the animals associated with water, birds are the most visible. While arguably the most beautiful riparian bird, the azure kingfisher (*Alcedo azurea*), is now extinct in the Adelaide area,[9] Adelaide remains home to a huge collection of bird species. However, it is not just the number of species that inhabit waterways that is spectacular, it is their diverse range of habits and body types. Also, water supports a huge variety of species with sedentary, migrant or nomadic lifestyles. Hence we enjoy pelicans, moorhens, coots, cormorants, darters, terns, waders, black (and occasionally white) swans, geese, and of course ducks.

Ducks belong to the family Anatidae, which includes geese and swans, and are found worldwide. All of Adelaide's ducks are puddle ducks, which means they bob in the water with their tails in the air. Adelaide has 11 native duck species, the most common natives being the Pacific black duck (*Anas superciliosa*), which roosts/nests on ground, and the wood duck (also called the maned duck, *Chenonetta jubata*), which roosts/nests in tree hollows. Two species of Teal are also relatively common in Adelaide, while feral domestic mallard (usually white with orange bills) and muscovy ducks also occur on the Torrens. The 'wild type' of mallard (*Anas platyrhynchos*) was introduced to Australia probably for hunting, and was first recorded in Adelaide in 1953 in the Botanical Gardens' duck pond. Hybridisation is known to occur between black and mallard ducks. Indeed, hybridisation between introduced mallards and close native relatives is of conservation concern throughout the world; it is a specific issue in Adelaide and has been the subject of considerable work. Ducks eat grasses, slugs, snails, weed, flowers, crustaceans and fish; while they seem to love bread, it is actually bad for them and causes bloating. It is better to feed ducks snails and slugs, which they crush using grooved-lamellae in their beaks.

Ducks have a high survival rate in urban areas because the chicks are precocial (reasonably independent), and the lack of predators enables numbers to build up rapidly. Pacific black ducks will make nests anywhere including on road median strips, centres of football ovals, or in gardens. The maned duck will nest in trees, often quite a distance from water, and will lead their brood along roads to the waterways. Luckily, drivers are incredibly good about stopping for ducks!

The estuarine communities also support large numbers of birds that feed on the abundant intertidal organisms. More than 120 species of birds have been identified as using the Barker Inlet Wetland, including threatened waterbirds such as the white-bellied sea eagle (*Haliaeetus leucogaster*), little egret (*Egretta garzetta*), and the slender-billed thornbill (*Acanthiza iredalei rosinae*).[1] In fact, the Barker Inlet and St Kilda Wetlands are a significant breeding habitat for waterbirds in general, with 18 species recorded as breeding there including cormorants, oystercatchers, terns and egrets.[1] Nearly 60 species of waterbird have been observed in the area, 12 of which are listed under international treaties such as the Japan–Australia and China–Australia Migratory Bird Agreements (JAMBA and CAMBA).[1] In addition, Barker Inlet contains breeding colonies of several species of local birds, and is also an important stopover point for migrating waders.

Because of its importance as a habitat for birds, the Barker Inlet/Port River estuary has a broad geographic significance.[1] On a local scale, silver gulls (*Larus novaehollandiae*) nest on islands in the estuary and forage over much of Adelaide during the day, returning to the breeding colony to roost at night. On a state scale, the two known breeding colonies of little egrets in South Australia are in the mangrove forests fringing the estuary, crucial for the persistence of this elegant white bird throughout the state.[1] On a national scale, Australian pelicans born at the breeding colony near Outer Harbor disperse throughout south-eastern Australia.

The estuary also has international significance. Two groups of wading birds use Barker Inlet/Port River estuary as a feeding site as part of their intercontinental migration strategy. One group, which includes the double-banded plover (*Charadrius bicinctus*), breed in New Zealand and then migrate to estuaries around the coast of south-eastern Australia in summer.[1] The other group includes about 15 species of wading bird that breed in Eurasia during the northern summer and migrate to the southern coast of Australia during the southern summer. Barker Inlet/Port River estuary is one of several wetlands critical to the continued existence of the migratory pathways for populations of species that move from Siberia and Mongolia, via Japan and the Indonesian archipelago, to southern Australia.[1] The coastal and marine wetlands of Barker Inlet and St Kilda are now recorded as wetlands of national importance.[1, 14]

Estuarine animal assemblages

In the estuary itself, there is generally a replacement of freshwater organisms by seawater organisms as one traverses the upper reaches of the estuary to the sea. This is because most freshwater species do not tolerate water with salinities greater than 10 parts per 1000 (ppt), whereas seawater organisms do not tolerate salinities less than 25 ppt.[1] Few organisms can cope with the variability of the salinity within estuaries. However, two types of organisms can be broadly distributed through the whole of the estuary. Species that are relatively sessile (e.g., whelks, periwinkles and other mollusks) may accommodate variations in temperature and salinity by possessing wide physiological tolerances.[1] More mobile organisms continually move to water with suitable characteristics. Fish traverse the estuary with the tide, and hence many people prefer to fish at the turn of the tide when fish are following bodies of water with different salinities.

BOX 24

Common waterbirds of Adelaide

Waterbirds are perhaps the most diverse groups of birds found in one ecological location. Members of up to 13 orders of birds can be found together in small strip of riverine or estuarine parkland.

Mallards are probably the most common ducks worldwide. The mallard belongs to the family Anatidae, a group of birds which feed on vegetable matter on the water surface, or by grazing on riverbanks. They rarely dive. Ducks can be seen in parks, on eutrophic lakes, woodland marshes and along seashores. The mallard is a medium-sized bird, 50 to 60 cm in length with a wingspan of 81 to 95 cm. The male in full plumage has a metallic-green head, narrow white collar, black rear end and a blue speculum edged with white. The bill of the male is yellow, whereas the female has a yellow-orange bill with blackish culmen. After the breeding season, the male displays an eclipse plumage similar to that of the female, although it still has a uniformly green-yellow bill. The juvenile mallard is similar to the adult female. Couples form in autumn and will be stable until the female lays eggs. During the spring mating season, it's not uncommon to see two or more males chase an isolated female duck, peck at her until she weakens, and then take turns copulating with her. Hens choose a wide variety of nest sites including under bushes, in tree holes, and under boats or buildings. The clutch comprises 9–13 eggs, and when they hatch, after 26–28 days, the hen leads the ducklings to water and does not return to the nest. Ducklings are precocial and can swim and feed themselves on insects as soon as they hatch, although they stay near the mother for protection. Young ducklings are not naturally waterproof. Mothers are very careful with the young, and watch over them tirelessly. When startled, the chicks start crying and calling the mother and only stop when they are reunited. While swimming or feeding, the hen calls out to the chicks constantly. Most mallard hens breed as yearlings, but they may not have much success in raising their young. Mallards are very good and fast flyers and can take off directly from the water surface. Mallards are very vocal, sometimes using 'soft' sounds resembling chick's calls, or a loud 'alert' sound to warn the group of possible danger. When they are very frightened they will repeat a series of loud 'quarr-quak' calls.

Stilts include the banded and the black-winged stilt and have a wide range, including Australia, Central and South America, Africa, southern and south-eastern Asia and parts of North America and Eurasia. They are widespread on the Australian mainland. Banded stilts are found mainly in saline and hypersaline (very salty) waters of the inland and coast, while black-winged stilts prefer freshater and saltwater marshes, mudflats and the shallow edges of lakes and rivers. Both banded and black-winged stilts, and their relatives the avocets, feed on crustaceans, molluscs, insects, vegetation, seeds and roots. They forage by picking, probing and scything (swinging their bill from side to side) on salt lakes, either by wading in shallow water or swimming, often some distance from the shore. However, unlike the banded stilt, the black-winged stilt rarely swims for food, preferring instead to wade in shallow water. Stilts are diurnal (feed by day), and are dependent on the availability of prey in ephemeral (appearing only after flooding) salt lakes. Stilts are dispersive and movements are complex and often erratic in response to availability of feeding and breeding habitat across the range. Populations may move to the coast or nearby when the arid inland is dry, returning inland to breed after rain.

Moorhens have a compact body and very long legs with long toes. They are shy birds and stay close to their hiding spots unless they feel very safe. Their long toes enable them to walk on floating vegetation or climb small branches. Moorhens are omnivorous and feed on plants, grasses, insects, snails and worms.

Sandpipers and terns are representatives of the Charadriiformes and are waders. Terns are widespread around the globe, and are a familiar sight in most coastal waters. The Arctic tern arrives in large numbers in Antarctica each summer to feed after a journey of 20,000 km from their nesting grounds in the Northern Hemisphere. The crested tern is the second largest Australian tern and one of the most commonly seen. It has a pale-yellow bill, scruffy black crest, grey wings and back, and a white neck and underparts. Although it is often observed on its own, the crested tern also frequently forms mixed flocks with other species.

The common sandpiper is grey-brown above and white below. There is an indistinct white supercilium (eyebrow) and white eye-ring. The bill is dark grey with yellow at the base and the legs vary from greyish-olive to a yellowish-brown. When at rest, the long tail projects well beyond the tips of the wings. This species is also known as the Eurasian sandpiper or summer snipe. It breeds in Europe and Asia. In Australasia it visits New Guinea and Australia, mainly in the north and west, and is less often seen in New Zealand.

Gulls represent the order Gaviiformes and include the silver gull (*Larus novaehollandae*), a common urban resident, breeding where appropriate safe places are found, and travelling in small flocks to the sea in the early morning and away at dusk. In general, they spend much of their time swimming on the water surface. Gulls do not often perch in trees, preferring to sit on the top of light poles, on walls, buildings and cranes. As with many other gull species, the silver gull has become a successful scavenger, readily pestering humans for handouts of scraps, pilfering from unattended food containers or searching for human refuse at tips. Other food includes worms, fish, insects and crustaceans. Silver gulls nest in large colonies on offshore islands. Often two broods will be raised in a year, and both adults share nest-building, incubation and feeding duties. Eggs are laid in a shallow nest scrape lined with vegetation.

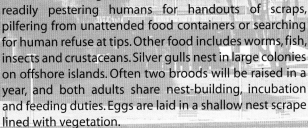

Cormorants and pelicans. A cormorant dives from the water surface and catches fish by propelling itself with its feet. It possesses a nictitating membrane (transparent third eyelid), protecting the eyes and allowing it to see underwater. The plumage is partially wettable in order to decrease buoyancy, but as a result it needs to dry its feathers on land, which it does by spreading its wings. Cormorants will sit on trees, often choosing the tallest branches to roost. When resting, preening or drying out they congregate in small groups, but otherwise they fly alone or in pairs. In Asia, fishermen use cormorants for catching fish and the cormorant species known as the guanay is the source of the fertiliser guano found along South America's Pacific coast.

Australian pelicans are large black and white birds found Australia wide. They have very long, pale-pink beaks with a small hook on the end and a stretching pouch underneath. When their wings are outstretched they measure over 2 m across. They have feet with four fully webbed toes. Australian pelicans feed on fish, yabbies and crabs. They sometimes swim in a group and drive fish into shallow areas. They call with pig-like grunting sounds. Australian pelicans make loose nests on the ground from sticks. Females lay two to four large white eggs. Both parents take turns to incubate eggs for up to 35 days.

Penguins (Order Sphenisciformes) The little penguin is 0.4–0.45 m high and, weighing about 1 kg, is the smallest of the world's 17 species of penguin. Also known as the fairy penguin, the little penguin is highly adapted for life in the sea. Its body is streamlined, its wings are modified as flippers, and its feathers are densely distributed over its body. Fairy penguins live along the southern edge of mainland Australia, Tasmania, New Zealand and the Chatham Islands, in temperate seas. The little penguin feeds mainly in inshore waters around the coast and breeding islands, and will swim out to the continental shelf. Little penguins come ashore at sunset in the breeding season. Breeding pairs live in colonies.

Spoonbills, storks and ibises are a group of large, long-legged wading birds in the family Threskiornithidae. Ibis have long, down-curved bills, and usually feed as a group, probing mud for food items, particularly crustaceans. Spoonbills have large, flat, spatulate bills and feed by wading through shallow water, sweeping the partly opened bill from side to side. Spoonbills generally prefer freshwater to saltwater. Spoonbills are monogamous, and most species nest in trees or reedbeds, often with ibises or herons. Nesting is colonial in ibises. Ibises have become urban pests in many Australian cities and will become aggressive if fed regularly by picnickers.

Chris Daniels

The biological communities can be conceptualised using an imaginary gradient from the landward end (above the high-tide level at the upper reaches of the estuary), to those communities living below the low water mark near the mouth of the estuary. Each region along the gradient has specific plant and animal assemblages. The upper reaches are dominated by freshwater species, the seaward end by marine species. Throughout the estuary, tolerant sessile organisms interact with mobile animals following the tides. However, the complex topography of the Barker Inlet/Port River estuary, and the extensive destruction of the natural vegetation surrounding the estuary above the high-tide mark, have complicated the basic pattern for the distribution of organisms.[1]

The mud crab (*Helograpsus haswellianus*) digs its burrow in the mud and plays an important role in cycling and aerating the sediments.[1, 6, 13] Indeed, these crabs may be termed 'ecological engineers' because they play a major role in determining the structure of the environment and fauna of the mangrove forests in temperate mangrove communities. Another conspicuous organism of the estuary in the razor 'fish' (*Pinna*),[1] a bivalve mollusc related to oysters and mussels and a feature at the low-tide level. *Pinna* are especially common where the bottom is soft and muddy. The gastropod snail, *Bembicium auratum*, common on the sandflats and mangroves, can be seen grazing on the surface of the sand when the tides are low and also on the trunks and pneumatophores of mangroves. Cockles (*Katelysia spp.*) are a common but inconspicuous species of the open sandflats of the estuary, so numerous that cockles in the Port River/Barker Inlet support a fishery valued at $AU1.5 million per year.[1] Their shells are washed up on local beaches in large numbers after they have died following predation by a small gastropod/snail called *Lepsiella*. These snails use their rasping radula to drill a hole into one valve of the cockle, gaining access to the cockle's soft parts, which they consume. The telltale small circular holes in many beach washed cockle shells is evidence of the enormous impact the predatory *Lepsiella* can have on the cockle population.[1]

Saline and brackish pools of water in the samphire and mangrove areas are the breeding sites of the saltmarsh mosquito (*Ochlerotatus vigilax*). A recent survey of mosquitoes in metropolitan Adelaide found this species to be quite rare in general, accounting for only 0.2% of all the mosquitoes found across the area.[5] However, the species is locally abundant, accounting for 84% of one sample of some 7000 mosquitoes caught in one night's collecting in the samphire and mangrove regions.[1, 26] Saltmarsh mosquitoes breed in small bodies of brackish water sheltered by vegetation,[26] rather than in larger, exposed bodies of water subject to wave action. The saltmarsh mosquito is a primary vector of Ross River virus in coastal temperate Australia.[27] Despite their abundance in the mangrove and samphire areas of the Barker Inlet/Port River and the Onkaparinga estuaries, there have been very few cases of locally acquired arbovirus infection in Adelaide.[27]

Threats to Adelaide's freshwater biodiversity

Freshwater habitat modification

The in-stream and near-stream habitats in Adelaide have been significantly modified since Europeans settlement in 1836. One of the most substantial changes has been through the realignment of a large proportion of the urban creeks. Many of the streams, such as the lower section of the Sturt River, have been concreted, and some (such as Second Creek) have been essentially converted to drainage pipes and are covered or built over. The lower section of the River Torrens was realigned in the 1930s following major flooding in 1931. A channel was cut to form a direct outlet through to the sea, and the surrounding embankments were built up to contain the expected maximum floodwaters. Before modification, the River Torrens ended in the Reedbeds, draining via the Port River or Patawalonga Creek.[28] The most significant threats to the biodiversity of riparian zones in Adelaide include:[1, 29]

1. The clearance of vegetation for agriculture, horticulture and urban development, resulting in the loss of native plant and animal species and causing erosion problems.
2. The intrusion of woody weeds (e.g., ash trees, *Fraxinus* spp., poplars, *Populus* spp., and willows, *Salix* spp.) and other exotic plants along riparian zones can result in a small number of these species dominating a whole stretch of river or lake, reducing the viable habitat for native aquatic and terrestrial species.
3. Unrestricted grazing along riparian zones can change the composition of plant species (promoting the spread of problematic weedy species), as well as reducing the survival and growth of the vegetation, leading again to bank erosion problems (damaging in-stream substrate habitat and increasing suspended particles in the water column).
4. The water runoff from industry, horticulture and agriculture near to riparian zones can contain high levels of nutrients, pesticides, herbicides, fertilisers, heavy metals, etc., which can severely damage an ecosystem.
5. Poor habitat management (e.g., removal of snags and native vegetation), introduced aquatic and riparian plants, animals and invertebrates, and inappropriate development (e.g., resulting in sedimentation).[17]
6. Unsuitable urban development of the floodplain and catchment area.
7. Increased water extraction and stormwater runoff.
8. Management of water temperature.

Some of the most important specific issues include loss of snags, altered streamflow, sedimentation, increased water extraction and water runoff, and temperature control. Each of these issues is discussed below, but for a more detailed analysis of the threats to our water, see Daniels and Tait, *Adelaide: Nature of a City* (2005).[1]

Loss of snags (large woody debris)

In an unmodified, unregulated river, most of the in-stream habitat available for aquatic animals comes from vegetation in the riparian zone. Large woody debris (LWD) (i.e., snags), such as logs, tree trunks and branches of fallen trees, are

Platypuses

BOX 25

Platypuses (*Ornithorhynchus anatinus*) occurred in the colony of South Australia at the time of its establishment. The creeks and rivers of the Fleurieu Peninsula were the western extent of their range. Around the turn of the century, a museum curator reported the species 'numerous' in the Torrens River. Specimen records in the South Australian Museum indicate that platypuses inhabited the lower Murray River, but most of the records come from the Riverlands. The most recent specimen recovered from this region was drowned in a drum net in 1975, while the most recent sighting was in 1990, near Renmark. Records also exist showing platypuses in the Mount Lofty Ranges, the Onkaparinga River and the Glenelg River near Mount Gambier in south-eastern South Australia, but no sightings have been made in these regions for over 50 years.

Photograph by Susan Ivory

South Australia seems to be the only state where the distribution of the platypus has changed considerably, being reported as 'rare' or locally 'extinct' in recent times. Low capture rates and the paucity of sightings has led biologists to speculate that platypuses in South Australia may not have ever been common and may represent transient non-breeding animals from New South Wales and Victoria. It has also been proposed that the occurrence of large Murray cod (*Oligorus macquariensis*) in the larger rivers of the plains may exclude platypuses.

A platypus floating on the surface would have no protection against these voracious predators.

Translocation of a number of animals from Tasmania and Victoria between the late 1920s and early 1940s resulted in the establishment of populations in several streams in the western end of Kangaroo Island, including Rocky and Ravine rivers and Breakneck Creek. Some of these populations continue to persist in ephemeral streams where permanent pools act as refuges during times when stream flows are reduced or absent due to drought conditions. A few platypuses from Rocky River were later relocated to the feral animal-proofed Warrawong Sanctuary in the Adelaide Hills where they continue to breed successfully.

Available evidence indicates that the platypus has probably always been uncommon in South Australia as it receives very little mention in published literature. Nevertheless, its distribution may be restricted or fragmented as a result of natural or human influence on the availability of suitable habitat. Its distribution has likely contracted so that it has now probably been extinct in the Mount Lofty Ranges and Fleurieu Peninsula area since the 1900s.

Kate Hutson

most often the major structural form of in-stream habitat.[30] Hence, some of Adelaide's original native fish species that use LWD as the preferred habitat are now extinct. De-snagging is the procedure used to remove LWD, which aims to improve the flow of streams, reduce the impact of flooding, allow safer navigation and recreation, and easier sand, gravel and gold extraction. The management practice of desnagging is now largely discontinued, however, the removal of riparian vegetation also leads to the loss of snags. Animals benefit from the presence of woody debris by being provided with:

1. Sites in which to spawn and rear juveniles. The freshwater blackfish (now locally extinct in the Adelaide area) attach their eggs to the inside of hollow logs; the removal of woody debris would have enhanced its decline and local extinction.
2. The vegetation provides cover and protection, especially during periods of flooding where individuals might easily be forced downstream out of their normal habitat.
3. Snags sticking out of the water can supply resting sites for waterbirds, reptiles and amphibians.
4. Protection from strong currents and sunlight.
5. Orientation points to identify habitat and territories.
6. Protection from predators; woody debris results in an assortment of flows within a single stream. Areas just downstream of a snag, or within the branches of a fallen tree or shrub, have slower currents, and are used by fish to rest or hide from predators.
7. Vantage points to help catch prey.
8. Increased productivity; LWD acts as a food source. On LWD, algae and bacteria growing on the surface or on other organic matter (e.g., leaves trapped by the debris) are eaten by invertebrates, which, in turn, are eaten by fish and waterbirds.

Although it is currently limited in quantity, woody debris provides important habitat in Adelaide's streams and rivers. In order to increase the amount of in-stream woody debris, more riparian vegetation is required. Fish and invertebrates have optimum water speeds (which differ for different stages in the lifecycle), and woody debris can provide shelter from fast currents as well as protection from terrestrial predators such as birds.[17]

Altered streamflow

In urban areas, rivers are modified for many reasons. Much of Adelaide's urban sprawl and agricultural production has historically taken place on floodplains previously susceptible to periodic flooding. By removing meanders (bends) or straightening a river channel, either by constructing a totally new path for the river to flow or cutting a flowpath across a bend, river modification can increase the physical and ecological damage to a waterway. One of the most common forms of river modification in urban areas is through the enlargement and straightening of a channel (followed by adding a concrete lining), to increase the river's water capacity. Almost all forms of river modification increase the velocity of a stream, thus increasing erosion and removing or altering in-stream habitat (e.g., rocks, logs and aquatic plants). The habitat is either swept away with the faster flows or covered in sediment by bank and streambed erosion and

deposition. The strength of a modified river flow can hinder the upstream migration of fish or other aquatic species (potentially forming a barrier to migration), or even prevent them from maintaining their position in the river without being washed away.

Some 70% of Australia's native fish species migrate at some point in their life in order to spawn;[31] barriers to migration can threaten their distribution and survival. Barriers can be small (e.g., a culvert under a road, funneling water at high speed) and are generally species-specific (e.g., *Galaxias* sp. are not able to cross the Breakout Creek weir while pouched lamprey, *Geotria australis*, can pass relatively easily). Access to marine or estuarine systems is required at some point in the lifecycle of many native fish species living in Adelaide's coastal streams (e.g., adult common galaxias move downstream between March and June to spawn in estuaries).[17]

Sedimentation

Sediment is a natural component of waterways, consisting of suspended particles in the water or on the bottom of rivers and streams. The artificially high level of in-stream sediments and nutrients in Adelaide result in the loss of suitable habitats, smothering of plant and animal species, increased turbidity and scouring, and changes to the freshwater community composition. The majority of sediment and nutrients come from the erosion of hill slopes, streambanks and gullies, as well as from groundwater. The presence of riparian vegetation can reduce runoff and sediment (by trapping it), thus protecting streams from rapid influxes of sediment and nutrients.[30] Sediment loads are regularly higher in cleared areas and where riverbanks are under frequent grazing pressure. Grazing in riparian zones, as in all other areas, causes erosion by disturbing the soil and striping the vegetation, resulting in slow regeneration of plants.

Suspended sediment in the water impacts on the resident plants and animals; an increase in turbidity occurs and water quality can also be affected (due to the heightened loads of nutrients and toxic substances attached to the sediment particles). 'Deposited sediment' is sediment that has settled out of the water onto the streambed, smothering in-stream habitat, reducing flow capacity, and filling pools (thus reducing the availability of spawning, resting and feeding habitats). An increase in turbidity significantly impacts on the physical, physiological and behavioural aspects of animal and plant ecology. For example, sediment damages the gills of fish, causing asphyxiation. Visibility is also reduced, diminishing the success of visual predators.

The importance of clean, clear water is obvious, especially for successful fish breeding. For example, the eggs of the river blackfish will only attach to clean, unsilted surfaces. Recent information highlights the high death rate of the eggs of several native fish species when they are only slightly covered in sediment.[32]

Increased water extraction and stormwater runoff

The high level of water extraction that has taken place in Adelaide since European settlement (and still occurs today) has left many of our rivers with substantially less water than they would have had naturally. The damming of several of Adelaide's waterways and altering of streamflows (e.g., the River Torrens Lake created by the construction of the City Weir) has also left some regions of rivers more or less permanently full of water, often with significantly reduced flows. Historically, these sections would have been ephemeral wetlands or floodplains (temporarily full of water depending on rainfall). In addition, stormwater flowing over impervious surfaces such as roads, buildings and paved areas has led to increased runoff (in both volume and peak-flow intensities), changing waterway environments. In Adelaide, as in most urban environments, water pollution is often a direct consequence of urban development. In particular, road runoff and associated pollutants from motor vehicles – as well as high loads of leaf litter, sediment and rubbish passing through drainage systems – enter our waterways at a great rate. Stormwater, and agricultural, horticultural and industrial runoff have resulted in litter, herbicides, fertilisers, pesticides, heavy metals, and many other pollutants entering Adelaide's waterways.[33]

Temperature control

Many species have temperature-specific survival ranges, particularly for breeding, and riparian vegetation can alter stream temperatures by providing shade.[17] The growth and development of the majority of aquatic organisms (e.g., algae, invertebrates, fish, reptiles and amphibians) are at least partly influenced by temperature. Changes in temperature are often critical for egg hatching and larval development. Temperature also influences the dissolved oxygen content of the waterway by decreasing the amount of oxygen in the water as temperature increases, limiting plant and animal life.[30]

Conclusion: the future of Adelaide's waterways from a biodiversity perspective

Aquatic, riparian, estuarine and coastal marine ecosystems on the Adelaide Plains, Gulf St Vincent and Barker Inlet, are the 'receiving' environment for most of the water flowing through Adelaide's four catchments.[1, 8] However, in many parts of the Adelaide Plains the creeks have been cleared and concreted so they maintain virtually none of their original ecological integrity. In other areas, the floodplains of the creeks and rivers have been turned into highly modified parks and public gardens, having varying impacts on the native wildlife inhabiting the waterways. The use of fertilisers and pesticides can have major local and downstream effects on a drainage system. Removal of 'untidy' fallen wood, rocky riffles and snags can simplify the physical structure of the riparian habitat, reducing the biodiversity of these highly managed areas. The deterioration of wetland and river condition reduces the ability of aquatic ecosystems to:

- improve water quality;
- refill subterranean aquifers, store water, reduce flooding;
- provide crucial habitat for native plants, animals and invertebrates during drought;
- supply valuable habitat for migratory birds (e.g., those species considered significant in the CAMBA and JAMBA treaties);

- reduce erosion by stabilising stream and riverbanks;
- sustain populations of threatened plant and animal species; and
- enhance recycling of nutrients.[10]

The native plants and animals living in or preferentially using these systems have been affected by the decline in water quality.[33] While there is now a push for the implementation of better, more environmentally conscious management strategies for Adelaide's waterways, there is a great deal of restoration that needs to occur. There are still the issues arising from multiple water-users with different, often opposing needs, occurring when so many people and industries are in such close proximity.[1]

Generally, watercourse management works should involve treating any erosion problems by addressing the causes of the erosion (e.g., halting vegetation removal and river path alteration); protecting all remaining remnant riparian vegetation (e.g., through active weed management and control of stock access for grazing); and avoiding activities that may initiate further erosion or decline in water quality (e.g., certain horticultural, agricultural, residential and industrial developments).[7] Where revegetation is attempted, local indigenous plants should be used whenever possible, and a diverse range of species and community structures (i.e., native aquatic plants, reeds, rushes, groundcovers, understorey plants, and larger shrubs and trees) should be planted. Weed control during revegetation projects and in general along Adelaide's waterways needs to be ongoing and part of a long-term weed removal program, rather than a one-off event.[1, 7] Woody weeds and exotic species can out-compete native plants and may provide inferior habitat value for native aquatic and terrestrial plants and animals using the freshwater environment. Adelaide's waterways are potentially the most important habitat linkages (or corridors) between patches of remnant vegetation that can encourage dispersal (of both populations and gene flow between populations) and migration of native plants and animals within and beyond the Adelaide metropolitan area. Thus, revegetation projects should attempt to link riparian zones with native vegetation in the catchment and on local landholders' properties.[7] Appropriate revegetation will enhance the habitat for use by native birds, mammals, reptiles, amphibians, freshwater fish and invertebrates.[1]

BOX 26

Reedbeds, sedgelands and vernal pools

Sitting in the niche between the salt marshes and the freshwater riparian wetlands, the reedbeds, sedgelands and vernal pools of the Adelaide Plains are ephemeral freshwater habitats of varying types. The *Phragmites*-dominated Reedbeds running south to Glenelg from Port Adelaide, which formed the discharge ends of the Port and Torrens rivers, were large, extensive wetlands, marked in the earliest maps of Adelaide. Other vegetation communities, such as cutting grass (*Gahnia filum*) sedgelands, intertidal-freshwater salt club rush (*Bolboschoenus caldwellii*) swamps, and swamp wallaby grass (*Amphibromus nervosus*) vernal pools, are smaller-patch habitats and many locations were never recorded.

The extensive Reedbeds wetlands were impacted early on, with many areas of loamy sediments cleared for horticulture by settlers within a few years of settlement. The smaller sedgelands suffered similar fates, and clearance and 'pasture improvement' for grazing left only one area of vernal pools on the Northern Adelaide Plains, while the cutting grass sedgelands and salt club rush swamps suffered the double jeopardy of clearance for agriculture and subsoil dehydration resulting from neighbouring drainage schemes.

The sedgelands and reedbed habitats vary quite dramatically in their preferred locations within the landscape, the plant species within them, and their hydrological requirements, but they are all marked by a dependence on water. Some of their characteristics are presented here:

Reedbeds were located on low-lying land at the end of the rivers. These wetlands accumulate soils and organic matter from the catchments and filter and slow the egress of water from the rivers to Gulf St Vincent. Edged with lignum (*Muehlenbeckia florulenta*) and swamp paperbarks (*Melaleuca halma-turorum*), they support dense stands of reeds (*Phragmites* sp.) with small, open areas of deeper water. Waterbirds, turtles and yabbies are plentiful, which would have made reedbeds a veritable larder for the local Kaurna people. Remnants of this habitat exist on land owned by the Adelaide Airport, in the coastal golfcourses, and near Buckland Park.

Cutting grass sedgelands occur on low-lying swampy land behind salt marshes on the Northern Adelaide Plains coastal strip. Freshwater lies on the surface around the tussocks of cutting grass (*Gahnia filum*) over the winter, drying out in summer. The groundwater table below these wetlands is very shallow and somewhat saline. Besides the dominant *Gahnia filum*, a range of native tussock grasses and low-growing shrubs make up this habitat. The yellowish sedge skipper (*Hesperilla flavescens flavia*) breeds only on *Gahnia*, and is possibly already extinct on the Adelaide Plains. A range of other threatened skipper and blue butterflies also breed on the grasses and sedges of these wetlands. Very small remnants of this habitat exist on the Dry Creek saltfields, SA Water's land at Bolivar, along Thompson Creek, and near Buckland Park. In almost all of these areas the sedgelands are in poor condition as horticultural drainage has denatured the soils.

Salt club rush intertidal freshwater swamps are essentially estuarine, occurring where the intermittently flowing rivers and creeks of the Northern Adelaide Plains debouch into the gulf. Dominated by *Bolboschoenus caldwellii*, these wetlands tolerate occasional inundation with the tide but do not persist if a seasonal source of fresher surfacewater is not available. Small remnants occur at the Little Para Estuary and at the Helps Road drain.

Swamp wallaby grass (*Amphibromus nervosus*) **vernal pools** occur slightly further inland on the more elevated alluvial fans of the Adelaide Plains. These deposits of shrink-swell clay develop a pattern of depressions (*gilgai*, or melon-holes) that hold water over winter and early spring. With very rare flows of sheet water across the surface, these tiny wetlands depend on the balance between rainfall, evaporation and seepage losses to support their mainly amphibious vegetation assemblages. With direct rainfall (the dominant water supply), the aquatic chemistry of these wetlands is unbuffered, resulting in wide daily pH excursions. The diurnal pH swings drive alterations in the soil chemistry, leading to a migration of clay within the profile and the development of a 'duripan' underlying the pools, which reduces groundwater seepage, gradually extending the period the pool remains wet each year. The vegetation association comprises species with a wide tolerance to pH variation and the resulting soluble metals concentrations. Vernal pools support small, drought-tolerant, shallow-water crustaceans including fairy shrimp, shield shrimp, and the daphnid *Simocephalus acutirostratus*.

They host numerous frogs and are a magnet to wading birds such as egrets. As they occur on land used for grazing, many of these small wetland patches were subject to deep ripping to improve drainage and seeding with 'improved pasture'. This has all but extirpated them. A small remnant exists along Elder Smith Drive at the southern end of Parafield Airport.

Altered hydrology will ultimately (possibly unintentionally) cause the extinction of these small elements of habitats. What actions do we need to take in order to preserve these special places?

- Return environmental flows to our rivers.

- Ensure tide and seasonal flows in coastal wetlands are not interrupted by embankments and crossings.

- Ensure drainage schemes do not lower shallow watertables near sedgelands unless a replacement water source is provided.

- Any remaining naturally 'panned' wetland habitats occurring in grazing areas should not be deep ripped.

Peri Coleman

References

1. C.B. Daniels and C.J. Tait, *Adelaide nature of a city: the ecology of a dynamic city from 1836 to 2036*, BioCity: Centre for Urban Habitats, Adelaide, 2005, chapter 12.
2. Onkaparinga Catchment Water Management Board, 'About the catchment area', 2005 (online).
3. D.N. Kraehenbuehl, '1836 vegetation map of the Adelaide metropolitan area', South Australian Department of Environment and Heritage, Adelaide, 1996.
4. A.C. Robinson, 'Reptiles and amphibians' in Wallace, H.R. (ed.) *The ecology of the forests and woodlands of South Australia*, D.J. Woolman, Government Printer, Adelaide, 1986, pp. 68–74.
5. M.S. Turner, *Conserving Adelaide's biodiversity: resources*, Urban Forestry Biodiversity Program, Adelaide, 2001.
6. P. Dillon, A. Hodson, J. Kochergen, A. Ciric and K. de Wett-Jones, 'Urrbrae Wetlands', City of Mitcham, 2004 (online).
7. Northern Adelaide and Barossa Catchment Water Management Board, 'The catchment overview', 2005 (online).
8. Patawalonga Catchment Water Management Board, 'Patawalonga Catchment Water Management Plan [2002–2007]', Adelaide, 2002, p. 184 (online).
9. C.J. Tait, C.B. Daniels and R.S. Hill, 'Changes in species assemblages within the Adelaide metropolitan area, Australia, 1836–2002', *Ecological Applications*, 15, 2005, pp. 346–359.
10. J.W. Holmes and M.B. Iversen, 'Hydrology of the Cowandilla Plains, Adelaide, before 1836' in Twidale, C.R., Tyler, M.J. and Webb, B.P. (eds) *Natural history of the Adelaide region*, Royal Society of South Australia, Adelaide, 1976, pp. 91–97.
11. Wetland Care Australia, 'Definitions of wetlands', 2004 (online).
12. T.G.B. Osborn and J.G. Wood, 'On the zonation of the vegetation in the Port Wakefield district, with special reference to the salinity of the soil', *Transactions Royal Society of South Australia*, 47, 1923, pp. 244–256.
13. K.S. Edyvane, 'Where forests meet the sea: mangroves in South Australia', South Australian Research and Development Institute (Aquatic Sciences), Adelaide, 1995.
14. T.F. Houston, *Reptiles of South Australia: a brief synopsis*, A.B. James, Government Printer, Adelaide, 1973, p. 11.
15. M. Hammer and G. Butler, 'Data sheet: freshwater fishes of the Mount Lofty Ranges, part (a): South Australian Gulf Division', Upper River Torrens Landcare Group Inc., Adelaide, 2000, p. 8.
16. T.D. Scott, C.J.M. Glover and R.V. Southcott, *The marine and freshwater fishes of South Australia*, 2nd edn, Handbook of the Flora and Fauna of South Australia, Government Printer, Adelaide, 1980.
17. G.R. Allen, S.H. Midgley and M. Allen, 'Field guide to the freshwater fishes of Australia', CSIRO Publishing, Collingwood, Victoria, 2002.
18. R. Wager and P. Jackson, 1993, *The action plan for Australian freshwater fishes*, Fisheries Division, Department of Primary Industries (Queensland) and Department of the Environment and Heritage (Queensland), Brisbane, 1993 (online).
19. H.G. Cogger, *Reptiles and amphibians of Australia*, 6th edn, New Holland Publishers (Australia) Pty Ltd, Sydney, 2000.
20. M.J. Tyler, G.F. Gross, C.E. Rix and R.W. Inns, 'Terrestrial fauna and aquatic vertebrates' in Twidale, C.R., et al (eds) *Natural history of the Adelaide region*, pp. 121–129.
21. R.J. Lynch and C.P. Catterall, 'Riparian wildlife and habitats' in Lovett, S. and Price, P. (eds) *Riparian land management technical guidelines, volume 1: Principles of sound management*, Land and Water Resources Research and Development Corporation, 1, 1999, Canberra, pp. 121–136.
22. S. Walker, 'Frog census 2002: community monitoring of water quality and habitat condition in South Australia using frogs as indicators in Frog census (series)', Environment Protection Authority, Adelaide, 2003.
23. M.J. Tyler, *Australian frogs*, Viking O'Neil, Penguin Books Australia Ltd, Melbourne, 1989.
24. M.S. Turner, 'Conserving Adelaide's biodiversity: resources', Urban Forestry Biodiversity Program, Adelaide, 2001.
25. M.J. Tyler, 'The action plan for Australian frogs', Wildlife Australia, Canberra, 1997 (online).
26. C.R. Williams, M.J. Kokkin, A.E. Snell, S.R. Fricker and E.L. Crossfield, 'Mosquitoes (*Diptera: Culicidae*) in metropolitan Adelaide, South Australia', *Transactions Royal Society of South Australia*, 125, 2001, pp. 115–121.
27. R.C. Russell, 'Arboviruses and their vectors in Australia: an update on the ecology and epidemiology of some mosquito-borne arboviruses', *Review of Medical and Veterinary Entomology*, 83, 1995, pp. 141–158.
28. R.P. Bourman and N. Harvey, 'chapter 4: landforms' in Nance, C. and Speight, D.L. (eds) *A land transformed: environmental change in South Australia*, Longman Cheshire Pty Ltd, Adelaide, 1986, pp. 78–125.
29. Onkaparinga Catchment Water Management Board, 'About the catchment area', 2005 (online).
30. S. Lovett and P. Price, 'Riparian land management technical guidelines, volume 1: Principles of sound management', Land and Water Resources Research and Development Corporation, Canberra, 1999.
31. Victorian Department of Natural Resources and Environment, 'In-stream barriers' in Victorian Department of Natural Resources and Environment (ed.) *Freshwater ecosystems: biodiversity management issues*, Department of Primary Industries Victoria, Melbourne, 9, 2001, p. 4.
32. Victorian Department of Natural Resources and Environment, 'Sedimentation of rivers and streams' in Victorian Department of Natural Resources and Environment (ed.) *Freshwater ecosystems: biodiversity management issues*, Department of Primary Industries Victoria, Melbourne, 4, 2001, p. 4.
33. C. Nicolson, J. Payne and J. Wallace, 'State of the environment report for South Australia: inland waters', Environment Protection Authority, Adelaide, 2003 (online).

Kaurna cultural heritage

Marni naa budni Kaurna yertaanna.
'Welcome to Kaurna country'
(Lewis Yerloburka O'Brien)[1]

Bukkibukkiunangko Kaurna meyunna yaintya pangkarilla tikketti
'Since ancient times Kaurna people have occupied this country'[2]

We would like to acknowledge that this land we meet on today is the traditional land of the Kaurna people and that we respect their spiritual relationship with their country. We also acknowledge the Kaurna people as the custodians of the Adelaide region and that their cultural and heritage beliefs are still as important to the living Kaurna people today.[3]

The above statements are traditional Kaurna protocol welcoming statements and expression.

Introduction

The Kaurna people have resided on *Mikawomma* – the Adelaide Plains – for over 40,000 years. To them, the *Mikawomma* is their 'country', which they care for as custodians on behalf of their Ancestors and Dreaming characters. Therefore, because of this long-standing relationship, a complex set of codes and moral narratives have evolved that explain the role of the Kaurna people and their responsibilities to the care, conservation and maintenance of the landscape. Within this semi-sedentary lifestyle, water was the lifeblood of existence and meaning. The landscape possessed a diversity of water sources and forms, of varying fresh and saline levels, which were linked to food, ritual, death, and often explained in myth, Dreaming story and nomenclature. Water was also carried to enable life for human, animal and spirit alike. The protection and care of water was essential, irrespective of the harshness of its absence during drought or its abundance during the rains or floods or its presence in narratives. While Jones[4] discussed the wider environmental associations of Kaurna occupancy of *Mikawomma*, this chapter focuses more specifically on the theme of water.

An important reference in understanding Kaurna consultation and engagement protocols was released in 2007, providing additional explanations of Kaurna cultural knowledge.[5]

The geographical character and form of *Mikawomma* has also changed considerably over the thousands of years the Kaura people have resided here. The Kaurna people are custodians of a landscape their Ancestors would first have encountered some tens of thousands of years ago. This is a landscape that has transformed from an ocean in the Pleistocene Period to a land that progressively dried, witnessing vegetation succession and megafaunal inhabitation and later extinction with the arrival of Indigenous peoples. The geographical and vegetative form of the landscape described by Colonel Light and other European settlers had not changed considerably for some 7000 years.[6]

As a consequence of the geographical transformation and evolution of the Australian continent, the Kaurna people are one representative of over 200–250 language groups or 'countries' (nations) of Aboriginal and Torres Strait peoples, which anthropologist Norman Tindale (1900–1993) identified and mapped as part of his research.[7] Through his work, substantial knowledge was accumulated about the natural systems of *Mikawomma*, including its water systems and patterns, vegetation, ecosystems and climate.

This chapter does not purport to be definitive, holistic or authoritative; rather it is offered as a guide only, using information accepted by researchers. It is a distillation of contemporary anthropological and environmental knowledge of the Kaurna people, providing a starting point to better understand and appreciate this landscape. This chapter considers the Kaurna 'history' of occupancy as it relates to water systems, and reviews their concept ('science', if you like) of nature and water systems. 'Water systems' in this chapter is treated holistically to not just mean the geographical forms of water drainage and evapotranspiration processes.

While Western knowledge is informed by positivist and objective forms of scientific experimentation – yes or no, black or white – Aboriginal knowledge is informed by oral traditions, vertical transferral of knowledge through story, practice and Indigenous research, and through the intellectual and experimental translation of knowledge at a place. Accordingly, our historical use of Western scientific documentation approaches explains why these landscape were not 'read', appreciated, understood or documented by early European immigrants to the Adelaide Plains landscape in the first instance, and secondly why 'truth' was apprehensively conveyed by the Kaurna people as it could result in loss of 'knowledge' and living with the surrounding water resources.

A complementary case study by Heyes considers the position of water in Indigenous notions of time and seasonality on *Mikawomma*.

Aboriginal landscapes

Aboriginal landscapes are ethnographic areas, often called 'countries', that stretch over the Australian continent and its waters. There were over 200–250 ethnographic 'countries' of varying sizes and populations prior to European settlement, with language groups loosely identifying them, and country size determined very much by environmental resources. Within each country were 'people', of which the Kaurna people were one such grouping. Kaurna country stretched from the Port Wakefield/Crystal Brook locality to

Cape Jervis, predominantly to the west of the 'forests' of the Mount Lofty Ranges. To the north-west were the Narangga people on Yorke Peninsula, to the north-east the Ngadjuri of the Mid North region, to the east were the Peramangk, and to the south-east were the Ngarrindjeri. Within each country people lived in clan groups, often involving extended families governed by strict moiety rules and narratives. On the *Mikawomma*, clan groups were associated with either the North Haven/Port Adelaide locality, the City of Adelaide precinct, to the north in the Prospect/Salisbury/Gawler area (*wirra meyunna*; the 'wood people'), the *Kawanda meyunna* ('north men') north of Gepps Cross, and to the south along the Sturt River/*Warriparri*. Connections between these clan groupings are also evident.[3]

Inter-clan communication and contact was common, as was inter-country communication and contact. Such connections enabled Dreaming journeys, trade and exchange, harvest and feasting, celebration and ritual semi-sedentary movement, according to strict protocols.

It is important to note that the Kaurna people, like other Aboriginal people, did not 'own' nor perceive 'ownership' of land in the Western sense; rather, they perceived that they were *part* of the land with a predetermined role to curate it. Thus, Aboriginal communities viewed themselves as custodians with a duty to care for and nurture the environment, including water, on behalf of and pending the return of their Dreaming Ancestors.

Anthropologist William E.H. Stanner[7] (1905–1981) stated:

A central meaning of The Dreaming is that of a sacred heroic time long ago when man and nature came to be as they are, but neither 'time' not 'history' as we understand them is involved in this meaning.

The interconnectedness relied upon stories, narratives and mythology. Unlike European inventory and documentation approaches, the Kaurna people obtained knowledge and information through progressive cultural immersion in the place with associated narratives explaining 'histories' and 'libraries' between the Kaurna people and their land, their Ancestors, their spiritual Ancestors, their practices and activities, their families, and their knowledge and stories. Instead of learning directly through schools and universities, learning was gained through experience of the land, their evolutionary journey through it, and life.

People talk about country in the same way they talk about a person; they speak to country, sing to country, visit country, worry about country, feel sorry for country and long for country. Country knows, hears, smells, takes notice, and is sorry or happy. Because of this richness, country is home. Peace and nourishment for the body, mind, spirit and eases the heart (Deborah Bird Rose).[3]

As Kaurna Elder Lewis O'Brien stated at the Adelaide Botanic Gardens master planning workshop in 2005, the Kaurna's spiritual attachment to the land was and continues to be explained through stories of the Dreaming, which 'aren't just fairytales; they speak about our country, our law and our knowledge'.

Central to this engagement with water in the landscape were six key traits: myth, cosmology, seasonality, partnership, death and name. But before moving into this review it is valuable to consider what *Mikawomma* landscape was like in 1836, as it relates to water.

The 1836 landscape: water and the Mikawomma (Adelaide Plains) landscape

To the Kaurna people, the *Mikawomma* biogeographical landscape was a rich domain of many animal and plant resources, supplied by a system of rivulets hosting cascades of irregular, semi-arid rains from the escarpment, feeding the tapestry of waterholes, swamps, marshes and streams. Many of the waterways were impeded in releasing their flows to Gulf St Vincent/*Wongayerlo* by the solid lines of sand dunes along the coastline. Resultant ponding of the waterways created the 'Reedbeds', near Fulham, and numerous shallow lagoons such as at Glenelg, West Lakes and in the south parklands, where rich silt from the ranges settled. Streams experienced dramatic waterflow fluctuations creating often deeply incised rivulet embankments and winding water passages, like along the River Torrens/*Karrawirra Parri* and the Sturt River/*Warriparri*.

Through different eyes, to the colonial surveyor and settler of 1836, here was an uncharted landscape that held the promise of rich, well-watered soils; a landscape that could be cleared to accommodate new crops, pastures and horticultural ventures. In achieving these colonial dreams, the landscape was transformed, felled, drained, harnessed and re-planted in the image of pastures and fields of Europe. In this act of transformation, waterbodies and routes were drained, channeled, filled and supplanted with alternate species and spatial configurations.

The floods of the River Torrens/*Karrawirra Parri*, for many years the main potable water supply for the city, resulted in various attempts to construct weirs and bridges. With success in the latter, the next objective was to create an artificial lake to both store the water and create a gardenesque vision of a 'city' on a lakeside. Harnessing this river, as well as the suite of creeks and streams that swept westwards from the hills, dramatically changed water flows, cyclical water regimes, silt deposition, flushing of swamp morasses, and water supplies to the aquatic vegetation and fauna that would otherwise be dependent upon these cyclical and seasonal flows, and their impacts on the plains.

George Hamilton, who arrived in 1839, described the scene:[8]

The land in the vicinity of this river was timbered with noble trees, and its banks sloped down to the water in gentle undulations thickly clothed in grass. The river itself meandered through a tangle of tea-tree [Melaleuca ssp], rushes, reeds and many flowering weeds here and there almost hidden by vegetation, but at intervals opening out into pretty ponds or tolerably large waterholes …

Helen Mantegani also shared a similar picture of the river in 1837:[8]

> The River Torrens as I first saw it in the winter of 1837 was very pretty and picturesque, high and steep banks on either side, closely covered with beautiful shrubs of all sorts; splendid gum trees also were growing on the banks, and in the stream, overhanging the water which was narrow and deep, small fish were plentiful, and that strange creature the platypus [Ornithorhynchus anatinus] was occasionally seen on its banks. But some stupid people cut away the shrubs and trees that still held the banks together. Consequently the soft alluvial soil fell away, and the river became broad and shallow and very ugly. After this the winter floods carried away the banks that remained, making it a most unsightly spot for many years, and entailing an enormous expense to restore it to anything like beauty, through it will never be as picturesque as nature made it.

Upstream, the river was more defined and enveloped in vegetation, as reported in the *South Australian Magazine* (1841–1843):[8]

> On either side of the river, but more especially on the northern one, the bank is in many cases exceedingly precipitous, and at least from 40 to 50 ft [12–15 m] above the bed of the stream although for many miles flowing through a flat and extremely depressed country. For the most part the bed of the river is of a sandy or gravelly formation, whilst the strata of the banks, through which it directs its course, are usually composed of a clayey or marlish substance, in many places ornamented with the Native Broom [Exocarpus sparteus], and myrtle [Myporum montanum], or honeysuckle [Banksia marginata], and in others overhung with the rustling Sheoak [Allocasuarina verticillata] and stately forest gum tree [Eucalyptus camaldulensis], or decorated with the 1000 plants and shrubs to which science has not yet offered a name.

To the east were the First to Sixth creeks, which brought water down from the Mount Lofty Ranges gullies and escarpment. To Nathaniel Hailes,[8] the locality of Norwood in 1839 possessed 'the sparkling clearness of the running water', with First Creek hosting 'big gums' felled for wharf timber. Waterfall Gully Creek, enveloped 'in natural shrubbery, the shade of which was most inviting', was inhabited by 'innumerable little niggers of crayfish as they darted about among the white and brown pebbles which formed the bottom of the stream' [sic].

Glen Osmond Creek originally flowed out into the Black Forests (hence *Kertaweena* – scrub with reeds) of northern Unley and Parkside before it was channeled. So too was the Brownhill Creek that fanned out its water-logged sediments between Ashford to Plympton before settling in swamps under the now land-filled airport area. The rich birdlife habitat of saline samphire flats and mangrove woodlands of Gillman-Barkers Creek Inlet/St Kilda were also drained, channeled and reconfigured for saltwater evaporation ponds. For both

the River Torrens/*Karrawirra Parri* and Sturt River/*Warriparri*, it wasn't until the 1970s – when extensive stormwater management schemes were implemented – that flooding largely ceased, thereby denying irregular flooding of the river red gums (*Eucalyptus camaldulensis*) and grey box (*Eucalyptus microcarpa*), both of which lined the watercourses.

With the harnessing of these watercourses came the transformation of the vegetation on their banks. Examples are simple turfing with exotic grasses, like in Heywood Park and Hazelwood Park, and the migration of pioneering weed species, including desert ash (*Fraxinus angustifolia*), caster oil plant (*Ricinus communis*), common prickly pear (*Opuntia stricta*), fennel (*Ferula communis*), bamboo (*Arundo donax*), and weeping willow (*Salix babylonica*).

The bountiful mangrove beds around Port Adelaide were extensively felled for timber for construction. Extensive stands of kangaroo honey myrtle (*Melaleuca halmaturorum*) lined the margins of Patawalonga Creek and Port River and the low shallows of the Le Fevre Peninsula; 'one of the most striking features of this creek is the fact that almost all the trees which line its banks consist of "paperbark trees"'.[8] Remnants of these honey myrtle-dominated swamps in the Camden–Grange locality were only land-filled in the 1950s–1960s, with the West Lakes estuary honey myrtles and samphire flats disappearing with redevelopment in the early 1970s. The Reedbeds of Fulham were extensively drained.

Helen Mantegani described the Reedbeds landscape in 1837,[8] recalling that:

> A little way inland were several large lagoons, formed probably from the overflow of the Sturt River, which was not far off. They were covered with tall reeds and splendid flooded gums grew on their margins, which were very swampy for a considerable distance. One of them was very lovely, just like a miniature lake, bordered by trees and reeds, and in its centre the bright water was the resort of wildfowl, making a beautiful picture. It was subsequently known as Hack's Lagoon, as Mr J.B. Hack had his camp there. I do not know if the lagoons are still in existence or dried up from the trees and reeds being cut down.

E.M. Waddy[8] verified a similar scene in his reminiscences:

> … the beautiful lagoon beside which Mr Hack has settled and built his cottage – known afterwards as Hacks Lagoon – was one of the largest and most beautiful. Giant gum trees cast their shadows upon its clear water; the graceful ibis, wild ducks and geese and several other waterfowl make an easy living from the numerous frogs and other food they found there in abundance … This extensive lagoon was 3 feet deep and was situated where the suburb of Helmsdale with its luscious gardens and flourishing trees is now situated.

To the north was mallee scrub vegetation, often lying on the poorly structured, sandy mallee soils. At Pine Forest, C.F. Folland settled in 1842, recording:[8]

… a densely-wooded spot, pleasantly situated on the spur of a hill commanding a beautiful view [westwards], and abounding in native birds, snakes and lizards … The air was full during the day of the cries of cockatoos, parrots and other birds, while at night wild dogs, curlews and mopokes, flies and mosquitoes, were strongly in evidence. … the land composed of a deep loam was very fertile …

Evident in these early colonial descriptions is the quality of water in the streams as well as the abundance of fauna and birdlife enabling a ready food resource. This was bountiful landscape for the Kaurna people to reside upon. They largely occupied the western portion during the summer period and shifted eastwards to higher ground as the rains came.

Mythological landscapes

The essential nature of a story is to convey a narrative. But within 'myths', a version of a story, are lores/laws, moral codes and rules, warnings and principles of human ethics. Aboriginal myths told stories of the passage of ancestral beings across the landscape, and their activities and events during this passage explaining biogeographical places, features, resources and spiritual meanings. This connectedness was multilayered, linking ancestral beings to 'Dreaming' places and tracks, to birthplaces and environs, to family relationships, and to cultural rituals and teachings.

The mythological journey of Tjilbruke (Tjirbruke) is the most well-known Dreaming story of *Mikawomma* (Figure 6.1). The story tells of the journey of Ancestor Tjilbruke along the Adelaide coastline to Cape Jervis/*Parewarangk* and backwards through the Currency Creek and Mount Barker regions to Brukunga. It conveys lessons about respect for life and human survival, explaining the creation of a series of geographical features and freshwater springs along the journey route.

Water is a key element in this story. The placement of potable water, the sacredness of quality drinking waterholes, the existence of a 'pathway' linking the waterholes, and the moiety associations of such waterholes were all explained.

The Tjilbruke Dreaming was one of several Dreaming stories enabling a spiritual connection to land. Other stories include the *Warriparinga*, the Emu Dreaming, the 'Seven Sisters'/ The Pleaides/*Mankamankarrana* Dreaming, the Kondoli Whale story (associated with fire mythology), *Tandanya* the 'Red Kangaroo' Dreaming, 'Tidley the Frog' Dreaming, the *Pootpobberrie* Dreaming story associated with the Waterfall Gully area, and the general teaching of totems.

Anthropologist Norman Tindale transcribed a version of the Tjilbruke story in detail from interviews with Milerum (Clarence Long), a Ngarrindjeri man, enabling it to be the only formally accessible Dreaming story associated with the Kaurna people and the Adelaide Plains today. There are various versions of the name of this ancestor including Tjilbruke and Tjirbruki, in period and contemporary literature. Tindale uses Tjilbruke, whereas Tjirbruke has been used on council internet page discussions about the 'Kaurna' and 'Indigenous histories'.

Mythological journey – Tjilbruke

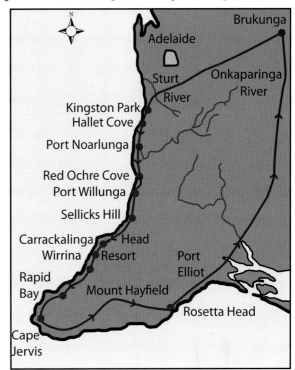

Figure 6.1 The mythological journey of Tjilbruke and the creation of coastal springs. The purple line shows the route taken by Tjilbruke as told in the Dreaming story.
Source: Seaford Rise and District Children's Centres website.

The following abridged version is readily accessible via the internet:

Tjirbruki [Tjilbruke] Dreaming

The Dreaming is a complex and multi layered story that tells of creation, the law and human relationships.

Tjirbruki was an ancestral being of the Kaurna people of the Adelaide Plains, whose lands extended from Parewarangk (Cape Jervis) in the south to Crystal Brook in the north. Tjirbruki's much loved nangari (nephew) Kulultuwi, his sister's son, killed a kari (emu) which was rightfully Tjirbruki's but he forgave him for this mistake. However, Kulultuwi was subsequently killed by his two part brothers [at Warriparinga], Jurawi and Tetjawi supposedly for breaking the law.

Tjirbruki, being a man of the law, had to decide if Kulultuwi had been lawfully killed. He determined Kulultuwi had been murdered. Tjirbruki avenged the crime by spearing and burning the two nephews, killing them. This happened in the vicinity of what is now called Warriparinga.

Tjirbruki then carried Kulultuwi's partly smoked dried body to Tulukudank (a fresh water spring at Kingston Park) to complete the smoking and then to Patparno (Rapid Bay) for burial in a perki (cave). Along his journey he stopped to rest and overwhelmed by sadness, he

BOX
27

The cultural landscape of Port Adelaide/Yerta bulti

In 2006 the City of Port Adelaide Enfield contracted GHD Pty Ltd, Hemisphere Design and Vivienne Wood, in consultation and collaboration with local Kaurna descendents 'to paint their own historical picture of the Port Adelaide region based on their memories, cultural stories/histories and their spiritual connections to family and "country"'.

One outcome of the Kaurna Cultural Heritage Survey (2007) was a diagram that depicted the 'cultural landscape'. Within this diagram are several places that possess direct associations to water by way of location, waterholes, spirit movement directions, fish and dolphins, summer and winter migration patterns, and seasonal food sources.

This painting is an evocative insight into the spatial and functional arrangement of a landscape through different eyes, and demonstrates the validity of alternative mapping techniques. The painting also shows that Kaurna biogeographical knowledge holds a valid place in our appreciation and reading of this landscape.

David Jones

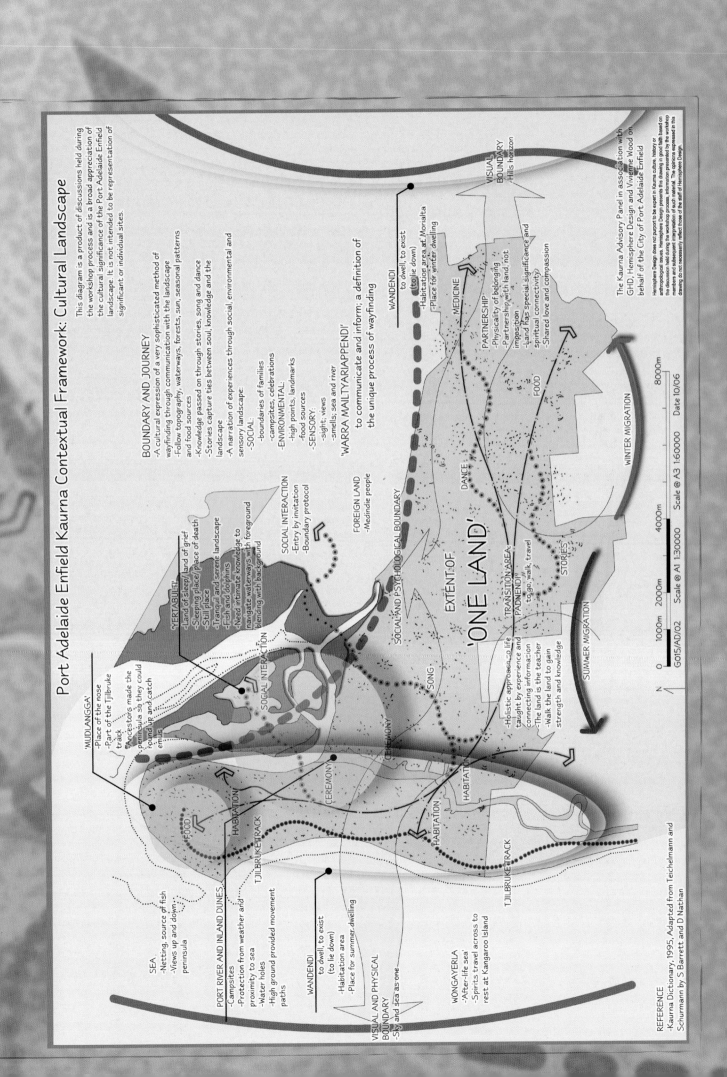

Port Adelaide Enfield Kaurna Contextual Framework: Cultural Landscape

This diagram is a product of discussions held during the workshop process and is a broad appreciation of the cultural significance of the Port Adelaide Enfield landscape. It is not intended to be representation of significant or individual sites.

BOUNDARY AND JOURNEY
- A cultural expression of a very sophisticated method of wayfinding through communication with the landscape
- Follow topography, waterways, forests, sun, seasonal patterns and food sources
- Knowledge passed on through stories, song and dance
- Stories capture ties between soul, knowledge and the landscape
- A narration of experiences through social, environmental and sensory landscape:
  - SOCIAL:
    - boundaries of families
    - campsites, celebrations
  - ENVIRONMENTAL:
    - high points, landmarks
    - food sources
  - SENSORY:
    - sight; views
    - smells; sea and river

'WARRA MAILTYARIAPPENDI' to communicate and inform; a definition of the unique process of wayfinding

'MUDLANGGA'
- Place of the nose
- Part of the Tjilbruke track
- "Ancestors made the peninsula so they could round up and catch emus

SEA
- Netting, source of fish
- Views up and down peninsula

PORT RIVER AND INLAND DUNES
- Campsites
- Protection from weather and proximity to sea
- Water holes
- High ground provided movement paths

WANDENDI
to dwell, to exist (to lie down)
- Habitation area
- Place for summer dwelling

VISUAL AND PHYSICAL BOUNDARY
- Sky and sea as one

WONGAYERLA
- 'After-life sea'
- Spirits travel across to rest at Kangaroo Island

'YERTABULTI'
- Land of sleep/ land of grief
- Sleeping place/ place of death
- Still place
- Tranquil and serene landscape
- Fish and dolphins
- Need intimate knowledge to navigate waterways with foreground blending with background

SOCIAL INTERACTION
- Entry by invitation
- Boundary protocol

FOREIGN LAND
- Medindie people

SOCIAL AND PSYCHOLOGICAL BOUNDARY

EXTENT OF 'ONE LAND'

'TRANSITION AREA' 'PADNENDI'
to go, walk, travel
- Holistic approach to life taught by experience and connecting information
- The land is the teacher
- Walk the land to gain strength and knowledge

WANDENDI
to dwell, to exist (to lie down)
- Habitation area at Morialta
- Place for winter dwelling

VISUAL BOUNDARY
- Hills horizon

PARTNERSHIP
- Physicality of belonging
- Partnership with land; not imposition
- Land has special significance and spiritual connectivity
- Shared love and compassion

MEDICINE
FOOD
DANCE
STORIES
SONG
CEREMONY
HABITATION
FOOD
TJILBRUKETRACK

WINTER MIGRATION
SUMMER MIGRATION

N

| 0 | 1000m | 2000m | 4000m | 8000m |

Scale @ A1 1:30000   Scale @ A3 1:60000

G015/AD/02   Date 10/06

REFERENCE
- Kaurna Dictionary, 1995, Adapted from Teichelmann and Schurmann by S Barrett and D Nathan

The Kaurna Advisory Panel in association with GHD, Hemisphere Design and Vivienne Wood on behalf of the City of Port Adelaide Enfield

Hemisphere Design does not purport to be expert in Kaurna culture, history or anthropological issues. Hemisphere Design presents this drawing in good faith based on the discussion held during the workshop process, information presented by the workshop members and subsequent interpretation of such material. The opinions expressed in this drawing do not necessarily reflect those of the staff of Hemisphere Design.

wept and his luki (tears) formed the freshwater springs along the coast at Ka'reildun (Hallett Cove), Tainba'rang (Port Noarlunga), Potartang (Red Ochre Cove), Ruwarung (Port Willunga), Witawali (Sellicks Beach), and Kongaratinga (near Wirrina Cove).

Saddened by these events Tjirbruki decided he no longer wished to live as a man. His spirit became a bird, the Tjirbruki (Glossy Ibis), and his body became a martowalan (memorial) in the form of the baruke (iron pyrites) outcrop at Barrukungga, the place of hidden fire (Brukunga – north of Nairne in the Adelaide Hills). Tjirbruki was a master at fire-making.

It is important to recognise that Tjirbruki was almost certainly not a glossy ibis as consistently used in these internet extracts. Milerum said, in another account, that Tjirbruke was like 'blue crane', and Clarke[9] has concluded that, consistent with cosmology, at the end of the Creation period Tjirbruke became the brolga, a species that became locally extinct after the arrival of Europeans to *Wikamomma*.

The Warriparinga Project in Marion is one of several that have given the Kaurna people an opportunity to protect and maintain significant sites associated with the Tjilbruke Dreaming as important places to celebrate the Dreaming.[10, 11, 12, 13]

Tarndanya refers to the 'Red Kangaroo' Dreaming and is associated with central Adelaide and the Festival Centre location.

Ivaritji (Amelia Savage) (1840s–1929) has provided a valuable insight into the significance of waterholes now converted into lakes in Adelaide Botanic Gardens.[14] In the *Advertiser* on 16 April 1864, William Cawthorne recorded the existence of a she-oak (*Allocasuarina cunninghamiana*) near the Frome Road bridge that may have held Dreaming significance:[15]

It is devoutly believed that a certain man was transformed into a sheoak tree, the one that stood a little way above the old Frome Bridge. In fact, every island, cape and point are transformations of one kind or another.

Cosmological landscapes

Aboriginal myths are multifaceted and multiplace in their expression. While they are bound together through reference to Ancestral beings, their journeys and activities, they explain the origins of cosmic beings and the (historical) role they have played out in the heavens and on the physical landscape. Thus, their cosmological or astronomological traditions (in other words, the 'night landscape') are equally as important as terrestrial and daytime traditions.

Colonial anthropologist and Moravian schoolteacher Christian Teichelmann (1807–1888) eloquently described this 'sky world' in 1841:[16]

… consider the firmament (heavens) with its bodies as a land similar to what they are living upon … It is their opinion that all the celestial bodies were formerly living upon earth, partly as animals, partly as men, and that they left this lower region to exchange for the higher one. Therefore all the names which apply to the beings on earth they apply to the celestial bodies, and believe themselves to be obnoxious to their influence, and ascribe to them malformation of the body, and other accidents.

Moravian schoolteachers Teichelmann and Clamor Schürmann (1815–1893) in 1840 recorded the term 'sky' as *Ngaiera* from their interviews with Kaurna people; but it was *Wodliparri*, or the Milky Way, that holds relevance here. *Wodliparri*, literally meaning 'hut-river' – perhaps being the home to *Yura* (a 'magnificent animal' or 'monster') – was a feature that dominated the Adelaide cosmic landscape, and was believed to be a river with reeds growing along its banks. Dotted along *Wodliparri* were dark spots representing waterholes or lagoons in which *Yura* lived; when *Yura* was visually predominant, in winter, rains were abundant.[17]

Seasons of the Adelaide Plains

As new residents to the Adelaide Plains, European settlers brought with them the notion of four seasons, which they were familiar with in the Northern Hemisphere, to explain general climatic and environmental patterns in our calendar year. They thought of summer, autumn (fall), winter, and spring, yet there is increasing debate whether these equal temporal periods and nomenclature are relevant to the Australian continent as well as the Adelaide Plains. While recent community and scientific debate has shifted towards grappling with climate change and its implications, the basic colonial premise is that the *Wikawomma* climate is shifting in conventional logic but is still contained within four three-month segments. This logic is enshrined in media presentations, and many of the calendars, diaries, recording systems, and information measurement and storage systems that associate time with four seasons.

These new residents are still trying to impose this four season system upon Australia, as well as their own lives. In truth, this *terra nullius* is a complex biosystem with its own temporal calendars, which are, and were, intimately 'read' and applied by its Aboriginal occupants. Therefore, these residents live within a false construction of time that hails from their European forebears, and have not embraced or engaged with the landscape and its environment.

Water is an essential feature of Aboriginal understandings of environmental calendars. As new residents, of only some 150 years residency on *Wikawomma*, we experience change in Adelaide's weather and the onslaught of evidence associated with climate change. We are measuring and accounting for time in small increments and not appraising the larger and longer perspective and understanding. This realm of information may hold some answers and demonstrate that change has been occurring not suddenly but incrementally over thousands of years. Therefore, what we are witnessing now is an acceleration of this change with an increasing plethora of scientific evidence that demonstrates immediate change patterns.

BOX
28

Princess Amelia and the Main Lake in the Adelaide Botanic Gardens

Ivaritji, or Amelia Savage, was born in the late 1840s near Port Adelaide and died in 1929. She is believed to be the last person of full Kaurna ancestry, contrary to previous colonial perspectives that 'the Adelaide tribe is extinct'.

Ivaritji was born to King Rodney (Ityamaitpinna) of the Kaurna and Tankaira, and Charlotte, from the Clare district. Ivaritji means a 'gentle, misty rain' in the Kaurna language. Ivaritji's father was one of the leading Elders of the Adelaide tribe of the Kaurna.

Ivaritji is important to the Adelaide Botanic Gardens because of her knowledge of places and the activities of the Kaurna. Anthropologist Daisy Bates interviewed Ivaritji in 1919 at Point Pearce. While the interview provided a list of Kaurna relationship terms, and elevated Ivaritji's status as the last person of full Kaurna ancestry, it also revealed some biogeographical insights.

Ivaritji pointed to the Main Lake of the Botanical Gardens as her father's principal waterhole, called *Kainka wira*. She also provided *Dharnda anya* (Tandanya) as a place name in the Adelaide area, along with a number of other place names. Botanist John McConnell Black, in 1919–1920 also interviewed Ivaritji and noted that *Kainka wira* meant literally 'eucalypt forest', and he recorded the term as relating to the North Adelaide area.

A reporter from the *Advertiser* interviewed Ivaritji at Moonta in 1927, and recorded her memories about the Adelaide Plains landscape:

She was born on the Adelaide Plains 72 years ago, and the tribe to which she belonged numbered thousands. Their headquarters were where the city of Adelaide now stands, with their central camp in or near Victoria Square … The city of Adelaide was being surveyed when she was a child, but it was very rough. The roads were dirt tracks. Gum trees and mallee trees grew in abundance where the city of Adelaide now stands. There were only a few houses dotted around the place (Advertiser, 8 December 1927, p. 13).

Ivaritji's knowledge has been accepted by contemporary anthropological researchers as providing the most authentic evidence of place names and meanings on the Adelaide Plains. Accordingly, her family association with the Main Lake in the Botanic Gardens points to an important Kaurna association.

Image of Princess Amelia reproduced with the permission of the South Australian Museum (Hale Collection AA124), and with the approval of the Kaurna Heritage Committee.

David Jones

Time and seasons are different conceptions for Aboriginal people. The former does not measure or structure life. Rather, it is subjective, multidimensional and determined by the nuances of ancestral beings, or 'nature'. There is a growing scientific and anthropological recognition that, because there are numerous biogeographical regions in Australia with different cyclic environmental patterns, we need to better understand Aboriginal conceptions of time and the landscape, and particularly Indigenous 'seasons' as they relate to a particular region or landscape.

For example, in managing the Kakadu and Uluru Kata Tjuta national parks in the Northern Territory, the Commonwealth Department of Environment, Water, Heritage and the Arts has adopted Indigenous season names, patterns and temporal periods as ways of structuring land management actions and Indigenous 'owner' and employee custodial activities. Interestingly water, rainfall and drought occurrences and the availability of water are chief determinants of seasonal names and the length of time of each season.

In terms of *Mikawomma*, Heyes has provided a major illustrated review that surveys the ethno-ecological patterns of *Mikawomma*, and proposed an alternate calendar.[18] This is discussed in detail in the accompanying case study in this chapter.

Partnerships with the landscape

A partnership implies a mutual arrangement to undertake a task or activity. To the Kaurna people, an integral part of their daily and seasonal activities was to undertake hunting, gathering and food harvesting activities within certain protocols, so as not to over-reap these resources and enable their continued availability, but also to care for the ecosystems these resources resided within. Thus firing the grasslands at particular times of the year, only hunting certain animals during certain seasons, harvesting tuberous root plants, or shifting encampments so as not to over-harvest resources within short walking distances, were very careful and premeditated land management practices, which might be naively interpreted today by new residents as simple routines in a semi-sedentary culture with no scientific or environmental strategies or rules as to land management.

In 1844, early settler William Cawthorne[19] observed the Indigenous diet, based upon his observations of Indigenous people encamped on the River Torrens/*Karrawirra Parri*, to consist of '… kangaroos, emus, opossums, birds, ants, grubs, frogs, lizards, guanos, gum roots … berries … white ants … wombats … wild ducks …'

Throughout the landscape were scattered physical attributes and places perceived to have been created by Ancestral beings to sustain the Kaurna people. Such features included waterholes, river systems, estuaries, sand dunes, mangrove forests, wetlands and swamplands, and the coastline. To destroy such places was destroy a relationship with and a work of an Ancestral being.

A central thread in this landscape was therefore water, of which the River Torrens/*Karrawirra Parri* was a key place.

It is likely that the river was directly associated with the 'Red Kangaroo Dreaming',[20] given its central role as a place for food harvesting, trade and exchange, encampments and proximity to 'neutral ground' at Victoria Square/ *Tarndanyangga*, and the *Tarnda Kanya* (the 'Red Kangaroo rock', reputedly located on the present site of the Adelaide Festival Theatre).

An early resident of Adelaide, Thomas Day, recollected Kaurna residency along the River Torrens/*Karrawirra Parri*:[15]

> *Women and children spent most of their time at the Torrens river – children bathing and practicing with spear and small waddy – Women crab[b]ing and going in the river with a net bag and picking up cockles. I have seen them go down – And I thought they would never rise again. They got many cockels And rose again on the other side after being under water A long time. The river torrens was A chain of water holes very deep When not in flood. It was full of timber Very dangerous to go amongst. Their time was also employed making mats, nets and rope clothes lines … They would then sell or exchange for food from the settlers … [sic].*

This account suggests that freshwater cockles (*Unio dubia*), *Ngampa* (edible roots of the yam daisy, *Microseris lanceolata*), and *kar'li* (crayfish or yabbies) were commonplace along the River Torrens/*Karrawirra Parri*. Further, the lagoons inland from the coast were a rich resource of freshwater, birds and fauna for the Kaurna people, until settlement and the sinking of wells filled and drained many of these swamps, and errant cattle and pigs grazed freely on the vegetation, or the vegetation was felled for construction, firewood and fencing purposes.

Culturally the Kaurna people continued to camp along the riverbanks, partake of the waterholes, and harvest the freshwater foods, resulting in considerable friction as to purposes between colonial authorities and settlers. An editorial in the *Register* in 1840[21] stated:

> *Something … ought to be done to keep the Natives out of the river. From the Botanic Gardens, downwards, it is full of them; not much, we should think, to the improvement of the water, which all the inhabitants who have not wells are obliged to use.*

Unmoved, William Cawthorne reported in 1843[22] that 'a policeman came with orders to burn all their wurlies or huts, which was done … the reason for them burning them out was because they swam and made the water dirty above the [water] hole.'

Movement shifts by Kaurna people from coastal to hinterland encampments was commonplace prior to European arrival. Early colonial records suggest that some of the Kaurna clans

Table 6.1 Select Kaurna names and their meanings, in metropolitan Adelaide, with associations with water. Sources: Amery (1997, 2002); Amery and Williams (2002); GHD *et al* (2007); Hemming (1998); Cooper (1966); and Howchin (1934) in Kraehenbuehl (1996), Four Nations NRM, 2007.

Table 6.1 Kaurna place names

Contemporary place name	Kaurna place name	Possible Kaurna meaning
Adelaide Plains	Mikawomma	'the plain'; at Kilkenny region; plain between Adelaide and Port Adelaide
Aldinga	Ngaltingga	Ngga = place; Ngalt is unclear in meaning.
Aldinga Plains	Mullawirra	
Barkers Creek Inlet	Medinda	'place of the Medinda people'
Black Forest	Kertaweeta	'scrub with reeds'; for the Black Forest – Unley – Parkside area
Brownhill Creek	Willa willa	
Cape Jervis	Parewarangk	
Carrickalinga Head	Karrikalinnga	'a camp near a swamp'
Cowandilla	Kauwandila	Kauwe = 'water'; full meaning is unclear; 'a place for drinking'; 'drinking water place'
Fulham, the Reedbeds	Witongga	Wito = 'reed'; reed place
Glenelg	Pattawilya	'eucalyptus tree scrub place'
Gulf St Vincent	Wongayerla	'after-life sea'
Hallett Cove	Ka'reildun	
Hahndorf	Bukartilla	'swimming place on the Onkaparinga River'
Hindmarsh	Karraundongga or Karra(k) undungga or Karawintanga	Redgum chest place
Le Fevre Peninsula	Medlangga or Mudlangga	'place of the nose'
Moana	Moana	'sea' from a Maori word
Morialta	Moriata	'ever-flowing permanent water'
Morphett Vale	Panangga or Parnangga	'place of autumn rains'
Munno Para	Munno-para	'creek with golden wattle trees'
Noarlunga	Ngurlongga	Nurlos = 'corner or curvature'; 'corner' on Horseshoe Bend on the Onkaparinga River
Onkaparinga River	Ngangkipari	Woman river place; women's river.
Para River district	Mulleaki	
Patawalonga	Pattawilyangga	Patta = a species of gum tree', possibly swamp gums; wilya = foliage, young branches
Pooraka	Pooraka	'dry waterhole'
Port Adelaide	Yertabulti	'the land of sleep or death'
Port Noarlunga	Tainba'rang	
Port Willunga	Ruwarung	
Rapid Bay	Patparno	
Red Ochre Cove	Potartang	
River Torrens	Karrawirra Parri	'river flowing through the red gum forest'
St Vincent Gulf	Wongayerlo	'western sea'
Sellicks Beach	Witawali	
Semaphore	Pu:lti	
Sturt River	Warriparri	Warri = wind; windy river
Taparoo	Taparoo	'air'
Tea Tree Gully	Kirraungdingga	
Toorak [Gardens]	Toorak	'swamp'; 'springs'
West Lakes	Witongga	
Willunga	Willangga	Willa = 'dust', +ngga = 'location'; real meaning is unclear
Yatala	Yertalla	?
Yatalavale	Yertalla	'water running by the side of a river'; inundation; cascade

frequented the coastal dunes and beaches in summer and shifted inland during winter to the foothills. As Tindale[23] has noted:

> ... *they wintered in the timbered foothill country, sheltering in the cover of gigantic and often hollow red gum trees along the riverbanks. Often as a basis for shelter they used a fallen tree, and hollowed out beneath it, a practice hazardous when floods and heavy rains destroyed their fires and flooded their sleeping places.*

Thus, certain places enabled and performed certain functions, but water was a key determinant. The coastline, the River Torrens/*Karrawirra Parri* outwash, and the Port Adelaide/*Yertabulti* mangrove swamplands provided ample saltwater and freshwater fish, shellfish and reeds for basket construction. The forests at Woodville provided shelter and timber, enabling a resting point between the wetlands and swamplands to the west before journeying eastwards prior to winter. The reliable waterholes along the River Torrens/*Karrawirra Parri* in the Adelaide/*Tandanya* and St Peter's/Tennyson Bridge localities provided summer freshwater sources, adjacent grassland hunting grounds, and wide expanses for encampments near to 'neutral ground' at Victoria Square/*Tarndanyangga*, which also served as a trading point because of its 'neutrality'. These concentrations of Kaurna activities and encampments, along the sides of watercourses and lagoons and at springs and swamps, is also corroborated in the 'concentrations of information' about the Kaurna people, as documented by Harris.[24] It also corresponds with anthropological research elsewhere which has observed that Aboriginal people tended to camp at the meeting points of often three eco-niches or ecosystems, thereby affording three different landscape types (for example: saline swamps, freshwater springs and open wooded grasslands), hosting often three separate foods resources.[25]

Death in the landscape

While birthplace was significant, so too was death place. While death place could often not be determined, it is clear that many death places and burial grounds on the Adelaide Plains are associated with water streamlines, maybe as a reflection of general life patterns and rituals. The River Torrens/*Karrawirra Parri* corridor features in many colonial records relating to individual burial sites and uncovering of human remains due to post-colonial earthworks or human/natural surface-changing activities. Other known sites include along the Sturt River/*Warriparri*, in the Gepps Cross/Dry Creek area, in the former Rosewater dunes area, in the Birkenhead Reserve area at Largs Bay and at Birkenhead, at Hindmarsh/*Karra(k)undungga*, and at St Peter's Billabong/Tennyson Bridge locality.[15, 23, 26]

Human remains were found during the construction of the new deer park in Adelaide Zoo in 1914, and human remains were also uncovered in the Botanic Gardens in 1856. Author Barbara Best recounts that 'On Sunday morning, Mr George Francis of the Botanical Gardens, was surprised by his children bringing into the house a skull and one of the arm bones of a human being. On making enquiries, he ascertained that they were found on the south bank, opposite the Old Garden, about three feet beneath the surface ... exposed ... by falling in ... [off] the bank for the river'.[15, 27, 28]

Table 6.2 Possible Kaurna place names

Park no. or place	Kaurna toponym proposed	Kaurna meaning or association	Contemporary place name	Origins of toponym
2	*Padipadinyilla*	Swimming place		*Padipadinya* = swimming; padendi = to swim + illa = locative
10	*Warnpangga*	Bulrush root place		*Warnpa* = bulrush root + ngga = locative
11	*Kainka Wirra*	*Eucalyptus* ssp forest	Botanic Park	*Kainka* –synonym, Karra = river red gum; wirra = forest
12	*Karra Wirra*	River red gum forest		*Karrawirraparri* = river red gum forest; River Torrens flowed through the Karra wirra red gum forest
22	*Wikaparndo*	Netball park		*Wika* = net (wallaby or fish net), parndo = possum skin ball (used as a football) + wirra = forest park.
River Torrens	*Karrawirra Parri*	River flowing through the river red gum forest		*Karrawirraparri* river red gum forest; River Torrens flowed through the *Karra wirra* River Red Gum forest

+= neologism (new term) constructed by Amery.
Table 6.2 Possible Kaurna place names (toponyms) for regions within the Adelaide parklands with references to water.
Source: Amery (1997); Amery and Williams (2002); and Jones (2007).

The story of Pootpobberrie

BOX 29

The story of Pootpobberrie, while purported as a Dreaming story as transcribed by an assistant to the colonial Protector of Aborigines, James Cronk, is believed by Kaurna Elder Lewis O'Brien to be a false narrative. Mr O'Brien concluded that it 'was deliberately told incorrectly' so that the real narrative was not imparted to Cronk. Notwithstanding this, waterholes play a strong role in the 'Story of Pootpobberrie'. The story also tells of the Kaurna arriving from the north-east to the Adelaide Plains, and warns of an evil spirit residing in waterholes – analogous to the colonial mythical 'bunyip'. The negative presence of the evil spirit at waterholes is a form of a precautionary story, demonstrating the need to take care at waterholes. The story also describes the careful harvesting of animals for food, an indication of prohibitions about killing and eating certain foods, and the potential role of Waterfall Gully as a spiritual place.

When Aboriginal people first arrived at the Adelaide Plains they came in small parties of three or four. They found plenty to eat and could have lived well. However they were annoyed by evil spirits who lived on the land and in the water. The chief of these spirits was Muldarpi ['of nonsense'], who had the power to change people into birds and animals. Paddling children were sometimes changed into ducks or fish. A baby left by its mother asleep in a wurly, might be a lizard when she came back and a boy climbing a tree for birds' eggs might be changed into a possum. A man snaring wombats was said to have been changed into a kangaroo and two women who quarrelled were changed, one into a parrot and the other into a hawk. As this could not continue, two of the men returned to their own country to ask the doctor, Warra Warra, to punish Muldarpi and stop his trouble making. Warra Warra Maldie collected a great many things such as yacca gum, hair, feathers and intestines. He then made a fire and burned them, causing a great smoke and horrible smell. This alarmed the spirits so much that they fled in terror, hiding in waterholes and creeks, where they have remained ever since, except on dark nights, when they came out just a little way. Warra Warra, however, could not restore the people who had been changed into birds and animals to their former shapes. He returned to his people and spoke so well of the country that they came in large numbers to occupy it.

There was now a problem about eating certain foods, for it was quite possible that any fish, lizards, possums or birds might be really a human being. It was against the law to eat human flesh. One man and his wife did eat part of a kangaroo they had killed and as a result their baby boy was born part human and part kangaroo. This was

Pootpobberrie, the first of the group of that name. He grew taller than any man, had long pointed paws for hands and fur covered his body. He was so strong that he could carry a large rock and leap across a gully with the rock in one bound.

As the years went by, Pootpobberrie, with his wife and children, lived in the Adelaide Hills. They lived on roots, fruits and wattlegum which they obtained by robbing from the land of their neighbours on the plain.

These thefts led to the undoing of the Pootpobberrie people as the people who had been robbed lay in wait for them and killed all they could. Thus, after many years, there were only three of the original Pootpobberrie people left, Meyu a man, Miminnie a woman and Moolarlar a child.

One day Meyu heard that men of the Wirra local group had stretched long nets, mindie munta, across the creek at Waterfall Gully, hoping to catch a large number of wallabies. Feeling sorry for the animals, he crept quietly along hoping to loosen the net before any of them were caught. The woman followed him at some distance and the child came after. Meyu was wounded by a spear. He threw a stone behind him to warn Miminnie of danger and she in turn threw a stone behind her to warn Moolarlar, but too late; many spears were thrown and the woman and child were killed.

A great number of wallabies was stopped by the net, some falling over into a large waterhole shaded by spreading gum trees where Muldarpi lived. He was so enraged by the intrusion of the animals and noise of rolling stones that he changed the only person visible, Meyu, into a Kookaburra. Meyu flew into a high tree carrying with him one end of the net which he made fast to a bough and broke out into a laughter at the strange things which had happened. He continued laughing so loudly that people at a distance wondered at the strange noise, some calling Gooeyanna, other Kerkoarta. The place where this happened was called Meyuru Marti and afterwards became known as Ngulta, a stage that Kaurna men went through to achieve full manhood.

Source: James Cronk, 'Pootpobberrie and Aboriginal legend', *Public Service Review*, February 1913.

David Jones

Naming and knowledge in the landscape

About 70% of 4,000,000 place names in use (in Australia) are Indigenous in origin or inspiration, and reflect Australia's record in giving official priority to Indigenous names. The use of Indigenous Australian place names acknowledges the worth of Aboriginal and Torres Strait Islander peoples, and fosters a sense of cultural identity. In some regions in Australia, place names are all that Indigenous communities have left of their language.[29] Every place name for the Kaurna people holds a meaning, a rationale, or a story. As new residents to the Adelaide Plains, early surveyors and settlers ascribed names to places to spatially locate themselves in this *terra nullius*, enabling intellectual security of location, ownership and direction. Names bred familiarity and enabled passage. Today, we do not question the history and meaning(s) behind contemporary place names because we have mental security in the 'history' of the place, trust in the familiarity of European transposed place names, faith in our past abstractions of folk etymology, knowing that the geographical meaning can be explained somewhere, somehow. Adelaideans call Adelaide 'Adelaide' but they do not appreciate that the naming occurred to honour Princess Adelaide before she rose to be Queen Adelaide, let alone who Queen Adelaide actually was. Geographical longitude and latitudinal references are expressed in Adelaide street directories, and knowledge that the Adelaide Hills are to the east and south and the sea to the west, enabling an authoritative guide map of the Adelaide Plains.

The Kaurna people have nomenclature for places. But, unlike conventional names, Kaurna names can be multifaceted. A place name disguises a 'library of knowledge' about a place, its natural history and resources, its 'history' and meaning in a Dreaming story, and its association with a Kaurna and/or Kaurna/colonial event. But it also may provide a geographical reference point.

Along with plants, animals and seasons, few colonial immigrants sought to understand and document Kaurna language, nomenclature and their meanings. Anthropologists and surveyors like James Cronk,[30] Christian Teichelmann and Clamor Schürmann often heard and abstracted place names into their journals, plans and writings, but the accuracy behind the nomenclature is now unclear.

Governor George Gawler (1795–1869, governor from 1838 to 1841) and Surveyor-General George Goyder (1826–1898, surveyor-general from 1861 to 1894) both encouraged the use and application of Aboriginal place names, as new place names and surveying information by colonists. Gawler instructed, in the *Government Gazette* (1839):[31]

> *In regard to the minor features of the country to which the natives may have given names, the governor would take the present opportunity of requesting the assistance of the colonists in discovering and carefully and precisely retaining these in all possible cases as most consistent with property and beauty of appellation.*

> *Furthermore, all information on this subject should be communicated in precise terms to the Surveyor-General who will cause memoranda to be made of it and native names, when clearly proven to be correct, to be inserted in the public maps.*

Recently, research has been undertaken to seek to document and reinstate Kaurna place names. Amery and Williams (2002) summarised much of this research undertaken for central Adelaide; GHD Pty Ltd *et al* (2007) for the Port Adelaide/Enfield region; and Amery (1997) reported this information as a series of recommendations to the Corporation of the City of Adelaide following an invitation from Kaurna woman Dot Davey and in collaboration with the Kaurna Aboriginal Community and Heritage Association (KACHA Inc).[1, 20, 32, 33] In proposing names, Amery applied four criterion: original names; nomenclature linked to a particular flora found or known to have existed on the site; an extant functional name; or a name that honours a Kaurna person. Water or the association of water was prevalent in several names.

Four names, *Piltawodli* (Park 1), *Karrawirra* (Park 12), *Tambawodli* (Park 24), and *Wirranendi* (Park 23), were officially endorsed by the corporation in March 2000, and *Karrawirra Parri* (River Torrens), being a geographical feature, gained approval from the Geographical Names Board of South Australia in November 2001. Some additional 19 names were approved for use by the corporation in May 2001, and in May 2002 *Tarndanyangga*/Victoria Square was dual named. The remaining city squares and named parks were dual named in March 2003. Amery has drawn a distinction between literal and connotative meanings in analysing place names.

Tables 6.1 and 6.2 provide a listing of names associated with water on the Adelaide Plains. The tabulation is inconclusive but provides a glimpse into the significance of Kaurna names to landscape and water.

Kaurna Elder, Georgina Yambo Williams, has invited the adoption of Kaurna toponyms for places on the Adelaide Plains to reinstate their Indigenous meaning:

> *I have grown up with oral traditions and understanding about places and their significance that do not necessarily appear in the historical record ... The placenames that have survived in an Anglicised form are part of the story, law and lore of the land formations and places in which they are situated. These placenames are the 'skeletal remains' of the historical surviving reality of Kaurna First nation peoples, once a peaceful and intact body of lore/law of the land ...*

> *By using our language reclamation in this way, we identify with the purposes served by the language in the reconstruction of identity from the roots remaining. Naming activity that is not rooted in the land and in the people of the land runs the risk of being a shallow and meaningless activity that misappropriates our language and culture.[32]*

Legacy of knowledge and spirit

It is often hard to express a legacy of one, or a group, which we know little about in terms of their full expertise, contribution, knowledge and activities. While we are still addressing reconciliation, it is increasingly obvious – through Indigenous workshops or guidance on land management activities, as well as the multilayer scope of evidence submitted as part of Native Title claims – that we are only scratching the surface of appreciating the Kaurna people and their regime of land management and land relationship protocols for the Adelaide Plains.

One clear trait is that freshwater was revered, respected, associated with spiritual and moiety connections; it was conserved and used carefully. The Kaurna people respected that it needed to be shared between people and animal alike. As present residents, given the context of climate change and our over-reliance upon River Murray water, we need to do the same, and better appreciate and respect this depleting resource.

References

1. University of Adelaide, 'Kaurna: Warra Pintandi', 2005 (online).
2. City of Onkaparinga, 'Aboriginal heritage and culture' (online).
3. Four Nations NRM Governance Group, 'Four Nations NRM governance group consultation and engagement protocols', Rural Solutions South Australia, Adelaide, 2007.
4. D.S. Jones, 'The ecological history of Adelaide 2: the Adelaide Plains environment and its people before 1836' in Daniels, C.B. and Tait, C.J. (eds) *Adelaide, nature of a city: the ecology of a dynamic city from 1836 to 2036*, BioCity: Centre for Urban Habitats, Adelaide, 2005.
5. P.A. Clarke, 'Early Aboriginal fishing technology in the Lower Murray, South Australia', *Records of the South Australian Museum*, 35 (2), 2002, pp. 147–167.
6. Australian Institute of Aboriginal and Torres Strait Islander Studies, 'Aboriginal Australia map', 2005 (online).
7. W.E.H. Stanner, *White man got no dreaming*, ANU Press, Canberra, 1979.
8. D.N. Kraehenbuehl, *Pre-European vegetation of Adelaide: a survey from the Gawler River to Hallett Cove*, Nature Conservation Society of South Australia, Adelaide, 1996.
9. P.A. Clarke, 'Adelaide as an Aboriginal landscape' in Chapman, V. and Read, P. (eds) *Terrible hard biscuits: a reader in Aboriginal history, journal of Aboriginal History*, Allen and Unwin, Sydney, 1996, pp. 69–93.
10. City of Onkaparinga, 'Tjilbruke Dreaming' (online).
11. City of Marion, 'Warriparinga', 2007 (online).
12. City of Holdfast Bay, 'Tjilbruke heritage and the Kaurna people' (online).
13. District Council of Yankalilla, 'Indigenous', 2005 (online).
14. T. Gara, 'The life of Invaritji ('Princess Amelia') of the Adelaide tribe', *Journal of the Anthropological Society of South Australia*, 28 (1), 1990, pp. 64–104.
15. S. Hemming and R. Harris, 'Tardanyungga Kaurna Yerta: a report on the Indigenous cultural significance of the Adelaide parklands', unpublished report, Adelaide City Council, Adelaide, 1998.
16. P.A. Clarke, 'The Aboriginal cosmic landscape of southern South Australia', *Records of the South Australian Museum*, 29 (2), 1997, pp. 125–145.
17. P.A. Clarke, 'Adelaide Aboriginal cosmology', *Journal Anthropological Society South Australia*, 28 (1), 1990, pp. 1–10.
18. S. Heyes, 'The Kaurna calendar: seasons of the Adelaide Plains', unpublished Honours thesis, University of Adelaide, Adelaide, 1999.
19. A.J. Butler, 'A survey of mangrove forests in South Australia', *South Australian Naturalist*, vol. 51, 1977, pp. 31–49.
20. R. Amery, 'Kaurna place names', unpublished report, Adelaide City Council, Adelaide, 1997.
21. Editorial, *Register*, 5 February 1840, pp. 5–6.
22. W. Cawthorne, diary entry, 27 January 1843.
23. GHD Pty Ltd, Hemisphere Design, Vivienne Wood and City of Port Adelaide Enfield, 'Kaurna cultural heritage survey', unpublished report, City of Port Adelaide Enfield Council, Port Adelaide, 2007.
24. R. Harris, 'Archaeology and post-contact Indigenous Adelaide', unpublished Honours thesis, Flinders University of South Australia, Adelaide, 1999.
25. D.S. Jones, 'Traces in the country of the white cockatoo (Chinna Junnak Cha Knæk Grugidj): a quest for landscape meaning in the Western District, Victoria, Australia', unpublished PhD thesis, University of Pennsylvania, Philadelphia, 1993.
26. N. Draper, J. Mollan, A. Maland, F. Pemberton and S. Hemming, 'Community land management plans: Adelaide parklands and city squares – Aboriginal heritage', unpublished report, Adelaide City Council, Adelaide, 2005.
27. Unknown author, *Advertiser*, 7 August 1914, p. 14.
28. B. Best, *George William Francis: first director of the Adelaide Botanic Garden*, Barbara Best, Adelaide, 1986, p. 66.
29. Council for Aboriginal Reconciliation, 'Sharing history: a sense of all Australians of a shared ownership of their history', Australian government, Canberra, 1994, p. 33.
30. B. Daniels, James Cronk: the first European to prepare a Kaurna-English dictionary' in Daniels, C.B. and Tait, C.J. (eds) *Adelaide, nature of a city*, p. 61.
31. R. Cockburn, *South Australia: what's in a name?*, Axion Publishing, Adelaide, 1990.
32. R. Amery and G.Y. Williams, 'Reclaiming through renaming: the reinstatement of Kaurna toponyms in Adelaide and the Adelaide Plains' in Hercus, L., Hodges, F. and Simpson, J. (eds) *The land is a map: placenames of Indigenous origin in Australia*, Pandanus Books, Canberra, 2002, pp. 255–276.
33. University of Adelaide, 'Kaurna place names within the City of Adelaide', 2005 (online).

CASE STUDY 1: Water: the lifeblood of a landscape, the heart of the seasons

Scott Heyes

The dry conditions southern Australia has experienced in recent years have drawn much attention to the appropriateness of the concept of four distinct seasons, and particularly how the essential element for life – water – has become a precious and valued commodity. In this era of increased awareness of climate change, the seasons have increasingly become defined and characterised by the presence or absence of water. Ironically, this burgeoning phenomenon has caused many otherwise non-environmental adherents to advocate that Australia's seasons be viewed and articulated from an Indigenous perspective; a way of viewing the land and its associated systems that has been disregarded, whether purposefully or unintentionally, since meteorological observations have been kept in southern Australia.

Indigenous notions of time and seasonality of the Adelaide Plains

The Kaurna people of the Adelaide Plains, the traditional custodians of this area, maintained a distinctively different perception of the seasons prior to European settlement in 1836. Unlike the conventional non-Indigenous calendar for southern Australia, the 'Kaurna calendar', reconstructed here by Heyes,[1] suggests that their notion of the seasons was determined by a combination of events and cultural beliefs including but not limited to Dreaming stories, myths and legends, cosmology, environmental systems, tidal fluctuations, constellations, and landscape modification practices (e.g. burning of native grasses). Their calendar had no start or end point, although perennial animal behavioural patterns, such as the arrival of grey teal ducks (*Anas gracilis*) after the diminishment of floodwaters in Outback Australia, suggest that time and the occurrence of major environmental events in distant regions was recognised.

The calendar presented here is not intended to speak of the seasons on behalf of the Kaurna people. At the time of generating the calendar, the Kaurna community was consolidating fragmented information on their people, history, language and culture from a variety of sources, with the assistance of anthropologists, geographers and historians. Attempts to gain Kaurna input on the nature of the seasons were sought by the researcher in 1999, although the community was unable to provide a knowledgeable person or spokesperson on the topic at the time. The calendar presented here has been previously reviewed but has not necessarily been endorsed by the Kaurna community.

The notion of four distinct seasons, a colonial construct imported by early settlers from Europe, has been used as a way to describe and distinguish between the physical, environmental and meteorological changes that take place in southern Australia on an annual basis. To non-Indigenous

Australians, the seasonal labels have become a measure of time; the yearly cycle commences and terminates at the start of the calendar year on 1 January, which falls within the middle of summer. Yet this calendar sits awkwardly upon a landscape that constantly reminds its inhabitants that natural systems are in a state of flux, and that the transition from one season to another is fluid and subtle rather than abrupt and immediate. From a temperature point of view for instance, the *actual* mid point of summer may be experienced in early December or in late March. Over the last few years, the summer season (based on non-Indigenous Australian definitions) has spanned from November to March rather than from December to February, the current accepted period for summer in southern Australia.

Precipitation, like the fluctuation of extreme temperatures throughout the course of a 'year', is a reminder that the earth is still spinning, that the annual cycle is on track. However, the association of large rainfalls with winter in southern Australia is tenuous. The notion of winter is an idea that has been superimposed upon the southern Australian context, transplanted by European migrants in the 18th, 19th and 20th centuries. On the ground, what does winter, as a seasonal construct, tell us about animal migration patterns, constellation patterns, wind patterns, and plant life for instance? It is inappropriate to assume that the same amount of information about the Australian landscape can be gleaned from the Gregorian-derived lexicons of summer, autumn, winter and spring as from the seasons identified in the Kaurna calendar.

Contextualising the Kaurna calendar

It is with lament that we consider the early settlers in the Adelaide Plains region did not take the time to understand the seasonal calendar already devised and in use for thousands of years by the Kaurna people. No precise calendar was recorded by the Kaurna people, nor the ethnographers and missionaries in the late 1800s and early 1900s. However, it has been possible to reconstruct, in visual form, what may have resembled the Kaurna seasonal calendar around the time of European settlement. This reconstruction has been drawn from a variety of ethnographic and linguistic sources.[2, 3, 4, 5, 6, 7, 8, 9] With respect to the visual representation of the calendar, historical documents on the natural systems of the Adelaide region, in addition to climatic data, were used to generate a picture of the landscape setting that the Kaurna people likely experienced around 1836.

With over 40,000 years of Indigenous occupation of the Adelaide Plains region, one presumes that the Kaurna calendar has metamorphosed over time to reflect the rise and fall of sea levels, shifting constellation patterns, as well as subtle and seismic changes to the cultural practices and migration patterns of the people and animals. The calendar presented here represents a 'moment' in time (1836) of Kaurna occupation of the Adelaide Plains.

The canvas of the Adelaide Plains presented a different picture in 1836 compared to the built-up regions and concrete-lined channels of today. To fully comprehend the calendar, one must cast the mind back to the period when

Kaurna calendar for the Adelaide Plains

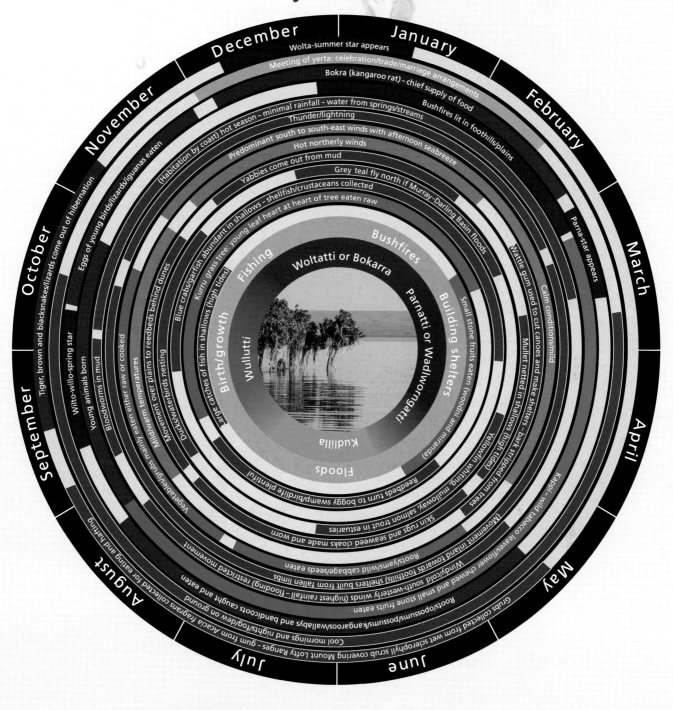

December
- Wolta–summer star appears
- Meeting of yerta: celebration/trade/marriage arrangements
- Bokra (kangaroo rat) – chief supply of food

January

November
- (Habitation by coast) hot season – minimal rainfall – water from springs/streams
- Thunder/lightning
- Predominant south to south-east winds with afternoon seabreeze
- Hot northerly winds
- Yabbies come out from mud
- Grey teal fly north if Murray-Darling Basin floods

February
- Bushfires lit in foothills/plains

- Eggs of young birds/lizards/iguanas eaten
- Snakes/lizards/iguanas come out of hibernation
- Blue crabs/garfish abundant in shallows – shellfish/crustaceans collected
- Kuru grass tree: young leaf heart at heart of tree eaten raw

October
- Tiger, brown and blacksnakes/lizards come out of hibernation
- Wilto-willo–spring star
- Young animals born
- Bloodworms in mud

March
- Parna-star appears
- Wattle gum used to cut canoes and make shelters – bark stripped from trees
- Calm conditions/mild
- Mullet netted in shallows (high tides)
- Small stone fruits eaten (woodn and miranda)

September
- Mild/warm temperatures
- Movement over plains to reedbeds behind dunes
- Large catches of fish in shallows (high tides)
- Ducks/waterbirds nesting

April
- Kaapi – wild tabacco leaves/flower eaten/chewed
- Yellowfin whiting, mulloway, salmon trout in estuaries
- (Movement inland towards foothills)

August
- Vegetables/grass mainly eaten either raw or cooked
- Grubs collected from wet sclerophyll scrub covering Mount Lofty Ranges – gum from Acacia collected for eating and hafting
- Roots/opossums/possums/kangaroos/wallabys and bandicoots caught and eaten

May
- Skin rugs and seaweed cloaks made and worn
- Reedbeds turn to boggy swamp/birdlife plentiful
- Roots/yams/wild cabbage/seeds eaten
- Windy/cold south-westerly winds (highest rainfall – flooding) restricted movement
- Cool mornings and nights/fog/dew on ground

July
- Acacia fragrans collected for eating and hafting

June
- Windy/cold and small stone fruits eaten
- Grubs collected from wet sclerophyll scrub covering Mount Lofty Ranges (flooding) shelters built from fallen limbs

Inner ring (seasons):
- Woltatti or Bokarra
- Bushfires
- Parnatti or Wadlworngatti
- Building shelters
- Fishing
- Birth/growth
- Wullutti
- Kudlilla
- Floods

The Kaurna conceptions of the seasons as reconstructed and interpreted by Dr Scott Heyes the site of Adelaide was a woodland populated by red kangaroos and bilbies. At this time the Adelaide Plains was characterised by expansive reedbeds and wetland systems that lay behind rolling sandunes from Semaphore to Glenelg. A series of parallel rivers, lined with large stands of river red gums (*Eucalyptus camaldulensis*), from the foothills to the reedbeds provided a permanent water supply. These river corridors and their tributaries, especially the Para, Torrens, Sturt and Onkaparinga rivers, were popular camping areas for the Kaurna. Much of their way of life centred on the water bodies of the Adelaide Plains, the water attracting animals hunted for subsistence. The calendar presented herein reflects the physical characteristics of the waterscapes of this period.

Kaurna nomenclature of 'seasons'

A survey of linguistic records revealed that the Kaurna possessed seven English-equivalent names for the seasons. Two English-equivalent names were used to describe summertime; two terms were used to describe both the autumn and winter periods; and one was used to encapsulate spring. A summary of each of the Kaurna seasons is provided below, and should be interpreted with reference to the graphic of the Kaurna calendar.

Woltatti[2] (hot season) or **Bokarra**[2] (hot north winds that blow in summer)

According to Teichelmann and Schürmann (1840), *Woltatti* was a time when '… friendly tribes visit each other and wander more about than in [*Kudlilla*] winter'.[9] They also reported that during the hotter months, possums and the gum of wattles were the chief supply of food.[10] It was a time of celebration, marriages and storytelling. The Kaurna lived behind the sand dunes near the coast during this season. Fishing was an active event and bushfires were lit to hunt out animals and to promote growth of the tall kangaroo grass (*Themeda triandra*).

Wadlworngatti (time of building huts against fallen trees) or **Parnatti** (autumn)

This is the shelter-building season. The Kaurna people move towards the foothills of the Mount Lofty Ranges and wooded plains away from the coast. Some species of fish are harvested, along with possum and the bark of trees. Seaweed and possum cloaks are made in anticipation of inclement weather. Small stone fruits and berries are eaten.

Kudlilla[2] (rainy season/winter) and **Kudlillurlo**[2] (in/during the winter)

A variety of roots, possums, kangaroos and wallabies contribute to the Kaurna diet during this season. More robust shelters are built to guard against the predominantly cool south-westerly winds. The reedbed area becomes flooded and, at times, impassable. The Kaurna are relatively confined to the gullies and gorges of the Mount Lofty Ranges and wooded areas on the plains.

Wullutti (variations: *Wiltutti* or *Wilutti*)[2] (spring)

During this period, the Kaurna lived chiefly on vegetables, grubs, insects, lizards and larger game. It was a season of birth and growth. Reptiles come out of hibernation, birdlife is abundant, and young animals are born. Eggs and snakes are easy to collect. The Kaurna gradually retreat from the plains and hills to the rich food source around the reedbeds as the temperature increases.[11]

An abridged version of the calendar has been provided to highlight the significance of water and water-related activities throughout the seasons, which is to remain consistent with the focus of other chapters in this publication. In so doing, however, the researcher cautions that the calendar should be viewed in a holistic sense, and should not be interpreted in terms of water necessarily being the dominant or overarching theme of the Kaurna notion of seasons or the construct of time. Water, like fire, the constellations, cosmology and natural systems are but one cultural feature, or environmental aspect, that constitutes the seasons. Further, it is imprudent to suggest that water in its various physical forms and manifestations (e.g. fog, vapour, steam, dew, ice) is regarded by the Kaurna in the same sense as non-Indigenous Australians. A complete interpretation of a water-related conception of the Kaurna seasons would require an exploration of how water binds the corporeal and spiritual worlds, as well as any Kaurna concepts relating to water and fertility, witchcraft, navigation, initiation ceremonies, and other traditional customs.

The acceptance and embracement of the Kaurna notion of the seasons by non-Indigenous Australians would allow for a better understanding of the natural systems of the Adelaide Plains, and could enrich current forecasting methods. The movement away from the conventional four seasons to a series of transitional seasons of irregular duration would reduce societal apprehension when uncharacteristic weather patterns are experienced, and may provide a useful device to understand climatic change. Australian society may become more attuned to the environment, its systems, and the sky if the Kaurna traditional knowledge that is embodied in the calendar, like other Indigenous calendars around Australia, is internalised and promulgated.

Acknowledgements

Thank you to Christine LaBond of Michigan State University for providing invaluable comments and feedback. This paper was generated with financial support from a University of Melbourne Early Career Researcher grant and a Canadian High Commission Faculty Research Program grant.

References

1. S. Heyes, 'The Kaurna calendar: seasons of the Adelaide Plains', unpublished Honours thesis, University of Adelaide, 1999.

2. R. Amery (ed.), *Warra Kaurna: a resource for Kaurna language programs*, State Print, Elizabeth, 1995.

3. J.M. Black, 'Vocabularies of four South Australian languages', *Transcripts in Royal Society of South Australia*, 44, 1920, pp. 76–93.

4. H.M. Cooper, *Aboriginal words and their meanings*, South Australian Museum and W.L. Hawes, Government Printer, Adelaide, 1966.

5. R.W. Ellis, 'The Aboriginal inhabitants and their environment' in Twidale, C.R., Tyler, M.J. and Webb, B.P. (eds), *Natural history of the Adelaide region*, Royal Society of South Australia, Adelaide, 1976, pp. 113–120.

6. J.W. Holmes and M.B. Iversen, 'Hydrology of the Cowandilla Plains, Adelaide, before 1836' in Twidale, Tyler and Webb, *Natural history of the Adelaide region*, pp. 91–97.

7. D.N. Kraehenbuehl, *Pre-European vegetation of Adelaide: a survey from the Gawler River to Hallett Cove*, Nature Conservation Society of South Australia, Adelaide, 1996.

8. C.G. Teichelmann, *Aborigines of South Australia: notes of the manners, customs, habits and superstitions of the natives of South Australia*, Committee of the South Australian Wesleyan Methodist Auxiliary Missionary Society, 1841.

9. C.G. Teichelmann and C.W. Schürmann, *Outlines of a grammar vocabulary and phraseology of the Aboriginal language of South Australia*, published by the authors, Adelaide, 1840.

10. No author, 'Transactions of the Statistical Society: report on the Aborigines of South Australia', *The Southern Australian,* 11 January 1842.

11. Education Department of South Australia, *The Kaurna people, Aboriginal people of the Adelaide Plains: an Aboriginal Studies course for secondary students in years 8–10*, Adelaide, 1989.

CASE STUDY 2
Norman Tindale and the Tjilbruke Dreaming story, an explorer of Aboriginal culture

David Jones

The body of this case study is from Norman B. Tindale's *The wanderings of Tjirbruki: a tale of the Kaurna people of Adelaide*[1] and is reprinted with the permission of the South Australian Museum Archives.

It is not common in our time to find men with the skills and insights which Mr Tindale has shown through his active and productive life. He is basically a scientist, while skills with language and human relations fit him for anthropology. His special interests in entomology, geology and botany broadened the scope of his ethnographic studies. As the use of film, tape, carbondating and blood grouping came into anthropological work he readily made use of these for the data that they could bring towards his final synthesis of the cultural history of the Aboriginals in time and space. In addition, Mr Tindale has shown throughout his work a tolerance, humility, honesty and adaptability which made it easy for him to find collaborators amongst both black and white men [sic.].

Professor W.V. MacFarlane[2]

Norman Tindale (1900–1993) was a prominent anthropologist who worked for some 49 years at the South Australian Museum producing a voluminous collection of manuscripts and articles that largely inform much of what we understand about Aboriginal culture in Australian today, including the significant 'Aboriginal Tribes of Australia' map (1974).

The Tjilbruke story has become the most accessible of the Kaurna Dreaming stories due to the transcription, and subsequent publication, by Tindale of this version as told by Milerum of the Jarildekald (Tanganekald) group of the Ngarrindjeri whose country the Dreaming story also traverses, together with some additional Ramindjeri informants. The substance and route of this Dreaming explains the origins and importance of a series of waterholes along the Adelaide–Cape Jervis coastline.

Within this story are places of water. There origins or general roles of the water places are explained within the story often implying the sacredness of the freshwater source/waterhole or the need to respect the food resource that resides within or is dependent upon that water source.

The following text is reprinted with the permission of the South Australian Museum.

The wanderings of Tjirbruki

Tjirbruki and his fellow Patpangga clansfolk were living at [Tankul'rawun] near Rapid Bay. Tankulrawun (its name has the meaning of the 'granite place') was one of their summer camping places near Witawatang]. Today Witawatang is known as Rapid Head.

There came an urge among some of the members of the band assembled there, including some young visitors, to go north and arrange a hunt for [kari] or emus. Many kari were to be seen in the ['ru:we], clan lands of the Tandanja people at Adelaide because that big bird was their ['ŋaitji], or totem. They did not kill them although they feasted on their eggs.

Tjirbruki, who was a hunter skilled in kangaroo spearing, did not wish to go, but his much loved [na:ŋari] or sister's son, named ['Kulultuwi], who was visiting with him along with several companions, did so wish. Kulultuwi called his mother's brother ['wan:u] or ['kawanu] as did two other younger lads whom he persuaded to go along with him. Both ['Jurawi] and ['Tetjawi] bore the same relationship to Tjirbruki although they were by different mothers. They departed hastily. (It may be assumed that their families accompanied them although the story, as told, often omits such details.)

Tjirbruki, not wishing to take part, shifted his camp more leisurely, moving through the *ruwe* of the ['Witjarluŋ] clan which began near ['Karika:liŋga], a name still on the map as Carrickalinga. He arrived at ['Wituwa'taŋk] now known as Brighton. He and his family were welcome visitors in the clan lands of ['Jatabiliŋ] at Wituwatangk, whose ['paŋkara] (hunting territory) extended northward along the coast beyond the place now know as Outer Harbour. Tjirbruki spent much of his time at Wituwatangk fishing for ['kurari], also called ['dar:awe] (beaked salmon, *Gonorhynchus greyi*). He used a special ['ŋe:re] or net, termed a ['dar:awenjeri 'ŋe:re], with which several persons helped in the haul.

Meanwhile Kulultuwi and his companions, travelling ahead, had sought out, and quietly were driving several emus ahead of them without revealing their presence, making their moves by holding up shields of branches of eucalyptus leaves. They moved across the middle of the ['Mikawom:a], the Adelaide Plains, because they needed to keep the birds close to the coast so as to corner them at ['Muldaŋ] on the northern tip of the Outer Harbour peninsula. Ancestors had made the Port River for them so that this could be done. Four male kari and four females, known as ['tartja], were caught up in their drive. By keeping on the coast-ward side of the plain, the hunters were avoiding trespass on Tandanja hunting grounds because they had not received permission to take emus there. The hunt was going well.

However there was a disturbance. Near ['Patawiljaŋk], now called Glenelg, some Jatabiling women were cooking herbs in their hot stone ovens. This caused the emus to turn away inland. Kulultuwi had to race around, going far into the Tandanja *ruwe* by way of ['Medaindi], now known as Medindie, to prevent the birds escaping from the trap. During this trespass Kulultuwi had killed a female bird. Some kari had escaped but others were successfully held over several days at Muldang while the men and their families fed on the body of the *tartja*.

While this was going on Tjirbruki and other people with him had shifted camp to ['Ṯulukudaŋk], now called Kingston Park. From there he made short excursions inland. He saw the old tracks of emus and their hunters going north but also the fresh tracks of one male bird. He decided that this would be his bird to hunt, since according to custom the first to sight the presence of game had the right to take it. For a while he continued to fish, taking further hauls of *kurari* for his journey.

Then Tjirbruki left, following the track of his *kari* along the coast to [Ka'reilḏuŋ] (Hallett Cove) and on to ['Tainba'raŋ], now Port Noarlunga, to ['Ru:waruŋ] (Port Willunga), and to ['Witawali] where the tacks turned inland. There, near Sellicks Hill, the old name of which has been forgotten, the tracks were lost.

Meanwhile the hunters decided to go back to the rest of their people. They arrived at Wituwarangk during a very heavy morning fog, found the camp empty and that Tjirbruki had left.

Tjirbruki, having lost all traces of tracks, and judging that the male bird would continue its movements southwards along the coast, turned inland on a path which took him through the valley at ['Maitpaŋ'ga] (which still bears the name as Myponga), travelling to ['Muta'pariŋa], a place where there are many blackwood trees, continuing down the Hindmarsh Valley ['Jaladula], and passing ['Jerlto'worti], to Victor Harbor at ['Lat:arŋ]. He still thought the emu might come around by the coast so he hid in ambush and watched for several days. No tracks appeared so he went back on his own trail and found a place where the old tracks had been covered by newer ones. There was good food for the bird here in the forest, far inland from his Witawatang camp at Rapid Head. In the distance he saw the smoke of a small fire and, heading in that direction, he heard the voice of Kulultuwi singing while one of the younger men was preparing a cooking fire for an emu Kulultuwi had killed.

This was the bird which Tjirbruki had been following and expecting to spear. He confronted Kulultuwi, claiming that his *nangari* had been wrong in killing his male bird. His own footprints should have indicated this to the younger man.

Kulultuwi said, 'Sorry, I did not know it was your *kari*. You saw the bird first. Cook it and take it home to your children'.

Tjirbruki replied, 'No! You killed it. You cook it and give us some of the meat'. He had some kangaroo meat and did not need the emu. Tjirbruki then departed.

Kulultwuli made ready the ['wintjimi] oven, making the bed of hot stones, placing the green herbs over them, putting the bird on and covering it with further herbs and earth, and pouring on water to make much steam. After waiting for it to cook Kulultuwi, as was customary, dug in and took out the head of the bird to see if it was ready when a sudden

burst of steam blinded him. Thereupon his part brothers, Tetjawi and Jurawi, taking advantage, rushed in, speared, and killed him.

The boys reasoned they had killed their ['juŋalia], or elder brother, because he had transgressed, having really known from reading the tracks that the bird belonged to their *wannu*. The youths cut off the meat from the bones of the bird and carried it to their own people of the Jatabiling clan. They left the body of Kulultuwi. They told their folk that Kulultuwi had done wrong: ['Ŋaitjau/'peindjəli] (in front of us/the emu). They used a northern word for emu, implying that the bird meat was evidence that Kulultuwi transgressed. Their people carried the body of Kulultuwi to ['War:pari] (Sturt Creek) on the Adelaide Plains near Marion where they continued the drying of the flexed body on a rack over a fire.

The youths made up a story that Kulultuwi, in fear of the anger of Tjirbruki, had gone away elsewhere to hunt further for emus. When this false story reached him at Rapid Bay, Tjirbruki asked several members of the Witjarlung clan living north of his country give a message of forgiveness to Kulultuwi. Although they knew of the death of Kulultuwi they, with malice, did not tell him the truth.

Searching for Kulultuwi, Tjirbruki went first to ['Loŋkowar] (Rosetta Head), the great bluff on Encounter Bay, then up ['Mu:lapari] the Inman River to ['Towara:nk], near Moon Hill, and on to ['Maikaba'naŋk] near the coast at Normanville. His family had gone with him. Then he began to wander about by himself, going as far as ['Nutaraŋ] (Lands End), at that time still in Kaurna country (according to informant Karlowan).

Heading north again, he came to the place near where he had seen Kulultuwi last and chanced to see some sugar ants on the track. He picked up some ants carrying human hair and others with blood and red ochre. Further on, he found more and knew in his thoughts that *nangari* was dead. He saw where the body had been, and where people had made a smoke fire. They had made a ['tirukati], or drying rack of poles tied together like a raft such as a man uses when fishing. On the third day they had, as was customary, covered the body with red ochre ['ta:u:we] from ['Potarta:ŋ]. They had carried the bier towards Adelaide.

Having made these discoveries Tjirbruki said, 'I have only one spear properly fixed. I am off!' He left the place in the *wita* (peppermint tree forest) and went towards ['Rāwarð 'ŋal] (Port Elliot). At Rawarangal he had opportunities, through his ['ŋiaŋi'ampe] trading partners, to obtain good spears which had come from the Tanganekald people on the Coorong. On the way, while walking along the ['Mu:lapardi] (Inman River), he met ['Jorlu] the red-backed kingfisher (*Halcyon pyrrhopygius*) man. On hearing the story Jorlu gave him a spear, as did another man, ['Joldi] of the black cormorant (*Phalacrocorax carbo*) totem at the Finniss. Tjirbruki, with his new weapons, chose to follow tracks along the eastern side of the Mount Lofty Ranges through Peramangk tribal country, keeping to their eastern boundary to avoid serious trespass. On the way he camped at ['Wiljau'a:r] near Strathalbyn, then at

['Peiera] (Woodchester Waterfall), then at ['Motoŋeŋal] (Mount Barker), and at [Bᵊrukuŋga], now the mining township of Brukunga. Travellling on through places now not remembered, he came to ['Kali:a] (Gawler) which was the beginning of Tandanja clan country. Keeping near the coast he travelled south. He had learned where a big camp ['taldamari] was gathering at Marion on Sturt Creek. He arrived, very weary, at Witawatangk.

Children saw him and cried out, 'Here is old ['muȚari]'. Old father's mother's brother soon was the centre of a gathering and he told them he would stay to rest only the one night. He saw that the two men, Jurawi and Tetjawi, were present. Acknowledging that Kulultuwi was dead, they deceived Tjirbruki about the real killers, blaming his death on strange people who might have been Peramangk tribes-folk who had come along the Mount Lofty Range.

Tjirbruki ignored their implications, knowing that they were lying. He practiced deception also, saying, 'Yes! I know!' Strange men came from the ['wir:a] (forest) country in the north'. He thus made out that he thought the young men were innocent. On the following day Jurawi and Tetjawi with their families made a part day's journey to Warrpari (the Sturt Creek at Marion) where they settled in at the big taldamari hut. The body of Kulultuwi was still being smoke-dried there on a rack. In the evening they began ['kuri] dancing for the old man and he initiated others. Then he sang the whole camp to sleep. He tested them by calling out, 'Come! Give help with a haul of kurari fish'. There was no response and the old man said, 'Ah! I've got you!'

Tjirbruki was a master at fire-making. He took powdered stringybark tree bark ['morθi] as tinder and set it around the taldamari with much grass, leaving only a small gap at the entrance. Then, using a [bᵊruke] (iron pyrites) stone and a piece of flintstone ['paldari] he started fires at each pile of morthi or tinder, telling the fire to blaze up quickly. He cried out loudly: 'Raŋkaŋ 'haŋand 'dol. You are getting burned! Camp on fire'.

The top of the taldamari began to fall in as it burned and all the people attempted to rush out. As children came out he kicked them with his foot and hit them with his club. Out came Jurawi whom he speared with a ['wundi] or dread-spear, one set with quartz chips in resin on its head. The spear entered Jurawi right up to the ['tuŋgi] or swelling of resin set on the spear to prevent its too ready removal from the wound.

Out came Tetjawi whom he speared also and held in the fire. Only when he felt no further kicking did he accept that 'they were done'. He pulled out the spears and waited until morning as the taldamari burned to the ground.

Tjirbruki took the dried body of his nangari to [Tulukudaŋk], a spring of good water on the beach of Kingston Park Reserve at Marion. There he completed the smoking of the body of Kulultuwi and an inquest was held, many people gathered for the ceremony. The names of the two killers were confirmed. Tjirbruki learned that his nangari had

indeed been struck down while raking the head of the emu from the fire, looking for the steam coming from its bill, indicating the bird was cooked.

Carrying his burden, now a dry compact parcel, Tjirbruki said, 'I go back now!' He departed, walking along the coast to [Ka'reilᵭuŋ], now called Hallett Cove, where he rested. As he reclined he began to think about his nephew and burst into crying [ka'reilᵭuŋ]. Tears ran down his face and where they fell to the ground a spring of water welled up (thus the spot became a camping place). Tjirbruki then journeyed to ['Tainba'ra:ŋ] (Port Noarlunga) where he burst into fresh tears. He went on to ['Porata:ŋ] (Red Ochre Cove), Willunga, where he cried again; yet another spring of water came up. He then walked to ['Ruwaruŋ] (several hundred metres south of Port Willunga jetty). The tide was out. He sat down on the beach and cried once more. The ['lu:ki] (tears) dropped on the sand, causing a spring to appear. At high tide the sea covered it, but when the tide fell again the fresh water could be obtained by scraping in the sand. It remains so today.

The old man then carried the body to ['Witawali] on the beach north of Sellicks Hill. He noticed that there was a fine bay which would serve at night as a good netting place for sea salmon. His tears were still flowing and brought a spring into being there (in the vicinity of Section 639, Hundred of Willunga).

While there, Tjirbruki began to think of further grudges and as he was passing through the pangkara of the Witjarlung families it disturbed him that they had failed to pass on his message of forgiveness to Kulultuwi and his other nephews. Instead of continuing along the beach he turned inland and climbed over Sellicks Hill. He kept Maitpanga on his left and climbed another high hill (it may have been Mount Jeffcott or Black Hill). There he made a smoke signal. White smoke went straight up. People who were camped as a place called ['Warabari] saw the smoke and began to interpret its meaning:

'Turtil garwand 'werati.' (Smoke plenty/going upwards).

Korn 'loro 'kutu 'malbur 'undul.' (Men/straight up/good news of killing.) (In this loosely translated remark ['malpuri] given as 'guilty of murder' in Jaralde.)

'Itji 'nel lund'. (He is coming home.) (In Tangane ['ŋerei'luna] has been translated as 'quickly; wasting no time'.)

Tjirbruki made other fires as he picked up the answering smoke, and continued to do so until he was close enough to hear the people shouting. It was the camps of the men ['Limi] and ['Ŋarak':ani].

'Naitj purntulunul.' (He is coming.)

Those who were still in their huts asking:

'Jaŋaleitj?' (How far away?)

'Niji teipuland.' (He is close.)

Tjirbruki heard their questioning. He untied his bundle of spears, taking as many as he could hold, and walked directly into the camp. A first spear he drove into Ngarakkani, another into ['Ŋeŋara'tawi], a third into Limi, and the last one into ['Tul:aki]. (Even in those days it was proper to spear people in the legs unless murder was the direct intention). The men saw that Tjirbruki meant mischief and all took headers into the water and turned into fish. Thus, in the sea off ['Ŋaldeŋga] today you will find ['ŋarak:ani] (the gummy shark, *Mustelus antarcticus*), ['limi] (the cobbler carpet shark, *Sutorectus tentaculatus*), also ['ŋeŋara'tawi] (the southern fiddler, *Trygonorrhina guanerius*), and ['tul:aki], 'the long thin shark with the flag on it' (which we have not identified, although it perhaps is the cocktail shark, *Carcharhinus brachyurus*). These fish became the *ngaitji* or totems of members of the Witjarlung clan of the Kaurna tribe. Any other people who were present when Tjirburke took his revenge fled and turned into birds, leaving only the old man there, alone. Satisfied, Tjirburke stayed there a while, and when his nephew's body was again dry enough to carry, he rolled it in a kangaroo skins and continued on his journey.

Tjirbruki came to ['Karika:liŋga] (Section 1018, Hundred of Yankalilla), just south of the place known to Europeans a Carrickalinga Head. Here there was (and is, for informant Karlowan had seen it himself) a little swamp flat where ['ŋuri] grows, very green like a reed. Rafts, called ['kundi] (not ['kandi] as misprinted in an earlier publication) were made of the dried stems of this plant (probably a *Typha*). Tied up bundles, they were also used along the Murray River.

Continuing his journey along the coast Tjirbruki went to ['Koŋarati'ŋga] where there is a ['perki] or cave. Just before he arrived at the *perki* he again sat down and cried: a small spring flowed there. He did not go into the cave but walked further on, a few hundred metres to the mouth of a small creek that is a camping place. He continued walking, sometimes on the shore and at other times above the cliffs, all the way to ['Parewar'aŋk] (now Cape Jervis). From Parewanangk he returned northwards along the foreshore below the cliffs and came to another *perki* (called [Ja'narwiŋ] by another informant). It is close to the place from which you *janarwing* (turn back) because the water is too deep for one to pass along the shore.

Tjirbruki left his nephew's body outside and, walking into the darkness, found a place where there was a suitable ledge of rock. He put sticks up, just as was done when the body was being smoked, carried the body in, placed it on the platform, and left it. He did not emerge from the cave but went on into the depths of the hills for a long way. He made the way wide enough for him to continue inside right up on top of the range at ['WaƗeira 'ŋeŋgal] (now Mount Hayfield). Emerging there he shut the 'airhole' where he came out. He 'fixed it up with gravel' to appear he had 'never come out there'. Going down to the foot of the hill he shook his body and dust came off. This became the ['mulkali] (yellow paint or ochre) which is used for decorating or 'making spears flash'. (A further comment from the informant: 'Gold has been found there; it may be from off him').

Tjirbruki arrived at ['Tjutju'gawi] (west of Mount Robinson), the camp of the Ramindjeri tribesman ['Ken'gori] of the ['wanma'rai] totem (ringtail possum, *Pseudocheirus peregrinus*). Kengori was a member of the Polumpindjeri clan and Tjirbruki received permission from him to take *wanmarai* so that he could make a skin rug for the coming winter. He was feeling old. He looked out and saw a swampy lagoon and said to himself, 'There is no use in my living like a man anymore'. However, he left the camp of Kengori (whose adventures, which became a separate story, took place after Tjirbruki departed). The old man walked along the southern shore of the Fleurieu Peninsula on land well above the sea until he came to the ['koiŋkanja] or 'high hill' called ['Loŋkowar] (Rosetta Head).

'This place will do for me', Tjirbruki thought. 'How will I do it?' The answer came. On a tree nearby there was a bird, a ['kelendi] (the grey currawong, *Strepera versicolor*). He stalked the bird, killed it, plucked the feathers, and then rubbed the bird's fat over his own body. He recalled that ['Kelendi], when he was still a man, was a great messenger who travelled around the country singing songs and telling people of the coming meetings for initiation of their young men. Tjirbruki tied the bird's tail feathers on his arms with hairstring. Then he split the flesh between his big toes, and the third and fourth ones, made a run, and 'straight away started to fly'. As a ['tji:rbruki], which white people today called the glossy ibis (*Plegadis falcinellus*), his spirit still appears in bird form where there are swampy areas. His body became a ['martowalən] (a memorial), a rocky outcrop at Barukungga (on Section 1887, Hundred of Kanmantoo), the place of 'hidden fire'.

References

1. N.B. Tindale, 'The wanderings of Tjirbruki: a tale of the Kaurna people of Adelaide', *Records of the South Australian Museum*, 20, 1987, pp. 5–13.
2. P.G. Jones, 'Obituary: Norman B. Tindale, 12 October 1900–19 November 1993', *Records of the South Australian Museum*, 28 (2), 1994, pp. 159–176.

CHAPTER 7

A history of water in the city

Martin Shanahan
David S. Jones
Sara Hughes

Introduction

Water has always been of fundamental importance to Adelaide. Indeed, the very location of Adelaide was determined, in large measure, by the perceived assurance of water being supplied by the River Torrens.[1] For many years since, meeting the demand for clean, safe water has been a focus of attention for governments at all levels. At various points a lack of water has threatened Adelaide's future, while floods have caused damage and loss of valuable assets. Mostly, however, it has been the relatively scarce supply of water that has been a limiting factor for the growth and development of the city of Adelaide.

The European settlers' approach to addressing these challenges was in stark contrast to that of the original inhabitants, the Kaurna people, who lived with the natural fluctuations in water supply.[2] Since European settlement, the history of water management has revolved around efforts to control, store, transport and clean water in order to sustain an ordered lifestyle and environment, rather than to live within natural limits. This chapter provides a brief historical perspective on various aspects of Adelaide's relationship with water resources.

Planning for settlement

Edward Gibbon Wakefield's land management ideas were central in the minds of the directors of the South Australian Land Company, constituted in 1831 to found and develop the new Province of South Australia.[3] Despite being formed with 'conscious purpose', in reality the Province was a real-estate venture, with Adelaide – 'a capital city' – the central product. Social reform, religious freedom, and equality of business opportunity, unfettered by the traditional social hierarchies of Europe, were key elements; as was the overall need for a profitable and viable company.

The 1834 prospectus for potential investors in and migrants to the colony of South Australia stated:

> It is, however, chiefly upon the PRINCIPLES to be pursued in the formation of this colony that the expectations of its success have been founded. It is of importance to know that the soil and climate are fitted for the residence and support of the colonists; but unless the plan upon which the colony is founded be a good one, the utmost fertility of soil and advantage of position may be rendered of little or no avail.[4]

Land fertility and, by necessity, soil quality and water access were critical to the plan's success.

In May 1835, the South Australian Colonisation Commission in London, began raising funds and recruiting individuals to establish and manage the new colony.[4] Colonel William Light was appointed Surveyor-General (on the recommendation of [Governor] Captain John Hindmarsh), and instructed on how to undertake land settlement choice and formulation,

Chapter photograph: *The Company's Bridge, Hackney,* 1864, James Shaw. Image courtesy of the Art Gallery of South Australia, Adelaide; bequest of Mary Ayling, 1961.

aided by his deputy, George Kingston. Land surveying and town planning were viewed as integral to the success of the venture.

Some information as to the qualities of the prospective colony was available to potential migrants and investors, including the reports of the explorations of Matthew Flinders and Nicolas Baudin in 1802, Captain Charles Sturt's publication *Two Expeditions into the Interior of Southern Australia*, in which he reported his explorations down the River Murray in 1830, and the ad hoc oral accounts of whalers and sealers. In Sturt's publication, there appear clues as to Light's subsequent nautical assessments:

> My eye never fell on a country of a more promising aspect or of more favourable position that that which occupies the space between the lake [Alexandrina] and the ranges of St Vincent's Gulf, and continuing northerly from Mount Barker stretches away without any visible boundary.[5]

The site of the capital and the search for water

Choosing the right site for a colony in a foreign location was a challenge. In Light's case it was made all the more difficult by the constraints of planning that had been designed a world away, with no experience of the local conditions but with great expectations for the Province's success.

Seven weeks before Light sailed to South Australia onboard the *Rapid*, he was issued with a set of instructions by the colonisation commissioners: undertake a fast comparative and environmental impact assessment of suitable town sites 'so that the different sites favourable for the erection of towns may be brought into exact comparison'. This was considered prudent to avoid repetition of the flawed decisions elsewhere around the Australian coastline, which 'led to an actual change, or to the desire for change, in the sites of certain towns after their first establishment'.[6] The commissioners were specific in their wishes, or their 'environment impact' performance criteria, when they required a site that 'combines in the highest degree the following advantages:

1. A commodious harbour, safe and accessible at all times of year.
2. A considerable tract of fertile land immediately adjoining.
3. An abundant supply of freshwater.
4. Facilities for internal communication.
5. Facilities for communication with other ports.
6. Distance from the limits of the colony, as a means of avoiding interference from without in the principle of colonisation.
7. The neighbourhood of extensive sheepwalks.

The above are of primary importance, the following of secondary value:

1. A supply of building materials, as timber, stone or brick, earth and lime.
2. Facilities for drainage.
3. Coal.'[6]

A letter by Colonel William Light to George Jones

BOX 30

In 1836, Coloniel Light penned a letter with an annotated drawing to his friend, British artist George Jones, in London. The letter, together with another to friend Henry Dixon, were acquired by the State Library of South Australia in early 2006.

Included in the drawing are references to two 'high roads' to the north and north-east, a 'canal' along the present Port Road alignment, several freshwater lakes denoted between the 'canal' and the coastline, delineation of the River Torrens/*Karrawirra Parri*, together with a discussion about the siting of 'the city of Adelaide'.

In the letter, Light wrote:

Nature has done so much that very little human labour and cost is requisite to make this one of the finest settlements in the whole world.[1]

David Jones

It is interesting that freshwater availability was placed so high (third) in the list of priorities and that drainage (placed ninth) was also considered a priority. Further, while harbour location (first) was important, the commissioners expressed that it should not override the overall site selection: 'it would be unwise for the sake of still superior advantages in the harbour to make any important sacrifice in the other essentials of a good site.'

With these criteria, Light set about the initial survey and assessment work while still in London. But he also had to contend with other opinions on the preferred site. Three days before Light's departure, Governor Hindmarsh directed him to establish the capital at Port Lincoln; later Hindmarsh sought to transfer the site to Encounter Bay. Fortunately, the commissioners had specified Light's decision as:

> … a duty which … you are hereby fully authorized and required to discharge … [and] should the governor [Hindmarsh] arrive sufficiently early in the colony … you will confer with him on the subject and pay due regard to his opinion and suggestions without, however, yielding to any influence which could have the effect of diverting you in any way of the whole responsibility of the decision.[6]

Light was in charge of the 'environmental impact assessment', as well as making the political decision as to its outcome. He was under no illusion as to the importance of the decision and his ultimate accountability for it.

Light arrived at Antichamber Bay (sic.), Kangaroo Island, on 19 August 1836. Respecting the subtleties of the commissioners' instructions, Light and his team spent some 11 weeks exploring the coast of Kangaroo Island and the east coast of Gulf St Vincent before spending one week at Port Lincoln.[7] The search for reliable water supplies featured prominently.

At Nepean Bay (Kangaroo Island), 'no fresh water was to be found' with 'lamentable accounts [about its] scarcity' by settlers. At Kingscote, Light found 'a well of fresh water [and perceived] that a settlement might be formed here at some future period, to great advantage'. At Rapid Bay he found 'a fine stream of fresh water' and at Yankalilla, 'a beautiful stream of fresh water running into the sea'.

He explored Jones' Harbour (Port Adelaide River and channel), recording that it 'will make an excellent harbour at some future period', but noting that 'no fresh water was seen' and the place was 'a great disappointment' with a landscape distinguished by mangroves, saltpans, and shallow, 'swampy ground' with little evidence of a stream divesting its 'fresh water stream from the [Mount Lofty] mountains'.

Port Lincoln disappointed him greatly:

> finding nothing but hard rocks and she oak … some pools of fresh water … [and less than] 1000 acres of tolerable land; … I must decidedly say, it cannot be thought of as a first settlement; some years hence it may be made a valuable sea port, but that can only be after the colony has increased considerably.

From Light's perspective, Encounter Bay lacked a safe, secure harbour, 'and its low sandy appearance, with its exposure to the whole Southern Ocean, was a sufficient danger that must attend the approach of it'.[7]

On 9 November, he explored the Patawalonga Creek/River Torrens/*Karrawirra Parri* estuary:

> We have this morning been looking for the mouth of the river [Torrens] and find it exhausts itself in the lagoons, these must either ooze through the sand into the sea, or be connected with the [Patawalonga] creek. I strongly suspect the latter, as the distance to the creek is small at this part, and the water in the upper part of the creek, where I grounded, was far from being salt. I feel more interested in this flat than ever, and have determined that a survey may be carried on here while I am in the other Gulf.[7]

Light perceived this area offered the best possibilities for water and settlement capability. While he journeyed to Port Lincoln to assess its capabilities, he set his deputy, Kingston, to review the plains to the east of Holdfast Bay. It appears that water was the overriding criteria used by Light in site selection, as evidenced in his evaluation of sites not just upon extant freshwater but also their propensity to obtain natural rainfall. In this, Light demonstrated a keen appreciation of Mediterranean climatology despite the otherwise foreign environment.

In mid December 1836, Light concluded a preference towards the east coast of Gulf St Vincent because:

> … all the vapours from the prevalent south westerly winds would rest on the [Mount Lofty] mountains here, and that we should, if we could locate this side of the gulf, be never in dread of those droughts so often experienced on the eastern coast of Australia.[7]

Site discovery and the assessment of water resources

Light appears to have made an intuitive assessment about the area ultimately to be the site of Adelaide. Even before he ventured onto the plains beyond the coastal sand dunes of Holdfast Bay, he noted:

> Looking generally at this place I am quite confident it will be one of the largest settlements, if not the capital of the colony, the [Patawalonga] Creek will be its Harbour. Six months labour would clear a road down to it … I am positive there is quite enough of good rich land for every purpose; the low parts of this plain are covered with fresh water lakes, many of which are full of rushes, and in the winter a great part of the plain may be covered with water, but the ground rises gradually towards the [Mount Lofty] mountains, and that part can never be flooded … Much remains to be done also by proper management of the waters that have hitherto run in natural courses, by collecting them with proper dams, and conducting them through more eligible channels. This will I am sure be one of the finest plains in the world.[7]

Water-carting

BOX 31

The distribution of water was difficult for Adelaide's residents, virtually from the first day. It is reported that the HMS *Buffalo*, standing at Holdfast shore, paid £100 to obtain 20 t (20,321 kg) of water from the Adelaide settlement – a large portion of the cost being for returning empty water casks.[1] The comparatively well-paid occupations of water-carter and well-digger quickly emerged in the new settlement. The early land allocations in the South Australian regions beyond Adelaide were dominated by settlers seeking reliable water sources; the best sites were quickly taken up.

Although a reason for settlement, the River Torrens was neither particularly reliable nor, after a short time, particularly clean. Initially, the settlers used the river (when it held sufficient water) for their drinking supply, for washing, and to dump refuse in. Human activity, both in the city and further along the river's banks, increased the opportunity for waterborne disease transmission. Outbreaks of cholera and dysentery were not uncommon.[2]

The Torrens and other nearby streams and water sources quickly proved inadequate for the needs of the growing population.

By the end of the 1840s, water supply and sanitation were both urgent problems for the new settlement's administration. Initially serviced by water-carters, by the mid 1850s private companies were pumping water to Adelaide households. This provided competition to the water-carters whose prices had sharply increased as drivers left for the Victorian goldfields. Nonetheless, the supply of clean, safe water to city residents was still problematic. One response was the foundation of the Waterworks and Drainage Commission in 1856, the department that was later to become the Engineering and Water Supply Department, and then SA Water. Addressing problems related to flooding also entered the city's agenda around this time, the year 1842 marking the first recorded River Torrens flood in Adelaide.[3]

Martin Shanahan and Sara Hughes

Here was a keen insight not only as to the potential the Adelaide Plains offered to settlement but, significantly, to its ready freshwater resources and the fact that the watercourses could be harnessed and drained to minimise flooding impacts on future development.

It was Kingston who actually 'discovered' the site of Adelaide. Kingston later recalled that he

> … had discovered a situation in the vicinity of the river [Torrens], about six miles north-east of the [Holdfast] Bay [at Glenelg] and the same distance south-east of the [Port Adelaide] harbour, which I felt assured he (Light) would finally select as well-suited for the site of the City of Adelaide. I informed (Light) that he would there find an abundant supply of fresh water; that the locality was moderately well wooded at an elevation of from 100 to 150 feet above the river … and that no other spot would be found possessing anything like the same advantages, as a site for the city.[8]

And at the time, 24 November 1836, Kingston reported by letter to Light, following his exploration of the River Torrens/*Karrawirra Parri* watercourse and the Adelaide Plains that:

> I examined the river and reaches again yesterday, and on my way … for about six miles was across a plain of exceedingly fine land; I again traced the plain and then kept to its edge, being all along able to trace the course of the river through the reeds, until I found it again running through a regular bed. The river, although in parts shallow and much obstructed by fallen tea-trees, would be navigable for flat-bottomed boats as far as the marshes … A very large body of water must come down the river in the winter, as in the upper part where the banks are thirty feet deep, there are evident marks of the floods reaching the top. I now feel assured, that we have obtained sufficient information to convince the most skeptical of the great value and eligibility of these plains – possessing as they do, abundance of fresh water, an excellent harbour, with at least one river running into it, which can easily be made eligible as a mode of communication between it and the plains.[7]

After reviewing site options at Port Lincoln, Light returned to Holdfast Bay and on 24 December 1836, 'walked over the plain to that part of the river where Mr. Kingston had pitched his tent … My first opinions with regard to this place, became still more confirmed by this trip'. On the 29th, he spent 'nearly all day examining the plain … I was delighted with the appearance of the county, and the supply of fresh water we were certain of possessing …'.[7]

Quality freshwater was imperative in the selection of Adelaide as a place for settlement, with its context on a potentially wide productive plain and a position with secure rivulets due to the western 'vapours' and rainfalls on the slope of the Mount Lofty Ranges.

Light's judgement was also supported by the diary entries of George Stevenson, incoming editor of the *South Australian Gazette and Colonial Register*:

> We were delighted with the aspect of this country and the rich soil of the Holdfast Plains … I have seen the Pickaway plains of Ohio and traversed the Prairie of Illinois and Indiana, but the best of them are not to be compared with the richness of the Holdfast Plains. The water is in plenty, and four or five feet of digging is required to obtain an abundant supply.[9]

Adelaide as a site with water

As a site for a settlement, it is still clear today that a supply of freshwater was available on the Adelaide Plains when Colonel Light was conducting his surveys. Water fed in from the hills down the Brownhill Creek and Waterfall Gully Creek drainage lines, along Botanic Creek, winding its way through the east parklands across Victoria Park Racecourse/*Bakkakkandi*/Park 16 and into the Adelaide Botanic Gardens' Main Lake and the now artificially restrained River Torrens/*Karrawirra Parri*. At the site of Adelaide, the River Torrens/*Karrawirra Parri* had gouged a series of sweeping bends distinguished by steep embankments opening into waterholes. The latter are still in evidence at *Tulya Wodli*/Bonython Park/Park 27 (formerly 'The Billabong'), where a series of low retaining barrages exist today, and also on the Torrens between the present Adelaide Oval and Adelaide Railway Station. Before the construction of rainwater tanks and bores, freshwater supplies were obtained directly from the River Torrens/*Karrawirra Parri* waterhole and from a smaller waterhole on a bend in the Torrens near today's Frome Road.[10, 11] Originally, the river exhibited a propensity to flood, providing water to aid the river red gums (*Eucalyptus camaldulensis*) that lined the original embankment edges.

A set of infrastructure and activities were established along the River Torrens/*Karrawirra Parri* to access the water for potable use, harvest it for the production of food, including vegetables and orchards and tree nurseries, and also as a conduit to flush away excrement and waste. It was because of the location of the principal water source, its geography, and its seasonality that:

- Light consciously included in his 1836–1837 survey a set of road access routes across the Torrens that were adjacent to, and thus conserved, the waterholes.
- 'The Billabong' was selected as the site for the public slaughterhouse; the river provided water for animals prior to butchering, and waste could be flushed away. The City of Adelaide constantly sought to dam or impound the Torrens waters at the present weir site.
- Governor Hindmarsh selected the 'Government Garden' site on the southern flank of the Torrens (now hidden beneath extensive building fill and rubble that created Elder Park).
- Horticultural proponent George Francis established a successful plant nursery and residence at a site on both banks of the Torrens to propagate specimen ornamental and horticulturally productive trees for sale and planting (now between Adelaide Zoo and the City of Adelaide Nursery).
- Prominent horticulturalist John Bailey established the most successful early colonial plant nursery in Hackney, 'Bailey's Garden', on the southern flanks of the Torrens,

The Mitcham Waterworks

BOX 32

The South Australian Company's promotion of Adelaide and the new colony of South Australia reported favourably on all aspects of the colony[1], and in 1837 John Morphett, assistant to the colony's surveyor-general, Colonel William Light, sent the following enthusiastic report home to England:

> Mount Lofty bears nearly east, and the whole of this side of the range is intersected with gullies, ravines and water courses, of the deepest kind, bearing evident marks of being acted on by powerful torrents. All of the hilly country along the coast has a similar character, but in no place is it so conspicuous as here. The facilities for damming up, and the creation of water power, are greater than I have seen in any country in an equal area …[2]

After only one decade, however, it was clear that the Company had misjudged the volume and reliability of natural flows from the ranges, and the advice to intending colonists described an almost desperate situation:

> … there is another circumstance equally important to internal comfort and to health, as to which Adelaide is singularly deficient – that is the want of good water – the river is of course, salt, and all the water available for drinking and for domestic purposes, is obtained either by catching the rain water, or from deep wells …[3]

In addition, the quality of water being supplied from the river was so contaminated that most water carriers placed warning signs on their carts: 'BEWARE OF DYSENTRY [sic] AND FEVER.'[4]

By 1847 the colonial engineer, Colonel Freeling, had surveyed the water catchments of the Mount Lofty Ranges and proposed the formation of a water company[5] to construct a dam on Brownhill Creek for a reservoir of 46 million gallons, sufficient for a town of 30,000 people.[6] Although this plan was soon discarded in favour of the Thorndon Park Reservoir on the River Torrens, George Green, a retired engineer, proposed a second plan in 1848 to supply water to Adelaide from Brownhill Creek:

> The reservoir at Brownhill Creek will be found just above the boundary of Mitcham, and will be 174 feet above the highest part of Adelaide. The conduit main will be 6 inches in the bore and will bring in 300,000 gallons of water daily to the closed cast iron receiver fixed in the centre of the reservoir …[7]

Although Green's proposal was not accepted, the need to supply water to the growing village of Mitcham was becoming increasingly urgent. Mitcham was at a higher elevation than Adelaide and therefore water could not be supplied by gravity flow from the new reservoir on the River Torrens; and by the 1860s the supply from Brownhill Creek was failing. In the late 1850s, the district council of Mitcham had constructed an access road to Section 1076 on Ellison Creek, a tributary of Brownhill Creek, in order to promote the selection of the site for the new reservoir.[8] When Mitcham was not selected, residents lobbied the state government for a permanent water supply.

Finally, in 1878 parliament agreed to construct a water storage and reticulation scheme on Ellison Creek. Construction began in 1879, using a revised version of Green's proposal. A brick-lined water storage well, a stone-lined valve well, a dam wall, and a settling pond were constructed in the bed of the creek. A six-inch cast-iron pipe, imported from Scotland, carried water along the Ellison Creek and Brownhill Creek valleys to the base of McElligott's Quarry. Here the pipeline ran up the side of the valley and across what is now the end of Fullarton Road to the Mitcham High Level Reservoir (now called the Mitcham Tank), with a 270,000 gallon capacity. From here the water was piped to the Mitcham village, and, in times of water shortages, to Unley and Adelaide.

The Mitcham Waterworks was officially opened on 25 November 1879. Water flowed by gravity, the pressure in the system being a function of the respective heights above sea level of the well on Ellison Creek and the Mitcham High Level Reservoir. The level in the reservoir was 150 ft lower than the level in the well. The system was officially discontinued in 1930, although it may have been out of use for several years prior to that date.[9]

Although references to the waterworks appeared in archival documents, the locations of many of the features associated with the Mitcham Waterworks had been forgotten. Between 2002 and 2004 each feature was identified by archaeological field survey and documented – a dam wall, settling pond, circular brick well, square stone-lined valve well, and sections of the cast-iron pipe that terminated at the Mitcham High Level Reservoir at the end of Fullarton Road.[9]

Pamela Smith

drawing irrigation water from the waterholes.

- George Francis, as inaugural director of the Botanic Gardens, crafted the first garden based upon the waterholes and watercourse of Botanic Creek.
- Henry Copas established one of the most extensive orchards and trickle-fed nurseries upstream at Campbelltown, which enabled him to successfully lay the first quality lawns on the Adelaide Oval.

Developments soon transformed the nature of water in the landscape and the watercourses and routes it bubbled up from or wandered across. By 1846, English surveyor and artist Frederick R. Nixon noted this environmental transformation:

> What is now called 'Hindley Street', was, seven years ago, in shape a mere idea, and in reality, the resort of the kangaroo and wallaby … The River Torrens itself then wore a very different aspect to what it now does. Then its banks were clothed with a rich carpet of grass and shrubs, and were enlivened by a variety of rude cottages of the early settlers. The trees, too, along its margin, had not then felt the devastating effect of the saw and the axe, and were as remarkable for their beauty as for the extent for their number. But all is now so altered as not to be longer recognizable. The drooping foliage is gone: the beautiful geraniums are destroyed; the stately gums have disappeared; and in a word, the whole character of this once pretty stream is so transformed, that a person who beheld it in its primitive state would not now believe it to be the same …[12]

Declining water supplies and expanding public works

With the growth of any settlement, the need to source quality water increases. Adelaide certainly fell into this category. With expansion, the water carters could not keep pace with demand nor source quality freshwater. The creeks and River Torrens were increasingly becoming polluted with night soil and other waste, while litter was being dumped into the waters and slaughterhouse excrement was emptied into the Torrens at *Tulya Woldi*/Bonython Park/Park 27. The increased fencing of land in the eastern and southern flanks of the plain, the construction of dams on creeks, and the sinking of bores further limited public access to quality freshwater.

A correspondent to the *Observer* in 1843 summarised the situation:

> It does appear strange to me that no steps are as yet taken to supply the town with cleaner water. It would be hard indeed to persuade me that the Torrens water is at all wholesome, if sheep washing with tobacco, tanning hides and so forth, are allowed on its banks … But besides all this, the cows, goats, bullocks and horse kept in the town are taken to the Old Bridge to water and to wash … In the neighbourhood of Currie Street, wells may be surrounded by water closets, with a few feet, and these wells are not dammed. This must be worse than drinking the Torrens water.[13]

One response was to construct a series of water reservoirs, in advance of the construction of an integrated reticulated system. In 1848, engineer George Green first proposed a supply from Brownhill Creek, based upon a weir and a six-inch main to a below-ground reservoir – a proposal that also prompted the formation of the unsuccessful Adelaide Water Company.

By 1848 the price of carted water was skyrocketing. Settlers could pay an average of £3 per week for carted water and the carriers become the most pandered-to tradesmen in the city. Colonial engineer Freeling evaluated the systematic waterbody option of Brownhill Creek in 1850, and proposed that the construction of a weir on the River Torrens some six miles upstream was the most viable option.

The first *Waterworks and Drainage Act 1856* authorised the raising of £280,000 to construct the Thorndon Reservoir, with a proposed capacity of 180 million gallons, and the Gorge Weir on the River Torrens/*Karrawirra Parri* – both to supply Adelaide with water. The Act also brought about the establishment of the Waterworks and Drainage Commission, headed by Sir Samuel Davenport. The public reaction to the passage of the legislation and authorised works was subject to comment by the editor of the *Register* in June 1856:

> A very extraordinary apathy characterises the public mind with regard to the proposed water supply and drainage of Adelaide. There are neither public meetings, public memorials, nor letters in the newspapers.[14]

The construction of Thorndon Reservoir and the life of the commission included the construction of the Kent Town valve-house, cottage and compound in *Kadlitpinna*/Park 13 on the corner of Hackney Road and North Terrace East, of which only the valve-house stands today. Reticulated water flowed into North Adelaide in 1859 with the completion of the Thorndon Park Reservoir, and this was soon extended to Port Adelaide and several other accessible suburbs. Of this supply, resident Frederick Sinnett wrote in 1862:

> Water is carried in pipes down all the streets, and is now introduced into most houses – an incalculable blessing to the inhabitants, though it has probably been obtained at an unnecessary expenditure of public money.[15]

A water gravitation point was established in *Kangattilla*/Park 4, on the corner of Jeffcott Street and Barton Terrace. The first semi-underground reservoir was built in the late 1870s on this site with a capacity of a million gallons, '104 ft square [9.66 m²], 19 ft [5.79 m] deep, with 8 ft 3 inches [2.53 m] of it above ground', with a roof built over brick columns and arches rendered with lime concrete. But its location was prone to flooding from the ponding of stormwater; particularly evident in times of high rains as evidenced in the 1856 flood when a 40 acre (16.2 ha) area lake formed, which could only be dispersed by the building of drains. Aspects of the system on the corner of O'Connell Street and Barton Terrace are still operational today.[16]

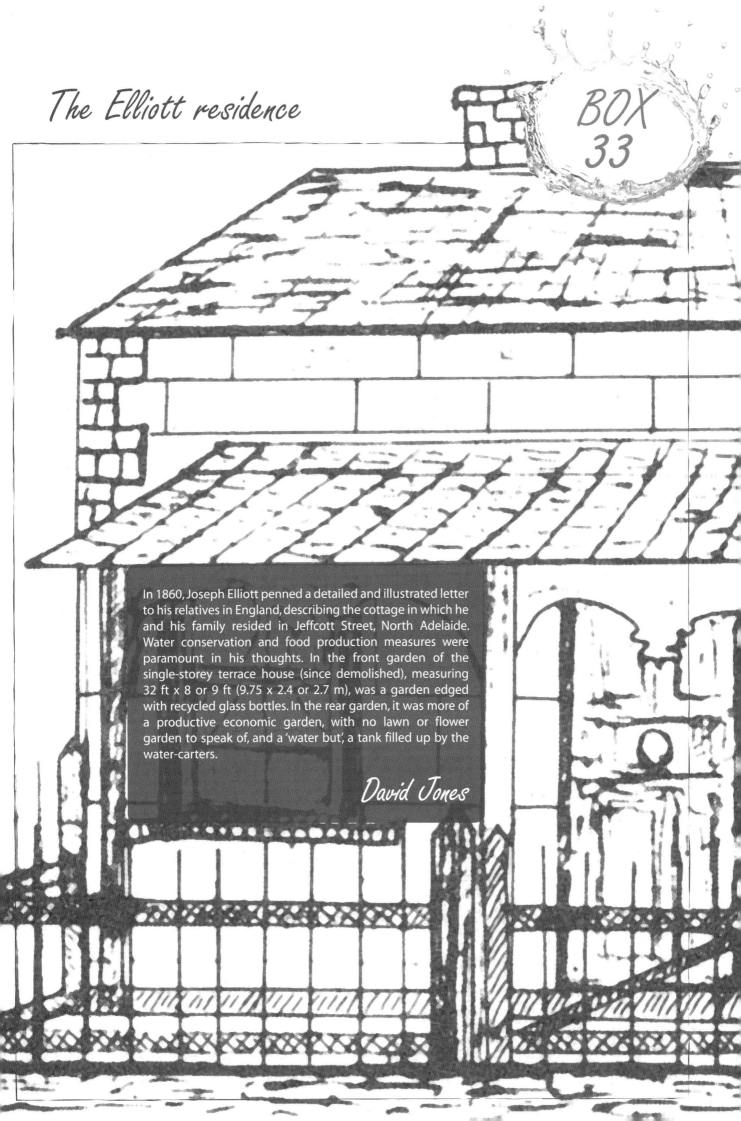

The Elliott residence

BOX
33

In 1860, Joseph Elliott penned a detailed and illustrated letter to his relatives in England, describing the cottage in which he and his family resided in Jeffcott Street, North Adelaide. Water conservation and food production measures were paramount in his thoughts. In the front garden of the single-storey terrace house (since demolished), measuring 32 ft x 8 or 9 ft (9.75 x 2.4 or 2.7 m), was a garden edged with recycled glass bottles. In the rear garden, it was more of a productive economic garden, with no lawn or flower garden to speak of, and a 'water but', a tank filled up by the water-carters.

David Jones

With the passage of the *Glenelg Waterworks Act No 173 1880*, the City of Adelaide agreed to construct a raised reservoir structure and mound in *Tuttangga*/Park 17, near the corner of South and East terraces. Constructed in 1880–1881, the tank was to service both Adelaide and Glenelg, and would hold a million gallons (4.5 million GL) of water. It is not operational today but its site is readily identifiable by the circular earth mound.[16]

One of the early developments that followed a more reliable supply of water was the construction of a city bath house on King William Road, owned by the Adelaide City Council and intended primarily for washing and cleaning rather than swimming. The baths were rebuilt in the early 1880s to include separate swimming pools for men and women, and redesigned in the early 20th century to include Turkish baths. The popularity of the bath house gradually began to decline, however, in part because of the requirement under the *Building Act 1923* for new residential dwellings to have their own baths. It is interesting to note that swimming continued in the Torrens with official permission well into the 20th century.[17] Nude bathing in the Torrens Lake was also common; the practice continued near the Jerningham Street sewer outlet well into the late 1880s.

Freely available, piped-in water with unfettered metering resulted in a boom of water use, most obviously in the 'greening of Adelaide' commencing in the late 1860s and blossoming in the late 1870s and 1880s when plant collecting and gardening became socially fashionable and councils instituted a scientifically based parks and gardens regime with qualified staff. The City of Adelaide, led by city gardener August Pelzer, was at the forefront of public garden development, using the latest advances in irrigation with each new garden. Many of the streetscapes, parks, gardens, and parklands we enjoy today owe their existence to reticulated water; but with reticulated water and excess use came a naivety about the availability of supply.

While most of Adelaide's early water supply came from surfacewater sources, a growing population and the accompanying demand for water prompted the exploitation of groundwater resources. Shallow wells were initially dug near watercourses but proved unreliable. Settlers began to construct deep wells into third-layer aquifers of the Adelaide Coastal Plain. The water in these aquifers was reliable and clean, and could be extracted at depths between 50–120 m.[18] At times, groundwater was used to augment the city's water supply as well as a local source of water for horticulture and irrigation. Since 1915 groundwater has been used six times to augment city supplies, most recently in the summer of 1967–1968.

The need for improved water quality

In the first 50 years of European settlement, not only were there difficulties in supplying drinkable water to the growing number of Adelaide residents, there were also significant problems in removing and treating wastewater. Early efforts to manage Adelaide's water and sewerage systems were inefficient, in part because the city lay adjacent to and higher than the river from which it received its water supplies,

which meant sewerage could easily, and naturally, find its way down to the river, polluting the remaining source water. Growth in the city's population exacerbated the situation; from fewer than 20,000 people in 1844, the population in metropolitan Adelaide roughly doubled by 1860 and was close to 100,000 by 1880.

Within 20 years of settlement, human waste was being disposed of via six absorbing grounds that lay in the parklands; the waste made its way to these sites via gravity feeds and earthenware drainpipes.[17] By the mid 1860s, however, many homes had their own water treatment centres: an outhouse linked to a cesspit. Poor repair of the pits and irregular clearing meant it was not uncommon for excrement to run down city gutters. Even when cleared by nightwatchmen, the contents of the cesspits were carried in open carts and dumped in the olive plantations near the city gaol. By the late 1870s Adelaide was being called 'the stench capital of Australia'.[17]

Adelaide's citizens were becoming increasingly aware of public sanitation issues and of the infrastructure modern and sophisticated cities such as London were developing.[17] In the 1850s Adelaide's main thoroughfares were being paved with the Macadam system of crushed stone;[17] if done poorly (as with West Terrace) the roads quickly became mud in winter and dust in summer. By the early 1880s, however, granite or blue-stone kerbing was connecting drainage systems to the street, as well as to sinks in houses and factory outlets. The systems were occasionally flushed by hydrants, or naturally by stormwater, to assist in the removal of manure, offal and other factory and transport waste from the street, and into the closest watercourses. While this removed the worst waste from the streets, it merely flushed it to sites where it was more concentrated and putrid, thus degrading water quality in surrounding areas.

By the 1870s it was clear that Adelaide needed a comprehensive reticulated water supply and sewerage system to address quality supply issues, and to clean up the domestic and industrial wastes being freely dumped throughout the city. Mayor Edwin Thomas Smith expressed the dilemma of water quality and waste pollution in Adelaide at the time, citing a city mortality rate of 188 for every 1000 babies born compared to a figure of 126 in rural South Australia.

> The causes at work are, in my opinion, the usual sanitary disadvantages of populous towns and especially the impure atmosphere arising from our defective drainage, our foul-smelling water tables, and the enormous accumulation within the limited area of the city of night soil.[19]

While two Health Acts had unsuccessfully sought to resolve this situation, it was not until 1878 that the work on a more sophisticated underground sewerage scheme really began. Adelaide's grid-like layout, and the fact that water pipes had already been laid, meant each pipe had to be broken to lay the deeper drainage pipes. Nonetheless, by 1881 the installation of sewer mains and a sewerage farm at Islington meant Adelaide was the first Australian city with a complete

August Pelzer and irrigating Adelaide's parklands

BOX 34

As one of the longest-serving City of Adelaide gardeners (1899–1932), August Pelzer (1862–1934) was largely responsible for the design, development and planting of many of the parks and gardens in Adelaide's parklands and squares. Over time many of these parks and gardens have been modified but their existence is owed to Pelzer's vision and tenacity in their creation.

Many of these gardens and parks were created in the 1910s to 1930s when reticulated water was becoming readily accessible throughout Adelaide. Pelzer sought, as part of his development works, to install irrigation in newly created parks and gardens to reduce the volume of hand-watering and water carting his staff were undertaking, especially during summer and drought periods, to water new street and garden ornamental trees, as well as shrubs and lawns.

It would appear that the first irrigation piping that Pelzer laid in the parklands was in Creswell Gardens in 1926.

He installed 'the "Elder" automatic irrigation system' in Adelaide, altering his previous watering program (City of Adelaide annual report 1926, p. 40). The system:

… does away with delivery hoses. The taps are so placed that the water is evenly distributed over a large area. The sprinklers have a flexible nozzle attachment, and various shaped areas can be watered. It is claimed that on a moderate pressure they will distribute water over a larger area than any other device known. The large-sized sprinklers deliver 25 gallons [95 L] of water a minute (City of Adelaide, annual report, 1926, p. 40).

The irrigation system was installed in Pennington Gardens West in 1929 (City of Adelaide, annual report, 1929, p. 33).

David Jones

and modern sewer system – a major advance for the city's residents and at the forefront of comparable developments in other cities around the world.

Over time, a second influence on water availability became more evident as the city's population grew. By the 1880s the catchment area of the River Torrens began to deteriorate as settlement in the Adelaide Hills spread into the river's watershed. Increased erosion and polluted runoff from houses and farms loaded the river with sediment, nutrients and chemicals. Cattle and sheep were often allowed to graze up to the edge of water storages and runoffs, damaging plants and soil and depositing faeces. Learning from this experience, Happy Valley Reservoir (opened 1896) was sited further to the south of the city than the older reservoirs. During the 1920s, plentiful rain and postwar reconstruction led to expansion of agriculture in South Australia, including in the Adelaide Hills. Other developments such as the Mount Bold Reservoir (1937), which almost doubled Adelaide's internal catchment water supply capacity, and the pipeline from Mannum to Adelaide (1954), have continued to use the Hills for catchment and storage.

By Federation, Adelaide considered itself to have a water reticulation and wastewater disposal system quite superior to those in other Australian cities, and on a par with the best overseas. Measured by the standards of the day, this was not as farfetched as it may seem; it was undoubtedly true that the climate was still quite severe and the summers hot, but many public health issues involving water had been solved by an active government and a professional public service. A comparatively small population well serviced by significant infrastructure projects meant that a lack of water was not seen as a constraint in metropolitan Adelaide. Outside the city, however, water availability was still a major factor influencing the nature of agriculture, industry location and rural settlement.

The engineering response to water scarcity

As described by Marianne Hammerton,[14] historically South Australia has responded to water scarcity through engineering works. Despite the projects of the late 1880s, 1890s and early 20th century, water was again in relatively short supply by the time of World War I. In 1918, several summers of water restrictions in Adelaide, together with minimal supplies to the northern suburbs, saw the completion of yet another reservoir in the Torrens Catchment region, this time at Milbrook. This relieved the pressure on existing water shortages but only temporarily. Adelaide's metropolitan population constituted 51% of the state's population by 1921, the largest proportion of any Australian state; by the 1930s Adelaide's population had increased to 315,000 and restrictions were again introduced.

The inter-war years saw a number of additional changes affecting water flows in and around Adelaide. For example, the Mount Bold Reservoir was completed (1938) and the *Waterworks Act 1932* restricted human activity on Crown lands surrounding water storages – an important attempt to ensure the water catchment areas remained clean and unpolluted. Human activity was also changing the water flows on the Adelaide Plains. As settlement expanded along creek and river floodplains, the increasing runoff from built-up areas was channelled into creeks, causing increased flooding. In response, water authorities cleared streams of vegetation and straightened their courses, and creeks were essentially turned into drains.

One water supply option, not exploited until after one hundred years of settlement had passed, was from the River Murray. This source initially appeared too distant from Adelaide to be useful, although as early as 1839 special surveys were conducted to promote farming along its course. Until the early 1900s, the river was primarily used as an inland highway, with riverboats bringing wool and produce to Goolwa. A new initiative saw South Australia press for the coordinated construction of weirs and locks to regulate one of the most variable rivers in the world. It was not until the River Murray Waters Agreement of 1914, however, that serious progress was made in this regard. By this time the river was no longer seen as an inland highway but rather a source of water for irrigation. For South Australia, the agreement was instrumental in establishing a minimum flow of water into the state in normal years, and an agreed distribution between three states in drought years. This 'guarantee' was, in fact, a political compromise between the states, made with little reference to environmental limitations. Nonetheless, it was to prove instrumental later in the century when Adelaide linked up to the river for its water supply.

By the end of the 1950s, Adelaide's population was around half a million people; postwar immigration and a focus on industrial development meant the city's demand for water grew rapidly. Between 1945 and 1965, water consumption rose from around 35,000 to 120,000 ML/annum. Unsurprisingly, in the 11 years after 1942, eight years included water restrictions.

The response to these constraints on growth and shortages of supply was two-fold. One was the construction of more reservoirs in the few remaining possible sites, this time at South Para and Myponga. The second response was to expand in a completely new direction: instead of building additional catchments the decision was made to use the supplies offered by the River Murray. The Mannum to Adelaide Pipeline was built in 1954 to augment supplies in the reservoirs and ensure water storages in the Adelaide Hills would always have enough supply for the city's demands. So successful was this approach that for the next 20 years the majority of policy and management attention was directed towards improving the *quality* of Adelaide's water rather than taking steps to manage the *quantities* of water demanded or supplied. Implementing filtration and fluoridation of water became important ongoing projects.

The 1950s saw the growth of more intensive market gardening on the plains north of Adelaide. While the plain around the city's square mile had for many decades been notable for its mixture of farm land, villages and then suburbs, the 1950s saw a reshaping of these concentrations. Land for farm use was progressively pushed to the outer suburban fringes, and housing concentration increased

The dams of Adelaide

BOX 35

The flanking Adelaide Hills proved to be a critical resource for the city to capture and store water. After several false starts, considerable public apathy, and notable engineering disasters (such as the destruction of the newly completed Torrens Weir in the winter flood of 1858), the construction of Adelaide's first water storage was finalised in 1860. Thorndon Park Reservoir was completed just after the first public water supply scheme at Port Elliot was constructed. Twelve years later, a second storage project was developed to form the Hope Valley Reservoir. Besides water for drinking and washing, these initially abundant supplies also permitted Adelaide's citizens to create a green garden city that was much admired by visitors and helped project the city's 'sense of difference'. As the population grew and spread, however, it soon became apparent that water from these reservoirs could not reach residents in the eastern suburbs. Adelaide's urban development was being constrained by a lack of reliable water. In response, the government undertook the construction of a much larger Happy Valley dam, in part sourced from the Onkaparinga River. This was completed in 1897 with significant engineering skill.

To respond to the recurring constraint caused by an expanding population straining the water supply limits of earlier infrastructure projects, dam building around the state became an important activity for the Engineering and Water Supply Department for over a century. Notable water storage projects around the state included reservoirs at Beetaloo (1890), the Barossa Reservoir (1902), Bundaleer (1903), Ullabidinie and Ulbana (1916), Baroota (1921), and Tod (1922).

Despite the reservoir projects of the late 1890s and early 20th century, water was again in relatively short supply by the time of World War I. In 1918, several summers of water restrictions in Adelaide, together with minimal supplies to the northern suburbs, saw the completion of yet another reservoir in the Torrens Catchment region, this time at Milbrook. This relieved the pressure on existing water shortages, but only temporarily. Adelaide's metropolitan population constituted 51% of the state's population by 1921 – the largest proportion of any Australian state. By the 1930s Adelaide's population had increased to 315,000 and restrictions were again introduced. The response was Mount Bold Reservoir, completed in 1938, along the Onkaparinga River and used to maintain flows to Clarendon Weir and then the Happy Valley Reservoir.

Other storages were built including South Para (1958), Myponga (1962), Kangaroo Creek (1969) and Little Para (1979). All the catchments built in South Australia suffered from a common problem – they did not have the advantage of being built in large mountain ranges or in areas of high precipitation. Rather, local engineers used comparatively low-lying ranges in areas of low rainfall to their maximum potential, in an effort to produce more reliable water supplies. By the 1950s it was clear that most of the feasible reservoir sites around Adelaide had been utilised; more recently, the limitations of South Australia's topography and natural rainfall have become even more evident as the existing storages have proved barely adequate to sustain Adelaide through longer periods of low rainfall. Even now, the water held in Adelaide's reservoirs is only enough to meet the needs of the city for one year.

Martin Shanahan and Sara Hughes

as new suburbs grew to 'fill in' the gaps closer to the city. Water supplies for the northern market gardens had been drawn from underground aquifers for many years, with little restriction. Increased population and the need to balance competing demands saw first a marked drop in aquifer water levels, which necessitated the prescribing of the *Underground Waters Preservation Act in 1967*. Restrictions on water use from this source began in the 1970s. Similar issues of competing use for water began to emerge in the southern suburbs, although not until the 1970s.

Engineering and environmental responses to water scarcity

The 1970s saw qualitative changes in the management of water issues in Adelaide. Increasing population pressure, ageing infrastructure, decreasing water quality, the need to control flooding, and the demands of a population seeking better standards in water quality and its management gradually altered the actions of the authorities. Through the 1980s and 1990s, increasing attention was given to managing water in ways more aligned with the local environment, more awareness was given to the watercycle, and steps began towards a more integrated approach to water management. While prescription of the region's water resources began in 1967 with groundwater in the Northern Adelaide Plains, by 2004 it included all water resources in the Western Mount Lofty Ranges.

The growth in population and expansion of the urban landform meant low-lying and flood-prone areas were turned over to housing, which was by now a serious issue for a significant portion of the metropolitan area. In the catchments of the River Torrens/*Karrawirra Parri*, Para, Onkaparinga, the 'five creeks' of the River Torrens/*Karrawirra Parri*, and Sturt River/*Warriparri*, irregular flooding and ponding continued, despite many of their watercourses being re-engineered into concrete drains. Many of the western suburbs between the CBD and the sea are located on low-lying and originally marshy land. Again, an engineering response to the flooding problem was initially favoured, with the construction of stormwater drains and the diversion of runoff into the sea. At West Lakes, the developers' immediate plan was to drain the land, like at Grange, Fulham and Henley Beach, and to create a recreational waterbody as part of an asset for the development project. Despite this development creation, an extensive series of saline and freshwater swamps, relatively untouched by development, were destroyed through drainage and landscape remodelling.[20] Still more streams were converted to drains, perhaps the most notable being the Sturt River in the south-western metropolitan area. While such schemes initially alleviated the most pressing problems of flood-prone areas, by the early 1980s it was deemed necessary to assign powers to local government to manage floodplain issues and watercourses in better ways.

Flood forecasts became part of the process by which flood-prone areas were managed, along with public works to raise bridges and structures, such as the Kangaroo Creek Dam wall. There was still more clearing of creek and river channels to widen the path for flowing water. Unlike during previous decades, however, there also began to emerge a heightened awareness of the impact such changes were having on the surrounding flora and fauna. Effort was put into ensuring a more 'natural' watercourse environment; for example, cement-lined channels started to be eschewed in favour of some degree of natural habitat restoration.

In the Para Catchment, erratic stormwater flooding was occuring even before the development of the Para suburbs. This prompted the Salisbury Council to search for engineering solutions to address and minimise the water damage. The scenario proposed was unconventional but has proved a precedent in Adelaide for stormwater management and water quality purification, and demonstrates the viability of a system that can harness stormwater and craft a recreational resource as well as store water. 'The Paddocks' wetlands stormwater management and park system model has shaped subsequent projects at Greenfields, Barkers Creek Inlet and St Andrews Farm wetlands, and at Mawson Lakes. Technical knowledge gained from these projects underpinned subsequent stormwater management projects in Adelaide, and represents a major shift in engineering thinking, with the environment being accommodated for rather than conquered.

Attention was also being turned to address erratic flooding in the River Torrens/*Karrawirra Parri* watercourse corridor. The eventual scheme was to recraft the watercourse into an integrated stormwater management system, and disguise this through the creation of a greenway or linear park and recreational corridor, the result of which remains today as the largest integrated metropolitan stormwater management system in Australia.

The city's increasing population, together with the presence of township settlements in the Adelaide Hills (the major catchment area), and reliance on River Murray water combined to make delivery of clean, clear water of acceptable standard a constant challenge. The use of fertilisers and pesticides in agriculture and unfiltered sewage from Hills' townships affected the runoff water quality, while the declining quality of Murray water meant water pumped into Hills' storages was often degraded. In the early 1970s the Waterworks Act was amended to preserve water catchment quality throughout the Adelaide Hills by imposing stricter controls on development and disposal activities. By the mid 1980s, continuing deterioration of water quality meant these controls were made substantially stricter, and by 1985 land division outside defined townships in the watershed of the Adelaide Hills was prohibited, helping to minimise runoff impacts on surrounding streams.

In the early 1970s, Adelaide held around 800,000 people and by the end of the millennium it contained over one million.[21, 22] The city's urban sprawl reached 80 km from north to south and around 25 km at its widest from east to west.[23] This footprint has a significant and dramatic impact on the catchment, distribution and disposal of water.

To meet the demand of these new water-users, the early 1970s saw the construction of a second major pipeline from the River Murray to Adelaide, this time from Murray Bridge

The idea of pumping water from the River Murray to Adelaide had been considered years before any work commenced. The period 1910–1920 saw rapid growth and drought put stress on Adelaide's water supplies. Several initiatives to conserve water were implemented, including appointment of inspectors to seek out prohibited water-use activities and stop new connections to the sewerage system. Installation of water meters increased and restrictions and penalties were imposed, resulting in much public debate, with one letter to the *Advertiser* in 1915 suggesting that given the example of Western Australia's Kalgoorlie Pipeline, the River Murray should be tapped for South Australia.

Administration procedures to allocate River Murray water were not yet established in 1915. Lock construction, necessary to regulate river levels and flows, had barely commenced. Even after the completion of the locks in the 1920s, the pool below Lock 1 at Blanchetown was unregulated; this section of the river is the closest to Adelaide and so the logical section to source water from.

It was recognised, however, that the lower reaches of the river would become extremely saline during drought, due to seawater intrusion. Moreover, River Murray droughts coincided with drought periods in the Mount Lofty Ranges. Barrages would prevent seawater intrusion to allow extraction of freshwater during such periods. This became part of the justification for the construction of the five barrages at the Murray Mouth in the 1930s.

The Mannum–Adelaide Pipeline

After World War II, Adelaide's water supply was becoming precarious. With post-war migration, the population grew from 379,000 in 1947 to 535,000 in 1958. Per capita usage was increasing rapidly and demand was approaching capacity of existing storages. Availability of resources and manpower were low, and there was concern about the cost of expanding supply. Materials such as cement and steel were scarce, coal-fired electricity expensive and restricted, and the cost of lost irrigation production due to withdrawal of water from the Murray-Darling Basin (opportunity cost) was a further concern.

Between 1947 and 1954 there were water shortages and restrictions. In 1947 the *Advertiser* reported that a number of buildings burnt down due to a lack of water to fight the fires. In 1949 water restrictions commenced and were applied every year thereafter until 1954. To compound matters, Hugh Angwin, engineer-in-chief of the relatively new Engineering and Water Supply Department, died in 1949, with Julian Dridan taking over

the role. Within two years the Korean War began, coinciding with high inflation and the Commonwealth limiting loan funds to the states. Nevertheless, construction of a reservoir on the South Para River was proposed to augment Lower North supplies. A pipeline from the River Murray to Adelaide was proposed as insurance against shortages. The success of the barrages in mitigating effects of the 1944–1945 drought was a crucial factor in this decision.

Work on the pipeline began in 1949 but progress was slow due to funding and resource constraints. The EWS 1953–1954 annual report noted 'the concurrent execution of too many works, with too great a spreading of the available resources and a consequent unsatisfactory rate of progress in each undertaking.'

Major construction of the Mannum–Adelaide Pipeline was completed in 1954, costing £1.4 m (~$40 m in current values) with a capacity of 66,000 ML/year. Officially opened by Premier Playford on 31 March 1955, the total length of the mains was 117 km, with 29 km of rising (pressure) main from Mannum to 'summit storage' at Tungkillo, 30 km of gravity main from summit storage to 'terminal storage' near Highbury, and a further 58 km of distribution mains to connect to the existing water network.

Three pumping stations with an initial combined power capacity of more than 19,000 kW (now 30,000 kW) were installed to lift the transferred water nearly 460 m. The major components of the pipeline are 1150–1450 mm in diameter, made of mild steel with a concrete inner lining. The pipeline is mainly above ground, supported on steel rocker supports on concrete pedestals, and incorporates expansion joints to promote flexibility. This design minimises forces on the pipeline supports, reducing their size and minimising the use of cement.

Supporting rural communities along the way, the pipeline's success was almost immediate, with Adelaide the only Australian capital city to avoid mandatory restrictions over the 1954–1955 summer. The current configuration for the pipeline flows through the Anstey Hill Water Filtration Plant; it can also discharge into Millbrook, Kangaroo Creek, Hope Valley, Little Para, Warren and South Para reservoirs. The pipeline's importance to Adelaide's water supply has increased with utilisation averaging around 40% of monthly capacity, reaching up to 95% on occasion.

John Murphy

to the Onkaparinga River near Hahndorf. With perceptions that water supply reliability had improved, attention was again turned to water quality. The first metropolitan water filtration project was constructed at Hope Valley (1977), followed by other large-scale filtration projects including Anstey Hill (1980), Barossa (1982), Little Para (1984), Happy Valley (1989), and Myponga (1993).[24]

The opening of the Bolivar treatment works in 1965/1966 meant that the sewerage system covered almost 100% of Adelaide; no other Australian city had more than 75% of its population linked to a sewerage system at this time.[25] Nonetheless, sewage effluent was still being disposed of in Gulf St Vincent, which, it was realised, was having a negative impact on the surrounding marine environment. (It has been estimated that up to 9000 ha of seabed grasses have been lost.[26]) Similarly, the use of drains and concrete channels to divert stormwater meant significant quantities of chemicals and pollutants are washed off roads and other hard urban surfaces to the sea, with no filtering taking place.

While some 'solutions' to water problems continued to damage the environment, other development proposals, although ultimately aborted, demonstrated that at least some South Australians were beginning to change their approach to water issues.

In the 1970s, concern over population growth turned planners' attention to the construction of a new development at Monarto – the semi-arid wheat fields between Mount Barker and Murray Bridge. From the outset, the designers and planners of the Monarto Development Commission concluded that any design had to be low-water tolerant. While the project was ultimately shelved, the essential research knowledge was not lost and was transferred to the design and development of arid settlements at Leigh Creek in the Flinders Ranges and at Yulara village at Uluru–Kata Tjuta National Park.

A second unfulfilled development was for a Multi-Functional Polis (MFP), a unique information technology-rich city. The proposal envisaged the design and development of an extensive suburb laid across 'a crescent of degraded and contaminated land between Gillman and Dry Creek', including Barkers Creek Inlet and the Cheetham saltpans. The design sought to work with the saline and freshwater drainage system, harnessing it to service the suburb and provide a unique aesthetic quality and character. It was also to provide an aquatic cooling and energy regime without draining or substantially modifying the landscape. Although ultimately unfulfilled, the notion of 'bridging' with water and engaging with this type of water-based landscape was translated into many of the design and planning ideals that underpinned the subsequent Mawson Lakes development.[27]

Despite such examples, other developments that did continue were less benign in their impact on water resources. Mount Barker as a settlement is an example uncontrolled peri-urban development, resulting in compromised dairy pasture land and detrimental runoff impacts on the catchment watercourse.[20]

Increasing environmental awareness as a response to water scarcity

Despite the reality that most commercial developments are conservative in approach and prefer established rather than innovative approaches to water problems, the 1990s saw an increasing awareness of the need to live with water scarcity and a willingness to consider environmental issues when dealing with water problems.

Following the success of some of the projects listed above, wetlands and 'soft' watercourses have begun to be favoured over cement-lined drains to control floods. Increasing awareness of the need to replenish underground aquifers meant 'open' areas were favoured over hard surfaces in some developments, in order to allow water to seep through the ground and replenish the aquifers. Most importantly, an increasing awareness of the need to tailor water expectations to the characteristics of the natural environment began to emerge in the population. Public gardens were more frequently planted with native species. Although initially restricted mostly to 'enthusiasts', native gardens began to be seen as a sensible option in a climate with low natural rainfall. At the same time, however, improved reliability of central water supplies meant that, paradoxically for most households, 'old fashioned' water conservation methods, such as rainwater tanks and the recycling of household water, fell from favour.

A good example of these changes has been the infrastructure improvements made to the treatment of wastewater. Four wastewater treatment plants (WWTPs) in Adelaide now handle approximately 95,000 ML of wastewater each year: Bolivar, Port Adelaide, Glenelg and Christies Beach supply treated water for a range of uses including growing vegetables, irrigating parks, and grape growing in McLaren Vale. Increased awareness of the toxicity of poorly treated effluent into the marine system has seen changes in the processes by which wastewater is treated and discharged, while stricter standards for the disposal of water and solid materials have been established, and research in the coastal areas of the city continues to improve management. Advances can also be seen in the way people use water in their homes.

Developments in the northern areas of Adelaide continue, with new approaches to water conservation and management being progressively introduced. For example, in a 4000 home development at Mawson Lakes, water from Bolivar and from stormwater collected in wetlands in Salisbury is recycled and treated to supply water for toilets, gardens, and car-washing as well as public parks and reserves. The water is delivered via a separate piping system. Such integrated systems, which also reduce the quantity of stormwater and pollutants flowing into the gulf, are seen as models for future water use.[28]

Many of these developments are the result of a 're-learning' of natural processes. The Paddocks in Salisbury was initially developed to control stormwater flows and to take advantage of under-utilised land. The impact on flora and fauna was soon appreciated, while the successful filtration of heavy metals and pollutants came as a bonus.

The Murray Bridge–Onkaparinga Pipeline

Following the completion of the Mannum–Adelaide Pipeline, work continued on the dam construction program and by 1962 the South Para and Myponga reservoirs had been completed.

By this time, reservoir construction for Adelaide was nearing full development. The spillway for Mount Bold was raised and a further reservoir was constructed at Kangaroo Creek in 1969, ultimately permitting the maximum use of the Mannum–Adelaide Pipeline. Demand was nevertheless expected to increase, and plans commenced in the early 1960s for an additional pipeline from the River Murray.

The government approved plans for the pipeline, following the recomendation of a parliamentary standing committee on public works in 1967. The Engineering and Water Supply Department undertook the design and most of the construction. Pipelaying commenced in 1969 and proceeded at a rate of nearly 300 m per week. Construction activity peaked over the period 1970 to 1972 with a combined workforce of 410 people.

The pipeline was constructed for $22 million, as reported on commissioning (about $150 m in current values), with a stated capacity of 164,000 ML of water per year. The pipeline was officially opened by the Minister of Works, Des Corcoran on 12 October 1973.

The pipeline extends some 48 km, using three pumping stations (at the time using some of the largest pumping units in Australia), with a total installed power of 45,000 kW. No. 1 pumping station is located over the River Murray just upstream of Murray Bridge; no. 2 is nearly 7 km to the west, 60 m above the river; and no. 3 is a further 19 km to the west (just north of Kanmantoo) and nearly 120 m higher again.

Water is lifted a total of 410 m over the first 38 km of the pipeline, and then gravitates nearly 2 km to 'summit storage' (a 490 ML earthen dam located to the north of Littlehampton). A buried pipe outlet from summit storage transfers water a further 8 km to discharge into the Onkaparinga River just west of Hahndorf. Water transferred to the Onkaparinga falls 110 m and has considerable energy when released, which is dissipated through specially designed valves. The original pipeline concept considered the possibility of installing a small-scale hydroelectric generator at this location to remove this energy and offset power costs. At the time, though, this was assessed as uneconomic.

Once released to the Onkaparinga River, water travels a further 10 km downstream to Mount Bold Reservoir. Water released from the reservoir flows to the Clarendon Weir where an aqueduct diverts it to the Happy Valley Reservoir, which in turn supplies the Happy Valley Water Filtration Plant. This plant provides nearly half of Adelaide's supply.

The pipeline is predominantly 1650 mm in diameter and is of mild steel with a concrete inner lining. Most of the first 38 km rising (pressure) main section of pipeline is constructed above ground, and, like the Mannum–Adelaide Pipeline, the above-ground section is non-rigid, with regular expansion joints along its length.

The pipeline is supported every 18 m by flexible slings mounted on concrete pedestals. The slings have bearings to allow for expansion and contraction of the pipe under normal temperature fluctuations. The pipeline design also allows for future expansion to bring the capacity to 200,000 ML per year.

Utilisation of this pipeline is not as high as that of the Mannum–Adelaide Pipeline and has averaged around 20% of monthly peak capacity over recent years, peaking in one month at nearly 70%.

The major pipelines to Adelaide are an essential element of the water supply system. Initially established as insurance against low yearly runoff to local catchments, the pipelines have become an essential and ongoing source of water, on occasions providing up to 90% of Adelaide supplies. The importance of these key assets will continue well into the future, only to be limited by the reliability of the supplies from the River Murray itself. Current work on the pipelines includes modifying the intakes to the pumping stations on the river to ensure water can be extracted during the low levels associated with the present drought.

John Murphy

The project was advanced in the 1990s by storing winter flows of stormwater in an underground aquifer for summer use; not only did this begin to restore aquifers depleted by agriculture, it increased awareness and management of significant underground water stores.

This council area now has more than 36 wetlands of approximately 250 ha, which serve to manage stormwater flows, restore underground water stores, filter pollutants, and encourage flora and fauna. By minimising polluted water flowing into the Barker Inlet (a nearby marine estuary), the wetlands projects have had a measurable impact on restoring marine wildlife and mangroves. The use of wetlands is now part of the development process required by this council for residential and industrial projects.[29]

Cooperation between government and industry saw Australia's largest wool processing company, G.H. Michell and Sons, integrate its heavy demand for water resources with a recycling process, involving reedbeds and an aquifer storage and recovery system. The result is a system that produces water of higher quality than that taken from the Murray.[30] Unfortunately, this project is still notable as being an exception rather than the rule.

Long-term depletion of aquifers in the Northern Adelaide Plains groundwater basin was also partially addressed by the construction of the Virginia Pipeline, which supplies 15,000 ML of reclaimed water for irrigation annually from the upgraded Bolivar WWTP. The scheme is one of the largest reclaimed water schemes of its type in Australia.[31]

While these projects are examples of significant and important changes in approach, and they set standards for future development, they are also noteworthy because they are still relatively novel. The need to significantly alter Adelaide's approach to gathering, distributing, using, cleaning, and recycling water has become more apparent in recent years. By the turn of the millennium, the supply of water became an issue once more as the city's residents became aware of the relative scarcity of clean drinking water. By 2004 it was clear that water was again a significant constraint on development. In the middle of a continuing drought, the worst since 1900, and in the shadow of predictions of significant climate change that would reduce Adelaide's water supplies further, the government undertook a major study, 'Water Proofing Adelaide'. This project, while marking a turning point in community awareness in the modern era, repeated a theme that has been recognised several times since European settlement. But while previous water crises sparked an engineering solution, this time the government looked for answers using both engineering solutions and public education. In combination with proposals for a desalination plant and upgraded water infrastructure, several education programs have been established to increase the community's awareness of water scarcity and the importance of the Murray River. Programs for school children and the construction of interpretive centres to demonstrate the feasibility and need to take an integrated approach to urban water have been instituted. Furthermore, the government has decided to use the price of water to reinforce the message of water scarcity and so

modify residents' demand. Recent increases in the sale of household rainwater tanks and dual-flush toilets, and restrictions to garden watering (although promoted by regulation), are external examples of residents' increasing acceptance of the scarcity of water and of the individual's responsibility to use it carefully.

Conclusion

Despite shifts in attitude and changes in approach, the issues surrounding water supply in Adelaide continue to pose difficult challenges. A example is the regular occurence of toxic algal blooms on the River Torrens/ *Karrawirra Parri*. The river, which remains close to the spiritual, historical and recreational heart of Adelaide's cultural life, is regularly poisoned by this growth in summer, a signal of a water supply under stress. For several years, technical responses, such as installing mechanical agitators or applying chemical treatments, have been tried. Slowly, after all engineering approaches have proved inadequate, attention has turned to the adoption of more environmentally sympathetic approaches.

At the current time, Adelaide is again returning to a period where the supply of water is perceived as a constraint on development. Unlike the previous periods in the 1840s and 1870s, and the 1930s and 1950s, there is now little scope to capture additional flows of freshwater. As the pressure on water resources increases, even the River Murray is seen to be a source with limited reliability. In 2007 it was announced that a desalination plant was to be built to supply a portion of Adelaide's future water needs. Importantly, however, this time the engineering solution was not viewed as the final answer; rather, the plant was seen as *part* of a more comprehensive and integrated approach. While this included efforts to educate the public about water use and to modify demand, it also saw the beginning of a far more significant debate on constructing better systems to take advantage of the watercycle – in effect, using water more than once.

Regardless of the outcome – whether freshwater is created from the sea, or if the capturing of storm flows and the recycling of wastewater is greatly expanded – it is clear that water will become more expensive. Its relative scarcity for Adelaide residents is a historical trend that will continue to be an issue into the foreseeable future. While this presents a challenge for all Adelaide residents, it may also allow for creative solutions.

In 2004, Adelaide's then Thinker in Residence, Professor Peter Cullen, leader of the Wentworth Group of Concerned Scientists, which produced the *Blueprint for a Living Continent*, noted that it was impossible to drought-proof Australia; extreme variability in rainfall, extensive droughts, and significant floods were characteristic of the continent. One of the Wentworth Group's recommendations, particularly pertinent to Adelaide, was that we need to learn to live *with* the landscape rather than continually fight against it. It would appear that it has taken us 150 years to learn this lesson – something the original Kaurna inhabitants along the banks of the River Torrens/*Karrawirra Parri* knew well.

Multi-Function Polis: a city in a waterscape

BOX 38

In 1989 the Commonwealth government invited state submission for the location and development of an Australian MFP city. The Multi-Function Polis (MFP) was envisaged as a 21st-century city integrating technopolis, biosphere and renaissance city themes to host a community comprising residential, business, educational and recreational elements. In May 1990, the South Australian government selected Sydney-based architects Edwards Madigan Torzillo Briggs to prepare a detailed submission using the Barkers Creek Inlet/Gillman/Dry Creek crescent landscape as a venue. A challenging brief sought the transformation of salt fields, marshlands and an urban wasteland, surrounded by visually detractive and polluting industrial activities and complexes into a polis that possessed environmental and social responsibility, and particularly environmental renewal, sustainability and replenishment.

A robust urban design strategy was prepared that enabled settlement organising principles draped across the landscape, including the adoption of field village edge installations and the visual dominance of three introduced thematic axes. Project leader Lionel Glendinning explained this different approach to city planning:

The difference is really an issue of process. Rather than 'planning thinking', we have worked from 'design thinking'. That is the difference between two-dimensional reality and three-dimensional reality. In doing this, we are accepting that complexity and plurality are essential and therefore must be addressed in design terms. The stimulus and interaction which is necessitated in the public realm is an enriching experience which, when achieved, provides layers of meaning and understanding as to how we live.

A core goal in this city proposition was the achievement of biosphere objectives. A biosphere community is one where natural resources are conserved and sustained by appropriate siting and development. Such communities seek to achieve high air and water quality standards; minimise air, water, solid waste and noise pollution; minimise natural resource, heritage and human-made feature impacts; improve degraded areas through innovative conservation and remedial techniques; conserve natural resources; and integrate human and nature in place.

It was a very challenging proposition to actually craft a polis on an ecologically sensitive saline wetland environment with high avifaunal habitat values. The South Australian government's brief inherently assumed that this landscape was a wasteland fit for transformation, had little biodiversity and water value, needed remedial attention to address flooding and water quality, and could support and accommodate an expression of 'authenticity and modernity through an application of technical solutions to environmental, infrastructure, built, energy, agricultural and horticultural challenges'.

Further feasibility research by consultants Kinhill and Delfin, over a development core of 800 ha, pointed to residential densities of 30 dwellings per hectare to accommodate approximately 42,000 people, in addition to some 416 ha being allocated to lakes and canals and 624 ha to forests and open space.

The combination of an economic recession, the collapse of the State Bank, and a change of Commonwealth government resulted in the project being abandoned in 1997. Despite this, the technical and design legacy of investigations for the MFP, bearing echoes of the Monarto city experiment, were re-translated by Delfin into the successful Mawson Lakes project, including much of the biodiversity ethos and objectives.

David Jones

References

1. M. Davies, 'Establishing South Australia' in Statham, P. (ed.) *The origins of Australia's capital cities*, Cambridge University Press, Cambridge, 1989, pp. 159–178.
2. R. Foster and T. Gara, 'Aboriginal culture in South Australia' in Richards, E. (ed.) *The Flinders history of South Australia: social history*, Wakefield Press, Adelaide, 1986.
3. A. Hutchings, 'With conscious purpose: a history of town planning in South Australia' (2nd edn), Planning Institute of Australia (South Australian Division), Adelaide, 2007.
4. South Australian Association, *Outline of the plan of a proposed colony*, Ridgway & Sons, London, 1834.
5. C. Sturt, *Two expeditions into the interior of Southern Australia, vol. 2*, Smith, Elder and Co., London, 1833.
6. Colonization Commissioners for South Australia, *First report*, London, 1836.
7. W. Light, *A brief journal of the proceedings of William Light, late surveyor-general of the province of South Australia: with a few remarks on some of the objections that have been made to them*, Archibald MacDougall, Adelaide, 1839.
8. G. Kingston, letter to the editor, *South Australian Register*, 21 May 1877.
9. Royal Geographical Society of Australasia, 'Extracts of the diary of G. Stevenson', *Proceedings of the Royal Geographical Society of Australasia (South Australian Branch)*, 30, n.d., pp. 21–23.
10. D. Jones, 'Adelaide parklands and squares cultural landscape assessment study', section 3: Tulya Wodli/Park 27 report, Adelaide Innovation and Research, Adelaide, 2007 (online).
11. D. Jones, 'Adelaide parklands and squares cultural landscape assessment study (Karrawirra report)', Adelaide Innovation and Research, Adelaide, 2007.
12. F.R. Nixon, *Adelaide Magazine*, 1 February 1846.
13. Anon, *Observer*, 25 May 1843.
14. M. Hammerton, *Water South Australia: a history of the Engineering and Water Supply Department*, Wakefield Press, Adelaide, 1986.
15. F. Sinnett, *An account of the Colony of South Australia*, W.C. Cox, Government Printer, Adelaide, 1862.
16. P. Nagel, *A social history of North Adelaide 1837–1901*, P. Nagel and T.J. Strehlow, Adelaide, 1971.
17. P. Morton, *After Light: a history of the City of Adelaide and its councils, 1878–1928*, Wakefield Press, Adelaide, 1996.
18. N. Gerges, 'Overview of the hydrogeology of the Adelaide metropolitan area', Department of Water, Land and Biodiversity Conservation report 2006/10, Adelaide, 2006.
19. E. Smith, 'Lord Mayor's annual report', City of Adelaide, 1880.
20. C. Forster and M. McCaskill, 'The modern period: managing metropolitan Adelaide' in Hutchings, A. (ed.) *With conscious purpose: A history of town planning in South Australia*, Planning Institute of Australia, South Australian Division, Adelaide, 2007.
21. J.B. Hirst, *Adelaide and the country 1870–1917: the social and political relationship*, Melbourne University Press, Melbourne, 1973.
22. T.L. Stevenson, 'Population change since 1836' in Richards, E. (ed.) *The Flinders history of South Australia*, Wakefield Press, 1986.
23. C.J. Tait and C.B. Daniels, 'Introduction' in Daniels, C.B. and Tait, C.J. (eds) *Adelaide, nature of a city: the ecology of a dynamic city from 1836 to 2036*, BioCity: Centre for Urban Habitats, Adelaide, 2005.
24. SA Water, 'Water treatment plants', 2008 (online).
25. SA Water, *SA Water: celebrating 150 years*, Government Printers, Adelaide, 2006.
26. Department of Environment and Natural Resources, 'South Australia: our water, our future. Sustainable management', Adelaide, 1995.
27. S. Hamnett and A. Hutchings, 'The era of strategic planning' in Hutchings, A. (ed.) *With conscious purpose: a history of town planning in South Australia*, Planning Institute of Australia, South Australian Division, Adelaide, 2007.
28. SA Water 2008, 'Mawson Lakes Recycled Water System', 2008 (online).
29. City of Salisbury (online).
30. I. Moyle, 'Water projects in the City of Salisbury', *SA Geographer*, 21 (3), 2006.
31. SA Water, 'Virginia Pipeline Scheme', 2008 (online).

Box references

Box 30: A letter by Colonel William Light to George Jones
1. Government of South Australia, 'Letter to George Jones RA', 2005 (online).

Box 31: Water-carting
1. H.T. Burgess, (ed.), *The cyclopedia of South Australia, vol. 1*, Adelaide, 1907.
2. Engineering and Water Supply Department, 'Water resources development 1836–1986: the historical background to the Water Resources Management Strategy', 1987, Adelaide, 1987.
3. D. McCarthy, T. Rogers and K. Casperson, *Floods in South Australia 1836–2005*, Australia Bureau of Meteorology, Canberra, 2006.

Box 32: The Mitcham Waterworks
1. G.F. Angas, collected papers, State Library of South Australia, Adelaide, n.d.
2. J. Morphett, 'South Australia: latest information from this colony', London, 1837.
3. J. Allen and D. Francis, 'Emigrant's friend or authentic guide to South Australia', London, 1848.
4. No author, *Adelaide Times*, 1 May 1850.
5. No author, *Register*, 13 November 1847.
6. State Library of South Australia, 'Manning index of South Australian History' (online).
7. No author, *Register*, 11 November 1848.
8. No author, *South Australian Government Gazette*, 9 December 1858.
9. D. Lane, P.A. Smith, M. Ragless and A. Ash, 'The Mitcham Water Works 1879–1930' in Smith, P.A., Pate, F.D. and Martin, R. (eds) *Valleys of stone: the archaeology and history of Adelaide's Hills Face*, Kōpi Books, Belair, South Australia, 2006, pp. 17–28.

CHAPTER 8

The River Torrens is also known by the Kaurna name *Karrawirra Parri* (see chapter 6 for a detailed discussion). However, this chapter is a discussion of the river since European settlement, and the author uses European nomenclature in order to remain concise and be consistent.

Introduction

The development of agricultural and urban settlement within the River Torrens Catchment since 1837 has progressively altered its hydrology, channel morphology and water quality. This chapter traces the 'unnatural history' of these alterations by documenting the events and processes that have converted the natural stream, chosen by Colonel William Light to be the centrepiece of the City of Adelaide, to the artefact of human impact and intervention it is today.

'A miserable dribbling current'

Australia's colonial capitals were founded during the age of the sail, and existed for decades without the benefit of water reticulation (Table 8.1). The imperatives for the sites of these settlements were the same as those described for the site of the City of Adelaide in the *Letter of Instructions by the Colonization Commissioners for South Australia to Colonel William Light, Surveyor General for the Province of South Australia, 9th March, 1836*:

> The commissioners are of opinion that the best site for the first town will be that which combines in the highest degree the following advantages:
> 1st: a commodious harbour, safe and accessible at all seasons of the year.
> 2nd: a considerable tract of fertile land immediately adjoining.
> 3rd: an abundant supply of fresh water ...

Colonel Light's reasons for choosing the site he did have been discussed and debated in numerous publications; the *results* of his choice were not what the colonization commissioners had envisioned. Light's plan for the city placed it on the north and south banks of a narrow, mainly shallow, intermittent stream, approximately 10 km inland from Holdfast Bay where Adelaide's first settlers had landed in 1836 (see Figure 1.1). The stream did not terminate in a 'commodious harbour, safe and accessible at all seasons of the year', but in the Reedbeds on the Cowandilla Plains, a low-lying area of winter-flooded wetlands, shallow open-water ponds, and scattered islands of higher ground separated from the sea by sand dunes on the coast of Gulf St Vincent.[3] The Cowandilla Plains drained northward into the river estuary that was to become the main port for Adelaide, although it was more than 12 km by road north-west of the city (see Figure 1.1).

Some of Adelaide's first settlers found the stream that Colonel Light named the River Torrens aesthetically pleasing; others disparagingly described it as a 'miserable dribbling current with an occasional waterhole'[1] and an 'unnavigable country brook ... deep in one part and shallow in another'.[2] Most settlers had expected Adelaide to be situated on a navigable river providing a perennial water supply, and were disappointed with the reality of the Torrens. The variability from season to season and year to year in rainfall, runoff and streamflow in the Mount Lofty Ranges and on the Adelaide and Cowandilla plains was baffling to colonists, who were conditioned to the regular progression of northern European seasons. In winter the River Torrens fords were often impassable; in summer the steep riverbanks within the city created a 'very formidable dry ditch', hindering passage from south Adelaide to North Adelaide and to the suburbs

Chapter photograph: River Torrens looking east, c. 1900.
Image courtesy of the State Library of South Australia, SLSA: B 26486.

Table 8.1 Australia's capitals – harbours and freshwater

Foundation	Site (initial freshwater source and navigable sea access)	Establishment of reticulated water supply
Sydney, NSW, 1788	Tank Stream Port Jackson (Sydney Harbour)	Lachlan Swamps via Busby's Bore (1826), limited reticulation 1844
Hobart, TAS, 1804	Hobart Rivulet Derwent River estuary	Lower Reservoir 1861
Perth, WA, 1829	Wetlands and Claise Brook Swan River estuary	Victoria Reservoir 1891
Melbourne, VIC, 1835	Yarra River Port Phillip Bay	Yan Yean Reservoir 1857
Adelaide, SA, 1836	River Torrens No navigable sea access	Thorndon Park Reservoir 1860
Brisbane, QLD, 1842 (free settlement)	Wheat Creek Brisbane River to Moreton Bay	Enoggera Dam 1866

Table 8.1 Foundation of Australia's colonial capitals – harbours and freshwater.

developing along the road to Port Adelaide.[4] For much of the year the supply of freshwater available from the Torrens was not sufficient to meet the needs of even Adelaide's first few thousand inhabitants; while those settlers who had purchased land with river-access for market gardens and orchards to the east and west of the city were dismayed by the Torrens' transportation and irrigation limitations.

The settlers' disdain for the Torrens was echoed by Adelaide's first newspaper. In February 1838, the *Register* observed

> *what the River Torrens may be capable of performing for a week or two of the rainy season beyond sweeping down to the swamp the summer filth of Adelaide we cannot guess; but the Torrens at other times is not a river at all, but merely a chain of fresh water pools. At the present moment, its running water may be spanned with the hand and sounded with the forefinger.*

In September 1844, the 'miserable dribbling' River Torrens displayed another side to its nature when late-winter rains caused the first of many floods on the Adelaide Plains, destroying a brewery to the east of the city and drowning two women attempting to ford the river at Klemzig. More significant floods occurred in July, August and September of 1844, marking the beginning of a pattern of repeated flooding, with loss of property and lives in Adelaide and the Torrens Catchment to the east of the city (see Figure 1.1). Fortunately, the most extensive and frequent flooding occurred downstream from the city in the western portion of the Torrens Catchment, which remained sparsely settled until the 1940s.[4]

During the early years of settlement, all the streams supplying freshwater to Australia's colonial capitals were also used as open sewers and waste dumps. Winter flows in the River Torrens were expected to sweep away the accumulated 'summer filth of Adelaide' dumped directly in the stream channel or fed into the channel by the city's (mainly open) drains and the river's tributary creeks (see Figure 1.1). The inadequacy of the Torrens in this respect earned Adelaide the title of the 'stench capital of Australia', which it retained until the completion of deep drainage and an effective sewerage system.[5] From January 1881, all the sewage that had previously run into the Torrens from south Adelaide was taken by main sewer to a sewage farm established near what was to become the location of the Islington Railway Workshops. All of North Adelaide was connected to this system by 1883.[6]

However, after the construction of Adelaide's sewerage system, agricultural pollution (e.g., runoff of manure from fields and from chicken and pig farms), seepage from cesspits, inflow from household drains, and direct dumping of night soil and other wastes (including, from 1862 till the 1970s, wastes from the Adelaide Zoo), continued for many decades in other parts of the river's catchment. Even following the extension of sewerage throughout the Adelaide metropolitan region, sewage continued to enter the Torrens at Marden and Gilberton until the early 1980s from pressure-release valves in the main trunk sewer that was frequently overwhelmed by heavy rains.[7]

Within a year of settlement, industries were established along the Torrens at Hindmarsh (the first subdivision established outside the city in 1838), and continued to concentrate along the Torrens in Hindmarsh, Thebarton and Torrensville during the following decades, although some industries were also located along the river and its tributaries upstream from the city. The main sources of industrial pollution (other than that associated with extractive industries) were sawmills, flour mills, brickyards, wineries, distilleries, breweries, slaughterhouses, fellmongeries, and tanneries – most of which were located next to the river and its tributaries in order to utilise water for cooling, cleaning and waste disposal. Hotels also favoured waterside locations so stable sweepings could be washed into the river.[8]

Given the negative impacts of repeated flooding, an unreliable and insufficient water supply, and deteriorating water quality on agricultural and urban development it is not surprising that most of Adelaide's early settlers were dissatisfied with the Torrens and set about remaking it to better serve their wants and needs. This process still continues, ironically at present with attempts to *restore* some of the pre-settlement character of the river that past interventions have eradicated.

'Stupid people cut away the shrubs and trees'

Light's survey of Adelaide was completed on 11 March 1837. By December, the city consisted of some 300 tents, huts and houses, most of which were located on North Terrace, Rundle Street and Hindley Street, as close as possible to the Torrens. Some residents sank wells on their blocks but the water obtained was often brackish. Until the advent of water reticulation in 1860, the majority of residents relied on water-carters who used buckets to fill barrels from the river then hauled the barrels through the streets of the city, filling householders' wooden casks and iron tanks. Stock were also driven to the Torrens to be watered. These practices, which followed the spread of settlement east along the river and its tributaries and downstream to the sea, were responsible for the breaking down and eroding of the channel banks at numerous locations.

Additional channel bank erosion resulted from the settlers' insatiable demand for fuel and construction materials. During the first few years of settlement, timber cutting and firewood collecting virtually denuded the vegetation cover of the Adelaide parklands, including the River Torrens channel and banks, and then spread throughout the Torrens Catchment in the following decades (Figure 8.2).

> *The River Torrens as I first saw it in the winter of 1837 [wrote Helen Mantegani] was very pretty and picturesque; high and steep banks on either side, closely covered with beautiful shrubs of all sorts; splendid gum trees also were growing on the banks, and in the streams, overhanging the water which was narrow and deep, small fish were plentiful, and that strange creature the platypus was occasionally seen on its banks. But some stupid people cut away the shrubs and trees that still held the banks together. Consequently the soft alluvial soil fell away, and the*

BOX
39

Cox Creek Nutrient Mitigation Program

The Cox Creek Nutrient Mitigation Program in the Mount Lofty Ranges watershed represents a collaboration aimed at improving water quality and ecological health, and building community awareness. The combined efforts of three South Australian government agencies, six technical consultants, tertiary institutions, members of the Piccadilly Valley community and Scouts Australia has produced major on-ground catchment works within three years. This collaborative and technically strong program has set a standard for catchment improvement initiatives aimed at improving water quality in the Mount Lofty Ranges.

The Mount Lofty Ranges watershed

The primary drinking water sources for metropolitan Adelaide are the River Murray and the Mount Lofty Ranges watershed. In an average year, the 1600 km^2 watershed provides 60% of Adelaide's drinking water, but it is more than just a drinking water supply – it supports intensive agriculture, rural urban development, and a fragmented natural ecosystem.

Upper Cox Creek is a multiuse sub-catchment within the Mount Lofty Ranges watershed, which demonstrates the conflicts associated with such a diversity of land uses. The headwaters of Cox Creek flow through the Piccadilly Valley, an area of intensive horticultural activity. Further downstream, Cox Creek joins the Onkaparinga River, which flows into Mount Bold Reservoir, and, downstream, into the Happy Valley Reservoir. A major challenge for public and private managers of the watershed is to satisfy often competing water quality and land use needs – often leading to a water resource under pressure.

The upper Cox Creek Catchment (Piccadilly Valley) area has long been recognised as exporting a disproportionate amount of nutrient and sediment loads into the Onkaparinga River, resulting in poor aquatic health and significant algae management issues at Happy Valley Reservoir. Various government agencies have run programs in the area, particularly in the late 1980s and early 1990s. The investigation and extension efforts of the Primary Industries Department of the time and the CSIRO helped improve land management practices on a small scale, didn't result in sustained change at the landscape level. As a consequence, the Cox Creek water quality and river health have remained poor, according to more recent surface water monitoring and analysis.

Over 2003–2005, SA Water formed a partnership with the Adelaide and Mount Lofty Ranges Natural Resources Management (AMLRNRM) Board, the Environment Protection Authority (EPA) and Australian Water Environments to fund and develop nutrient management options for the area.

Feasibility studies

Several studies were completed to determine the most effective nutrient management option, including:

- a 'nutrient budget' for Happy Valley Reservoir (Ingleton, 2003); and
- a feasibility study for nutrient mitigation (Australian Water Environments, 2005, available online.

A 'triple bottom line' feasibility study assessed the most effective and efficient nutrient load mitigation solutions. The program partners engaged a consultancy team to identify the source and fate of nutrient load contributions, social and environmental benefits, the social and economic drivers influencing the adoption of mitigation options, and a detailed cost/benefit assessment of each option. The social aspects of the feasibility study included detailed community consultation, to help ensure the resultant recommendations were acceptable to the community and would provide ongoing social benefits.

The study concluded that engineering works (an in-stream wetland plus a sedimentation basin upstream of the wetland), along with improvements in land and watercourse management, could be implemented to reduce nutrient and sediment loads. No single solution was found to be fully effective, rather a range of water quality improvement initiatives needed to be implemented – thus the Cox Creek Nutrient Mitigation Program was born.

The Cox Creek Wetland

The concept design for the wetland and sedimentation pond was based on best practice management attributes of wetland design, including the provision of fish ladders, monitoring stations (to enable post implementation review), and comprehensive environmental construction methods, with project costs refined and anticipated water quality improvements modelled and verified.

The Cox Creek Nutrient Mitigation Program

The aims of the program overall are to:

- Derive and implement mitigation works to reduce nutrient exports from Cox Creek in order to reduce the frequency and intensity of cyanobacteria blooms at Happy Valley Reservoir. The program will result in the reduction of copper dosing with only one to two

dosings required prior to Christmas (rather than three or four), thus meeting the strategic management aim for the reservoir.

- To improve the general water quality of Cox Creek and thereby the aquatic ecosystem.
- To improve community awareness of water quality issues in the Mount Lofty Ranges watershed and develop a consultative process for determining the most effective long-term catchment management strategies.

The collaborative process used to achieve these aims was and remains a critical element of the success of the program.

Environmental benefits – water quality

Happy Valley Reservoir experiences severe annual episodes of cyanobacterial growth, requiring ongoing monitoring and recurrent operational intervention in the form of dosing with copper sulphate. This treatment is not only expensive – with significant primary costs, plus costs associated with disposal of the resultant contaminated alum sludge – it is not always effective. Importantly, the EPA, South Australia's environmental regulator, has reviewed the use of copper sulphate as an intervention for cyanobacterial growth and indicated they would like to see a move away from the practice in the long term. Therefore in this instance, a 'business as usual' approach can be seen as generally unsustainable.

Ecosystem benefits

The EPA and the AMLRNRM Board were concerned that nutrient and turbidity levels in the waterway were well above the Australia and New Zealand Environmental Conservation Council (ANZECC) guidelines and state environmental thresholds for inland waters, with the potential to impair ecosystem function and river health. The aquatic ecosystem downstream of the Piccadilly Valley will benefit from this program, by improving water quality and by slowing down water flows, thus reducing the eroding force of the catchment runoff.

Additional environmental benefits have also been achieved through this program. The AMLRNRM Board are investing both their own funds and those sought from the federal government to stabilise watercourses in the area. The value of this work was approximately \$200,000 over the 2005/2006 financial year, and work will continue into the future until the remaining priority actions identified in the feasibility study have taken place.

Community benefits

Since its inception, comprehensive community and stakeholder focus has been a significant feature of the program. This has been demonstrated through media releases, public flyers, public presentations, local landholder consultation, and developing positive relations with local members of parliament, the local council (Adelaide Hills Council), and Scouts Australia. Key educational and social elements have been included into the program, for example the Woodhouse Wetland design, which includes interpretative signage, walking tracks, contributions towards an environmental resource centre, wetland website links, and Waterwatch sampling.

'[The program provides] a model for cooperation between a major public sector instrumentality and a private Association … over 30,000 school children, visitors and Scouts will visit Woodhouse in the next 12 months – all will [have] access to the wetland. The wetland not only has significant water quality benefits for the State, but is also magnificently positioned as a "hands-on" educational resource for young people, families and the corporate community.' – Dan Ryan, CEO, Scouts Australia (SA branch)

Measures of success – watch this space!

The success of the wetland and the program overall will be determined based on water quality monitoring data, specifically collected for the program, in comparison with historic data and modelled predictions. Impacts at Happy Valley Reservoir will also be closely assessed, as will improvements in biodiversity and aquatic ecosystem health through ongoing monitoring by the EPA and Waterwatch participants.

By implementing the Cox Creek Nutrient Mitigation Program, SA Water is showing a move away from reactive 'intervention' management strategies and towards a proactive, longer-term solution to algal management. The project also provides benefits in terms of environmental outcomes within the catchment, community awareness of water quality issues, and shows commitment to investment in the long-term sustainability of source waters. It is intended that similar programs can be rolled out in other priority subcatchments in the Mount Lofty Ranges watershed.

Karla Billington

Figure 8.1: The River Torrens in the 1880s. Source: L. Henn & Co., Adelaide, 1882? Print: lithograph, hand col., 24.7 x 32.8 cm. Series: Rex Nan Kivell Collection NK9990 nla.pic-an 7832080v, National Library of Australia.

river became broad and shallow and very ugly. After this the winter floods carried away the banks that remained, making it a most unsightly spot for many years, and entailing an enormous expense to restore it to anything like beauty, though it will never be as picturesque as nature made it.[9]

Sandstone quarried from the south bank of the Torrens, within the city, was used in the construction of some of Adelaide's early public buildings. The river remained the main source of gravels, sands and clays for construction in Adelaide and its developing suburbs until the 1930s when there were 16 gravel-crushing and sand-washing plants discharging sludge into the Torrens within the eastern urbanised portion of its catchment upstream from the city. All of these plants had closed by the 1970s; in their later years they mainly treated material from quarries in the Mount Lofty Ranges and elsewhere, having exhausted the River Torrens' supply.[7]

Excavation of construction materials was concentrated on the inside of meander bends, where the Torrens channel was bordered by alluvial flats. The holes left when the excavations were abandoned made ideal rubbish dumps for the developing suburbs along the river. Dumps filled to capacity were normally capped with earth to form flat-topped mounds rising several metres above the river level. Channel banks then required reconstruction to at

least the angle of repose of the new mounds. This resulted in substantial narrowing of the river channel at several locations, causing accelerated erosion of the channel banks on the outside of meander bends.[7]

Because the normal process of erosion on the outside of meander bends usually creates a high bank bordering a deep pool, the conversion of gravel, sand and clay pits to rubbish dumps probably had a greater impact on public concern about swimming in the Torrens waterholes than it did on public concern about pollution of the river.

Last week I visited my old boyhood swimming resorts [wrote Mr A.E. Kenny, founder of the South Australian Swimming Association, in a letter to the Editor, Advertiser, 25 February 1937]. What did I find? The Star [at the end of Gilbert Street, Gilberton] which used to be 150 yards [137.2 m] long and five to fifteen feet [1.5–4.5 m] deep almost silted up and a rubbish tip. The spoil from Parliament House excavations [construction of the East Wing 1936–1939] has been deposited here, and at the first flood will be washed back to Adelaide, palm bulbs and all. The Blue Bell and the Eureka [in and near what is now St Peters Billabong] are also displaced by a rubbish tip. The Gilberton and Walkerville pools [near the end of Severn Street, Gilberton] are likely to be silted up with sand washing establishments which have almost entirely stopped the flow of water necessary

BOX 40

Improving water quality in the upper Torrens Catchment

SA Water is working towards improving water quality within the upper Torrens Catchment through partnerships with the Upper Torrens Land Management Project and the Adelaide and Mount Lofty Ranges Natural Resources Management (AMLRNRM) Board. SA Water aims to decrease nutrient and pathogen loads from this area of the watershed, and has developed water quality models to estimate pollution sources and predict improvements based on property management initiatives and watercourse fencing.

Property planning
Property plans have been completed for several of the major primary producers (focusing on dairy farms) and seven smaller, part-time producers. These landholders are considered to have the greatest potential impact on the water quality in the upper Torrens Catchment.

Property planning has concentrated on the four dairy farmers in the Birdwood area and two large sheep/cattle graziers near Gumeracha to develop property and work plans involving pathogen and nutrient management. Integral to these plans are: buffering of riparian zones:

establishment of offstream water points; improved internal fencing; and strategic grazing management to keep young stock away from runoff zones within the property.

Stock management options to reduce pathogen loads are evaluated in the property planning process, and in combination with a number of other programs such as the control of Bovine Johnes disease, where young stock are segregated from older animals and in general removed from the floodplain.

In addition, the property planning consultant, project manager and Dairy SA industry officer have discussed the development of nutrient management plans with the priority landholders to ensure that application of fertiliser or disposal of effluent onto pasture does not degrade the water quality on the farm.

The implementation of these work plans for priority properties is nearing completion. The project partners are now developing a larger program based on the success of this initial work.

Karla Billington

to carry away silt. N.A. District [at Warwick Sreet, Walkerville] and other pools are similarly doomed. Upon whom does the blame lie? A natural watercourse turned into a rubbish tip. I fear it is too late to restore the river to even a semblance of its former state without enormous expense.

Concerned citizens were aware of many of the inadvertent ways the establishment and growth of Adelaide were directly impacting the River Torrens. However, the most widespread and significant of the impacts were *indirect* and the result of urban development in the river's catchment below the Torrens Gorge (marked by 'Gorge Weir' in Figure 1.1). The impacts of urbanisation were largely disregarded or attributed to natural processes, partly due to the general lack of understanding in Australia of Mediterranean environments, which prevailed throughout the colonial period in South Australia and into the early years of the last century. In addition, the impacts of urbanisation on river systems were poorly documented until the 1950s when they became the explicit subject of numerous scientific investigations in different localities across the world.[10]

In 1967, Wolman developed a conceptual model of the impacts of land use change in the catchment of a river system[11] (Figure 8.2), which has since been confirmed by empirical observations. The model can be used to reconstruct the changes most likely to have occurred in the hydrology and channel morphology of the River Torrens and its tributaries from the beginning of settlement in 1837 to the more or less complete urbanisation of the eastern and western portions of the Torrens Catchment, below Gorge Weir, after 1940. As indicated by the Wolman Model, during the pre-settlement (or equilibrium) phase, the Torrens River system channels would have been metastable, with high-volume winter flows ponding on the Cowandilla floodplains and spilling overbank onto alluvial flats in localities upstream on the Adelaide Plains. Settlement would have destabilised the river system and initiated a period of significantly increased sediment yield and channel aggradation (i.e.,

infill due to sediment deposition). As the native vegetation cover was cleared and land was ploughed to establish crops, there would have been substantial increases in sheet and gully erosion, erosion in agricultural drains, channel bank collapse, and channel blockage by fallen trees and landslips. The beginning of urbanisation would have caused a massive increase in sediment yield and channel aggradation as new areas of land were repeatedly cleared for construction, and bare ground exposed to erosion for a few months to a few years until it was covered by buildings and other impervious surfaces. In turn, channel aggradation would have caused more frequent and severe occurrences of winter flooding on the Adelaide and Cowandilla plains, where, as predicted by the Wolman Model, the Torrens and it tributaries inundated some settled areas virtually every year from 1841 until the completion of the Breakout Creek flood mitigation scheme in 1938.

The rapid urbanisation after 1940 of the Torrens Catchment below the Torrens Gorge and continued clearing for agriculture in the rural portion of the catchment (above Gorge Weir) renewed concerns about flooding on the Adelaide and Cowandilla plains that had not been regarded as a significant problem since 1938.[12] As shown by Figure 8.2, with urbanisation of its catchment, sediment yields to a river system decline to levels at or below those experienced during the pre-settlement phase, but inflows to the system greatly increase due to the runoff from impervious urban surfaces fed to gutters and stormwater drains. In metropolitan Adelaide, urban runoff to the Torrens River system is mainly the result of winter rains.

However, beginning with the supply of reticulated water to part of the city of Adelaide in 1860, and continuing until the current imposition of watering restrictions, urban runoff from the metropolitan area has also included 'leakage' to groundwater and stormwater drains of substantial quantities of the water used to irrigate household gardens (particularly lawns) with unattended sprinklers. Since 1955, much of this leakage has been water imported by pipeline from the River Murray beyond the Torrens Catchment.[13]

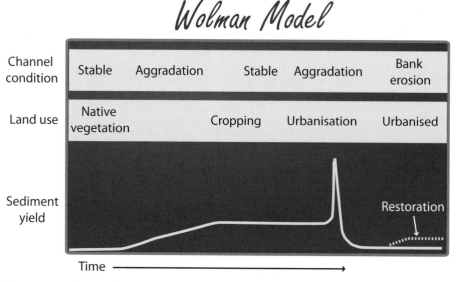

Wolman Model

Channel condition	Stable	Aggradation	Stable	Aggradation	Bank erosion
Land use	Native vegetation		Cropping	Urbanisation	Urbanised

Sediment yield

Restoration

Time ⟶

Figure 8.2 Wolman Model of the effects of land use change in the catchment of a river system. Source: adapted from K.J. Gregory, 'The human role in changing river channels', *Geomorphology*, 79/3–4, 2006, pp. 172–191. Copyright (2006), with permission from Elsevier.

The decreased sediment yields and increased flows associated with the urbanisation of a river system's catchment normally cause bank erosion, channel erosion (or scour), and channel widening (Figure 8.2). However, where river channels have been extensively modified by engineering works, sediments produced by bank erosion and channel scour may not be flushed from the river system and instead may accumulate upstream in constructions that impede flows (dams, weirs, bridges, etc.). Also, increased flows may undermine constructions that inhibit channel widening or may cause overbank flow and flooding of the catchment, upstream of constructions

Table 8.2 Human impacts on rivers

Cause of impact	Processes: hydrology (H) sediment (S)	Effect of impact: River channel adjustment
Impacts on a channel cross-section/point		
Dam construction	H−, S−, then S+	Scour below dam. Downstream may have reduced channel capacity, but variations are possible depending on sediment trapped
Weirs	H−, S−, then S+	Scour below weir
Outflow to reservoir	H−	
Return flow, drains, outfalls	H+	Scour at inflow point
Bridge crossings	H+	Scour around bridge piers
Culverts under road crossings	H+	Scour at downstream end
Impacts on a channel reach		
Desnagging and clearing	S+	Sediment transport increase and there may be accelerated erosion
Grazing	S+	Localised bank and channel erosion may occur
Embankments and levee construction	H+	May have effects downstream
Channelisation:		
Bank protection and stabilisation	H+, S+	
Resectioning, dredging	H+, S−	Effects downstream may arise from increased flow velocities
Channel straightening, cut offs	H+	Knickpoint recession may occur upstream from channelisation
Clearance of riparian trees and vegetation	S+	Localised bank and channel erosion may occur
Sediment removal/gravel, sand, clay mining	S+	Sediment transfer downstream or scour may affect channel morphology
Sediment addition/mining spoil	S+	Localised aggradation possible. May affect aggradation downstream and upstream as well
Invasion by exotic plants	S−	Stabilise sections of channel and banks. May reduce sediment transfer downstream
Impacts on catchment/river system		
Agricultural drains	H+	Localised effects possible where drainage reaches channel
Irrigation networks	H−	Channel adjustment possible where flow extracted. If many locations of flow extraction, can have significant effects
Stormwater drains	H+	Scour around outfalls. If numerous outfalls, can increased peak flows and induce erosion downstream
Deforestation	H+, S+	Gully development may occur with knickpoint recession upstream. Downstream may have channel change and often increased capacities, but depending upon sediment availability
Grazing	S+	Local effects by degradation of channel banks
Fire, burning off	H+, S+	Channel change after fire events
Agriculture, ploughing	H+, S+	Localised effects, often where tributaries join main streams
Building construction	S+	Channel affected by large sediment load locally
Urbanisation	H+, S−	Scour may occur at stormwater outfalls. Stormwater drains augment drainage network

Notes: Discharge is shown as H, with increased flows designated as H+ and decreased flow as H−. Similarly, increased sediment transport is shown as S+ and decreased sediment transport as S−.

Table 8.2 Human impacts on river channels – causes and effects. Source: adapted from Gregory, 'The human role in changing river channels', 2006. Copyright (2006), with permission from Elsevier.

preventing channel widening. This was the case with the River Torrens, which, by the 1980s, was not 'as nature made it' but as engineering works had made it, intentionally and unintentionally (Table 8.2).

Reservoirs: impounding the Torrens' flow

The Wolman Model of the impacts on a river system of land use change in its catchment includes a number of simplifying assumptions, including that the river system is not being directly modified by engineering works. From 1860, the hydrology and channel morphology of the Torrens River system *was* significantly modified by engineering works intended to provide the city and then the suburbs with reticulated water.

The first of these engineering works was the construction of the Gorge Weir (where a 'weir' is a dam designed to be overtopped by high-volume flows) to impound flows from Sixth Creek and the rural portion of the Torrens Catchment in the Mount Lofty Ranges. Work on the weir began in 1857 but the first weir was destroyed in the relatively minor flood of 1858; the replacement weir of squared rubble masonry was completed in 1860 and used to feed the newly constructed Thorndon Park Reservoir via a 21 inch (53.3 cm) pipe. Reticulated water finally flowed into the city from Thorndon Park Reservoir along Main North East Road to a water works in the east parklands on 28 December 1860.[14]

Within a decade of the construction of Thorndon Park Reservoir there was urgent demand for a second reservoir to meet the water needs of the now more than 50,000 inhabitants of Adelaide and its expanding suburbs. By 1872, the much larger Hope Valley Reservoir had been constructed and filled via an aqueduct channel from the Gorge Weir, to supply reticulated water for Port Adelaide, the city, and 20 of its neighbouring suburbs.[15]

The drought years of 1913 to 1916 motivated further impoundment of the discharge from the rural portion of the Torrens Catchment in the Mount Lofty Ranges by the construction of the Gumeracha Weir and a tunnel to carry water from the weir to the Millbrook Reservoir in 1918.

Finally, in 1969, the Kangaroo Creek Dam, the only structure to dam the flow of the River Torrens itself, and the onstream Kangaroo Creek Reservoir were commissioned to impound all the discharge from the rural portion of the Torrens Catchment in the Mount Lofty Ranges above Sixth Creek. Thereafter, flows to the River Torrens and its first five tributary creeks were largely limited to stormwater runoff from the urbanised portion of the river's catchment, with some supplementary flows from the rural portion of the catchment when the Gorge Weir is overtopped, the reservoirs fed by the Torrens River system are full, and/or water is released from the Kangaroo Creek Dam.

The Torrens Lake: 'Muddy and noisome ditch' to 'magnificent sheet of water'

The engineering work that did the most to transform the section of the River Torrens within the city of Adelaide was the construction of a weir near the western edge of the city where the river passes through a high-walled narrow gully.

By the 1860s, the part of the River Torrens located within the Adelaide parklands was not only an aesthetic embarrassment to the more than 35,000 inhabitants of this rapidly growing Victorian city, it was increasingly being viewed as a public health hazard. Although the river had ceased to be the main source of drinking water for the city in 1860, public health concern was not, at this time, focused on bacterial contamination of drinking water (Pasteur and Koch did not begin to promote the germ theory of disease in Europe until the late 1870s). Accordingly, during the 1860s the focus of public health concern in Adelaide, the 'stench capital of Australia', was the supposedly disease-causing 'miasma' of foetid gases rising from the rotting organic matter dumped in the Torrens and flowing into it from city drains.

The river Light had seen was a pleasant watercourse lined with tall trees, but, 40 years after settlement, had turned it into a muddy and noisome ditch. Most of the trees had been grubbed out for building materials and the constant procession of water carts had beaten down the banks. Every hot summer reduced it to a chain of evil-smelling waterholes.[16]

To the Victorian mind, the obvious solution for these aesthetic and atmospheric problems was to convert the 'chain of evil-smelling waterholes' to a picturesque lake, with landscaped surrounds promoting 'elite recreation' in the parklands, and to cover the source of the miasma with several metres of water. The creation of Torrens Lake was first attempted in 1862 when a wooden weir was constructed by 'private enterprise' at Hindmarsh, for unrecorded manufacturing purposes. This dam was destroyed by the floods of 1866 and succeeded by a cement weir (the Corporation Dam), constructed, using prison labour, by the Corporation of the City of Adelaide near the Adelaide Goal in 1867. The Corporation Dam was also washed away almost immediately in the floods of that year and not replaced until 1881. This was a fortunate public health outcome (given the prevalence of bathing in Torrens Lake), as construction of the third and final weir (the Adelaide City Weir) was quickly followed by an end to the practice of draining sewage from south Adelaide into this section of the Torrens.[14]

The Adelaide City Weir was a 6 m high mass-concrete gravity-overflow dam with a wrought-iron, timber-floored footbridge running along the top. Its base was holed by six pipes, each a metre in diameter and fitted with sluice gates. Closed in summer to impound Torrens Lake, the sluice gates were opened during heavy winter flows, supposedly to prevent overflow of the lake and the accumulation of sediment upstream from the weir. However, the sluice gates failed 'to perform either function effectively', and flooding and sediment accumulation became major problems by the end of the decade.[4]

Have you ever noticed a class of students exploring a local river? Or seen people on the weekend collecting a bucket of water out of the pond at the end of the street? There's a good chance they are one of the 155 school or 78 community Waterwatch groups in Adelaide. Waterwatch is a national community water monitoring and education program that encourages Australians to become actively involved in the protection and management of local catchments. It has been operating in Adelaide for 15 years.

What do Waterwatch groups do? They are caretakers of our catchments. Having chosen a special local waterway, their regular visits mean they become experts in the health of these areas. Participants collect regular water samples as part of the regional water quality monitoring program, and report back on their results and on impacts such as chemical spills, animal deaths and illegal dumping. Groups monitoring in areas such as the lower reaches of the Onkaparinga River and Hindmarsh River use specialised estuarine equipment. Groups expand on their monitoring by cleaning up litter, planting local native species, stencilling stormwater messages on drains, constructing frog ponds, and sharing their discoveries with the community. Here are a few of our groups' achievements:

- Nuriootpa High School found tadpole shrimp when biologically monitoring their site; before this there were no other known records of the tadpole shrimp in the region.

- An Adelaide Hills Waterwatch group alerted the council and the Environment Protection Authority (EPA) to a sewage leak in a local watercourse.

- Burton Primary School students joined forces with AV Jennings to monitor the establishment of the storm-water-treatment wetlands in an adjacent housing subdivision. When litter was identified as an issue, they embarked on a builder education program.

- While assisting Waterwatch with fish monitoring, members of the North East Hills Environmental Conservation Association recorded galaxid species at their site.

- North Haven Primary School sells alternatives to plastic bags at the Torrens Island markets as a direct way to protect their Port River from dangerous discarded shopping litter.

- Friends of Brownhill Creek use their water testing results to monitor the success of their revegetation works, as well as keeping an eye out for other polluting activities in their catchment.

- St Francis School at Lockleys divert stormwater from several classrooms to fill their frog pond.

In Adelaide, these groups are supported by Natural Resources Management (NRM) Education staff based at Mawson Lakes, Nuriootpa Flinders Park, Stirling and Victor Harbor. Major funding for the program is provided by the Adelaide and Mount Lofty Ranges Natural Resources Management (AMLRNRM) Board, with support from Adelaide Hills Council, City of Victor Harbor, City of Salisbury, and KESAB environmental solutions. NRM Education works to build the natural resources management capacity of a diverse cross-section of the community including individuals, schools, local government, businesses, service clubs, and community groups and organisations. Among this diverse audience, schools have been recognised as an important sector to focus effort on, particularly primary schools. NRM Education plays an important role in encouraging and supporting teachers to incorporate and integrate sustainability across their school activities. Staff provide students with hands-on experiences that allow them to develop a connection with their environment, develop their value system, and participate in natural resource management activities. Educators are supported through professional development, networking, resources and equipment loans. In 2007–2008, NRM Education worked with over 24,000 students and teachers across Adelaide. The community can take part in talks and tours to build awareness of local issues and find out how they can improve their local environment. NRM Education has also produced a range of resources on local wetlands, estuaries and aquatic life to guide personal discovery.

From its early days, which saw the program focus primarily on water-related issues, NRM Education in the Adelaide and Mount Lofty Ranges region has significantly expanded its charter to include issues such as soils, native plants, pest plants and animals, and matters surrounding biodiversity preservation. NRM Education is now considered the main flagship environmental education program of the AMLRNRM Board. This program is seen by the board as an important tool in moving people from awareness to behaviour change, and as the platform for building good environmental citizens for today and tomorrow.

Amy Blaylock

According to the *Register* (7 July 1881), those who watched the mayor of Adelaide close the sluice gates of the new weir for the first time were enthusiastic in their admiration of 'the magnificent sheet of water … thrown back by the dam' and 'the wonderful transformation that had been effected' of the 'muddy and noisome ditch'. This enthusiasm was short-lived, however, as the defects of the weir soon became apparent. During the decades following its construction, the weir sluice pipes were repeatedly overwhelmed by high-volume flows and jammed by logs and other debris, causing flooding upstream and along First Creek. Every year, whole trees, uprooted from the river channel and catchment, piled up against the weir, together with a tangled mass of 'immense logs of dry timber, fruit trees, reeds, and pieces of broken boats, tables, chairs, lockers, pumpkins, marrows and melons', which had to be removed at great expense and occasional loss of life (*Register*, 19 April 1889).[4]

In addition, inadequate flows through the sluice gates caused sediment to accumulate in the Torrens Lake each winter. The great flood of April 1889 deposited so much sediment behind the weir that gravel and sand banks, more than 2 m high and covered in black mud washed down from the market gardens and orchards upstream from the city, became a permanent feature of the lake in all seasons.

> *Measurements showed that over the six years from 1889 … the width of the lake was halved by the luxuriant growth of rushes on the permanent mud banks near the Albert Bridge. Even in summer boating [and bathing, the main recreational uses of the lake] had become difficult and unpleasant, and there was no boating at all between June and September because the lake was emptied every winter in a vain attempt to increase the scouring action. Most of the waterbirds were abandoning the muddy and dreary flats, leaving behind a few tame tired black swans and one white swan [the sole survivor of a group of five birds imported from London and introduced to the lake in 1885], which had to be cared for by the parklands ranger until the lake filled up and partially submerged the mud banks in the spring.*[17]

Despite continuing efforts to remove sediment accumulation with a steam dredger, Torrens Lake remained an expensive aesthetic embarrassment to the City of Adelaide until the weir was rebuilt in 1929. This involved removing a section of the weir to install more effective sluice gates, permitting the occasional flushing of the lake and scouring of sediment accumulation. At the same time, work was done downstream to dissipate the energy of the water discharged from the weir, which had caused erosion at its base. This work included the installation of a stilling basin, concrete cribbing, and an expanse of rubble in the stream course.[18]

The creation of Torrens Lake was an ecological disaster for the river, even apart from the introduction of the ill-fated 'white swans' (i.e., Mute swans, *Cygnus olor*), goldfish or golden carp (*Carrassius auratus*), brown trout (*Salmo trutta*), and the planting of bamboo (sp. or spp. unknown) and willows (*Salix* spp.).[17] The lake has arguably remained an ecological disaster, with sediment accumulation a continuing problem and the proliferation of European carp (*Cyprinus*

carpio), together with high nutrient levels associated with stormwater pollution, causing repeated blue-green algae (cyanoprokaryote) blooms during summer hot spells.

However, reduced flows and sediment load downstream from the weir did provide a temporary reprieve for the Reedbeds, which had been impacted by the effects on the of land clearance for agriculture and urbanisation. As noted by the *Register* (8 July 1884), with approval of these impacts:

> *before the dam was erected the sediment now filling the lake was carried down and deposited in the Reedbed swamps converting them gradually into beautifully fertile fields, and in course of time the whole of the great swamp at the back of Henley Beach, covering about 300 acres [121.4 ha], would have ceased to exist.*

Improving the Torrens: bridges, weirs and drains

For Adelaide's first few inhabitants, the section of the River Torrens separating the northern and southern parts of the city was more a nuisance than a formidable barrier to passage. In summer the settlers walked with their animals across the 'dry ditch', skirting the deeper waterholes or crossing on a fallen tree. In winter the river was crossed on a punt located near the present Adelaide City Weir, pulled back and forth on a rope tied between two trees. The first attempt at bridging the river within the city was a wooden footbridge built near the modern-day Adelaide Oval in 1837.[4, 19]

By 1839 these makeshift measures were inadequate to serve the growing population; a sequence of bridges was built at easy crossing points with established fords connecting major streets in south Adelaide to North Adelaide and Port Road (see Table 8.3). The first of these was built in 1839 by Alfred Hardy, later the town surveyor for Adelaide, and located some 500 m west of the present King William Street crossing. Typical of bridge construction until the mid 1850s, Hardy's Bridge was a small, low structure built of local timber and employing under-strutting girders to minimise the number of piers required in the river. In 1842 a second timber bridge was built near the present Frome Road crossing, and in 1843 Hardy's Bridge was replaced by a more substantial structure, the City Bridge, using timber and stone from the parklands.[19]

All of the bridges built in City of Adelaide, and indeed all of the bridges built on the River Torrens and its tributaries from the beginning of settlement until the mid 1850s, were repeatedly breached or washed away by both 'normal' winter flows within the river system channels and 'abnormal' winter flows that overtopped the banks (both referred to as 'floods' by the settlers). This was partly due to the limited engineering capability in the early colony and the reliance on locally available construction materials (i.e., timber and some stone for masonry as opposed to the concrete and imported wrought iron that became increasingly available after the mid 1850s, and which began to be manufactured in Australia after 1915). However, an additional problem was the settlers' understandable ignorance of the Torrens River system.[4]

Table 8.3 Adelaide's main bridges

YEAR	Present river crossing			
	Morphett Street	**King William Street**	**Frome Road**	**Hackney Road**
	Bridges contructed of timber and stone			
1839	Ford and footbridge in the vicinity of Morphett Street provides access to Port Road	Hardy's Bridge built		
1842	Footbridge washed away by flood		'Old' Frome Rd Bridge built	
1843		City Bridge replaces Hardy's Bridge		
1844		Bridge breached by flood	Bridge damaged by flood	
1845				SA Company Bridge built at Prescott's Crossing
1847		Repaired bridge breached by flood	Repaired bridge breached by flood	Only bridge in the city to survive flood and remain open, but only to pedestrians
1850		Bridge repaired with masonry arch		
1851		Masonry arch bridge breached by flood; timber footbridge constructed on the ruined abutments	Repaired bridge damaged by flood	Repaired bridge breached by flood
1855		Footbridge washed away	Bridge destroyed by flood and replaced by ford	Repaired bridge breached by flood
	Bridges begin to be built with wrought iron and concrete (imported from England until 1915)			
1856	Railway to Port Adelaide opens blocking access from Morphett Street to River	Second City Bridge opened; wrought iron girders with masonry abutments		
1860	Level crossing established at railway			New bridge opened; four span, trussed timber
1863			Footbridge built beside ford	
1867			Footbridge swept away by flood and rebuilt	
1868	Overway Bridge built to span railway			
1871	Victoria Bridge opened; wrought iron girders			
1877		Third wider City Bridge (or Adelaide Bridge) opened		
1878			Footbridge damaged by flood; Albert Bridge foundations laid	
1879			Albert Bridge opened; cantilevered span of wrought iron girders	
1881		Torrens Weir and Torrens Lake established		
1885				New concrete Hackney Bridge opened

Table 8.3 Timeline for Adelaide's main bridges from 1839 to 1885. Source: adapted from Stacey and Venus[19] with additional information from Smith and Twidale (1987).[4]

BOX
42

*Politics and the problems
of the River Torrens*

Although there is a long history of local and state government responses to the environmental problems of the River Torrens, the last few years have seen an almost unprecedented level of government intervention in this area. The repeated closure of Torrens Lake through the 1990s and the early years of the 21st century as a consequence of algal blooms may have prepared the ground, but the catalyst for this outburst of political activity was the publication in May 2006 of a report on the inorganic pollution of the sediments of the River Torrens. This revealed the extent of the degradation of the river, particularly along its urban reaches. The story was broken by the *Australian*, and the headline 'River metal "enough to mine"'[1] ensured that for over a month not a day passed without the problems of the Torrens being debated in local, state and national media.[2, 3, 4, 5] The *Advertiser* threw its weight behind the campaign. The political response was rapid. Two organisations were charged with solving the problems of pollution in the Torrens. The Torrens Taskforce was formed in August 2006 at the request of Gail Gago, the South Australian Minister of Environment and Conservation. This was quickly followed by the Urban Rivers Symposium, held in Adelaide on 23–24 November 2006, and established under the umbrella of the Adelaide City Council.

The aim of the symposium was to bring together community and professional views on the present state and future recovery of the River Torrens. The vast body of information, expertise and ideas generated during the symposium was collated on a single website. To supplement this, Adelaide City Council launched a Torrens Lake water quality monitoring website with the intention of providing the community with daily (and ultimately hourly) information on conditions in the Torrens.

The Torrens Taskforce, by contrast, was set the task of providing concrete solutions to the environmental problems of the river. Its final report[6] recommended that:

1. A set of water quality targets should be established for the river, and regular monitoring should be undertaken to assess progress in meeting these targets. In order to gain the support and enthusiasm of the community for the project, the Taskforce proposed as its objective the re-establishment of the platypus in the river.

2. More rigorous attention should be paid to catchment management. In rural parts of the catchment, watercourses should be fenced off to exclude stock, stock management practices should be improved to reduce the chance of pathogens reaching the river, acidic soils should be managed to improve the retention of nutrients and organic matter in the soil, and poorly maintained septic tanks should be remediated. In the urban zone, gross pollutant traps should be installed across 95% of the catchment. Throughout the entire basin, stricter approaches should be adopted to ensure compliance with existing catchment management regulations.

3. Several recommendations were concerned with modifications to the current flow regime of the river:

   (i) The periodic transfer of water from the Murray to the upper reaches of the river should be managed to avoid the problems associated with steeply rising artificially-generated flood pulses. These include bank collapse and the injection of sediment into the channel.

   (ii) The proliferation of farm reservoirs in the rural part of the catchment should be halted. These have significantly reduced base flows in the river by diverting water into storage during low flow conditions.

   (iii) Summer base flows along the urban Torrens should be maintained to enhance water quality, particularly in Torrens Lake, to maintain hydrological connections along the channel and to improve the appearance of the system.

   (iv) The flow of water along the channel under high-stage conditions should be reduced to increase the opportunity for natural filtering and decomposition of the pollutant load, and to improve the ecological status of the river. There are various ways in which this might be achieved, by the use of water sensitive urban design, for example, or by the construction of wetlands and other storm water detention features.

4. The water destratification devices that have been installed in Torrens Lake should be maintained and refined, and a biological filtration system for the lake's waters should be tested. Water currently pumped from the lake should be taken solely from the

downstream end in an effort to increase water turnover. Aquatic vegetation should be established in the lake to stabilise the bed and to remove nutrients from the sediments and water column. Carp should be removed from the lake to reduce their impact on sediment disturbance and artificial feeding of the bird community should cease to reduce the bird population and thus faecal input to the lake.

5. The report stressed the critical role of community support in solving the environmental problems of the river and argued that the duties of the agencies responsible for the river should be clarified to avoid the potential for conflict, overlap and gaps in responsibilities.

The proposal that the recolonisation of the river by platypuses provide a test of the success of efforts to rehabilitate the state's urban rivers is certain to be seized on by the community as a tangible symbol of the success of the strategies devised to save the Torrens. It is now many decades since the last specimen of platypus swam and foraged along the banks of the river, its fate sealed by a combination of pollution and habitat destruction. By the late 1980s a similar situation existed along the Yarra, and it was widely believed that the platypus had vanished entirely from metropolitan Melbourne. Yet in less than two decades, improvements to the environmental health of Melbourne's rivers have seen such a dramatic turnaround in platypus populations so that the platypus may now be found within five km of the city centre. The question remains, does Adelaide have the political will to emulate Melbourne's success?

The Taskforce's recommendations are not cheap. Unfortunately, too, there is no guarantee that many of the proposals will solve the problems the river faces. In dealing with the algal blooms on Torrens Lake, for example, $7 million is proposed. Yet this expense is to be incurred with no real understanding of the derivation of the nutrients that give rise to the blooms. If, instead, we were able to determine the origin of the nutrients (and this could be achieved at a cost of thousands rather than millions of dollars), we might be able to tackle the problem at its source rather than facing the never-ending task of dealing with the downstream consequences of problems whose origin lies elsewhere in the catchment.

Perhaps more importantly, the overwhelming focus of the Taskforce's report is on water quality. Yet this preoccupation overlooks the fact that the major source of pollution in the river is its sediments. Concentrations of lead in the sediments of the Torrens, for example, are about a thousand times greater than those in the river's waters. Even worse are concentrations of phosphorus, which are around 10,000 times more than those in the water. Indeed, the lead and zinc concentrations in the river's sediments are among the highest recorded in any Australian aquatic environment. This problem cannot simply be ignored. The assertion of the Taskforce that these toxicants are not bioavailable has been comprehensively discredited.[7,8] At the very least, the remarkably high levels of pollution that have been identified justify a precautionary approach to the problems of the river. What is certain, however, is that unless the current levels of lead, zinc and cadmium in the sediments are reduced, the platypus will never again be seen in the Torrens.[9]

What have been the long-term results of these political initiatives? Information from the Urban Rivers Symposium was meant to have been passed on to the Torrens Taskforce. Beyond that, its legacy is difficult to discern. The consolidated report of the symposium's findings is no longer available on the internet, and there is no evidence that any of the vast corpus of information that it contained was exploited by the Taskforce during its deliberations. Similarly, the website containing daily water quality data from Torrens Lake has yielded no information since January 2008.

By contrast, at least one of the initiatives of the Torrens Taskforce has been implemented. On 13 February 2008, a $300,000 biological filter was installed in Torrens Lake in the first stage of a trial to filter nutrients from the lake's waters. Despite this advance, most of the measures recommended by the Taskforce remain unimplemented. Indeed, even short-term schemes such as setting new water quality targets, installing gross pollutant traps, and fencing rural watercourses have been ignored. It is difficult to escape the conclusion that the end result of all this activity will be little more than a string of reports destined to gather dust on the shelves of the state's archives.

Hilary P.M. Winchester

Initially, the settlers did not know what even the 'normal' water levels might be during winter flows in the Torrens, let alone to what heights the river system might rise during floods (overbank flows) – so they could not judge how high above the channel bed they should build their bridges. In any case, winter flows could not have been predicted during the early years of settlement and are still difficult to predict due to rainfall variability in the catchment. Then, too, from the beginning of settlement to the present day, the hydrology of the Torrens River system has been adjusting and will continue to adjust to land use changes in its catchment and to direct human impacts on its hydrology and channel morphology, both inadvertent and as a result of engineering works deliberately intended to 'improve' the system for human use. (Table 8.2 describes the main impacts.) Finally, each impact on a river system has a lag time (reaction time) from the commencement of the impact to the beginning of hydrological and morphological change, as well as a lag time (relaxation time) from the beginning to the end of hydrological and morphological change (not represented in Figure 8.2, due to the simplifying assumptions of the Wolman Model). The duration of these lag times is even now poorly understood for most forms of impact. This means impacts that commenced years or even decades ago may still be working through the system, which can be assumed to have been unstable more or less from the beginning of settlement with no new equilibrium in sight, particularly due to the predicted external impact of climate change on rainfall patterns in the Torrens Catchment.[10]

The most immediate impacts of bridging the River Torrens within the city of Adelaide are now obscured by Torrens Lake. Within the urbanised eastern and western portions of the Torrens Catchment, continuing proliferation of the road network (including construction of Stage 1 of the O-Bahn busway from the city to the Paradise Interchange in the period 1983–1985) has required the construction of numerous bridges, large and small, across the river and its tributaries. Most of the modern bridges do not have the problems of the earlier ones, but many of the earlier ones are still in use or have only been repaired rather than remodelled since their construction.

In the relatively narrow, mostly high-banked River Torrens channel upstream from the city, water levels have often reached the underside of a bridge there. In these circumstances, the bridge becomes an enclosed conduit, which can cause overtopping of the bridge and overbank flow (flooding) upstream (Table 8.2). This problem is exacerbated when debris becomes trapped against the upstream side of the bridge, blocking flow and often causing damage that, together with the force of water, may breach the bridge or wash it away, as happened with all of Adelaide's early bridges on a number of occasions. The force of the water passing under the bridge can also undermine its footings and cause channel scour.[20]

Downstream from the city, where the river has more gentle gradients and meanders between levee banks somewhat higher than the adjacent floodplain, bridges are normally approached by embankments across the floodplain on either side of the channel. Here, bridges have often acted as

closed conduits during high water flows, while the bridge embankments have acted as dams or weirs inhibiting overbank flow on the floodplain.

By the 1980s, close to 20 'intentional weirs' had been constructed on the River Torrens between the Gorge Weir and the sea, and numerous 'unintentional' weirs had been created by bank collapse, fallen trees trapping other debris, and rubbish dumping (including a number of abandoned cars in the upper reaches of the river). During the early years of settlement, weirs were constructed mainly to create ponds for watering stock, irrigating market gardens and orchards, collecting household water, and various industrial uses. Later, weirs were used to create swimming pools and landscape effects, as well as to assist flood mitigation efforts. As described for the case of the Adelaide City Weir, the impacts of weirs are mainly sediment accumulation (aggradation) in the channel above the weir and scour of the channel below it (Table 8.2).

The closed conduit effect of bridges at road crossings is also common on the creeks tributary to the River Torrens; a highly visible and seemingly intractable example being the frequent flooding of Rymill Park where First Creek passes into the Adelaide Botanic Gardens under Torrens Road. In addition, lengthy sections of First, Second and Third creeks have been converted to closed conduits; while all five of the creeks tributary to the River Torrens in the eastern urbanised catchment have become little more than stormwater drains, at least in their lower reaches, usually running in open concrete conduits where not entirely enclosed.

More than 200 other smaller and mainly enclosed stormwater drains discharge into the River Torrens, mainly in the eastern urbanised portion of its catchment.[21] The velocity of the water exiting these drains following precipitation events is high, and various measures have been used to dissipate this energy, including directing the flow of water over a cascade of rock or down a stepped concrete structure to create turbulence and minimise scour in the river channel.[7] Recently, trash racks and silt basins/wetlands have been installed on some drain outlets (all of which can be observed at the outlet of Second Creek into St Peters Billabong and the outlet of the billabong into the River Torrens). The effects of these measures have been to narrow the river channel and, except for the use of wetlands, replace the earth channel bank with synthetic materials.

Taming the Torrens: flood mitigation

Flood mitigation has been a 'work in progress' on the Torrens since the early years of settlement. Until the 1930s, and with the exception of the Adelaide City Weir and the Gorge Weir, both of which had some flood mitigation function, most of this work involved mere 'tinkering' with the River Torrens channel upstream from Adelaide, by both government agencies and private landholders. This tinkering included widening the river channel in some parts, straightening, deepening and narrowing it in others, shoring up the banks with timber, masonry and concrete, and other relatively minor engineering activities that were often counterproductive, tending to accelerate flows and increased flooding downstream.

Following extensive flooding and property damage in the Torrens Catchment during 1931 and again in 1933, the South Australian government passed the *Metropolitan Drainage Act 1933* to take control of the whole river and its banks. In 1935, in the midst of the Great Depression, using the labour of unemployed men and a grant from the federal government, work began on a major flood mitigation scheme that would transform the western portion of the Torrens Catchment downstream from the city.[12]

The scheme involved shoring up the river channel banks on the outside of meander bends, some minor channel straightening, and the installation of five water-retarding and drop weirs to the west of Port Road, as well as increasing the height of the natural levee banks further downstream. However, the most significant feature of the scheme was the blocking of the old terminus of the River Torrens channel and, by 1938, the construction of a new concrete-lined channel (Breakout Creek) for some 3.5 km through the Reedbeds and coastal dunes to the sea (Figure 1.1).[12]

These flood mitigation works were specifically designed to alleviate flooding in the western portion of the Torrens Catchment, and had little impact on flooding upstream from Adelaide. Hence, the largely ineffective or counter-productive tinkering with the River Torrens channel, and the conversion of its tributaries to stormwater drains, continued after 1938. Most notably, a steep-sided earth-walled drain was constructed at Marden in 1968 to straighten the River Torrens channel by cutting through a meander bend and infilling the old channel for residential development. In 1978 a similar channel was constructed creating the St Peters Billabong, which is filled by the Second Creek drain in the old meander bend.[12]

Restoring the Torrens

By the 1980s it was recognised that most of the suburbs in the western urbanised portion of the Torrens Catchment, and some of the suburbs in the eastern urbanised portion, were still flood prone after more than one hundred years of piecemeal flood mitigation work. Through the 1980s and 1990s, the River Torrens Linear Park Project attempted, in an integrated and systematic manner, to address not only flood mitigation but also degradation of the river channel and its associated riparian landscapes. An additional and, to some degree, overriding objective of the project was the creation of a linear walkway/bikeway within a landscaped, public-access parkland that would eventually extend along the River Torrens some 50 km from the sea to the Kangaroo Creek Dam in the Mount Lofty Ranges.[22]

Since the 1990s, ecological restoration has become an important management objective for urbanised river systems, as it has for environmental management in general. In this context, critics of the Linear Park Project have described it as an ecological disaster akin to the creation of Torrens Lake, criticism which is, at the very least, unfair. As shown in Figure 8.3, ecological restoration includes a range of management options depending upon the ecological integrity of the landscape being restored. Many critics of the Linear Park Project have equated ecological restoration

with returning the Torrens to 'the river which Light had seen'. Given the level of degradation experienced by the river and its associated riparian landscapes, the only management option for the implementation of this objective would have been *reconstruction* using indigenous or introduced materials to *replace* some elements of the form and function of a pre-existing native landscape (Figure 8.3).

However, reconstruction of the River Torrens and its associated riparian landscapes was never a management objective of the Linear Park Project. Instead, its objectives for the river channel were mainly utilitarian, intended to assist flood mitigation by

> *altering the cross-section and shape of the channel so as to increase flood capacity, improving energy dissipation during peak flows, reducing erosion to background levels to reduce sediment transportation downstream, along with enhanced deposition on channel beaches and subsequent vegetation regeneration.*[22]

An additional objective was to remove introduced vegetation from the river channel, where this would not compromise flood mitigation, a process that has continued as the Linear Park has been extended upstream and downstream. Critics of this process have argued that it destroyed valuable aquatic and wetland habitat without replacement with indigenous vegetation that would have performed the same functions.

In terms of the riparian landscapes bordering the river channel, the project's management objectives were also mainly utilitarian, intended to create an aesthetic environment suitable for both active and passive recreation rather than to restore anything like the structure and composition of the pre-settlement vegetation. Nevertheless, the project was not intended to 'create an open-grassed park, but rather a river environment within metropolitan Adelaide which consists of natural areas interspersed with more accessible passive and active recreation areas'.[23] The creation of these so-called natural areas was to involve removing some introduced vegetation, retaining some remnant native trees and shrubs, and planting groups of indigenous trees and shrubs – in order to:

- create a variety of densities of both upper- and lower-storey planting to provide different wildlife habitats;
- create interest by defining a variety of spaces, providing seclusion or framing views;
- screen unsightly features;
- provide buffers between incompatible uses;
- emphasise existing significant features;
- not increase the level of potential flooding due to planting; and
- create a strong visual identity for this 'linear spine' of open-space and the river, which may also serve as a wildlife corridor to which species (uncommon within the metropolitan area) might be attracted.[24]

Another criticism of the Linear Park Project is related to the last landscape design objective cited above – the value of the park as a wildlife corridor. Prior to settlement, the

Ecological restoration options

Rehabilitation (help for recovery		Reconstruction (making anew)	
Repair	**Reintroduction**	**Replication**	**Replacement**
Repairing damage to an existing native landscape (mainly by removal of damaging intrusions and impacts)	Reintroducing original elements of an existing native or relict landscape	Replicating the form and function of a pre-existing native landscape using indigenous materials	Replacing some elements of the form and function of a pre-existing native landscape using indigenous or introduced materials

High	Level of ecological integrity before restoration	Low
Low	Level of ecological artificiality after restoration	High

Figure 8.3 Types of ecological restoration (returning a landscape, in some measure, to a pre-existing ecological state).

River Torrens Valley would have been a major wildlife and Aboriginal cultural corridor extending from the Mount Lofty Ranges to the Cowandilla Plains. This corridor has been extensively fragmented below the Kangaroo Creek Dam and Reservoir by the construction of bridges, weirs, drainage outlets, Torrens Lake, and numerous other minor engineering works, as well as by a host of other human impacts and interventions including vegetation clearance and replanting during the Linear Park Project. Fragmentation of riverine wildlife corridors by physical barriers to movement and the disruption of habitat gradients can affect both aquatic and terrestrial species, but will affect each species differently. Consequently, while much of the ecological function of the River Torrens wildlife corridor has been lost through fragmentation, the Linear Park still functions as a greenway for the Adelaide metropolitan region, from an aesthetic and recreational perspective, and as a movement corridor for a number of wildlife species, unfortunately including foxes (*Vulpes vulpes*) and koalas (*Phascolarctos cinereus*) (both introduced species), but also including native bats, birds and reptiles.

In defence of the Linear Park Project, it can be argued that ecological restoration of the River Torrens and its associated riparian landscapes to their pre-settlement condition is neither practical nor, in fact, possible, because ecological restoration of any urbanised river system must accommodate a number of limitations, most insurmountable. This chapter has attempted to demonstrate that the hydrology, channel morphology and water quality of the River Torrens is now almost entirely an artefact of a history and will probably continue with increasing frequency and intensity, since the river's catchment is still being urbanised, and is unlikely to be a reversible process.

Moreover, the notion that even part of an urbanised river system can be restored to some pre-existing state, and maintained in that state, is a fallacy that ignores the way impacts on a river system at a given locality propagate spatially and temporarily upstream and downstream for distances and over periods of time that are extremely difficult to predict and therefore to manage. The notion of restoration to a steady state also ignores the impacts on the whole river system of changes in the external physical environment of its catchment; particularly climatic fluctuations (e.g., drought) and directional climatic changes, such as those predicted to occur through global warming that may decrease rainfall in the river's catchment by as much as 10% by 2030 and cause more extreme rainfall events, with consequent reductions in average flows but increasingly frequent flooding.[25]

Finally, and perhaps most importantly, urbanised river systems exist in an external environment that is social as well as physical. It is difficult to imagine an ecological restoration project for the Torrens that would gain public support for decommissioning the reservoirs fed by the river catchment, removing the bridges and road crossings over the river channel, draining Torrens Lake, infilling Breakout Creek, and undoing decades of other extremely costly flood mitigation work. Even an attempt to restore some semblance of the native woodland vegetation of the Torrens Valley for any considerable portion of the Linear Park would be likely to generate strong public opposition, especially given safety concerns about dense understorey plantings in public parks. The Breakout Creek Wetlands and St Peters Billabong do, however, demonstrate what can be accomplished using replacement techniques to restore ecological value to small areas.

The River Torrens will never again be 'as nature made it', due to decades of human exploitation, deliberate abuse, accidental damage, neglect, and ill-informed management. Nevertheless, current and future management interventions informed by understanding of the impacts of urbanisation on river form and function, by the development of improved ecological restoration techniques, and increasing appreciation of the social value of city-flowing rivers can create an ecologically healthier and more sustainable River Torrens.

References

1. J.C. Hawker, *Early experiences in South Australia (Australiana facsimile edition, 1975)*, E.S. Wigg & Son, Adelaide, 1879, p. 7.
2. D.N. Kraehenbuehl, *Pre-European vegetation of Adelaide: a survey from the Gawler River to Hallett Cove*, Nature Conservation Society of South Australia, Adelaide, 1996, p. 78.
3. J.W. Holmes, and M.B. Iversen, 'Hydrology of the Cowandilla Plains, Adelaide, before 1836' in Twidale, C.R., Tyler, M.J. and Webb, B.P. (eds) *Natural history of the Adelaide region*, Royal Society of South Australia, Adelaide, 1976, pp. 91–97.
4. D. Smith and C.R. Twidale, 'An historical account of flooding and related events in the Torrens River System from first settlement to 1986, volume 1: 1876–1899', Engineering and Water Supply Department, Adelaide, 1987.
5. P. Morton, *After Light: a history of the City of Adelaide and its councils, 1878–1928*, Wakefield Press, Adelaide, 1996, p. 123.
6. M. Hammerton, *Water South Australia: a history of the Engineering and Water Supply Department*, Wakefield Press, Adelaide, 1986, p. 73.
7. Hassell and Partners Pty Ltd, 'River Torrens study: a co-ordinated development scheme', Government of South Australia, Adelaide, 1979.
8. Morton, 1996.
9. Kraehenbuehl, 1996, p. 76.
10. A. Chin, 'Urban transformation of river landscapes in a global context', *Geomorphology*, 79, 2006, pp. 460–487.
11. M.G. Wolman, 'A cycle of sedimentation and erosion in urban river channels', *Geografiska Annaler*, 49A, 1967, pp. 2–4.
12. D. Smith and C.R. Twidale, 'An historical account of flooding and related events in the Torrens River System from first settlement to 1986, volume 4: 1931–1988', Engineering and Water Supply Department of South Australia, Adelaide, 1989.
13. E. Woolmington and J.S. Burgess, 'Hedonistic water use and low-flow runoff in Australia's national capital', *Urban Ecology*, 7, 1983, pp. 215–227.
14. Hammerton, 1986.
15. M. Williams, *The making of the South Australian landscape*, Academic Press, London, 1974.
16. Morton, 1996, p. 160.
17. ibid., pp. 162–163.
18. Aussie Heritage, 'Torrens Lake weir and footbridge', *Aussie Heritage*, 2007 (online).
19. B. Stacy and R. Venus, 'Bridging Adelaide's River Torrens: pre- and post-Federation technologies' in Baker, K. (ed.) *Proceedings of the Eleventh National Conference on Engineering Heritage: Federation Engineering a Nation*, 2001, Institution of Engineers (Australia), Barton, ACT, pp. 37–42 (online).
20. K.J. Gregory, 'The human role in changing river channels', *Geomorphology*, 79, 2006, pp. 172–191.
21. Torrens Taskforce, 'River Torrens Stormwater Drainage Inlet GPT Opportunities', *Memorandum from Andrew Thomas to Alan Ockenden*, 2007 (online).
22. D. Mugavin, 'Adelaide's greenway: River Torrens Linear Park', *Landscape and Urban Planning*, 68, 2004, pp. 223–240.
23. Hassell and Partners Pty Ltd, 'River Torrens study: summary report', Government of South Australia, Adelaide, 1979, p. 25.
24. ibid., pp. 26–27.
25. D. Bardsley, 'There's a change on the way: an initial integrated assessment of projected climate change impacts and adaptation options for natural resource management in the Adelaide Mount Lofty Ranges region', Department of Water, Land and Biodiversity Conservation report 2006/06, Adelaide, 2006.

Box references

Box 42: Politics and the problems of the Torrens

1. J. Roberts, 'River metal "enough to mine"', *Australian*, 22 March 2006, p. 3.
2. G. Hunt, 'Dear premier: clean up the filthy Torrens', *Advertiser*, 18 May 2006, p. 18.
3. B. Hurrell, 'Red faces grow over polluted Torrens', *Advertiser*, 7 April 2006, pp. 1, 8.
4. C. Jenkin, 'Mind change urged to turn trash into a city treasure', *Advertiser*, 13 May 2006, p. 28.
5. H.P.M. Winchester, S.J. Gale, and R.J.B. Gale, 'River Torrens on the critical list', *Independent Weekly*, 1–7 April 2006, p. 4.
6. Torrens Taskforce, 'Torrens Taskforce summary of findings final report', report to the Minister for Environment and Conservation, Government of South Australia, Adelaide, 2007.
7. S.J. Gale, R.J.B. Gale and H.P.M. Winchester, 'Contaminated sediments in the River Torrens, South Australia', *South Australian Geographical Journal*, 105, 2006, pp. 78–105.
8. S. J. Gale, R.J.B. Gale, H.P. M. Winchester, N.J. Dorrington and N.F. Cano, 'The River Torrens 2: contaminated sediments in the river' in Daniels, C.B. (ed.) *Adelaide: water of a city*, Wakefield Press, Adelaide, 2009.
9. M. Serena and V. Pettigrove, 'Relationship of sediment toxicants and water quality to the distribution of platypus populations in urban streams', *Journal of the North American Benthological Society*, 24, 2005, pp. 679–689.

CHAPTER 9
The River Torrens 2: contaminated sediments in the river

Stephen J. Gale
Richard J.B. Gale
Hilary P.M. Winchester
Neill J. Dorrington
Nelson F. Cano

Introduction

Most attempts to assess the pollution of rivers have been based on analyses of the dissolved component of river waters. This is usually because of concerns about drinking water supplies or because water quality data may be directly related to potentially toxic effects in organisms. However, audits of water quality are likely to be inconclusive unless measurements are made over long periods of time and under a range of conditions. Variations in river discharge and pollutant source areas, irregular pollutant emissions, and changes in temperature and pH and Eh, can all result in large, short-term fluctuations in water quality. Unfortunately, long-term and temporally detailed monitoring of river waters is both costly and time consuming. In Australia in particular, the highly variable nature of river regimes means that long records of discharge and water quality are required before reliable interpretations are possible. Moreover, water quality data alone rarely reflect all the possible adverse effects of pollution on aquatic biota. One reason for this is that the interaction of organisms with the river environment is not restricted to the direct uptake of pollutants from the water. In particular, contamination of the channel bed and toxicity of suspended particulate matter may both have serious consequences for the ecology of aquatic systems.

There have been numerous studies of the water quality of the River Torrens, with investigations continuing at several sites.[1, 2] Unfortunately, many of the locations studied have been monitored only discontinuously and, of the remainder, most are sampled at intervals too infrequent to establish the detail of variations in water quality over time, particularly through individual flood and drought events. In addition, much of the sampling is composite, consisting of integrated samples of between two and six weeks of flow; information on short-term variations and temporal trends in water quality is thus lost. Even the longest records are still too short to be confident that they reliably portray trends in water quality. Finally, a rather narrow range of water properties has typically been monitored and, with a few notable exceptions, much of the resultant data is not easily accessible despite having been collected by publicly funded enterprises.

The focus on the dissolved load of rivers neglects the fact that it is the sediments that represent the major carriers of pollutants along the length of the channel. Concentrations of trace metals in river sediments, for example, may be between 10^4 and 10^5 times higher than those in the corresponding river waters,[3] making sediments the major repository of contaminants in river systems. Furthermore, those studies that concentrate their attention on water quality tend to overlook the fact that channel sediments may represent a source of bioavailable contaminants to benthic biota, and hence potentially to the aquatic food chain. Sediments may also provide an important secondary source of dissolved and suspended contamination of the river water. By this means, sediment quality may exert a significant influence on water quality and indirectly on all aquatic organisms.

Chapter photograph: The Torrens Weir.

Because sediment quality is a major determinant of the biodiversity and ecological health of rivers, sediment analysis provides a rapid and cost-effective way of characterising channel pollution and its ecological consequences. It can also play a key role in the detection of pollution sources and dispersion pathways, and may be useful in the selection of critical sites for routine water sampling. Moreover, by contrast with the temporally variable and difficult to interpret record of river contamination offered by water analysis, fluvial bed sediments provide a composite picture of pollution integrated over long periods of time. Bed sediments also integrate the effects of pollution over space, with contaminants from the contributing catchment being reworked into and along the channel. Finally, although the bedload most obviously acts as a sink for coarser particulate pollutants, fluvial bed sediments also incorporate materials transported in solution and suspension, deposited by mechanisms such as adsorption, chelation and precipitation. In this way, they unify, at least in part, the dissolved contaminant load, the suspended pollutant load, and the pollutant bedload.

The polluted river

Recent press reports have propagated the notion that Adelaide lies on the dirtiest river of any capital city in Australia.[4] On the evidence of a single water sample, it was claimed that the Torrens has the highest levels of copper, lead and faecal coliforms, and the second highest levels of nitrogen and phosphorus of any river in any state capital, and that '… generous traces of heavy metals remain in the sediments of all our city waterways'.

Until recently, few data were available to test such claims. In the last couple of years, however, systematic studies of the pollution of the sediments have been undertaken.[5, 6] These have established the regional pattern of inorganic pollution in the sediments of the Torrens and its tributaries, providing baseline data for the assessment of the impact of past and current land uses on the river, and for the development of future environmental strategies for the catchment.

The pattern of sediment contamination

The trunk stream of the Torrens is fed by a series of tributaries, all of which rise in the Adelaide Hills. The headwaters of one of these branches, First Creek, lie within the Cleland Conservation Park and are thus likely to be the least affected of any tributary by the impacts of European land use. The branch may consequently provide a control against which to judge the effects of land use changes downstream. A series of sediment samples was collected along the course of First Creek and along the River Torrens, downstream off its junction with First Creek (Figure 9.1). The samples were analysed to determine their cadmium, chromium, copper, lead, phosphorus and zinc concentrations.

Analyses were conducted on the <2.00 mm fraction of the sediments. With the exception of chromium, the concentration of every metal investigated in this study lies above the national trigger value for sediment quality (Box 43) at some point along the course of the river. In general, the rural sites (FC1–FC8) show little cause for concern, and the

BOX 43

Contaminant	Interim sediment quality guideline	
	Low (trigger value) ($\mu g\ g^{-1}$)	High ($\mu g\ g^{-1}$)
Cadmium	1.5	10
Chromium	80	370
Copper	65	270
Lead	50	220
Zinc	200	410

Table 1: Interim guidelines for sediment quality in aquatic ecosystems in Australia and New Zealand.[1] Only those guidelines applicable to the metals investigated in studies of sediment pollution in the Torrens[2,3] are listed.

National guidelines have only recently been established for sediment quality in Australian aquatic ecosystems.[1] It is important to stress that these are based on disturbance to biological indicators rather than on consequences for human health. Those guidelines relevant to the contaminants investigated in studies of sediment pollution in the Torrens[2,3] are listed in Table 1. The 'high' values are those beyond which adverse biological effects are likely to occur frequently. However, as it is important to intervene before damage takes place, 'trigger' values have also been established as prompts for further action.

In employing these procedures, a number of considerations must be taken into account. Firstly, no specific guidelines have been established for nutrients such as phosphorus. In part, this is because it is difficult to predict how these are released from the sediment to the water column. Nevertheless, nutrients may have significant impacts on stream systems. In particular, they may stimulate blooms of cyanobacteria, algae and macrophytes. These may result in the displacement of endemic species, the clogging of waterways, and the smothering of the streambed, altering aquatic habitats and threatening benthic communities. Some cyanobacteria and algae release toxins into the water, while blooms may alter stream pH and deplete dissolved oxygen, killing aquatic organisms.

Secondly, there are few reliable data on the biological effects of sediment toxicity in Australia. The guidelines are therefore based on the best available overseas data, largely employing the database of Long, et al.[4]

Thirdly, the guidelines are largely based on tests undertaken on one or two species, principally amphipods.

Finally, the guidelines should be regarded as providing only a first step in the determination of ecological risk. If possible, they should be supplemented and refined by site-specific studies of physical and chemical stressors (such as the use of background reference data), and by local biological studies.

Stephen J. Gale

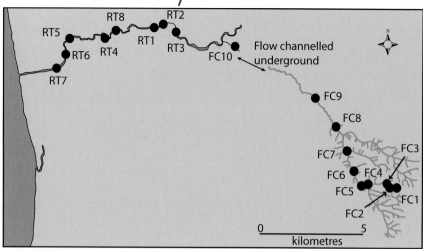

Figure 9.1 Sediment sampling sites along First Creek and the River Torrens.

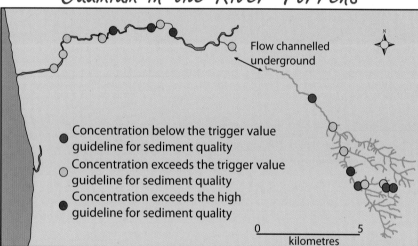

Figure 9.2 Cadmium in the mud and plant organic fractions of the bed sediment along First Creek and the River Torrens, by comparison with the Australian guidelines for sediment quality.

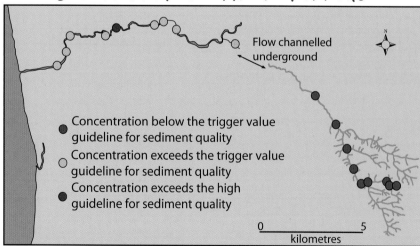

Figure 9.3 Chromium in the mud and plant organic fractions of the bed sediment along First Creek and the River Torrens, by comparison with the Australian guidelines for sediment quality.

high values of copper, lead and zinc are probably a consequence of the extensive copper–lead–zinc mineralisation of the underlying Proterozoic rocks. By contrast, all the downstream residential and industrial sites exceed either the trigger value or the high guideline (Box 43) for one or more of the measured toxicants.

Importantly, the process of fluvial deposition is likely to result in high silt and clay concentrations at particular sites. Since most toxicants have a particular affinity for the fine and plant organic fractions of the sediments, this may result in the guideline concentrations for the total sediment being exceeded. Under such circumstances the guidelines recommend that the sediment be assessed on the basis of analyses of the <63 μm fraction.[7] If this is done, a rather different picture emerges (figures 9.2 –9.6). First, every metal lies above the high guideline for sediment quality at some point along the river. Second, for each metal analysed every site from the city centre to the mouth of the river lies above the relevant trigger value. Third, the analyses reveal a belt of high toxicity running through the city with almost every site from FC10 to RT8 possessing concentrations exceeding the high guideline for every metal analysed.

The source of the contaminants in the sediments

In order to understand the origins of the pollutants in the sediments, we shall focus on perhaps the most significant of the trace metal contaminants, lead, and the most important of the nutrient elements, phosphorus.

The mean concentration of lead at the rural sites is 104 μg g^{-1}, somewhat higher than would normally be anticipated under natural conditions (the mean concentrations of lead in uncontaminated Australian regolith materials typically lie below 20 μg g^{-1}). This is likely to be largely a consequence of the extensive lead mineralisation of the underlying Proterozoic rocks between sites FC1 and FC8 rather than the product of widespread rural pollution.

As the stream enters the residential sprawl of the Adelaide Plains, however, mean lead concentrations increase by a factor of almost eight (Figure 9.7). The widespread use of lead in Western society over the last two centuries has been well documented; most important has been the use of tetraethyl lead as an additive to petrol. By 1998–1999, 98% of lead fallout from the atmosphere in Adelaide came either directly or indirectly from leaded fuels.[8] This was despite the prohibition on the use of leaded petrol by new vehicles in South Australia in 1986. Although the sale of leaded petrol ceased in the state in 2000, this can have had little effect on existing reservoirs of lead in the environment, which are likely to continue to be reworked into the river system for many decades.

With the passage of the river into the industrial zone, downstream of the city centre, there is a further substantial increase in lead levels. Average concentrations are almost 20 times those found at rural sites, with a maximum concentration of around 4800 µg g^{-1} at site RT2.

The lead may be derived from a number of possible sources. First, site RT2 lies immediately downstream of one of the busiest road junctions in Adelaide at Hindmarsh Bridge; the river sediments are thus likely to have received a combination of direct fallout of tetraethyl lead from vehicle exhausts and lead reworked from roads and other surfaces into the river. Second, the increase in lead pollution along this stretch of the river may be related to the former presence of a string of metal-processing industries along the reach downstream of RT2. In addition, a dozen or more metal-working concerns were located in the streets back from the river. Those alongside the river included the metal foundry of Mason & Cox, established in 1917 adjacent to site RT2, where it remained until the firm was progressively removed to a new location during 1966–1976. The Union Engineering Company's heavy engineering works shifted to a riverside location in the first decade of the 20th century while F.W. Hercus, an engineering and machine tool company, moved to premises in Anderson Street, Southwark, in 1938, remaining there until 1994. Although numbers have declined in the last few decades, several small-scale metallurgical concerns still operate in the area. Third, the Adelaide Chemical and Fertiliser Company, located upstream of site RT8, may have been responsible for the high sedimentary lead levels downstream of the plant.

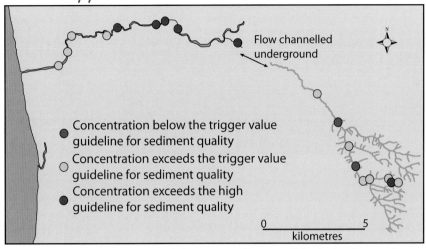

Figure 9.4 Copper in the mud and plant organic fractions of the bed sediment along First Creek and the River Torrens, South Australia by comparison with the Australian guidelines for sediment quality.

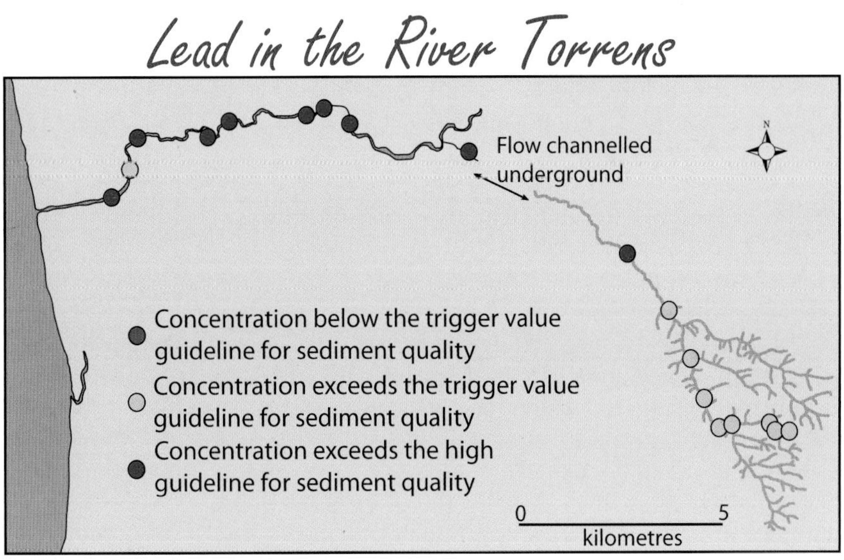

Figure 9.5 Lead in the mud and plant organic fractions of the bed sediment along First Creek and the River Torrens, South Australia by comparison with the Australian guidelines for sediment quality.

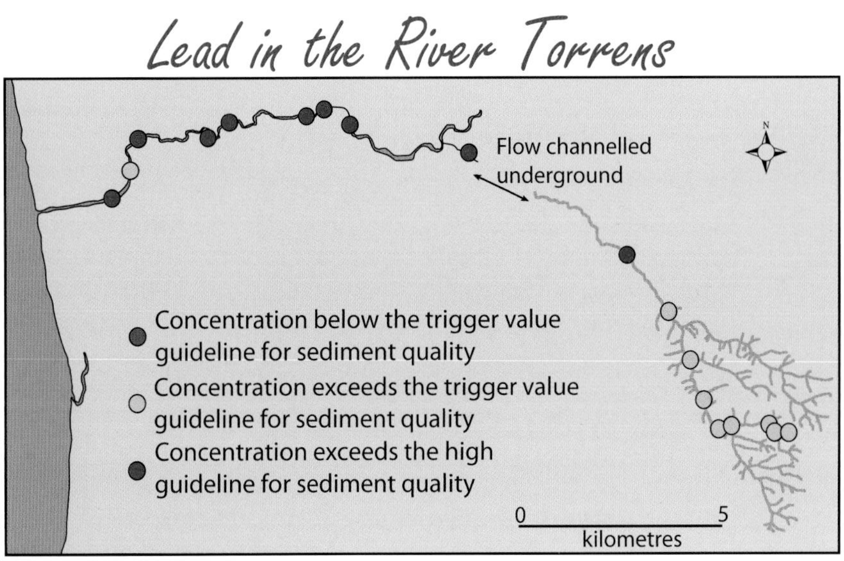

Figure 9.6 Zinc in the mud and plant organic fractions of the bed sediment along First Creek and the River Torrens, by comparison with the Australian guidelines for sediment quality.

Lead vs distance downstream

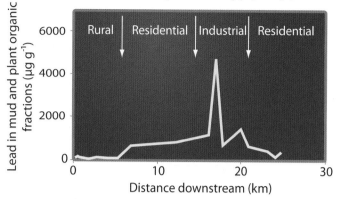

Figure 9.7 The concentration of lead in the mud and plant organic fractions of bed sediment with distance downstream along First Creek and the River Torrens.

The factory was established in 1881 or 1882 and remained in operation until at least 1956. A range of chemicals was manufactured including sulphuric acid; since this was produced using the lead-chamber process, in which the reactions forming the acid take place in large lead-lined chambers and towers, lead pollution may have been a by-product of the operation.

With the change to largely residential land use downstream of the city, lead concentrations in the river sediments return to levels characteristic of the residential areas that lie at the foot of the Adelaide Hills.

Phosphorus concentrations at sites in the Cleland Conservation Park lie within the range expected in uncontaminated Australian regolith materials. Downstream of Waterfall Gully (site FC6), however, phosphorus concentrations more than double. This is unlikely to be explicable other than by reference to human activities. The presence of phosphorus in a rural environment is likely to indicate inputs from fertiliser, sewage, animal excrement and detergent. During the 19th century, numerous market gardens and orchards lay along the reach between FC6 and FC8, and it is probable that manures were applied intensively and extensively on the hillslopes bordering the stream. It is equally probable that the newly cleared slopes experienced high rates of erosion, carrying both soil and manure into the channel. Interestingly, the highest rural values of phosphorus lie immediately downslope of a vineyard at site FC8. In this case at least, the phosphorus may come largely from the erosion of superphosphate-enriched soils.

There are continuing high levels of phosphorus at the residential sites (Figure 9.8). These may be a consequence of the intensive domestic use of artificial fertilisers, although the use of phosphatic detergents may also be important (and observation of the public park at site FC9 points to animal excrement as a major source of phosphorus).

There are further increases in phosphorus concentrations within the industrial zone (Figure 9.8). The most important source is likely to have been effluent from the Adelaide Chemical and Fertiliser Company. The company was the first in South Australia (and only the second in the country) to produce phosphatic fertilisers, which formed the mainstay of its business for decades. The factory was located upstream of site RT8, where phosphorus concentrations reach their highest value along the entire length of the river, 7300 µg g^{-1}, over twice the concentration at any other site.

The pharmaceutical company F.H. Faulding and Co. acquired a riverside site just upstream of RT1 in 1910, though new buildings were not erected until 1915. In addition to manufacturing a wide range of pharmaceutical products, the factory made fertilisers, insecticides and a range of poisons, and may have contributed to the high levels of many of the pollutants found downstream of the plant, particularly given the evidence of poor practice in waste disposal at the site. The company operated until 1990, when it moved to a new location to the north of the city.

Downstream of the city, phosphorus concentrations return to levels at and below those characteristic of the rural and residential areas adjacent to the Adelaide Hills.

The environmental impact of inorganic pollution of the river sediments

With the exception of chromium, the concentration of every metal investigated lies above the national trigger value for sediment quality at some point along the course of the river. In general, the rural sites show little cause for concern and the high values of copper and zinc probably reflect natural background conditions rather than pollution. By contrast, the lead content of every residential site exceeds either the trigger value or (at site RT4) the high guideline value for aquatic sediments. Similarly, the zinc content of every residential site but one lies either above the relevant trigger value or (at sites RT4, RT5 and RT6) above the value at which adverse biological effects are likely to occur. The cadmium content of residential sites RT4, RT5 and RT6, immediately downstream of the industrial zone, is also above that at which intervention is judged to be required.

Phosphorus vs distance downstream

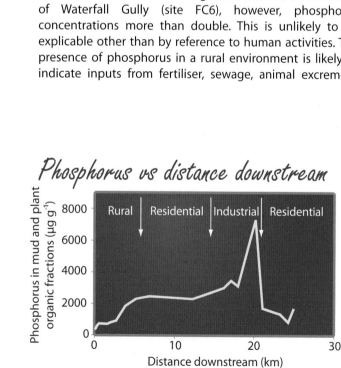

Figure 9.8 The concentration of phosphorus in the mud and plant organic fractions of bed sediment with distance downstream along First Creek and the River Torrens.

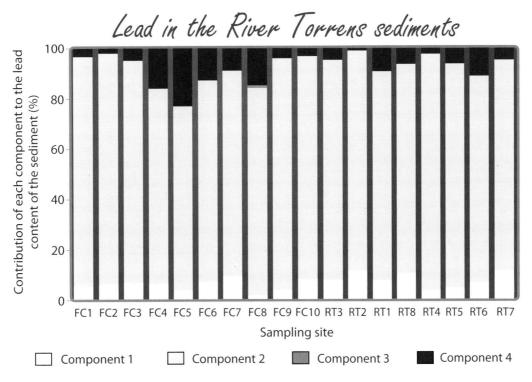

Figure 9.9 The relative proportion of the lead content of the <2.00 mm fraction of bed sediment of the River Torrens released into solution under specific environmental conditions. The sampling sites are plotted as a function of distance downstream along First Creek and the River Torrens. Component 1 represents the most mobile forms of lead, while components 2, 3 and 4 represent progressively less abile forms.

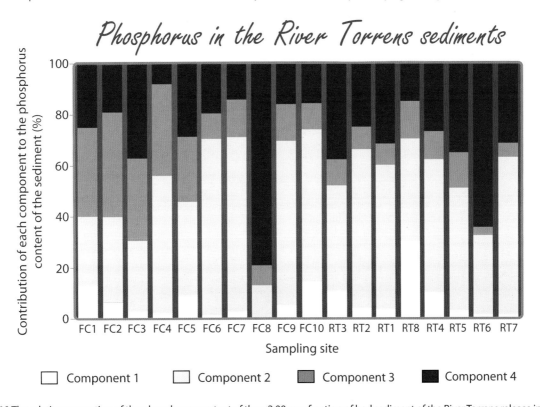

Figure 9.10 The relative proportion of the phosphorus content of the <2.00 mm fraction of bed sediment of the River Torrens release into solution under specific environmental conditions. The sampling sites are plotted as a function of distance downstream along First Creek and the River Torrens. Component 1 represents the most mobile forms of phosphorus whilst components 2, 3 and 4 represent progressively less labile forms.

Within the industrial zone, every site exceeds the trigger values for lead and zinc. Site RT2 (in the case of lead) and sites RT1 and RT2 (in the case of zinc) lie far beyond the high guidelines for biological impact. Two sites (RT1 and RT2) exceed the trigger value for cadmium, and one site (RT1) exceeds that for copper.

These threshold values are based on analyses made on the entire <2.00 mm fraction of the sediments. However, if the sediments are assessed on the basis of analyses of the <63 μm fraction, as recommended in the national guidelines,[7] the picture is far more serious. For every metal analysed, every site from the start of the industrial zone to the mouth of the river lies above the relevant trigger value. In the case of lead and zinc, every industrial site lies above the high threshold for aquatic sediments. For all the metals analysed, at least one of the industrial sites lies above this threshold.

In the residential zone, the zinc concentration of all the sites and the lead concentration of all but one of the sites lie above the high guidelines for sediment quality. The situation is not quite so bleak for the other metals, though every residential site downstream of the industrial area and most of those above it exceeds the relevant trigger value.

The rural sites present a more favourable picture, though site FC6 is a focus for cadmium and zinc, perhaps a consequence of overindulgent application of fertiliser in the neighbouring gardens. Most of the relatively high metal values in this zone are likely to reflect the background mineralisation of the bedrock, and it is possible that the ecology of the river is little affected as a consequence.

Although there are no national guidelines against which to assess the phosphorus content of the sediments, there is strong evidence that human activities have had a significant impact on phosphorus levels in the river. There have been major cyanobacterial blooms along the lower Torrens over the past decade, which have been linked to the release of nutrients from the sediments.[9] Phosphorus levels in the water have also reached dramatic levels. The South Australian Environment Protection Authority reported a mean total phosphorus content of 0.439 mg l^{-1} in the River Torrens during the period June 1995 to January 1997.[10] This is far above the relevant default trigger value for South Australian lowland rivers.[7]

The biological availability of toxicants in the sediments

The chemical analyses of the sediments of the Torrens provide values that approximate the total concentration of each pollutant in the sediment. However, not every component of each toxicant is necessarily available to organisms. A greater or lesser fraction may instead be chemically inert under the environmental conditions likely to prevail in rivers, and may thus play no part in contributing to the toxicity of the waterbody. The thresholds listed in Box 43 are based on a wide range of work that considers the response of aquatic organisms to 'total' (acid extractable) toxicant concentrations. This is the approach employed in almost all attempts to assess sediment quality in aquatic ecosystems, both in Australia and worldwide. Nevertheless, the application of more sophisticated procedures to establish the mobility of each pollutant element under a range of environmental conditions would clearly be of considerable value in efforts to understand the environmental impact of these toxicants. The mobility of each pollutant element in the river's sediments was therefore determined by employing a stepwise extraction procedure to establish the proportion of each pollutant released into solution under specific environmental conditions.

The first extraction liberates the most highly mobile component. This is the fraction that is adsorbed to carbonates, or is soluble in water or weak acids. Step two targets the reducible constituent, that bound to iron and manganese oxides. The third step aims to recover the oxidisable components, the fraction bound to organic matter and sulphides. The final step yields the remaining constituents of the sediment.

The results of the application of this procedure to the analysis of lead and phosphorus are shown in figures 9.9–9.10. It is clear that the largest component of both elements becomes mobilised during the second stage of extraction, that which replicates the occurrence of reducing conditions in the river. We should therefore anticipate that a significant proportion of the lead and the phosphorus in the sediments would be released into solution under reduced, anoxic conditions. Monitoring of the river's waters close to the sediment interface in Torrens Lake reveals that anaerobic conditions occur frequently during the warmer half of the year (Figure 9.11), often lasting for days at a time and may result in the remobilisation of contaminants into the water column and/or into aquatic food chains, with potentially significant implications for the ecology of the river.

In an attempt to break down the thermal stratification thought to contribute to the existence of anoxic conditions in the lake, Adelaide City Council has installed a series of aerators in the lake (the first

Dissolved O₂ in the River Torrens

Dissolved oxygen (mg l⁻¹)

12
8
4
0

2000.0   2000.2   2000.4   2000.6   2000.8   2001.0

Year

Figure 9.11 The variation in the dissolved oxygen content of the waters of the River Torrens at a depth of 5 m at City Weir, Torrens Lake, from 21 December 1999 to 27 January 2001. The measurements were taken <1 m above the bed of the lake. The data were collected by the Australian Water Quality Centre.

was commissioned in 2001). These may have reduced the incidence of anoxic events in this part of the system, though the continued occurrence of cyanobacterial blooms in the lake suggests that conditions under which nutrients (and thus potentially other contaminants) are released from the sediments to the waterbody may still occur.

But what of the remainder of the river? Are reducing conditions experienced in other parts of the Torrens? The absence of sites at which frequent long-term monitoring of dissolved oxygen levels has been undertaken makes this question difficult to answer. Nevertheless, the Torrens Catchment Water Management Board[11] has acknowledged that low levels of dissolved oxygen occur at many points along the river. In addition, intermittent monitoring of the upstream reach of the river at Mount Pleasant suggests that, even away from the lake, low dissolved oxygen conditions are repeatedly experienced in the river's waters. As in Torrens Lake, these events all occur in the warmer half of the year. Such evidence that exists, therefore, points to the repeated occurrence of conditions under which contaminants may be mobilised from the sediment body to the water column.

Conclusions

The fact that pollutant concentrations in the sediments of the River Torrens reach values that are amongst the highest recorded in any Australian aquatic environment is hard to ignore. The concentrations of lead in the river's sediments are 10^3 times those of the river's waters, while the sedimentary concentrations of phosphorus are 10^2–10^4 times those of the water, making the sediments the major repositories of pollutants in the system. It is likely that the sediments provide the major source for much of the pollution in the waters of the river. Indeed, the Patawalonga and Torrens Catchment Water Management boards[9] have already linked major cyanobacterial blooms along the lower Torrens to the release of nutrients from the sediments. Despite frequent assertions to the contrary,[12, 13] there is compelling evidence that conditions under which the pollutants may be released to the water column occur periodically throughout the year. The environmental implications of this should be the subject of careful and immediate investigation. At the very least, the remarkably high levels of pollution identified in many parts of the river justify a precautionary approach.

The claim that Adelaide lies on the dirtiest river of any capital city in Australia[4] is difficult to assess. Some of the enrichment in the headwater streams is likely to be natural, the product of copper–lead–zinc mineralisation of the underlying rocks. It is clear, however, that the sediments of the Torrens downstream of its junction with First Creek are contaminated by cadmium, lead, phosphorus and zinc. Levels of chromium and copper are also of concern. Pollution at every site along the trunk stream is such that immediate investigation is required to ensure that ecological damage is prevented. In the case of lead and zinc, several sites far exceed the likely threshold for frequent adverse biological impact. It is probable that much of this contamination is a response to past pollution practices, whether the result of industries intentionally or unintentionally discharging waste into the river, or the product of long-term reservoirs of contaminants being reworked into the channel. Unfortunately, these problems

are likely to persist; there is little evidence that bed sediments are being moved downstream and flushed out of the system, or that they are being diluted by mixing with relatively uncontaminated sediments. Indeed, the close association between pollution peaks in the river and likely sources of pollution would suggest little mobilisation of sediment has taken place since deposition first occurred. This is not surprising. The chains of weirs, sluices and reservoirs along the river must have reduced the magnitude of the flood flows that would have entrained the sediments under natural conditions, while simultaneously acting as sediment traps to prevent pollutants from being transported downstream.

References

1. Eco Management Services Pty Ltd, 'Revised Torrens Catchment water management plan working paper on water quality, riverine habitat and aquatic biodiversity', Torrens Catchment Water Management Board background report, Adelaide, 2000.
2. R. Kawalec and S. Roberts, 'Review of South Australian state agency water monitoring activities in the Torrens Catchment', Department of Water, Land and Biodiversity Conservation report 2006/12, 2006, p. 110.
3. A.J. Horowitz, 'A primer on trace metal-sediment chemistry', *United States geological survey water supply paper*, 2277, 1985, p. 67.
4. A. Hodge, 'Waterways of abuse', *Australian*, 18 August 2003, p. 10.
5. R.J.B. Gale, S.J. Gale and H.P.M. Winchester, 'Inorganic pollution of the sediments of the River Torrens, South Australia', *Environmental Geology*, 50, 2006, pp. 62–75.
6. S.J. Gale, R.J.B. Gale and H.P.M. Winchester, 'Contaminated sediments in the River Torrens, South Australia', *South Australian Geographical Journal*, 105, 2006, pp. 78–105.
7. Australian and New Zealand Environment and Conservation Council and Agriculture and Resource Management Council of Australia and New Zealand, 'National water quality management strategy', paper no. 4, *Australian and New Zealand guidelines for fresh and marine water quality*, vol. 1, (chapters 1–7), Canberra, 2000.
8. F. Adeeb, R. Mitchell and L. Hope, 'Future air quality monitoring for lead in metropolitan Adelaide: a report to the National Environment Protection Council', Environment Protection Authority, Adelaide, 2003.
9. Patawalonga and Torrens Catchment Water Management boards, 'Blue-green algal blooms: a challenge for Torrens Lake', *Pat and Torrens water: news from the Patawalonga and Torrens Catchment water management boards*, December 2002, p. 11.
10. Environment Protection Authority, 'State of the environment report for South Australia 1998, summary report', Department for Environment, Heritage and Aboriginal Affairs, Adelaide, 1998.
11. Torrens Catchment Water Management Board, 'Torrens Comprehensive Catchment water management plan 1997–2001', Adelaide, 1997.
12. A. Ockenden to H.P.M. Winchester, 28 April 2006, University of South Australia archives, Adelaide.
13. Torrens Taskforce, 'Torrens Taskforce summary of findings final report', Report to the Minister for Environment and Conservation, Government of South Australia, Adelaide, 2007.

Box references

Box 43: Australian sediment quality guidelines
1. Australian and New Zealand Environment and Conservation Council and Agriculture and Resource Management Council of Australia and New Zealand, 'National water quality management strategy', paper no. 4, *Australian and New Zealand guidelines for fresh and marine water quality*, vol. 1 (chapters 1–7), Canberra, 2000.
2. R.J.B. Gale, S.J. Gale and H.P.M. Winchester, 'Inorganic pollution of the sediments of the River Torrens', 2006.
3. S.J. Gale, R.J.B. Gale and H.P.M. Winchester, 'Contaminated sediments in the River Torrens', 2006.
4. E.R. Long, D.D. MacDonald, S.L. Smith and F.D. Calder, 'Incidence of adverse biological effects within ranges of chemical concentrations in marine and estuarine sediments', *Environmental Management*, 19, 1995, pp. 81–97.

CHAPTER 10

The River Torrens 3:
creating a new riverscape

David S. Jones

The River Torrens/*Karrawirra Parri* Linear Park disguises the fact that it is the largest integrated stormwater management project undertaken in Australia to produce an aesthetic greenway which has become a core recreational venue for the people of Adelaide. A major recreational corridor, little thought is now given to the fact that the historically regular floods and pondings along the length of the River Torrens/*Karrawirra Parri* now do not occur due to this major piece of infrastructure. What was a naturally semi-arid watercourse subject to semi-arid-characteristic flash flooding is now a calm streamlet regulated with artificially deposited addition water flows and a sequence of water retention devices that visually imply that the ephemeral stream is now perennial.[1]

Introduction

The demise and transformation of the River Torrens/*Karrawirra Parri* from a major ephemeral ecological corridor occurred quickly with the advance of post-1836 European settlement. As vegetation was felled, watercourse alignments changed, land was furrowed, and the waters were pumped to enable irrigation and water supply – and biodiversity was lost. Intensive agriculture claimed the adjacent alluvium soil flats with industry in others, and the watercourse became a dumping venue for stormwater, pollutants and hard wastes.

Narratives quickly abounded about the despoliation of the watercourse, and what it contained. These stories were made worse by the fact that despite the stench and condition of the River Torrens/*Karrawirra Parri*, it remained the principal potable water source for many Adelaideans:

> A loathsome gutter may be seen … which conveys filth into the creek, thence into the river and thence into the cisterns of the cisterns. Let anyone who doubts … visit the spot and look about him. Let him also trace the creek from St. John's Church down to the Lunatic Asylum and say whether what is now a filthy ditch might not, by embankment and other careful management, be converted into a refreshing stream.[2]

The City of Adelaide Council's slaughterhouse (1835–1910), originally situated in *Tulya Wodli*/Bonython Park/Park 27, was consciously sited adjacent to the River Torrens/*Karrawirra Parri* watercourse, at the 'The Billabong', before it shifted to Gepps Cross.[3, 4] The slaughterhouse was infamous in its condition and the way it dumped waste into the watercourse:

> There the cattle were supposed to be killed under regulated conditions in the presence of the superintendent, W.J. Hemsley, and the quality of the meat checked by men with no personal interest in getting all of it to market. But no one who knew anything about the slaughterhouse or heard the ugly rumours which were circulating in the 1890s could place much trust there. The building itself dated back to the early days of the colony and originally has been a barracks. The roof was lofty, unceilinged, and made of crumbling old slate covered with iron. The place where the meat was hung was a mass of dirty, stained

timber posts and beams, impossible to keep clean. The floors were in a disgraceful state; a patchy mixture of absorptive brick which soaked up the filth and worn concrete on which pools of blood and water lay until hosed away. There was only one pitching pen, so all the cattle, diseased and healthy alike, had to be killed in the same place. Up to ten were poleaxed at once, and bled where they fell, so that the floor of the pen was 'all blood and bullocks', as an observer described it. No veterinary surgeon was on hand to check the cattle. A mere two% of the carcases were condemned – an impossibly low figure – and they were disposed of by being sprinkled with kerosene and buried in the nearby olive plantation, a procedure fairly described by Mayor [Charles] Tucker in the corporations' own public report as 'cumbersome, inelegant, and from a sanitary and health point of view utterly ineffective [sic.].[4, 5, 6]

'The Billabong' was to the south-east of the slaughterhouse site, the source of stone for the construction of the slaughterhouse and possibly the Adelaide Gaol, thereby increasing the river's width and the apparent sharpness of its embankments. With its abandonment as a source of stone, parklands rangers and city gardeners appear to have appropriated it as a venue for rubbish and hard waste dumping. This function operated from the 1860s to 1889 when it was reputedly filled in and covered over.[6]

Parklands ranger William Campbell reported over 1887–1889:

> The work of filling in the billabong on the south side of the Slaughterhouse with household refuse still continues.[7]

> The work of filling in the billabong on the south and east sides of the slaughterhouse with household refuse still continues, and which will, when completed, form a good road to the entrance of the cattle market.[8]

> The work of filling in the billabong on the South and East sides of the Slaughteryards, with household refuse has been completed, and now forms a good road to the entrance of the Cattle Market.[9]

Rubbish tipping continued to be a constant management practice over the years as River Torrens/*Karrawirra Parri* swimmer A.E. Kenney recorded in 1937:

> Last week I visited my old boyhood swimming resorts. What did I find? The Star which used to be 150 yards [137 m] long and from five to fifteen feet [1.5 m–4.5 m] deep almost silted up and a rubbish tip. The spoil from Parliament House excavations has been deposited here, and at the first flood will be washed back to Adelaide, palm bulbs and all. The Blue-Bell and Eureka are also displaced by a rubbish tip. The Gilberton and Walkerville pools are likely to be silted up with sand washing establishments which have almost entirely stopped the flow of water necessary to carry away silt. N.A. [Northern Area] district and other pools are similarly doomed. Upon whom does the blame lie? A natural watercourse turned into a rubbish tip. I fear that it is too late to rescue the river to even a semblance of its former state without enormous expense.[10]

Greenways and corridors

In this context of riverine pollution, deterioration and neglect, it is important to understand the theoretical concept that drove attempts to rejuvenate and renovate the riverine corridor in the 1960s and 1970s. It was not simply that the river now existed as a disorganised, neglected stormwater corridor, but that many people recalled the natural qualities it used to possess and questioned the ethics of why such a transformation had occurred, in contrast to other cities that had respected their core watercourse as a major recreational corridor and a symbol of city pride and cultural advancement. This cultural advancement was the very point Edwin Thomas Smith used in 1880 when, as City of Adelaide mayor, he successfully argued for a master plan to green the Adelaide parklands[11] and to create Elder Park and the 'serene waterbody' formed by the River Torrens Weir.[4, 12]

The River Torrens/*Karrawirra Parri* is an ephemeral watercourse that runs east–west through the metropolitan area of Adelaide and through the city of Adelaide. It is a major topographical feature, literally dividing the city in two portions. Embracing these two portions is the larger Adelaide parklands, making it an unique international town planning and landscape architecture achievement. The parklands are considered an international town planning precedent, used by 'City Beautiful' proponent Ebenezer Howard (1850–1928) in his *Gardens Cities of To-morrow* (1946),[13, 14, 15, 16] which first put forward the notion that a city needs a healthy green landscape to service and renew its citizens, but also to distinguish suburbia from rural productive land. In this context, the Adelaide parklands is an exemplar of a 'green belt' or 'greenway' encircling the primary city, whereas the River Torrens/*Karrawirra Parri* is today a lineal 'greenway', linking hill to sea and interconnecting with the parklands. Both coexist and are dependent upon each, and both define and support the biological and hydraulic fabric and character of the Adelaide metropolitan area. In addition, within the reaches of the River Torrens/*Karrawirra Parri*, on the plains/*Mikawomma*, are the 'five creeks' of the River Torrens that stretch across the Burnside region creating minor 'greenways' or ecological corridors in their own right.[17]

Christine Garnaut has summarised the City Beautiful/green belt theory:

> *Garden City ideology shaped local thinking [in Australia] by promoting the green belt as a means of redressing the absence of metropolitan open space arising from uncontrolled nineteenth-century expansion, and as a tool for structuring and organising metropolitan space … The function of the green belt – sometimes called a 'lung', 'girdle', or 'ring' – was variously perceived to contain expansion, frame and separate new from existing development, reserves natural bushland, and provide contact with nature.[18]*

The 'green belt' theory underpins and characterises town planning and landscape architecture projects in South Australia. The green belt concept was adopted by surveyor-general George Goyder (1826–1898) in directing future settlement survey and planning in South Australia.[19, 20, 21]

Goyder's surveying performance criteria directed the expansion of the 'park land' town concept throughout South Australia in the 1860s–1890s; inaugural government town planner Charles Reade (1880–1933) continued this agenda, moulding the notion in City Beautiful language and design configurations to craft a suite of new towns in the 1910s–1920s, including Colonel Light Gardens, which his successors carried forward in towns and subdivisions.[22, 23]

Accordingly, the River Torrens Linear Park is a continuation of the green belt theory, creating a multifunctional 'green lung' and a major lineal venue for recreation, biodiversity and water quality enhancement. This philosophy was articulated by United States landscape architect Frederick Law Olmsted (1822–1903) in his proposition statement with his successful Central Park design competition entry for the City of New York.[1, 24]

Natural systems

Metropolitan Adelaide lies approximately on latitude 35° south on a semi-arid, low, sloping plain, characterised by a Mediterranean climate and ephemeral watercourses. Average rainfall in the 1970s for Adelaide was 533 mm with a majority falling in April–October with some 50% in May–August, and with the majority of rain being associated with winds moving south-west to north-west across Adelaide. Summer temperatures in the 1970s averaged between 27°–29°C, with winter averages of between 18°–21°C. There are some 2500 hours of sunshine annually.

The River Torrens flows for some 30 km through metropolitan Adelaide, effectively dividing the region into northern and southern suburbs. It originates on heights of 230–250 m and quickly descends into a major valley system before seeking out an old meandering course and flooding pattern.

The geological structure of the Torrens Valley consists largely of quartzite and slates in the upper reaches with recent Torrens alluvium deposits westward past the Burnside Fault, which crosses near Athelstone. Accordingly, soils along the watercourse are mainly silty, clayey alluvium deposits in the upper reaches or sand and gravel deposits in the lower reaches.

The original vegetation structure, prior to European colonisation (see Figure 10.1), was characterised by a ribbon of river red gum (*Eucalyptus camaldulensis*) riverine community with a middle-storey of *Bursaria spinosa*, *Callistemon salignus*, *Acacia rotundifolia* and *Acacia retinodes*, and patches of *Eucalyptus leucoxylon* and *Acacia melanoxylon*. In the Reedbeds of Fulham and Henley Beach, the vegetation was originally dominated by *Phragmites communis* and *Typha augustifolia* with patches of *Triglochin procera* in open water.[25]

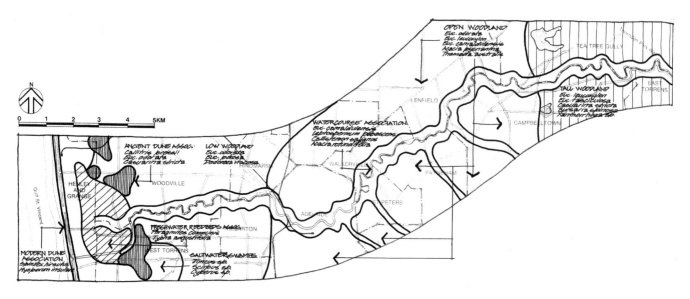

Figure 10.1 Vegetation prior to European settlement. Source: HASSELL, 'River Torrens study: a coordinated development scheme', 1979, p. 42.

Flooding

The River Torrens/*Karrawirra Parri* is essentially a semi-arid ephemeral watercourse fed by headwaters that flow from the rains and springs in the Mount Lofty Ranges. Artist George Fife Angas (1789–1879) expressed this situation in a note accompanying his *Lower Fall of Glen Stuart, on the Morialta Rivulet in the hills near Adelaide*:

> *Although South Australia is generally considered a dry country, it is watered by numberless small streams or 'creeks' as they are styled by the settlers … During the summer months these 'creeks' cease to run, whilst their course is marked by occasional ponds or natural reservoirs here and there, and the rank and luxuriant vegetation that arises within the influence of their moisture.*[26]

Prior to colonisation, the River Torrens/*Karrawirra Parri* functioned as an intermittent conduit that was not constant. The Torrens has a steep longitudinal gradient over a short distance and peak flows occur after short, intense storm events. These flows have increased exponentially with suburbanisation, as a consequence of increased non-porous surfaces and human attempts to remove water as quickly as possible through conventional stormwater and watercourse engineering measures.

Historically, land development and suburbanisation has engulfed the former marshlands and swamps at the morass near the mouth of the River Torrens/*Karrawirra Parri*, necessitating major drainage and water channel measures to allow the waters to exit into the Gulf St Vincent artificially through a dredged Torrens channel and mouth at Henley Beach and the Patawalonga Creek at Glenelg. But as suburbanisation increased upstream in the catchment, with peak flows being generated from its upper reaches, it became glaringly obvious that the Torrens watercourse and its 'five creeks' tributaries could not handle these increasing flows, and that hydraulic engineering channeling was inadequate to cater for erratic flows.

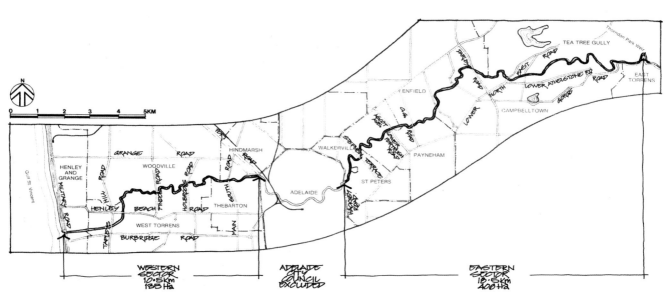

Figure 10.2 River Torrens study area. Source: HASSELL, 'River Torrens study', 1979, p. 2.

Linear parks as travel spaces

BOX 44

In the aftermath of Adelaide's freeway debates, the South Australian government emerged with proposals to improve cycling facilities and build a light rail line from the central city to the rapidly expanding north-eastern suburbs. The government's preferred light rail route was to run alongside the Torrens Valley, criss-crossing the river at several points. Respondents to the government's consultation process generally supported the need for an upgraded public transport system but many were opposed to the use of the river for this purpose. There was an over-riding concern that the Adelaide parklands and parks along the route would be destroyed and construction would harm remaining native vegetation and birdlife. Submissions to the North East Area Public Transport Review, including from the River Torrens Committee, variously implored government to desist from the plan or take the opportunity to implement a planting program to enhance native vegetation and habitat in the area. A change in government saw plans for the light rail line shelved but a guided busway, the O-Bahn, was built in its place. More than 20 years on, the busway is one of the most highly patronised public transport routes in metropolitan Adelaide – an achievement attributable, in part, to the speed of the journey on a dedicated route.

This public transport initiative added impetus to an idea, suggested by Adelaide's first town planner, Charles Reade, early in the 20th century, to create a park along the River Torrens. Although the transport corridor would be 12 km long, the Linear Park would be 35 km, almost the entire length of the river. The Linear Park proposal did not emerge directly from the light rail/busway scheme, but, as a 'reservoir for native flora and fauna', it addressed some public concerns about using the Torrens Valley for public transport. Apart from its environmental and recreational roles, one of the most important social attributes of this riverside park is mobility.

The social role of travel is generally reduced to a person's ability to access destinations. The social benefits of the journey itself are ignored, with the emphasis being placed on rapid movement from origin to destination. But journeys are comprised of a range of possibilities from exercise, socialising, clearing our minds of a busy day, to reaching a destination. We can combine, recombine and add new possibilities to each journey we make. The cycling, walking and public transport facilities incorporated into the Torrens Valley bring to the fore the multiple permutations of travel. Early morning walkers and runners are not only exercising, they are also building their relationships with their walking/running partners, the people they see regularly, and building a relationship with the context in which they move. Cyclists out for a family bike ride are also working to knit family members together in the shared experience and space of mobility. The Linear Park tends to be thought of in terms of recreation and leisure but the paths are also used by commuters. For commuter cyclists seeking to avoid motor vehicle traffic, the route along the Torrens makes cycling to work a possibility and allows them to combine transport and health. The pathways, designed to be shared by cyclists and pedestrians, recommend the journey at a slower pace. The sensuousness of travelling on foot or by bicycle is accentuated in the travel spaces along the river. Slowing the journey shifts emphasis to the bodily experiences (especially sound, smell, touch) of being *in* rather than moving *through* a place.

Jennifer Bonham

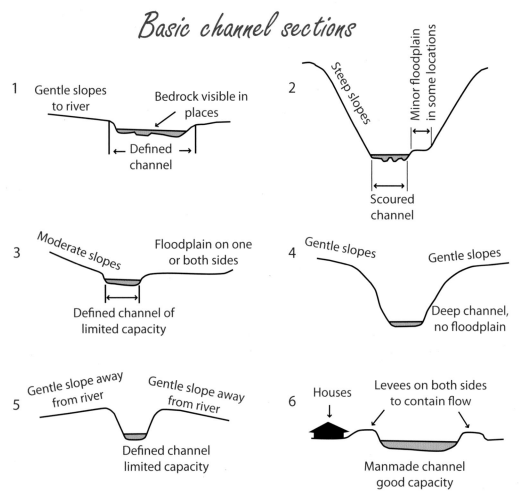

Basic channel sections

1 Gentle slopes to river
Bedrock visible in places
← Defined → channel

2 Steep slopes
Minor floodplain in some locations
Scoured channel

3 Moderate slopes
Floodplain on one or both sides
Defined channel of limited capacity

4 Gentle slopes
Gentle slopes
Deep channel, no floodplain

5 Gentle slope away from river
Gentle slope away from river
Defined channel limited capacity

6 Houses
Levees on both sides to contain flow
Manmade channel good capacity

Diagram 10.1 Six basic watercourse channel types. Source: HASSELL, 'The River Torrens study', 1979.

Flooding was a central issue when the Torrens was discussed by Adelaideans in the 1870s–1970s. Clear felling for agriculture and subsequent suburbanisation only assisted in adding to waterflow volumes. A generation of engineers, with the use of concrete and excavations in their training, added further to the problem by increasing the velocity of flows and thereby the frequency and intensity of damage. With these ingredients, the Torrens experienced major floods in 1844, 1847, 1852, 1889, 1897, 1889, 1898, 1917, 1923, 1931, 1937, 1965 and 1968. The 1931 floods, as an example, resulted in some 1845 ha of western Adelaide being inundated including most of the suburb of Henley Beach, with the waters at the Albert Bridge on Frome Road being some 60 m wide.[27]

Seeking a design solution

By the 1950s it had become very clear to a number of councils adjacent to the River Torrens/*Karrawirra Parri* that the watercourse had become severely degraded and presented a major problem in its management. In 1959, some 13 councils, together with state government representatives, formed a 'watchdog committee', the River Torrens Improvement Standing Committee. Initially lacking political and planning mechanisms to enable action, the committee expended its deliberations upon general environmental matters including landfill, flooding and sand mining. In 1964, the committee was reconstituted as the River Torrens Committee to

formally advise the minister of works 'in relation to the conservation and enhancement of the River Torrens'.[27]

A state town planning committee in 1962 recommended the area be developed as a major open space corridor. However, this recommendation gained no legislative weight until the passing of the *Planning and Development Act 1967*, which enabled this policy to be pursued albeit in a fragmented, uncoordinated manner by councils. This was except in terms of drainage, where strategic objectives authorised a multilevel planning approach in the corridor to address community use, land use suitability, transport, flood mitigation, water pollution, and transportation.

Released in 1968, the state government's Metropolitan Area Transport Strategy (MATS) proposal provoked more debate and a need for coordinated planning. MATS proposed freeway arterial routes down the valley corridor, especially to service the growing north-eastern suburbs, as well as a light rail option. These controversial transportation proposals prompted considerable debate in the Adelaide but, thankfully, were not implemented, except the recommendation to construct a dedicated guided busway, the O-Bahn, along the corridor, which became the catalyst for a major environmental planning investigation of the river corridor.

The O-Bahn involved the construction of a concrete-track guided busway route from Hackney Road, near the Adelaide Botanic Gardens, to Tea Tree Gully, running roughly parallel to the river watercourse. Such works anticipated extensive earthworks, numerous watercourse crossings, and strategic flood mitigation measures, and prompted a need to address overall environmental and aesthetic concerns in the scheme and a need for integrated planning and design.

Community and governmental debates resulted in the commissioning of Hassell & Partners in June 1975, in conjunction with a multidisciplinary team, to prepare an outline report, 'River Torrens Study Stage One Report' (1976), which analysed the key issues and set forth possible strategies for debate and consultation, later redrafted into the 'River Torrens Study: A Coordinated Development Scheme' (1979).[27]

Hassell devised a three-stage process of overall investigation and planning, which drew much inspiration from the frameworks set out in *Design with Nature* (1969) by Ian McHarg.[28] These processes involved:

1. Collecting and collating the available information.
2. Transplanting the complex ecological processes and issues involved into a manageable form so that decisions, based upon the best available information, could be made.
3. Preliminary consideration of the potential environment, social and economic impacts of alternative forms of development.
4. Public participation in the decision-making process.
5. Providing the pubic with adequate information about the process, plans and findings, so that they could make reasonable value judgements.[27]

Within this investigation, the definition of 'development' was interpreted as a generic term embracing all forms of development whether recreational, building, active or passive conservation, etc.

A formal public consultation process was employed in 1976 that included a major public questionnaire and a series of five structured workshops. Approximately 20,000 questionnaires were distributed in the catchment area, and 5076 (25.4%) were returned. The questionnaires sought the opinions of the public as to values, activities and uses. A series of five public workshops, with 200 attendees, promoted discussion as to 'the River's existing use and character' and 'planning your section of the River'. The process resulted in the following themes:

1. Most people used the river and environs for passive recreation: walking, relaxing and enjoying nature.
2. Parts of the river were not used, being polluted, overgrown, poorly accessible and potentially dangerous.
3. The river should be reserved for public recreation, nature reserves, and bushland retained.
4. Less emphasis on intensive development and more on the promotion of activities such as walking, cycling, fishing, informal picnics and games.

Arising from these investigations, the 1979 River Torrens Study realised a state government objective to:

> … *prepare a plan and strategy to co-ordinate development and conservation of the River Torrens and environs to ensure the rational use of resources at both a regional and local level.*[27]

The objectives of the scheme sought to:

1. Give due regard to the riverine ecosystem and surrounding environment.
2. Conserve, make accessible or apparent, areas of high social value, whether they be of historic, scientific, or education value.

3. Restore or rejuvenate degraded areas and those areas which are incompatible with the river and its intended uses.
4. Identify areas suitable for various forms of development or use.
5. Determine appropriate recreational uses for the land suitable for such development.
6. Provide a variety of recreation facilities in accordance with the needs and aspirations of the community.
7. Capitalise upon the linear quality of the river and provide a continuous link between the developable portions of land and if possible other community recreation facilities.
8. Coordinate all planned and imminent development.
9. Provide a planning framework which serves to identify short and long-term development proposals and guidelines, but is sufficiently flexible to take into account changing social needs.[27]

Key components of the scheme, in order of priority, included:

1. Ecosystem conservation: the retention of remnant native vegetation and aquatic habitats, including recovery of degraded areas.
2. Geomorphology: emphasis on and provision for stream accretion and erosion, as against constructed 'prevention' along with flood detention in the upper catchment, so as to regain water levels and flow velocity within stream capacity.
3. Protection of cultural landscape values: provision to protect items and areas of historical or social value, such as swimming holes.
4. Restoration of despoiled resources: a concerted program to improve water quality, re-vegetate and halt inappropriate activity such as rubbish dumping.
5. Identification of development potentials: within a development suitability framework, areas available for a range of uses, both public and private, were identified.
6. Recreation: existing and potential recreation areas were identified and the general type of activities nominated, such as sports field, camping, or picnic area.
7. Linear parkway: the alignment for a walkway/cycle path network, along and across the full length of the river and linking to local and regional facilities, including an outline of remedies to specific impediments, such as road crossings.
8. Landscape guidelines: the setting of standards by provision of outlines for construction techniques, clearing of river channels, structures, standards for trail systems, vehicular access and carparks, recreation facilities and buildings, park furniture and revegetation regeneration techniques.

Assessing and designing

As part of the study, the physical and ecological characteristics of the watercourse length were mapped and analysed. Arising from this approach, and associated community engagement and additional technical investigations, a suite of design proposals were drafted that depicted works to be undertaken along the length of the watercourse.

In terms of the evaluation or assessments, the entire watercourse was mapped out in series of annotated drawings that considered: topography and slope; soils and geological features; existing vegetation; hydrology; water quality and aquatic biology; wildlife; transport and access; land use; land ownership; zoning; recreation facilities; historic sites; river structures; visual resources; recent developments and activities; and development and conservation proposals. Several examples of the annotated maps depicting this information are included in this chapter.

In terms of design proposals, a series of interconnected maps detailed the extent of the project's scope; in particular new pedestrian, cycle and equestrian trails, existing and proposed tree planting, contours, existing historic features, together with annotated proposals as to quarry rehabilitation, trail renovations and strategies, and new recreational facilities including their siting and associated facilities and earthworks, new bridges and carparks.

A typology of six basic watercourse channel types were identified and mapped (Diagram 10.1),[27] including:

1. Upper Rural: for some 30 km of the river headwaters, with steep grades and exposed bedrock in parts;
2. Lower Rural: for some 20 km principally through the Torrens Gorge, with steep grades, high velocities, and scouring over bedrock;
3. Urban Zone 1: between Athelstone and Windsor Gardens, a constrained watercourse channel with extensive adjunct floodplains, with natural perennial waterholes and some human-made intrusions;
4. Urban Zone 2: from Windsor Gardens to Hackney Road, a deep and wide channel with no adjunct floodplains;
5. Urban Zone 3: from Port Road westwards with reductions in channel size and width and gently sloping adjacent land; and
6. Urban Zone 4: a human-constructed channel replacing nature.

Along some sections of the river, steep banks were of concern. Diagram 10.2 demonstrates why designers sought to 'lay back' these banks as part of the project.

Water quality

Historically the Torrens had been treated as a surrogate stormwater and waste removal system, provided by nature in lieu of engineering construction. Thus, solids and liquids were both deliberately and unconsciously dumped into the watercourse and its tributaries and found their way to the main watercourse, thereby contributing to the general pollution of the watercourse.

Urban stormwater quality is dependent upon the types of activities in its catchment, together with the effectiveness of the sewerage system and the effectiveness and frequency of street sweeping and flushing events. While a large proportion of pollutants are reputedly flushed away in the first flush cycle following initial rainfall, it has been recorded that these pollutants can be at levels equivalent to raw sewerage. Summative potential pollutant sources, perceived for the River Torrens/*Karrawirra Parri*, are set out in Table 10.1.

In reviewing water quality, the study found that a number of patterns were occurring. Non-point sources (e.g., stormwater) were identified as the major source of pollutants to the river. Faecal contamination, on the basis of physical and chemical water quality testing, was the most significant form of pollutant. Nitrogen and phosphorous levels, of nutrient sources overall, were very high and often accompanied by algal blooms and aquatic macrophytes, according to seasonal eutrophication patterns. High levels of lead and zinc were recorded in some locations and significant levels of chlordane pesticides were also recorded. Biological evidence validated that the watercourse system could support macroinvertebrate fauna and higher forms of animals including fish and birds.

Laying back the banks will ...

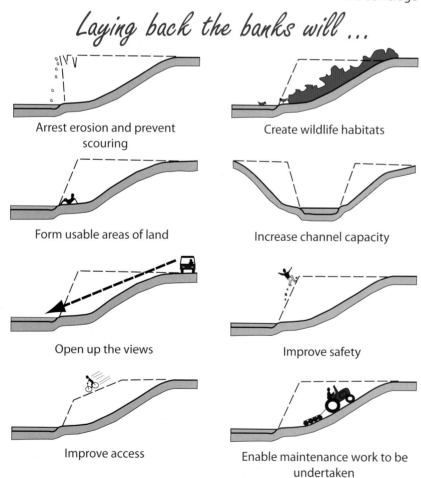

Arrest erosion and prevent scouring

Create wildlife habitats

Form usable areas of land

Increase channel capacity

Open up the views

Improve safety

Improve access

Enable maintenance work to be undertaken

Diagram 10.2 The positive effects of laying back the banks of the River Torrens. Source: HASSELL, 'The River Torrens study', 1979, p. 215.

Table 10.1 Urban runoff pollutants

Constituent	Point sources	Non-point sources	Possible effects
Solid wastes Debris, domestic and industrial wastes	Construction projects	General litter from household, commercial and industrial sources	Unsightly and reduction in stormwater channel carrying capacity
Suspended solids Silts, dust, colloidal substances	Construction projects, sanitary landfills in the form of leachate	Street surface contaminants, erosion from pervious areas as well as unlined channels	Provides a major transport mechanism for a number of more dangerous pollutants and reduces the stormwater system capacity
Algal nutrients Nitrogen, phosphorous	Fertilisers	Atmospheric fallout, mineral leaching and the decomposition of organic matter, animal excretions	Increase in aquatic plant growth leading to the possibilities of algal blooms and eutrophication. Reduction in recreational amenity
Bacteria and viruses Total coliform, faecal coliform, faecal streptococci	Illegal connections of household and commercial sullage units to stormwater systems, sewer overflows, septic system overflows		Health hazard due to pathogenic bacteria and viruses
Toxic materials Lead, DDT, etc	Pesticides, household, commercial and industrial	Automobile pollutants (leakage of fuel, asbestos, lead, etc.)	Possible health hazards with excess concentrations
Oxygen demanding substances		Degradation of vegetation, organic matter, oil and grease, animal excretions and human waste products	Reduction in the health of waterways, can lead to fish kills, foul odours and slime growth

Table 10.1 Typical urban runoff pollutants – sources and possible effects. Source: HASSELL, 'The River Torrens study', figures 2–18, 1979, p. 25.

Implementation

The design, based on the 'River Torrens Flood Mitigation Study' (1980), the River Torrens Study (1979), and agreements arising thereto, was approved by the state government's Parliamentary Standing Committee on Public Works in 1981. At the same time, proposals for the O-Bahn busway were integrated into the River Torrens Study (1979), which had envisaged a light-rail transportation system and not specifically the present guided bus route.

Key elements of the agreement by the state government, were:

1. The acquisition of the necessary land.
2. The modification of the Kangaroo Creek Dam to mitigate the catchment flows.

3. The undertaking all flood mitigation works identified for the urban areas of Adelaide in the above reports.
4. The construction of the basic Linear Park proposal.[1]

Key elements of this agreement, by respective local governments, were:

1. Assuming responsibility for landscape management of the Linear Park.
2. Constructing pedestrian and bicycle trail systems
3. Developing other passive and active recreational facilities considered to be appropriate to their resources and identified in the above reports.

Conclusion

In summing up, a simple question can suffice, 'Have you witnessed flooding on the Torrens?' To the majority of Adelaideans the answer is 'no'. Thus, we take for granted a major design and engineering endeavour and have progressively forgotten what used to occur regularly. A further point rotates around biodiversity enhancement. Over the last 30 years we have witnessed the greening of the Torrens watercourse with the renewal of maturing vegetation and indigenous vegetation. We have also witnessed an increase in animal, reptile, fish and bird habitat residency, demonstrating the natural environmental renewal of the area as a major ecological corridor in the metropolitan area. At the same time, for us humans, Linear Park has become an integral recreational activity and movement corridor.

It was once observed that 'the most successful design is that which we do not see and which we eagerly inhabit'. This is a clear conclusion from the works that have created the River Torrens Linear Park. Accordingly, we have created an abstract and highly modified version of nature's green corridor, which once existed along the River Torrens/*Karrawirra Parri*. It is a version far from the pre-settlement one but still valuable in contributing to our aesthetic and psychological health and that of our city.

References

1. T. Dexter, 'Adelaide creates a great asset: River Torrens Linear Park', *Landscape Australia*, 4, 1997, pp. 343–348.
2. No Author, 'Summer Odours', *Register*, 8 December 1855.
3. R. Maurovic, *The meat game: a history of the Gepps Cross abattoirs and livestock markets*, Wakefield Press, Kent Town, 2007.
4. D. Jones, 'Adelaide parklands and squares cultural landscape assessment study', section 3: Tulya Wodli/Park 27 report, 2007, Adelaide Innovation and Research, Adelaide (online).
5. City of Adelaide, 'Mayor's annual report', City of Adelaide, 1896.
6. P. Morton, *After Light: a history of the City of Adelaide and its councils, 1878–1928*, Wakefield Press, Adelaide, 1996.
7. City of Adelaide, 'Annual report 1886–1887'.
8. Ibid., 1887–1888.
9. Ibid., 1888–1889.
10. A.E. Kenney, letter to the editor, *Advertiser*, 25 February 1937.
11. J.E. Brown, *A report on a system of planting the Adelaide parklands*, Government Printer, Adelaide, 1880.
12. D. Jones, 'Adelaide parklands and squares cultural landscape assessment study, section 3.1 25: Tarndanya Womma/Park 26 report, 2007 (online).
13. F.J. Osborn, *Garden cities of to-morrow [original text by E. Howard (1898) Tomorrow: a peaceful path to real reform]*, Faber, London, 1946.
14. R. Freestone, *Model communities: the garden city movement in Australia*, Nelson, Melbourne, 1989.
15. S. Hamnett and R. Freestone, *The Australian metropolis: a planning history*, Allen & Unwin, Sydney, 2000.
16. R. Freestone, 'City beautiful movement' in Aitken, R. and Looker, M. (eds) *The Oxford companion to Australian gardens*, Oxford University Press, Melbourne, 2002, pp. 140–141.
17. J.W. Warburton, *Five metropolitan creeks of the River Torrens, South Australia: an environmental and historical study*, Civic Trust of South Australia and Department of Adult Education, University of Adelaide, 1977.
18. C. Garnaut, 'Green belts' in Aitken *et al The Oxford companion to Australian gardens*.
19. D. Jones, 'Goyder, George Woodroofe' in Aitken *et al The Oxford companion*.
20. M. Williams, *George Woodroofe Goyder: a practical geographer*, Wakefield Press, Adelaide, 1982.
21. R. Bunker, 'Town planning at the frontier of settlement' in Hutchings, A. (ed.) *With conscious purpose: a history of town planning in South Australia*, Planning Institute of South Australia, 2007, pp. 45–60.
22. C. Garnaut, 'Reade, Charles Compton' in Aitken *et al The Oxford companion to Australian gardens*, pp. 502–503.
23. J. Tregenza and C. Reade, 'Town planning missionary' in Hutchings *With conscious purpose*, pp. 21–33.
24. C.E. Beveridge and D. Schuyler, *Creating Central Park, 1857–1861*, Johns Hopkins University Press, Baltimore, 1983.
25. D.N. Kraehenbuehl, 'Pre-European vegetation of Adelaide: a survey from the Gawler River to Hallett Cove', Nature Conservation Society of South Australia, Adelaide, 1996.
26. G.F. Angas, *Lower fall of Glen Stuart, on the Morialta rivulet in the hills near Adelaide*, McLean, London, 1847.
27. HASSELL and Partners Pty Ltd, 'River Torrens study: a coordinated development scheme', Government of South Australia, Adelaide, 1979.
28. I.L. McHarg, *Design with nature*, Doubleday/Natural History Press, Garden City, New York, 1969.

CHAPTER 11
Wastewater treatment and recycling

Grant Lewis

Introduction

Adelaide, like all cities, generates many different types of waste products including human solid and liquid wastes (sewage), which poses a risk to human health and the environment unless properly managed. The process of collecting, conveying, treating and disposing of wastewaters underlies the sustainability of a city.

Good public sanitation is the cornerstone of good public health. Sanitation engineering has had a hugely positive impact on community health over the last one hundred years. Before sewerage infrastructure was commonplace, when human waste was still being emptied into the street, a variety of diseases were prevalent in cities. People today generally have little appreciation of how conditions have improved over time. In the main, wastewater management occurs out of sight of the general community and is very much taken for granted.

As many Australians learn to live with drought, treated wastewater is now increasingly being seen as a resource rather than a waste product; recycling is now an increasingly important source of water for non-potable use in Adelaide.

Adelaide's sewer system

Adelaide's sewer system has a number of components. All houses within metropolitan Adelaide are required to be connected to the sewer system. The sewer network efficiently removes wastewater from residences, businesses, commercial facilities and industrial sites, and minimises contact between the population and raw sewage.

Sewage collection and treatment is centralised through a network of buried pipes and underground pumping stations, which collect and convey sewage to wastewater treatment plants (WWTPs). Houses typically discharge domestic wastewater through 80 to 100 mm diameter pipes, which then connect with a 100 to 150 mm diameter sewer pipe buried underground in the adjacent road. The majority of wastewater discharged flows under gravity.

Sewer manholes are provided at strategic locations along the sewer pipes laid in roadways, allowing access to the sewer pipes in order to clear blockages and provide for the intersection of pipes from different directions.

The Adelaide metropolitan area sewerage system contains more than 300 pumping stations and 6900 km of buried sewer pipes, with more than 455,000 house and business connections and serving about one million people.[1]

The nature of the terrain dictates the number and location of sewer pumping stations. A number of gravity sewer mains will feed a larger pipe that terminates at a pumping station. These stations typically consist of a 'sump', containing submersible pumps located at the bottom of the sump. Pumps convey wastewater through pipelines over hills or through long distances either to other pumping stations or,

Chapter photograph: Christies Beach wastewater treatment plant.

eventually, to one of the four large WWTPs located near the coast.

Metropolitan Adelaide is segregated into four major drainage areas for wastewater collection, each of which discharges to a separate WWTP. There are two plants at Bolivar to the north, a plant at Glenelg on the western side of the city, and another at Christies Beach in the south. All employ variations of the activated sludge treatment process and discharge treated wastewater into the coastal marine waters of Gulf St Vincent through outfalls, with increasing quantities being recycled (see Figure 11.1). A small WWTP at Aldinga serves some of the population in the expanding southern fringe of the metropolitan area.

The WWTPs produce good-quality treated wastewater that has limited environmental impacts on receiving waters. Some of the treated wastewater is being reused for irrigation of parks and gardens, market gardens, vegetables and vineyards. This trend is likely to continue, particularly if drought conditions persist.

What is sewage?

Sewage is wastewater generated by the community and typically has domestic, commercial and industrial components. Household contributions arise from toilet flushing ('blackwater'), and from the kitchen sink, shower and washing machine ('greywater').

In most parts of Australia, about 40% of the water entering residential properties becomes sewage. The rest is used mainly to water gardens. Quantities of sewage vary, but a typical volume is about 200 L per person per day. A town with a population of 10,000 therefore produces about 2 ML of sewage per day (2000 t/day). The volume of sewage can increase markedly if significant quantities of groundwater and stormwater infiltrate into the sewers through cracked pipes, poorly fitted manhole covers, or leaky joints.

Commercial enterprises, such as restaurants, butchers, supermarkets and retail outlets, all contribute wastewater containing a range of different contaminants. Larger-scale industrial business, including wool scourers, manufacturers, chemical and petroleum processors, paint factories, abattoirs, and food processing factories, also contribute a variety of wastewaters and a range of other processes. Some of these pollutants are difficult for the conventional wastewater treatment process to deal with, and these trade wastes are regulated. Discharges to sewer may only be permitted if the wastewater is of a similar strength to domestic sewage, otherwise special onsite treatment may be required.

Wastewater is more than 99.9% water, with a small amount of putrescible solid matter and other pollutants. These contaminants are categorised as physical, chemical and biological in nature.

1. **Physical contaminants** include solids ranging from cans, pieces of wood, twigs, plastics, small toys, cardboard packaging, leaves, gravel, sand, grit and clay.

220

Table 11.1 Wastewater treatment plant capacities

Plant	2006/2007 inflows	Design capacity
Bolivar activated sludge	147 ML/d	165 ML/d
Bolivar high salinity	23 ML/d	32 ML/d
Glenelg	46 ML/d	60 ML/d
Christies Beach	26 ML/d	31 ML/d
Totals	**242 ML/d**	**287 ML/d**

ML per day of wastewater. 1 ML = 1,000,000 L.
Table 11.1 Metropolitan Adelaide wastewater treatment plant capacities. Source: SA Water.

2. **Dissolved and particulate putrescible organic (carbonaceous) matter** includes faecal material, waste meat and vegetable products, and nutrients (nitrogen and phosphorous) in various forms. Organic matter and ammonia are 'oxygen demanding' wastes that are converted by bacteria to harmless substances if there is sufficient free oxygen present.

3. **Chemical contaminants** include inorganic trace metals (e.g., copper, lead, chromium, aluminium, tin, zinc and others), and a range of natural and synthetic organic chemicals.

4. **Microbiological contaminants**. Many disease-causing organisms (pathogens) can be waterborne. Waterborne diseases that used to be common in human populations include typhoid, cholera, dysentery, poliomyelitis, hepatitis, and a large number of gastroenteric and other diseases.

If sewage is released untreated into the environment, all of these contaminants can pose a risk to public health and also cause damage to ecosystems. Raw sewage released into a stream can cause bacteria and other microorganisms to use all of the available dissolved oxygen in the water, resulting in the death of aquatic lifeforms and extensive ecosystem damage. Excessive nutrients in treated wastewater and the right weather conditions may also result in uncontrolled algal blooms in waterbodies contaminated with sewage. Algal blooms can block sunlight from reaching aquatic plant species and 'choke' them, as well as depleting oxygen overnight.

WWTPs are employed to reduce or remove the putrescible solid and dissolved wastes in sewage by physical and biochemical processes, which mimic natural processes but are concentrated in space and time through the application of engineering technology. The resultant treated wastewater (sometimes also referred to as 'effluent') contains low quantities of biological and chemical pollutants, nutrients, and, depending on the technology employed at the plant, the discharge poses little or no risk to the environment.

Adelaide's wastewater treatment plants

After Adelaide was founded and settled in 1836, the growing population relied on the River Torrens as a source of water and for disposal of human wastes. Night carts were employed to remove wastes from residences from 1848 and households also employed 'soakage trenches' with outside 'privies'. Noxious odours were commonplace and disease outbreaks, notably the waterborne diseases cholera and typhoid, occurred periodically.[2] Parliament finally acted, and from 1871 sewage was piped from the city to a sewage treatment farm at Islington to the north-west. The subsequent reduction in the mortality rate of Adelaide's inhabitants was significant; by 1901 two out of three houses in the entire urban area were connected to sewer and by 1965 Adelaide was nearly completely sewered.[2]

WWTPs were constructed to serve the Port Adelaide and Glenelg residential areas in the 1930s. Port Adelaide, at the time, was the most advanced plant in the southern hemisphere and one of the first to utilise the 'activated sludge process'. In 1966, the Bolivar Sewage Treatment Works opened and replaced the old Islington Sewage Farm. The new Bolivar plant was designed to cater for a population of 600,000 persons and consisted of secondary treatment and stabilisation lagoons, a major advance from practices at Islington. The last major metropolitan facility, Christies Beach WWTP, was commissioned in 1971 to serve the rapidly expanding southern suburbs. Current WWTP capacities are shown in Table 11.1.

All of these plants discharge treated wastewater into coastal marine waters. In the case of Glenelg and Christies Beach, the discharge pipes are near popular metropolitan beaches. At Bolivar, treated wastewater is discharged into coastal mangroves near St Kilda, and Port Adelaide WWTP formerly discharged into the Port Adelaide River but was shut down in 2004. Treated wastewater discharges contain nutrients (nitrogen and phosphorous), and there are concerns that the discharges may be causing harm to marine and estuarine ecosystems along the metropolitan coastline.

The introduction of the *Environment Protection Act 1992* and the establishment of an independent Environment Protection Authority (EPA) provided a comprehensive framework for the regulation and management of all wastewater discharges in South Australia. Assessment of environmental impacts of projects became more important and there was recognition that decisions involving environmental consequences could no longer be wholly driven by cost or political will. The Act provided for the preparation and implementation of Environment Improvement Programs (EIPs) for WWTPs. An EIP is a list of actions or activities to a timetable, designed to manage environmental impacts of a particular activity. The initial licences for the metropolitan WWTPs granted by the EPA in the mid 1990s contained a requirement for EIPs to be developed and implemented.

A major focus of environmental concern then, and now, was the degree of adverse impacts of treated wastewater and stormwater, which could happen when were discharged on the seagrass beds along the metropolitan Adelaide coastline into Gulf St Vincent. The thrust of the EIPs was to reduce the quantities of nitrogen discharged to coastal waters through a combination of treatment process upgrades and an increase in the reuse of treated wastewater, wherever feasible. A total of approximately $260 million was spent on EIPs between 1995 and 2004. The EIPs included the following projects:

1. At Bolivar it was decided to replace the existing trickling filters with an 'activated sludge process'. The upgraded plant was commissioned in February 2001.[3] Just prior to this, reuse was to be increased with the implementation of the Virginia Pipeline Scheme (VPS).
2. The Christies Beach WWTP was upgraded to enhance nitrogen removal using the 'intermittent fixed film activated sludge process'. This adds small, free-floating plastic media to aeration tanks in order to increase the biomass present and thus enhance treatment capacity without building new aeration tanks. The upgraded plant was commissioned in August 2002.
3. The Glenelg WWTP 'A' plant was decommissioned, 'B' and 'C' plants converted to the 'intermittent fixed film activated sludge process', and a conventional activated sludge 'D' plant was constructed. The upgraded plant was commissioned in October 2002.
4. The Port Adelaide WWTP is no longer in operation. Sewage from parts of the former Port Adelaide drainage area is now pumped to the new Bolivar high-salinity WWTP. The lower-salinity wastewater from the drainage area is pumped into the main Bolivar sewage system, which conveys it to the Bolivar activated sludge WWTP, where the treated wastewater is available for reuse. The Bolivar high-salinity WWTP was commissioned in October 2004.

All three major metropolitan WWTPs employ various forms of the activated sludge process (Diagram 11.1) to treat the influent sewage. The basic function of these plants is to speed up the natural process through which wastewaters are purified. A typical activated sludge plant includes the following components:

- The first stage is called 'preliminary treatment', where incoming sewage is screened, grit is removed and primary settling occurs.
- Mechanical screens remove large objects such as rags, cans, bottles, packaging, pieces of wood, leaves, etc from the sewage stream.

- The sewage often but not always then passes through a grit removal facility where sand and coarse particulate inorganic matter settles to the bottom of the tank and is removed. If not removed, this material may damage pumps and other machinery.
- The sewage then passes into a primary sedimentation tank where the organic solids settle under gravity and can be removed for further processing.
- In the secondary treatment stage, the settled wastewater (which contains fine and soluble organic material) is mixed with recycled biomass and then enters the main bioreactor where the mixture is aerated. Microorganisms biochemically oxidise the fine particulate and dissolved putrescible organic wastes and ammonia to eventually produce additional biomass, carbon dioxide and nitrogen gas. Thus the oxygen-demanding wastes are stabilised.
- The mixture of biomass and wastewater is then passed to a secondary setting tank or clarifier, where the biomass settles to the bottom of the tank and the clear supernatant water is discharged.
- Settled biomass (activated sludge) is pumped from the bottom of the clarifier and recycled to the head of the aeration tank to 'seed' the incoming sewage. Some is also 'wasted' from the system to the digesters.
- The biomass from the primary sedimentation tank and wasted biomass from the secondary clarifier are still biologically active and need to be broken down. These wastes are typically discharged to an anaerobic digester, when the waste products undergo further treatment and are stabilised.
- The digesters are heated and produce digester gas, mainly methane. This is usually used to generate electrical power which is used in the WWTP, with waste heat from the engines being used to heat the digesters.
- The digested sludge is then mechanically dewatered using a centrifuge or air-dried in drying beds. After a withholding period of up to three years, the dried sludge (or biosolids) may be used as a soil conditioner on farmland. A successful program has supplied biosolids to a number of farmers in country South Australia.
- The treated wastewater from the clarifiers is disinfected by adding chlorine and then discharged to coastal marine waters via an outfall structure. Disinfection by ultraviolet light is an alternative.
- A proportion of the treated wastewater is recycled from the Bolivar, Glenelg and Christies Beach plants. Uses include irrigation of market gardens, winery vineyards, parks and gardens.

The EIPs focused on nutrient reduction through plant upgrades, and it was recognised that increased reuse would also reduce the quantities of treated wastewater, and hence nutrient loads, discharged to Gulf St Vincent. A combination of the plant upgrades and increased reuse have reduced the nitrogen discharges from all four metropolitan WWTPs by 60 to 70% since 1996/1997, as shown in Figure 11.1. This represents a significant improvement. The decommissioning of the Port Adelaide plant in late 2004 has also resulted in a halt to all treated wastewater discharges into the Port Adelaide River, another significant environmental

An activated sludge wastewater treatment plant

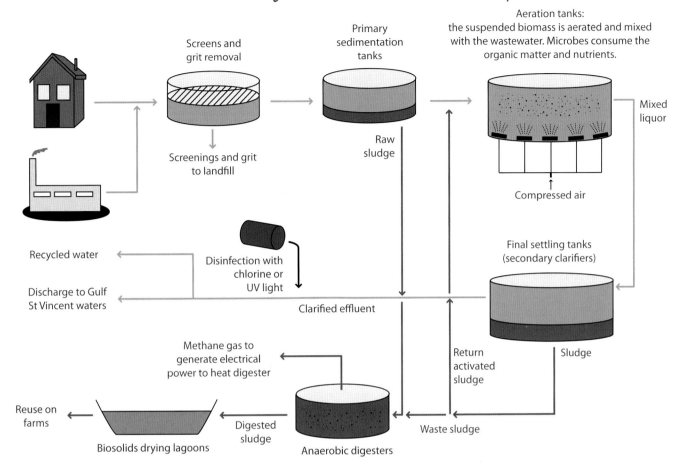

Diagram 11.1 The stages of treatment for industrial and household sewage at a typical activated sludge wastewater treatment plant. Source: SA Water.

achievement. Further reductions in nitrogen discharges will occur in the future as reuse increases at Virginia and in the Willunga Basin.

Reuse of treated wastewater

Treated wastewater may be reused, reclaimed or recycled (it means the same thing) in a range of ways. For example:

- municipal uses – irrigation of playing fields, ovals, recreation reserves, flower beds, ornamental plants, etc;
- agricultural uses – irrigation of pasture, vegetable crops, grape vines, tree plantations (e.g., pines and blue gums), turf production;
- industrial uses – boiler feed, process water;
- dual reticulation – use to water household gardens and for toilet flushing (e.g., Mawson Lakes).

In South Australia, water recycling schemes require the formal approval of the Department of Health, and the EPA is also involved in the assessment of schemes. The regulation of recycled water use became more straightforward in 1999 when the *South Australian Reclaimed Water Guidelines (Treated Effluent)* were published,[4] providing clear direction for both regulators and potential users regarding what quality of recycled water was need for safe use in particular

applications. Treated wastewater was classified as shown in table 11.2. Note that the highest-quality recycled water (class A) has the widest range of applications.

The main determinant of recycled water quality is the microbiological characteristics. Class A recycled water, for example, is required to have median *E. coli* levels less than 10 organisms/100 mL, whereas class C recycled water has less than 1000 *E. coli* organisms/100 mL. *E. coli* is a coliform microorganism present in the human gut and its presence serves as an indicator of faecal contamination. If *E. coli* levels are high it is likely that other harmful or pathogenic microorganisms may also be present. In an approval for the use of class A recycled water, there may also be requirements to monitor other microorganisms of concern such as adenovirus, rotavirus, giardia and cryptosporidium. There are also criteria for chemical and physical recycled water properties.

Metropolitan recycling schemes were developed when the 1999 state guidelines were in place. In 2006, new national guidelines, *Australian Guidelines for Water Recycling*, were released, and it is anticipated that the state guidelines will be revised to be compatible.[5] The national guidelines were developed in response to recognition that there was a need to protect the health of the public and the environment

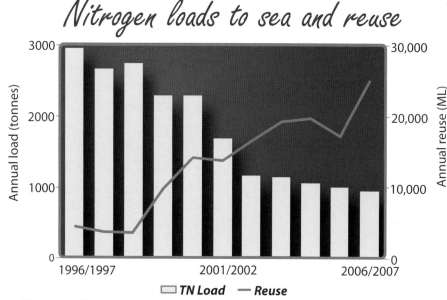

Nitrogen loads to sea and reuse

Figure 11.1 Wastewater treatment plant total annual nitrogen loads to sea and annual reuse during the period from 1996 to 2007. Source: SA Water.

while taking a consistent and Australia-wide approach. The national guidelines take a different approach to the 1999 state guidelines by applying a comprehensive risk management assessment to recycled water use. This is potentially a more rigorous, systematic and complex management technique, providing a higher degree of protection to the community, albeit at a higher cost for water managers as more reuse schemes are developed. The risk management approach requires the reuse scheme proponent to:

- identify the hazards of recycled water;
- estimate risk from hazards;
- identify preventative measures;
- monitor the ongoing situation; and
- verify that the management system is working.

The 2006 national guidelines have replaced the nomenclature of class A, class B, etc., with a more detailed description focused on reuse applications, as shown in Table 11.3. (The class A, B terminology is used below as a number of water recycling schemes discussed here were developed before the introduction of the 2006 guidelines.)

A key component is the degree of microorganism-removal required, expressed as 'log virus reduction'. There are also log-removal criteria for bacteria and protozoa, however, viral-reduction tends to require the greatest degree of treatment. A 6.0 log-removal means that 99.9999% of viruses are removed from the treated wastewater, which is a very stringent criteria.

'Municipal use with unrestricted access' is equivalent to the 1999 class A guideline and 'municipal use with restricted access' is the equivalent to the previous class B. The dual reticulation uses are of a higher quality than the previous class A and require very high reductions in viruses. In practice, this may require both ultraviolet (UV) irradiation and chlorine for disinfection to ensure that a wide range of microorganisms are inactivated. This reflects the new, more stringent approach to guaranteeing public health in recycled water applications.

The quality of the recycled water in the 2006 national guidelines is defined by the treatment processes used to produce it. For example, if treated wastewater is passed through a sand filter and chlorinated appropriately, it may be classified as 'suitable for unrestricted reuse'. If more chlorine is added and possibly combined with UV irradiation, the recycled water may be classed as 'suitable for dual reticulation'. The emphasis is on ensuring reliable removal of particulates (which may shield microbes from disinfectants) and inactivating pathogenic microorganisms by putting appropriate, robust treatment filtration and disinfection barriers in place.

Table 11.2 Classification of reclaimed water

Application	Class A	Class B	Class C	Class D
Primary contact	✓			
Non-potable use in homes	✓			
Unrestricted crop irrigation	✓			
Secondary contact	✓	✓		
Stock watering	✓	✓		
Golf courses and playing fields	✓	✓	✓	
Pasture and fodder	✓	✓	✓	
Restricted crop irrigation	✓	✓	✓	✓
Woodlots	✓	✓	✓	✓
Turf irrigation	✓	✓	✓	✓

Table 11.2 Classification of reclaimed water in 1999 guidelines. Source: SA Water.

Table 11.3 Water recycling guidelines

2006 water recycling guidelines	Log virus red'n required	Required treatment processes
Dual reticulation – toilet flushing, washing machines, garden use	6.5	Secondary treatment, coagulation, filtration, disinfection
Dual reticulation – outdoor use only or indoor use only	6.0	Secondary treatment, coagulation, filtration, disinfection
Municipal use – open spaces, sportsgrounds, golf courses, dust suppression or unrestricted access and application	5.0	Secondary treatment, coagulation, filtration, disinfection
Municipal use with restricted access and application	n/a	Secondary treatment with disinfection
Municipal use with enhanced restrictions on access and application	n/a	Secondary treatment with >25 days lagoon detention

Table 11.3 *Water recycling guidelines: recycled water quality classifications*, 2006. Source: SA Water.

This approach has the advantage that long-term, costly recycled water sampling and analysis of recycled water for a range of different pathogenic microbes is not necessarily needed to establish that the required performance is being achieved by the treatment process. However, the treatment processes used must have their performance confirmed by a qualified laboratory in accordance with the United States' Environmental Protection Agency or American Waterworks Association testing standards, a process referred to as 'validation'.

The 1999 state guidelines required irrigation management plans to be developed and implemented with recycled water use to ensure long-term environmental sustainability. Irrigation management plans are documents which typically outline the location and areas to be irrigated, the types of crops, grassed areas or vegetation to be irrigated, watering times and monthly quantities applied, and estimated nutrient and salt loadings. Monitoring programs are also detailed. Nitrogen and phosphorous can build up in soils and become a problem for maintaining healthy vegetation; salt can also build up in plant root zones and inhibit plant growth. Groundwater also needs to be monitored. Irrigation management plans are in place for all metropolitan reuse schemes. The 2006 guidelines require the development of a recycled water management plan, incorporating all elements of an irrigation management plan, plus additional material.

In South Australia, the lead agency in approving recycled water schemes is the Department of Health. SA Water and other treated wastewater suppliers (e.g., local councils) must obtain an 'approval to supply' from the Department of Health to provide recycled water to customers. Approval conditions may include the minimum recycled water quality required, reporting requirements, and incident reporting criteria.

Users (customers) of recycled water require an 'approval to use' recycled water, and must provide information to the Department of Health regarding the nature of the reuse application, irrigation methods, buffer zones, types of crops or plants to be irrigated, and information on locations where irrigation takes place including extent of public access. Approval conditions may include restrictions on watering times, irrigation application methods, reporting requirements and incident reporting criteria. The EPA may also impose conditions relating to the monitoring of long-term impacts of recycled water on soils and groundwater.

A history of metropolitan recycling

The reuse or recycling of treated wastewater from the metropolitan WWTPs has been undertaken in limited ways for many years. In 1933, treated wastewater from the Glenelg plant began to be used to irrigate lawns and gardens onsite. In 1958, the then Engineering and Water Supply (EWS) Department encouraged the West Beach Recreation Reserve Trust to pioneer the use of treated wastewater on playing fields south-east of the Glenelg plant. The success of the scheme encouraged the establishment of a major reticulation scheme to irrigate West Beach Trust land north of the works. A reclaimed water supply was made available to the Glenelg Council in 1962 for irrigating the fringe areas along the Patawalonga Basin foreshore. In February 1968, the government approved funds to reticulate reclaimed water to Adelaide Airport, Kooyonga and Glenelg golf courses, Lockleys Primary School and Lockleys Oval. In the late 1960s, market gardeners near the St Kilda outfall channel from Bolivar successfully lobbied for access to the treated wastewater, establishing their own privately owned schemes, with restrictions on use and crops grown to safeguard public health. Crops irrigated included pasture and fodder crops, others included potatoes, carrots, onions, tomatoes, grapes and olives. Annual reuse quantities were small compared with the amounts of treated wastewater discharged into Gulf St Vincent. At the Christies Beach plant in the mid 1970s, the local council began to use small quantities of treated wastewater to irrigate recreational areas and parklands adjacent to the plant. No reuse was possible from the Port Adelaide WWTP, due to elevated salinity levels in the treated wastewater as a result of infiltration of highly saline groundwater into the sewer pipes in low-lying parts of the drainage area.

Support for the increasing reuse of treated wastewater gathered momentum in the late 1990s due to demand from potential users. It was also recognised that diversion of treated wastewater to reuse would address concerns about the impacts that nutrients in the discharges from the metropolitan plants could possibly have on the ecology of coastal marine waters. There is evidence that the loss of seagrasses along the metropolitan coastline results from the impacts of discharges of stormwater and treated wastewater. In addition, wastewater discharges into the Port Adelaide River have increased the occurrence of algal blooms including noxious 'red tides'.

At the same time, the water industry was becoming aware that many of the state's surface and groundwaters were approaching or had exceeded their sustainable yields. Amendments to the *Water Resources Act 1976* had established new arrangements for the management of water resources with the establishment of Catchment Water Management boards, embracing the 'whole-of-catchment' approach and emphasising sustainability.

A major impetus came when the state government released a policy, *A Cleaner South Australia Towards 2000 and Beyond: Statement on the Environment' in 1995*, which supports ceasing marine treated wastewater discharges and encourages the reuse of treated wastewater where economically and environmentally sustainable.

Demand for water was increasing at the time, due to the commercial and economic agendas of private industry and government. For example, the wine industry was undergoing expansion as a result of successful penetration into overseas markets. Recycled water represents a relatively stable water supply and is not subject to limitations associated with extending the reticulated water network, changes to groundwater supply regulations limiting availability, or the natural variability of streamflows.

The community has become increasingly supportive of recycling in general. There are environmental and social advantages to recycling, which have now become part of the public consciousness, whereas it was once considered to be the domain of the conservation movement. However, the true cost of recycled water usually exceeds that of potable water.

Awareness of the advantages of recycling has come into sharp focus in recent years, as drought conditions have continued and River Murray flows have declined. Tough water restrictions were introduced in 2006 and 2007 in many Australian cities. These factors have served to encourage public debate on the importance and sustainability of the country's water resources.

All of these factors have contributed towards the development of two significant reuse schemes, one at Virginia in the north, the other at Willunga Basin in the south.

The Virginia Pipeline Scheme

The Northern Adelaide Plains (NAP) is a horticultural region 30 km north of Adelaide city centre, adjacent to the semi-rural suburb of Bolivar. It has been described as the 'vegetable

bowl' of South Australia and also referred to as the Virginia Vegetable Triangle. There are approximately 1200 growers in the area with an estimated annual turnover in fresh and processed horticultural products of $160 million.[6] Produce is supplied to local, interstate and international markets.

Prior to the availability of large-scale supplies of recycled water, further development and ongoing sustainability of horticultural production in the area had been hampered by the lessening availability of suitable quality groundwater for irrigation. Put simply, the groundwater beneath the NAP was being mined. Consumption was estimated to be about 28 GL/year, far exceeding the estimated annual replenishment of only 6 to 10 GL/year.[6] The overuse of groundwater in the NAP saw a decline in water quality to the extent that, in some areas, groundwater became unsuitable for irrigation of horticultural crops. Power costs were increasing as the watertable dropped and water had to be pumped from deeper in the aquifer.

In the mid 1990s, the EPA issued the initial licences for the operation of the metropolitan WWTPs. These licences contained a clause requiring negotiation of an EIP to reduce the impacts of marine discharges of treated wastewater. Reuse offered an opportunity to achieve this, provided that funding contributions could be obtained from the federal government and involvement of private industry could be secured. The alternative was very costly upgrading of the Bolivar WWTP process to maximise nitrogen removal.

After several years of negotiations, a project was developed for $52 million of new infrastructure to further process treated wastewater from Bolivar to a standard suitable for reuse on market gardens, and to pump and distribute the recycled water to growers throughout the Virginia area. The Commonwealth government provided $10.8 million from the Building Better Cities program, funds which were critical in establishing the scheme.

The state government offered the private sector an opportunity to construct the pipeline through a build, own, operate and transfer (BOOT) arrangement. The successful company, Euratech Pty Ltd, constructed the $22 million Virginia Pipeline Scheme (VPS), which now distributes recycled water to the growers. A public company, Water Reticulation Systems Virginia Pty Ltd (WRSV) was formed by Euratech to manage the pipeline and access to the supply of recycled water, and to deal directly with the irrigators.

The VPS has been a cooperative effort between SA Water, the private company WRSV, and the Virginia Irrigation Association (VIA), which represents the growers. WRSV is also required to comply with the VPS irrigation management plan, ensuring the scheme will remain ecologically sustainable. The VPS pipe network is about 105 kms long and includes pipes from 100 mm to 826 mm in diameter. Funds to construct the VPS were contributed by Euratech ($7 million), SA Water ($6.7 million), and the majority of the Commonwealth grant ($8.3 million).

Recycled water supplied to VPS is sourced from the lagoons at the Bolivar WWTP. Treated wastewater is detained for a minimum of 16 days in shallow, rock-lined lagoons before

Glenelg to Adelaide Parklands Recycled Water Scheme

BOX 45

In 2008 the state government approved the construction of the Glenelg to Adelaide Parklands Recycled Water Scheme that will provide more than 3.8 billion L of high-quality recycled water annually to irrigate the parklands.

The scheme seeks to ensure the sustainable long-term water supply for the parklands. In particular it will seek to reduce treated wastewater discharges to Gulf St Vincent, enable the tripling of wastewater reuse from the upgraded Glenelg wastewater treatment plant, and seek to improve the overall health of the River Torrens/*Karrawirra Parri* and quality of water in the Torrens Lake.

Essentially the scheme includes: construction of a new wastewater treatment facility at Glenelg; construction of new recycled water storage tanks and a pump station at Glenelg; provision of an underground pipeline, encircling Adelaide through the parklands; a new pump station and underground recycled water storage tank in the parklands; and a new underground pipeline from Glenelg to the parklands.

David Jones

Water recycling process

Treated wastewater	Feed storage tanks	Ultrafiltration membranes	UV disinfection	Chlorine disinfection	Recycled water storage basins	Pump station	Parklands irrigation

The water recycling process for the Glenelg to Adelaide Parklands Recycled Water Scheme. Source: SA Water.

being discharged into the 13 km outfall channel and discharged to coastal waters at St Kilda. Natural disinfection of pathogenic microorganisms takes place as the treated wastewater slowly passes through the lagoons. However, environmental conditions are also favourable for the growth of algae. The presence of algae, suspended particulate material and the remnant microorganisms in the wastewater discharged from the lagoons, posed problems for the use of the water on crops for human consumption.

Class A recycled water quality is achieved by tertiary treatment of the Bolivar lagoon treated wastewater at the St Kilda dissolved air flotation and filtration (DAFF) plant. The plant was designed to produce up to 120 ML/day of filtered recycled water.[7] The DAFF process is described in more detail later in this chapter and is often referred to as 'tertiary treatment'. A similar form of treatment process is employed at Myponga, south of Adelaide, for treatment of drinking water sourced from an adjacent reservoir.

When it was commissioned, recycled water produced by the St Kilda DAFF plant met some of the most stringent Australian and international guidelines. The product water is suitable for 'unrestricted use for agricultural irrigation' (as classified by the Department of Health and EPA), in particular for 'unrestricted spray irrigation' on all fruit and vegetables including leafy salad vegetables. The Department of Health based their requirements for ensuring the safety of the scheme, in parts, on the Monterey Reclaimed Water scheme in California and the *Californian Title 22 Regulations*.[6] These were then generally considered to have set the benchmark for development of reclaimed water schemes dedicated to the irrigation of vegetables, which may be eaten raw.

Filtration provides a barrier to remove protozoa and some viruses, and is the only treatment process capable of removing cryptosporidium. The reduction in fine particulates and algae by the St Kilda DAFF plant allows chlorine to more effectively disinfect the water, producing a product with very low levels of microorganisms, which is safe to use where humans may be exposed to it (e.g., spray drift or through

consumption of vegetables such as carrots which have been irrigated with recycled water). A pumping station extracts the recycled water from the storage basin and pumps it to farm properties via the VPS pipeline network (see Figure 1.1).

The approval for the scheme from the South Australian Department of Health includes stringent requirements for monitoring and regular reporting on a range of recycled water quality parameters and, in particular, a number of microorganisms. Very low limits on key viruses and bacteria must be maintained for continued supply of recycled water for irrigation in the NAP. The key recycled water quality requirements were set as follows:

- originally turbidity <10 nephelometric turbidity units (NTU) mean and 15 NTU maximum, but this has been reduced to <1.5 NTU;
- originally mean faecal coliform concentration less than 10 organisms/100 ml, now reduced to a target of median E coli ≤1 organism/100 ml; and
- pathogenic microorganisms <1/50 L, with an objective of nil.

The VPS has the design capability to supply up to 30 GL/year to the over 200 growers in the Virginia area who signed up for recycled water. However, less than 20 GL/annum has been utilised by growers in recent years. The VPS began operations in November 1999.

The VPS pipework has a capacity of 120 ML/day and SA Water is contracted to supply up to 32 GL/year of feed water to the St Kilda DAFF plant, and 12.5 GL of that during summer. Consumption totalled 1.63 GL in 1998/1999 and rose to 16·30 GL in 2006/2007. WRSV originally had contracts for up to 19.27 GL/year, with the growers having the option of taking an additional 15% without penalty. Recycled water is available in autumn, winter and spring (19,500 ML), and has not yet been fully committed under grower annual contracts. Water demand is very low in winter when rainfall typically satisfies most of the crop demand for water.

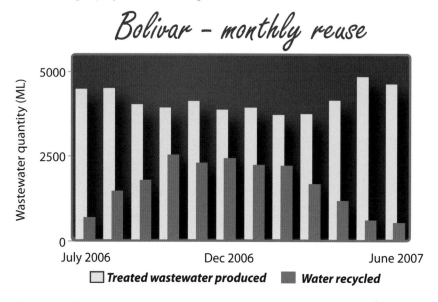

Bolivar – monthly reuse

☐ *Treated wastewater produced*   ■ *Water recycled*

Figure 11.2 The monthly reuse from the Bolivar wastewater treatment plant from July 2006 to June 2007. Source: Daryl Stevens, 'Growing crops with reclaimed wastewater', CSIRO Publishing, 2006.

Summer treated wastewater discharges from the Bolivar plant would be almost fully committed for recycling if the ultimate demand from customers occurs (see Figure 11.2). During the peak demand month in 2006/2007, nearly 65% of the treated wastewater produced by Bolivar was recycled including treated wastewater used around the plant as well as water pumped to VPS and Mawson Lakes.

Recycled water is supplied by SA Water from the St Kilda DAFF plant to WRSV at no cost. Reclaimed water supply contracts with WRSV require that growers construct storages at their farms. There are two main reasons for this: firstly, to supply a break in the supply chain to prevent backflow of recycled water into the primary distribution pipework; secondly, in the event of major problems with the DAFF plant, on-farm storage must have the capacity to store a minimum volume equivalent to one day at the guaranteed supply rate.

An extension of the VPS distribution infrastructure to serve the Angle Vale area to the west of Virginia was approved in 2007. Recycled water for the Mawson Lakes housing development is also supplied from the VPS (see Diagram 11.2).

The implementation of large-scale recycling of water from the Bolvar WWTP has provided an excellent solution to two significant environmental problems. This has been achieved by combining government and private sector funding initiatives, by selecting relatively novel treatment processes, and by delivering the project at minimum capital costs to taxpayers and irrigators alike. The St Kilda DAFF plant and VPS have been successfully operating for eight years, consistently delivering a high standard of recycled water. This project provides a good example for recycled water solutions that have been proposed in other areas of Australia.

Mawson Lakes Recycled Water Scheme

Mawson Lakes is a new residential subdivision north of Adelaide and to the east of Bolivar. From concept, it was designed to be an innovative 'green' development to reduce environmental impacts. When complete in 2010, the Mawson Lakes development will incorporate 3900 allotments accommodating about 10,000 residents.

Mawson Lakes is home to a recycled water system which was a first for South Australia and has established a benchmark for future sustainable urban developments in the state. All homes, businesses and organisations at Mawson Lakes will be connected to a recycled water system in addition to the normal drinking water supply and sewage collection network. Treated stormwater and treated wastewater are recycled for use in watering gardens and parks, washing cars and toilet flushing. This recycled water is delivered to properties through a recycled water system that is *separate* from the drinking water supply and is sometimes referred to as a 'third pipe system'. The system is designed to reduce usage of drinking water (sourced from the River Murray or Mount Lofty Ranges) in Mawson Lakes by 50% as compared to the Adelaide average.

This system was developed for Mawson Lakes by Delfin Lend Lease, the state government's Land Management Corporation, SA Water, and the City of Salisbury. The project

Mawson Lakes water systems

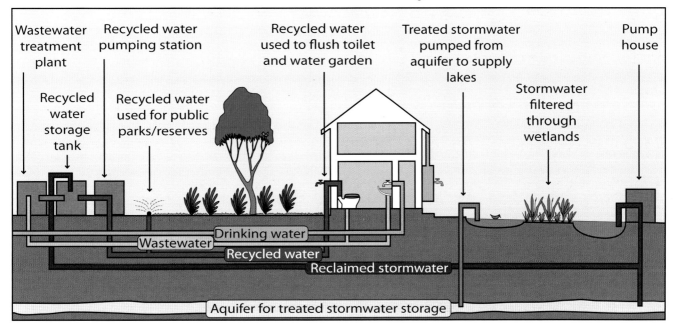

Diagram 11.2 The recycled water system at Mawson Lakes. The coloured underground pipes show the different types of water that are distributed to and from a Mawson Lakes' household. Source: adapted from Mawson Lakes community website.

cost $16 million to construct. The treated wastewater component of the project consists of a below-ground pumping station at Bolivar, a chlorination station for additional disinfection, approximately 12 km of pumping mains (see Figure 1.1), a 2.6 ML reinforced-concrete tank, and a pump station at Greenfields to transfer the recycled water to the Mawson Lakes reticulation network. The recycled water system became operational in late March 2005, from which point SA Water became the owner of the system and responsible for its operation.

The system works by collecting and distributing recycled water from two sources:

- SA Water's Bolivar WTTP (treated wastewater); and
- the City of Salisbury's wetlands/aquifer storage and recovery (ASR) (treated stormwater).

The treated wastewater and treated stormwater are mixed together and disinfected. The recycled water is then held in a large storage tank before being pumped to Mawson Lakes properties via a separate recycled water pipe network (see Diagram 11.2). (The recycled water network system is identified by purple- or lilac-coloured pipes and fittings.)

Recycled water is subject to stringent quality standards and is almost visually indistinguishable from reticulated drinking water. While recycled water is more saline than drinking water, it is suitable for toilet flushing, managed irrigation, washing cars, and other household garden uses. It is not suitable for drinking, swimming pools/spas, washing clothes, personal washing, indoor cleaning, or recreational activities involving water. The Department of Health approval for the Mawson Lakes Recycled Water Scheme required stringent quality control of recycled water from all sources, comprehensive community education, and routine auditing of the recycled water supply system.

The department required additional disinfection of the treated wastewater from Bolivar to ensure that very low levels (less than 1 organism/50 L) of potentially pathogenic bacteria, protozoa and viruses were achieved. This required additional chlorination and a minimum Ct >60 mg.min/L. (Ct is the free chlorine residual concentration in recycled water after it passes through a chlorine contact tank, multiplied by the detention time in the contact tank.) In terms of the 2006 national guidelines for water recycling, this is a recycled water suitable for 'dual reticulation – indoor and outdoor use'. Routine audits are undertaken to:

- ensure the potable and recycled water supplies are connected to the correct supply points within properties; and
- the two supplies remain separated from one another (i.e. that no cross connections exist between the two sources at the supply end).

Prior to the connection of the recycled water scheme, on-property audits are carried out in domestic dwellings and commercial properties to ensure that there are no interconnections between the potable water plumbing and the recycled water plumbing. Network audits are undertaken

to ensure that the correct water source is connected to the potable and recycled meters; these audits are carried out on all properties including vacant land. Audits are carried out at critical points during the construction of new houses, at change of property ownership, or at a maximum interval of five years.

Recycled water is currently set at 75% of the price of drinking water above 125 kL/annum consumption. The cost of recycled water was initially a flat rate of 77 cents per kilolitre (2004/2005 financial year). When compared with the average household water consumption outside Mawson Lakes, an average household in Mawson Lakes could save approximately $30 each year, an annual saving of approximately 110 kL of mains water per household. Commercial properties will save in accordance with property size and use.

Cleaning treated wastewater for recycling – the St Kilda DAFF plant

A vital component of the VPS is the $30 million St Kilda DAFF plant, which began operations in 1999 to provide class A recycled water for irrigators in the NAP. The plant was designed to produce up to 120 ML/day of recycled water to a high standard, suitable for spray irrigation of market gardens. The plant is owned by the SA Water and is managed, operated and maintained by United Water Pty Ltd under contract. The plant is operated in accordance with international standards ISO 9002 for operations and ISO 14001 for environmental management.

The recycled water produced by the plant has to meet stringent guidelines issued by the Department of Health, EPA and SA Water. Ongoing, regular monitoring of a range of physical, chemical and microbiological recycled water quality is undertaken, with the results reported to these agencies.

Treated wastewater is pumped to the St Kilda DAFF plant from the Bolivar lagoons numbers 3 and 6, where the water has been held for a minimum of 16 days. At times, the water can be rich in algae and may be turbid. In order to produce recycled water safe to use on crops destined for human consumption, it is necessary to remove the algae and suspended particles, so that later disinfection by chlorine can be effective in killing any pathogenic microorganisms. (Particulate matter can shield microbes from the action of disinfectants.) Safeguarding human health when using recycled water is the key to successful reuse and gaining Department of Health approval for reuse.

The DAFF treatment process is composed of the following stages (also see Diagram 11.3):

- Aluminium sulphate is mixed with the lagoon-treated wastewater stream. This is a coagulant that helps bind together the very small organic and inorganic particles.
- The next stage involves flocculation, in which a polyelectrolyte (polymeric flocculant) chemical is added, causing the coagulated particulates to flocculate

The St Kilda DAFF plant

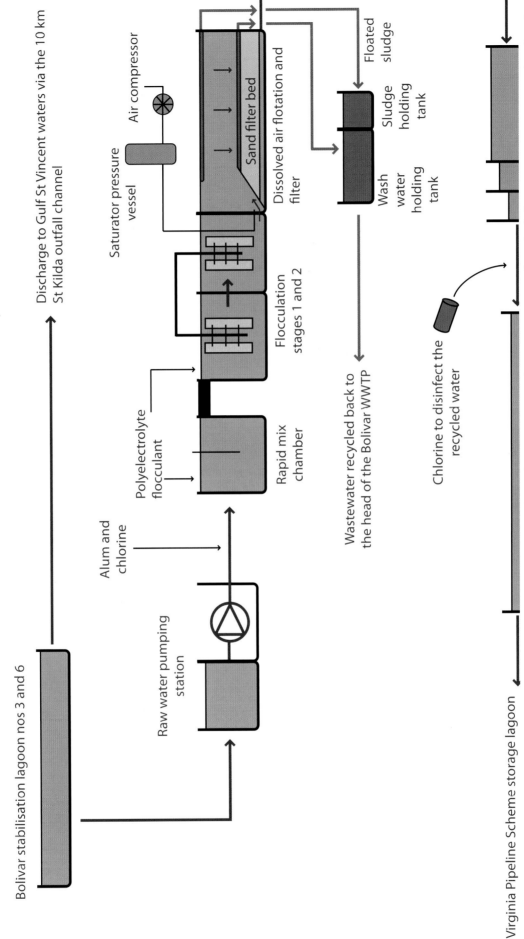

Diagram 11.3 The St Kilda dissolved air flotation and filtration plant, showing the additional processes required to treat wastewater from the Bolivar wastewater treatment plant, in order to provide class A recycled water for irrigation through the Virginia Pipeline Scheme. Source: SA Water.

Bolivar stabilisation lagoon nos 3 and 6

Discharge to Gulf St Vincent waters via the 10 km St Kilda outfall channel

Air compressor

Saturator pressure vessel

Sand filter bed

Dissolved air flotation and filter

Floated sludge

Sludge holding tank

Wash water holding tank

Flocculation stages 1 and 2

Polyelectrolyte flocculant

Rapid mix chamber

Wastewater recycled back to the head of the Bolivar WWTP

Chlorine to disinfect the recycled water

Alum and chlorine

Raw water pumping station

Virginia Pipeline Scheme storage lagoon

together under gentle agitation in tanks stirred slowly with paddles. The resultant 'floc' particles are large enough to be individually visible and be separated from the water stream by the next stage.

- The coagulated and flocculated wastewater is next mixed with a stream of clean water, into which air has been dissolved at high pressure, until it becomes saturated. The air-saturated water stream is then injected into the wastewater flow. The pressure decreases when the water mixture is released into the tank. Billions of microscopic air bubbles form and rise to the water surface. These rising air bubbles lift the light floc particles to the surface of the tank, forming a thick, brown scum layer. This is removed, collected and pumped back to the Bolivar WWTP sludge digesters. Most of the algae and suspended solids particles are removed by this process.
- At the bottom of the tank, where the floated scum is removed, the treated wastewater passes downwards through a multimedia filter bed. This comprises layers of anthracite (filter coal), sand and gravel, which removes most of the remaining fine particles and microbiological pathogens. The filters are periodically cleaned to remove accumulated solids. The filter beds are taken offline and compressed air is passed upwards through them. Clean water then flows upwards through the filter and is collected in troughs. The resultant waste washwater is pumped back to the Bolivar WWTP for treatment.
- The filtered water is then disinfected with chlorine to kill any remaining pathogenic microbes before it flows through a covered lagoon to a storage basin. Pumps extract the class A recycled water and deliver it to growers through the VPS.

The plant is fully automated, although there are usually operators on site during working hours. Key operating parameters are continuously monitored and displayed on computer screens in the control room. Process problems are automatically detected by the control system and alarms are displayed. The control system may call out an operator if a serious problem develops out of normal working hours. During winter there is little demand for recycled water and Bolivar lagoon treated wastewater bypasses the St Kilda DAFF plant and is discharged to sea via an open channel. The St Kilda plant currently treats about 17,000 ML of recycled water annually.

During the design stage of the St Kilda DAFF plant, a small-scale pilot plant was constructed and successfully operated to prove that the treatment process proposed would deliver the performance required to consistently produce class A recycled water. The sand filters adopted for the plant were 'state of the art' when the concept for the plant was developed, but, increasingly, reuse schemes are turning to newer, more effective technologies such as disk filters and, particularly, membrane filters.

The Willunga Basin Pipeline Scheme

The Willunga Basin is a rural area south of metropolitan Adelaide, approximately 320 km² in extent. It is a significant vine-growing region and includes McLaren Vale, which is home to a number of premium wineries. The region is also noted for recreation and tourism associated with the vineyards, orchards and popular beaches. Its close proximity to Adelaide is also a factor in its popularity.

The Willunga Basin area is defined by surfacewater catchment boundaries, with drainage westwards towards the coast. The area is bounded by the Onkaparinga River to the north and the Sellicks Hill Range to the south and east. In 1996 it was estimated that farming activities took place over an area of about 13,300 ha.[8] Of this, about 4100 ha was licensed for irrigation in 1990, with about 70% associated with wine grape cultivation.

Despite reliable rainfall, irrigation viability was dependent on the availability of groundwater. Surfacewater resources were not adequate and reticulated water supplies limited. Typical groundwater demand for irrigation was 7.5 GL/year and 1.0 GL/year of water from streams was diverted for irrigation. An additional 0.2 GL/year of potable reticulated water was also utilised.

In 1990 the groundwater supply was proclaimed under the *Water Resources Act 1990*. The licensing system restricted the areas to be irrigated, but, initially, not the volumes of groundwater that could be utilised. However, groundwater levels were falling, indicating that exploitation of the groundwater supply had exceeded the sustainable yield of the aquifer.

Salinity was also becoming an issue; some wineries were resorting to carting in potable water by tanker, at enormous cost. Recycling wastewater from towns within the catchment would have yielded insufficient quantities of water to make any difference to the situation. Commercial opportunities and an increasing demand for Australian wine were resulting in increased plantings of wine grapes elsewhere, however, this was out of the question in the Willunga Basin. Water-users thus focused on augmenting supplies from outside the catchment; recycled water from the Christies Beach WWTP was an obvious possibility for investigation.

The Christies Beach WWTP is located about 10 km to the north of Willunga Basin and typically produces about 10 GL/year of treated wastewater, which is discharged to the sea via a 400 m outfall pipe near Port Stanvac. This quantity of water exceeded the typical annual abstraction of groundwater at Willunga Basin in the 1990s. As noted above, a licensing system for South Australian WWTPs was introduced by the EPA in the mid 1990s. The license for Christies Beach included a requirement for the development and implementation of an EIP at the plant.

Following negotiations with the EPA, an EIP was approved that included upgrading the treatment capability of the existing plant, which would reduce the load of nitrogen discharged to coastal waters in order to ameliorate offshore environmental impacts, including seagrass losses. Local potential demand for recycled water was extremely limited as the plant was surrounded by residential areas. SA Water commissioned a consultancy into the feasibility of limited reuse in the Willunga Basin in 1996.

A users group of interested growers was formed to facilitate gaining access to recycled water for Willunga Basin. After a year of negotiations with various agencies and interest groups, the government called tenders for the right to negotiate for the treated wastewater, on the basis that no government funds would be available. In October 1997, the users group was successful in gaining the exclusive right to negotiate for uncommitted Christies Beach treated wastewater. A 40 year license agreement was signed in January 1998. The users group formalised their joint venture by forming the Willunga Basin Water Company (WBWC) to build and operate the recycled water supply.

Planning approval followed, together with preparation of an irrigation management plan to ensure environmentally sustainable management of irrigation using recycled water. Infrastructure, consisting of pipelines, pumping stations and balancing storages was designed, and tenders were sought for construction by private contractors. The resultant Willunga Basin Pipeline Scheme (WBPS) was commissioned in April 1999 and officially opened in August 1999.

The final capital cost of the WBPS was $7.2 million. The scheme was wholly funded, owned and operated by the water-users. The original consortium of 24 growers were allocated only 25% (2100 ML) of the water the scheme is capable of delivering, which corresponds approximately to the summer flows from Christies Beach.[8] This allows for further expansion of reuse within Willunga Basin, without augmenting the pipeline. Extension of the pipeline would be to the east and south, and storage of winter flows would be needed as well.

In Stage 1 of the scheme, the main underground 450 mm mPVC-diameter lilac-coloured pipeline extends approxi-mately 20 km south from the Christies Beach WWTP (see Figure 1.1). There are about 4 km of spur mains in diameters including 300, 200 and 150 mm.[8] With the implementation of Stage 2, there is now a total of 70 km of pipelines serving 80 irrigators and about 1500 ha of land.[9]

Outlets are installed at each user's property and recycled water is available on demand; there is enough capacity to supply all 17 initial customers simultaneously. The users do not require on-farm pumps or storage basins (unlike in the VPS). Pressure is sufficient to directly feed drip-irrigation systems. Users provide their own equipment downstream of the outlets including automatic filtration to prevent potential dripper blockage from algae or particulate matter.[8]

There is a risk to human health from the use of recycled water, so the conditions and applications using recycled water are manipulated in order to minimise the risk. Any restrictions are detailed in the Department of Health approval for the particular use. At Willunga, class B recycled water is being used mainly for grape vines but a number of measures must be implemented to manage the health risk. Biological agents, including potentially infectious microorganisms,

may be present in the reclaimed water. Measures adopted at Willunga Basin include:[9]

- warning signs at every property utilising or able to utilise recycled water;
- the use of lilac-coloured pipes to distribute reclaimed water (this is the standard recognition device to identify recycled treated wastewater pipes);
- limiting irrigation methods to drip or subsurface irrigation only, with spray or sprinkler application of recycled water not permitted;
- growers are not permitted to use the recycled water for non-irrigation activities (e.g., mixing and diluting chemicals); and
- growers must observe personal hygiene at all times and all users are issued with a booklet containing information on compulsory health and safety measures when they are connected to the scheme.

Pipeline capacity is 24 ML/day, which is the typical minimum estimated discharge from the Christies Beach plant in summer. Recycled water from Christies Beach WWTP is supplied to WBWC at no cost. Users are licensed by WBWC to draw a certain amount of water each year. There is no requirement for users to take all of their allocation, but annual charges apply for water used and for water not used. The licence limits the flow rate at the farm outlet to that requested by the customer.

A 6 ML polyethylene membrane-lined basin was constructed at the Christies Beach plant on SA Water land, as a balancing storage. Pumps extract water and pump to a 12 ML open storage near Old Noarlunga. The pumping stations at both locations automatically respond to grower demands for water using variable speed controls, all monitored by a computer. Each farm outlet has a flow meter and data logger so that consumption can be continuously monitored to better manage the scheme. The WBWC employs a full-time water supply officer to maintain and monitor the system.

The adoption of reclaimed water in the Willunga Basin has reinvigorated the horticultural industries in the area, particularly vineyards. In contrast to Virginia, the advent of a guaranteed recycled water supply at Willunga was used to open up about 600 ha of additional plantings of vines for wine production. It is estimated that this represents about $30 million/year of added value through increased wine sales.[9] Another benefit is that land formerly only suitable for broadacre crops became more valuable and the original owners reaped the benefits when the land was sold.[10]

In the first growing season (1999–2000) consumption totalled 950 ML; this rose to 3200 ML in 2006/2007. Highest demand occurs in the summer months, and in the peak month (December) 80% of the available recycled water produced by the Christies Beach plant is reused (Figure 11.3). This is a limitation to further growth in reuse. More water could be reused if storages were constructed to capture winter discharges for subsequent use in summer. This will require very large storages (1 to 2 GL in size) to be effective. Additional recycled water may be used through initiatives as part of the Water Proofing the South project.

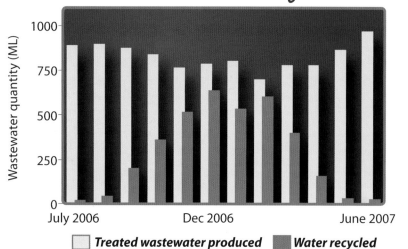

Christies Beach – monthly reuse

Wastewater quantity (ML)

July 2006   Dec 2006   June 2007

☐ **Treated wastewater produced**   ■ **Water recycled**

Figure 11.3 Christies Beach wastewater treatment plant monthly reuse during the period from July 2006 to June 2007. Source: SA Water.

- expansion of the existing WBWC recycled water supply network to transfer greater volumes of recycled water for viticulture;
- dual reticulation of recycled water in a new housing development at Seaford for toilet flushing and watering gardens;
- more storage so that recycled water produced in winter can be utilised in summer for irrigation;
- the integration of Aldinga WWTP into the WBWC network;
- irrigation of council parks and gardens with recycled water and some commercial reuse; and
- localised stormwater capture and reuse.

This strategy has the potential to increase recycled water use from 4.4 GL/year to 8.8 GL/year by 2011, with the potential to provide further expansion as population increases.

A key difference between the Willunga Basin and Virginia recycled water supplies is the quality of the water supplied. At Virginia, the St Kilda DAFF plant filters and disinfects the water to achieve class A standard. At Willunga Basin, recycled water is extracted after treatment (including chlorination) at Christies Beach WWTP but it is not filtered, so it is classified as class B. There are restrictions on the application of the class B recycled water, mainly to ensure that exposure of the general public to the water is minimised. Measures must be taken to control spray drift, to irrigate at night if necessary, and to provide appropriate buffer zones to roadways and residences. These constraints do not present a problem for rural areas irrigating vines using mainly drippers.

The WBPS has been a very successful project and provided an example of how private enterprise can initiate and carry through recycling projects. It also provides an illustration of the relationship between land value and water rights.

Water Proofing the South

Water Proofing the South is a project that aims to diversify southern Adelaide's water resources and to reduce reliance on the River Murray and current groundwater supplies. Alternative water sources, such as treated wastewater and stormwater, will be developed. Increased recycling of treated wastewater will also reduce the environmental impacts of discharges to coastal waters from Christies Beach WWTP. Project partners include the City of Onkaparinga Council, SA Water, Adelaide and Mount Lofty Ranges Natural Resources Management Board, and WBWC, with a contribution from Flinders University. The project includes a number of elements, including:

- accessing more treated wastewater from the Christies Beach WWTP, which is about to have a capacity upgrade;

A key part of the strategy is the Southern Urban Reuse Project (SURP), which aims to provide recycled water for dual reticulation, and municipal irrigation in new housing developments south of the Onkaparinga River, including Seaford. This is similar in concept to the Mawson Lakes scheme north of Adelaide. Planning and concept design work for the SURP has commenced. Treated wastewater from the Christies Beach and Aldinga WWTPs will require further treatment to attain the high standard needed for dual residential use. Membrane filtration, UV disinfection and chlorination at a new tertiary treatment plant are likely to be required. Additional pipelines and balancing storages will also be needed. The Commonwealth government is providing $34.5 million to the Water Proofing the South project through the National Water Commission (NWC). Total investment from all sources, including the Onkaparinga Council and the state government, is estimated at $119.3 million for the entire range of infrastructure.[11]

Glenelg reuse

Treated wastewater from the Glenelg WWTP has been reused since the 1930s. Since 1964, customers have included the local council and the West Beach Trust – using the water to water recreational areas, grassed areas along the Patawalonga foreshore, a caravan park and playing fields. Several local golf courses (Kooyoonga and Patawalonga) have also benefited from recycled water. In recent years, the limitations and restrictions associated with using class B recycled water reduced its attractiveness as an alternative to reticulated water and groundwater supplies. In response, SA Water and United Water installed a tertiary treatment plant to improve the recycled water quality to class A, for all those customers who needed it. Commonwealth government funding through the Coast and Clean Seas Program contributed to the capital cost of the project. The 3.5 ML/day-capacity Holdfast Bay Water Reclamation Plant, located at the Glenelg WWTP, was commissioned in 2004 and consists of four tri-media (anthracite-sand-gravel) pressure filters and UV light irradiation for disinfection. Adelaide

A wash with recycled water

BOX 46

One of the most confronting issues for Adelaide is how to source and supply potable and irrigation-quality water to its suburbs. The media and politicians debate desalination plants, the need to increase water storage reservoirs in the Adelaide Hills, the fragility of the River Murray flows and ecology, and the effectiveness of rainwater tanks – but these are metropolitan-scale debates and initiatives that do not address the problem from the bottom up.

Can we not better capture and reuse water within a residential development and make that development sustainable?

The $400 million Point Boston Peninsula project on South Australia's Eyre Peninsula demonstrates a future vision as to where we could go in Adelaide. Representing the state's largest coastal residential development, the project aims to maximise its use of rainwater by extensive recycling, with the aim that the development will be 90% independent of mains water. The Port Lincoln area, like other Eyre Peninsula towns, depends primarily upon mains water pumped from the depleting Uley Basin aquifer.

Point Boston has adopted the WaterCress (water–community resource evaluation and simulation system) stormwater-harvesting model, which reduces the need for tank capacity to as little as 50% of the traditional 'farmhouse' model.

Using the Water Cress model, rainwater will be captured from roofs and cleansed by first-flush diverter systems, then by carbon filters and ultraviolet light to enable use for drinking, bathing and showers. All other water will be obtained from recycled water, purified using a combination of an onsite Biolytix mini-treatment plant for each allotment, together with a centrally located recycled water treatment plant.

The Biolytix wastewater system uses worms and micro-organisms to break down waste. Following treatment, the Biolytix water is pumped to a central recycled water treatment plant where it is filtered and disinfected using power supplemented by photovoltaics. The treated water is then pumped back to houses for use in toilets, car washing and gardens, and stored in tanks for fire-fighting and landscaping. Excess water will be piped to revegetated land planted with native plants, which absorb nitrogen and phosphorous. This vegetation will be harvested every few years and composted for garden use. A series of water-retention swales will also adjoin road reserves so that stormwater will flow naturally to water-common areas.

This approach is so technically advanced that recycled water will used in clothes washing machines –a level never before achieved for household use in Australia.

Overall, the project has five qualities of water: rainwater and mains supplementation for drinking, pressure effluent from household onsite treatment units, domestic recycled non-drinking water, and treated effluent irrigation water.

David Jones

Airport Ltd negotiated for the supply of higher-quality recycled water (requiring chlorination after the filtration and UV processes) for non-potable use onsite, including toilet flushing. Only the Patawalonga golf course currently receives class B (unfiltered) recycled water from this plant.

Glenelg to Adelaide Parklands Recycled Water Scheme

The Glenelg to Adelaide Parklands Recycled Water Scheme (GAPRWS) is a project that will result in high-quality recycled water being used to irrigate the parklands surrounding the centre of Adelaide. The Adelaide parklands, with an area of about 690 ha, is one of the largest urban parks anywhere in the world, and nearly three times the combined size of London's Hyde Park and Kensington Gardens.[12]

For many years, the Adelaide City Council was reimbursed for mains water used in irrigating the parklands by the state government through SA Water,[13] so there was little pressure to develop alternative sources of water supply or, for that matter, to practice efficient irrigation. However, water restrictions in response to recent drought conditions have resulted in the City Council severely reducing watering of the parks and gardens surrounding the CBD.

The potential for using recycled water from Glenelg WWTP on the parklands was raised in a 1993 report and investigated in detail in 1999.[13] At the time, the state government concluded that the project was not financially viable. In 2004, the proposal was again investigated, this time within an environment of even greater pressure on water resources.

A funding contribution from the Commonwealth government was sought through the NWC. In May 2006, a memorandum of understanding with the Adelaide City Council was signed, setting the recycled water maximum price at 75% of the top-tier statewide price of potable water. The agreement is subject to mutually acceptable commercial terms and state government and NWC funding. The project has been approved in principal for 50% subsidy funding from the NWC, with the remaining 50% funded by the state government, based on an initial estimated project capital cost of about $60 million.

The GAPRWS will treat and convey 2000 ML and more of class A recycled water from the Glenelg WWTP to irrigate the parklands. City buildings may also be supplied with recycled water (for toilet flushing). There is also the potential for discharging recycled water into the River Torrens for environmental flows, if this gains community acceptance and EPA approval. Infrastructure to be constructed includes:

- A 34.5 ML/day-capacity filtration plant on the site of the existing Glenelg WWTP, utilising membrane technology and both UV and chlorination for disinfection;
- a 10 km, 750 mm-diameter underground pipeline to convey recycled water to the parklands (see Figure 1.1);
- approximately 10 km of 525 mm to 100 mm diameter underground pipelines laid in a ring around the parklands; and
- a major pumping station at Glenelg.

Recycled water will replace the approximately 1000 ML/year of mains water sourced from the River Murray and Adelaide Hills catchments currently used for parklands irrigation. A further 500 ML/year of River Torrens water and approximately 500 ML/year of groundwater currently used for irrigation purposes will also be replaced. The scheme is planned to be operational by late 2009 or early 2010. Another benefit will be a further reduction in the quantity of treated wastewater and nutrient loads discharged to Gulf St Vincent marine environment from the Glenelg WWTP.

Increased use of recycled water through the proposed GAPRWS will also increase public awareness of recycled water use and the technology that is in place to create highly treated wastewater suitable for a range of purposes. This in turn supports the principles of water conservation and, more generally, integrated water resource management, both of which are foundation elements of the Water Proofing Adelaide strategy.[14]

Increasing reuse – the challenges ahead

In Adelaide, the peak reuse month in 2006–2007 was December 2006, in which 49% of the 6.8 GL of treated wastewater produced by the metropolitan WWTPs (including Aldinga) was recycled – an outstanding outcome. If the quantities of treated wastewater discharged from the Bolivar High Salinity WWTP are excluded (the wastewater is too saline for reuse), the reuse percentage increases to 54%. This is the highest proportion of treated wastewater currently recycled in any Australian capital city. On an annual basis, 30% of the treated wastewater produced by the metropolitan WWTPs was recycled in 2006–2007, another outstanding outcome for a city of well over one million people. Table 11.4 provides reuse figures for Australian state capital cities; in percentage terms, Adelaide is currently leading the country.

In 2002 the Water Proofing Adelaide project was established by the state government to develop a strategic framework for the management, conservation and development of Adelaide's water resources to 2025. The work of the Water Proofing Adelaide project indicated that the way forward was in the implementation of a large number of smaller goals working together, including increasing water recycling. The final report[14] indicated that:

- 'Further opportunities for large-scale recycled projects including the expansion of existing schemes will be implemented where they are viable according to an economic, environmental and social impact assessment;
- localised reuse of recycled water and/or stormwater will be considered for new land divisions as part of water-sensitive urban development requirements; and
- regulations relating to sewer mining and greywater reuse systems will be reviewed to ensure that restrictions on reuse are justified by the public health and urban amenity criteria.'

A key outcome of implementation of the strategy was a target to increase the use of recycled water from 14,000 ML/annum in 2002 to more than 30,000 ML/annum by 2025. A range of projects will help meet this target. The Angle Vale

Table 11.4 Recycling in Australian cities

Capital	Recycling 2001–2002[1] %	Recycling 2005–2006[2] %	Recycling 2006–2007[3] %	Future targets (2003 aspirations)
Sydney	2.3	3.5	4.3	35% reduced consumption by 2011, 10% wastewater recycling by 2020
Melbourne	2.0	14.3*	23.0*	15% reduced water consumption, 20% wastewater recycling by 2010
Brisbane	6.0	4.8	7.2	Increase recycling to 17% by 2010
Adelaide	11.1	18.1	29.6	30,000 ML recycled water/y (33% by 2025)
Perth	3.3	5.3	6.0	20% recycling by 2012
Hobart	0.1	3.1	?	10% reduction in water consumption

1. J.C. Radcliffe, 'Water recycling in Australia', Australian Academy of Technological Sciences and Engineering, Melbourne, 2004, p. 233.
2. J.C. Radcliffe, 'Advances in water recycling 2003–2007', 3rd Australasian Water Recycling Conference, University of New South Wales, Sydney, July 2007.
3. WSAA/NWC national performance report 2006–2007.
* Almost all from Werribee STP on beef pastures in lieu of previous ground spread of sewage on beef pastures.
Table 11.4 Recycling of treated wastewater in Australian capital cities. Source: SA Water.

extension of the VPS, further expansion of Mawson Lakes Dual Reticulation Scheme, and the GAPRWS are examples of reuse projects that have been implemented or are in the process of being implemented. There may also be opportunities to expand reuse from Bolivar, which currently provides the majority of recycled water in Adelaide (see Figure 11.4). In addition, plans are being made to undertake further expansion of reuse from the Christies Beach WWTP as part of the Water Proofing the South project.

The major impediment to substantial reuse expansion is the understandable lack of demand during the winter months from horticultural users. The potential solution requires storage of winter water for reuse during the next summer. Alternatively, a large industrial user with a relatively constant year-round recycled water demand would be ideal for expanding reuse. The former would be extremely expensive to implement and the latter is unlikely to eventuate, although the mining industry north of Port Augusta could utilise large quantities of recycled water. Studies have been done into the feasibility of pumping Bolivar water northwards, but high cost has been a deterrent.

Substantial increases in water recycling are likely to require the inclusion of some form of bulk recycled water storage, so that treated wastewater produced during winter can be stored for subsequent reuse during the summer peak demand period. Earth-lined open storages are costly and recontamination may be a problem unless the storages are covered, incurring further costs. Locations suitable for storages in the developing rural fringes of the city are also limited. Aquifer storage and recovery (ASR) offers an alternative to open storages. ASR is a technique whereby treated wastewater is injected down bores to be stored in an underground aquifer until needed. The water is pumped out of the same bore or nearby wells. Successful trials have been conducted at Bolivar over recent years,[15] but the technique has yet to be implemented on a large scale. ASR offers a number of advantages over the alternative of surface storages and needs to be vigorously pursued.

Another issue of concern is a slow rise in treated wastewater salinity over time. Salt in wastewater is contributed from various household cleaning products and the infiltration of saline groundwater into some sewer networks. A salinity of up to about 1500 mg/L is generally acceptable for most horticultural applications, although a lower figure is typically preferred. In 2006–2007, average Bolivar recycled water salinity total dissolved salts (TDS) was 983 mg/L, Christies Beach 730 mg/L and Glenelg 1186 mg/L. At Bolivar, potable water can be added into the VPS recycled water supply to control salinity. Growers have been guaranteed water with

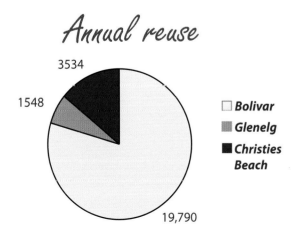

Annual reuse

3534
1548
19,790

☐ **Bolivar**
▩ **Glenelg**
■ **Christies Beach**

Figure 11.4 The annual volume in megalitres of recycled water, provided in reuse projects, from the Bolivar, Glenelg and Christies Beach wastewater treatment plants. Source: SA Water.

a TDS of less than 1500 mg/L, but this use of potable water under current water restrictions may not be popular with the general public. Long-term salinisation of the land irrigated with recycled water (and reticulated water) is also an issue, and there is a potential risk of problems from the long-term accumulation of nitrogen and phosphorous in soils from the application of both recycled water and fertilisers. Good irrigation management and ongoing soil monitoring is the key to addressing these risks. Recycled water is an important element in the range of strategies available to address Adelaide's water supply problems. However, its use must be carefully managed to minimise the potential impacts on the health of the general public and the environment.

References

1. SA Water, 'SA Water annual report 2005/2006', 2006 (online).
2. M. Hammerton, *Water South Australia: a history of the Engineering and Water Supply Department*, Wakefield Press, Adelaide, 1986.
3. B. Kracman, R. Martin, R. and P. Sztajnbok, 'The Virginia Pipeline: Australia's largest water recycling project', *Water Science and Technology*, 43 (10), 2001, pp. 35–42.
4. Department of Human Services and the Environment Protection Authority, *South Australian reclaimed water guidelines: treated effluent*, Adelaide, 1999.
5. Natural Resource Management Ministerial Council, Environment Protection and Heritage Council, *Australian guidelines for water recycling: managing health and environmental risks (phase 1)*, Australian Health Ministers' Conference, Canberra, 2006.
6. J. Kelly and D. Stevens, 'From problem to profit: wastewater reclamation and reuse on the Northern Adelaide Plains', *Water*, 27 (5), 2000, pp. 37–41.
7. C. Huijbregsen, K. Yerrell, C. Bosher and L. Sickerdick, 'Bolivar DAFF plant: Australia's largest reclaimed water treatment plant', *Australian Water Association 18th Federal Convention*, NWC, Adelaide, 1999, pp. 65–69.
8. J. Gransbury, 'The Willunga Basin pipeline: stage one, *Water*, 27 (5), 2000, pp. 27–31.
9. M. Carter, 'An investigation into agricultural use of reclaimed water in the Willunga Basin, South Australia, 2001', *GEOView*, 2004.
10. Frey Water Engineering and Optimatics Pty Ltd, 'Private reclaimed water scheme', *The Optimatics Letter*, 13, 2001.
11. National Water Commission, 'Driving national water reform', 2008 (online).
12. H. Girardet, *Creating a sustainable Adelaide: a green city*, Government of South Australia, Adelaide, 2003.
13. BCT/SKM Alliance, 'Glenelg WWTP effluent reuse report', Tonkin and Associates and Sinclair Knight Merz, Adelaide, 1999.
14. Government of South Australia, 'Water proofing Adelaide: a thirst for change 2005–2025', Adelaide, 2005 (online).
15. P. Dillon, S. Toze, P. Pavelic, S. Ragusa, M. Wright, P. Peter, R. Martin, N. Ganges and S. Pfeiffer, 'Storing recycled water in an aquifer at Bolivar: benefits and risk', *Water*, 26 (5), 1999, pp. 21–29.

CASE STUDY 3
Water treatment in Adelaide

Thorsten Mosisch

This case study is adapted from SA Water's Drinking Water Quality Report[1]

Introduction

The Adelaide metropolitan area is supplied with water collected from streams within the Mount Lofty Ranges, and supplemented with water from the River Murray. In average rainfall years, from June to November about 700 to 1000 mm of rain falls in the catchment areas of the ranges. Once soils in the catchment are wet, water runs off the land and into streams, which flow into reservoirs where the water is stored and pumped, or gravity-fed to water treatment plants (WTPs). Ten reservoirs supply Adelaide's water supply systems: Barossa, Happy Valley, Hope Valley, Kangaroo Creek, Little Para, Millbrook, Mount Bold, Myponga, South Para and Warren reservoirs. Greater Adelaide is supplied by six WTPs – Anstey Hill, Barossa, Happy Valley, Hope Valley, Little Para and Myponga – where water is filtered, disinfected and sent into the distribution network.

All metropolitan reservoirs except Myponga Reservoir have their catchment runoff supplemented by water piped from the River Murray. The percentage of water supplied from the Murray varies from year to year, with the river providing about 40% of metropolitan Adelaide's water in an average year, although this can increase dramatically during dry years to as much as 90%.

Water treatment in Adelaide

Adelaide's drinking water supplies are managed from catchment to tap, with systems and expertise ensuring ongoing provision of high-quality drinking water to customers, and to protect public health. This is achieved through a multiple-barrier approach to managing water quality to ensure hazards are reduced at each step of the management process. Hazards in the water can take many forms, generally categorised into three types: biological, physical and chemical. Typical examples of hazards found in South Australian raw waters include:

- biological – algal by-products, pathogens (e.g., *Cryptosporidium*, *Giardia*, *E. coli*);
- physical – sediments (turbidity), colour; and
- chemical – pesticides, hydrocarbons, iron, manganese.

Barriers to protect Adelaide's water supplies include:

- The water supply catchment (minimising the introduction of hazards into source water).
- Water storage reservoirs (minimising the introduction of hazards and removal of some hazards).
- Water treatment plants (removal of most hazards).
- Disinfection of water (neutralising microbiological hazards and algal by-products).
- Maintaining a chlorine residual in the distribution systems (neutralising microbiological hazards).
- Closed distribution systems (to prevent the introduction of hazards).
- Backflow prevention (to prevent the introduction of hazards).

Hazards are identified at each barrier so that appropriate actions can be implemented to prevent issues in the water supply system.

To ensure drinking water meets health and aesthetic guidelines, water quality from catchment to tap is continuously monitored. During 2006–2007 water samples were routinely collected from a total of 278 sample points within the Adelaide metropolitan area. Results of tests performed on the samples were compared and assessed against the *Australian Drinking Water Guidelines*.

Conventional water treatment process

Conventional WTPs use the following processes to deliver safe drinking water (refer to Diagram A):

Step 1. Coagulation: collects small particles and dissolved organic matter. A chemical (coagulant) is added to untreated water and reacts with impurities. The coagulant is usually a metal hydroxide (alum or ferric) that enmeshes and separates most of the suspended particles from the water.

Step 2. Flocculation: the coagulant combined with the captured particles is called a 'floc', some of which can be up to 5 mm in diameter. Flocculation is a gentle mixing process to bring together the flocs formed in the coagulation step to form a larger floc that will settle more easily. Water remains in the flocculation tanks for approximately 20 to 30 minutes.

Step 3. Sedimentation: the water and suspended flocs pass through to sedimentation basins where, after several hours, most of the floc settles to the bottom of the basins and forms sludge. The water – now containing only a small amount of very fine floc particles – continues on to the filters. The sludge is removed from the basins for further treatment and disposal. Another technique called dissolved air flotation and filtration (DAFF) uses the addition of air, causing the floc to float to the surface for removal by overflow or skimming.

Step 4. Filtration: the separation of any remaining solids from the waterflow by straining them through a filter. The most common filters are deep beds of sand or a sand/anthracite combination. As the water passes through the filter bed, any particles remaining from the sedimentation process are trapped in the fine spaces within the media, resulting in clear water. More modern filter types include membranes.

Step 5. Disinfection: deactivation of any disease-causing microorganisms such as bacteria, viruses or protozoa. The most common types of chemical disinfectant are chlorine compounds – chlorine, chlorine dioxide and chloramines. Other disinfection methods include ozone and ultraviolet light (UV). Chlorine is generally added at a point between the filters and the filtered water storage tank, to destroy any microorganisms that may not have been removed in the earlier flocculation and filtration stages. Chloramination (a combination of chlorine and ammonia) is often used for disinfection in longer water pipelines, such as those in country areas where it is more effective and longer lasting than chlorine alone.

The water treatment process

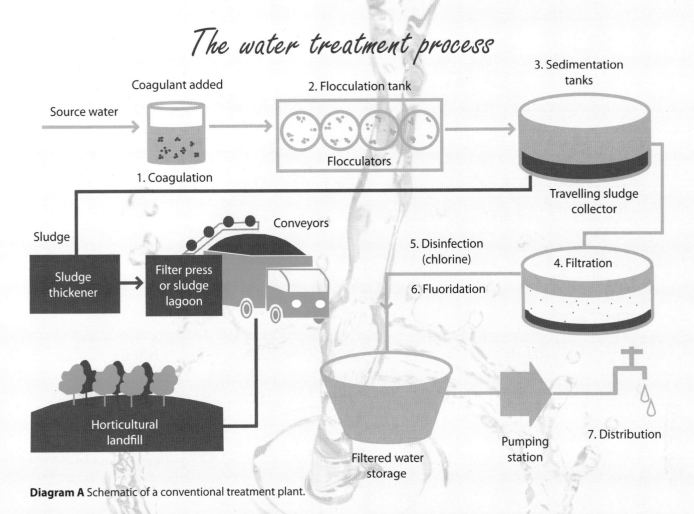

Diagram A Schematic of a conventional treatment plant.

Step 6. Fluoridation: in metropolitan Adelaide supplies, fluoride is added to the treated drinking water to help prevent tooth decay in the population.

Step 7. Storage and distribution: after disinfection the finished water is transferred to covered water storage tanks, ready for distribution to customers.

Other water treatment processes

Ultraviolet light disinfection
UV light at a wavelength of 253 nm has the ability to alter the DNA of microorganisms so that they cannot reproduce, rendering bacteria, viruses and protozoa non-pathogenic to humans. In the UV disinfection process, water passes through reactors containing sufficient UV lamps to produce the required dose. The power required to drive this process is affected by different organisms requiring different doses, the clarity of the water, and the dissolved compounds in the water being treated.

Magnetic Ion Exchange (MIEX®)
MIEX® resin is a reusable ion exchange resin designed to remove dissolved organic carbon (DOC) from water supplies. DOC is found in all natural water sources and is the result of the decomposition of organic material, causing colour, taste and odour in drinking water. The orange/brown colour of many surfacewaters is attributed to DOC compounds. The MIEX® resin works like a magnet, attracting DOC in the water. The small size of the resin particles (<200 micron) facilitates rapid exchange reaction with the DOC. The magnetic

component allows for efficient removal of the resin particles from the bulk water stream, taking with it the DOC.

Pre-treating with the MIEX® process results in a significant reduction in chemical usage, sludge generation, and the amount of chlorine required for effective disinfection and public health protection.

MIEX® was developed after years of research. The process was launched at SA Water's Mount Pleasant WTP in 2001 and is now being used at 10 WTPs in Australia, the United States and Europe.

Desalination
SA Water's desalination plant on Kangaroo Island uses conventional reverse osmosis technology to convert seawater into freshwater. Seawater is drawn into the plant through an intake pipe and prescreened to remove seaweed and other objects. UV disinfection is used to minimise biological growth and filters remove most of the particles. The filtered seawater is forced under high pressure through reverse osmosis membranes, allowing freshwater to pass through but very little salt. About 30 L of freshwater is produced from every 100 L of saltwater.

References
1. SA Water, 'Drinking water quality report 2007', 2008 (online).

CHAPTER 12

Who is responsible for water management?

Jennifer M. McKay

'Ecosystems may not only be more complex than we think, they may be more complex than we can think.' – Frank Egler[1]

Introduction

Defining *our water* and *who is responsible for it* are two vexed concepts in many parts of the world, especially in Australia, and Adelaide in particular. For South Australia the picture looks like this: there are many types of *our water*, multiple types of water stakeholders and human users, and a mosaic of laws and policies that have created organisations *responsible for our water*. Laws, policies and organisations have evolved over time in response to stakeholder understandings of the issues and demands for different economic and ecological outcomes from water use. It is not a static picture. Indeed, it is an interactive and highly interrelated picture; if you change one aspect, all others need to change in response.

This chapter will focus on defining *our water* and mapping the *responsibility* for it, looking at the laws, policies and organisations responsible for water management. Today's is a complicated situation, heavily influenced by laws and policies from earlier times when water demands were different. The current scenario involves a prolonged drought, conflict between the states over River Murray water, water quantity and quality issues, and community demands for ecological and social sustainability.

Our water

'Our water' for Adelaide means all of the water that all of the approximately 1 million people, animals and biota use in the Adelaide Mount Lofty Ranges Natural Resources Management (AMLRNRM) region. The largest human use of water is irrigation, at about 75%, with urban and industrial use accounting for the remainder. The sources of Adelaide's water are diverse, with most coming from the River Murray and from dams in the Mount Lofty Ranges (MLR), which is filtered and reticulated to households and industrial premises. Other, newer, sources of water are, for example, recycled water for agriculture in Virginia, and for toilet flushing and garden watering at Mawson Lakes. Stormwater is harvested and groundwater is used for irrigation of golf courses, and many domestic householders have recently put down wells. Desalinated water will become available after 2011 from the new plant at Port Stanvac.

In the early days, the River Torrens was vitally important to the city of Adelaide for water supply, food (fishing and agriculture), and sanitation. The following two excerpts[2] demonstrate that early residents did all of these, and that fishing in the river was adequate to supplement the colonial diet:

In the first decade following European settlement large numbers of sweet little 'slipperies'? [Galaias sp.], which seldom ran beyond three inches in length, were caught in the River Torrens which then consisted of continuous deep pools; also 'scalers'?, a species of cod, which were from 8 to 10 inches long and lurked beneath logs. Boys in

those days were too poor to buy hooks so they resorted to bent pins, strong thread and a stout reed, and when they got a bite the fish was snatched up and swung around the head until the young fisherman could vacate the log, from whence he was fishing, and dash his prey upon the shore before it could wriggle off the barbless pin.

It was a paradise for the children of Adelaide, Hindmarsh and Thebarton: secreted in deep holes were yabbies, which, when cooked, were hawked for a profit of a few pennies. For small recompense, native women were persuaded to dive for mussels. Such offerings from the river were a welcome addition to the frugal dinner tables of the working classes of Adelaide.

As time moved on the most important aspect of the Torrens in the city was for recreation, both passive and active. Public baths were built in 1882 on King William Street (where the Festival Theatre is now) and were supplied with water from the Torrens. The baths were remodeled in 1940 and finally demolished in 1970.[3, 4] The weir damming the Torrens today, which created Torrens Lake, was built in 1881, replacing an earlier timber weir built in 1862 that had washed away. The lake opened up leisure possibilities for the River Torrens, Elder Park was created, and the riverbanks were landscaped for recreational use.[5]

In 1960, Adelaide resident Robin Boyd was critical of recent attempts to alter the river features around the lake,[6] lamenting the loss of 'the shady timbered, untidy natural banks of the river that recently ambled unmolested past the University being bulldozed and terraced with trim rubble walls and neat plants'. In 1985, Derek Whitelock,[7] wrote

… where it flows through the central belt of the Parklands and the heart of the city, the Torrens was later improved and beautified effectively too, its 'Wind in the Willows' ambience has become a popular inner city 'Pleasuance for Adelaideans', however out of public and genteel sight, the river became and largely remains an ecological disaster area.

As the city grew, more water and water storages were required. Two large pipelines, the Mannum Pipeline (135 km, built in 1955) and the Onkaparinga Pipeline (48 km, built between 1960 and 1973), were constructed to import water from the River Murray. Today, in an average year the Murray supplies about 40% of the state's urban water needs; in dry years this can increase to 90%. Ten reservoirs and four weirs can store 198,400 ML of water, which is derived from local streams, augmented by the water pumped from the Murray. Other sources of water include groundwater, stormwater and recycled water; desalinated water is soon to be available.

In 1992 it was recognised that the volume of stormwater discharged to sea was equal to the average quantity of water consumed in Adelaide.[8] Many local governments set up cooperative schemes to manage stormwater flooding. Salisbury Council has been extremely active in reusing its stormwater by collecting and filtering it through wetlands for storage in aquifers. This water is then able to be used to irrigate local parks and ovals and the lake at Mawson Lakes.

Chapter photograph: Parliament House, Adelaide.

The Torrens Catchment Water Management Board was established under Section 53 of the *Water Resources Act 1997*. On 3 September 1999, the board issued two separate notices to Macag Holdings Pty Ltd, the owner of land situated adjacent to Duncan Road, Beaumont, pursuant to sections 14 and 17 of the Act. The notices required Macag to take action 'to maintain the watercourse situated on the land in good condition', and 'to take reasonable steps to prevent damage to the bed and banks of the watercourse situated on the land'. Macag appealed. The issue for the court to determine was if the topographical features con- stituted a watercourse as defined in the Act. 'Watercourse means a river, creek or other natural watercourse (whether modified or not) and includes: (a) a dam or reservoir that collects water flowing in a watercourse; (b) a lake through which water flows; (c) a channel (but not a channel declared by regulation to be excluded from the ambit of this definition) into which the water of a watercourse has been diverted; and (d) part of a watercourse.'

The stream in question was an unnamed tributary to First Creek. While it is beyond doubt that the River Torrens and First Creek are watercourses within the meaning of the Act, the question was whether the unnamed, third-order tributary stream flowing through the appellant's land was also such a watercourse. The common law definition of watercourse had already been applied in Victoria in Lyons v Winter (1899) 25 VLR 464.

'To constitute a watercourse … there must be a stream of water flowing in a defined channel or between something in the nature of banks. The stream may be very small, and need not always run, nor need the banks be clearly or sharply defined. But there must be a course, marked on the earth by visible signs, along which water usually flows …'

The Environment Resources and Development (ERD) court reviewed a number of Australian and UK decisions and concluded: 'A consideration … leads us to the conclusion that a channel through which water passes will constitute a watercourse at common law if it exhibits the following characteristics: a defined channel having both bed and banks, which channel must exhibit features of continuity, permanence and unity sufficient to distinguish it from a mere drain or drainage depression; water in it should usually flow in the same direction; water must flow through it, although such flow may be seasonal or intermittent.'

The ERD court reported that 'When walking up the valley floor, we saw clear signs that water still runs along it. Along or adjacent to much of the path, and in some parts of the old creek channel, we saw evidence of water erosion. The greater part of this evidence consisted of a small but clearly defined channel of varying depths and widths, running through the valley.' Overall the channel was ~200–300 mm wide and ~100–150 mm deep. Engineering witnesses agreed that the creation of such a channel was the result of water erosion. The evidence suggested that there would be 20 occasions when water would flow along the channel at flow rates of 400–700 L/sec at least once a year. These were considered significant. The court concluded: 'This is a watercourse … a very small part … of a series of tributaries and creeks which drain the River Torrens Catchment. That it is small and almost insignificant is not surprising, for every watercourse must start somewhere.'

Upon appeal, the Supreme Court held: 'The last of the characteristics identified by the ERD court is incorrect namely that … Water must flow through it, although such flow may be seasonal or intermittent.' On the facts of the flow rates of 20 days in each year, the Supreme Court found that this was not within the definition of intermittent and concluded: 'The ERD court has therefore erred in its definition of watercourse and in applying the common law definition to the facts.' Hence there was no watercourse on the land, and the appellant won.

Jennifer McKay

The Adelaide region produces nearly 100 GL of wastewater from major wastewater treatment plants (WWTPs) – Bolivar, Christies Beach and Glenelg – with some contributions from local government-operated community wastewater management schemes (CWMS) and septic tank effluent drainage scheme (STEDS). Adelaide reuses the highest proportion of this water compared to other Australian cities,[9] from 20% in 2004 to 29% in 2006/2007.[10] The wastewater is used to irrigate horticulture in the Virginia Triangle area, in McLaren Vale (in the Willunga Basin) for grape-growing, and in residential areas for irrigation and toilet-flushing. The water used in the Willunga Basin is sold to a private company that distributes it through a network of lilac pipes to growers, who have paid for the pipes; all such uses are strictly regulated and the water must be tested and the quality monitored. There are new national guidelines for water recycling, which the Department of Health in South Australia administers throughout the state.

Desalination of water is a recent initiative. Stage 1 of a new plant will be built at a cost of $3 billion, providing 50 ML per annum of drinking water for the metropolitan area. It will be owned by SA Water and is expected to commence operations in 2011. The drawbacks are the enormous energy costs and potential environmental impacts of the discharge on the marine environment, which will be monitored by the Environment Protection Authority; the operation of the Western Australian desalination plant will provide guidance.

There are moves for each new homeowner to capture rainwater. As from 1 July 2006, rainwater tanks must be installed in new houses and be plumbed to a toilet, to a water heater, or to all cold water outlets in the laundry of the home. The requirement for a (minimum) 1 kL plumbed rainwater tank is additional to any other water storage tank requirements (e.g. other tanks are required in some areas for bushfire-fighting purposes). Where a number of dwellings contribute to a communal rainwater storage tank, each dwelling must contribute rainwater from 50 m² of its roof area.

Finally, there is another type of 'our' water. Adelaide (like all cities) imports water in the form of food, fibre and manufactured goods from all over the world. As such, our water includes a substantial 'virtual' water component. Virtual water is the amount of water embedded in food or other products needed for its production. For example, to produce 1 kg of wheat we need about 1000 L of water, i.e., the virtual water of this kilogram of wheat is 1000 L.[11] The term virtual water was coined many years ago in the United Kingdom by Tony Allan, and places water at the centre stage of world trade.

The management of virtual water occurs in the place of origin. Australia as a whole tends to be net exporter of virtual water because of exports of grain and meat. In order to achieve long-term sustainability, water losses need to be rationalised. In many cases in South Australia, older water allocation policies allocated too much water, and there have been cut-backs. For example, for a decade there has been a limit of 50% on the runoff a grower can capture in a farm dam in the MLR. The present AMLRNRM plan has a number of water plans connected to it, and these will change the allocations of water to various users over coming years.

Influences on South Australian water laws, policies and organisations responsible for managing our water

Table 12.1 identifies the key major legal and policy events at the local, state, national and international level, which have created the present arrangements (see Figure 12.1). There is complexity in the distribution of power between state, local and federal governments, which has many implications.

The administrative arrangements in South Australia are overwhelmingly complex (but arguably more streamlined than in other states). There are many inter-departmental committees and other organisations providing services, and many groups that make substantial contributions in the administration of government obligations. For example, after the CoAG reforms of 1994, SA Water contracted United Water to manage the water supply pipes, while SA Water owns the pipes on behalf of the state.

Ecologically sustainable development

The adoption of the concept of ecologically sustainable development (ESD) has affected water management. Of importance is the Rio summit of 1992, where ESD was put forward. As a consequence, there were state-based approaches under the first State Water Plan to achieve integrated water management, and a host of other local initiatives.

SA Water[12] has described sustainability as:

decisions and actions which meet the needs of the community today without compromising the needs of the future, expressed in terms of social, environmental and economic considerations about the way we operate". For water utility to be acting sustainable it needs it needs to be designed and managed to fully contribute to the objectives of society now and in the future while maintaining the ecological, environmental and hydrological integrity of its natural resource base.

Another example is the statement from the Department of Water, Land and Biodiversity Conservation (DWLBC):

Water is a vital resource for South Australia's future prosperity. Its sustainable use within an integrated natural resources management framework for economic, social and environmental gains underpins activity in the government, industry and community sectors.

At a national level, the *Water Act 2007* requires the state to refer some of it powers over water to the federal government. This takes place under section 51(37) of the constitution, and the new Act and the referred powers will change the shape of water plans in the future, with the federal government having powers to accredit state plans.

National policies to manage River Murray water

Since 1955 Adelaide has relied on water piped from the River Murray, therefore the management of the whole river system is crucial for the city. While only a small proportion of the river flows through South Australia, there are long-standing arrangements in place guaranting our access to water.

Adelaide and Mount Lofty Ranges Natural Resources Management (AMLRNRM) Board – creating a sustainable future.

The AMLNRM Board was established in December 2004 and replaced 14 existing boards responsible for animal and plant control, catchment water management and soil conservation.

From its very first meeting, the board's priority has been to develop an overarching regional Natural Resources Management (NRM) plan that will achieve significant, tangible results through genuine community partnerships.

The plan covers a broad spectrum of natural resources management issues such as land and biodiversity conservation, water quality and resources, education and community capacity building, sustainable development and coast and marine programs.

The document has a 20-year timeframe and includes:

- vision and goals;
- state of the region report;
- resource condition targets;
- management action targets;
- strategic plan for the region;
- three-year business plan; and
- community engagement report.

The Adelaide and Mount Lofty Ranges regional NRM plan is a plan for the region, not just the board. To succeed, it must be owned and implemented by the entire NRM community. This extended family includes community and Landcare groups, businesses, government agencies, local councils, schools, research institutions and individual landholders. Hence, engaging with the community has been at the heart of the development process.

Our community engagement program has been comprehensive and far-reaching. It is designed to work hand-in-glove with the development of the plan, and we are excited and encouraged by the success we have achieved to date. Activities to engage the community and stakeholders have included:

- workshops with experts in fields of biodiversity, land and soil, water, marine, estuaries and coastal resources;
- meetings with major organisations, particularly government departments, 26 local councils, peak community and environment bodies, and business;
- personal contact and electronic surveys of 250 community organisations, industry groups and research bodies;
- local government forums;
- nine community workshops attended by 450 people from community organisations and the public;
- four follow-up community workshops, which attracted 120 attendees;
- 'open house' sessions at board offices; and
- informal and formal feedback provided throughout development and implementation of the plan.

Every care has been taken by the board to ensure that those involved in the process have been well informed about local issues and know what needs to be done. The challenge now is to harness this energy to ensure we create the right partnerships and generate the right investments for good environmental outcomes. Through the implementation of our regional NRM plan, and the partnerships created along the way, we are creating a strong base for an improved future for our precious natural resources.

Christel Mex

Like many downstream users of a major river water source, South Australia has long maintained that its upstream neighbours (New South Wales, Queensland and Victoria) have over-allocated water. State-based water allocations have proven difficult to reconcile because of the cross-border implications. The federal Water Act 2007 takes a whole-of-catchment approach.

Up to 7 March 2008, there was 2640 GL available to be shared between New South Wales, Victoria and South Australia. Of this, South Australia has 350 GL, New South Wales 580 GL, and Victoria 990 GL. South Australia has 720 GL for dilution flow, which is non-allocatable. Of the state's 350 GL allocation, 176 GL is allocated to River Murray licence holders. Additionally, 30 GL is allocated for carry over, 31 GL for country towns, 20 GL for private stock and domestic, and 90 GL for metropolitan Adelaide and associated country areas, including industry.[12]

The Commonwealth government's Water for the Future program was announced in July 2008. This 10 year package seeks to secure river health by reducing water abstractions, through modernising irrigation technology, and 'buying-back' water.

Water plans

ESD has been endorsed (in a non-enforceable way) through major international agreements. The main mechanism to achieve ESD in South Australia are water plans, dating from 1997 and reinforced in the *Natural Resources Management Act 2004*.[14] This Act seeks to promote sustainable and integrated management of the state's natural resources and to make provision for the protection of the state's natural resources.

The Act covers more than water. Indeed, according to section 3 (14), 'natural resources' include:

1. Soil
2. Water resources
3. Geological features and landscapes
4. Native vegetation, native animals and other native organisms
5. Ecosystems

The Act aims to bring together government and non-government organisations, and engage the community in the process of NRM planning – creating an integrated vision for the management of natural resources, addressing environmental, social and economic issues. Water planning is an example of this cooperation, where local communities agree on water-sharing arrangements, within the context of ESD. The water planning process is part-funded by levies paid on both urban and rural landholders.

The South Australian government has expressed its strategic direction for water management in a series of cascading plans starting with the South Australia's Strategic Plan,[16] and then the State NRM Plan,[17] which establishes a 50 year vision:

South Australia, a capable and prosperous community, managing natural resources for a good quality of life within the capacity of our environment for the long term.

The plan charts four major goals:

1. Landscape scale management that maintains healthy natural systems and is adaptive to climate change
2. Prosperous communities and industries using and managing natural resources within ecologically sustainable limits
3. Communities, governments and industries with the capability, commitment and connections to manage natural resources in an integrated way
4. Integrated management of biological threats to minimise risks to natural systems, communities and industry

The State NRM Plan was launched in February 2006, providing for state-level policy and strategy over the next 10 years, and must be evaluated every five years. The implementation strategy was finalised in September 2007, creating a reporting and accountability framework to meet the statutory obligations in the NRM plan. The framework consists of networks of organisations shown in Figure 12.1, in collaboration with non-government bodies including the South Australian Farmers Federation and Conservation Council of South Australia.

Underneath the State NRM Plan is a series of regional plans for the state's eight NRM regions (each managed by its own NRM board).[18] For Adelaide, the regional plan is known as the Adelaide Mount Lofty Ranges Regional NRM plan. Regional plans are created by the relevant board and must be consistent with the four goals of the State NRM Plan. The *Natural Resources Management Act 2004* establishes extensive consultation processes that these regional bodies must follow when drafting their plans.

The AMLRNRM plan is a very long document, agreed to after extensive public consultation. Its four volumes report on the condition of resources, strategies, monitoring activities, and other Acts the plan must refer to. Local government and its subsidiaries must conform with the plan when performing functions or exercising powers under the *Local Government Act 1999* or any other Act.

Within the Adelaide and Mount Lofty Ranges region there are five 'prescribed areas' for water resources management: Western Mount Lofty Ranges, McLaren Vale Prescribed Wells area, Northern Adelaide Plains Prescribed Wells area, Barossa Prescribed Water Resources area, and Central Adelaide Prescribed Wells area. Where a water resource is prescribed, the NRM Act requires that a water allocation plan (WAP) be prepared (some of the plans were established under the *Water Resources Act 1997* and have been adapted under transition arrangements to the NRM Act). Once completed, WAPs become government policy. Licensed users are limited in the volume of water they may take and use from a prescribed area.

Prescription of water a resource follows extensive consultation with the community and consideration of economic and environmental consequences. Where a water source is not prescribed it is less protected. However, if there is a mention of water in development plan then it is possible, under some decisions of the South Australian Supreme Court, that the impacts on the water source may be regulated.

BOX
49

Community knowledge of Natural Resources Management boards

The *Natural Resources Management Act 2004* (NRM Act) came into operation in 2005. The new Act replaced the previous *Water Resources Act 1997*, the *Animal and Plant Control Act 1986* and the *Soil Conservation and Land Care Act 1989* – bringing catchment water management, animal and plant control, and soil conservation under one umbrella. Under the Act, eight regional Natural Resources Management (NRM) boards were established and have taken up the work of previous catchment water management boards, animal and plant control boards, and soil conservation boards across South Australia.

Before 1 July 2006, local councils collected catchment water management levies in the former Catchment Water Management Board areas on behalf of the relevant regional boards. Likewise, councils contributed from general rate revenue for animal and plant control work, and made these payments to relevant regional Animal and Plant Control boards. Under the NRM Act 2004, since 1 July 2006 these contributions have been consolidated into a regional NRM levy, which is presented to the public as a separate line located on their council rate. Councils continuously collect the NRM levies and subsequently pay them to the relevant regional NRM board(s).

During the study period (2006–2007), the amount South Australians have paid for the NRM levy has varied: from as high as $76.10 for the district council of Le Hunte to as low as $6.50 for the City of Port Augusta (subject to a fixed charge); and from as high as 0.012 cents in the dollar for Adelaide Hills Council (which is in the NRM area of the Adelaide and Mount Lofty Ranges NRM Board (AMLRNRM Board), to as low as 0.000067 cents in the dollar for City of Norwood Payneham and St Peters.

A survey was conducted in May 2007 to gather information on community knowledge of NRM boards. Results indicated that only a quarter of survey respondents knew they had a local NRM board. The result was similar to the finding of an investigation conducted in May 2006 by the AMLRNRM Board that reported the percentage of awareness of the NRM Board was 29%. It was further discovered that of those who knew about the board, local government was the highest source of this awareness. Newspapers and brochures from local NRM boards followed.

Information source	Responses	per cent
Local government	90	31.3
Newspaper	65	22.6
Brochures from NRM board	56	19.4
TV	42	14.6
Radio	22	7.6
Internet	13	4.5
Total	*288*	*100.0*

Table 1: Information source of participants' awareness of the existence of NRM boards.

The interaction of government and community should lie at the very heart of a democratic society. Community feels unconfident in processes in which public authorities alone undertake and validate assessments of natural resource management policy. Therefore, the need for community and their governments to interact more effectively in making and implementing policies and legislations for natural resource management must be recognised.

The high proportion of unawareness suggests that the community does not feel well informed on this institutional arrangement or change. Therefore, continuation of the work of increasing community awareness is needed, through a number of effective information and education approaches.

Zhifang Wu and Jennifer McKay

Table 12.1 Timeline of sustainable development

Date	Event
From 1839	An emphasis on development and water supply, sanitation , recreational and agriculture use of water.
1901	Section 100 of the Australian Constitution guarantees state control of water management.
1915	The River Murray Waters Agreement (RMWA) signed by the governments of Australia, New South Wales, Victoria, and South Australia. It took a further two years to establish the River Murray Commission (RMC), which had the task of putting the Agreement into effect.
1987	World Commission on Environment and Development (WCED), *Our Common future*, launched the idea of 'environmentally sustainable development' (ESD), characterising it as development that meets the needs of the present without compromising the ability of future generations to meet their own needs.
1988	Mount Lofty Ranges strategy and River Murray strategy.
1989	Response to WCED by Hawke federal government spelt out a range of measures for dealing with environmental problems. There were conflicts between developers and environmentalists, and so a broad-ranging, less-confrontational strategy occurred in the shape of an Inter-Departmental Committee (IDC) to produce a discussion paper on ESD.
1990	IDC report released for public comment. The IDC was set up with nine sectors where environmental concerns were deemed most important: agriculture, fisheries, forestry, mining, manufacturing, energy production, energy use, transport and tourism.
1992	Rio Earth Summit outlined the 'precautionary principle' and the concept of 'intergenerational equity'.
1992	National Strategy for ESD constructed by a state and Commonwealth committee and accepted by the Council of Australian Governments (CoAG).
1992	Intergovernmental agreement on the environment, which created seven Australian themes for promote ESD and the National Greenhouse strategy.
1993	Mount Lofty Ranges Catchment Program established, focused on on-ground action that would lead towards sustainable use and development of natural resources in the ranges. The program is jointly funded by the community, and the local, state and Commonwealth governments.
1994	First CoAG reforms prescribe ESD and encourage: • private sector participation in water planning; • consumption based on two part tariffs: urban 1998, rural 2001; • full cost recovery by water suppliers; • separate identification and funding of community service obligations; • trading in rural water entitlements by separating water from land titles; • allocation of water for the environment; and • broader social values to be embraced.
1995	An interim 'cap' was imposed in June 1995 on the River Murray waters that limited the amount of water able to be diverted for consumptive uses to those being diverted as at 30 June 1994. There was an independent review of equity issues and this cap was made permanent for New South Wales, Victoria and South Australia from 1 July 1997. The cap has been described as '… the most monumental decision in resource management ever undertaken in Australia'. Yet environmental deterioration continues to threaten the sustainability of the nation's most important water resource.
1996	State of Environment Advisory Council find little evidence that ESD and related strategies have any impact on decision-making.
1997	*South Australia Water Resources Act* provides that Catchment Management Boards must create catchment management plans to manage the water resources of a region and ecosystems in close cooperation with the community, local government and state and federal governments.

Date	Event
1998	Final Final Murray-Darling cap on water diversions to 1993–1994 diversion levels. There is much community debate, with some sectors saying this level still demands too much of the river. Under the cap arrangements, each state is required to monitor and report to the Murray-Darling Basin Commission (MDBC) on diversions, water entitlements announced, allocations, trading of water within, to and and from the state, and to report on compliance with the target. The state must also report on measures undertaken or proposed to ensure that the water taken does not exceed the annual diversion target for every ensuing year. The MDBC appoints an Independent Audit Group (IAG), which annually audits the performance of each state government. There is also power to order a special audit if: • the diversion for water supply to metropolitan Adelaide has exceeded 650 GL; or • if the cumulative debit recorded in the register exceeds 20% of the annual average for a particular river. Many special audits have been conducted on the basis of the the latter. The IAG reports provide a vehicle for each state to voice concerns about the actions of other states. A key concern of South Australia has always been that New South Wales has exceeded is cap in a number of rivers, notably the Border, Wilder, Lachlan and the Baron Darling rivers. Further concerns in 2000/2001 were that New South Wales had not finalised its cap compliance through an independent verification process and South Australian remained concerned about the state's 'ability to bring diversions back into balance with cap limits into the future' (IAG, 2002).
2000	The State Water Plan addresses statewide resource issues and all regional plans must be consistent with it.
2002	The MDBC, a group responsible for managing the country's most significant source of drinking and irrigation water, initiated The Living Murray restoration program whose aim is to recover 500 GL of water for the Murray River (see MDBC 2006).
2002	Johannesburg Summit – world summit on sustainable development. This attempted to further clarify the concept of ESD.
2004	NRM Act reinforces waterplans and incorporates other acts to create a more comprehensive management scheme for natural resources.
2004	National Water Initiative (NWI) reforms provide more detailed protocols to the states on over 80 issues in water planning (e.g., requiring harmonisation of water trading rules).
2005	Water Proofing Adelaide strategy released
2006	State NRM and strategic plans drafted to implement the NRM Act 2004.
2007	Commonwealth Water Act attempts to get all states to refer powers over Murray-Darling Basin water to the Commonwealth. The Act comes into force in 2008. Despite difficulty in gaining consensus over the Act, it is unlikely to be challenged for constitutional validity. The Act requires a basin plan (operational in two years), which gives consideration to international agreements and impacts on biodiversity with a basin cap on extraction from rivers and aquifers. A national plan for water security creates Murray-Darling Basin Authority, which has a role in water trading and pricing and gives the Bureau of Meteorology functions regarding water information and standards.
2008	Office for Water Security within DWLBDC designed to be a single policy hub for developing and coordinating water security policy with SA government on issues of local regional and national significance.

Table 12.1 Timeline of sustainable development from 1839 to 2008.

South Australian water management

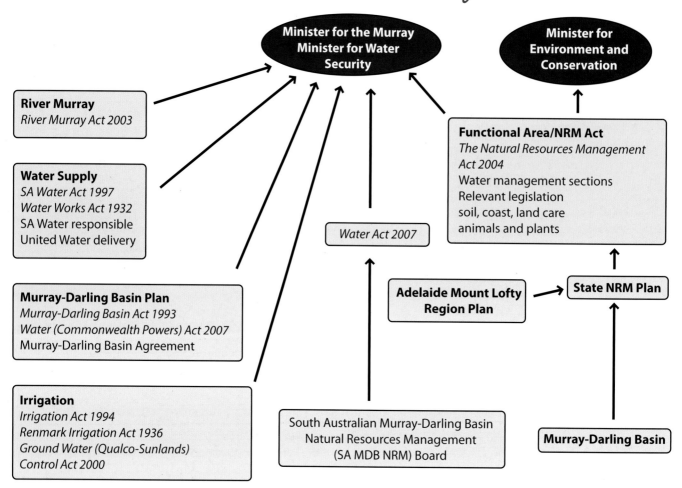

Figure 12.1 Organisations responsible for water management in South Australia at present.

The wording of WAPs has created some ambiguity, and it is desirable in the long term to standardise their wording to create certainty for all stakeholders. There have been many appeals to the Environment Resources and Development Court regarding this issue.

In addition to WAPs, there are community activities under federally funded schemes which take some responsibility for monitoring the health of water. These often become part of the local process of consultation for the plans. For example, Waterwatch[19] is a national community water monitoring program, which aims to increase community awareness, education and involvement in catchment care. Waterwatch in South Australia began in 1993 with about ten groups. It has since grown to over 490 school and community groups monitoring water quality at about 810 sites. Projects operating under the Waterwatch program include:

- Saltwatch;
- gutter guardians;
- our patch; and
- catchment carers.

Saltwatch[20] occurs annually during the second week of May and involves schools and community groups learning about the effects of salinity on water quality in their catchments. Sixty-eight groups from around South Australia participated in Saltwatch 2002, an increase of 13 groups from the previous year.

Users of our water in Adelaide

The most up-to-date statement on this topic comes from the plan for the Adelaide and Mount Lofty Ranges region. The problems with reconciling data collection tools held in various departments means these figures are patchy. The new scheme under the Water Act (Commonwealth) will provide a new role for the Bureau of Meteorology to coordinate and require the states to provide data on water.

Environment

Under the AMLRNRM plan, the environment (including both terrestrial and aquatic ecosystems) is now recognised as the 'first user' of water. The plan[18] states:

BOX 50

How the Natural Resources Management levy should be calcualted

Currently, the Adelaide and Mount Lofty Ranges Natural Resources Management (AMLRNRM) Board mainly bases the Natural Resources Management (NRM) levy on the capital value of rateable properties; a fixed charge is applied to a few council areas. However, the Natural Resources Management (NRM) Act sets out a number of alternative ways in which the levies can be determined:

1. The value of rateable land.
2. A fixed charge of the same amount on all rateable land.
3. A fixed charge of an amount that depends on purpose for which rateable land is used.
4. The area of rateable land.
5. The purpose for which rateable land is used and the area of the land.
6. The location of rateable land.

This list highlights that the Act does not relate the NRM levy to the volume of water or any other natural resources used, and therefore does not provide incentive for responsible and sustainable water management. Where a fixed charge applies, high- and low-valued property owners, and high- and low-level water-users pay the same rates. Where the value of rateable land is charged, high-valued property owners have to pay more regardless of their water consumption. Furthermore, the NRM levy is charged to every rateable property in South Australia such that some property owners pay more than once, irrespective of their impact on their local environment or local natural resources.

Departing from the Act, a survey was conducted to investigate how the Adelaide urban community thinks about the levy base method. Respondents were asked to answer an open-ended question: 'How do you think the Natural Resources Management levies should be calculated?' Over half of the participants indicated that they thought the NRM levy should be calculated on the volume of water used. Only 4.7% of respondents agreed with the current levy base method using property value.

Further, the survey investigated the preferences of levy payers to the levy base method, between percentage of property value and percentage of water bill. Based on an average property value of $300,000 and an average water bill of $600, the value of 1/10000 of the property value is equivalent to the value of 5/100 of the annual water bill, viz. $30. Four scenarios (see table below) were then presented to participants, with two options per scenario.

The responses clearly indicate that regardless of the scenario respondents were more willing to accept an NRM levy calculated on the volume of water consumption, despite the fact that in scenario 3, the value based on the water bill was double the amount of that based on property value.

However, re-establishing a levy method is not easy, as it is constrained by the federal constitution (Section 90), which restricts state powers to levy taxation: 'the power of the [Commonwealth] Parliament to impose duties of custom and excise, and to grant bounties on the production or export of goods, shall become exclusive.' A levy on water use is likely to be considered an excise. The only solution is to amend Section 90 or, alternatively, to permit the states to levy excise taxes as approved by a resolution of both houses of federal parliament.

Participants' preferences to pay: calculated on water bill or on property value?

	When levy amount =	Response %	When levy amount =	Response %
Scenario 1	1/10000 of property value	28.23	5/100 of annual water bill	71.77
Scenario 2	2/10000 of property value	31.79	10/100 of annual water bill	68.21
Scenario 3	1/10000 of property value	42.01	10/100 of annual water bill	57.99
Scenario 4	2/10000 of property value	24.64	5/100 of annual water bill	75.36

Zhifang Wu and Jennifer McKay

The biodiversity of the region is of state, national and global significance. The woodland and forest habitats of the Region are effectively an island surrounded by ocean to the south and west and more arid habitats to the north and east. As a result, the Region supports a number of species which occur nowhere else on the planet. Because of this isolation, the Region is also particularly susceptible to the impacts of climate change. These facts, along with the wide range of threats to biodiversity, contributed to the Region (with Kangaroo Island) being identified as one of 15 national Biodiversity Hotspots by the Commonwealth Government.

Urban residents

Urban residents use filtered and chlorinated water, reticulated by SA Water, for all purposes inside and outside the house. There are 556,000 household meters in South Australia, and water is priced using a fixed supply charge and a consumption charge. A three-tiered process is used to determine consumption charges (see Table 12.2). Some of the funds raised will be used to pay for infrastructure for the desalination plant.

There are variable figures on water use per household. It has been reported that mains water use per household fell from 271 L/day in 2000–2001 to 245 L/day in 2003–2004, and now is put at 235 L/day.[21] The outside use of water on gardens has been restricted over the last few years by SA Water, in order to reduce our take from the Murray. These water restrictions have been enforced by a well-established community information program, with fines and prosecutions.

Agriculture

The main user of water is agriculture. Water is used for irrigation and stock watering. Sources of water include rainfall and water captured in farm dams, pumping from aquifers, and imported water from the River Murray. The 15,300 farm dams in the Mount Lofty Ranges capture 40,000 ML each year. In some places, recycled water is used such the Northern Adelaide Plains and the Willunga Basin to water horticultural crops and grapes. The Bolivar WWTP supplies 14–18 GL to the Virginia Triangle, supplementing declining groundwater

supplies, to about 200 growers in an area of 200 km^2.[22] Customers in some recent recycled water schemes are being charged 75% of potable water process. In older schemes, established before the drought, recycled water prices were based on recouping annual pumping costs. Elsewhere, for example the Virginia Pipeline Scheme (VPS), recycled water is supplied at no cost, but growers are charged by the private company distributing the water, to recover infrastructure establishment costs. This use of recycled water close to the regions producing food reduces our need to use Murray River water and local water.

Agricultural users do pay for the licence to use surface or groundwater, but not a price for the original allocation for the raw water. If the grower used water provided by SA Water, the price is as above. Growers can trade water in regions close to each other, and the price paid is fixed by the market. The rules for market trading of water are set by WAPs.

Industry

SA Water supplies water under special contracts to big private users and also to public facilities. There are specific rules relating to these contracts that ensure full cost recovery.

Innovative state strategies in organisations responsible for our water

Levies funding the plans

The NRM Act gives the power to impose regional levies on all properties in order to address issues identified in the State NRM Plan. This levy is one of the few of its type in Australia and in the world. In 2004, the levy became known as a 'natural resources management levy'. The levy is collected by local government along with the collection of council rates.

The average amount paid per household by urban residents was between $16 to $32 in 2006/2007.[23] The levy is struck on the value of urban rateable land in the ALMRNRM Board region. It is also possible to strike a levy in relation to the amount of water taken by licensed water-users, which is also

Table 12.2 Water pricing

Tiers	Water use prices (pre 2007) per kL	Water use prices (2007–2008) per kL	Water use prices (July 2009) per kL
Tier 1	(0–125 kL) $0.50	(0–120 kL) $0.71	(0–120 kL) $0.97
Tier 2	(>125 kL) $1.16	(120–520 kL) $1.38	(120–520 kL) $1.88
Tier 3	No Tier 3	(> 520 kL) $1.65	(> 520 kL) $2.26

Table 12.2 Water pricing for urban domestic users. Source: SA Water, 'Your account' (online).

Regulated water reuse

BOX 51

Water is a precious resource, and the options for its storage and reuse should form a key element of any conservation strategy. Its reuse also addresses the undesirable impacts of stormwater discharge into coastal waters. But reuse and recycling can also introduce their own health and environmental issues. The key national principles that guide the management of water reuse and the legislative priorities have been identified in the *Australian Guidelines for Water Recycling: Managing Health and Environmental Risks*:

- The protection of public and environmental health is of paramount importance and should never be compromised.
- The protection of public and environmental health depends on implementing a preventive risk management approach.
- The application of preventive measures and requirements for water quality should be commensurate with the source of recycled water and the intended uses.

The agencies involved in regulating the quality of recycled water are Health SA (Department of Health) and the Environment Protection Authority (EPA). Other agencies, e.g., Primary Industries and Resources SA, have specific roles where statutes impose particular obligations.

Regulating water reuse

The documents that govern the reuse of water in South Australia are the *South Australian Reclaimed Water Guidelines 1999*, and the *Australian Guidelines for Water Recycling*, which, while not imposing mandatory requirements, establish the circumstances in which reclaimed water can be used sustainably and without risks to human health. The guidelines cover water reuse for industrial, agricultural and recreational purposes. Potentially, they have a legal effect since they can be called up as a mandatory condition of compliance for activities licensed under the *Environment Protection Act 1993* (EP Act), or for an approval issued under the *Public and Environmental Health (Waste Control) Regulations*. These regulations prescribe codes with which wastewater systems must comply. With respect to agricultural irrigation, additional requirements may apply, such as Section 32 of the *Livestock Act 1997*, which prevents the disposal of potentially hazardous material where grazing animals might be exposed. Other uses covered by the guidelines include irrigation of parks and gardens with reclaimed water, its non-potable use in residential premises, environmental uses (in wetlands), and industrial uses such as boiler or washdown water.

More generally, water reuse and recycling has the potential for adverse environmental impacts; operators whose activities might pollute the environment have a general duty under the EP Act to take all reasonable and practicable measures to minimise pollution. In addition to this general duty, the *Environmental Protection (Water Quality) Policy 2003* (EP Policy) imposes specific and general obligations in relation to the protection of water, and calls up the guidelines in relation to the collection for reuse of waters from septic tank effluent disposal or sewerage systems. Though not mandatory, it envisages that operators will comply. In both cases, failure to comply may result in an Environment Protection Order being issued that requires the operator to take specified action or to comply with the guidelines. Furthermore, were the process of recycling or the reuse of wastewater to amount to an insanitary condition, as defined by the *Public and Environmental Health Act 1987* (most obviously a 'risk to health'), a local council could impose a remediation order and the person responsible may also be prosecuted for an offence.

Aquifer storage and recovery

The process of aquifer storage and recovery (ASR) or managed aquifer recharge, where stormwater is pumped into an aquifer and then extracted for reuse, is an important method of harvesting superfluous water and having it available during dry periods. But it can have adverse impacts on groundwater and therefore is regulated through the EP Act and the EP Policy. In particular, the operation of an ASR scheme requires a licence under the Act if it is within the City of Mount Gambier or in the Adelaide metropolitan area. More generally, the EP Policy provides that a person who operates an ASR scheme must comply with the EPA Code of Practice for ASR 2004. Failure to comply may lead to the EPA issuing an Environment Protection Order.

Chris Reynolds

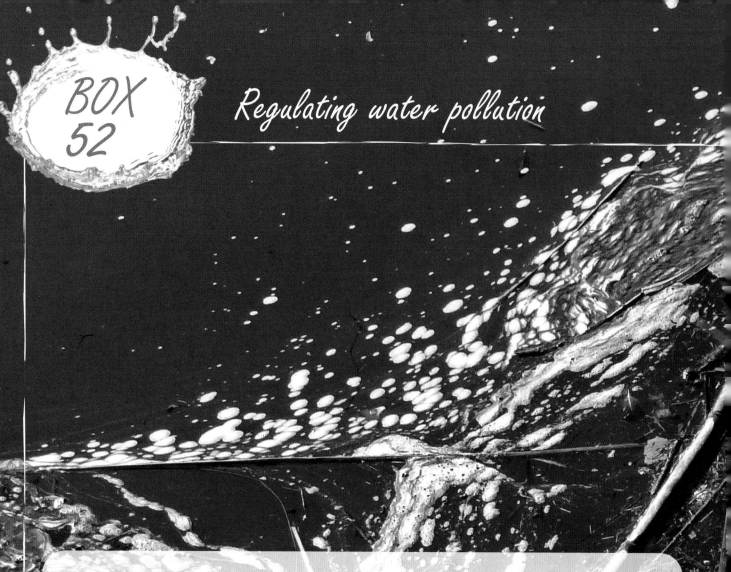

BOX
52

Regulating water pollution

The origins

South Australia regulates both the access to water (its quantity) and its purity (its quality). Regulation of the latter was among our first laws and our earliest pollution controls. In 1839, after the death of five children from dysentery, persons were prohibited from bathing or throwing animal carcasses into the Torrens within one mile of the city. Controls became more sophisticated and laws prohibiting water pollution became the cornerstone of 19th-century public health legislation. The *Public and Environmental Health Act 1987* continued that tradition with provisions for sanitation and specific penalties (ss 21 and 22).

The *Environment Protection Act 1993*

More recently, though, pollution controls have been part of environment protection functions, and the quality of South Australia's underground and surface water is also regulated under the general laws relating to pollution, in particular by the *Environment Protection Act 1993* (EP Act) and the *Environment Protection (Water Quality) Policy 2003* (EP Policy). Both the Act and the Policy are administered by the Environment Protection Authority (EPA), a statewide entity, and in some respects also by local councils as part of their stormwater management programmes. The key purpose of the EP Act is to prevent environmental harm, which is not defined in the Act other than to specify that harm occurs by pollution, with 'pollutants' defined very widely. The Act seeks to protect the environment in general and the aquatic environment in particular in the following ways:

- In the case of incidents amounting to actual harm, by prosecution for either serious environmental harm (where the harm or potential harm is of high impact or on a wide scale) or material environmental harm (where the impacts are significant in the sense that they are more than 'trivial'). In the most serious cases penalties of up to $2 million or four years' imprisonment can be imposed for deliberate or reckless actions that may cause harm (ss 79 and 80).

- In cases where the incident amounts to an environmental nuisance, which is defined as an act of pollution that does or may 'unreasonably interfere' with the rights of others exposed to the nuisance. This is also an offence, attracting a penalty of up to $60,000 (s 82).

- By imposing a general duty on persons undertaking activities that pollute, or might pollute, the environment to take 'all reasonable and practicable measures to prevent or minimise' any environmental harm that might result from those activities. It is not an offence to be in breach of this general duty but it is open to the EPA to issue an environment protection order, requiring the person to comply with their general duty (ss 25 and 93).

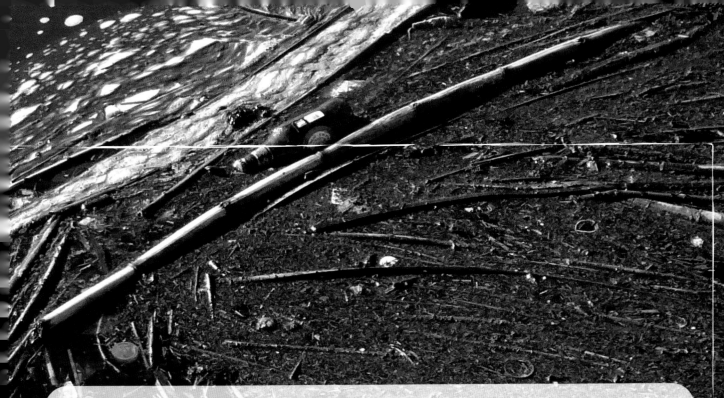

The *Environment Protection (Water Quality) Policy 2003*
The EP Act also allows for the creation of policies (broadly similar to regulations) through a process of consultation specified under the Act (s 28). These policies cover particular areas of the environment, regulated in general terms by the Act (water, air, noise and waste). Specifically, the EP Policy relates to all waters in the state other than reticulated waters and sewage systems (cl 4(1)). The Policy mirrors the Act by creating a general duty to take 'all reasonable and practicable measures … to avoid the discharge or deposit of waste' into any waters or places where it is reasonably likely to enter waters (cl 11(1)). The Policy also contains a series of mandatory requirements, breaches of which are typically punished by penalties of up to $4000. The Policy specifies offences for harm to the aquatic environment, defined to include: loss of seagrass or other native aquatic vegetation; a reduction in numbers of any native species of aquatic animal or insect; an increase in algal or aquatic plant growth; the water becoming harmful or offensive to humans, livestock or native animals; or increased turbidity or sediment levels (cl 12). As in other states, general water quality objectives are established and water quality criteria set to satisfy those objectives (see schedule 2). However, South Australia is unique insofar as it seeks to enforce its objectives by a mandatory requirement (cl 13). Other specific obligations in the Policy include end-of-pipe limits that can be imposed for particular industries or pollutants. These apply over and above any discharge limits that might be imposed by a condition of licence (cl 16). It is also an offence to discharge a range of pollutants (listed in schedule 4) into waters or places where they might be washed into waters (cl 17). Mainly these are gross pollutants: building material, wash-down water, and domestic chemicals. This provision is used regularly by local councils as part of stormwater management.

The Policy also regulates specific or 'point source' activities that impact on water quality. In Part 4, Division 2, a range of activities are listed. These include piggeries, cattle feed lots and wineries, whose effluents can all adversely affect water quality. Activities that directly impact on the marine environment, such as the use of antifoulants or the discharge of waste from vessels, are specifically regulated. In many cases these activities have mandatory requirements and are subject to a range of codes of practice called up by the Policy. Failure to comply with a code is likely to be dealt with by the issuing of an environment protection order, and failure to comply with the order can lead to an offence. Diffuse activities are also regulated under the Policy (Part 5), principally by tying activities such as road and building construction and the management of urban stormwater into codes of practice called up under the Policy. These are also enforced by way of an environment protection order.

Planning controls – 'prevention is better than cure'
The protection of water quality is best done before problems have occurred, and therefore in the planning stages of a development rather than afterwards when the impacts are felt. For this reason there are important links between the development control system and the EPA. In particular, developments of environmental significance as listed in the Development Regulations (schedules 21 and 22) must be referred to the EPA for comment or direction. Where these impact on waters, the EP Policy requires the EPA to 'take into account the potential of the activity or development to cause pollution of any waters', and 'consider the requirements that should, in the event of an authorisation being granted, be imposed on all relevant persons for the purposes of preventing or minimising the pollution or its harmful effects' (cl 43, see also cll 18(1), 23, 30). There are also special referral provisions for developments that might impact on the Adelaide Dolphin Sanctuary and the River Murray, both of which are protected by special legislation.

Chris Reynolds

done in the region. A meter must be installed on the well or pump. If that is inaccurate, other means may be used to estimate the volume used (NRM Act, Section 101).14 The levy can be payable even if the water is not used. There is provision to authorise special purpose levies in the relevant regional plan, and this can be used to pay for special projects.

The Water Proofing Adelaide strategy

The aim of this strategy,[24] which is housed in DWLBC, is:

> Working with the people of South Australia to secure our future water supply. Creating the blueprint for the management, conservation and development of water resources in Adelaide and semi-rural areas to 2025.

Specific projects include:

- Irrigating Adelaide's parklands using recycled water piped from the Glenelg WWTP – saving 1 GL of River Murray water used now for irrigation, and stopping up to 50% of wastewater from being pumped from the treatment plant out to sea every summer.
- Water Proofing the North – 24 stormwater recycling projects, two wastewater recycling projects and an experimental project to inject, store and cleanse stormwater in underground aquifers for recovery as potable water suitable for human consumption. Saving up to 30 GL annually.
- Water Proofing the South – extending the Willunga Pipeline from Christies Beach WWTP into the southern suburbs to irrigate parks, gardens, school grounds and for use in public toilets, saving 1 GL of mains water annually.
- Extending the VPS to pipe up to 3 GL of class recycled water from Bolivar WWTP to Angle Vale for market gardens.[25]

The Water Proofing Adelaide project has 63 strategies related to local surfacewater, groundwater, stormwater and recycled water, as well as water imported from the River Murray . It also has a focus on fostering innovation.

Federal innovations since 1994 and the Water Act 2007

Until March 2007, federal responsibility over freshwater has been limited by Section 100 of the constitution,[26] which provides that the power over water is to remain with the states. The former colonies viewed water as a key stumbling block to Federation, and, hence, inserted a prohibition clause in the new Constitution. Section 100[26] states:

> The Commonwealth shall not, by any law or regulation of trade or commerce, abridge the rights of the State or of the residents therein to the reasonable use of waters of rivers from conservation or irrigation.

The parliamentary debates reveal that Section 100 was inserted because New South Wales, Victoria and South Australia feared that Commonwealth laws under Section 51 might affect their common interest in water for irrigation.[27] The contest was really between the Commonwealth power over water for navigation and the state desire to use the water for irrigation. The object of Section 100 is thus twofold:

to limit Section 51; and to limit the inherent paramountcy of Commonwealth navigation power.[28] This has now been changed: in March 2008 in the Murray-Darling Basin area, by the process of all relevant states referring some power to the Commonwealth over water in the basin. States still retain water allocation power. This is the last significant reform for some time, and builds on three earlier periods of reforms in 1994, 2000 and 2004 (see Table 12.1).

Federal reforms implemented by the Council of Australian Governments (CoAG) in 1994 separated land and water rights, promoted ESD, and encouraged private sector involvement in water management. In November 2000, CoAG agreed to a regional model for the delivery of the National Action Plan on Salinity and Water Quality.[29] Following this, the NRM Ministerial Council adopted a regional delivery model for funding environmental activities at a regional level, leading to the integrated implementation of both programs, based on regional needs.

> The principal driver underpinning the regional delivery model for NRM is to harness the capacity of those closest to the problem on the ground', building on local knowledge, experience and expertise and enabling flexible and responsive solutions to local NRM challenges.[30]

There are 56 NRM regions in Australia, which have been agreed to by the states and the Commonwealth. The role of regional bodies at per the Commonwealth is expressed as:

> The key role of regional bodies involved undertaking regional natural resource management planning, prioritising regional level investments, coordinating actions at the landscape scale, getting community ownership in decision-making and reporting on progress.

The key features of the regional delivery model include:

- The development of a framework that sets out the respective NRM roles for Commonwealth, state/territory and local governments, and the community.
- A shift from funding individual projects to funding outcomes determined through regional NRM strategic planning.
- Devolution of decision-making to a regional level: a dispersed rather than centralist approach that allows for flexible decision-making, tailored to local conditions and needs.
- Introduction of national standards and targets to guide and provide direction for investment in NRM.
- A comprehensive accreditation, monitoring and evaluation framework to achieve consistent and acceptable standards of program delivery.
- Encouragement of community capacity-building through involvement in local NRM.[30]

In light of the continued complexity and lack of coherence between jurisdictions, CoAG announced a second major series of water reforms, the National Water Initiative (NWI), in 2004.[31] The NWI is a product of the CoAG process and aims to achieve national compatibility in the markets, and regulatory and planning schemes to achieve sustainable management of surface and groundwater. The NWI agreement[32] was signed by all governments at the CoAG meeting on 29 June 2004, with the exceptions of Tasmania and Western Australia, who signed in mid 2005 and early 2006 respectively.

The NWI specified that consumptive use of water requires a water-access entitlement to be described in legislation as a perpetual share of the consumptive pool of a water resource (NWI, paragraph 28).[33] The NWI represents a shared commitment by the Australian Commonwealth government and state/territory governments to ESD principles insomuch as it recognises:

The NWI is a comprehensive reform agreement containing objectives, outcomes and agreed actions to be undertaken by governments across eight inter-related elements of water management covering: water access entitlements and planning; water markets and trading; best practice water pricing; integrated management of water for environmental and other public benefit outcomes; water resource accounting; urban water reform; knowledge and capacity building, and; community partnerships and adjustment.[33]

Key areas for attention were the Murray-Darling Basin Agreement, and adopting a common lexicon for water terms in legislation and between the states:

Just under half of the NWI's 70 or so actions involve national actions or other action by governments working together. This reflects not just the emphasis in the Agreement on greater national compatibility in the way Australia measures, plans for, prices, and trades water. It also represents a greater level of cooperation between governments to achieve this end.[33]

This process will be driven by the new National Water Commission, and the $2 billion over six years to be invested through the Australian Water Fund.[33]

The National Water Commission is established under Commonwealth Government legislation (i.e., the National Water Commission Act 2004). It is an independent statutory authority reporting to the Prime Minister and, on some water reform matters, through the Prime Minister to CoAG.[33]

The commission consists of seven commissioners – four (including the chairman) nominated by the Commonwealth government, and three nominated jointly by the states and territories. Unique among Australian intergovernmental institutions, commissioners are appointed for their expertise in a range of water-related fields (including freshwater ecology, hydrology, resource economics, and public sector management), rather than as representatives of sectoral or government interests. The commission is supported by a staff of just over 40.[33]

The National Water Commission has three main functions:

1. Assess governments' progress in implementing the NWI (e.g., through biennial assessments of progress commencing in 2006–2007).

2. Help governments to implement the NWI (e.g. by acting as lead facilitator on certain actions under the NWI, such as nationally compatible registers of water entitlements and trades, and nationally consistent approaches to pricing).

3. Administer two programs under the Australian Government Water Fund (including recommending projects for decision by the Commonwealth government on financial assistance from the Water Smart Australia program, and the Raising National Water Standards program).[33]

The NWI also requires that water provided to meet environmental and other public benefits is to have statutory recognition, and have at least the same degree of security as water-access entitlements for consumptive use (NWI, paragraph 35). This is to ensure that water for environmental outcomes is not made less secure considering the need for greater security for consumptive water entitlements.[33]

The states agreed to refer powers to the Commonwealth, which then had power to operate the Water Act 2007.

In 2008, a Memorandum of Understanding was signed to create a new body: the Murray-Darling Basin Authority (MDBA).[34] This authority will create and implement a basin plan, including a sustainable cap on surface and groundwater diversions. The plan places critical human needs for water first, then sustainable industry and enhanced environmental outcomes. The basin plan is to be completed by 2011. There will be influences of this Act on the administrative arrangements (set out in Figure 12.1), evolving over the next few years.

Conclusions

Water supply is an emotive and complex issue in Adelaide and other parts of the world. Adelaide uses local water but has a heavy reliance on water imported from the River Murray; hence we have an extremely complex set of administrative arrangements. River Murray water supplies drinking water, and, once treated, water for irrigation of crops in the region. Residents in the city of Adelaide also use virtual water when importing food and other commodities.

As Adelaide is at the end of the River Murray system, national arrangements were put in place in the 1950s to pipe water to Adelaide. The state initially managed this through a government unit (Engineering and Water Supply Department), but since 1994 was obliged to alter this, creating SA Water and separate pricing and environmental regulators. These changes were the results of Commonwealth influence.

Up to 1994, and to the present day, South Australia remains in control of the management of local and imported water, however, since 1994 there has been increasing federal influence pushing the state (and the other states) to adopt certain principles such as ESD, full cost recovery, separation of water and land titles, and private sector provision of some aspects of water supply. There have been other special purpose funded schemes as well. The ESD concept was the international influence driving many of these changes.

BOX 53

Water restrictions

Drought conditions and declining water resources led to the introduction of Level 2 water restrictions in Adelaide in 2006, which were increased to the more stringent Level 3 restrictions in 2007. These restrictions were effective in helping to reduce Adelaide's water consumption, which fell from 216 GL in 2005 to 140 GL in 2007.

However, plants suffered, lawns began to turn brown, and there was widespread community concern about the health of gardens. Many people flouted the new rules and used extra water on their gardens. Other people reported community members who were over-watering. Water conservation officers issued over 7000 friendly reminders, warning notices and expiations for restriction breaches in a one-year period.[1]

Adelaide's Level 3 water restrictions may seem moderate compared to those enforced in other Australian cities, such as Brisbane, which has reached Level 6 restrictions. However, Brisbane's climate is very different to that of Adelaide – Brisbane's average annual rainfall is nearly double and it experiences hot, wet summers.[2] In comparison, Adelaide summers are dry and plants suffer from increasing water stress. Due to its climate, Adelaide cannot afford to implement higher, tighter levels of water restrictions.

There are pros and cons in using water restrictions to help solve Adelaide's water crisis. On the positive side, the tough water restrictions and consequent plant die off have made people more aware of gardens and plants better suited to Adelaide's climate. People now think more seriously about plant choice, and nurseries have seen an increase in sales of native and low water-use plants.[3] Selecting the right plants can result in significant water savings. A garden with native or drought-tolerant plants and grasses requires three–quarters less water than one composed of exotic species.[4] Restrictions have also stimulated people to become more involved in finding solutions to current water problems. For example, more people are now considering alternative water sources, such as installing rainwater tanks and the use of greywater.

The downside to water restrictions is that urban vegetation is now deteriorating. The decline of this green infrastructure affects the people of Adelaide and their quality of life, as urban vegetation provides a wide range of benefits to the community.

The most obvious of these benefits is that gardens are aesthetically pleasing and gardening is a popular and enjoyable pastime. Urban vegetation also plays an important role in moderating the local climate, including the 'urban heat island' effect.

An urban heat island describes the phenomenon where heat from the sun is reflected and re-radiated off buildings and urban structures, thereby increasing the temperature in cities.[5]

Urban vegetation creates shade and a cooling effect around homes and buildings, thus mitigaing the heat island effect. A decline in vegetation would lead to increased city temperatures and, therefore, an increase in the use of energy for cooling. Trees also moderate the local climate by creating windbreaks, which reduce the amount of heating needed in the cooler months.[6]

Urban vegetation improves the air quality in cities by taking up carbon dioxide (a greenhouse gas), producing oxygen, and filtering air-borne pollution and particulate matter.[6] Plants and trees also slow water runoff, helping to reduce erosion. Without this service, the quality of stormwater runoff would decline, as would the water quality in creeks and waterways.[6] Finally, urban vegetation creates habitats for wildlife such as birds, insects, bats, reptiles and small mammals.

Green infrastructure provides important ecological services and enhances the quality of life for the people of Adelaide. As Level 3 water restrictions continue in 2009, we need to consider how best to preserve our trees, plants and gardens without depleting our water resources. Taking responsibility for our personal water use by creating waterwise gardens is an important step in the right direction.

Carolyn Herbert

	Permanent water conservation measures	Level 3 – enhanced
Sprinklers	After 5 pm and before 10 am on any day (or, when day light saving is in force, after 6 pm and before 10 am)	Not allowed
Hand-held hoses fitted with trigger nozzles and dripper systems	Anytime	Maximum of 3 hours a week during the following times: • Even numbered houses Tuesday and Saturday, 7–10 am or 4–7 pm • Odd numbered houses Wednesday and Sunday 7–10 am or 4–7 pm
Buckets and watering cans	Anytime	Anytime

Residential water restrictions for garden watering introduced by SA Water. Source: SA Water.

These national and international influences, along with local concerns, have created a complex set of organisational arrangements (outlined in Figure 12.1). This arrangement is executed now by the series of cascading plans from the state strategic plan to the State NRM Plan, and local plans such as the AMLRNRM plan. The content and operations of the NRM plans and regional plans are found in the *Natural Resources Management Act 2004*. The plans were drafted after extensive community consultation. They have the potential to change past water-use practices. This Act, and the plans, are a charter to achieve ESD; the Act and plans are in the process of being fully implemented by assignment of responsibility to the organisations.

The task confronting the community of Adelaide is to reduce dependence on the River Murray, and to manage all water assets in an effective way. The plans look at long-term strategies for water use. The reliability of water supply is a crucial issue under climate change scenarios. The planning for this is a major activity of all governments.

One key lesson for the future is that responsibility for *our water* means making sustainable water use decisions that are in keeping with long-term availability of supply. There are many more stakeholders in the equation now than ever before, so decisions on sustainability are more complex. The system of locally based, community-drafted plans is the best option to ensure fair, transparent decisions; however, the processes are still evolving.

References

1. W. Dietrich, *The final forest: the battle for the last great trees of the Pacific Northwest*, Penguin Books, New York, 1992.
2. State Library of South Australia, 'Place names of South Australia', no date (online).
3. J.W. Warburton, 'Five metropolitan creeks of the River Torrens, South Australia: an environmental and historical study', Civic Trust of South Australia and Department of Adult Education, University of Adelaide, 1977.
4. W. Prest, K. Round and C.S. Fort, *The Wakefield companion to South Australian history*, Wakefield Press, Adelaide, 2001.
5. State Library of South Australia, 'River Torrens', 2006 (online).
6. R. Boyd, *The Australian ugliness*, Cheshire, Melbourne, 1960.
7. D. Whitelock, *Conquest to conservation: history of human impact on the South Australian environment*, Wakefield Press, Adelaide, 1985.
8. R.D.S. Clark, 'Multi-objective management of urban stormwater in Adelaide part 1: technical considerations', *International Symposium on Urban Stormwater Management: preprints of papers*, Barton, ACT, 1992, pp. 298–304; Institution of Engineers (Australia) (online).
9. J.C. Radcliffe, 'Water recycling in Australia: review report', Australian Academy of Technological Sciences and Engineering, Victoria, Australia, 2004.
10. Victorian State Government, 'Media release: Melbourne's water use lowest, water bills cheapest of major Australian cities', 2008 (online).
11. World Water Council, 'Virtual water', 2005 (online).
12. SA Water, home page, 2007 (online).
13. World Commission on Environment and Development, 'The Brundtland report (our common future)', Rio, 1992.
14. Government of South Australia, *Natural Resources Management Act, 2004* (online).
15. Department of Water, Land and Biodiversity Conservation, 'Water resources overview', 2008 (online).
16. Government of South Australia, *South Australia's strategic plan, 2007*.
17. Department of Water, Land and Biodiversity Conservation, *State Natural Resources Management Plan 2006*, Adelaide, 2006 (online).
18. Adelaide and Mount Lofty Ranges Natural Resources Management Board, 'Creating a sustainable future: an integrated Natural Resources Management Plan for the Adelaide and Mount Lofty Ranges region', Adelaide, 2008 (online).
19. Waterwatch SA, homepage (online).
20. Waterwatch SA, 'What is saltwatch?', 2005 (online).
21. Australian Bureau of Statistics, 'Feature article: household water use and effects of drought', catalogue no. 1350.0, Australian Bureau of Statistics, Canberra, 2005 (online).
22. B. Kracman, R. Martin and P. Sztajnbok, 'The Virginia Pipeline: Australia's largest water recycling project', *Water Science and Technology*, 43 (10), 2001, pp. 35–42.
23. Z. Wu, J.M. McKay and E. Hemphill, 'Attitudes to the Natural Resources Management Levy in Adelaide', *Water*, 32 (2), 2008, pp. 154–157.
24. Government of South Australia, 'Water Proofing Adelaide: a thirst for change 2005–2025', 2005 (online).
25. M. Rann, J. Hill and M. Wright, (2005) 'Water Proofing Adelaide news: Adelaide's water supply secured', 2005 (online).
26. Commonwealth of Australia, 'Commonwealth of Australia Constitution Act', 1900 (online).
27. P.H. Lane, *Commentary on the Australian Constitution*, Law Book Company, Sydney, 1986.
28. J. Quick and R. R. Garran, *The annotated constitution of the Australian Commonwealth*, Legal Books, Sydney, 1976.
29. Commonwealth of Australia, 'National action plan for salinity and water quality', Commonwealth of Australia (online).
30. Commonwealth of Australia, 'Living with salinity: a report on progress', Senate Standing Committee on Environment, Communications and the Arts, Canberra 2006 (online).
31. Commonwealth of Australia, 'National Water Initiative', National Water Commission, 2008 (online).
32. Commonwealth of Australia, 'Intergovernmental Agreement on the National Water Initiative', 2004 (online).
33. M. Thompson, 'National water initiative: the economics of water management in Australia: an overview', OECD workshop on Agriculture and Water: Sustainability, Markets and Policies, Adelaide and Barmera, 2005 (online).
34. Department of the Environment, Water, Heritage and the Arts, 'The Water Act 2007 and Water Amendment Act 2008' (online).

Box references

Box 53: Water restrictions

1. South Australian Water Corporation, 'Change to enforcement process for restriction breaches', 2007 (online).
2. National Water Commission, 'Australian water resources 2005: regional water resource assessment: capital city Brisbane Water Supply Area', 2005 (online).
3. Nursery and Garden Industry, 'Australian Garden Market Monitor report for the spring period ending 31 December 2006', 2006 (online).
4. J. Argue and A. Barton, 'Water-sustainability for Adelaide in 2020 based on stormwater: options, opportunities and challenges', *Proceedings of the International Conference on Water Sensitive Urban Design*, Adelaide, 2005.
5. A.H. Rizwan, L.Y.C. Dennis and C. Liu, 'A review on the generation, determination and mitigation of Urban Heat Island', *Journal of Environmental Science*, 20 (1), 2008, pp. 120–128.
6. C.L. Brack, 'Pollution mitigation and carbon sequestration by an urban forest', *Environmental Pollution*, 116, 2002, pp. S195–S200.
7. Wikipedia, 'Water restrictions in Australia', 2009 (online).

CASE STUDY 4: The politics of water: a brief history of the governance of the Murray River and Murray-Darling Basin

Martin Shanahan

Disputes over the River Murray, and later the Murray-Darling Basin (MDB), have loomed large in Australia's history. This case study briefly examines the context and outcomes from some of the more notable instances of conventions, commissions, or enquiries that have been held into the river and/or MDB over the past 150 years.

The management of rivers and river systems often prove difficult for humans. The desire to access, possess and own water quickly reveals the limits of our ability to manage wisely and plan thoughtfully. No better example of this can be found than the history of Europeans' efforts to govern, first the River Murray, and subsequently the MDB.

An important reason for the highly political nature of the disputes over the Murray, the Darling and the associated rivers of the MDB is that these watercourses represent far more than a water supply; they embody, or have embodied at different times, visions of settlement, land ownership, boundary demarcation, agricultural development, transportation, environmental amenity, tourism, nation building, trade, and economic prosperity. Thus, disputants have concerns with not just the water but its ownership, flow, storage, control, distribution, and future use.

After an aborted attempted at a conference in 1857, the first inter-colonial conference on the navigation and management of the River Murray between representatives from New South Wales (NSW), Victoria (Vic.) and South Australia (SA) occurred in 1863. The conference was not a success – mainly because of the different objectives held by each colony. SA sought developments that advanced river navigation, an activity it heavily supported. Vic. sought the right to develop irrigation and divert water within its territories. While NSW argued that under the declarations of the Imperial Parliament establishing the colonies, it had exclusive use of the water above SA, and its border with Vic. extended to the southern bank.[1] The claims of other colonies, it argued, had no impact on its sovereign rights to use the water, as it saw fit, within its boundaries. Unable to come to an agreement, another attempt to convene a convention in 1865 also failed.

Given the large differences in colonial objectives, it is not surprising that by 1882 SA's efforts to call another meeting would only produce formal acknowledgements from the other colonies that her correspondence had been received. Unable to work collectively, the colonies individually conducted royal commissions into the River Murray and its tributaries – NSW in 1884/1885, Vic. in 1885/1886 and SA in 1887–1894.

Common interests in development opportunities saw NSW and Vic. focus heavily on irrigation, and they began to work more closely, passing common resolutions declaring the rivers belonged to them and agreements to divert the waters that flowed through their territories. SA protested violently and '[her] bitter protests reached screaming point with the signing of the first Chaffey agreement' – an agreement that followed the NSW/Vic. decision to grant land to private irrigator developers.[1] Invited by Vic. to form its own royal commission and investigate how both navigation and irrigation could be promoted, SA was unable to convince either NSW or Vic. to even participate in its deliberations.

It was only the stirrings of Federation and appeals to a 'national sprit' that brought the colonies to discussion. The distance between their views on water, however, can be seen in the following exchange from the official record of the debates of the Australian Constitutional Convention (1891–1898):[2]

> Barton (New South Wales): 'What is the difference between taking away some of our water and taking away some of our land?'
> Deakin (Victoria): 'Exactly.'
> Kingston (South Australia): 'But it is not your water.'

In the convention debates of the 1890s, NSW asserted: (i) she owned the water; (ii) Vic. enjoyed coequal rights of navigation to the SA border; (iii) SA had no statutory claim; (iv) NSW and Vic. had moral rights to the water as they contributed to the Murray's flow; SA had none; (v) NSW had ceded certain rights to some Vic. tributaries to Vic., but only if she allowed NSW to build dams on the southern bank; and (vi) NSW had superior rights to the river but would agree to treat the matter as a federal issue if Vic. and SA agreed to submit all plans to NSW.[1]

The final compromise was section 100 of the Constitution – a piece of drafting that left all the crucial matters unresolved. It states: 'The Commonwealth shall not, by any law or regulation of trade or commerce, abridge the right of a State or of the residents therein to the reasonable use of the waters of rivers for conservation or irrigation.' The Constitution, in sections 73 (3) and 101–104 had made provision for an inter-state commission. However, in 1915 this commission was declared unconstitutional; it never was considered a possible solution for inter-state disputes regarding the River Murray or MDB. The opportunity for the Commonwealth to be involved in conservation was missed, being unanimously rejected by the delegates.[1]

It was not politicians but members of the River Murray Main Channel League in Cowra who in 1902 produced some movement to the political impasse. They persuaded each state to participate in a royal commission with equal membership. The result was the creation of permanent administration for the development of the river system – although the rights of irrigation and general supply were to be put ahead of navigation (this was, by 1902, a far less contentious issue; railways were usurping riverboats as the preeminent mode of inland transportation). Concurrently, an agreement was made limiting the amount of water NSW and Vic. could divert for a period of five years. This compromise was heavily criticised by all sides as either too great a concession or too little. Nonetheless, from these inauspicious beginnings, the 1915 River Murray Waters Agreement would eventually emerge. Undoubtedly one of the most critical elements in breaking the impasse was the Federation drought – until recent times the most severe drought in Australia's recorded history.

Moving from the Cowra conference of 1902 to the agreement of 1915 was not simple. In the interim, SA considered legal action against NSW and Vic. to protect its right to water. Four premiers' conferences and a Victorian Royal Commission on the Murray waters were held. Assisting the eventual compromise agreement, however, state demands too began to change. SA now began to emphasise the need for water for irrigation rather than navigation, and NSW and Vic. agreed to allow SA to construct storages on Lake Victoria to improve her consistency of supply. Geographically, this location, while just inside NSW, was comparatively isolated from NSW's and Vic's development and population interests.

The interstate conference of engineers in 1913 was crucial to the establishment of the River Murray Commission and the subsequent agreement of 1915. Equal representation on the commission, and a technical and engineering focus on water gauging, verifying water deliveries and constructing works, gave the participants a common objective – to use and develop the river – while downplaying ownership disputes. The commission assisted cooperation by allowing state water authorities and government to undertake most of the work in their jurisdictions. A scheme for sharing water was also set out, with all the optimism engendered by the hope of nation building and with only some regard to the reality of scarcity. It is still telling, perhaps, that while an arbitration process was to determine disputes, where the parties could not agree to a suitable arbiter, the Chief Justice of Tasmania would appoint one![1]

Having refocused the issue from the distribution of water to enabling the expansion of its supply, the next 70 years saw relatively minor modifications to the basic agreement of 1915 (amendments have been made to the agreement approximately every 7 to 10 years since 1920). This is not to say there were no disputes. SA regularly felt exposed to the threat of insufficient water and at various times sought to further improve certainty of supply. For example, SA ensured the Snowy Mountains Agreement did not override its right to receive its original allocations; SA was also heavily involved in disputes surrounding the Chowilla and Dartmouth dams. The commission, however, worked to ensure cooperation, and remained heavily focused on the construction of infrastructure and its maintenance.

Nature, however, does not take kindly to being ignored. While humans built all manner of constructions, turned rivers and permanently changed the flow of billions of litres of water in ways that suited them, and to which they agreed, they did so with little regard for the long-term sustainability of the river system. Thus the 1960s and 1970s saw the emergence of salinity as a major issue. For a commission focused on water supply, this problem sharply revealed its limitations. The result, in 1982, was a broadened focus on water quality, environment and recreation matters via amendments to the River Murray Agreement. By 1987 the growing awareness of viewing the system as a whole saw the emergence of a new body, the Murray-Darling Basin Commission.

But even as governments and communities began to appreciate more fully the complex and interconnected nature of river systems, and the impacts of human intervention, the environmental problems rapidly deteriorated. By 1992 a new agreement was forged, and by 1995 the first agreements to cap water diversions were initiated. Within two years the caps were made permanent, and the ACT and Queensland were officially involved in governance decisions. Within a few years of the initial agreements to cap water, it was clear that they were inadequate to deal with the conflicting demands of excess diversions, drought, economic and environmental sustainability, and predicted long-term reductions in rainfall.

In the face of rapidly mounting problems, the past decade has seen numerous inter-governmental meetings, national initiatives, plans and blueprints – all providing 'solutions' but each falling short of achieving long-term sustainability. While national bodies are now heavily involved in governance, the Commonwealth is still not ultimately and fully responsible for the MDB system.

Conclusion

It is no coincidence that the conventions and commissions that have had the most impact on altering the politics and governance of the River Murray and MDB have coincided with the two periods of longest drought in Australia's recorded history. Rather than serving as testament to Australian intelligence and farsightedness, the political history of Australia's MDB waterways is mostly a history of opportunism and short-term solutions. In recent years the cumulative effects are beginning to emerge of individuals, states and regions putting their interests before those of the entire system. It is to be hoped that although we are clearly slow learners, we will learn, and that we still have time to rectify the worst of our mistakes.

References

1. S.D. Clark, 'The River Murray question: part I colonial days', *Melbourne University Law Review*, 8, 1971, pp. 11–38.
2. D. Connell, *Water politics in the Murray-Darling Basin*, The Federation Press, Canberra, 2007.

CHAPTER 13

Climate change and water management

Tim Kelly
Peter Gell
Jim Gehling
Kelly Westell
Greg Ingleton

Introduction

To understand the risks of climate change and variability on Adelaide's water supply system, it is useful to consider the changes that have occurred throughout the earth's history, and changes due to natural climate variability that occur across decades and centuries. In this chapter we first take a look at climate change in earth's history that has helped shape the natural systems of the Murray-Darling Basin (MDB) and the Mount Lofty Ranges (MLR), which provide Adelaide with its drinking water. We consider the forces that drive natural climate variability. In the second part, we introduce the new climate change force of increasing greenhouse gases, caused by human civilisations. The chapter explores future global scenarios of human behaviour and examines SA Water's role in managing its greenhouse gas emissions as it adapts to climate change, seeks to reduce its emissions, and supports necessary research.

Climate change in earth's history

Climate change has become a common part of the popular lexicon on account of its link to anthropogenic global warming, a function of the increase in the efficiency of the lower atmosphere in trapping long-wave radiation emitted from heated surfaces of the earth. The increased concentration of greenhouse gases in the atmosphere, derived from fossil fuel burning and other activites, drives this warming, or what is effectively a *reduction in cooling*. However, climate change is not restricted to the post-industrial period and is not only a function of the changes in the concentration of carbon dioxide and other gases.

In the 1970s, concern was raised in popular magazines about the consequences of a likely cooling of the earth's climate, and the problems that might be expected if we entered another ice age. The idea of climate change is not new. References to viticulture in Britain during Roman occupation, and the freezing over of the Thames during Elizabethan winters were reminders then of a natural cyclicity to climate that was not just on a geological scale. While there is lip service to the fact that the early human migrations out of Africa into distant continents were enabled by sea level lows, allowing either terrestrial access to new lands or, at most, short aquatic passage, there is a naive idea abounding that the global climate averages of the last six to eight thousand years are normal. This period of relatively stable sea level and climate allowed humans to abandon hunter-gathering and settle down to permanently occupy suitable sites, with the consequent evolution of agriculture and animal husbandry. We consider that this development of civilisation is a demonstration of the superior intellect of home-loving humans over their distant nomadic ancestors, when arguably it was a short-lived opportunity made possible by feed-back loops in global climate.

Whether future climates will shift to some cooler or warmer steady state, or begin to resume a more oscillatory pattern, cannot be predicted. The study of gases and their isotopic characteristics in ice cores from continental ice sheets in Greenland and Antarctica suggest that climatic instability, of the kind that would not allow seasons to be predicted, has been the most common pattern in the last few hundred thousand years. The total period of human settlement may be just a brief hiatus in this geological era of climatic instability.[1]

The geological span of our genus, *Homo*, is the Quaternary Period, which began 2.6 Ma ago with the establishment of the circum-polar current around Antarctica and the sinking of cold, salty surface ocean water in the north and south Atlantic.[2] We evolved during the Pleistocene Epoch – an ice age that featured waxing and waning of polar ice sheets over Europe and North America with consequent fluctuations in sea level of up to 200 m. The Holocene Epoch of the last 10,000 years is a tiny time of stasis in the greater sinusoidal graph of climatic fluctuation.

Climate change across hundreds of millions of years

On a geological scale of hundreds or even thousands of millions of years, it is arguable that the mean climatic condition of the earth has been one of considerably warmer climates, minimal polar ice, higher rainfall, and higher sea levels. For much of the last 150 million years, the Australian continent had barely one third of its present land surface; the other part was either shallow marine, like the Gulf of Carpentaria today, or involved large freshwater lakes and swamps. During the Mesozoic Era, when Australia was still docked with Antarctica near the South Pole, shallow seas separated western Australia from the highlands along the east coast. For much of the Cenozoic Era, of which the Quaternary Period is a short, final chapter, Australia was en route from Antarctica, at latitude 50–70° south, on its way north to meet Asia. Forty Ma ago, rainforests, like those in north Queensland today, occupied much of southern South Australia. The fossil leaves of these forests have been dug up in Golden Grove and Maslin Beach sand quarries, or lie buried beneath the Adelaide Plains. In the late Miocene Epoch (5–20 Ma ago), the salt lakes of Eyre, Frome and Callabonna were freshwater habitats occupied by flamingos, crocodiles and freshwater dolphins.

Further back in time, about 300 Ma ago during the early Permian Period, southern Australia was the melting ground of ice sheets sliding north off Antarctica (when our two continents were joined as the super continent, Gondwana). Today, the islands off Victor Harbor are the remnant landforms scoured by these Permian ice sheets. Backstairs Passage and the Inman and Hindmarsh valleys of the Fleurieu Peninsula were cut by glaciers and filled with the debris as the ice melted. Today, as the Fleurieu Peninsula and the MLR are being squeezed upward, these valleys are being exhumed. Backstairs Passage, between Kangaroo Island and Cape Jarvis, is being scoured out by tidal currents. The coastal cliffs south of Adelaide have only recently formed, as the MLR rise and Gulf St Vincent sinks. As the cliffs of Hallett Cove are progressively stripped of their sediment cover, we can see the telltale evidence of ice sheets moving west-north-west. In Permian times, ice sheets, which scraped and polished the ancient rock beneath, left scratch marks covered with ice-borne debris. However, this Permian ice age, although 300 Ma older than our current Quaternary ice age, was by no means either the oldest or greatest ice age the earth has suffered.

Snowball Earth

BOX 54

Early in the 20th century, geologists like Sir Douglas Mawson realised that there were rock formations ('tillites'), composed of ice-borne deposits sandwiched within the layered Proterozoic eon rock formations that make up the foundations of South Australia from the Mount Lofty Ranges through the Flinders Ranges and into Central Australia. The evidence is plain to see on the southern doorstep of Adelaide in Sturt Creek, by Flinders University and Flagstaff Hill. The Sturt Tillite is a massive deposit of boulders, gravel, sand and mud – a sort of 'pudding stone' that can only have originated from the melting of massive ice sheets.

Evidence of a slightly younger ice age, near the coast at Marino Rocks, is even more prominent in the Flinders Ranges. Near Arkaroola, in the northern Flinders Ranges, these ancient tillites are several thousands of metres thick. This ice age, represented by both the Sturt and Marinoan glacial record, can be recognised in the Proterozoic Eon rocks of every continent except Antarctica. Current dating of these rocks suggests a period of refrigeration that lasted from about 720 to 630 Ma before the present. Considering the age of such rocks, and the chance that many areas are covered beneath younger rocks, the global coverage was quite extraordinary.

However, evidence from palaeomagnetic studies of these rocks near Hallett Cove and south of Quorn has shown that the tillites and associated melt-water sediments were deposited when Australia straddled the equator.[1,2,3] The magnetisation of iron particles in these sedimentary rocks shows that the lines of force of the magnetic field were almost parallel to the sea floor. Today that only happens at or near the equator. Furthermore, there is recent evidence of low-latitude tillites of the same general age, from the Arabian Peninsula and Namibia.[4]

The implication of the global distribution of these ancient tillites, and the evidence of their equatorial position at the time, led to the idea of 'Snowball Earth'.[5] It was not until the study of carbon isotopes from these 'cryogenian' rocks that a model was put forward arguing for a progressive freezing over of the earth's oceans due to extraordinary low carbon dioxide levels in the atmosphere at the time.[6,7] This ice-encrusted era would have all but extinguished the simple microbial lifeforms that inhabited the earth at the time. However, since we know that microbial life did survive this icy world, it is likely that the oceans never completely iced over, especially near the equator. The ice might have become permanent had it not been for the fact that the our planet has an active interior, powered by radioactive decay, which has resulted in volcanic exhalation through most of earth's history. With low levels of photosynthetic microbes during the Snowball time, there was little recycling of the greenhouse gases emitted by volcanism; consequently, carbon dioxide built up in the atmosphere to such an extent that the earth was rapidly reheated despite the reflective effects of ice sheets. The Snowball theory argues that there was a rapid meltdown of ice, with acid rain stripping the continents of limestone such that a widespread layer of carbonate rocks deposited over the last deposited tillites (ice debris). This 'cap carbonate' has some peculiar characteristics that allow it to be recognised in many regions. Locally, the Nuccaleena Formation forms the 'cap' that can be traced from Hallett Cove into the northern Flinders Ranges.

It may not be a coincidence that within a few millions of years after the last of the Snowball ice ages, the first large fossils of complex marine animals appeared. These are known as the Ediacara biota – segmented mats, fronds and sac-shaped organisms that lived on the sea floor before predators evolved. They include the earliest known sponges, cnidarians, arthropods, molluscs and chordates – the ancestors of most living animal phyla.[8] They were first recognised in the Flinders Ranges by Sprigg,[9,10] and later gave their name to the first new geological period to be named in 120 years: the Ediacaran Period.

Jim Gehling

Pre-industrial climate change (past 2 million years)

So, as geological time has progressed there have been considerable changes in the earth's climate, varying from an almost completely snow- and ice-covered planet to other hot periods with tropical conditions extending towards the poles.

In recent geological history (over the past 2 million years), before the era of industry, and before the population of humans on earth began their inexorable rise towards the 6.3 billion souls of the 21st century, global climates continued to go through variations of great amplitude, across a range of return times. Driving these changes were not necessarily the biota, although the forest cover/carbon dioxide/greenhouse nexus has always been a feature, and not necessarily particulate emissions, although volcanic explosions are known to have impacted upon the atmosphere enough to cause short-term climatic aberrations, but a combination of the shape of the earth's orbit relative to the sun, and the intensity of the sun's rays, each of which vary predictably over a range of timescales.

The earth's orbit of the sun varies over long periods of time. These variations were identified as early as 1842, but it was Milutin Milankovich, a Serbian geophysicist, who in the 1920s/1930s substantially advanced the knowledge of orbital and axial changes to reveal drivers of climate change in what are now termed 'Milankovich Cycles'.[3] The first of these describes a combination of the tilt of the earth and the elliptic shape of the earth's orbit that changes through a full cycle of hotter northern summers to hotter southern ones every 21,000 years (Diagram 13.1). The second describes the angle of the earth's tilt relative to the plane of the ecliptic, moderating the seasonal insolation over a period of 41,000 years. Lastly, the narrowing and broadening of this path brings the earth closer to and further from the sun at different key parts of the year and influences the amount of incoming solar radiation by up to 3.5% during different seasons, on a recurring cycle of 100,000 years. In the past, these three cycles have interacted, leading to periods of increased and decreased solar warming. Since the emergence of hominids, two cycles have dominated. For much of the last 2 Ma small amplitude cycles driven by the obliquity of the ecliptic have occurred with a frequency of ~ 41,000 years. However, from around 700,000 years ago, the eccentricity of the earth's orbit has come to dominate and the earth has come under the influence of large, amplitude glacial-interglacial cycles generating cold, dry ice ages and warm and relatively wet interglacial phases.

It is highly fortuitous that industrialised people find themselves in an interglacial phase, as the alternative would spell a climate regime that would challenge the technological capacity of even the wealthiest nation today. It is, however, rather humbling to ponder the challenges such climatic extremes must have posed for Australia and Adelaide's original people, for it is certain that the Kaurna people experienced both the extremes of the ice age as well as the more recent time of warmth and plenty.

To reconstruct the climate of the past, scientists draw on a rich and varied research tool kit. They seek out sites that have archived, optimally in a continuous sequence of time, evidence from the land, air, water, or the biota – which can be used to infer particular climatic regimes. In many instances these records are archived in locations permanently inundated by water or where sediment accumulations of sand dunes have buried evidence of the past. Alternatively, rings from long-living trees or the continuous record of atmospheric conditions tied up in cave carbonate deposits can be used. The climates of South Australia, even through these cycles of such contrast, did not lend themselves well to preserving such evidence. A reconstruction of the climate of Adelaide over millennial timescales therefore requires reference to evidence from sites that are widely spread. So a local view can only be generated by extrapolation between sites and then with a degree of inference based on topography and position relative to the shifting coastline.

If we accept the arrival of people in Australia to be at around 50,000 years ago, with the oldest, well-dated evidence being for human occupation of Lake Mungo, north-east of Mildura, at 46,000 years ago, we can contemplate people living in the vicinity of Adelaide from around 40,000 years ago. However, if there were people living in the Adelaide area from that time they would not have been a coastal people as the sea was 87 ms lower than today, and retreating. The nearest sea would have been up to 300 km to the south-west, disconnecting the site from the rain-bearing maritime winds that prevail through a modern Adelaide winter. This drying effect was exacerbated globally by the progressive locking up of vast volumes of water as ice on the poles and at altitude. This phase represented the descent into the last ice age, which reached maximum effect around 20,000 years ago as sea level dropped to 138 m lower than present (Diagram 13.2).

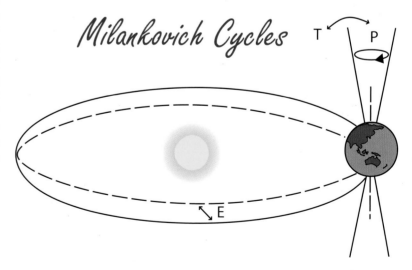

Diagram 13.1 Milankovich Cycles. 'T' shows the changes in the tilt of the earth's axis; 'E' shows the changes in the eccentricity or shape of the earth's orbit around the sun; and 'P' shows the changes in the direction of the earth's axis of rotation. Source: Rahmstorf and Schellnhuber, Der Klimawandel, Beck Verlag, Munich, 2006, p. 144.

What climate change means for Adelaide

BOX 55

The term 'global warming' says it all – a heating of the atmosphere right across the world. But that does not mean that the warming, or its impacts, will be the same everywhere. Regional and local differences can cause things to be worse, or better, depending on where you are.

One example of this unevenness is in the Arctic. Snow and ice melt over progressively larger areas and for longer periods as the temperature rises, causing the earth's surface to be duller. Bare rock, soil, vegetation and the open ocean are all much darker than bright ice, and so, just like the dark panels on solar hotwater systems, they absorb substantially more sunlight. This leads to greater heating, more melting, and so on – just one example of an amplifying feedback that can make global warming worse that it would otherwise be. There are many other such feedbacks, some of which remain poorly understood and could lead to more severe and more rapid warming than expected.

Perhaps the biggest regional impact of climate change that Adelaide faces is a shift in equatorial weather systems. Global warming causes the overturning tropical air masses that circulate in giant loops (called Hadley Cells and the Walker Circulation) to expand north and south. This has been recently shown to have happened already – up to 2° of latitudinal expansion over the last 30 years. Atmospheric heating also causes polar winds to whip around the Southern Ocean more rapidly. Together, these effects of global warming act to push rain-bearing mid-westerly weather systems further north and south. So, instead of being doused in rainfall brought in from the Indian and Southern oceans, progressively more of this rain will be dumped uselessly over the sea, below the Australian continental margin. This means less rainfall for Adelaide and South Australia's agricultural areas, as well as the south-west of Western Australia and other mid-latitude regions such as South Africa, southern Europe, Mexico and the western United States.

With less rain, the vegetation and soils around Adelaide and the Mount Lofty Ranges will dry. In combination with higher temperatures, the risk of bushfires intensifies. Heatwaves are the most dangerous culprits in this relationship. The 15 day March 2008 heatwave in Adelaide was, on the basis of the 20th-century temperature record, a staggering 1 in 3000 year event. Yet under a mid-range projection of global warming (should no action be taken to quickly curtail carbon emissions), such an event would be an expected part of an average summer. Such heatwaves, and regular intensive fires, also cause great stress to most species, leading to higher mortality, failed reproduction and reduced body condition. These synergies – between water availability, hotter temperatures and changed fire regimes – are some of the primary reasons why unrestrained climate change is anticipated to lead to the extinction of an appallingly large fraction of our biodiversity.

Sea level rise is a more universal threat. When salty seawater mixes with fresh groundwater, the result is diluted seawater. A once useable water resource becomes worthless, with obvious impacts on coastal drinking and irrigation water supplies, as well as ecosystems that tap into aquifers. Severe storm surge events occasionally result in this exchange, but if these events are rare and do not encroach too far up the shoreline the impacts are generally minor and localised. But what if the frequency of flooding events from the sea were to increase dramatically, and do so across the entire coastline of South Australia? That ominous threat is just what is anticipated due to climate change, and should therefore be a major concern to coastal planners and beachside residents alike.

There is clear evidence that sea levels have risen over the past century. Long-term records from a globally distributed network of reference tidal gauges show that sea levels rose about 20 cm from 1870 to 2004, correlating with a globally averaged rise in temperature of about 0.8°C. Since 1992, a satellite monitoring system has made regular and precise measurements of sea level, and shows an accelerating rise over the last decade. If the Greenland and West Antarctic ice sheets hold together, the most recent estimates suggest another 50 to 140 cm of sea level rise this century. A worst-case scenario, now being predicted by some eminent scientists, is 3 to 5 m by 2100, should the polar melt accelerate. Yet 50 cm would be enough to make a 1 in 100 year storm surge event a yearly occurrence.

The need for action is urgent and our window of opportunity for avoiding severe impacts is rapidly closing. Yet the obstacles to change are not technical or economic, they are political and social.

Barry Brook

Effective rainfall was probably only half that of today, and the early inhabitants may have had to withstand average temperatures 8–10°C cooler than present. Conditions for many millennia were very cool and dry – a climate with very few analogues in modern Australasia. Reduced evaporation on account of the lower temperatures may have allowed groundwaters to rise and surfacewater to persist, but, by and large, Adelaide was semi-arid to desertic.

The climate warmed as the eccentricity driver switched. Global warming began and sea levels rose in response. At around 14,000 years ago the rise may have occurred as quickly at 50 cm/decade. At this time, increased tropical monsoon activity is evident and, soon after 10,000 years ago, Backstairs Passage was being flooded and the gulfs were beginning to refill. Again, the maritime air masses began to reach through to the MLR where sheoak woodlands were able to recolonise due to increased effective rainfall in the now warm interglacial. Through this phase, effective rainfall probably increased for surfacewater to be a regular feature of the Adelaide Plains landscape. This continued to maximal conditions around 7000 years ago when uplands became saturated, eucalypts invaded the hills, and Adelaide beaches were reestablished. The stabilising of sea level from this time and the increased surface moisture allowed for the reestablishment of rivers and estuaries and the development of the tea-tree swamp in the lower Torrens where West Lakes is today. This period of plenty was relatively recent in the Indigenous settlement history of Adelaide and lasted for more than 1000 years – Adelaide was probably at its most rainy for over 100,000 years. From this time to today the regional climate has become more variable although there is little evidence for this in the palaeoclimatic record. There is evidence of a certain resilience to these lower-amplitude changes in recent centuries.

Australia 20,000 years ago

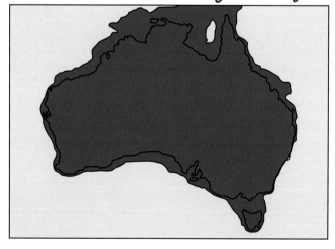

Diagram 13.2 The Australian continent 20,000 years ago during the last ice age. The inner line shows the coastline of Australia as it is today. Source: Explore SahulTime, Monash University (online).

Understanding climate variability and climate change since European settlement

Climate change evidence embodied in landscape evidence after Europeans settled Australia is unreliable, as many indicators of change have been overwhelmed by the stronger forces of catchment clearance, riverflow extraction, and the pollution of air and water. Fortunately, however, the instrumental record can take over and, as elsewhere, attests to considerable warming over recent decades, and reduced effective precipitation. While this is consistent with projections of the future climate over Adelaide under greenhouse warming, it remains difficult to partition the influence of industrial emissions from the prevailing cycles of climate change much better known to the earth through time.

Regardless of the percentage contributions, Adelaide's water supplies from the River Murray and MLR are vulnerable to a drying climate whether this is caused through natural variability in Australia's climate or anthropogenic climate change, or both. In considering the risks associated with water scarcity, it is important to consider how robust Adelaide's water sources are against natural climate variation first, and then consider the additional risks associated with climate change.

Post-European settlement climate variability

Instrumental records of Murray-Darling System inflows recorded since the early 1900s show that since European settlement, climate shifted to a drier, drought-dominated period from the 1890s to around 1950, followed by a postwar wetter period lasting into the 1990s. Since the mid 1990s, however, Australia's climate in the southern part of the continent appears to have switched once again into a drier, drought-dominated period leading to reduced streamflows in the Murray River and the drawdown of water reserves. See Figure 13.1.

Adelaide's local MLR water storage reservoirs have been considered as semi-independent water resources that can provide 60% of Adelaide's water supply in favourable years, filling local reservoirs. Under local drought conditions, however, local catchments may only provide for 10% of Adelaide's water requirements. The full water storage capacity in the MLR provides approximately 12 months' water supply but is not sufficient to cover longer periods should the River Murray supply become unavailable due to insufficient waterflows or poor water quality.

Anthropogenic climate change

The term 'anthropogenic climate change' refers to additional climate forcing caused by human activities, or changes in the climate in addition to natural changes and variations, such as the burning of fossil fuels and increased methane from livestock and land clearing. When considering water availability, water industry planners must now consider additional impacts of anthropogenic climate change. Several reports, such as the 2003 CSIRO's 'Climate Change in South Australia',[4] indicate that warming and drier conditions in mid-latitude zone areas, particularly in winter and spring, are likely to result in higher temperatures and reduced rainfall on average; however, extreme weather events also include risks of severe flooding as well as longer periods of drought.

While it is not possible to separate changes in short-term rainfall in southern Australia into natural and anthropogenic proportions, increasing temperatures of approximately three quarters of a degree Celsius are apparent in Bureau of Meteorology historic datasets, which are consistent with global warming trends. Figure 2.6 (chapter 2), showing the annual mean T for the Adelaide and Mount Lofty region, is typical of such trends. In very simple terms, water availability for

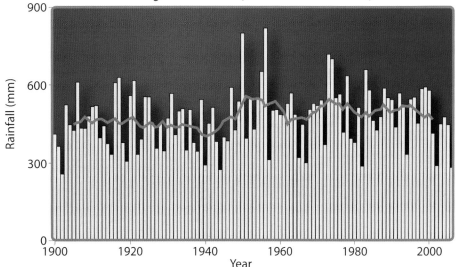

Figure 13.1 Murray-Darling Basin annual rainfall (mm) from the early 1900s. Source: Bureau of Meteorology, 'Timeseries: Australian climate variability and change, Murray-Darling Basin annual rainfall', 2009, copyright Commonwealth of Australia, reproduced by permission.

human settlements and the environment is equal to rainfall minus re-evaporation, either directly or from vegetation. Hotter conditions, such as those experienced over the past century and particularly over the past decade, drive greater rates of evaporation from land surfaces, rivers streams and water storages, causing significant system losses.

Informing decisions through climate science

Quantifying the risks from climate variability and change will involve ongoing research and modelling, which has commenced at the global scale and is progressively being downscaled to catchments and regions.

The River Murray Sustainable Yields project, being undertaken by the CSIRO, is an example of the ongoing research effort. It involves reviewing each subcatchment in the MDB to estimate current and likely future water availability in each catchment/aquifer, and for the entire MDB, considering:

- climate change and other risks;
- surface/groundwater interactions; and
- current and future water availability compared to that required to meet the current levels of extractive use.

The Department of Water, Land and Biodiversity Conservation (DWLBC) is also preparing computer models that can show the combined impacts of natural variability and different climate change emission scenarios in the MLR.

Plausible climate impacts

In the immediate coming decades it is likely that natural climate variability may continue to be the dominating influence on rainfall and streamflows. As time goes on, the impact of climate change will continue to materialise. Future water infrastructure for Adelaide will need to provide water

security during drought-dominated periods, combined with drying conditions from global warming – in conditions of an ever-growing water deficit.

Long-life infrastructure, such as reservoirs and even pipelines, are built to last in excess of 100 years, when climate variability and change may well combine (at times) to cause a severe water deficit, beyond previous experience. It is important, therefore, to plan ahead using the right future scenarios that correctly identify plausible risks.

The natural variability risk can be assessed by modelling available water supply during drought-dominated decadal periods, such as 1890–1950 or 1995–2007, rather than using long-term average flow conditions. The additional impact of climate change then needs to be added into the models.

Firstly, the current impacts of climate change must be factored in. Globally there has been an increase of approximately three quarters of a degree in temperature, attributed mostly to anthropogenic climate change. Over inland areas such as in the MDB, conditions are predominantly drier and can lead to the regional temperature increase being greater than the global average. The observed temperature increase – acknowledged as being anthropogenic – should be included because of the impact on evaporation.

Secondly, the impacts of a continuation and even increase of climate change must be factored in. Using Intergovernmental Panel on Climate Change (IPCC) emission scenarios as the starting point for national and regional downscaled models, it is important for modellers to select the most plausible scenario or several scenarios that will ensure risks are managed for potential impacts. The IPCC Special Report on Emissions Scenarios (SRES) are about projecting human behaviour. The modelling reflects what happens as nations pump more or less greenhouse

BOX 56

What is climate change and why are we to blame?

Earth's climate has always been dynamic and changeable. In the distant past there have been bouts of intense volcanic activity, periods when vast deserts spanned much of the globe, warm epochs when forests covered Antarctica, and glacial ages when much of Europe and North America were entombed under miles of ice. When large climatic changes occurred rapidly, a mass extinction of species was the result. Life later recovered, but this process took millions of years.

Just one species – humans – are now the agent of global climate change. As we develop our modern economies and settlements at a frantic rate, we have caused deforestation and fragmentation of natural habitats, over-hunting of wild species we use for food, chemical pollution of waterways, and massive draw-downs of rivers, lakes and groundwater. These patently unsustainable human impacts are operating worldwide, accelerating, and clearly constitute an environmental crisis. Yet the threat now posed by human-caused global warming is so severe that it may soon outpace all others.

Recent global warming is caused principally by the release of long-buried fossil carbon through burning oil, natural gas and coal. Since the furnaces of the industrial revolution were ignited in the late-18th century, we have dumped more than a trillion tonnes of carbon dioxide (CO_2) into the atmosphere, as well as other heat-trapping greenhouse gases such as methane, nitrous oxide and ozone-destroying chlorofluorocarbons. The airborne concentration of CO_2 is now 38% higher than at any time over the past million years (and perhaps much longer – information beyond that time is too sketchy to be sure). Average global temperature has risen about 0.8°C in the last two centuries, with almost two-thirds of that warming having occurred in just the last 50 years.

Complex computer simulation models of the atmosphere have been developed and refined for over 40 years. They are now sufficiently advanced that they can reproduce most of the major features of climate change observed over the last 150 years. Under a business-as-usual scenario, which assumes a continued reliance on fossil fuels as our primary energy source, these models predict 1.8°C to 6.4°C of further global warming during the 21st century. There is also a real danger that we have reached or will soon reach tipping points that will cascade uncontrollably, taking the future out of our hands. But much of the uncertainty represented in this wide range of possibilities relates to our inability to forecast the probable economic and technological development pathway global societies will take over the next few decades.

It remains within our power to anticipate the many impacts of future global warming, and make the key economic and technological choices required to substantially mitigate our carbon emissions. But will we act in time, and will it be with sufficient effort to avoid dangerous climate change?

2008, and rising at 3 parts per million per year), we would still only have a roughly 50/50 chance of averting dangerous climate change. This will require a global cut in emissions of 50–85% by 2050 and certainly by more than 90% for developed nations like Australia.

Peak oil, global warming and long-term sustainability all require that we reduce energy needs and switch to nuclear and renewable energy sources. Many credible studies show we can quickly and cost-effectively reduce greenhouse gas emissions through dramatic improvements in energy efficiency. For details of the mitigation options, see the 2007 Intergovernment Panel on Climate Change (IPCC) Fourth Assessment Report and the Millennium Ecosystem Assessment.

Unfortunately, there is little evidence so far that we are taking meaningful action. Indeed, energy efficiency (expressed as gross domestic product per tonne of carbon) in developed nations such as the United States and Australia has actually decreased over the last decade, the global rate of emissions growth has risen from 1% to 3% a year, and total carbon emissions from all sources now exceed 9 billion t a year. China overtook the United States in 2006 as the single biggest greenhouse polluter and will be producing twice as much CO_2 within little more than another decade at its present rate of economic activity.

This exponential growth in carbon-based energy, if sustained, will mean that over the next 25 years, humans will emit into the atmosphere a volume of carbon that exceeds the total amount emitted during the 150-year industrial period of 1750 to 2000. Of particular concern is that long-lived greenhouse gases, such as CO_2, will continue to amplify global warming for centuries to come. For every five tonnes added during a year in which we dither about reducing emissions, one tonne will still be trapping heat in 1000 years. It is a bleak endowment to future generations.

Should we choose to take no effective action, we can expect increasingly severe consequences. For instance, beyond about 2°C of further warming, the Great Barrier Reef will be devastated. Extreme events will become much more frequent, such as storm surges, adding to rising sea levels of many metres and threatening coastal cities. There is the possibility that a semi-persistent or more intense El Niño will set in, leading to frequent failures of tropical monsoonal rains which provide the water required to feed billions of people. Above 3°C, up to half of all species may be consigned to extinction because of their inability to cope with such rapid and extreme changes.

Worryingly, even if we can manage to stabilise CO_2 concentrations at 450 parts per million (it was 385 in

Barry Brook

gas emissions into the atmosphere (Figure 13.2). Box 57 provides an explanation of the IPCC emission storylines.

As a basis for policy and planning for the water industry, the question is about considering the four different scenario families, A1, A2, B1 and B2, and then considering the scenario group that matches likely human behaviour in probabilistic terms. It would not be appropriate to 'pick' any particular scenario or range of scenarios without defining the logic behind such a decision.

This area of discussion is highly contested by a number of interest groups. Some assert that higher emission scenarios, as used in the United Kingdom's Stern Review on the Economics of Climate Change,[5] are extreme and don't take into account economic changes and technology improvements, while others suggest that global behaviour is tracking to the A1FI fossil fuel intensive scenario. A similar economic impact assessment has commenced in Australia under economist Ross Garnaut. In media interviews, Garnaut has raised concerns similar to those of Stern – that the rate of growth in global greenhouse gas emissions is exceeding even the highest IPCC A1FI emissions scenario. Adelaide water planners and climate modellers will therefore need to:

1. Consider which is the most likely scenario group and scenario family to plan towards.
2. Consider planning for scenarios where the probability is high enough and the consequences are severe enough to take further action and/or develop contingency plans.
3. Decide on the signals that might take place in the political/policy arena that would show a global shift from one scenario to another.

Fossil-intensive behavior (A1FI)

In looking at the different emission scenario storylines,[6, 7] it can be argued that current global behaviour aligns best to the A1FI emission scenario. There may be some regions of the globe that may not increase emissions to the high rates of First World countries, or in the same way that China, India and Asia are rapidly increasing their emissions, however, the big population regions are, and these dominate by exceeding previous emission growth estimates.

It is hoped that human behaviour will shift away from the A1FI behaviour scenario at some point in the future, and that it could move towards an A1B scenario if there is global progress in implementing alternative technology scenarios on scales that might negate the impacts of fossil fuel usage. When such a change may take place is hard to predict.

Risks of abrupt climate change steps and escalating feedback loops are also present. While difficult to quantify, it is logical to anticipate that these risks are increased with fossil intensive behaviour.

Why not the A2 scenario?

The A2 storyline and scenario family may eventuate in part to be heterogeneous world, however, the convergence of the big population countries appears to be dominating.

It is the heterogeneous economic and emissions growth aspect of this scenario that might give the greenhouse benefit, where emissions growth won't be as great in some global regions. The low emissions growth regions may include Africa, South America, and smaller states may be slower; however, the risk is that many might catch up to First World emissions.

It does not appear that nations are on a path of self-reliance and preservation of local identities, or that economic development is primarily regionally oriented. Continuously increasing populations would push even this scenario to higher emissions.

Why not the B1 scenario?

The B1 scenario is that there has not yet been a signal to suggest a global rapid change in economic structures towards a service and information economy that would replace or substitute resource-based industries. There is still a passion for the cycle of mining, logging, manufacturing and disposal, etc.

To base the planning of long-life infrastructure on a B1 scenario, visible progress would need to be underway (rather than in planning) towards achieving cooperation for global solutions to economic, social and environmental sustainability, including improved equity, with or without additional climate initiatives.

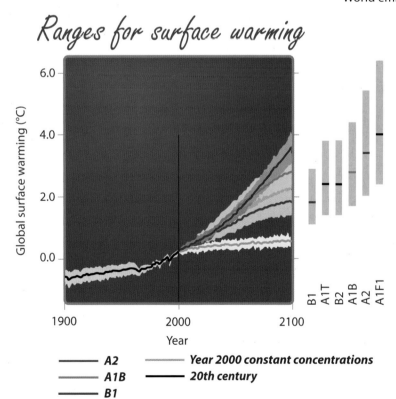

Figure 13.2 Multi-model averages and assessed ranges for global surface warming. Source: Intergovernmental Panel on Climate Change, climate change 2007: synthesis report. Contribution of working groups I, II and III to the Fourth Assessment Report, 2007.

BOX
57

Emission scenarios of the IPCC Special Report on Emissions

The international body, the Intergovernmental Panel on Climate Change (IPCC), modelled families of different scenarios reflecting potential outcomes for the world in response to climate change; these scenarios have been termed A1, A2, B1 and B2.

A1. The A1 storyline and scenario family describes a future world of very rapid economic growth, global population peaking in mid century, and declines thereafter, and the rapid introduction of new and more efficient technologies. Major underlying themes are convergence among regions, capacity building and increased cultural and social interactions, with a substantial reduction in regional differences in per capita income. The A1 scenario family develops into three groups that describe alternative directions of technological change in the energy system. The three A1 groups are distinguished by their technological emphasis: fossil intensive (A1FI), non-fossil energy sources (A1T), or a balance across all sources (A1B), where balanced is defined as not relying too heavily on one particular energy source, on the assumption that similar improvement rates apply to all energy supply and end-use technologies.

A2. The A2 storyline and scenario family describes a very heterogeneous world. The underlying theme is self-reliance and preservation of local identities. Fertility patterns across regions converge very slowly, resulting in a continuously increasing population. Economic development is primarily regionally oriented, and per capita economic growth and technological change are more fragmented and slower than other storylines.

B1. The B1 storyline and scenario family describes a convergent world with the same global population, peaking in mid century and declining thereafter, as in the A1 storyline, but with rapid change in economic structures towards a service and information economy, with reductions in material intensity and the introduction of clean and resource-efficient technologies. The emphasis is on global solutions to economic, social and environmental sustainability, including improved equity, but without additional climate initiatives.

B2. The B2 storyline and scenario family describes a world in which the emphasis is on local solutions to economic, social and environmental sustainability. It is a world with a continuously increasing global population (at a rate lower than A2), intermediate levels of economic development, and less rapid and more diverse technological change than in the A1 and B1 storylines. While the scenario is also oriented towards environmental protection and social equity, it focuses on local and regional levels.

Tim Kelly

This material is a direct extract from the Intergovernmental Panel on Climate Change Special Report Emission Scenarios, Summary for Policy Makers, 2000, p. 4–5.

Why not the B2 scenario?

To be confident that the B2 scenario could be used as a basis for planning, visible progress would also need to be towards a world of 'environmental protection and social equity with a focus on local and regional levels'.

Reduced water supply availability

Once the impacts of natural climate variability, recent climate change, and plausible scenarios of future climate change have been factored into regional and catchment downscaled models, planners of Adelaide's future water supply will have the best chance of planning robust, long-term water solutions for Adelaide.

Without pre-empting the extent of any change to water supply availability, there is enough evidence to suggest that River Murray flows and local flows from the MLR will decrease into the future. As this takes place there will also be additional demand for water to adapt to climate change, to grow (population) in line with South Australia's strategic plan, and to allocate water for environmental flows.

Other climate change impacts

Water quality impacts

Climate change poses particular risks to water quality, magnifying existing risks during periods of drought and higher temperatures.

Pathogens and cyanobacteria

It is considered likely that climate change will provide conditions more frequented with events involving pathogens and cyanobacteria (often referred to as blue-green algae) Such events may threaten health, livestock and the greater environment, and may require additional water treatment. Recently, certain strains of cyanobacteria have been on the increase, causing concern among water scientists and managers. Such events are difficult to manage through water treatment.

Cylindrospermopsis is a strain of cyanobacteria that has not been evident in South Australia. For the period 1991–2002, only two of the 127 positive samples along the South Australian stretch of the Murray contained *Cylindrospermopsis* in cell counts greater than 100 cells/ml. However, during the past four years (to December 2007), one hundred of 350 positive samples contained *Cylindrospermopsis* cell counts greater than 100 cells/ml, with 16 samples containing more than 1000 cells/ml. In addition, *Cylindrospermopsis* has recently been detected in the Warren Reservoir for the first time.

During late 2007 and early 2008, over a three-month period when the River Murray was amidst low flows and high temperatures, 26 positive samples contained more than 1000 cells/ml, with the highest detected at Swan Reach. Research has shown that chlorine will destroy the toxin produced by *Cylindrospermopsis*, however, for other river users and private water harvesters, toxins may still be present. The emergence of *Cylindrospermopsis* may be typical of how the risks from cyanobacteria may change as they respond to climate change.

A range of pathogens are known to become more problematic with higher temperatures in freshwater bodies and in some treated water systems, particularly where long-distance pumping is required. One example is *Naegleria fowleri*, a single-celled amoeba that may cause amoebic meningitis if high numbers enter the nasal passages under force. The pathogen is more prevalent in hot weather and it is believed that the risk of contracting amoebic meningitis from contact with untreated freshwater or from under-maintained swimming pools will increase as average rising temperatures result in hotter summers with longer heatwaves. The current high-treatment and monitoring standards for drinking water will also become even more important.

Salinity

The risk of increasing salinity in the River Murray translates to potentially higher costs for providing Adelaide's drinking water. Typically, the salinity of the Murray is well below levels that would impact on health; however, there is a nearer-term risk that salinity levels could approach the threshold for desirable drinking water quality.

Prolonged drought and higher evaporation in the MDB has reduced the flushing capacity to remove and transport salt in all water pathways, including through soil, streams and tributaries, from backwaters, and in the main river channel. As a result, salt has been accumulating since the early 1990s, without effective flushing periods. In the worst contributing areas, salt interception schemes are used to remove saline groundwater to surface basins away from the river. When higher rainfall events do take place in the future, it is anticipated that there will be a number of salinity spikes as accumulated salt travels into the main river channel and cannot be intercepted from all sources. In the periods that follow significant rainfall events, mobilised salt will continue to enter the main channel and may also cause a further delayed pulse of salinity.

In the past, Adelaide Hills catchments that are supplemented with River Murray water also provided excellent dilution for water supplies to remain well within the threshold for desirable water quality. Quantifying any future risk will involve research based on plausible emission scenarios and how these would impact on the River Murray and Adelaide's catchments.

Bushfires

While bushfires are a natural part of the Australian landscape that water quality managers have to deal with, they do cause a number of significant water quality problems when they occur. After a bushfire, the landscape is left with a great deal of light, potential surface pollutants, and, in addition, the soil is destabilised and becomes extremely vulnerable to erosion from wind, land slippages and flash floods. The soil itself becomes a pollutant when washed into waterways in high volumes because of the particulates it contains – dissolved organic carbon, phosphorous, iron and manganese. Also, sediment build up and debris cause other management issues.

Groundwater and climate change

Adelaide has a variety of groundwater aquifers extending from the MLR to beneath the Adelaide Plains, and adjacent to coastal areas. Adelaide's Tertiary aquifers (uppermost being T1 and T2) are important yet have long recharge pathways such that climate change impacts would not show for thousands of years. Cracked rock aquifers in the MLR and shallow aquifers on the Adelaide Plains have much faster recharge pathways and could be more easily affected by climate variability and change.

In the Northern Adelaide Plains, the T1 and T2 aquifers are used for irrigation of market gardens, particularly in the Virginia Triangle, with supplementary recycled water provided from the Bolivar WTTP. The T1 aquifer is diminished in a zone of depression caused by high abstraction rates for irrigation in the region. Salisbury Council has established large-scale stormwater collection and detention wetlands to harvest water for use, and has utilised aquifer storage and recovery (ASR) techniques to store water from winter rainfall and wet weather events, for recovery and use in dry periods. There are also golf courses, industry and private bores across the Adelaide region.

In the Willunga Basin, groundwater is used for irrigation of vineyards, almonds and other crops, and is also being supplemented with recycled water.

Climate change will impact on Adelaide's groundwater systems; it is envisaged that research will be undertaken to assist in quantifying changes to recharge rates and sustainable yields. More complex impacts that will affect irrigated landscapes may involve changes in evaporation rates that increase the potential for irrigation salinity. Climate change is also likely to drive growth in the use of groundwater systems because they provide the opportunity to store water from rainfall events without significant evaporation losses, and without the need for above-ground land and water storage infrastructure.

Flooding

Despite the risk of hotter and drier conditions in southern Australia anticipated from climate change, the risk of greater extremes is higher for flooding rains, causing the need for reservoir spillways, flood mitigation schemes and planning approvals to protect communities from the risks of greater flash flooding.

Adapting to climate change

Water Proofing Adelaide

'Water Proofing Adelaide' is the state government's blueprint for the management of Adelaide's water resources until 2025.[8] It was finalised in 2005 and work began shortly after on planning and implementing its 63 strategies.

Water Proofing Adelaide is based on:

1. Better managing existing water resources.
2. Increasing water use efficiency.
3. Investigating alternative water sources.

It aims to protect Adelaide's domestic water supply and includes strategies to achieve responsible water use, additional supplies, and to foster innovation. Key focus areas will be the expanded use of recycled water, and community education and incentives. Specific initiatives include:

- Water Proofing the North includes 24 stormwater recycling projects, two wastewater recycling projects, and an experimental project to inject and store stormwater in underground aquifers for recovery as potable water suitable for human consumption. The project aims to harvest and utilise up to 30 GL/annum.
- Water Proofing the South is an initiative to extend the Willunga Pipeline from Christies Beach wastewater treatment plant (WWTP) into the southern suburbs, and will integrate further sources of reuse water from the Willunga Septic Tank Effluent Disposal Scheme, the Aldinga WTTP, and stormwater sources. The scheme will expand opportunities to supply water for viticulture, parks, gardens, school grounds and public toilets. The initial project will provide the backbone infrastructure to double the current use of recycled water.
- A major expansion of the Virginia Pipeline is being undertaken to deliver approximately 3000 ML/annum of additional treated wastewater from Bolivar WTTP for irrigation of Angle Vale crops. The Commonwealth government has pledged a 50% financial contribution.
- A Glenelg to Adelaide Parklands Recycled Water Scheme (GAPRWS) is being undertaken to irrigate Adelaide's parklands and provide opportunities for business to use recycled water supplied from the Glenelg WTTP. The scheme will provide the potential to recycle up to 4000 GL/year of treated wastewater and reduce the amount of treated wastewater entering Gulf St Vincent. The GAPRWS has financial support from the Commonwealth government under the National Urban Water and Desalination Plan.
- A range of education and incentive measures have been established to promote water efficiency to householders towards achieving a reduction in household water consumption by 31 GL/year by 2025. Incentives include a state government rebate scheme to plumb rainwater tanks into existing homes, and a $200 rebate for water-efficient front-loading washing machines.

Water Proofing Adelaide also made provision for environmental flow trials in the MLR as part of the process for prescribing surface and groundwater supplies in the western area of the ranges. The project is being managed by the Adelaide Mount Lofty Natural Resources Management Board. Trial flows will determine the benefits of controlled water releases to a range of watercourses and their environments, and will enable long-term planning for sustainable water management. SA Water is the largest user of surfacewater within the catchment and has therefore made the greatest contribution to providing water for environmental flows.

More information about Water Proofing Adelaide can be found online.

Water security projects and further enhancements to Water Proofing Adelaide

Since the Water Proofing Adelaide strategy was released, drought conditions associated with natural variability and climate change have continued, while water storages in the River Murray continued to be drawn down due to the nature of water entitlements that provide no adjustment or linkages to system inflows. Greater vulnerability to reduced flows in the River Murray is now better understood, and research into system-wide management and sustainable yields is being undertaken by the CSIRO.

To manage the risks of extremely low flows into the River Murray, the South Australian government in October 2007 announced the 4-Way Water Security Strategy, to ensure South Australia's long-term water security. The strategy builds on Water Proofing Adelaide to secure water under the pillars of desalination, recycling, managing water use, and improving catchment management. The strategy has been underpinned by a major 'Water Matters' campaign to raise public awareness on water issues and further encourage communities to conserve water. Under each of the strategy pillars, major projects have been announced and considerable work is underway to secure the state's water supplies for the future. Key elements are outlined below.

Desalination

Work has already begun on a $1.37 billion desalination project for Adelaide. In December 2007 the government announced Port Stanvac as the preferred site for the plant, which will initially supply 50 GL/year for Adelaide. This decision was based on investigations by the Desalination Working Group, established in March 2007, which examined the impact of the drought on water supply, the feasibility of desalination, the preferred size and location of the plant, integration into the existing supply network, and cost implications. Environmental studies on the possible impacts of the plant are continuing.

Catchments

The state government is exploring options to increase future storage capacity in the MLR that would double capacity from the current approximate one year's reserve. South Australia was the first state to put in place a strategy for environmental flows for the River Murray, to make best use of environmental water. The health of the River Murray remains a top government priority, with many actions implemented to improve the health of the river and to protect water supply and quality within state and beyond state borders, including:

- the introduction of the *River Murray Act 2003* that give the government powers to limit activities that damage our precious Murray;

- state-based and basin-wide salinity strategies including revegetation, improved irrigation practices, better planning for new irrigation developments, and construction of the salt-interception schemes; and
- construction of fishways to improve the ability of native fish species to move along the river, while removing carp.

Recycling

As already described, a number of projects are being undertaken to make for greater use of treated wastewater and stormwater.

Managing water use

The South Australian community has made significants effort to save water, with total Adelaide metropolitan consumption for 2007 being just under 140 GL compared to 194 GL in the 2002 drought year and 173 GL for the 10 year average. To assist, in 2007 the government introduced the $24 million H2OME rebate scheme to encourage South Australian households to achieve greater water savings inside and outside the home. This includes a home water audits scheme and rebates on selected garden goods, low-flow showerheads, efficient washing machines, dual-flush toilets, and rainwater tanks. Industry is also playing a role, with major commercial and industrial water-users implementing water efficiency plans in order to identify and prioritise water-saving opportunities.

SA Water has introduced a water audit service for commercial, industrial and some institutional water-users, offering advice on water savings including promotion of the financial benefits for the industrial user, along with a Business Water Saver program to assist industrial and commercial customers to save water and minimise waste. More information about the 4-Way Strategy can be found on SA Water's website.

Development of SA Water's long-term climate change strategy

SA Water is developing a long-term climate change strategy to 2100 that will build on the work undertaken for drought response and the 4-Way Strategy. The strategy will be SA Water's adaptive management framework for climate change adaptation, mitigation, and supporting necessary research.

The long-term strategy aims enhance SA Water's current long-term planning mechanisms and climate change vulnerability assessments, and will strive to incorporate the best climate science and climate scenarios for long-term planning of water supply regions, catchments, coastal regions and infrastructure. The strategy will include actions to manage priority risks, identified by SA Water business units and through a high-level climate change risk-assessment workshop assisted by the CSIRO. A long-term approach is necessary particularly when planning long-life infrastructure and systems that will need to perform under plausible emission and climate scenarios. Risks to be addressed in SA Water's long-term climate change strategy will include:

BOX
58

SA Water greenhouse achievements, 2006–2007

- Achieved 10% renewable energy use through topping-up generated renewables use with 32,000 MWh of accredited GreenPower.

- Achieved 6000 t CO_2 sequestration from post-1990 revegetation.

- Established a further 97 ha of revegetation on principles to maximise water quality, biodiversity and improve land management in our catchments.

- Created and sold 50% share of 5200 MWh of accredited GreenPower for other electricity users to reduce their emissions from the Hope Valley mini hydro joint venture with Hydro Tasmania. SA Water is not able to keep and use the renewable benefit from this facility under current rules preventing the system owner from using the GreenPower benefit where the electricity is not used onsite.

- Achieved independent verification of SA Water's greenhouse reporting in the Greenhouse Challenge Plus Program.

- Maintained pump performance efficiency programs, which help save in the order of 10,000 MWh of electricity each year.

Tim Kelly

- water supply quantity and source water quality;
- low-lying infrastructure exposure to sea level rise and increased storm surges;
- increased potential for bushfires;
- increased potential for extreme weather events such as longer, more extreme heatwaves and more extreme flash flooding; and
- increased costs to provide necessary infrastructure and to adjust to a carbon-constrained economy.

The role of water catchments in improving habitat, and helping species adapt to climate change

When considering water and adapting to climate change, there is an opportunity to enhance natural habitats and ecosystems when managing catchments and reservoir buffer zones. The MLR are an 'island' of relatively high rainfall forest and woodland, isolated from similar areas in eastern Australia by much drier mallee habitat.[9] However, the MLR 'island' now contains only about 10% of its original 500,000 ha of native vegetation.

Possingham and Field[9] explain that while there may be a substantial time lag between the loss of habitat and the consequent loss of species (likely to be hundreds of years for long-lived species of birds), the destruction of 90% of native vegetation has already incurred an extinction debt (the future loss of species) as a consequence of past actions. In the case of the MLR, this may result in the loss of another 40 species of terrestrial birds.

The impact of climate change is expected to compound the difficulties for populations and species to survive. Animal species will seek to migrate within and beyond habitats to seek favourable conditions for survival. Where habitats are limited and fragmented, ecosystems will have difficulty adapting to climate change. Those species vulnerable to past habitat destruction and climate change can, however, be given a helping hand through a concerted effort to restore approximately 30% native vegetation to the MLR, enabling the expansion of populations to safer levels.

To this end, the South Australian government has established its Naturelinks Implementation Strategy, to connect habitats across South Australia for species and ecosystems to survive, evolve and adapt to environmental change. The program identifies the need for core protected areas, buffered and linked by lands managed for conservation objectives. More information about Naturelinks can be found online. Many private landholders are also undertaking revegetation activities, particularly along watercourses, in an effort to reduce erosion and improve water quality.

In the context of water management and climate change, reservoir reserves in the MLR, and even WWTP lagoons such as at the Bolivar plant, already provide significant core biodiversity refuges similar in scale to government conservation parks and reserves. SA Water has also established a five-year program to enhance biodiversity on its landholdings through revegetation of up to 135 ha/year, adding to core areas and buffer zones.

The state government has established the SA Urban Forests Million Trees Program, managed by the Department for Environment and Heritage, to establish 3 million local native trees and associated understorey species across the Adelaide metropolitan area, from Gawler to Sellicks Beach, by 2014. The program has support from local government, industry and the community.

The challenge now is for government agencies, SA Water, communities, and biodiversity experts to support the Naturelinks Implementation Strategy in a coordinated way for the maximum benefit to species and ecosystems. This process is being assisted by a University of Adelaide project to restore approximately 100 ha of native vegetation on the recently acquired Glenthorne Farm to become a model system in restoring and linking biodiversity. The university objectives include establishing a research centre for scientists, technicians, teachers and managers. The Glenthorne Farm project will show how this site and others can be enhanced to support the ultimate objective of restoring 30% native vegetation across the landscape.

Glenthorne Farm has the potential to be linked to Gulf St Vincent, along watercourse reserves to be protected and improved. Extending landward, the closest potential biodiversity refuge is SA Water's Happy Valley Reservoir. Scattered smaller islands of natural habitat exist on the hillsface towards Clarendon, where SA Water's Mount Bold Reservoir Reserve and the adjoining Scott Creek Conservation Park provide one of the MLR's largest and most important core biodiversity refuge areas. Nearby there are more scattered islands of biodiversity, towards the Cleland Wildlife Park and across catchment farmlands to Jupiter Creek and Kuipto Forest. Towards the Murray and Lake Alexandrina, and northwards towards Mount Crawford, Clare and the Flinders Ranges, habitats exist mostly as scattered remnants within farmland.

Maximising biodiversity benefits will involve ensuring a diversity of habitats within corridors, better managing introduced pests and predators, linking across barriers such as major arterial roads, and finding better ways to reduce road kill. The opportunity to combine efforts to protect Adelaide's water catchments, while at the same time delivering the maximum benefits for Naturelinks, will help to define Adelaide as a city that is connected to its surrounding landscape.

Mitigating greenhouse gas emissions from water use
Greenhouse gas emissions and trends

As the state's major water provider, SA Water has the primary responsibility for greenhouse gas emissions associated with providing water and wastewater services for Adelaide. SA Water's direct and indirect greenhouse gas emissions (before offsets are deducted) currently range between 400,000 t CO_2e in a favourable water catchment year to approximately 700,000 t in a severe drought year such as 2006–2007, equivalent to the emissions from electricity use of 100,000 households.

The largest source of SA Water's greenhouse gas emissions is from electricity use (Figure 13.3), and mostly from major pumping activities. Fugitive emissions of nitrous oxide and methane from wastewater treatment processes are also significant. Climate variability has a significant impact in greenhouse gas when more water is required to be pumped from the River Murray. Greenhouse gas emissions from local government-owned WWTPs, water treatment facilities and stormwater ASR schemes are smaller in scale compared with SA Water's activities, and are not currently reported towards an aggregated data set.

Figure 13.4 shows SA Water's greenhouse gas emissions made up from records of (major pumping) electricity use, extending back to 1986, and estimates of emissions from other sources. On trend, SA Water's current greenhouse gas emissions (before deducting offsets) are approximately 550,000 t CO_2e per year. Approximately 360,000 t CO_2e (65%) are attributed to the Adelaide region.

Water activities and emissions intensity

From SA Water's data and estimates, Table 13.1 provides an indicative summary of greenhouse gas emissions for various water supply and wastewater treatment activities. Individual circumstances may result in different values.

Reducing greenhouse gas emissions

SA Water is working towards achieving the climate change targets and objectives of South Australia's Tackling Climate Change strategy and the new South Australian *Climate Change and Greenhouse Emissions Reduction Act 2007*. Exact targets are to be confirmed in a Climate Change Sector Agreement between the SA Water and the state government. The following targets are intended, subject to confirmation. Firstly, limiting greenhouse gas emissions to the Kyoto target within 108% of 1990 levels on average between 1 July 2008 and 30 June 2012. This could also include:

- using renewable electricity for 20% of energy requirements by 2008 (state government target);
- new desalination plants (including headworks/ networks integration) to be powered using accredited renewable electricity; and
- new infrastructure to be undertaken in a way that does not add to SA Water's annual emissions.

Secondly, managing greenhouse gas emissions beyond 2012 on a defined pathway towards 60% reductions by 2050 (Figure 13.5).

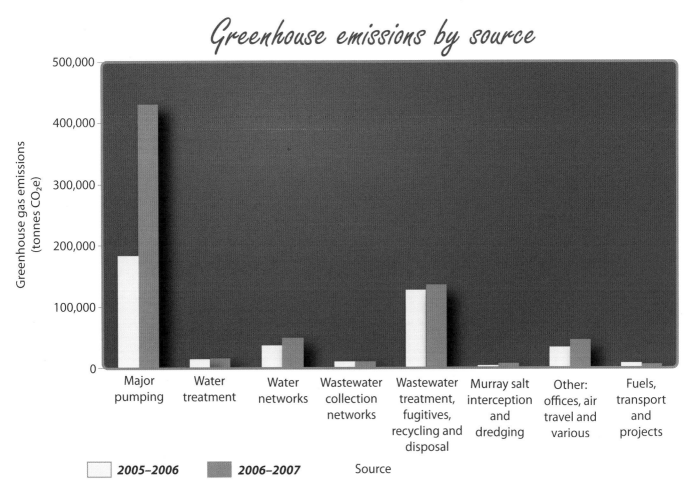

Figure 13.3 SA Water greenhouse gas emissions by source in 2005–2006 compared to 2006–2007. Source: SA Water.

In forecasting future emissions scenarios for SA Water the impact of water security projects, including the planned 50 GL desalination plant and establishing larger MLR storages, shows up as a significant challenge. The costs of dealing with such a challenge through purchasing accredited renewable energy and offsets are in the tens of millions of dollars per year. For comparison, however, the costs of providing additional water security infrastructure are currently forecast at $2.5 billion (Australian dollars, 2007).

To provide water infrastructure robust enough to deal with climate variability, climate change and population growth, significant investment will be required, and even higher investment if the climate change problem is not addressed comprehensively.

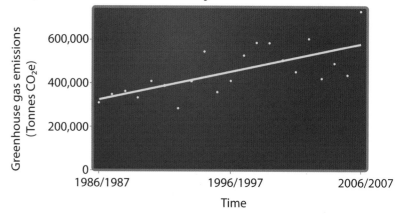

Greenhouse trend for SA Water

Figure 13.4 Greenhouse gas emissions trend for SA Water (before offsets deducted). Based on electricity use, fugitive emissions and fuel use, using AGP revised historic electricity factors (December 2005), and changing fugitive emission factors. Source: SA Water.

Responsibilities for managing water in the context of climate change impacts

As responsibilities for water supply in South Australia rest with different agencies, the responsibility to adapt to climate change impacts on water resources also rests with a number of agencies:

- The DWLBC has responsibility for the allocation of water from prescribed water supply sources, and therefore has a responsibility to consider the impact of climate change when planning allocations.
- The Murray-Darling Basin Authority (recently transferred from the Murray-Darling Basin Commission) has charge of administering agreements covering water allocations and flow management in the Murray-Darling Catchment, and therefore has a responsibility to consider the impact of climate change when planning the future management of the Murray and allocations.
- The South Australian Environment Protection Authority (EPA) regulates licensed wastewater treatment discharges of treated effluent to the environment and irrigation management plans in connection with water

allocation to ensure the sustainability of irrigation for projects such as the Clare Valley Irrigation Scheme. The EPA has a role to play in considering climate change impacts on proposed projects and the greenhouse consequences when higher standards for wastewater disposal are being considered.
- SA Water is responsible for providing water and wastewater services to approximately 1.4 million customers in line with its charter, licence conditions, its commitment to supporting South Australia's strategic plan policies, and its share strategies such as the Water Proofing Adelaide strategy and the Tackling Climate Change strategy. SA Water has direct responsibility for ensuring that it is able to manage the risks associated with climate change in order to provide reliable water and wastewater services to customers. SA Water also has the direct responsibility to manage its greenhouse gas emissions in line with agreed performance targets.
- Natural Resource Management (NRM) boards have a role in implementing NRM plans that include

Table 13.1 Water activities and emissions

Water activity	Indicative greenhouse emission intensity (MWh/ML unless marked otherwise)
Elevated catchment and gravity water filtration	0.1
Murray pumping to Adelaide and gravity filtration	1.6
Small scale desalination and pumping to headworks	10
Large scale desalination and pumping to headworks	5–6
Wastewater treatment, recycling or disposal (current mix)	0.4
Fugitive CO₂e emissions from wastewater – nitrous oxide and methane	0.1 t CO₂e per person

Table 13.1 Greenhouse gas emissions intensity of various water supply and wastewater treatment activities.

The economics of water and climate change

BOX 59

Internalising climate externalities

Addressing the climate externalities of water poses unique challenges, which require attention. Internalising externalities of water supply, use, and disposal is a key requirement of the National Water Initiative (NWI). When one considers this requirement carefully, it becomes clear that it requires much more than ensuring sufficient water is provided for environmental needs and maintaining riparian ecosystems.

Internalising externalities is about acknowledging the wider social and environmental costs of activities that are not captured in the price of products and services. Astute management of these costs requires the development of regimes that, as well as signalling their presence, reveal opportunities to reduce them.

The greenhouse impacts of providing and using water are often the forgotten externalities, yet there is growing acceptance that dealing with these impacts must be addressed. SA Water is already facing the challenge of reducing its greenhouse gas emissions to support South Australia's Strategic Plan and Tackling Climate Change targets, and facing higher prices and carbon costs with the proposed Carbon Pollution Reduction Scheme (CPRS). The result is upward pressure on the prices charged for water supply and wastewater management services.

Approaches to internalising externalities

There are different viewpoints on how the climate change externalities should be addressed. Professor Mike Young from the University of Adelaide believes that direct affirmative intervention by the water industry is not sufficient. As an economist, Young argues that the firm who supplies SA Water with energy should have to acquit emissions permits equivalent to the emissions it makes as it produces electricity, and then pass the costs of doing this to SA Water. SA Water would then, in turn, pass these costs on to customers in the normal manner.

The rule from Young's viewpoint is that the producer of the externalities pays for them and is given an incentive to manage them. In SA Water's case, it would be liable for its direct emissions from nitrous oxide, methane and fuel burning, and would have to find permits for such emissions.

Difficulties with such an economic approach are that the necessary framework, such as the proposed CPRS (a form of emissions permit and trading framework), are not yet established. Further, despite the fact that a CPRS is planned to commence in 2010, the gradual implementation of carbon constraints and the grandfathering of emission permits will not automatically internalise the externality costs nor reduce emissions fast enough or sufficiently enough for SA Water to achieve the state's greenhouse targets.

There is a driver for the water industry to tackle its climate change impact faster and harder than some other sectors because of its dependency on climate and its rapid transition to energy-intensive technologies such as desalination and membrane treatment for recycling wastewater. Indeed, for almost every desalination plant either installed or under planning in Australia, there is either a commitment or consideration for these to be powered using renewable energy.

In such an environment, the water industry will be operating at a performance standard to prevent or offset climate externalities beyond mandated requirements. This creates a potential conflict between those who see that tackling climate change within the water industry business and recovering the cost is breaking the rules of pricing policy because it is not a separate instrument, and others that view greenhouse mitigation as a cost of doing business in a carbon-constrained economy.

Complimentary mechanisms and standards for business that choose to act in advance of the mandated requirements of the proposed CPRS are likely to be clarified during 2009 and 2010 prior to commencement of the scheme.

Tim Kelly

the sustainable use of water, and therefore have a responsibility to consider the impact of climate change when preparing NRM plans.

- Universities and research organisations, such as the SA Water Centre for Sustainable Water Systems, play a valuable role in researching and demonstrating sustainable water management in ways that can be adopted into urban systems, and assisting communities with research findings and solutions to help adapt to climate change.
- The South Australian government is responsible for water legislation and ensuring that South Australia's Strategic Plan, policies and other strategies are achieved.

An effective global response to tackling climate change will require the contribution of all nations, supported by their government agencies, businesses and individuals. In this sense, the agencies and organisations listed above, and every consumer, can play a part by directly or indirectly supporting water conservation, by supporting the implementation of sustainable water technologies.

Conclusion

In this chapter we have considered the interactions between climate variability and climate change on water resources, and Adelaide's vulnerability to climate change impacts on water supply, infrastructure and biodiversity.

It can be argued that the IPCC A1FI scenario (based on fossil fuel intensive behaviour and high growth) most resembles current global behaviour, and has the worst-case potential to result in global temperature increases above 6 °C by 2100. To play its part in avoiding such consequences, Adelaide can develop its water and wastewater infrastructure in a greenhouse-friendly way – to achieve deep, absolute reductions in greenhouse gas emissions by mid century, in line with South Australia's Tackling Climate Change strategy. Should Adelaide and the global community be successful in avoiding worst-case emission scenarios and limiting anthropogenic climate change, there will still be an huge task to adapt to climate change that is already inevitable. Through the development of climate change strategies by all levels of government, working with business and communities, Adelaide can prepare for direct and indirect climate change impacts, particularly as they affect water supplies.

As governments, corporations, such as SA Water, and communities undertake actions to prepare for climate change, there are opportunities to assist species and ecosystems in having the best chance of adapting to climate change by protecting and enhancing core biodiversity refuges, expanding buffer zones, and linking islands and fragments with habitat corridors on a landscape scale.

Recommendations

Adelaide already has a comprehensive set of strategies and program initiatives that will enable it to respond to climate change, while securing water for human settlements and environmental needs. Adelaide also has its Naturelinks Implementation Strategy that can be supported at the same time as improvements are made in catchments to protect and enhance water quality, both on public and private land. Recommendations of this chapter are about overcoming the barriers to implementation. It is recommended that:

- The South Australian government, and its agencies and business enterprises including the Office of Water Security and SA Water, should continue and extend their initiatives on drought response and water security to prepare for the longer-term impacts of climate change on water resources and infrastructure needs.

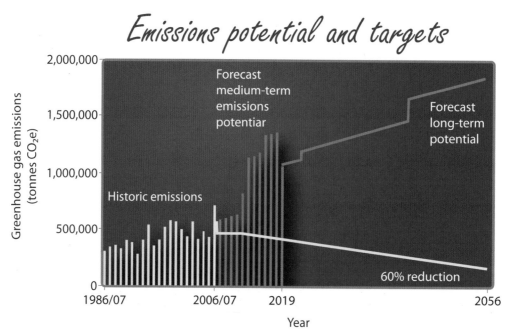

Figure 13.5 SA Water's historic production of greenhouse gas emissions as well as emissions potential and targets up to the year 2056. Source: SA Water.

- South Australia's new water infrastructure, such as desalination plants, wastewater and stormwater schemes, be constructed and operated in a way that does not cause even greater levels of greenhouse gas emissions.
- The Naturelinks initiative requires collaboration between government agencies and communities as well as funding and incentives on higher scale than previous biodiversity initiatives undertaken in South Australia. It will be important to support and build upon demonstration sites, such as the Glenthorne Farm project, as program managers find different ways to plan, design, fund and implement such projects.

Acknowledgements

The authors would like to thank Professor Mike Young and Peter Dillon for sharing ideas during the development of this chapter.

References

1. J.M. Diamond, *Guns, germs and steel*, Norton, New York, 1997.
2. M.A.J. Williams, 'Quaternary climatic changes in Australia and their environmental effects' in Gostin, V.A. (ed.) *Gondwana to greenhouse: Australian environmental geoscience*, Geological Society of Australia, 2001, pp. 3–11.
3. M. Bell and M.J.C. Walker, *Late Quaternary environmental change*, Pearson-Hall, London, 2005.
4. K.L. McInnes, R. Suppiah, P.H. Whetton, K.J. Hennessy and R.N. Jones, 'Climate change in South Australia: assessment of climate change, impacts and possible adaptation strategies relevant to South Australia', CSIRO, 2003 (online)
5. N. Stern, 'Stern review on the economics of climate change', 2006 (online).
6. Intergovernmental Panel on Climate Change working group I, 'The scientific basis', 2001 (online).
7. Intergovernmental Panel on Climate Change working group III, 'Special report on emissions scenarios', 2000 (online).
8. Government of South Australia, 'Water proofing Adelaide: a thirst for change 2005–2025', 2005 (online).
9. H.P. Possingham and S. Field, 'Regional bird extinctions and their implications for vegetation clearing policy', 2001 (online).

Box references

Box 54: Snowball Earth

1. P.W. Schmidt and G.E. Williams, 'The Neoproterozoic climatic paradox: equatorial paleolatitude for Marinoan glaciation near sea level in South Australia', *Earth and Planetary Science Letters*, 134, 1995, pp. 107–124.
2. B.J.J. Embleton and G.E. Williams 'Low latitude of deposition for late Precambrian periglacial varvites in South Australia: implications for palaeoclimatology', *Earth and Planetary Science Letters*, 79, 1986, pp. 419–430.
3. G.E. Williams and P. Schmidt, 'Proterozoic equatorial glaciation: has 'snowball earth' a snowball's chance?', *The Australian Geologist*, 117, 2000, pp. 21–25.
4. D.A.D. Evans, 'Proterozoic low orbital obliquity and axial-dipolar geomagnetic field from evaporite palaeolatitudes', *Nature*, 444, 2006, pp. 51–55.
5. J.L. Kirschvink, 'Proterozoic low-latitude glaciation: the snowball earth' in Schopf, J.W. and Klein, C. (eds) *The Proterozoic biosphere*, Cambridge University Press, Cambridge, 1992, pp. 51–52.
6. P.F. Hoffman and D.P. Schrag, 'Snowball earth', *Scientific American*, 282, 2000, pp. 68–75.
7. D.P. Schrag and P.F. Hoffman, 'Life, geology and snowball earth', *Nature*, 409, 2001, p. 306.
8. M.A. Fedonkin, J.G. Gehling, K. Grey, G.M. Narbonne and P. Vickers-Rich, *The rise of animals: evolution and diversification of the Kingdom Animalia*, Johns Hopkins Press, Washington, 2008, p. 302.
9. R.C. Sprigg, 'Early Cambrian jellyfishes from the Flinders Ranges, South Australia', *Transactions of the Royal Society of South Australia*, 71, 1947, pp. 212–224.
10. R.C. Sprigg, 'Early Cambrian 'jellyfishes' of Ediacara, South Australia, and Mt John, Kimberley District, Western Australia', *Transactions of the Royal Society of South Australia*, 73, 1949, pp. 72–99.

Box 57: Emissions scenarios of the IPCC Special Report on Emissions Scenarios

1. Intergovernmental Panel on Climate Change Special Report Emission Scenarios, 'Summary for policy makers', 2000, pp. 4–5 (online).

CHAPTER 14

What price water?

Martin Shanahan

'For only what is rare is valuable; and "water", which, as Pindar says, is the "best of all things", is also the cheapest.'[1]

Introduction

These words, written by Plato over two thousand years ago, have strong resonance for residents of a city that sees itself to be the 'Athens of the south'. Faced with a dry Mediterranean climate of long, hot summers and cool winters, similar to Athens, the citizens of Adelaide are aware that water is not as plentiful as in many other cities. Nonetheless, local residents have, over the past 30 years, felt reasonably secure in their access to freshwater. Decades of engineering work on reservoirs, pipelines and filtration systems by government authorities, coupled with the ultimate 'back-up' supply of the River Murray, meant city residents were confident that water would always be available. Water was relatively plentiful and comparatively cheap.

The recent drought, one of the most extensive since Federation, has shaken this confidence. The fear of a permanently drier climate has increased anxiety about the certainty of water supply. The River Murray, despite a century of dam building and regulation, is not the absolute guarantee of water it appeared to be. Reservoirs that were formerly sufficient are now perceived as too shallow; pipelines and distribution systems are antiquated and frequently wasteful; and stormwater catchments, recycling infrastructure and greywater systems are seen as significantly underdeveloped. Confidence in the centralised water system has been shaken as more residents install rainwater tanks and drill bores. The imposition of water restrictions has signalled to many that water – something which once was cheap because it was plentiful – is actually quite rare and valuable.

Of course, the truth lies somewhere between the extremes. Adelaide's residents have been comparatively well served by water supply authorities over many years. Extensive distribution systems have supplied reliable, clean (albeit somewhat 'hard') potable water to city residents over hundreds of square kilometres of suburbs. Stormwater works have been established in high-risk areas to minimise flooding; extensive sewerage systems have served many suburbs. But it is also true that under governmental and political pressure, the water distribution and collection system has been constructed with cost effectiveness and cost minimisation as higher priorities than 'drought proofing' or environmental concerns.

The current drought and predictions of significant declines in average rainfall and increased evaporation across southern Australia suggests the need for a different approach. Water Proofing Adelaide became a public issue in 2004 as Adelaide's increasing population and associated increasing water demand revealed the balance between plenty and scarcity was much closer than many previously thought.[2] Recent short-term responses to scarcity have included partial bans on water use, the promotion of water-saving devices, and plans for desalination plants and enhanced water supply capacity.

Chapter photograph: by David Radzikiewicz, A7Designs.

An important mechanism that can be used to respond to a situation of decreasing supply and increasing demand is the market process. Already partially in use, the market, and more particularly, the price mechanism, if used appropriately, can complement other devices to help balance the demand for and supply of water. This chapter will examine some of the issues surrounding the price of water in Adelaide and suggest some ways forward.

The price mechanism

Water is a complex resource to manage. It supplies benefits to individuals, households, agriculture, industry and the environment – both directly and indirectly. It has characteristics that make it suitable for the application of a 'market allocation mechanism' (such as where my consumption of a glass of water prevents another person's consumption of that same glass of water). It also has characteristics that make it difficult to apply market answers (such as my use of a river does not necessarily preclude your use of the same river). It has characteristics that make it a common resource (a lake), and it has characteristics that mean the same water can be used more than once (via the watercycle). Changing the relative price of water will not only affect the market for water, it will have impacts external to that market. Despite or perhaps because of these complexities, it is important that water is managed well, with care, and with a view to sustainability, sufficiency and equity. This is a difficult task.

The 'price mechanism' is perhaps one of the most efficient ways of providing suppliers and users with information about the relative scarcity of a good. Price information is conveyed almost instantaneously and users can use this to judge the relative 'worth' of the good against other goods. Similarly, suppliers can quickly assess whether there is money to be made in supplying such a good. In well-functioning markets high prices signal that something is either relatively costly to produce or in relatively low supply on the one hand and in relatively high demand on the other, or some combination of both. Similarly, low prices signal a good is either relatively cheap to produce or in relatively great supply and in relatively low demand, or a combination of both. As the supply or demand for water changes, therefore, so a change in prices will signal this, and induce responses by either suppliers, consumers, or both.

Some may argue that it is 'wrong' to put a price on water, as in one respect water is priceless. Without water, life cannot exist. Thus, the argument continues: water should be supplied to all as an essential human right along with fresh air, and perhaps basic shelter. But this overlooks the need to manage and coordinate the use of water, and the effort and resources it takes to supply reliable, potable water to individuals. Such resources are not costless and must be paid for, either via the price mechanism or taxation. By supplying water to many individuals, and diverting large amounts of infrastructure and human effort to do this, it has been possible to reduce the price of water to an extremely low level – mere fractions of a cent per litre. It is becoming increasingly clear, however, that the current low price does not accurately signal the relative scarcity of water, or the true cost of supplying it.

Nor does the low price (set by regulation rather than market forces) currently respond quickly to changes in supply.

It is also true that in a market economy, things that are relatively highly priced are generally used more carefully. Goods or services that are perceived as 'expensive' or 'precious' tend to be used judiciously in the knowledge that they are valuable and not to be wasted. Increasing the relative price of water will cause consumers to use it more thoughtfully and with an increased awareness of its value.

Accurately and clearly signalling the 'true' cost of water allows for better decision-making across a range of areas. For example, if water becomes relatively more expensive (relative to previous periods and/or relative to other goods) should we put resources into building additional storages, or in reducing wastage? Calculating which option is the relatively more efficient and cost-effective is far easier if water is correctly priced rather than undervalued. Similarly, recognising the 'true' cost of water causes consumers to reconsider their behaviour. If water is costly it makes no sense to waste it through inappropriate watering or irrigation practices, or to use it only once when recapture and reuse is possible. Importantly, the market process means each consumer has a personal incentive to modify their behaviour. Regulations and penalties are less effective than 'self interest' in enforcing behaviour modification.

Price incentives may also cause 'new' markets to develop. For example, a market for water of different qualities (and hence at different prices) may be used for different purposes. Further, once water is recognised as 'relatively expensive', the value of other goods and services that contain water also change. Thus freshwater that provides a recreational service, for example, increases in value as do other goods such as food or wine. Accurate pricing of water provides a personal incentive to all potential suppliers to discover 'new' markets or supply possibilities, as water is traded, recycled or reclaimed in ways previously overlooked or considered uneconomic. Pricing water accurately permits a direct comparison of such choices. The true cost of implicit and explicit subsidies to large water-users are more starkly revealed, and the cost of such subsidies may be brought into question.

Nevertheless, it is also true that the price mechanism requires a large supporting framework of institutional arrangements to deliver the major benefit of the market process – efficiency. For example, to operate effectively and accurately, the market requires well-defined property rights. Who actually *owns* the water being traded? Such a question has been contentious for South Australians ever since inter-colonial discussions on the regulation of the River Murray began over a hundred years ago. Is there sufficient information about the good to allow all market participants to make informed decisions? For example, what is the true cost of supplying water to a metropolitan household (as opposed to a regulated price set with political considerations in mind)? To have a well-functioning water market also requires complementary markets in other areas. For example, to determine the true cost of supplying fresh potable water should mean that markets for recaptured and recycled water, or for supplying water of a lower quality, are also well functioning.

Not only are these necessary complementary elements frequently missing or impaired, but society often has other objectives, such as equity, that it also wishes to achieve. It also takes time to develop markets. Given the importance of water in sustaining life in all its forms, and our still comparatively simple understanding of the natural environment, what constitutes an efficient allocation in human terms may well be at extremely high cost to the environment. For all these reasons, price should be viewed as only one of a range of mechanisms by which to signal scarcity, alter behaviour, and make allocation decisions. But it should be used, consistently, and with a view to producing long-term shifts in demand and supply. The market process and the price mechanism are extremely powerful tools by which to produce change at both the individual and aggregate level.

The situation in South Australia

What price should people in Adelaide pay for their water?

In Adelaide, a premium bottle of imported mineral water in a restaurant can easily cost $8/L. Local mineral water bought in a supermarket may cost $2/L. People can pay $7.50 to swim in a megalitre of water at the local public swimming pool. These prices are a long way from the price charged by SA Water: less than $1.50/1000 L ($1.50/kL or 0.15 cents/L). This suggests that people are prepared to pay a wide range of prices for different 'types' of water, or to use or consume it in different circumstances.

Approximately 45% of the water used in Adelaide is for residential purposes, 28% for primary production, 17% for public purposes, and 10% for commercial and industrial use.[2, 3, 4] Among residences, 60% of water is used inside (with 31% in the bathroom and toilet, and 16% in the laundry), and 40% in the garden. Over the past few years, annual consumption of water in Adelaide has fallen to around 151 kL/person (for all users) and to around 99 kL/capita for households. Total demand for water has grown, however, as the population has increased and the demand for water for agriculture has increased.

In 2007/2008, residential users in South Australia paid a service charge of around $160 a year regardless of the amount of water used, and a two-step charge for water use: 50 c/1000 L for the first 125,000 L and $1.16/1000 L thereafter.[5] In late 2007, the government announced a new three-tier system: in addition to the service charge, consumers would pay 71 c/ kL for using quantities under 120,000 L, $1.38 for amounts between 120,000 and 520,000 L, and $1.65 for amounts over 520 kL.[6]

This approach attempts to meet two objectives – equity and efficiency. In theory, paying more for using more is an appropriate way to price water. By setting the pricing policy in 'three steps' the government is attempting to meet this principle, with the lower-priced first step recognising that lower-income households may require assistance to access water. An initial allocation under 120 kL is an acknowledgement that some minimum quantity should be available at nominal cost so that an individual or

BOX 60

Pricing your water: is there a smart way to do it?

'For anything worth having, one must pay the price ...'
– John Burrows, American author, 1837–1921.

The issue

A recent National Water Commission (NWC) stocktake revealed an amazing array of charging regimes for household water use. The stocktake of 57 of Australia's urban supply systems found that:

- 25 set a fixed service charge and then add an 'inclining block' charging regime on top of this fixed charge, increasing the charge per kilolitre in a number of steps;
- four use an 'inclining block' regime without a fixed service charge;
- one uses a 'declining' block regime;
- 24 use a 'two-part' tariff regime that superimposes a volumetric charge on a fixed service charge; and
- three apply a service charge only, and don't charge for the amount of water used.

The record for the maximum number of 'inclining blocks' goes to Busselton Water with an eight-block regime. You pay $0.48/kL for the first 150 kL, $0.62 for the next 150 kL, etc. Over 1950 kL per annum, you pay $2.53/kL. Busselton, however, does not set a fixed service charge.

Lower Murray Water is the only water supplier with a seasonal charge, and water is cheaper in winter.

Given the state of our water supply systems, what is the best way to charge for and ration household water? Have any water suppliers of the 57 supply systems got it right, or have they all got it wrong?

Clarification of objectives

Unfortunately, governments tend to use water-pricing regimes to achieve equity, environmental, revenue and economic efficiency objectives simultaneously. This approach violates a golden rule in policy development to avoid conflicts – use a separate instrument to achieve every objective and, once an instrument is assigned to one objective, don't try to use it to achieve another objective.

Economic efficiency when there's lots of water

What we pay influences what we do and what we buy. If water is abundant, the efficient price to set is the long-run marginal cost of supplying one more kilolitre – including management costs, the costs of being the supplier of last resort, and providing a return on capital. The next step in pursuing efficiency is to charge according to the actual costs of delivering water to each suburb in each season.

Further, it is also necessary for the cost of upstream environmental and other externalities to be reflected in your water supply bill. To encourage you to manage for downstream externalities, however, these need to be charged separately and in proportion to their extent. Once built, the cost of maintaining and depreciating all infrastructure becomes part of the long-run marginal cost of water supply. The more supply reliability you want, the more you have to pay per kilolitre. Desalination plants, for example, are expensive and once built have to be paid for. Great when there is a water shortage, but an expensive white elephant if there is lots of water around.

Economic efficiency when water is scarce

When it unexpectedly gets or stays dry, water supplies have to be rationed. There are two ways to ration water use. One is to introduce water restrictions, which impose indirect costs on many people; the other way is to increase the price.

Economic research keeps pointing to the fact that water-users respond to price increases. Pragmatic as ever, Quentin Grafton recommends the best way to set a scarcity price is to estimate the amount of water in storage every quarter and charge accordingly. As dam storage goes down, the price goes up. To drive home the scarcity message, meters need to be read and bills sent at least quarterly. In the United States, many utilities read every meter every month.

Another way of achieving the same outcome is to allow urban water trading. Set the maximum amount of water that an average household can use in a quarter and let those who really want water buy it from those prepared to sell. In times of scarcity, the water supplier collects more money than is needed to cover costs. Some people think this money should be returned to users, others think it should be used to finance new infrastructure. Either way, it is quite clear that there is a need to change the way we charge for water.

Equity

Many people think that water, especially non-discretionary water (water used inside houses), should be supplied at an 'affordable' price. This is why there is so much interest in inclining block tariff regimes. 'Affordable' is code for not having to pay for the full cost of the water delivered. The idea is that the first amount of water you use should be cheap. Those who use lots of 'discretionary' water (gardens, pools, etc) should have to pay more for it. The result is a cross-subsidy from large water-using households to small water-using ones. At first glance, this may seem reasonable.

But when you dig a bit deeper, it becomes clear that inclining block tariff regimes transfer money from disadvantaged households to richer ones, which, as a result of the block regime, gain access to cheap water. Concerned that inclining block systems are inequitable, John Quiggin has shown that if you want to help disadvantaged households, it is better to set a uniform charge and then pay rebates to everyone or only to those in need. In short, use a separate policy instrument to chase each objective you are interested in. Remember, however, that a typical person uses around 46 kL/year. At current prices, the cost of water used per person is less than the cost of running an old fridge in your garage.

Inclining block tariffs are inequitable also because most of them are implemented on top of a fixed service charge. For the 25 NWC's water supplier utilities who combine an inclining block tariff with a fixed service charge, the average fixed service charge is $124 per household. If you use 100 kL/year and are charged $0.50/kL for this first block of water, the real cost per kilolitre delivered to you is $1.74/kL. This is not cheap water.

Revenue

The real reason water supply utilities set fixed charges is that it guarantees them a revenue base. These utilities are monopolies, but it is hard to argue that they should not be subject to the same pricing disciplines as other businesses. In summary, inclining block tariff systems represent a clumsy attempt to achieve efficiency and equity objectives simultaneously. We believe they should not be used.

Where to from here

With all these arrangements in place, and if we leave sewage connection charging arrangements for another day, several guidelines for household water pricing emerge.

1. Send an efficient price signal to everyone by charging them the same for every kilolitre of water they use.
2. Send a scarcity signal to all water-users. Read meters and send out a bill quarterly. Expect un-metered apartments to start applying for meters.
3. Inclining block tariff systems should be phased out – they are very inequitable.
4. Fixed water service charges should be phased out – for a monopoly, revenue protection is unnecessary.
5. Only help those in need and use targeted programs to do this. Consider increasing Centrelink and pension payments instead.
6. In times of abundance, supply water at the long-run marginal cost of securing an additional unit of water. Plan well but recognise that the cost of building excess supply capacity can be high. Take some risk and use scarcity pricing and/or trading to get out of short-term trouble.
7. In times of scarcity, change the price every quarter according to a formula, use an independent price regulator to do the same thing, or give households the option to trade water.
8. Keep water restrictions to a minimum and contemplate using them only after the scarcity price has risen by several orders of magnitude.

Mike Young and James McColl

household's income does not preclude them from accessing sufficient water for basic human wellbeing. Various pensioner concessions and exemptions for other specified organisations reinforce this equity aspect.

The government establishes water prices under the *Waterworks Act 1932* (urban prices) and *Sewerage Act 1929*. It does this based on a series of economic principles influenced by the 1994 Council of Australian Governments' strategic framework and the National Water Initiative (NWI) – both designed in part to promote efficient pricing policies. Once gazetted, the prices are binding for SA Water. The Essential Services Commission of South Australia conducts regular independent reviews of the processes by water prices, and sewerage charges are set.

While the government needs to be mindful of the equity considerations associated with pricing water, current scarcity suggests there is a strong argument that the three-tiered system is still too simplistic and the price of water still too low. It can be argued that higher prices should begin at lower water volumes, and that prices should become significantly higher for larger users. Under the scheme introduced in 2008, low water-users actually pay *more* per litre than some larger water-users. For example, a household of four, who consume the average of 99 kL/head, pay fractionally more per kilolitre over the year ($1.58/kL) than a household of four who consume twice the average quantity of water per person ($1.57/kL). The flat service charge that produces this result is essentially regressive in effect, and for this reason alone works against attempts to have an equitable pricing scheme in place. Thus, it has been argued, the flat service charge should be removed and all users should be charged the same rates, with equity being achieved by increasing rebates to targeted groups.[7] As selected groups already receive concessions, such a process would not appear to be particularly difficult to implement.

Many different water pricing schemes are possible. Some suggest a flat charge, others argue for progressive prices that increase with water use. It has also been suggested that water should be cheaper in winter and more expensive in summer.[8] All these suggestions, while feasible, involve additional administrative costs such as the introduction of more finely calibrated meters or more frequent meter readings. In many cases they make the pricing, charging and collection system more complex, and distract users from price signal itself. They may also result in unintended consequences as the relative price of water changes in comparison to other goods.

At least one author has suggested an approach that combines both more accurate pricing with equity.[9] This scheme would see water priced at around $1.50/kL for all water. This price signals its scarcity (and is comparatively expensive compared to current schemes), and approaches the current marginal cost of supply. This price would be combined with an initial free or low-cost allocation equal to around 50 kL/person. There would be no service or access charges.

Regardless of the final scheme, a balance must be struck between the need to signal scarcity and still allow access to affordable, clean drinking water. Nonetheless, there would appear to be considerable scope to increase prices further and in a manner much more designed to alter consumer behaviour.

A further equity consideration concerns geographic location. The price of water to residential users is the same regardless of where they live in metropolitan Adelaide, or the state. This decision places equity above efficiency in determining water allocations. One justification is that it considers the possible detrimental impact of variable water pricing on development: high-priced water in outer suburban, remote or difficult locations makes those regions harder to develop. Nonetheless, it is arguable such an approach has also hampered the development of a more diverse range of water sources, and increased the state's dependence on the Murray. For example, if water were more expensive in the Mid and Far North of the state, and cheaper in the South-East, there might already be water desalination plants for urban populations on the York and Eyre peninsulas, and more advanced water storages online in the South East. Raising the price of water to residential users will send this signal, and differentiating prices by location would also result in a more diverse range of water sources and less dependence on single sources. Thus it may also increase the supply available to the city of Adelaide.

Commercial users (wholesale, retail, finance and service sectors) buy water under a slightly different pricing structure. They too have an annual supply charge, of at least $174.60 (in 2007/2008), but which increases with property value. Thus purchasers based in the city's CBD pay more, as their properties are generally more highly valued than equivalent commercial purchasers in outer areas. In 2007/2008, the consumption charges for commercial users were, however, exactly the same as for residential users. Given that some individual commercial users are likely to be large consumers of water, such a simple approach would appear to undervalue water and provide only a modest incentive to avoid large consumption. It also ignores the cost of supplying commercial water to new outlying commerical users.

Even here, equity consideration must be balanced against the need to send appropriate scarcity signals to business. Corner stores and other small business owners need time to change their water infrastructure in response to price signals. While recognising the investment necessary to lower water use in business, there would still appear to be significant scope to:

1. Increase the incentive for businesses to lower water use via a sharply increased pricing regime.
2. Increase the incentive for business to invest in water saving technology via a combination of increased prices plus government assistance and subsidies.
3. Establish world-best practice in water use in businesses via an integrated program of targeted investment and assistance in identifying the best technology.
4. Remove government subsidies for inefficient technologies and practices.

While previous governments have a long history in encouraging businesses to locate in South Australia – by providing financial and infrastructure assistance (including major assistance to access water) – future policies in this field should have a strong requirement to price water correctly and to develop sustainable water use practices.

Another category of users charged by SA Water are categorised as 'non-residential properties' and include commercial properties engaged in industry, mining, quarrying, medical and health services, land used for primary production, and some other categories. These pay a flat annual access charge of $174.60 with no charge related to the cost of their property. There is no effort in this charging scheme to increase access charges in line with capacity to pay, or to counter the regressive impact of this charge for very small users. In 2007/2008 these users also paid the same step rate and charges as other commercial users. Such a scheme, while administratively simple, contains little incentive to promote the conservation of water, particularly where it is used in very large quantities. Here, too, government investment assistance should be refocused to promote sustainable use of water and water recycling, and to phase out explicit and implicit subsidies. Water prices that more accurately reflect the true scarcity of the good for these users would be a significant step in this process.

Equally important for the market mechanism to work effectively, government subsidies to business that artificially lower their water costs should be eliminated. Purpose-built, business-specific infrastructure, individual business 'exceptions' to water charges, and other forms of subsidies, at all levels of government, should be quickly identified and removed. While such a process would be politically difficult to achieve, it is likely to have an important impact on the aggregate demand for water.

Although governments can distort the price signal downwards by providing direct subsidies to water-users, they also frequently distort prices indirectly, by regulation. For example, regulation that prevents individuals from recapturing and recycling water – or which raises the cost of doing so by subjecting these processes to higher than average scrutiny – make using clean water relatively cheaper by comparison thereby inhibiting the adoption of water-saving processes. Raising the relative cost of potable water, while immediately making it more attractive to use less, could be even more effective if regulations inhibiting alternative methods of water saving and recycling were modified to lower their relative cost.

Steps to lower Adelaide's aggregate demand for water have already begun. In part a response to the increased awareness of scarcity and reinforced by relatively small price changes, many Adelaide residents and businesses are already attempting to reduce their water consumption. Water Proofing Adelaide[10] identified the cost of installing water-saving devices as amongst the cheapest initial option. Installation of water-efficient washing machines, tap timers, low-flow showerheads, low-flow taps, dual-flush toilets, and water-efficient garden practices with native gardens had the potential to produce aggregate savings of over 40 GL/year (compared to Adelaide's aggregate use of around 300 GL/year), at a cost of generally less than $1/kL. Aggregate savings were based on 100% uptake of devices and savings based on a $1.03/kL. The majority of the 40 GL savings occurred when native gardens and efficient water practices were established. The other methods listed produced around 16 GL in savings.

In December 2007, the government's new three tiered pricing scheme raised the price of water in real terms. Combined with standing restrictions on water use for residential users, the effect was to send a clear signal that water was, indeed, scarcer and more valuable than previously believed. Nonetheless, critics argued, with some justification, the price rise was insufficient to signal the *true* extent of the water shortage. While water restrictions can modify consumers' behaviour in the short run, they tend to become less effective in the longer term. Price rises on the other hand, if they signal a real and fundamental shift in the relative price of a good and not a temporary price 'spike', can produce more permanent changes in behaviour. They do, however, take longer to work. It is thus arguable that the government missed the opportunity in late 2007 to signal properly the true scarcity of water to Adelaide users.

Complementary to the studies on 'water proofing' the state and raising prices directly, legislative changes have progressively been made to raise the price of water indirectly. Legislation now requires rainwater tanks or dual-flush toilets be installed in new residences. Local governments have begun to expect developers to include wetland areas and other forms of stormwater capture in their proposals. Other regulations could prohibit farmers from allowing fertiliser-contaminated water to re-enter the watercycle or require businesses to filter and restore water to potable quality after use. These legislative changes effectively raise the cost of using water. Provided these costs are at or below the price of purchasing water, individuals have an economic as well as legal incentive to adopt water-saving behaviours. Raising the cost of water directly *immediately* makes these alternatives more economically viable. Creating consistent economic incentives is one of the most reliable and effective ways of ensuring compliance as it becomes in the individual's own interest to behave appropriately. If water prices across a range of water markets are out of alignment (for example, if it is far cheaper to purchase water and discharge it uncleaned or to purchase water rather than conserve it), then monitoring and enforcement costs rise and the quantity of water saved via individual behaviour is lessened.

Ensuring consistency in policies between markets is one of the more difficult aspects of regulation. While raising water prices directly signals its relative scarcity, unless other areas affected by water markets complement this signal, perverse incentives may be established that create unanticipated outcomes. Increasing water prices to primary producers, for example, could easily make dam building, bore drilling and other water-capturing activities more financially attractive – the result may be less runoff into reservoirs or seriously diminished underwater aquifers. While such outcomes are less likely in the current regulatory regime (for example, licences are required in sensitive water catchment areas before individuals may build dams), it is still important that there is coordination, particularly such that the costs of obeying regulations are clearly less than the benefits of

BOX 61

Urban water trading

'What's good for the goose, is good for the gander.'
– a proverb of unknown origin.

The issue
In response to increasing water scarcity, River Murray irrigators have watched the cost of buying a water allocation on the temporary market rise from $44/ML in January last year to $380/ML – a 764% increase. Over the last year, the value of a permanent water entitlement rose by 52%.

Contrast this 'scarcity' price signal with that given to households in cities like Adelaide. These households face a fixed two-tier pricing structure. The first tier, up to 125 kL per annum, costs $0.50/kL. The second tier – for any amount above 125 kL – costs $1.16/kL. Recently, the charge for second-tier water was increased a paltry 6.4%.

What would happen if urban and rural Australia were subject to the same price disciplines for water? Should the old proverb 'what is good for the goose is good for the gander' apply? In a drought, should, like farmers, urban households expect to see annual water prices go up seven fold?

Well-designed rural water trading arrangements have the potential to bring significant improvements in water use. What would happen if urban households and industrial water-users were able to buy and sell water entitlements and water allocations – just like irrigators? Could it be made to work? Is it worthwhile?

In this box, we search for a way to make household and industrial water rights tradeable – to expose urban and rural Australia to the same price discipline. Can a pragmatic, low-cost way be found to do this?

Valuing and charging for water
Essentially, there are two ways water supplies can be managed: a quantity limit (a cap) can be placed on the water supply and the market left to help decide who gets to use water; or the charge per kilolitre is fixed and restrictions are used to reduce consumption in times of scarcity. Under the former approach, prices rise with scarcity; under the latter, increasingly severe restrictions are introduced as scarcity increases. You pay the same but get less access to water, and experience considerable

inconvenience as the level of restrictions increase. Conceptually, the more that the price per kilolitre is allowed to increase in times of water scarcity, the less the need for restrictions on when and how people can use water.

Experience
The idea of linking urban and rural water markets is not new and is happening. In Queensland, the shire of Gayndah currently holds an entitlement to more water than it needs, and sells the surplus back to irrigators on an annual basis. SA Water has been buying water entitlements from River Murray irrigators.

In Arizona, in the United States, urban developers in cities like Phoenix and Tucson are required to certify whether or not any block of land they sell has guaranteed access to a water supply for the next 100 years. Given this requirement, most developers choose to buy enough water to guarantee supply to any future development.

In Beijing, each person in a household is allocated a quota. When the metered amount of water they use exceeds this amount by more than 20%, they are required to pay double for this water.

Allocating water to commercial and industrial users
Many businesses, like irrigation farms, use large amounts of water. Arguably there is little reason why a large commercial business cannot be given a water entitlement and then be required to enter into a water supply contract with a water utility. A price for delivery of a kilolitre of water would be struck in a manner similar to that set by rural water supply companies. There are many different ways of deciding upon the initial entitlement to be given to each business. One option is to give each business an entitlement equal to the maximum amount they have used in any of the last three years.

Household entitlements and allocations
One of the simplest ways to introduce household trading would be to make the first 200 kL tier of water given to each household tradeable. Families and households would then be able to sell off any savings they make. For water use above 200 kL, there are two choices: set a scarcity price for the second tier; or 'cap' the volume of the second-tier pool.

Under a scarcity pricing regime, if the charge for the first tier was $1.00/kL and the second-tier charge was regulated to rise as reservoir levels fall, one might expect a scarcity price to vary from $3.00 to $5.00/kL. Facing charges like these, there would be an incentive for low water-users to sell off part of their first tier to large water users – especially if first-tier water was not subject to restrictions in all but exceptional circumstances.

An alternative to the scarcity pricing approach is to cap access to second-tier water. Once 'capped', second-tier entitlement shares could be sold using a tender process, where anyone – including water utilities – could bid for a share of the available pool. Every quarter, shareholders would be notified of the tradable allocation, which they could either use or sell.

At the household level, many variants are possible. To protect people under financial pressure, the first 100 kL entitlement issued to each house could be made non-tradeable.

Urban household water trading
With low-cost internet trading platforms like eBay and a choice of water brokers, trading in household water entitlements and quarterly allocations is conceivable. The average household bill is around $330 per annum and most of these costs are unavoidable.

If first-tier water was available at $1.00/kL, a household with a large, water-dependent garden and a swimming pool would have a strong incentive to buy a low-cost first-tier entitlement to secure low-cost access to the water they need every year.

If any household exceeded its first-tier allocation, their nominated broker, or, by default, their water supplier, would be required to restore balance to their account by purchasing unused allocations and adding the cost of doing this to their next water bill. Households who use less than their allocation would have the opportunity to sell this saving.

It would also be possible to make urban subdivision approval conditional upon the purchase of a water entitlement sufficient to supply the proposed development – as already happens in Arizona. New or expanding industries would also be required to buy water entitlements sufficient to cover their needs.

Water industry implications
From a water utility perspective, the introduction of urban trading raises many questions. Utilities would still have to charge for delivery and infrastructure maintenance. Among other things, part of each water utility's bulk water entitlement would be broken up and transferred to those that buy access to second-tier water.

There may be lessons to learn from recent reforms in the electricity and gas industries. Sections of businesses that manage water infrastructure may need to be separated from retailing activities. Access to delivery capacity may need to be rationed.

As households and industry become exposed to the value of scarce water resources, competition from privately managed sewage recycling, stormwater capture and desalination businesses would increase. Water saving and alternative source development might become profitable. If the annual price saving per kilolitre was $2.00/year, one can imagine a 100 kL tier one entitlement rising in value to over $2000.

Many buildings are managed collectively and many houses are rented. In each case, someone will need to decide who should pay the water bill and who would hold any entitlement issued.

Where to from here?
We think the concept is worth serious evaluation, and, for large industries, could be readily implemented. A broader pilot trial, with no long-term guarantees, would be worthwhile. One option would be to test urban trading in a large regional centre which is on extreme water restrictions.

Mike Young, James McColl and Tim Fisher

not doing so. Charging differential prices between summer and winter may induce householders and businesses to store water inappropriately. This could increase the strain on underground storages or change the stability or salinity in soils. Again, the framework of regulations that complements price changes needs to be considered along with price rises.

A further undesirable but predictable outcome caused by raising water prices is an increase in water theft. In the longer term, as markets differentiate, water fraud (as providers pass off lower-quality water for higher-quality, higher-priced water) is likely to occur, as is the possibility of compromised health outcomes from water that has not been treated to the required standard (cost cutting). Theft of water from environmental stores is also likely to increase. It is also possible to predict that there will be a rise in a range of behaviours such as individuals devising their 'own' (possibly dangerous) solutions for reusing water, increased disputes over water sources (particularly in the Adelaide Hills), attempts to illegally tap underground springs (as opposed to sinking bores), and so on.

One response to this in the past has been to centralise control for water quality and security in government agencies. While this removes the influence of the profit motive, if these organisations are under-funded they will struggle to police water standards and security. As the price of water increases, therefore, it will be necessary for governments to devote more resources to constructing and maintaining water inspection offices and health compliance inspectors and the like to enforce government regulations and minimise the possible detrimental impacts caused by people attempting to circumvent health and water standards.

Focusing exclusively on the price mechanism to 'solve' issues of water scarcity also tends to overlook the integrated nature of the watercycle, and the integrated nature of markets. For example, in 2007/2008, sewerage rates were calculated as a percentage of the property value of a site, with differential prices between metropolitan and country properties and between residential and non-residential users. Rates were payable if the property *could* be connected to the sewer main, not if it was actually connected. It was also possible for some industrial users to dispose of their waste legally, via the sewerage system, so long as a higher levy was paid. Given the total lack of relationship between price and quantity in this part of the watercycle, there would appear to be significant and useful pricing options that could be implemented if a fully integrated water pricing system was adopted. For example, unless the discharge levy is correctly priced, it will 'pay' users to dispose of industrial waste via the sewerage system. It is then the environment and other users who ultimately pay the full cost of cleaning such effluents. On the other hand, businesses could actually be paid for the quantity of water they discharged from their property, if its quality was higher than that which they purchased. This would provide increased incentives to install filtration and filtering systems and, in the longer term, could lessen the aggregate demand on large, centralised filtration plants. Similarly, residential users could be offered incentives to either pay for discharging untreated wastes via the sewer system or receive payment for discharging treated effluent.

The price of water not only impacts on demand, it also affects supply. Historically, the size of the infrastructure necessary to supply whole populations with drinkable water, and the economies of scale that can be derived from such projects, has placed the responsibility of supplying water in government hands. This has also meant they have ultimate control over its price. Critical to the pricing decision, therefore, must be the cost of supplying water; the price must reflect the true (rather than subsidised) cost of providing adequate water supplies. Part of the solution to the problem of insufficient supply is to permit more suppliers to enter the market. Previous prices did not adequately cover the cost of supply. As prices increase, so will the incentives for more suppliers to produce water supply solutions.

Undercharging for water, particularly over the longer term, can mean inadequate infrastructure provision and insufficient investment in future supplies. Indeed, by under-pricing water and decreasing the amount of investment in supply infrastructure, it can be argued that the less well off receive less access to water, and at a higher per unit charge than would have occurred if appropriate charges had been set originally. Current deficiencies in infrastructure are in part a legacy of under-investment in the past. The price rises announced in late 2007 were closely tied to the need to provide additional infrastructure. This included a desalination plant with an initial capacity of 50 GL, a doubling of dam capacity, and the construction of pipes to connect northern and southern regions of Adelaide.[6] A recent study estimated that to recover the full cost of supplying water to Adelaide residents would require an increase in water revenues of around 22% per annum.[11] Note too that this was below the national average of 33% for all capital cities. Such an increase would mean that the full cost of supplying water to Adelaide each year would be met through revenues. The price rises in late 2007 amounted to an increase in real terms of around 12.7%.[6] Expressed as a 'flat charge', a 22% increase would be the equivalent of charging $1.54/kL.[11] In comparative terms (and estimating full cost recovery charges in all the capital cities), this placed Adelaide's estimated price slightly behind Perth ($1.59/kL) and Canberra ($1.56/kL), and slightly ahead of Sydney and Melbourne ($1.47/kL). There would appear, therefore, to be a strong argument for an increase in volumetric charges to approximate more closely the full cost of supplying water.

Expressing the price of water per kilolitre also allows us to compare the costs of providing alternative water supply options. Compared to other cities in Australia, Adelaide has relatively low storage capacity, high evaporation, and is heavily dependent on the Murray – a source that is rapidly deteriorating in quality. This suggests that a range of alternative supply options need to be investigated. In studies done on this issue, the word 'supply' includes 'finding' water via demand savings. Thus, studies have ranked water supply/demand options by cost per kilolitre, depending on the exact circumstances of individual cities.[11] Alternatives estimated to cost less than $3/kL were (in increasing order of cost): catchment thinning (removal of trees in catchment areas to increase runoff); purchase of irrigation water; demand management (including subsidised showerheads, toilet cisterns, etc); stormwater reuse; groundwater exploitation;

A stormwater impact offset system would require councils and/or a stormwater management authority to allow developers to have their development assessed on a 'net' basis that includes credit for reductions in stormwater load elsewhere.

A tradeable stormwater credit system would mean that any person who reduces runoff could receive credit for action that is in the community interest, and then hold it until someone else wants it. In effect, stormwater credits would become bankable.

The question that has to be asked is whether or not a practicable way to allow people to generate and bank stormwater credits can be found. A considerable degree of pragmatism is necessary; the system does not have to be perfect.

Many proposed market-based instrument systems have failed to get up because they cost too much to implement or are too complex. Probability of adoption increases with simplicity and pragmatism.

Elements of a stormwater credit system

One way of starting up a stormwater credit system would be for a stormwater management authority to raise the money to fund the construction and maintenance of infrastructure, and to guide the development of stormwater policies. South Australia is in the process of doing this. This stormwater management authority would empower local government authorities to issue and collect stormwater credits as and when appropriate. The existence of a credit would be recorded on an internet-accessible register similar to that used by banks. The holder of a stormwater credit could decide, at any stage, to offer it for sale on an internet-based auction site.

Anyone, including a developer, could earn stormwater credits by acting to reduce expected peak stormwater load. For most investments in stormwater reduction, a standardised set of look-up tables would be used, for example: connect a 500 L rainwater tank to a toilet and receive, say, 300 credits; reduce runoff from a supermarket carpark by 100,000 L and receive 100,000 credits.

When planning approval is given for a new development, it would be accompanied by a statement indicating how many credits would need to be surrendered in order to obtain building approval. Credits for surrender could be obtained by: generating them; buying them from someone who has generated them; or buying them from the stormwater management authority.

The cost of purchasing credits from the stormwater management authority would be seen as a last resort option and priced accordingly. There is a qualifier, however: such an authority may find it is cheaper to buy credits from existing households and parking areas rather than build drains! Considerable savings may be possible.

Outstanding issues

When designing the detail behind such a system, it needs to be remembered that the aim is to reduce the need for ratepayers and the government to build and then manage more and more stormwater infrastructure, and that such a system could grow through time.

Mike Young and James McColl

BOX 63

Thinking like an accountant

'You can not manage what you cannot measure.'
– Malcolm Turnbull quoting Sextus Julius Frontinus, the chief executive of 'Rome Water' in the late 1st century

The issue – why it is important

The business world is full of accounts and accountants. The financial accounting profession has many well-established standards, protocols and conventions. All are designed to help managers manage and give investors confidence. Accounting places very strict disciplines on those responsible for governance of a corporation.

The purpose of accounts is to enable people to make better decisions. If the accounts are flawed, inappropriate decisions can be expected. Misleading accounts can lead to unreasonable optimism and, as a consequence, adverse outcomes.

We suspect that very few of the water accounts available to Australia's water managers would pass even the simplest of accounting tests. What would an accountant expect to find in a set of water accounts?

Double-entry accounting

Double-entry accounting began to be widely adopted throughout Western Europe some 500 years ago and is thought to have played a pivotal role in facilitating European dominance of the world's economy.

Developed by Muslims in the 10th century, double-entry bookkeeping requires that for every credit there is a debit, forcing assessment of where gains and losses reside.

Applied to water, double-entry accounting would require managers to ensure that whenever one person takes more water, someone else takes less. If environmental allocations were managed in a similar way, then a manager would be able to rule off the accounts at any point in time and, using a set of well-established performance ratios, assess the health of any river or aquifer system.

A robust double-entry accounting system, supported by the equivalent of a balance sheet, could play an important role in revealing whether or not a river system is healthy. Summaries of assets and liabilities could be used to reveal the effects water management decisions are having on third parties.

Account consolidation

In the world of commerce, it is common for businesses to be at least partially connected with one another. To deal with connectivity, consolidation principles and conventions have been developed to prevent double counting and to ensure the accounts submitted to a board are not misleading. No such principles and conventions have been developed for water resource management.

When consolidated accounts are prepared, transfers within the system are netted out and all data adjusted and reclassified for consistency. An overall picture of the health of the consolidated business is presented. In addition, there is always a group of people responsible and accountable for ensuring the consolidated business remains healthy.

Water accounting protocols and conventions necessary to prevent double counting of surface and groundwater resources have yet to be developed. A well-designed, interconnected water account should reveal how groundwater management and use is affecting surface water resources, and vice versa. It should also account for the significant time lag between groundwater use and surface water impact.

Solvency

Another financial convention is that all accounts are prepared on the assumption that the business is viable and will continue to be so. This means that assets must be depreciated and liabilities accounted for. When prepared in this manner, it becomes obvious when a business is no longer sustainable and is trading while it is insolvent. This concept is so important that independent auditors are required to confirm that accounts present a true and

much water is to be released from the dam rather than what must be delivered downstream?' He wanted to know why the Murray-Darling Basin Agreement does not include a set of rules designed to ensure that reasonable amounts of water always flow through the Murray Mouth.

What would happen if we reversed the way allocation rules in plans are written, and put downstream obligations first?

A national water accounting framework?
The National Water Initiative proposes to develop water accounting arrangements that are 'able to meet the information needs of different water systems in respect to planning, monitoring, trading, environmental management and on-farm management'. Australia could prepare a national water accounting framework that:

- Defines scope – a definition of water accounting and water systems that accounts describe.
- Establishes a conceptual framework – objectives, structure and key elements to be included.
- Guides preparation – what can be counted, how amounts should be measured, and techniques of measurement.
- Guides presentation of balances, storage, flows, consumption and compliance with protocols.
- Sets out processes for reviewing and changing agreed policy conventions, audit arrangements, etc.
- Establishes monitoring and enforcement provisions.

The accounting profession is in the process of developing a new framework to guide account preparation and auditing. Water managers could decide to build upon the knowledge that this profession has taken centuries to develop.

correct record of what has happened. If a business is caught trading while insolvent, the board and management can be held responsible, and typically an administrator is called in to solve the problem. Counting the same resource twice – double counting – is a classic example of a practice that can lead to insolvency.

Well-designed water accounts could reveal whenever a river or aquifer is being managed unsustainably. While it may not be possible to call in an administrator, well-designed water accounts might speed up restoration efforts.

What would happen if all water accounts had to be audited and 'qualified' reports were not tolerated? The auditing profession would insist that all accounts – with a statement as to whether or not the river is being managed sustainably – be available to the public.

An end-of-system focus
While many may criticise accountants and management's focus on profit, arguably this is one of the reasons why the west European approach to business has been so successful. At each stage in the account preparation process, accounts are designed to reveal how much each part of the process is contributing to the 'bottom line'. In fact, what is the bottom line is an important question in water management.

Reflecting upon this, a professor of accounting asked us why all water allocation plans are back-to-front. In his view, all river water allocation rules should be written in terms of obligations to deliver water downstream and ultimately to the mouth of any river system. He asked: 'Why are most allocation plans written in terms of how

Mike Young and James McColl

BOX 64

New water for old: speeding up the reform process

'The laws relating to the transfer and encumbrance of freehold and other interests in land are complex, cumbrous and unsuited to the requirements of the said inhabitants.' *Torrens Title Act, South Australia 1857–1858.*

The issue

There is an important relationship among water resource plans, entitlements and registers – all three are of equal, essential and vital importance. Good registers do not fix bad plans; you can't have a good plan unless it gives effect to a good entitlement system. Like Romeo and Juliet, plans and entitlements go hand in hand.

Knowing what we know today, to produce a good water resource plan, one would expect the planning process to begin by rigorously identifying the separate bodies or pools of water to be managed, and then establish the rules for assigning water to each pool. The effect of one pool on another would be defined in a way that has hydrological integrity. Water could be assigned only to one pool at a time.

Entitlements to share access to the water assigned to each pool would then be defined and recorded on a register that guarantees ownership security and, through trade, facilitates efficient use and adjustment. This is clearly not where we are today. Many entitlements were first issued in a developmental era when water resources were relatively abundant and were not defined to manage scarcity, interconnectivity and climate change. Today, many water allocation systems are under considerable stress, so much so that some plans have been suspended. Entitlement trade, especially among states, remains cumbersome.

Given the circumstances Australia now finds itself in, we wonder whether or not there may be a need to adjust much more quickly than envisaged when existing water resource plans were put together.

What features would a new planning, entitlement and register system have to have to be deemed future proof? Could it be developed so that most entitlement holders would prefer it?

A way forward

In this box, we explore the proposition that it may be advantageous to consider improving plans, entitlement systems and entitlement registers simultaneously. And, that this could be done in such a way that most water-users would be keen to transfer their entitlements into this 'new' system.

The approach we have in mind draws upon the New South Wales experience in persuading land holders to convert 'old' system land titles into a 'new' Torrens land title system. Conversion was voluntary and implemented over a number of years. The approach also draws upon experience gained when company share registers were moved from individual state registers to a single national register.

Water resource planning and entitlements

Good plans start by rigorously defining the relationship among the pools of water to be made available for environmental use, consumptive use and system maintenance.

Under the 'new' system, and consistent with the National Water Initiative, plans and entitlements would be aligned in a manner that guaranteed that the water supply reliability would be a function of climate and nothing else. Entitlements would be defined as shares in the water assigned to each pool. Under this system, whenever one person's shareholding is increased, another's must be decreased. The system would have hydrological integrity. Among other things, each plan would require the offset of any adverse effects of land use change or other similar processes on entitlement reliability.

An 'indefeasible' register

Good registers define ownership unequivocally. In the past, many of the processes used to issue water entitlements lacked consistency. Typically, entitlements were issued as licences. At the time, no one envisaged that these licences would be separated from land titles and be used to define assets that could be traded across large distances. When licences were first issued, it was always assumed that they could be changed as and when necessary. As a result of this history, even today some governments remain reluctant to unequivocally guarantee the integrity of their water registers.

One of the most desirable features of the new register would be 'indefeasibility'. Whoever is named on the register as the owner of an entitlement is guaranteed to be its owner. The only way that ownership could be transferred to someone else would be to change the entry on the register. To provide for investment security, changes are made only with the consent of all registered

interests. Anyone who suffers a loss as a result of fraud, administrative error, etc. would be entitled to just compensation.

Financial risk

Throughout the world banks are required to hold a proportion of their assets in extremely safe asset classes. Moreover, as a general rule, those who offer to mortgage these safer assets can borrow at a lower interest rate than everyone else. While this may not seem important, if an entitlement register was defined so that it and its underlying assets were of the highest security, then irrigators should be able to use their entitlement to borrow money at cheaper rates than is presently the case. For this level of security, entitlements would need to be defined in a manner that prevented an entitlement from being forfeited or cancelled. There are many ways to penalise bad water-users – the threat of forfeiture of a water entitlement or, more seriously, the actual forfeiture of an entitlement need not be one of them.

Mortgageability and other interests

Well-defined registers record third-party interests and guarantee that these interests will be protected. On the 'new' register, electronic access could be given to banks and other similar bodies so that they could clear part or all of a mortgage themselves. When loan payments are seriously in arrears, mortgagees (lenders) would have rights to foreclose their registered interest and sell enough entitlements to enable them to recover their interest. Mortgagors (borrowers) would have an equitable right of redemption. A mortgagee would not be able to recover more than that owed to them.

Trading

Imagine a register that provides maximum investment security and facilitates trading at very low cost. If the new register was built from scratch and designed for electronic trade, it should be possible to complete an entitlement trade in a few minutes. Enter the name of the person to trade to, let the computing system check whether or not the proposed trade is possible, press the confirm button, and the deal is done – irrevocably! Partnerships with the banking system and brokers could be used to make execution conditional upon payment. No trade should cost more than the $60 or so currently charged for the electronic sale of shares in an ASX-registered company.

Where to from here

If we had more space in this box we would add a lot more detail and also point to all the good work that states and the National Water Commission are doing to build state-of-the-art water resource plans, registers and trading systems. We also recognise the reality of the many water-sharing plans and the constraints of plan review timelines and commitments among the various states and the Commonwealth.

The CSIRO Sustainable Yield Project for the Murray-Darling Basin aims to identify over-committed water resource planning areas. One reason for over-commitment is that some plans do not adequately account for groundsurface water interconnectivity and the consequences of a long drought. When the results of this project are considered alongside the impacts of the current drought, for some areas it may be worth considering simultaneous improvement of the plan, the entitlement and the register system in a way that benefits all.

In areas where simultaneous change might be in the interests of all, an offer could be made to prepare a new resource area plan that properly accounts for system interaction and change, and then offer to issue new entitlements recorded on a new state-of-the-art register. While we recognise that states may be reluctant to allow this to occur, we can see merit in letting the new Murray-Darling Basin Authority (MDBA) offer to do this, and then invite all those with interests in a water resource planning area to choose between the 'old' and the proposed 'new' system. For irrigators, this approach would involve no downside risk. There would, however, be a strong incentive for the MDBA to get it right – reputations would depend upon it.

If a majority want the new system then this process could set the standard for all to follow and provide a pathway for the progressive transfer of all water resource area plans and entitlements in the Murray-Darling Basin to a single basin-wide system.

Mike Young and James McColl

BOX
65

The value of South Australian

wetlands

South Australia is second only to Queensland in the number and area of wetlands recognised by the Commonwealth government as being of national importance. The 69 nationally important wetlands in South Australia cover an area of over 4 million ha. Approximately half (47%) of this wetland area is located along the Murray River, from the Victorian border to the sea; 43% is in the arid north; nearly 9% in the south-east of the state; and 18 of the 69 wetlands are in the vicinity of the city of Adelaide.

Wetlands provide a range of important environmental services to society, for free. They play critical roles in the regulation and movement of water within watersheds as well as in the global water cycle. They have important hydrological functions including aquifer recharge/discharge and water storage, act as a nutrient source/transformer/sink, and provide habitat for a multitude of species. One of their most important functions is water purification. The water filtration process includes sedimentation, adsorption, bacterial metabolism, plant metabolism, passive plant absorption, and natural die-off. The net result is a substantial improvement in water quality as water passes through wetlands. Surprisingly, even though wetlands are often referred to as 'the kidneys of our rivers', very little information exists in Australia on the ability of natural wetlands to filter water. This has resulted in wetlands being seriously undervalued, thereby providing misleading grounds for their destruction.

1. Wetland destruction

While significant areas of wetlands still remain in South Australia, large, important areas have already been destroyed – the Lower Murray dairy swamps were once part of a series of freshwater wetlands stretching from Mannum along the Murray to the Coorong. Of the original 5700 ha of wetlands only 500 ha remain today. The destruction of this wetland area is typical of wetland losses that have occurred across the country; it is estimated that over half of Australia's wetlands have been destroyed since European settlement.[1]

Human-induced restrictions to water flow along the Murray River (such as the construction of barrages, weirs and locks to assist with boat passage) have not only destroyed wetlands but resulted in a change in wetland type. The cutting off of wetlands from the sea or from environmental flows has resulted in some freshwater wetlands becoming saline, and some previ-

ously saline wetlands becoming fresh. Further, many wetlands once temporal in nature are now permanent, resulting in significant changes to the flora and fauna of the region.

Further causes of wetland destruction include the conversion of wetlands to farming land, changes in wetland wetting and drying cycles as water is diverted to farming, and pollution from farming and industrial activities. Because the environmental services of wetlands are so poorly understood, the impact on the environment of the loss of these wetland areas is little recognised and has seldom been valued.

2. Wetland valuation

Few Australian studies have been undertaken to specifically target the value of wetlands to society; of those that have, most have generally relied on willingness-to-pay methodologies. Thus, valuation of wetlands in Australia has focused on recreation/aesthetic aspects. Effectively this means that what people 'see, feel and hear' is being valued rather than what wetlands 'do'. Unfortunately such valuation methods are unable to capture wetland functions, such as water filtration, because the public generally has little idea of what such functions are worth. Estimates of wetland values based on willingness-to-pay methodologies can only be useful when considered in conjunction with valuation methodologies based on environmental functions.

One reason why environmental functions have not been explicitly valued in Australia is the lack of available information. Environmental functions are difficult to quantify and few wetland studies have been undertaken. This problem is compounded by the interdisciplinary skills required to assess the value of environmental services, the variability in functions between wetlands, and the temporal nature of many wetland complexes.

Perhaps the most significant reason for the lack of recognition of the value of wetland environmental functions is the lack of protective legislation. Unlike the United States, Australia has no specific legislation to ensure that a developer who destroys a wetland must replace the wetland function they destroy. The legal requirement that wetland functions be maintained removes the problem of relying on studies involving people's willingness to pay for wetlands – the cost of

providing the wetland function now becomes the important factor drawing wetlands into the commercial market. Thus, if Australian legislation required those who destroy natural wetlands to replace the ecosystem services wetlands once provided, the value of those services would be more accurately revealed and recognised.

3. Water filtration

The conversion of wetlands along the Lower Murray river to dairy farms (mentioned earlier) removed the natural filtration system wetlands had always provided for water consumers in the region; by the 1990s water quality had became a political issue. To address this problem, in 1999 the state government contracted the construction of four Water Filtration Plants (WFPs) at a cost of approximately $AUD10 million.

A recent economic analysis by Schmidt[2] examined how much of the WFP costs could have been saved had dairy swamps in the vicinity of the WFPs been turned back into constructed wetlands to undertake all or part of the water filtration process. In this way, the commercial value of wetlands for water filtration was determined.

The value of constructed wetlands for domestic filtration was found to be a minimum of $14,000 per hectare per year, and possibly up to $28,000 per hectare per year (depending on how much of the domestic water filtration system wetlands could replace). These figures are consistent with those from New York, where the city is spending US$1.5 billion in restoration costs to protect 80,000 acres (approx 32,400 ha) of watershed lands (including wetlands) required to ensure the quality of upstate drinking water supplies. If the upstate watersheds were not restored, the cost of installing an artificial filtration plant would be US$6 to 8 billion plus annual operating costs of $300 million.[3]

The Lower Murray study further determined that if legislation existed in Australia requiring those who destroy natural wetlands to replace the water filtration process that such wetlands once provided, permanent wetlands would be worth in excess of $7100 per hectare per year, even where the water filtered is not used for human consumption. Temporal wetlands in the region would be worth less (approximately $3000 per hectare per year), as typically they are only connected to the river (and thus can filter water) for five months of the year.

4. Mitigation banking

Following the example set in the United States, Australia needs to consider the introduction of a wetland mitigation banking system whereby businesses (or individuals) who enhance or restore wetlands are provided with wetland 'credits', which they are able to sell to needy developers. Such a system would help to ensure the protection of the important functions wetlands provide such as water filtration – a very valuable function. Such a system could halt the destruction of wetland areas, or provide the incentive to rehabilitate degraded areas, not only yielding environmental benefits but providing tangible economic benefits based on market prices.

A mitigation banking system would also allow for long-term land use planning, with farmers able to rehabilitate land and then 'bank' the credits in anticipation of future development needs. Given the additional benefits that wetlands can provide (e.g., flood mitigation and habitat), the government would be wise to consider this important option.

5. Summary

Wetlands provide a range of very important services to the community and are a very valuable asset. The set of factors that have caused the valuation of wetland functions to be overlooked – the 'invisibility' of the functions performed, the complex nature of these functions, and the need for more interaction between economists and others involved in the research and development process – means governments are poorly equipped to make investment decisions where wetlands are concerned. This has probably been partially or predominately responsible for the lack of legislation to ensure accountability and safeguards against ill-judged wetland destruction.

If South Australia's remaining wetland areas are to be protected, they must be viewed in the same way as other important public resources (such as education and defence). This means they must receive government funding and water rights based on the important services they provide, regardless of whether the benefits are clearly visible or understood by the public.

Carmel Schmidt

BOX 66

Icebergs in Adelaide?

The limited freshwater resources in South Australia have encouraged an enduring search for supplementary supplies. One commonly touted plan is to exploit the water frozen in icebergs that break away from Antarctica. This concept sounds intuitively fanciful, but there is an astonishing amount of freshwater stored in Antarctica – around 70% of the earth's reserves and 1.25 million GL floats away from the continent each year as icebergs.[1] As difficult as it is to ignore such a resource, technical, economic and environmental concerns may mean it is impractical and irresponsible to ever make use of it.

Harvesting icebergs requires long-distance haulage and processing techniques that have not yet been developed. Proposals to simply tow large bergs to where freshwater is needed are fanciful. Impediments include the extraordinary weight of big bergs, the lack of vessels capable of towing them (and their potential fuel consumption), and the loss of water as the ice melts.[1] If a large iceberg could be towed to Australia, its draft would prevent it getting closer than the continental shelf – around 50 km offshore from Robe or Kangaroo Island.[2] New infrastructure would then be required to mine the iceberg for its water and transfer it to land. Environmental concerns raised over the hauling of icebergs include the

increased ship traffic near Antarctica, the largely unknown ecological effects of removing icebergs from Antarctic waters (or introducing them to South Australian waters), and that any iceberg collisions with the ocean floor would devastate that environment.[3]

Some novel schemes have been devised to overcome the technical problems of towing icebergs. Firstly, to deal with the melting, a plastic wrap has been proposed. As the wrapped berg melts it will float on the heavier seawater and reduce the depth of its draft (allowing it closer to shore).[4] Environmental concerns have been raised over the production of such large sheets of a plastic durable enough to withstand the operation, and over the potential for it to be accidentally ditched at sea.[5] Secondly, to reduce operational costs, there has been suggestion that haulage of big bergs could be assisted by ocean currents and the rotation of the earth.[2] However, neither of these schemes have been properly developed or trialled.

During the development of the Water Proofing Adelaide strategy, the South Australian government assessed the viability of harvesting icebergs and presented it among other strategies to bring more freshwater to Adelaide.

A detailed review estimated it would cost $800 million to develop the required infrastructure and a further $17 per kilolitre to deliver 4.2 GL of freshwater to Adelaide per year.[6] In terms of ongoing costs, this amount represented the most expensive option. Importantly, both ongoing and establishment costs were calculated for a scheme that has never been attempted. The value of water would need to be considerably higher to stimulate the development of appropriate technology and outweigh the environmental risks.

It is not currently viable to harvest icebergs to supply freshwater in Adelaide, nor does it seem that it will be in the near future. However, it is difficult to rule out that it will ever happen; human ingenuity has overcome more complex technical, environmental and economic challenges. Nevertheless, it will serve us better to direct our ingenuity and economy to work with natural processes and use more readily available resources efficiently and sustainably rather than altering the structure and function of systems we do not fully understand.

Philip Roetman

References

1. Plato, *Euthydemus [online]*, eBooks @ Adelaide, 2003 (online).
2. Government of South Australia, 'Water proofing Adelaide: exploring the issues – a discussion paper', 2004 (online).
3. M. Rann, 'Water proofing Adelaide', Government of South Australia, 2008 (online).
4. National Water Commission, 'Australian water resources 2005: a baseline assessment of water resources for the National Water Initiative', Commonwealth of Australia, Canberra, 2007.
5. SA Water, homepage (online).
6. M. Rann and K. Maywald, K., 'News release', Government of South Australia, 2007 (online).
7. M. Young, 'Pricing your water: is there a smart way to do it? Droplet no. 10', University of Adelaide, 2007 (online).
8. Q. Grafton and T. Kompas, 'Sydney water: pricing for sustainability. Government working paper 06–10', Crawford School of Economics, Australian National University, 2006.
9. J. Quiggin, 'Issues in Australian water policy', Committee for Economic Development of Australia, Melbourne, 2007.
10. Government of South Australia, 'Water proofing Adelaide: a thirst for change 2005–2025', 2005 (online).
11. J. Marsden and P. Pickering, 'Securing Australia's urban water supplies: opportunities and impediments. A discussion paper prepared for the Department of the Prime Minister and Cabinet', Marsden Jacob Associates, Melbourne, 2006 (online).
12. W.D. Shaw, *Water resource economics and policy: an introduction*, Edward Elgar, Cheltenham, 2005.
13. No author, 'Don Juan: Canto the Second' online.
14. No author, 'T. Macci Plavti Mostellaria' (online).

Box references

Box 65: The value of South Australian wetlands

1. Australian Nature Conservation Agency, 'Wetlands are important', Canberra, 1996.
2. C.E. Schmidt, 'The valuation of South Australian wetlands and their water filtering function: a cost benefit analysis', PhD thesis, University of Adelaide, Adelaide, 2007.
3. Commission on Geosciences Environment and Resources, 'Watershed management for potable water supply: assessing the New York City Strategy Committee to review the New York City watershed management strategy', National Research Council, 2000.

Box 66: Icebergs in Adelaide?

1. P.G. Quilty, 'Icebergs as a water source? Looking south: managing technology, opportunities and the global environment', Australian Academy of Technological Sciences and Engineering, 2001 (online).
2. Engineering and Water Supply Department, 'South Australian water futures: 21 options for the 21st century', Adelaide, 1989.
3. M. Osborne and C. Dunn, *Talking water: an Australian guidebook for the 21st century*, Farmhand Foundation, Sydney, 2003.
4. Government of South Australia, 'Water proofing Adelaide: beyond the drought', 2003 (online).

Introduction

'Food glorious food, we're anxious to try it,' so sang Oliver Twist in the famous tale by Charles Dickens. While food was difficult to come by for Oliver, and no doubt for some early Adelaidians, today we take for granted a ready supply of fresh foods, rarely questioning the underlying elements essential for producing it. Critical to plant growth is water, along with air, light, space, nutrients, and a medium for growth, generally soil. Supply of any of these can be limiting in a city environment; so often food production is pushed out to make way for other, higher-value land uses. Yet, in Greater Adelaide, the inherent suitability and diversity of land for food production, the relatively low population density, and the market demands of 1.2 million people for fresh foods, all ensure agricultural production remains an important element of our city. With growing demands and reducing supplies, water and how we manage it is fundamental to the critical issue of food supply for Adelaide.

Agriculture is an important feature of the greater city of Adelaide, which stretches from the Barossa in the north to the Fleurieu Peninsula in the south, and from the Gulf St Vincent across to the Mount Lofty Ranges in the west. This includes seven of the South Australian government regions prescribed by Planning SA: southern, western, eastern and northern Adelaide, plus the Adelaide Hills, Barossa, Fleurieu Peninsula and Kangaroo Island. Forty per cent of this area is zoned agricultural, with much of the remainder once used for agriculture but now developed for housing, parks and gardens, industry or commerce. While the trend to build on fertile agricultural land continues, agriculture remains an important industry in the near-metropolitan regions.[1]

The diverse environments across the Adelaide and Mount Lofty Ranges catchments provide for the cultivation and supply of a wide spectrum of produce for a large cosmopolitan market, living and working in close proximity. This produce ranges from stone fruits (requiring chilling and high rainfall) in the Adelaide Hills, to market gardens in the north, or grapes and almonds in the south where altitudes are lower and seasons are drier yet where water may still be provided for irrigation. While Australian agriculture overall is strongly export-oriented, with over two thirds of annual farm produce exported (ranging from 98% and 51% for wool and dairy products respectively), this dominance is reversed for near-city producers, such as those around Adelaide, who are generally geared to provide for the ready market on their doorsteps, with an estimated 80% of market garden produce sold locally.

At the state level, South Australia's horticultural industries contribute around a quarter of total food revenue – $10 billion in 2006/2007. Of the state's horticultural food, 63% is sold on retail markets, 8% into the hospitality trade, 24% is exported interstate, and the remaining 5% was exported overseas.[2] Regionally, gross food revenue for the Adelaide Hills region in 2005/2006 was valued at $552 m. The major component of this was livestock ($200 m) followed by horticulture and wine (each some $120 m in value), field crops ($60 m), and dairy ($35 m). The figures for the Fleurieu region at the same time show a similar total

($526 m), but with wine representing 40% of this or $231 m in gross food revenue, and livestock back in second place at around $100 m. The Northern Adelaide region earns the greatest gross food revenue, estimated at $858 m for 2005/2006. Livestock and horticulture are the equal largest contributors at around $300 m each, followed by field crops at around $100 m, with wine only a small component.

With the trend to warmer and drier conditions across all areas but the Adelaide Hills, changes are being forced on established agricultural practices. This chapter explores these changes and the ongoing challenge of ensuring agricultural production continues to supply the many and varied needs of a hungry city.

Home gardens were once a dominant feature of life in Adelaide, and original subdivisions and home and garden layouts were specifically aligned for the provision of a productive vegetable patch and accompanying small orchard. The trend to subdividing blocks and doubling the size of dwellings combines to all but rule out backyard food production – an unfortunate situation in the face of a looming global food crisis and rising food prices. Yet for those able, home gardens can provide a productive source of fresh vegetables and fruit, and a healthy and rewarding pastime – backyard vegetable production is starting to make a comeback. Managing water about the home, be it main, rain or greywater, can be integrated with backyard food production in a microcosm of what is happening at the greater city level.

Water supplies

In a year of average rainfall, most (44.2%) of Adelaide's water is sourced from the catchments of the Adelaide Hills (Diagram 15.1). Water is also sourced from the River Murray (28.4%), groundwater (20.0%), recycled water (7.1%), and a small amount is collected in rainwater tanks (0.3%).[3] In a dry year, however, demand for water from the River Murray increases to 54.9%, as the overall demand for water also increases (Diagram 15.2). The Mount Lofty Ranges watershed contains significant water resources that support agriculture, industry and rural towns, along with important wetland and riverine environments. Increasing development in recent years has put pressure on these resources.

SA Water has primary responsibility for water provision in and around Adelaide and has been heavily involved in the Water Proofing Adelaide strategy,[3] which identifies sustainable sources of water for the future while encouraging water conservation and the responsible use of our water supplies.[4]

Reservoirs

Early European settlers to Adelaide initially relied on the River Torrens and nearby streams for water, and some inventive irrigation schemes resulted (see Box 67). Yet irregular supplies and a growing population soon highlighted the need for a more stable supply. A series of reservoirs were constructed throughout the Adelaide Hills, commencing in 1860 with the now decommissioned Thorndon Park Reservoir, followed by the Hope Valley Reservoir (1872) and a further eight reservoirs constructed between then and

Colonial irrigation (Eagle Terraces)

BOX 67

Market gardeners and orchardists in the Adelaide Hills during the 19th century depended on rain and springwater to irrigate their crops and, prior to the introduction of affordable metal pipes in the second half of the century, used centuries-old irrigation technologies.[1] A recent archaeological field survey identified a rare example of colonial irrigation technology in the valley of Ellison Creek in the Brownhill Creek Catchment.[2] This terraced irrigation system, now named the Eagle Terraces, is described here.

The Brownhill Creek Catchment in the Adelaide Hills had been identified by the first European settlers as one of the most fertile catchments close to Adelaide. Today, the historic market garden areas along Brownhill Creek are well known, but there was a market garden area at the headwaters of Ellison Creek and Brownhill Creek, below Crafers, that was also highly regarded; Ellison Creek was considered to be one of the most reliable sources of permanent water in the 19th century.

Only hints of colonial irrigation systems were described by 19th-century chroniclers of horticultural endeavours. A market garden and orchard visited and described by E.H. Hallack in 1893 was Mount Eagle, the well-established property of Mr John Mack, the then secretary of the SA Market Gardeners' Cooperative Society and the Market Gardeners' Horticultural Society:

'The area under cultivation, taking the undulating character of the country into consideration, does him great credit … There is also a spring on the southern slope of the hill, the water from which is used for tomato plots in the spring and vegetables in the autumn. Heavy crops of the latter are invariably obtainable by Mr Mack …'[3]

On this property, archaeological research has recorded one of the best preserved examples of colonial irrigation technology in the Adelaide Hills, and identified the spring referred to by Hallack. Title searches confirmed that Section 304 (title volume 5407, folio 605) belonged to John Mack. Extensive quarrying has destroyed much of the adjacent landscape – the farm house is now buried under sediment eroding from the quarry and the area below the terraces is overgrown. This damage was not apparent, however, when visited in 1974 by Claude Thorpe, a descendant of John Mack:

'Ancient cherry trees still survived on the steep slopes visible through the kitchen window. Across the creek I saw where wooden fluming was used to bring water from up the valley somewhere to water fruit trees or vegetables on the gentler slope.'[4]

A cluster of features was identified on a north-east facing slope including terraces cut into the hillside. Following a site survey and the excavation of a stone-lined water race, it was possible to understand the function of each feature as a part of an irrigation system. The first element of the irrigation complex to be identified was a wet area where a small creek ran across the pathway. Also recorded was a series of seven agricultural terraces covered by regenerating acacia and eucalypt species and a dry stone wall.

The terraces followed the contours on the slope and each terrace was between 50 and 60 m long. The width of the terraces ranged from <550 mm to 1200 mm. They were protected from heavy water runoff by a dry-stone wall and drainage ditch above terrace 1. The south-eastern end terminated in a small, steep gully down which the water was gravity-fed to the terraces.

Timber fluming and an in-ground water race appear to have carried water from the spring to the terraces. The western end of the race terminated in the same small gully as the eastern end of the ditch and dry-stone wall. The stonework associated with the water reticulation system that must have distributed water to the terraces had been disturbed and it was difficult to reconstruct.

The eastern end of the water race terminated within 10 m of the spring. The terrain at this point is steep and the underlying rockface is exposed. In these situations it was common to use timber fluming, however no evidence of the wooden race remains. The spring described by Hallack in 1893 is now choked by dense vegetation and is inaccessible.

The Eagle Terraces complex is a rare example of an early colonial landscape in which agricultural terraces, complete with a dry-stone wall drainage system and irrigation system, have survived for approximately 150 years. Its preservation is largely an outcome of the protection afforded by the Hills Face Zone legislation, and the location of the Eagle Terraces within the buffer zone around the former Eagle Quarry.

Pamela Smith

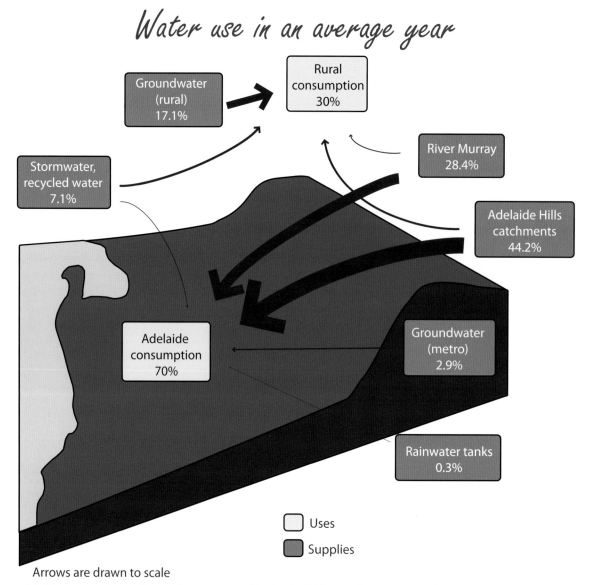

Water use in an average year

Groundwater (rural) 17.1%

Rural consumption 30%

Stormwater, recycled water 7.1%

River Murray 28.4%

Adelaide Hills catchments 44.2%

Adelaide consumption 70%

Groundwater (metro) 2.9%

Rainwater tanks 0.3%

Uses

Supplies

Arrows are drawn to scale

Diagram 15.1 Sources and uses of water (in percentages) in the Adelaide area during a year with average rainfall.
Source: adapted from Government of South Australia, 'Water proofing Adelaide: a thirst for change 2005–2025', 2005.

1977. At full capacity these reservoirs can hold 200,000 ML of water, almost a full year's supply for Adelaide. On average, however, they capture 120,000 ML while farm dams capture some 10,000 to 12,000 ML per year.

River Murray and filtration

To supplement Adelaide's steadily growing needs, pipelines from the River Murray were also required. The first was the Mannum–Adelaide Pipeline (completed in 1955), which flows to a terminal storage facility at Modbury and holds 136 ML. From there water is filtered at the Anstey Hill Water Filtration Plant (WFP) for provision into six separate reservoirs, as well as directly supplying some Hills residents. Adelaide's second pipeline, the Murray Bridge–Onkaparinga Pipeline, constructed in 1973 and designed to deliver 163,100 ML/year, discharges into the Onkaparinga River system, which carries water down to the Mount Bold Reservoir. Again, water supplied from here is filtered. WFPs now treat 98% of Adelaide water through six main plants. With more than half of Adelaide's water currently sourced from the River Murray in dry years, these pipelines and filtration plants are critical in meeting our needs.

Yet the Murray supply is now at an all-time low, recent seasons have been drier and demand continues to rise, albeit tempered by strong water restrictions. In response, SA Water is looking to innovative methods of supplying water, such as desalination, and reducing reliance on water supplies through greater recycling[5] and groundwater use.

Groundwater, recycled water and rainwater

In an average year, groundwater accounts for around 20% of the total water supply; however in some areas resources are stretched. This has resulted in much closer management in recent years and is continuing to develop. The groundwater resources of the Northern Adelaide Plains, for example, were first prescribed in 1976, and the current water allocation plan (WAP) for the region was adopted in December 2000. The process of prescription and water allocation planning is both time and resource intensive, involving significant discussions and consultation with the community, technical investigations, and other work. However, this is vital to ensure the best outcome for all water-users and the sustainability of the resource into the future.

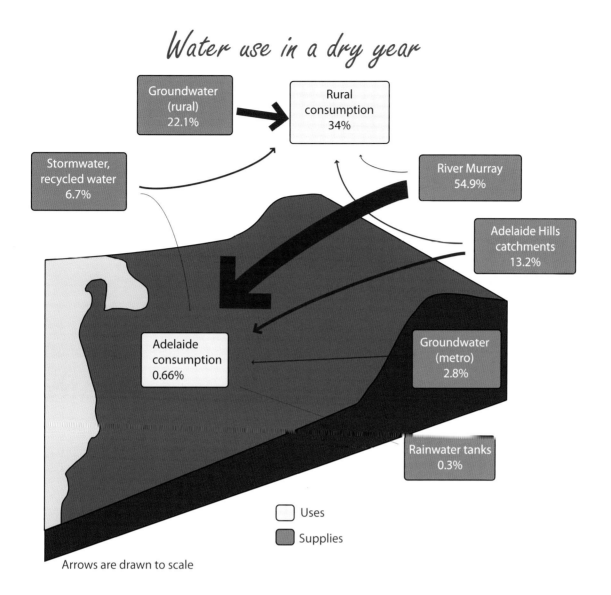

Water use in a dry year

Groundwater (rural) 22.1%

Rural consumption 34%

Stormwater, recycled water 6.7%

River Murray 54.9%

Adelaide Hills catchments 13.2%

Adelaide consumption 0.66%

Groundwater (metro) 2.8%

Rainwater tanks 0.3%

Uses
Supplies

Arrows are drawn to scale

Diagram 15.2 Sources and uses of water (in percentages) in the Adelaide area during a year with below average rainfall.
Source: adapted from Government of South Australia, 'Water proofing Adelaide', 2005.

Urban runoff and treated wastewater totalling over 100,000 ML is collected across Adelaide each year. Currently, the vast majority of this is discharged to Gulf St Vincent, with the remainder being used for agriculture. Some 18% of all effluent discharged from metropolitan treatment plants is now reused for agricultural irrigation. Several examples of this are described later in this chapter and in the associated boxes. Large-scale reuse is potentially very expensive but opportunities remain to use some of this water to replace mains water at reasonable cost for watering public gardens and industrial use. Recycled wastewater use has continued to increase in the past few years. Several small stormwater capture schemes are being used to supply industry and for watering parks and gardens. While some scope exists for expanding these systems, the major challenge will be to utilise, or store, winter flows when water volumes are high but demands are low.

Water availability
Water availability is an issue of great importance and interest to all water-users in and around Adelaide, and the activities of all users are interrelated. For example, the ongoing development of farm dams and increased groundwater extraction in upstream catchments can reduce the availability of water for Adelaide's water supply reservoirs. There are indications that increasing diversion of water to dams or pumping and increasing use of groundwater in the Adelaide Hills impacts not only on other users but on the environment. This is evident from the number of important wetlands, permanent waterbodies and native flora and fauna species under stress. Farm dams and reservoirs also reduce the amount of water available for environmental flows, having an impact on watercourses downstream of dam walls. This issue is discussed in the SA Water information sheet 'New Reservoirs in the Adelaide Hills'.[6]

Future development and the implications of predicted climate change may further reduce runoff to Adelaide's supply reservoirs in the long term. Equitable water resource management will be vital and can be achieved more effectively through controlling water extractions. If uncontrolled taking of water from all sources in the Adelaide Hills increases, there will not be enough to meet the long-term demand of existing users. This could result in insufficient supplies for human consumption, stock, irrigation, commercial use, and/or for the environment. With the gross value of food production in the regions around the City of Adelaide some $2 billion per annum, there is clearly a lot at stake.

Many private land and homeowners are looking to greater self-sufficiency in water supply, evidenced by a surge in rainwater tank installations in recent years. Scaling this philosophy up has seen at least one farming family 150 km from Adelaide taking water security into their own hands and installing a private desalination plant (see Box 71). Other agricultural-based production and processing businesses are also investing heavily to secure the supplies of quality water required for their operations, for example securing recycled water and boosting quality with their own filtration, as illustrated in Box 68.

Water quality

Areas with concentrated, mixed land use, which occur in and around cities, almost always will have issues around maintaining water quality. Inappropriate development or land use practices, for example, increase the risk of water pollution and decrease water quality. Risks may include overflow of human waste, spray drift of herbicide applications, and nutrient leaching from intensive agriculture. Also, with intensive land use there may be a reduction in the natural ability of watercourses and catchments to process some of the pollutants.

One body responsible for improving water quality through appropriate land management around Adelaide is the Adelaide and Mount Lofty Ranges Natural Resources Management (AMLRNRM) Board. Their water quality activities are focused on a 'whole-of-catchment' basis, and they support a range of projects to improve water quality and restore watercourses. These activities include protection of riparian zones through revegetation, bank stabilisation, fencing, weed removal and stock watering point control. In urban areas, their major efforts are focused on surfacewater degradation through stormwater runoff and litter management. A comprehensive review of surface and groundwater use is being undertaken and a range of measures are being implemented to help achieve responsible management of water use, including maintenance of water quality. Other groups with an interest in or responsibility for water quality management include the three tiers of government, as well as catchment management boards, industry and community groups.

Water use

In an average year, metropolitan Adelaide uses some 216,000 ML of water, with rural areas consuming an additional 94,000 ML, according to the 'Water Proofing Adelaide: a thirst for change 2005–2025' blueprint.[3] Water use in Adelaide and surrounds is split between primary production at 28%, residential 45%, public 17%, and commercial or industrial 10%.[7] (There is some variation in the breakdown of water use reported in Water Proofing Adelaide documents; primary production is stated as consuming 28%[7] whereas rural consumption is reported at 30%.[3])

Primary production includes market gardens, orchards, vineyards, pastures and livestock production. Livestock production may be intensive (mainly pigs and poultry) or extensive (dairy, beef or lamb). The residential component of water use includes water used for food production from home gardens as a minor component.

Water demand is increasing and is quickly approaching the levels of available supply. Based on current trends, it is estimated that by 2025 Adelaide's total water demand will be approaching 350,000 ML/year, the equivalent of 350,000 Olympic-sized swimming pools – exceeding our expected supply in dry years.[3] Moves are underway to address this, for example, with the construction of a water desalination plant to the south of Adelaide.

Production

Agricultural industries within the Water Proofing Adelaide boundary use some 94,000 ML/year, obtained from various sources, as shown in Table 15.1. This table highlights the importance of groundwater (56%), along with surface and recycled water, as compared to River Murray water (9%) for Adelaide's agriculture. It also shows how the Northern Adelaide Plains are making good use of recycled wastewater.

Processing

Water is used not only in primary production but is critical in most food- and fibre-processing systems where large volumes may be required. Food processing in the Adelaide Hills region, for example, in 2005/2006 was valued at $300 m wholesale value – the major components of which were wine (36%), livestock (35%) and horticulture (27%). In the neighbouring Fleurieu region, food processing was valued at $191 m for the same period, with wine representing 73%.[2]

Water use in processing varies greatly depending on the food product. For example, confectionary manufacturers will often work to exclude water from the processing area to avoid contamination of their products, using vacuum or other dry systems in their cleaning. At the other end of the spectrum, juice or dairy manufacturers with 'wet plants' will use water to move fluid products through the processing system and also to clean the plant at regular intervals. There is generally excess water produced which is either discarded to waste, diverted to suitable alternatives, such as watering a golf course, or treated for reuse. Some factories are well located to divert wastewater to alternative uses but many are not. In all cases it is important to look carefully at the full manufacturing and cleaning processes in order to minimise water use and maximise opportunities for reuse.

Water allocation

Water is a finite, precious and valuable resource; as such it is important to provide certainty of supply to current and future water-users, particularly those whose livelihoods depend on it. This can be provided through a WAP, which give consideration to environmental water requirements and economic development, as well as identifying areas where development places pressure on existing resources. WAPs aim to strike a balance to ensure that the allocation of the area's water resources is done fairly, taking into

The global average household water consumption per person per day is approximately 250 L, less if you live in a developing country, a little more if you live in a developed country. In Adelaide the figure is 388 L.

On top of this is approximately a further 5000 L per day per person to provide your food, about half of that if you eat a vegetarian diet.

CSIRO research indicates that if you eat a kilogram of steak once a week, that will require almost 1 million L of water per year, the equivalent of what four average Adelaide homes would consume in a full year.

Household water consumption is almost insignificant compared to the amount we consume in producing food. If the world is going to solve our water problems it needs to consider food production, making it efficient and optimising water use. Food is a necessity and the water to grow it is a necessity, and we need that food to be produced as efficiently as possible.

Some 75% of South Australia's water goes into agriculture and more than half of that (54%) is used to grow pasture and grains. While some farmers, particularly in the wine industry, have invested in technology and irrigation techniques to become water efficient, there are many that haven't.

The key point is that what food we eat, as well as how and where we grow it, makes a far bigger impact on our water usage than the water we use at home.

We have the knowledge and the technology to build an irrigation sector that can double the wealth obtained from using half the water. This may involve switching the crops we grow to those that South Australia has a relative advantage in growing.

An example of precision irrigation is an Adelaide-based company who produce and export beautiful, luscious, vine-ripened tomatoes. The precision irrigation story behind these tomatoes is amazing.

The company reuses city wastewater (blackwater that has been treated to a level where it is safe), purify it further by passing it through reverse osmosis, then add in a specific mineral diet customised precisely for tomatoes, which is fed to the tomato plants in a glasshouse where soil moisture monitoring technology delivers the precise amount of water precisely when it is needed to optimise root growth and plant yield.

Water is further supplemented by 30 ML of rainwater harvested off the roof every year. The glasshouse is more like an operating theatre of a hospital than a typical glasshouse; it is heated to a constant 25°C by using the waste heat of a onsite gas-fired cogeneration electricity plant. The exhaust from the cogeneration plant is harvested and the CO_2 is injected into the glasshouse, lifting CO_2 levels from 380 ppm (what they are out on the street today) to about 1200 ppm. As well as reducing global greenhouse gas emissions, the harvested CO_2 nourishes the tomatoes, who just love CO_2.

Now, the amazing part: this business has 12 times the normal plant yield, using 1/10 of the water, and sells year-round, consistent-quality tomatoes that consumers prefer over the field-grown product!

There are probably 30 or so South Australian technology and management suppliers that could have been involved in this tomato growing project, including companies that specialise in wastewater reuse treatment, desalination, glass-fibre pipelines, as well as soil moisture monitoring technology invented here in Adelaide.

Let me emphasise. Precision irrigation may or may not involve glasshouse production. The essential ingredient is optimising water use for efficient food production.

So, what could the law do to encourage efficient food production? Imagine if there was a regulated water-efficiency public labelling system – just like the star-rating for energy and water efficiency – that conveyed how efficiently food had been produced? What a great combination to add to South Australia's world-renowned food produce. Maybe a 'SA Great' label on food could become synonymous right around the world with water-efficient produce. It would give the public the knowledge to create a demand for water-efficient produce, while rewarding efficient farmers and irrigators.

Joe Flynn

Table 15.1 Agricultural water use (GL)

Geographic area	Ground-water	Surface-water	Reused waste-water	River Murray imports	Annual use (GL) by water use area	
Northern Adelaide Plains Prescribed Wells Area	18	0	13	0	31	33%
Barossa Prescribed Water Resources Area	4.4	2.8	0.1	8+	15.3	16%
McLaren Vale Prescribed Wells Area	6.0	1.6 *	2.5	0	10.1	11%
Mount Lofty Ranges watershed	25*	12*	~1	0	38	40%
Annual use by water source	53 (56%)	16 (17%)	17 (18%)	8 (9%)	94	100%

* Estimated consumption from unmetered water sources.
+ Some surface and River Murray water is supplied through the mains water supply system. The largest user of River Murray water for irrigation is the private Barossa Infrastructure Limited scheme, which currently provides 5 GL of water for irrigation in the Barossa Valley. SA Water also supplies about 3 GL of off-peak water to irrigators.
Table 15.1 Agricultural water use by region showing a breakdown of sources. Annual use in shown in gigalitres.
Source: adapted from Government of South Australia, 'Water proofing Adelaide', 2005.

account the needs of all water-users and the environment. This involves placing limits on how much water can be taken from each groundwater aquifer and river system. WAPs have been or are in the process of being developed for the following areas: Western Mount Lofty Ranges (WMLR); McLaren Vale Prescribed Wells Area (MVPWA); Northern Adelaide Plains Prescribed Wells Area (NAPPWA); the Barossa Prescribed Water Resources Area (BPWRA); and the Central Adelaide Groundwater Area (CAGA).

On completion of a WAP, allocations will be determined for the use of each groundwater aquifer and river system. These allocations will be determined using scientific means, taking into account available water supplies. In setting the allocations, the community and peak industry and consumer groups will be consulted to ensure that allocations are fair and equitable. Allocations will define the overall limits or caps on usage.

Regional water allocation plans
The AMLRNRM Board has responsibility for developing WAPs for the WMLR, the CAGA and the Northern Adelaide Plains (NAP). From October 2004, the WMLR, including the catchments of the five metropolitan Adelaide water supply reservoirs, was placed under a Notice of Prohibition and a Notice of Intent to Prescribe. These notices placed a freeze on any new or additional diversion of surfacewater and groundwater, except under specific circumstances. From late 2005, the minister for environment and conservation went on to prescribe the water resources of this area and commenced the process of formulating a WAP, through the AMLRNRM Board.[8]

The adjoining CAGA also supplies water for industry, agriculture, and the irrigation of recreational open space. Over time there has been increasing development and pressure on the underground water resources. In order to

protect them, the state government prescribed the wells in the CAGA in June 2007, following an extensive period of consultation and investigations into the economic and environmental implications. The AMLRNRM Board then began to develop a WAP for the CAGA, to implement change in a positive way for people living and working in the region. The intention is for underground water resources to be managed and developed in a sustainable manner for the benefit of irrigated agriculture, industry, the environment, and for future generations. This means that groundwater in the CAGA will be better protected and water-users will have greater security of supply. Prescription establishes a water allocation and licensing system to protect the resource from overuse and to ensure water-users are not adversely affected by other water-users. It also ensures that the environment gets enough water to maintain biodiversity and water-dependent ecosystems for generations to come.[9]

The current WAP for the northern region of Adelaide has been reviewed and a new plan is being developed incorporating a review of groundwater trends covering water levels, salinity and use. The new plan will then include a revised numerical groundwater model which will be used to define a sustainable groundwater yield for the region. The sustainable yield is likely to be based on managing the underground water resources to a defined range of pressures or water levels. Community consultation on the draft WAP is an important step before final adoption by the relevant state government minister. Further detail on the process involved and on progress towards developing the WAP can be found via the NAPPWA water allocation plan weblink,[10] or on similar sites for other regions. In general, water allocations can be traded between users, encouraging uptake of higher value and/or more water-efficient crops, as discussed in the next section (also see Box 72).

BOX
69

The hidden water in food 2

Virtual or embedded water is water used in the production of various foods and other goods, including water used in all parts of their production.

Generally, livestock products have higher virtual water content than crop products. This is because an animal uses water both directly and indirectly. In addition to drinking, water is used to produce animal feed (crops and pasture) and in other related farm activities.[1, 2] For example, 1 kg of beef produced in Australia requires more than 16,000 L of water.[1,2]

Examples of water embedded in other products are shown in the table below. (Reproduced from Chapagain and Hoekstra, 2004.)

Jennifer McKay

Embedded water in products

Product	Virtual water content (L)
1 glass of beer (250 ml)	75
1 glass of milk (200 ml)	200
1 cup of coffee (125 ml)	140
1 cup of tea (250 ml)	35
1 slice of bread (30 g)	40
1 slice of bread (30 mg) with cheese (10 g)	90
1 potato (100 g)	25
1 apple (100 g)	70
1 cotton T-shirt (medium sized, 500 g)	4100
1 sheet of A4 paper (80 g/m²)	10
1 glass of wine (125 ml)	120
1 glass of apple juice (200 ml)	190
1 glass of orange juice (200 ml)	170
1 bag of potato crisps (200 g)	185
1 egg (40 g)	135
1 hamburger (150 g)	2400
1 tomato (70g)	13
1 orange (100 g)	50
1 pair of shoes (bovine leather)	8000
1 microchip (92 g)	32

BOX 70

Irrigation water: use it or trade it because you can't save it

'He who controls the past, commands the future.'
– George Orwell

The issue
Large dams reduce water supply variability and provide access to water when we need it. Surprisingly, there has been little research on when to release water and when to store it. In most rural systems, water-users accept all the water that is given to them. The old adage was 'use it or lose it'; with trading, the adage has changed to 'use it or trade it as the government won't let you save it!'

Would Australia be better off if all irrigators were allowed to decide when to use water, when to trade water, and when to leave unused water on their account and carry it forward to the next season?

In the River Murray system, and as a once-off drought response measure, this year Victorian irrigators are being allowed to 'carry forward' any unused water. For many years, New South Wales general security irrigators have been allowed to carry forward water. For the first time ever, and as an emergency response measure, New South Wales high security entitlement holders have been allowed to carry forward unused water.

Should any irrigator be allowed to save water for the future, by leaving it on their account and having it automatically carried forward and made available to them in the next season? We think the answer is 'yes'.

Should carry forward accounting rules be consistent among states and among entitlements? We think the answer is 'yes'.

Allocation risk
In the past, storage management and inter-temporal decisions have been taken in the public arena. In the Murray-Darling Basin, states first agree on how much water to keep in an unallocated reserve and how much to allocate to each state's account, via a process that allows quite a large amount of flexibility. Each state then decides how much water to keep on their account and how much to allocate.

Recently several states made seasonal allocation announcements they could not honour. One of those not honoured was water that some New South Wales irrigators had carried forward from the previous irrigation season. We recognise that it has been announced that as soon as feasible water accounts will be re-credited with this water, but this decision has significantly reduced irrigator interest in carry forward.

In South Australia, provided there is storage space in the system, we assume that policy will be changed to define any seasonal allocation carried forward to be highly secure. Bank water, and after adjustment for storage losses you can use it whenever you like. If irrigators are to be encouraged to make astute water-saving decisions, they need to be confident that their storage decisions will not be undermined by government.

The economics of carry forward
The community impacts of not allowing irrigators to carry forward water are considerable. Donna Brennan has estimated that lack of carry forward in the 2002/2003 drought year cost Victoria's Goulburn Valley over $100 million. She also found that the maximum price paid for a seasonal water allocation was three times higher than it would otherwise have been if irrigators had been allowed to carry forward water in previous years. With less price volatility, an economist might predict that the value of water entitlements would rise.

Expanded trade versus supply reliability
In another paper, Brennan points out that expansion of interstate trade into New South Wales, where some carry forward is possible, will mean that less water will remain in Victoria's storage account. Pressing alarm bells, she observes that the expansion of interstate trade without the introduction of carry forward arrangements will make the Goulburn Valley worse off. Water previously left in the system and shared among Victorian irrigators will now be transferred to another account. This means that subsequent allocations to Goulburn Valley irrigators will be less than they otherwise would have been.

The solution, Brennan argues, is to give each individual irrigator the opportunity to choose between using, carrying forward, or trading water. If irrigators are allowed to trade water across space, and, with adjustment for losses, store water, both irrigators and the community should be better off.

We believe that all water-users in all states should be given the opportunity to optimise all water management decisions, including those concerning how much water is carried forward from year to year.

Who owns the airspace?

Just as it makes sense for irrigators to carry forward water, it also makes sense for an environmental manager to carry forward water. Given this observation, in a situation where a dam is full it is critical to work out whose water gets spilled first.

One simple way of providing access to airspace is to allow carry forward, or, as some states prefer to call it, 'carry over', on a first-up, best-dressed basis. This is fine while there is a lot of space in the dam, but when it gets close to full, there may not be enough space to give everyone their full allocation.

When a dam is nearly full, and if one thinks that storage priority should be issued in proportion to entitlement, it is possible to have a rule that carry-forward water is always the first to spill. This means that when storages approach full capacity, irrigators have an incentive to use or trade water rather than to save it.

Another way of allocating access to storage space in a dam is to set a limit on the maximum amount of water that may be carried forward. While it would be possible to unbundle this maximum storage right from an entitlement and make it tradeable, the benefits are likely to be minimal. In a free trading environment, irrigators can be expected to sell their water to someone who has access to unused storage in order to avoid losing an opportunity to profit from saving it.

One of Australia's most sophisticated carry-forward systems is the St George System in Queensland. Following trials, a continuous sharing approach has been developed with entitlement holders allocated a maximum storage right, and daily adjustments for evaporation losses are made to all unused allocations irrespective of when they were issued. To prevent carry forward causing delivery problems, the right to have water delivered within a season is capped. Customers who 'cap out' but still have water available within their storage share can purchase seasonal delivery rights from others.

Where to from here?

Failure to optimise storage management and water supply decisions is costing Australia a lot of money. Just as it makes sense to allow water trade within a season, it also makes sense to allow trade from one season to the next.

Obviously, all would benefit from more research on the question of how to optimise the storage and release of water, recognising the complexity of connected multiple storages and tributary inflow systems. To us, however, it seems obvious that the costs of not allowing individual participation in the process of deciding how to optimise the amount of water carried forward from year to year, and how much is used, are very high – too high.

Once the opportunity for private carry forward is introduced it will be necessary to review government storage and allocation policies. In particular, it may be necessary to ensure the introduction of carry forward does not increase the total amount of water used. In particular, some entitlement reliabilities and some allocation rules may have to be redefined.

From a Murray-Darling Basin perspective, this box raises an important question: 'Should the proposed new Murray Darling Basin Agreement be drafted in a manner that allows permanently for carry forward of unused allocations, or should this privilege remain limited to the few?'

Finally, under the current Murray-Darling Basin Agreement it is possible for South Australia to request access to Victorian and New South Wales dams. With the new commitment to managing the Murray-Darling Basin System as one, all irrigators – including those in South Australia – could be given an immediate opportunity to carry forward water.

Mike Young and James McColl

Table 15.2 Costs of water

Water source		Estimated costs per kilolitre
Farm dams	•	$3.00–$5.00
Groundwater (bore)	•	$1.00–$2.00
Mains	•	$1.03
Stormwater and wastewater	•	$0.40–$0.80

Table 15.2 Estimated costs of water per kilolitre from four different sources. Source: Water Proofing Adelaide information sheet: 'Sustainable water use for agricultural industries' (online).

Sustainable water use for agricultural industries

As stated in the 'Sustainable water use for agricultural industries' in 'Water Proofing Adelaide':

Sustainable water management is about ensuring that the volume of water extracted from a resource and how it is used, does not degrade the resource or the land. To maximise water savings, water efficient crops suited to our soils and climate need to be planted and a range of sustainable water management practices adopted. These practices include; irrigation scheduling, mulching and leakage minimisation and monitoring of rainfall, soil or plant moisture. In addition, greater utilisation of stormwater and wastewater will help to reduce the pressure placed on traditional water sources.

To read more about this topic see the Water Proofing Adelaide website.

Sustainable water management

The South Australian Department of Water, Land and Biodiversity Conservation's (DWLBC) primary objective is to improve sustainability through the integration and management of the state's natural resources, and to achieve improved health and productivity of our biodiversity, water, land and marine resources. A DWLBC information sheet[11] looks at potential water savings if more agricultural industries adopted sustainable water management, discussing:

the need to: ensure water resources are used within sustainable limits; implement irrigation management plans; adopt water saving technology and water efficient practices; and source new sustainable water.

Agricultural industries for this information sheet include cropping, floriculture, horticulture, viticulture, dairying and animal husbandry.

A large proportion of water in the Adelaide region is used for irrigated agriculture. It is estimated that water savings of up to 13,000 ML could be made by continuing improvements in water-use efficiency based on choice of crop, irrigation scheduling, behaviour modification, and control of leakage and evaporation in private watering systems.

Most agricultural water-users have in the past been able to take large quantities of groundwater or surfacewater without authorities charging them for the volume used. However, as Table 15.2 shows, such landholders do pay in part for the water, due to costs in establishing the bore or dam and in the costs of pumping. Currently, landowners are not required to measure how much they extract nor monitor changes in water quality or groundwater level. Further, riparian rights allow some landowners to pump water from creeks and streams with little or no restrictions or requirements to monitor volumes taken. However, landholders are discouraged from significantly over-watering when this would cause waterlogging, drainage problems, fertiliser wastage, and increased pest and disease problems. Water application rates may be compounded by salinity; where salts are present in water, high leaching fractions may be required to satisfy plant water needs.

There are both opportunities and limitations – by industry and region – to increase water-use efficiency in agriculture. Agriculture is continually changing and new, higher-value, water-efficient land uses continue to replace less profitable enterprises. For example, the replacement of some irrigated pastures with viticulture. Relative to viticulture, irrigated pasture is a high water-using crop with a low monetary return. However, at lower limits of availability, when water allocations are cut below what will sustain viticulture, dryland crops and pastures become the only viable agricultural option. Deep-rooted perennial plants such as lucerne have the potential to grow whenever water is available, and to persist in a semi-dormant state during periods when water is not available – making this plant a valuable tool for livestock or hay producers.

Critical to ensuring the long-term viability of the agricultural industries is ensuring that the water resources are used efficiently and managed within sustainable limits. This is achieved by gaining an understanding of the needs of a resource (e.g., river system) to remain healthy, and ensuring that the available water is appropriately allocated to all water-users. The process must include an effective consultation process and the collection of suitable data. Managing water resources in this way helps ensure that everyone receives their share and that the environment is protected.

Desalination for agriculture

BOX 71

Agricultural producers depending on the lower Murray River for water have, for the last two generations, believed that they were part of a highly dependable water source. Experience over the last five years has severely eroded this outlook to the extent that farmers depending on the Lower Lakes of the Murray River for 'farm water' have a real prospect of having no water available for stock or irrigation purposes. To illustrate the severity of the problem to users of the Lower Lakes, the maximum salinity suggested for human consumption is 800 EC units; most of the Lower Lakes is currently hovering around 3000 EC units to 5000 EC units. Salinity levels above this are entering the critical range for livestock.

Desalinating seawater is an expensive, high-energy process. However, some producers are now utilising brackish underground water for desalination, of which there are many benefits (as compared to seawater):

- much lower capital cost;
- much lower energy consumption and operating cost;
- much lower quantity of saline waste discharge; and
- the saline discharge is not hyper-saline (as is the case for seawater desalination). As such it could be used for aquaculture or reintroduced to a saline environment with little impact.

'Tauwitchere' is a 2300 ha property on the Narrung Peninsula that normally runs 350 milking cows, 350 beef cows, and has 600 ha of cropping. There is no mains water supply to this region. Prior to the 2002 drought, we were already canvassing options to either move our operation to a more secure water supply area or make our water supply independent of the River Murray. We initiated a strategy of reduced water use on irrigation in 2006. The solution for stock water proved more problematical. Bore holes yielded water but it was not suitable for large

numbers of livestock. So, what next? Desalination ('desal') became the only remaining option.

We do have an adequate supply of poor-quality water from either Lake Alexandrina or underground. At the time of writing (January 2008), our lake water is 4000 EC units, and we have underground water at 8000 to 10,000 EC units. We investigated the possibilities of desalinating this water via the reverse osmosis process, and it proved surprisingly affordable.

In our case, only a 7.5 kW multistage centrifugal pump operating at 900 kPa is required to produce 4500 L/hour of good-quality water from brackish input water. This quality water is produced at a fraction of the cost of mains water.

It is our hope that we will not be required to use the desalination plant on a permanent basis. Rather, it will provide a certainty of water supply into the future.

In the future, successful and sustainable agriculture will develop around those systems that:

a) use less water in total; and
b) can also effectively use alternative water supplies.

To further complicate agriculture's future, agriculture relies heavily on a healthy environment. The success of whatever we do on our farm at Tauwitchere is still heavily dependant on the macro water decisions taken at state and federal level. Water, currently diverted – for a multitude of purposes – out of ecosystems must be allowed to follow its natural path. This will mean sacrifices being made by all.

David Harvey

On an individual farm level, irrigation management plans are useful in reducing water wastage by ensuring that the right quantity of water is applied to a crop at the right time, in the right way. Without following such a plan there is the potential for landholders to over- or under-water their crops. Best-practice irrigation minimises drainage beyond the root zone while maximising yield. To implement a successful plan the landholder requires an understanding of how much water is being applied to a crop and how much water the crop needs. Additional research is needed to improve this knowledge. Some landholders may need assistance and/ or information to develop and implement an irrigation management plan.

Irrigation management plans can be integrated into an environmental management system, which provides a framework for businesses, including primary producers, to implement best management practices to reduce the impact of their operations on natural resources and ecosystems. An environmental management system strives for continuous environmental improvement through a 'plan, do, check, act' cycle, and has the potential to improve the financial and competitive position of an enterprise. Incentives for some primary producers to implement an environmental management system are available from the federal Department of Agriculture, Fisheries and Forestry.

Water-efficient practices

Water-saving technology, such as soil or plant moisture meters, can be expensive but can provide more accurate information on a crop's water requirements. Rain gauges can be connected to irrigation scheduling devices to prevent over-watering. Many water-efficient practices are simple and effective and can be adopted by existing water-users:

- Dairy: water-cooled heat exchangers used in dairies could potentially reuse the water several times before discarding it.
- Viticulture: adopt partial stressed vine irrigation, cover crop, mulch, increase the organic matter in the soil and monitor plant moisture.
- Horticulture: cover crop, mulch, increase the organic matter in soil, and monitor fertiliser and plant moisture.

Many of the required changes are already starting to occur within the industry as part of natural business processes, and some organisations are capitalising on a 'green' marketing image. It has been suggested that any involvement from government could be directed at supporting these businesses, and making the process of change easier for others. This could be achieved by developing or expanding on initiatives such as water-efficiency rebates, water audits, educational materials, changeover financial packages, or setting an enforceable industry water standard.

Increasing the use of new, sustainable water supplies, such as treated wastewater and stormwater and previously untapped groundwater, may allow new agricultural opportunities while still improving sustainable water management. Barriers to reuse include ensuring recycled water is of appropriate quality and the need for new infrastructure.[12, 13]

Environmental, social and economic issues around water for irrigation

There are several key environmental issues around the use of water for irrigated agriculture. On the downside, poor equipment and practices such as over-irrigating are not only wasteful but result in farm runoff and water polluted with salts and nutrients draining into creeks and local groundwater. Thankfully, however, improved irrigation technology and practices are becoming more commonplace, greatly increasing the efficiency of irrigation and substantially reducing farm runoff, as well as reducing energy use through pumping and groundwater extractions.

Another issue flagged is that the adoption of water-efficient crops may result in a monoculture regime, which may lead to increased disease or pest spread and increasing economic risk through a lack of diversity. Mixed crop regions offer greater diversity and therefore reduce the impact of pests and market fluctuations.

There are social implications around changes to water-use patterns. For example, if water-saving practices were mandated before a business was financially or operationally able to adopt them, there may be significant impacts on agricultural industries. This is occurring around the lower lakes of the Murray River where low river levels and increasing salinity have completely disrupted access to freshwater for any agricultural enterprise in the region – from dairying to vines. Training and education are important for all affected producers so they may consider their options and adopt efficient water use or other measures to improve their businesses. Increased efficiencies should make for a more competitive and viable industry, resulting in greater security for employees and investors. However, changes in crop or irrigation type to increase water efficiency may not be supported by some parts of the community; for example where the aesthetic value of one crop may be preferred over another, or when management practices impact on neighbours such as noise and spray drift.

There are economic implications in improving water use. In particular, mandating water-saving practices may result in short-term costs to landowners even though it is necessary to ensure the long-term viability of agriculture. Consideration could be given to those landholders who cannot afford to undertake a conversion to more water-efficient operations. In these situations, government financial support for agricultural businesses to improve water efficiency could result in significant costs and it may be difficult to design a water allocation system that ensures equity to all parties.

Planning for agriculture and the environment

BOX 72

Proper water planning is key to delivering improvements in agricultural productivity and environmental outcomes. The demands on water resources in the Adelaide and Mount Lofty Ranges Natural Resources Management (AMLRNRM) region to service a diverse array of social, economic and environmental benefits is arguably unique. Historically, these demands were particularly directed to delivering economic benefits to the community and the state; social (recreational) benefits were also important outcomes – while the needs of the environment were typically ignored.

In hindsight, decisions made in progressing economic and social development at the expense of the environment were naïve at best, with no complete understanding of aquifer behaviour and based on an assumption that the water resource was infinite. Knowledge of groundwater systems is still evolving, but we better understand that our catchments are finely balanced. Environmental benefits provided by water resources are not only vital to protect and enhance biodiversity, the health of the environment (water resources) is critical to the sustainable maintenance and development of irrigation-based industries, plantation forestry, industrial needs and urban consumption.

Improved productivity is built on and can coexist with good environmental outcomes – so long as planning processes are established that recognise and encourage the simultaneous delivery of multiple outcomes. This premise is central to the *South Australian Natural Resources Management Act 2004*, which is consistent with and delivers on sustainability targets in South Australia's Strategic Plan. The Act is underpinned by the National Water Initiative, a fundamental element in implementing sustainable and viable use of Australia's water resources.

Effective and sustainable management of water is dependent on a holistic approach to water allocation planning, based on the best available scientific knowledge of aquifer behaviour and water flow, connectivity between surface and groundwater, forestry impacts (recharge, extraction), urban impacts, agricultural impacts (dams, extraction), and environmental requirements, to set sustainable levels of extraction and use. Management processes must also be able to adjust to new information and be future-proofed to meet the challenge of climate change.

Agriculture and horticulture are a significant social and economic resource in the AMLRNRM region, occupying almost 50% of the region and producing a diverse array of agricultural products worth over $600 million in gross value. Despite the region only occupying around 1% of the state's land mass, the value of agriculture is approximately 14% of statewide gross value.

Enormous pressure is being placed on the viability and sustainability of agriculture in the AMLR – from urban and peri-urban expansion, plantation forestry, and the perception that agriculture is the primary culprit for the degradation of land and water resources. The pressure from urban development and plantation expansion not only impact on the area of valuable agricultural land available for food and fibre production, it also has consequences for the region's water resources. Proper water planning should account for impacts on recharge and direct extraction (plantation forestry), and interception of rainfall and surface water flows (housing and 'lifestyle' dams).

As a key catchment for water supply to the Adelaide metropolitan area, the Mount Lofty Ranges' capacity to supply Adelaide will dictate how much water is diverted from the River Murray. Urban development of all currently vacant rural allotments in the Mount Lofty Ranges Catchment, of which there are 3800 plus, would have serious implications to the integrity and sustainability of Adelaide's drinking water.

Agriculture is a legitimate water-user and has an entitlement to a reliable and long-term supply of good-quality water. Agriculture has recognised the need to improve its effective use of water. Irrigators accessing River Murray water have been moving towards 80% water-use efficiency, and similar efficiencies are being delivered by irrigation systems drawing from groundwater reserves. This has been aided by significant advances in irrigation technology, sustainable production systems, and crop varieties. Water allocation plans are also putting in place tighter controls on the diversion of surface water and the extraction of groundwater by agriculture.

Agriculture's long-term sustainability and viability in the AMLRNRM region is dependent on sound planning of water resources ensuring accessibility, sustainability and quality for all users in the region. Plantation forestry and urban development must also be a part of this integrated approach.

Nigel Long

BOX 73

Urban wastewater reuse and sustainable development

The concept of sustainable development is interpreted differently by specialists in different disciplines. Social scientists usually focus on social sustainability, economists deal with economic sustainability, and environmentalists deal with environmental sustainability. However, a holistic approach to understand sustainability involves dealing with all three dimensions. According to the Brundtland report (1987), sustainable development is the development that meets the needs of the present without compromising the ability of future generations to meet their own needs. While it is difficult to quantify sustainability, some indicators, such as economic profit, social benefits to the community, and environmental conservation, can be used to indirectly measure economic, social and environmental sustainability respectively. By extending these concepts to urban wastewater reuse in agriculture, it is plausible that a successful water reuse scheme can contribute to the sustainable development of the region in which it operates, as demonstrated in the case of the Virginia Pipeline Scheme (VPS) operating in the Northern Adelaide Plains.

Declining groundwater resources in the region and realisation by growers to use reclaimed water, along with the social, economic, and environmental drivers, led to the commissioning of the VPS in 1999. The scheme is built on the build, own, operate, transfer (BOOT) model and is the largest of its type in Australia. It is a cooperative undertaking of the Virginia Irrigation Association (VIA) representing the market gardeners, SA Water (public sector), and a private company (Water Reticulation Systems Virginia, WRSV). The scheme has successfully operated and supplied

Class A reclaimed water to approximately 250 growers operating within an area of 200 km² of the Virginia Triangle, also contributing to the sustainable development of the region.

The reclaimed water use data for the region indicates that there was an increase of 6100 ML over a period of five years, attributed to the growth of horticultural industry in the region. Horticulture comprises glasshouse-based enterprises and broad-acre vegetable concerns. Data indicate that the area under 'glasshouse' (an industry that supports approximately 600 growers) is expanding at a rate of 8 to 10% per year within the Northern Adelaide Plains. As the horticulture industry continues to expand, reclaimed water use will further increase.

In October 2005, local, state and federal governments announced funding of $4 million for the Virginia Pipeline extension from Bolivar to Angle Vale, ultimately increasing the reuse from Bolivar wastewater treatment plant by a further 3000 ML per annum.

Generally, water allocations add value to the land; good land for horticulture with water allocation fetches double the price of the land without water and improvements. The growers in the region expressed the view that reclaimed water has the same influence on land values as groundwater. Further, growers agreed that the nutrients in reclaimed water provide fertiliser value in crop production, helping to reduce the need for the purchase of off-farm fertilisers. Regarding the market for the produce grown using reclaimed water, the growers reported that the acceptance level of the produce grown

with reclaimed water was encouraging at all levels of the retail chain. This was made possible through communication campaigns carried out at different levels to train and educate the key stakeholders – industry, retailers, and the public. In addition, the wholesalers were kept informed of the development of the scheme and reassured that product quality would not be compromised. Endorsement of the scheme by the South Australian Department of Human Services and the Environment Protection Agency was also helpful in building up the confidence level of consumers regarding product quality.

Currently, with the domestic market being well supplied with fresh vegetables, and increased water supplies leading to increased production, there is scope for developing export markets in the area. All the indicators – expanding the horticulture industry, reduced off-farm fertiliser purchases, increased land values, and acceptable markets – indicate economic sustainability.

Social sustainability means maintaining social capital, which involves investments and services that create the basic framework for society, lowering the cost of working together and facilitating cooperation. Good communication, trust, mutual support and community cohesiveness are central to social sustainability. It is argued that division of irrigators by cultural and/or other social differences affects their capacity to communicate with one another. The VPS case refutes this argument; despite variations in ethnicity and cultural background, the irrigators were able to act collectively to find a solution to their problem of groundwater depletion, indicating the presence of a high degree of social capital within the community. Besides, the horticulture industry supports almost all the families in the community and, given that the industry will expand in the future, more people will take up farming. The expansion of the horticulture industry in turn has resulted in the creation of more job opportunities in the region. Horticulture being a labour-intensive industry has resulted in increased labour force for the agricultural sector in the region. The Australian Bureau of Statistics census data (2001) shows a 7.5% increase in the labour force after the implementation of the VPS. Apart from agricultural labour, more jobs will be created downstream in the packing, processing and marketing industries.

The commissioning of the VPS has resulted in reduced use of groundwater, evidenced by increased water levels of the two aquifers beneath the region. Furthermore, by using reclaimed water from Bolivar WWTP, the VPS has decreased the volume of nutrient-rich treated effluent entering St Vincent Gulf. Currently, the scheme has reduced the outfall by around 30% since 12,100 ML of flow from Bolivar is used for irrigation. Given the scheme capacity of 23,000 ML, the outfall can be further reduced to more than 50%. The scheme has helped to improve the marine environment and reduced the impact of the nutrients discharged on the gulf's seagrass. An issue faced with use of reclaimed water in agriculture in the long run is that it might result in adverse environmental hazards including soil salinisation caused by high sodium content in the irrigation water, which reduces soil permeability and creates an unsustainable environment for plant growth. Developing an irrigation management plan, which includes specific items such as water balance, subsurface drainage, and overall irrigation strategy, helped address this issue in case of the VPS. The risk of human exposure to pathogenic micro and macroorganisms present in the reclaimed water is minimised via training and awareness programs and an agreement with the growers to pump reclaimed water in lilac pipes, displaying sign boards on their properties to warn about the use of reclaimed water. These measures have reduced the health hazards to a large extent.

Use of reclaimed water for non-potable use (agricultural irrigation) has emerged as an alternative option to augment continuously depleting groundwater resources in the Virginia region. In addition, the VPS demonstrates that a reuse scheme can be instrumental in achieving sustainability of a region with its economic, social and environmental dimensions through the provision of knowledge and information on the 'current best practices', and communicating this information in a form that is understandable to the key stakeholder groups.

Ganesh Keremane

Research and extension to improve water use in agriculture

A variety of research and technology transfer activities aimed at improving water use in agriculture are underway in organisations such as the Department of Primary Industries and Resources SA (PIRSA).

Improved means to monitor and predict water requirements

Knowing that Australia's water resources are being tested by record drought and are likely to come under increased pressure with future climate change, the South Australian Research and Development Institute (SARDI) is undertaking a project to measure crop water use in a Riverland vineyard. The aim is to assess and improve upon the water-use monitoring and prediction methodologies currently available to irrigators. The findings will provide information for irrigation scheduling strategies to optimise irrigation efficiency and prevent deep drainage, while maintaining crop productivity. Amongst the findings, researchers have learnt that 40% of the water applied by drippers to a vineyard in the Riverland was lost to evaporation from the soil surface.

Advice for producers

PIRSA Rural Solutions SA consultants work with commercial operations to achieve a win/win for their business and the environment. In the case of vineyards, this means getting optimal results from their location, soil and water in order to create a plan to help grow their business. Better environmental management of the soils or water means better grape yields, better water use, and better profits. Rainfall data, lab tests, pesticides, sprays, and a host of environmental impacts can all be advised on with in-depth knowledge from a professional team. Rural Solutions SA has also been involved in stormwater capture and reuse design not only for agriculture but for public facilities such as the Oaklands railway station.

Improving irrigation on the Adelaide Plains

Adelaide Plains irrigators have been actively looking at ways to improve their irrigation practices to sustain their natural resources base as part of a national landcare project. Through collaboration between the Virginia Horticulture Centre and with combined funding from PIRSA Horticulture, the National Landcare Program and the AMLRNRM Board, the project has encouraged growers to monitor their irrigation practices and identify strategies that work best for their circumstances. This included testing new monitoring devices in highly saline areas and identifying what practices were most effective for particular crops, soil types and drainage conditions. Growers traditionally find it hard to invest in technology because their cost structure is so tight – so any new practices and monitoring devices must be affordable. This project has been important in demonstrating the benefits of new practices and convincing growers that such practices are good for their business.

Agricultural phosphorus and water

When considering the impact of agriculture on surface and groundwater, one of the elements of interest is phosphorus. Large quantities of phosphorus are often applied to irrigated or high-rainfall pastures in the Adelaide Hills and can result not only in strong pasture growth but in the unwanted export of phosphorus. This can lead to reduced water quality in streams as a result of increased algal growth. Warwick Dougherty in his PhD studies[14] looked at phosphorus accumulation, and the processes associated with its runoff in groundwater, in two soils, one at Flaxley near Mount Barker in South Australia and the other in New South Wales. He found that rainfall intensity and soil properties influence phosphorus levels in runoff – confirming the need to reduce the levels of phosphorus in the immediate topsoil in order to reduce runoff concentrations.

Improved water-use efficiency through differential watering

Improved water-use efficiency, early maturity in crops, and differential watering are all techniques that can reduce water use. SARDI scientists have investigated a technique known as partial root-zone drying, aimed at halving the quantity of water used to irrigate vines by watering only half of the vine then alternating to the other half the next time irrigation is applied, thus improving plant water-use efficiency.

New plants

Water conversion in agriculture can be improved with the development and adoption of new plant varieties, either through the introduction of promising new plants or plant breeding to develop new varieties of established plants. SARDI is investigating drought tolerance in lucerne and related plants to improve establishment and persistence of plant varieties under drought conditions. This should reduce the risk of poor establishment and performance of these fodder plants and may reduce the requirement for water in irrigated plantings. However, gains within a plant type are often small and hard-earned in terms of research investment, requiring careful consideration before projects are undertaken. Greater gains may arise from trying new plants altogether, and native plants are showing promise in research underway through the Future Farm Industries Cooperative Research Centre and related programs.

International conference

Improving agricultural plant performance under drought is a major theme for international agricultural science. Great advances have been made in recent years in understanding the molecular basis of plant response and tolerance to drought stress. Hundreds of drought-responsive genes have been identified and the cellular function of many has been resolved. Still, a huge gap remains between the findings at the molecular level and the application of this knowledge in the field. Excellent meetings have been dedicated to the cellular and molecular aspects of drought stress. However, there is a need in both public and private research sectors for crosstalk between disciplines involved with the molecular sciences and those seeking practical solutions to improve crop performance under drought conditions.

Tarac Technologies was established in 1929 by ex-CSIRO scientist Alfred Allen. One of Tarac's missions is to provide environmental solutions to the wine industry. Tarac has four processing plants, two in Nuriootpa (South Australia), one in Griffith (New South Wales) and one in Berri (South Australia). These plants process approximately 100,000 tonnes of grape marc, 35 million L of distillation wine, lees and de-sludge, and 5000 t of filter cake to recover grape alcohol, brandy, tartaric acid, grape seed and skin extracts/tannins. Tarac's closed-loop process ensures maximum utilisation of wine industry residuals.

In addition to providing waste management solutions to the wine industry, Tarac also owns a 30% share in North Para Environment Control (NPEC), who run a Waste Water Treatment Plant in the Barossa Valley. NPEC was established in 1975 with the aim of managing the effluent generated from wine industry businesses.

As the effluent streams increased and odour generation became a community and environmental issue, NPEC purchased a property west of Nuriootpa to dispose of untreated wastewater. In the early 1990s, a covered anaerobic lagoon (CAL) was built to provide some treatment of the waste stream prior to disposal.

The rapid expansion of the wine industry over the last 15 years and resultant increases in winery effluent has been matched by constant review and tightening by the Environmental Protection Agency of environmental standards and monitoring requirements. To keep pace, in 1999 the NPEC board decided to upgrade the effluent management process in an environmentally sustainable manner. The influent waste streams were segregated based on their strength and potential for reuse into two separate plants.

The 50 ML of concentrated streams produced each year from Tarac's manufacturing processes were diverted to the CAL. An activated sludge process was commissioned to complete the biological treatment of the organic components. The treated water retained its very high salt concentration and could not be reused. In 2005, Tarac installed a Microfiltration/Reverse Osmosis plant to harvest 30 ML per annum of this water for reuse as boiler feed.

An upgraded low biochemical oxygen demand treatment plant was commissioned in 2002. Since then, the plant has successfully treated increasing quantities of winery wastewater to a class A standard. Stormwater harvesting has added an additional 60 ML of dilution water to raise it to irrigation quality. The plant treats and stores 250 ML of wastewater annually and has the capacity to handle up to 600 ML annually.

To study the long-term impact on grape quality of irrigation with treated winery wastewater, in 2004 a long-term trial was initiated with the South Australian Research and Development Institute (SARDI). A block of shiraz grapes irrigated with NPEC water is being compared with a similar block being irrigated with River Murray water. Results to date are encouraging.

Recent water shortages have seen an increased demand for the treated water for irrigation purposes. However, because vineyards use drip irrigation, further media filtration is required. In addition there was a need to establish a distribution network. A federal government community water grant assisted NPEC in completing these projects in 2008. In the 2007/2008 irrigation season, seven vineyards used 95 ML of treated water.

Tarac is confident that ultimately all winery waste processed at NPEC will be returned to the vineyard in the form of high-quality irrigation water to achieve the optimal closed-loop solution, and minimise the impact of wine production on the environment.

Ira Pant

Water recycling process for the wine industry

Embedded water

'Embedded water' is a term to describe the volume of water required to produce a given product (see boxes 68 and 69). The water used in the production process of an agricultural or industrial product is called the 'virtual water' of the product. In general, livestock products have a higher virtual water content than crop products, as a live animal consumes a large quantity of feed and water in its lifetime before it produces some output. One table in Box 69 shows that 1 glass of milk requires 200 L of water while 1 hamburger takes 2400 L, while beer only requires 75 L of water – suggesting to some perhaps that we should consume more beer than milk!

In contrast to livestock, producing 1 kg of grain requires for instance 1000–2000 kg of water, equivalent to 1–2 m,^{3} while producing 1 kg of cheese requires 5000–5500 kg of water, and 1 kg of beef needs an average 16000 kg.[15] According to a recent study,[16] the production of a 32 megabyte computer chip of 2 g requires 32 kg of water.

Food miles is another concept, and addresses the transport costs associated with food production. This is not discussed in this chapter other than to say that food produced and consumed locally, in Adelaide or elsewhere, generally has a low transport cost and hence an energy-use advantage over distantly sourced foods.

Irrigation efficiency

Irrigators are being encouraged to become more efficient in an effort to reduce the levels of environmental damage due to irrigation, and to conserve Australia's precious water resources. PIRSA and SARDI scientists in the Riverland have developed a suite of indicators of irrigation performance, and assessed and compared sample groups of irrigation units. Levels of performance were compared within sample groups and with benchmark information from other sources, including international sources. Case studies of the best performers were used to develop best management practices for irrigation. The performance indicators took into account the volumes of irrigation applied and the drainage produced, and included measures of water cost, productivity and return from the irrigation enterprise. This enabled the overall efficiency of different irrigation enterprises to be compared, given appropriate relative values.

The benchmark performance indicators and best management practices developed provide the knowledge base for irrigators to assess their own irrigation efficiency, to compare themselves to other irrigators, and to make adjustments to their practices in order to improve their efficiency. The process of benchmarking, where irrigators compare themselves against other irrigators, can act as a powerful motivation in itself.

Irrigating with reclaimed water

Recycled wastewater or reclaimed water is proving to be very sustainable for irrigation, demonstrated by the Willunga Basin Water Company (WBWC) (see Box 76), which is even looking at a proposal to water parks and reserves in the Seaford area. The Adelaide Plains use reclaimed water to irrigate horticultural crops (through the Virginia Pipeline, serviced by SA Water's Bolivar wastewater treatment plant, WWTP).[17]

Large quantities of wastewater are generated in Australia each year, yet only a small proportion of this valuable resource is reused in agriculture and industry. For example, the reuse of treated sewage wastewater for irrigation ranges from 5.2% in the Northern Territory to 15.1% in South Australia.

Use of reclaimed water alleviates pressure on valuable water resources and reduces the amount of discharge of nutrients from sewage into the sea. One of the main concerns in the widespread adoption of reclaimed water is the impact on the soil environment and influence on soil pathogens. A study is reported in Box 76 that investigated the influence of reclaimed water on soil microbial activity and soil pathogens to support its use as a safe and viable alternative water source. Through studying the soils under vines irrigated with reclaimed water, it was found that reclaimed irrigation water did not adversely impact microbial activity of soils nor did it increase infection by potential root pathogens. There were actually less pathogens in the soil and indeed higher levels of microbial activity – providing a net benefit through nutrient transfer to the vine. Reclaimed water appears to be a viable means of alternative irrigation.

Reclaimed water first became available in the Willunga Basin region through the WBWC in 1999. Another SARDI study has shown that there is no difference in yield between vines irrigated with this reclaimed water and vines irrigated with mains water.

The WBWC takes treated water from SA Water's Christies Beach WWTP, 10 km north of the Willunga Basin, and pumps it via 70 km of pipeline to more than 90 users whose properties cover more than 1500 ha. The Christies Beach WWTP treats about 10,000 ML of wastewater a year and about a third of that is being used by the WBWC for irrigators. The remaining treated wastewater is pumped out to sea. The WBWC will eventually have the capacity to take most of the wastewater from the WWTP. Better still, the wastewater issuing from wineries can also be collected, treated and recycled back to the wineries for reuse (see Box 74).

Tarac Technologies is another company with water-processing plants. It has four, strategically positioned in the key wine-growing regions of Australia – two in Nuriootpa in Barossa Valley and one each in Griffith in New South Wales and Berri in South Australia's Riverland. Through these, Tarac services approximately two thirds (by volume) of the Australian wine industry. Tarac currently processes approximately 100,000 t of grape marc, in excess of 35 million L of distillation wine, lees and de-sludge, and 5000 t of filter cake to recover grape alcohol, brandy, tartaric acid and grape seed, and skin extracts/tannins.

Gemtree Wetlands

BOX 75

Gemtree Wetlands is situated in the McLaren Vale wine region, and is a community asset being developed by Gemtree Vineyards with the assistance of Greening Australia. Gemtree's aim for the wetland is for it to serve as an example of effective land management by the wine industry while creating much needed habitat for native fauna. The ultimate goal is to engage the local community in environmental rehabilitation and rebuild the habitat back to its original condition, prior to European settlement, thereby attracting native fauna including endemic frog species .

Four local native frog species have been recorded in the Onkaparinga Catchment area: the common froglet (*Crinia signifera*); brown tree frog (*Litoria ewingi*); eastern banjo frog (*Limnodynastes dumerili*); and the spotted grass frog (*Limnodynastes tasmaniensis*).

The specific habitat requirements of these frogs vary, but all require waterbodies with good water quality, that persist for at least three months of the year. Most also require good-quality native streamside vegetation.

Gemtree Vineyards had the environmental foresight to establish the wetlands site in 1998. The site sits adjacent to vineyards within the Pedler Creek Catchment, with the creekline extending through the property. Once a degraded area, the 10 ha site has been rejuvenated with the establishment of six interlinking dams and the planting of over 25,000 native trees and plants.

The wetlands are nestled in the heart of the Gemtree Vineyards, and viticulturist Melissa Brown believes the native vegetation enhances the viticulture by providing alternative habitats for beneficial predators, decreasing the need for chemical sprays on the vines.

The creekline had become severely eroded, choked with woody weeds, and had been used as a rubbish dump under previous ownership. Gemtree, together with the assistance of Greening Australia and the Natural Resources Management Board, have undertaken extensive works on this creekline to remove the rubbish, eradicate the weeds, and revegetate the area with natives. This has facilitated unimpeded water flow through to the Pedler Creek Catchment area.

In 2008, Gemtree will be introducing recycled water to the vineyards in order to reduce the reliance on groundwater resources. The use of recycled water on the site fits well with environmental initiatives the company is undertaking. Using treated wastewater on the vineyards will take pressure off the aquifer and reduce the amount of waste being released into the ocean.

Melissa Brown

Water Proofing Northern Adelaide

Water Proofing Northern Adelaide is a joint project to improve urban water management in Adelaide's northern region. The cities of Playford, Salisbury and Tea Tree Gully are working together with the help of a $41.8 million grant from the federal government's Water Smart Australia program, as well as funding from the state government, the Natural Resource Management board and private partners. The goal of the project is to complete a system to sustainably manage the catchments and water in the urban areas, and demonstrate additional innovations on current stormwater management best practice.

In the City of Playford, Water Proofing Northern Adelaide will help provide up to 80% of irrigation water needs. Key projects include developing major wetland sites and associated aquifer storage and recovery (ASR) facilities at Munno Para West, Andrews Farm, Andrews Farm South and Adams Creek, along with a separate network of community bores adjacent to irrigation sites to extract water from the aquifer.

The project integrates into deliverable form the proposals of the state water plan, the catchment management plan and Water Proofing Adelaide, while delivering infrastructure to implement the National Water Initiative's objectives. Specific objectives include producing a sustained stormwater harvest of 18 GL/annum, fit for use in irrigation of community open spaces, industrial process water, residential non-potable uses and commercial irrigation. They also include sustaining riverine linear parks, and injecting water into aquifers for storage and reuse.

The Andrews Farm ASR site has been operating since late 2000, injecting and recycling between 80 and 120 ML of stormwater annually. A schematic on ASR from the SA Water website is provided to illustrate the principles (Diagram 15.3).

The Virginia Pipeline Scheme

Declining groundwater resources in the region led to the commissioning of the Virginia Pipeline Scheme (VPS) in 1999. The scheme is built on the build, own, operate, transfer (BOOT) model and is the largest of its type in Australia. It is a cooperative undertaking of the Virginia Irrigation Association (VIA) representing the market gardeners, SA Water (public sector), and a private company (Water Reticulation Systems Virginia, WRSV). The VPS has successfully operated and supplied class A reclaimed water to approximately 250 growers operating within an area of 200 km^2 of the Virginia Triangle, and has contributed to the sustainable development of the region.

Horticulture comprises glasshouse-based enterprises and broadacre vegetable concerns. Data indicate that the area under 'glasshouse' (an industry which supports approximately 600 growers) is expanding at a rate of 8 to 10% per year within the Northern Adelaide Plains. As the horticulture industry continues to expand, reclaimed water use will further increase. Local, state and federal governments announced funding of $4 million in October 2005 for the Virginia Pipeline extension from Bolivar to Angle Vale, ultimately increasing reuse from Bolivar WWTP by a further 3000 ML/annum.

The use of reclaimed water for potable and non-potable use has emerged as an alternative option to augment continuously depleting freshwater supplies in the Virginia region. In addition, the VPS demonstrates that a reuse scheme can be instrumental in achieving sustainability of a region with its economic, social and environmental dimensions through provision and communication of knowledge and information on the current best practices, understandable to the key stakeholder groups.

Aquifer storage and recovery

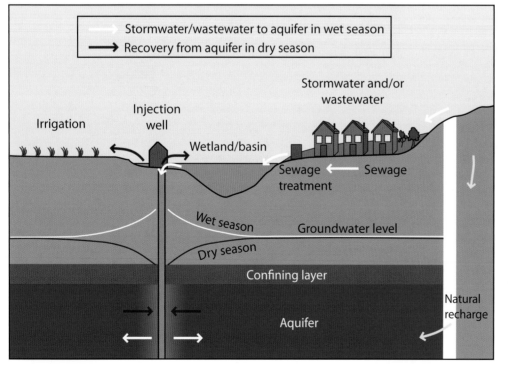

Diagram 15.3 Aquifer storage and recovery during wet and dry seasons. Source: adapted from P. Dillon and P. Pavelic, 'Guidelines on the quality of stormwater for injection into aquifers for storage and reuse', Urban Water Research Association of Australia research report no. 109, 1996.

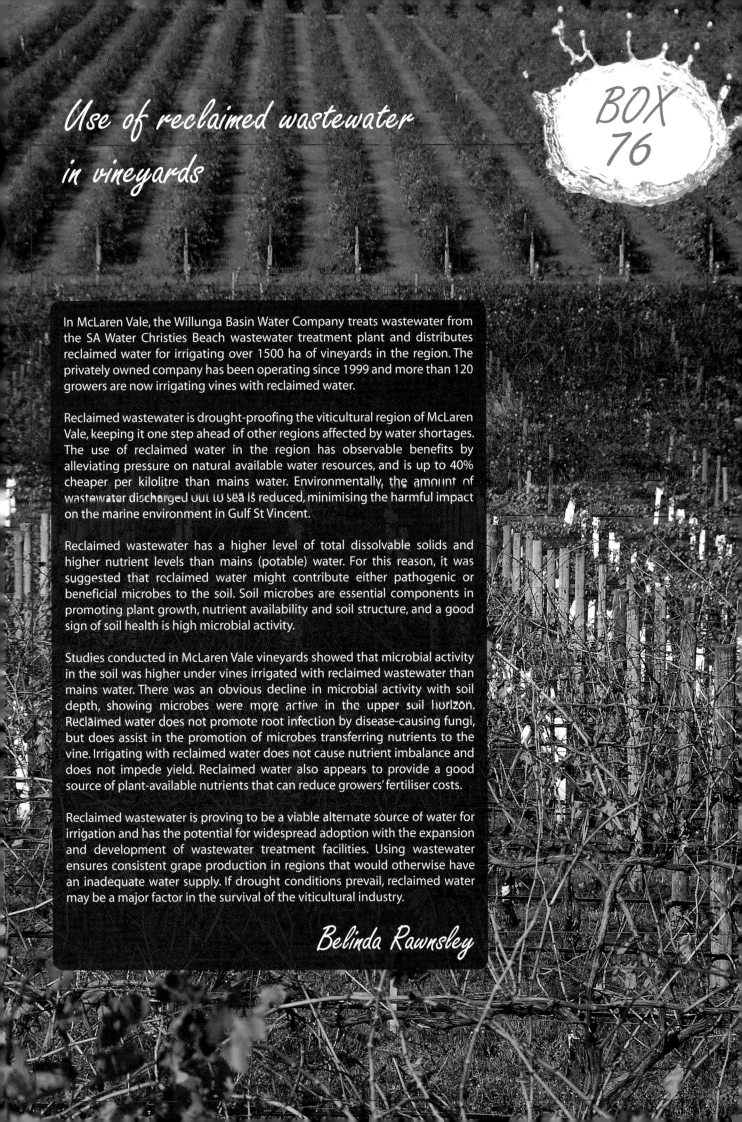

Use of reclaimed wastewater in vineyards

BOX 76

In McLaren Vale, the Willunga Basin Water Company treats wastewater from the SA Water Christies Beach wastewater treatment plant and distributes reclaimed water for irrigating over 1500 ha of vineyards in the region. The privately owned company has been operating since 1999 and more than 120 growers are now irrigating vines with reclaimed water.

Reclaimed wastewater is drought-proofing the viticultural region of McLaren Vale, keeping it one step ahead of other regions affected by water shortages. The use of reclaimed water in the region has observable benefits by alleviating pressure on natural available water resources, and is up to 40% cheaper per kilolitre than mains water. Environmentally, the amount of wastewater discharged out to sea is reduced, minimising the harmful impact on the marine environment in Gulf St Vincent.

Reclaimed wastewater has a higher level of total dissolvable solids and higher nutrient levels than mains (potable) water. For this reason, it was suggested that reclaimed water might contribute either pathogenic or beneficial microbes to the soil. Soil microbes are essential components in promoting plant growth, nutrient availability and soil structure, and a good sign of soil health is high microbial activity.

Studies conducted in McLaren Vale vineyards showed that microbial activity in the soil was higher under vines irrigated with reclaimed wastewater than mains water. There was an obvious decline in microbial activity with soil depth, showing microbes were more active in the upper soil horizon. Reclaimed water does not promote root infection by disease-causing fungi, but does assist in the promotion of microbes transferring nutrients to the vine. Irrigating with reclaimed water does not cause nutrient imbalance and does not impede yield. Reclaimed water also appears to provide a good source of plant-available nutrients that can reduce growers' fertiliser costs.

Reclaimed wastewater is proving to be a viable alternate source of water for irrigation and has the potential for widespread adoption with the expansion and development of wastewater treatment facilities. Using wastewater ensures consistent grape production in regions that would otherwise have an inadequate water supply. If drought conditions prevail, reclaimed water may be a major factor in the survival of the viticultural industry.

Belinda Rawnsley

The Gemtree Wetlands project

Gemtree Wetlands is situated in the McLaren Vale wine region and is a community asset being developed by Gemtree Vineyards with the assistance of Greening Australia (see Box 75).

Stormwater management

There are a number of active water-related associations in South Australia (the Australian Water Association, the Stormwater Industry Association, Engineers Australia Environmental Society, the Hydrological Society of South Australia, and the Limnology Society of Australia), with one specifically focused on stormwater management, the Stormwater Industry Association in South Australia. Their website explains:

> Stormwater practice is going through a period of rapid and fundamental change brought about by advances in technology, environmental awareness and the need for water conservation. Whereas previously, stormwater was perceived only as a flood threat and a waste product to be conveyed as quickly as possible beyond the urban area, it is now increasingly recognised as a valuable resource which is best managed by slowing its flow and capturing and retaining its resource values within the urban area. This change has spawned a new set of stormwater industries which are still in their initial stages of development and which have great potential for bringing a new range of services to urban dwellers in water supply, flood control, pollution reduction, amenity and biodiversity.[18]

Tools for water management

Developing tools for water management is a growth area; one body active in this area is the eWater Cooperative Research Centre (CRC). This is a technology development initiative set up by the Australia's water resource management and research sector where the core business is building water management tools for partners and bringing those tools to Australian and international markets.

The eWater CRC product portfolio includes tools for integrated systems for efficient urban water management, tools for developing monitoring programs, and models for joint management of surface and groundwater. The CRC is a joint venture made up of 45 of Australia's leading watercycle management, consulting and research organisations, supported by the Australian government's Cooperative Research Centres program. Urban water management tools will allow users to consider concurrently stormwater, wastewater, rainfall, rainwater and stormwater harvesting, infrastructure, wetland filters, groundwater and the ecology of waterways. Water authorities and businesses, working on new and existing developments, will be able to implement sustainable urban development from a local to a whole-of-catchment context, minimising costs and maximising environmental benefits. Planners will be able to use these tools to evaluate the costs and benefits of major infrastructure proposals.

Conclusion

This chapter aims to provide food for thought for readers – such food wouldn't have helped Oliver Twist's hunger pains, but then we are fortunate we live in more plentiful times! Water and its wise use is fundamental to those good times and we are all challenged now to maintain and improve our circumstances through actively engaging in the topic of water.

Water is a shared community resource where the activities of one group of users impact on others downstream. Adelaide is in a more perilous position than most major cities, being highly reliant on the tail-end of an overstretched and undersupplied River Murray, especially during dry years (Diagram 15.2). The finite limitations of this river and our other water sources, and the restrictions imposed by climate and rainfall, are challenges for us all.

Readers are urged to be aware of the big picture; issues surrounding not only our individual home water supplies but also those which surround water use by the broader community, be it for agriculture, the environment, commerce or industry. Be aware also of where our water is sourced, how it is used, and how it can best be conserved. Consider the relationship between water, domestic and industrial use, food production and an aesthetic environment in Greater Adelaide.

It is then easy to support and encourage those who work hard to adopt sound water management practices, to educate others, and lead by example wherever possible. Children can be taught the basis of agriculture and responsible and productive water use at home, as well as some fundamental recycling of greywater and the conservation and use of rainwater.

It is interesting to see the flurry of rainwater tank installation around Adelaide and surrounds. What is less clear is how the collected water will be utilised about the home. Once full, a rainwater tank collects no further water, therefore it needs to be in constant use so long as it contains water. Vegetable production is a great way to put some of this water to good use.

Enjoy the rewards of water awareness, conservation and sound governance, as, used wisely, water will maintain our fresh and productive city of Adelaide.

References

1. N. Schellhorn and G. Auricht, 'Utilising urban land 4: urban agriculture' in Daniels, C.B. and Tait, C.J. (eds) *Adelaide, nature of a city: the ecology of a dynamic city from 1836 to 2036*, BioCity: Centre for Urban Habitats, Adelaide, 2005.
2. SA Food Centre, 'South Australian regional score cards', 2008 (online).
3. Government of South Australia, 'Water proofing Adelaide: a thirst for change 2005–2025', 2005 (online).
4. M. Rann, (2008) 'Water proofing Adelaide', South Australian Government, 2008 (online).
5. South Australian Water Corporation, 'SA Water annual report 2005/2006', 2006 (online).
6. Government of South Australia, 'New reservoirs in the Adelaide Hills, information sheet', 2007 (online).
7. Government of South Australia, 'Water proofing Adelaide: exploring the issues, a discussion paper', 2004 (online).
8. Adelaide and Mount Lofty Ranges Natural Resources Management Board, 'Fact sheet: groundwater trends in the Northern Adelaide Plains 2004–2007', 2007 (online).
9. Department of Water, Land and Biodiversity Conservation, 'Fact sheet 79: central Adelaide groundwater area: prescription of groundwater resources', 2007 (online).
10. Adelaide and Mount Lofty Ranges Natural Resources Management Board, 'Northern Adelaide Plains Prescribed Wells Area: water allocation plan', 2008 (online).
11. Government of South Australia, 'Sustainable water use for agricultural industries, information sheet', 2007 (online).
12. Government of South Australia, 'Stormwater reuse, information sheet', 2007 (online).
13. Government of South Australia, 'Large-scale wastewater reuse, information sheet', 2007 (online).
14. W.J. Dougherty, 'The mobilisation of soil phosphorus in surface runoff from intensively manged pastures in south-east Australia', PhD thesis University of Adelaide, Adelaide, 2006.
15. A.Y. Hoekstra, 'Virtual water: an introduction' in Hoekstra, A.Y. (ed.) *Virtual water trade: proceedings of the International Expert Meeting on Virtual Water Trade*, Value of Water Research report series no. 12, 2003, pp. 13–23 (online).
16. E. Williams, R. Ayres and M. Heller, 'The 1.7 kilogram microchip: energy and material use in the production of semiconductor devices', *Environmental Science and Technology*, 36 (24), 2002, pp. 5504–5510.
17. Willunga Basin Water Company, 'Welcome', 2008 (online).
18. Stormwater Industry Association South Australia, 'About SIA(SA)' (online).

Box references

Box 67: Colonial irrigation (Eagle Terraces)

1. R. Cunningham, *Hints for Australian emigrants: with engravings and explanatory descriptions of the water-raising wheels, and modes of irrigating*, T. and W. Boone, London, 1841.
2. P.A. Smith, 'Dry stone walls and water races' in Smith, P.A., Pate, F.D. and Martin, R. (eds) *Valleys of stone: the archaeology and history of Adelaide's Hills Face*, Kōpi Books, Belair, South Australia, 2006, pp. 69–92.
3. No author, *Observer*, 13 May 1893, p. 3.
4. C. Thorpe, 'The Thorpe family in South Australia 1840–1930 and beyond', unpublished memoirs of Claude Thorpe held by Joy McDonald, n.d.

Box 69: The hidden water in food 2

1. A.K. Chapagain and A.Y. Hoekstra, 'Water footprints of nations, vol. 1', main report, *Value of Water research series*, no 16, UNESCO IHE, 2004.
2. Lenntech, 'Use of water in food and agriculture', Lenntech Water Treatment and Air Purification (online).

Box 73: Urban wastewater reuse and sustainable development

1. R. Goodland, 'Sustainability: human, social, economic and environmental' in Timmerman, P. (ed.) *Social and economic dimensions of global environmental change, Encyclopedia of Global Environmental Change*, John Wiley and Sons Ltd, New York, 2002, pp. 489–491.

Further reading

1. Adelaide and Mount Lofty Ranges Natural Resources Management Board, 'Adelaide and Mount Lofty Ranges water allocation plans', 2008 (online).
2. Department of Environment, Water, Heritage and the Arts, 'Virigina Pipeline Scheme extension', 2009 (online).
3. eWater CRC, homepage (online).
4. Interdrought III, conference homepage (online).
5. Department of Primary Industries and Resources of South Australia, Rural Solutions SA homepage (online).
6. Department of Primary Industries and Resources of South Australia, SA drought e-news (online).
7. Department of Primary Industries and Resources of South Australia Horticulture, 'Improving irrigation on the Adelaide Plains', homepage.
8. Playford Council, 'Water Proofing Northern Adelaide', 2008 (online).
9. South Australian Research and Development Institute, 'Viticulture, root zone drying', 2009 (online).
10. South Australian Water Corporation, 'Our water systems: history', 2004 (online).
11. South Australian Water Corporation, 'Virginia Pipeline Scheme', 2009 (online).
12. Government of South Australia, 'Water proofing Adelaide: sustainable water use for agricultural industries' (online).

CASE STUDY 5
Recycling urban wastewater, an alternative source for agricultural irrigation

Ganesh Keremane

Exponential growth of population, rapid industrialisation and urbanisation, higher cultivation intensities, and poor water management practices have made freshwater availability a limiting factor in agricultural development.[1,2] Also, the options for increasing supply have become expensive and often environmentally damaging.[3] Therefore, we need to manage the available freshwater resources effectively and use water based on fitness-for-purpose criteria. Our actions to counter water scarcity challenges should be sustainable, without depleting the natural resources or harming the environment. In such situations, 'source substitution' appears to be the solution, as it allows higher-quality water to be reserved for domestic supply and poor-quality water for less critical uses.[4] Consequently, urban wastewater (treated) is now considered a reliable alternative water source without compromising public health, and wastewater management is prominent in the water management agenda of many countries.[5,6]

Recycling water in Australia

The scope for Australia to recycle water was first identified during 1977–1978 in a report commissioned for the Victorian Government.[7] However, this failed to attract the attention of policymakers and hence had little impact until the 1980s, when issues of environmental health, sustainability, water availability and water quality for consumptive uses emerged as significant political issues.[8]

Australia is currently experiencing the highest ever amount of pressure on its water resources. It has been stated that 'substitution of water used in agriculture and urban irrigation with reclaimed water will free up water and help make appropriate allocations to the environment, thus ensuring good environmental condition for stressed water supplies'.[9] This means that reclaimed water can definitely become a major resource for the agriculture sector (irrigated agriculture accounts for around 67% of Australia's total water usage).[10]

Water recycling was given impetus in the early 1990s when the states established Environment Protection Authorities, which imposed compositional standards on the discharge of treated effluents from wastewater treatment plants (WTTPs) to the oceans. As a result, interest increased in recycling for productive purposes on land as an alternative to installing expensive biological nutrient-removal plants. Further, the water sector reforms of 1994, in the form of Council of Australian Governments reforms, provided policy direction and gave momentum to water recycling in Australia. The droughts of 2001–2003 reinforced the need for more effective water management, with recycled wastewater, urban stormwater and rainwater being seen as resources rather than problems. A recent review on water recycling in Australia by the Australian Academy of Technological Sciences and Engineering reported that by 2001–2002 over 500 WWTPs were recycling some or all of their treated wastewater.[11]

Water recycling was brought within the National Water Reform Framework in 2003, an intergovernmental agreement aimed at encouraging water conservation in cities through better use of stormwater and recycled water.[12] The subsequent signing of the Intergovernmental Agreement on the National Water Initiative, and the creation of the Australian Government Water Fund, laid the foundation for encouraging innovation and the use of recycled water in Australia's cities and towns. Thus, a series of events in the late 1990s provided powerful incentives for cities and town to consider including water recycling in their water development plans, and ultimately converged to accelerate implementation of water recycling.[13] Given that the current freshwater resource status in the country is bleak, water recycling certainly has a role to play.

Annually, large quantities of wastewater are generated in Australia (Table A) but only a small proportion is reused in agriculture and industries. The reuse of treated sewage wastewater for irrigation ranges from 5.2% in the Northern Territory to 15.1% in South Australia.[11] Nevertheless, Australia is a world leader in the use of treated wastewater (Dimitriadis, 2005). According to Australian Bureau of Statistics (2004) data for 2000–2001, around 27.8% (511.3 GL) of the total volume of effluent produced (1837.2 GL) was reclaimed. (Because of differences in definitions, estimates of reclaimed water use can vary considerably.[9]) A number of successful wastewater irrigation schemes have been developed in Australia (Table A).

Apart from the schemes mentioned in Table A, Australia has several other schemes operating successfully, as most state governments and water authorities have policies on reuse and are devoting efforts to develop new applications. Most wastewater irrigation schemes in Australia are formal; two examples are described below.

Brighton irrigation scheme, Hobart[11]

Brighton is a dry area near Hobart, with significant broad-acre farming. In an attempt to drought-proof the area, Brighton Council is now recycling all its wastewater (800 to 900 ML/year), which is reclaimed from residential properties connected to sewerage in the areas Old Beach, Gagebrook, Bridgewater, and the Brighton township itself.

The Brighton lagoon recycling system was established in 1996 and since then treated effluent has been used for irrigation of poppies, cereals and pasture on the neighbouring farm. In 1997, the council formed a joint venture with the local pulp mill at Boyer to establish a 17 ha irrigated pine plantation. The success of the Brighton lagoon recycling system enabled council and local farmers to secure funding to establish infrastructure required to also recycle treated effluent from the Green Point WWTP. Participating farmers agreed to install and pay for suitable storage and irrigation infrastructure. The Brighton Council, with the help of a National Heritage Trust Coasts and Clean Seas program grant of $788,000, paid for the recycled water distribution network. Farmers were irrigating with recycling water 12 months after project funding had been announced, leading to an initial reduction of the demand for potable water by 20%. Apart from the recycled water being an important and affordable source of water to local farmers, the scheme also led to a significantly reduced discharge of nutrients (nitrogen and phosphorus) into the Derwent River, thereby helping farmers save on fertiliser.

Table A Water reuse projects

Project	Annual volume (ML)	Water quality*	Application
Virginia (SA)	22,000	A	Unrestricted irrigation of horticultural crops
South-east Queensland	100,000+	A and C	Class A to horticultural crops, class C to cotton and cereal farms
Hunter Water	Up to 3000	C and B	Coal washing, electricity generator cooling
Eastern Irrigation Scheme	10,000	C	Horticulture, public spaces, golf courses, household gardens and toilet flushing
Barwon Water Sewer Mining	Up to 1000	-	Agricultural and industrial uses
McLaren Vale	Up to 8000	C	Application to vines for producing premium-quality grapes
Rouse Hill	Up to 1500	A	Dual reticulation system to 15,000 households
Georges River Program	15,000 to 30,000	Varying standards based on application	To serve existing potable water customers and new residential developments
Other projects**	-	All class B or A	Application includes horticulture, wine grapes, pasture, fodder, public spaces, golf courses, household gardens and recreational area

Table A Water reuse projects operating in Australia. Source: US Environmental Protection Agency, 2004.
Note: *states have to treat the water to standards that meet Australian Environment Protection Authority and Department of Health requirements.
** includes more than 50 schemes.

Hawkesbury Water Recycling Scheme, Richmond, NSW[14]

The scheme builds upon an established partnership with Sydney Water Corporation, and a project to harvest and treat stormwater supported by the New South Wales Stormwater Trust and the Natural Heritage Trust, in collaboration with Hawkesbury City Council, Clean Up Australia, and a number of industry supporters. Situated near Richmond on the Hawkesbury campus of the University of Western Sydney, the Hawkesbury Water Recycling Scheme is a practical demonstration of the environmental and economic benefits of water recycling.

Two resource streams, treated effluent and harvested stormwater, are strategically utilised. The upgraded Richmond STP supplies approximately 600 ML annually. Advanced biological tertiary treatment is used to produce high-quality recycled water suitable for a range of non-potable applications. In addition, approximately 250 ML/year of stormwater is harvested from the urban and peri-urban areas of Richmond, cleaned in constructed wetlands, and utilised as a complementary resource.

While historically the university's dairy was the main user of recycled water, a range of water-users now draw upon this resource, and new commercial agreements are being established. Current uses include pasture for cattle, sheep, deer and a commercial horse unit (as part of the university's outdoor laboratories and grazing unit), and a range of horticultural crops for commercial, research and teaching purposes. The multifaceted aspects of the scheme are coordinated through management systems that incorporate developing approaches to entitlements, differential securities, and risk communication strategies; approaches of great potential relevance to emerging water recycling schemes in western Sydney and elsewhere.

Wastewater treatment schemes in South Australia

With around 18% of water being directly reused, South Australia stands second after Victoria (45.2%) in the country,[10] and, per capita, has the highest level of wastewater reuse in the country.[15] In South Australia, SA Water treats annually 95,000 ML of wastewater across four WWTPs, serving around 1.1 million people within metropolitan Adelaide.

- **Bolivar** treats about 45,000 ML wastewater annually and has resulted in high-quality class A treated wastewater being piped to vegetable growers on the Northern Adelaide Plains.
- **Port Adelaide** treats about 18,000 ML wastewater annually, resulting in the elimination of discharge of treated wastewater into the Port River and providing water for irrigation.
- **Glenelg** treats about 21,000 ML wastewater annually, which is being used to irrigate parks.
- **Christies Beach** treats about 11,000 ML annually and provides treated wastewater to the important wine-growing region of McLaren Vale.

Table B Classification of reclaimed water

Class	A	B	C	D
Uses	Residential non-potable (garden watering, toilet flushing, car washing, path/wall washing); municipal use with public access; unrestricted crop irrigation	Municipal use with restricted access; restricted crop irrigation; irrigation of pasture and fodder for fodder animals	Municipal use with restricted access; restricted crop irrigation; irrigation of pasture and fodder for fodder animals	Restricted crop irrigation; irrigation for turf production; silviculture and non food chain aquaculture
Microbiological criteria (E. coli/100 ml)	<10; specific removal of viruses, protozoa and helminths may be required	<100; specific removal of viruses, protozoa and helminths may be required	<1000; specific removal of viruses, protozoa and helminths may be required	<10,000; helminths need to considered for pasture and fodder
Chemical/physical criteria	Turbidity ≤2NTU; BOD <20mg/L chemical content to match the use	BOD <20mg/L; SS <30mg/L chemical content to match the use	BOD <20mg/L; SS <30mg/L chemical content to match the use	Chemical content to match the use
Typical treatment process	Full secondary plus tertiary filtration plus disinfection coagulation may be required to meet water quality requirements	Full secondary plus disinfection	Primary sedimentation plus lagooning OR full secondary (disinfection if required to meet microbial criteria only)	Primary sedimentation plus lagooning OR full secondary
Best irrigation practices	Plants with large surface area on or near ground = drip or furrow; Root crops consumed raw = spray, drip, flood or furrow; Crops without ground contact = spray	Plants with large surface area on or near ground = drip or furrow; Root crops consumed raw = subsurface; Crops without ground contact = flood; Crops with or without ground contact and skin that is removed before consumption = spray	Plants with large surface area on or near ground = drip or furrow; Crops without ground contact = drip or furrow; Crops without ground contact and skin that is removed before consumption = flood; Crops with ground contact and skin that is removed before consumption = drip, flood or furrow; Root crops processed before harvesting = spray, drip, furrow or subsurface; Root crops processed before harvesting = spray, drip, furrow or flood	Crops without ground contact = subsurface; Crops without ground contact and skin that is removed before consumption = drip, furrow or subsurface; Crops with contact with ground and skin that is removed before consumption = subsurface; Surface crops processed before harvest = subsurface; Crops not for human consumption = unrestricted

Table B Classification of reclaimed water for use in South Australia. Source: Department of Human Services and the Environment Protection Authority, 'South Australian reclaimed water guidelines (treated effluent), 1999.

Note: NTU = nephelometric turbidity unit; BOD = biochemical oxygen demand; SS = suspended solids.

Regarding quality specifications for reclaimed water reuse for irrigation, in South Australia recycled water quality is classified as either class A, B, C or D (Table B). When recycled water is used for irrigation, the level of treatment (i.e. class of the treated water) determines the crops suitable to receive recycled water irrigation and the method by which it can be applied. Recycled water of class A, for example, has received a level of treatment greater than classes B to D and is suitable for unrestricted irrigation to all crop and fodder types. The use of class C recycled water, however, is restricted to a more limited selection of crops.

References

1. A. Dupont, 'Will there be water wars?', *Development Bulletin*, 63, November 2003, pp. 16–20.

2. I. Ray and S. Gul, 'More from less: policy options and farmer choice under water scarcity', *Irrigation and Drainage Systems*, 13, 1999, pp. 361–383.

3. K.D. Frederick, 'Water marketing: obstacles and opportunities', Forum for Applied Research and Public Policy, 16 (1), 2001, pp. 54–62.

4. I. Hespanhol 'Wastewater as a resource' in Helmer, R. and Hespanhol, I. (eds) *Water pollution control: a guide to use of water quality management principles*, E. and F. Spon on behalf of WHO/UNEP, 1997.

5. P. Cullen, (2004) 'Water challenges for South Australia in the 21st century', Department of the Premier and Cabinet, Adelaide, online.

6. T. Asano 'Water from (waste) water: the dependable water resource', 11th Stockholm Water Symposium, Stockholm, Sweden, 12–18 August 2001.

7. GHD Pty Ltd, 'Planning for the use of reclaimed water in Victoria', Reclaimed Water Committee, Ministry of Water Resources and Water Supply, Melbourne, 1978.

8. M. Taylor and R. Dalton 'Water resources: managing the future', Australian Water Association 20th convention, Perth, Western Australia, April 2003.

9. A.J. Hamilton, A.M. Boland, D. Stevens, J. Kelly, J. Radcliffe, A. Ziehrl, P. Dillon and B. Paulin, 'Position of the Australian Horticultural Industry with respect to the use of reclaimed water', *Agricultural Water Management*, 71 (3), 2005, pp. 181–209.

10. Australian Bureau of Statistics, 'Water account, Australia, 2001–2002', 4610.0, 2004, online.

11. J.C. Radcliffe, 'Water recycling in Australia', review report, Australian Academy of Technological Sciences and Engineering, Victoria, 2004.

12. A.C. Hurlimann and J.M. McKay, 'What attributes of recycled water make it fit for residential purposes?: the Mawson Lakes experience', *Desalination*, 187, 2006, pp. 167–177

13. P.J. Dillon, 'Water reuse in Australia: current status, projections and research' in Dillon, P.J. (ed.) *Proceedings of Water Recycling Australia 2000, Adelaide, 19–20 October 2000*, 2000, pp. 99–104.

14. R. Attwater, J. Aiken, S. Booth, C. Derry and J. Stewart, 'Adaptive systems of management and risk communication in the Hawkesbury Water Recycling Scheme', electronic publication (o5148), *Ozwater watershed: the turning point for water*, AWA, Brisbane, 2005, online.

15. S. Dimitriadis, 'Issues encountered in advancing Australia's water recycling schemes' research brief, no. 2, Department of Parliamentary Services, Commonwealth of Australia, Canberra, 2005.

CASE STUDY 6: Farm dam policy: Adelaide and Mount Lofty Ranges

Jennifer McKay

The erection, construction, modification, enlargement or removal of a dam, wall or other structure that collects or diverts water in the Adelaide and Mount Lofty Ranges region is controlled by the *Development Act 1993* and underlying individual local government development plans, the *Natural Resources Management Act 2004* (NRM Act) and the underlying Adelaide and Mount Lofty Ranges Natural Resources Management (NRM) plan (volume D), and prescribed water resources water allocation plans (WAPs).

Works relating to dams and water diversion are defined as a 'water-affecting activity', and require a permit under the NRM Act. However, if dams have a wall height greater than 3 m or a volume greater than 5 ML, they require development approval and not a permit. Development approval is issued by local councils, and applications are referred to the Department for Water, Land and Biodiversity Conservation (DWLBC), which will consider the objectives and principles of the NRM plan in providing advice. Where a WAP applies, the objectives and principles contained within the WAP are used as the basis for assessment.

WAPs for prescribed water resources form part of the Adelaide and Mount Lofty Ranges NRM Plan and contain more detailed policies for water-affecting activities. The surfacewater resources of the Barossa and the Western Mount Lofty Ranges are prescribed, and hence it is the policies in the WAPs that must be applied. The Barossa Prescribed Water Resources Area has a current WAP, which has recently been reviewed with a new plan due to be adopted in the near future. The first WAP for the water resources of the Western Mount Lofty Ranges was being drafted for community consultation in late 2008.

Policies in WAPs are more specific to the particular resource, with extensive investigation having been undertaken to determine the characteristics of the resource, including resource capacity, environmental water requirements, and management zones for appropriate policy application.

For policy related to dams, a WAP will determine, among other things:

- the maximum storage volume (total dam volume) allowable in a management zone;
- the location of new dams;
- threshold flow rates for diversion of water into dams from watercourses; and
- requirements for low flow passes.

Policy extracted from the Adelaide and Mount Lofty Ranges NRM plan, volume D

The Adelaide and Mount Lofty Ranges NRM Plan sets out the following policy relating to dams, with section numbers referring to the NRM Act.

Dam construction (Section 127(3)(d) and 127(5)(a))

The objectives and principles that follow apply specifically to an activity under:

- section 127(3)(d) – the erection, construction, modification, enlargement or removal of a dam, wall or other structure that will collect or divert, or collects or diverts (i) water flowing in a prescribed watercourse; or (ii) water flowing in a watercourse in the Mount Lofty Ranges watershed that is not prescribed; or (iii) surfacewater flowing over land in a surfacewater prescribed area or in the Mount Lofty Ranges watershed; and
- 127(5)(a) – the erection, construction, modification, enlargement or removal of a dam, wall or other structure that will collect or divert, or collects or diverts, water flowing in a watercourse that is not in the Mount Lofty Ranges watershed and that is not prescribed, or flowing over any other land that is not in a surfacewater prescribed area or in the Mount Lofty Ranges watershed.

Policy objectives

1. Maintain and improve the quality and quantity of water flowing in the region.
2. Ensure that dams, walls or other water collection or diversion mechanisms in watercourses and drainage paths are constructed and managed in a manner which:
   a. protects the needs of downstream users;
   b. protects water quality and quantity; and
   c. protects ecosystems dependent on these resources.

Principles

1. The combined capacity of all dams in a catchment within an allotment shall not exceed 50% of the annual runoff for that catchment within the allotment.

Note: For the purposes of principle 1, 'annual runoff' is a volume derived from 10% of the mean annual rainfall for the allotment, multiplied by the area (of the allotment).

Note: For the purposes of principle 1, the term 'allotment' means an allotment delineated on a certificate of title under the *Real Property Act 1886* and includes two or more contiguous allotments owned or occupied by the same person and operated as a single unit for the purpose of primary production.

2. In order to minimise impacts on downstream water-dependent ecosystems:

   a. dams must not be located onstream for third order or higher streams;
   b. water should be diverted to an offstream dam wherever possible for first- and second-order streams;
   c. any onstream dam must incorporate a low-flow bypass mechanism;

d. the low-flow bypass mechanism must be of an approved design and remain operational; and
e. any overflow from a dam or water diverted by a low-flow bypass mechanism must not be recaptured or diverted.

Note: For the purpose of principle 2, an 'onstream dam' means a dam, wall or other structure placed on, or constructed across, a watercourse or drainage path for the purpose of holding back and storing the natural flow of that watercourse, or the surface runoff flowing along that drainage path.

Note: For the purpose of principle 2, an 'offstream dam' means a dam, wall or other structure that is not constructed across a watercourse or drainage path and is designed to hold water diverted, or pumped, from a watercourse, a drainage path or aquifer, or from another source. Offstream dams may capture a limited volume of surfacewater from the catchment above the dam, but may not take an amount of surfacewater, from the catchment above the dam, in excess of 5% of its total volume.

3. Provision shall be made for flow to pass the dam as follows:

a. a diversion structure shall include a device that prevents the diversion of water from the watercourse or drainage line during periods of flow at, or below, the threshold rate; and
b. An onstream dam, wall or structure shall include a device that regulates the diversion of any flow at, or below, the threshold rate, away from the dam and returning it back to the same watercourse or drainage line below the dam, wall or structure.

Note: For the purposes of principle 3, the threshold flow rate (litres/second) means:

c. The flow rate of a watercourse or drainage line (litres/sec) determined by multiplying the unit threshold rate (litres/sec/km^2) by the area of catchment (square kilometre) that contribute to the watercourse or drainage line, that is above the point where the water is diverted from the watercourse or drainage line: or
d. 1 L/second, whichever is the greater.
e. For the purposes of (a), the unit threshold flow rate of a subcatchment can be determined by dividing the 10th percentile flow rate (litres/second) for a subcatchment (square kilometres). Where the 10th percentile flow rate is the flow rate (litres/second) obtained from a time weighted annual flow duration curve (with the time step being 1 day – mean flow), which is greater than or equal to 10% of all flows during that period.

4. Collection or diversion of water flowing in a watercourse, or over land, must not adversely affect downstream water dependent ecosystems by causing reduced streamflow duration, lengthened periods of no or low flow, or other such impacts, unless it is part of a regional Natural Resources Management plan project of the board (eg., constructed wetland).

5. Dams shall not adversely affect the environmental flow requirements of ecosystems dependent on surfacewater or watercourses.

6. A dam shall only be constructed if the capacity of the dam is no more than two times the volume of the proposed extraction from that dam.

Stock and domestic dams

Dams for stock and/or domestic purposes shall only be constructed if there is insufficient or inadequate water available on the property, such that:

a. there is no capacity to connect to SA Water supply; or
b. the flow rate of water from wells is less than 0.1 L/sec; or
c. the salinity of the water from the wells is greater than:
   i. 1500 mg/L for general domestic purposes;
   ii. 1000 mg/L if the water is used for drinking purposes; or
   iii. 3000 mg/L for stock purposes.

Location

Dams, including dam walls and spillways must not be located:

a. in, immediately upstream, or immediately downstream of, ecologically sensitive areas;
b. in areas prone to erosion;
c. onstream for third order streams of higher; or
d. where it is likely to adversely affect the migration of aquatic biota.

Dam construction and design

Dams should be sited and constructed to:

a. minimise the loss of soil from the site through soil erosion and siltation; and
b. minimise the removal or destruction of in-stream or riparian vegetation.

Removal of a dam

Removal of a dam shall not result in:

a. increased erosion;
b. increased flooding;
c. bed and bank instability;
d. downstream sedimentation;
e. loss of riparian vegetation;
f. decline in water quality; or
g. alteration to the natural flow regime of a watercourse.

The site of the dam should be remediated and revegetated so that there are no ongoing impacts on the downstream environment.

CHAPTER 16

Water and population

Graeme Hugo

Introduction

Any comprehensive treatment of the relationship between the Adelaide metropolitan area and water must include a full consideration of the people of Adelaide. Often, water is considered purely in environmental terms, or for the needs of agricultural production, but the population of Adelaide is an important element when evaluating the availability and quality of water. This chapter traces the development of the population of Adelaide and shows how water has been an important influence, and discusses some of the implications of this relationship into the future. The South Australian government has committed the state to a population policy that sees the population increasing from its current level (1,584,513 in 2007) to 2 million in 2050. In light of the water shortages experienced during the drought of 2007–2008, and the realisation of climate change, it is relevant that the policy envisages a 26.2% increase in the state's population over the next four decades, with the bulk of this increase in the Adelaide region. Is the policy to achieve environmental sustainability compatible with this population policy? This question will be addressed later in this chapter.

We begin with a consideration of the complex relationship between population and water, exploring the Australian context, in which there is an increasing mismatch between the national distribution of population and runoff. We then move on to analyse the historical, contemporary, and likely future patterns of growth of Adelaide and its population. Some of the features of this growth that have impinged on water consumption are discussed, such as the nature of urban development and the changing size and nature of households and housing. This is then related to the availability of water resources. The options for the future are then discussed within the context of the state's population policy.

The water/population relationship

The relationship between water and population is multilayered and complex. At its simplest level water availability clearly places an environmental constraint on the limit of human settlement. In Australia, the driest continent on earth, the significance of water constraining close settlement has long been appreciated and was forcibly and clearly demonstrated in the work of Griffith Taylor in 1947.[1] He argued that by 1860:

> ... the contemporary margins of settlement in Australia already closely approximated the limits which had been set by the very nature of the physical environment: whether people, plants or animals were considered the appropriate environmental 'controls' could be ignored only at a cost.

Hence at one level it is apparent that water, along with other environment elements, has placed a limit on development. At another level, however, the relationship is more complex as is suggested by Equation 16.1.

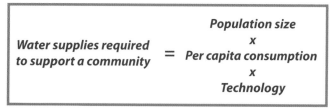

$$\text{Water supplies required to support a community} = \frac{\begin{array}{c}\textit{Population size} \\ \textit{x} \\ \textit{Per capita consumption} \\ \textit{x} \\ \textit{Technology}\end{array}}{}$$

Equation 16.1 Water supplies required to support a community.

The amount of water needed to support a community is not simply a function of the number of people in an area but also their per capita consumption of water and the technology they employ to capture and maintain the sustainability of that water resource. Per capita consumption is influenced by many variables. Firstly the way populations group themselves into households is significant, since some domestic consumption is household based rather than individual based. The size of the houses and the allotments they are built on are also significant influences, and the mix of land use has an impact on demand for water. Figure 16.1 shows how agriculture is far and away the major user of water in Australia, accounting more than four times the amount of water used by the nation's 8 million households. Agriculture also has the lowest value of output per megalitre of water input. Some industries are more water intensive than others.

The way in which a community uses its water resource (the technology element) also has an impact. On the one hand, a community that is able to capture a high proportion of runoff can be larger, given a water resource of a particular size, than one that pollutes its water system or does not capture a high proportion of runoff. The fundamental issue evident in the equation is that there is no simple relationship between a volume of water and a specific number of people it can support. This must be recognised in any discussion of Adelaide's future population, and the role of water in influencing this. Technology also influences the efficiency of water supply systems through, for example, the control of evaporation and leakage.

Australia: water consumption

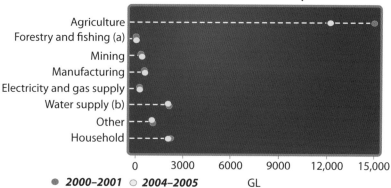

● *2000–2001* ○ *2004–2005*   GL

(a) Includes services to agriculture; hunting and trapping.
(b) Includes sewerage and drainage services.
Figure 16.1 Annual water consumption in Australia, 2000–2001 and 2004–2005. Source: Australian Bureau of Statistics, 'Water account: Australia, 2004–2005', catalogue no. 4610.0, data used with permission from the Australian Bureau of Statistics.

In examining the water/population relationship, however, it cannot be assumed that the water side of the equation remains a fixed quantum. Of course, there is seasonal and annual variation in rainfall but of increasing interest are the long-term shifts occurring in rainfall (and runoff) levels associated with climate change. The CSIRO has delineated the parts of Australia (shown in Figure 16.2) that have experienced an overall increase in rainfall and those areas that have recorded a long-term decline in rainfall and runoff.

A clear spatial pattern is in evidence whereby the south-east of the continent is experiencing long-term decline and the north-west an increase in rainfall. The former covers 38% of the continent, but at the 2006 census was home to 89.6% of the national population. Two decades ago Nix[2] drew attention to the distinct spatial mismatch in Australia in the distribution of runoff on the one hand and that of population on the other. Table 16.1 shows that the Far North had 52% of the national annual surface runoff and only 2% of the population, while in the south the relevant proportions were 27% and 82%. With some prescience, Pittock and Nix[3] speculated about the potential impact of climate change.

By far the largest volumes of uncommitted water are in northern Australia and western Tasmania. In the most heavily populated regions of south-western and south-eastern Australia, surfacewaters are committed to a high degree and the consequences of climate change are potentially most serious.

The significance for Adelaide is twofold. Firstly, Adelaide is squarely located in the area that is experiencing a long-term decline in rainfall and runoff; hence it can expect a secular decline in water runoff in its immediate catchment area over coming decades. Secondly, Adelaide has become increasingly reliant since the 1950s on water being pumped via the Mannum–Adelaide Pipeline so that increasingly its water has been drawn from the wider Murray-Darling Basin (MDB) and not just from its immediate catchment. However,

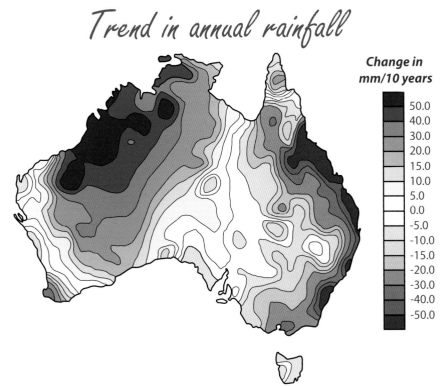

Trend in annual rainfall

Change in mm/10 years

50.0
40.0
30.0
20.0
15.0
10.0
5.0
0.0
-5.0
-10.0
-15.0
-20.0
-30.0
-40.0
-50.0

Figure 16.2 Trend in annual total rainfall in Australia from 1950 to 2006 (mm/10 years). The blue areas have seen an overall increase in rainfall, whereas the yellow, orange and red areas have seen a decrease in rainfall. Source: Bureau of Meteorology, 'Trend in annual total rainfall, 1950–2006', copyright Commonwealth of Australia, reproduced with permission.

the MDB also lies largely within the declining rainfall and runoff region so that the amount of water flowing into that system will decrease over time.

The growth of Adelaide's population

The past and projected future growth of Adelaide is shown in Figure 16.3. The graph shows a clear break at the end of World War II. There was steady population increase up to the onset of the Great Depression and World War II, which saw a plateauing of growth for over a decade. At the 1947 population census, Adelaide's population stood at 382,454. Thereafter were several decades of steep growth, with the population reaching one million in 1986. The national economic downturn and State Bank crisis of the early 1990s saw a significant reduction in growth levels due to the combined effects of reduced fertility, reduced immigration, and increasing net interstate migration loss.

Over the next decade the population of Adelaide grew at less than 0.5% per annum, less than half the national population

Table 16.1 Water and population

	Far North Australia (%)	Southern Australia (%)
Population	2	82
Potentially arable land	4	65
Annual mean surface runoff	52	27

Table 16.1 Australia's mismatch between water and population. Source: H.A. Nix, 'Australia's renewable resources', 1988.

Adelaide's population

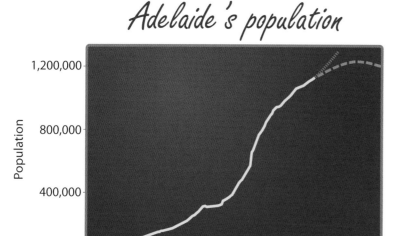

ABS ERP --- **ABS projections** ⋯⋯ **Planning SA projections**

Figure 16.3 Adelaide's population, 1846–2006, and projected population, 2007–2051. Source: CBCS (SA office), 'South Australian year book', various issues; Australian Bureau of Statistics, historical population statistics; 2005 projections, series B; Planning SA, projections, high series.

Table 16.2 presents the ABS projections, which show that the Adelaide Statistical Division (ASD) population is likely to continue to increase from 1.132 million in 2006, reaching a peak in 2031 of 1.229 million, and thereafter declining to 1.203 million in 2051. However, over the same period the population aged over 65 in the ASD would double and the proportion they make up of the total population will also almost double.

This spectre of population decline and ageing has been a concern for recent South Australian governments, both Labor and Liberal. The Rann Government responded to the ABS projections by developing a South Australian population policy[4] – the first jurisdiction in Australia to develop and implement such a policy. The centrepiece of the policy was to reverse the trends used as a basis for the ABS projections to the extent that the state's population would increase to 2 million in 2051 rather than decline to less than it is today (under the ABS projection scenario). To achieve this objective, the policy sets a number of targets within South Australia's Strategic Plan:[5]

growth rate. These trends convinced the Australian Bureau of Statistics (ABS), in developing their 2001 census-based population projections, that Adelaide would continue to experience low levels of population growth and indeed its population would begin to decline in 2036. However, it was not only the projected overall population decline in South Australia and in Adelaide which were of concern. Within the population there was significant ageing anticipated. Figure 16.4 shows that in the ABS projections, virtually all of the population growth anticipated in South Australia would be occurring in the older age groups while there is anticipated to be a decline in most ages under 40 years.

- Target 1.22: total population: 'increase South Australia's population to 2 million by 2050, with an interim target of 1.64 million by 2014.'
- Target 1.23: interstate migration: 'reduce annual net interstate migration loss to zero by 2010, with a net inflow thereafter to be sustained through to 2014.'
- Target 1.24: overseas migration: 'increase net overseas migration gain to 8500 per annum by 2014.'
- Target 1.25: fertility: 'maintain a rate of at least 1.7 births per woman.'
- Target 5.9: regional populations: 'maintain regional South Australia's share of the state's population (18%).'

Population change by age

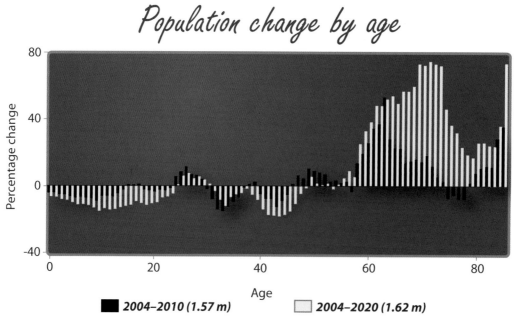

■ **2004–2010 (1.57 m)**   □ **2004–2020 (1.62 m)**

Figure 16.4 South Australia's population change by age, 2004–2010 and 2004–2020. Source: Australian Bureau of Statistics, 2005 projections, series B; based on Jackson, 2004.

Table 16.2 Adelaide population projections

Year	ABS Series B projections			Planning SA High Series projections		
	65+	Total	% 65+	65+	Total	% 65+
2006	173,984	1,132,569	15.4	173,620	1,138,089	15.3
2011	193,164	1,157,813	16.7	190,088	1,186,378	16.0
2016	226,322	1,180,815	19.2	220,728	1,237,703	17.8
2021	256,695	1,201,252	21.4	250,179	1,292,651	19.4
2031	314,394	1,229,043	25.6			
2041	340,743	1,226,919	27.8			
2051	346,516	1,203,937	28.8			

Table 16.2 Adelaide statistical division: alternative projections of total population and population aged 65+, 2006–2051.
Source: Australian Bureau of Statistics, 2005 projections, series B; Planning SA, projections, high series.

The population policy has achieved a significant degree of success. If the 2005–2006 population annual growth rate was maintained, South Australia's population would reach 2 million by 2034, and if the 2006–2007 rate continued this would happen in 2030. South Australia in 2007 grew by 1.02%, more than twice the rate of increase at the time of the introduction of the population policy. This has largely been achieved by a spectacular increase in international migration settlement in the state, although the fertility and natural increase levels have also increased.

Figure 16.5 shows the striking change which has occurred in immigration to South Australia since the initiation of the population policy. The state created an office of immigration (Immigration SA) and became active in lobbying the federal government to obtain concessions for the state through the State Specific and Regional Migration Scheme[6] and through active recruiting of skilled migrants in countries like the United Kingdom and India. International migration now contributes 80.6% of current population growth in South

Australia and this proportion is likely to increase over time. The majority of these new immigrants are settling in the Adelaide region. Of immigrants counted in South Australia at the 2006 census who had arrived between 2001 and 2006, 90.4% (36,367 persons) were in the ASD and 2.8 (1110 persons) were in the Outer Adelaide Statistical Divisions (OASD).

The ASD and OASD have increased their share of the state's total population from 77.3% in 1976 to 81.4% in 2006.

Although the continuous built-up area of Adelaide is largely contained within the ASD, it is important to include the OASD in consideration of the environmental impact of Adelaide. Much of the economic activity of the OASD is related to the city and 30.2% of the workforce commute daily to the ASD. Figure 16.6 shows how the Adelaide region has increasingly dominated the distribution of the state's population. Despite the substantial non-metropolitan based growth of the mining industry in South Australia, Adelaide will remain the home of the majority of the state's population. Table 16.2 shows projections undertaken by Planning SA, which see the ASD population increasing from 1.138 million in 2006 to 1.292 million in 2021.

This increase of over 150,000 in Adelaide's population over the next 15 years will influence future water needs in Adelaide, but the changing composition of the population is also significant; in particular the way in which the population groups itself into households is significant. This is because some household consumption of water is regardless of the number of people living in that house. As Australian households become smaller,[7] and other things being equal, we can expect an increase in water consumption because of the household consumption factor. In this context we need to note a long-term decline in the average size of households in Adelaide and in South Australia. The average number of persons

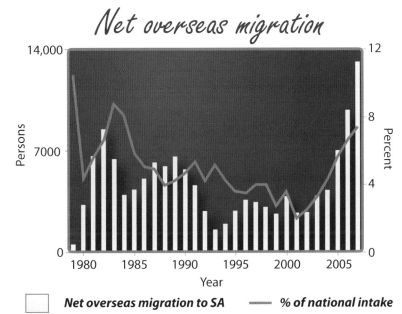

Net overseas migration

Figure 16.5 South Australia: net overseas migration, 1979–2007.
Source: Australian Bureau of Statistics, Australian demographic statistics, various issues.

Legend: Net overseas migration to SA — % of national intake

BOX 77

How much water does a person need?

It is important to know how much water a person needs because it has a bearing on how we deliver water, our water pricing regimes, and how many people a city's water resources can support. But the calculation is not simple. The United Nations, in its Agenda 21 documentation, considers that, 'In developing and using water resources, priority has to be given to the satisfaction of basic needs and the safeguarding of ecosystems'.[1] Some authors have gone further and suggested that 'basic needs' include the safeguarding of the local ecosystems on which we depend. I will discuss the figures available to help us determine how much water a person needs, the volumes of water needed for people and households in Adelaide, and touch on the implications of this calculation.

We need water for ourselves. The World Health Organization[2] states that adults have a basic physiological need for around 2 L of water per day (not including water in food). This is, however, a minimum requirement that should be adjusted to account for the climate, how active a person is, their size and diet.[2, 3, 4] For example, on a hot day in Adelaide, an active adult will need more than 2 L of water. In addition, each day (on average) we require a minimum volume of water for hygiene and sanitation (around 20 L), bathing (around 15 L, enough water to run a water-efficient showerhead for 1.5 minutes), and food preparation (around 10 L).[4] Thus, Peter Gleick[4] proposes that everyone has a 'basic water right' of 50 L/day. This volume of water satisfies our physiological needs, allows us to stay clean (which is necessary to maintain our social status), and is important for the prevention of disease.

We must also maintain the land we live on. We depend on the local environment and the city's green infrastructure to provide clean air and water, to shelter us (and our houses) from heat and wind, and we gain psychological benefits from experiencing the natural environment. Watering our gardens is also important in the maintenance of our built infrastructure. Watering during dry periods can reduce ground movement and minimise damage such as cracks in walls.

How we maintain 'our' patch of land is also relevant to our relationships with our neighbours and our social status. Furthermore, there are health considerations: if we don't water our gardens we risk a dusty existence that threatens the health of people, pets and wildlife. Therefore, some water should be allocated for us to maintain our gardens as part of our basic water needs.

I am not suggesting that this entitlement should allow us to establish large, lush lawns and gardens, but it should cover a sensible, water-sensitive garden design. It must be remembered that supplementary water may be required in urban gardens, even where native species of local provenance are planted, because of the changed hydrologic conditions inherent in cities. The volume of water appropriate for this need is difficult to determine. It will depend on local rainfall, soils, plantings and the size of the garden.

Very little work has been published regarding how much water we need to maintain the land we live on. Hillel Shuval, in work on the Middle East, has proposed that we require around 95 L/day to satisfy our personal needs, including some water for garden use.[4] I suggest that cohabitation diminishes the need for personal entitlements for garden water (unless the water is dedicated to food production). Therefore, the 45 L that is allocated over-and-above personal needs in Shuval's example can be allocated to households rather than individuals. To demonstrate how much water that is – if it was used over the driest four months of the year – it would allow the household to water their garden for around one hour a week using a hose that delivered water at 15 L/hour.

Using these numbers, and with an average household of 2.4 residents, the volume of water they 'need' would be 165 L/day (50 x 2.4 + 45), or around 60 kL/year. That represents less than a quarter of the water we actually use (an average Adelaide household uses 265–300 kL/year[5]).

In addition to household use, we use water indirectly, and the calculations thus far have not included major water uses such as agriculture, industry, public use and power generation (these are difficult to include in a regional analysis as they are often imported). When all delivered water uses are included, Australians' per capita water consumption blows out to 598 kL each year![6] Australians are among the heaviest water-users in the world.

The information presented above highlights that we do use a lot of water, and that there must be potential for us to reduce our water use. However, the situation should not be viewed simplistically. In terms of our household water use, there are a number of factors that *allow* us to use more water than other countries. First, we do not directly export water. We must balance our local water budget and reduce our water use to ensure that we are using our resources sustainably, but we do not necessarily need to go as low as the basic requirements I have described above. Second, we have water to use; many developing countries lack water resources, and, therefore, their water use is lower. In addition, water consumption increases with access to water; households in developed countries, with piped water, use more water than households that need to transport their own water.[4] Third, we should not unnecessarily reduce our water consumption so much that we increase associated health risks. Fourth, we should maintain the green infrastructure of the city. These caveats by no means mean that we should continue to use water the way we do now, but they must be considered in setting reasonable targets for reductions in water use. The use of water-saving technologies, capturing our own water, and recycling water can reduce the volume of water we need to import into our houses. Such strategies will contribute to a city of sustainable water use. Overall, we need to balance the volume of water we use with the resources available to the population of the city – without placing excessive pressure on the natural environment.

Philip Roetman

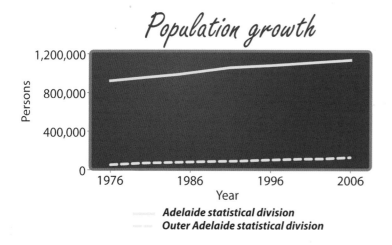

Population growth

Persons (y-axis): 1,200,000 / 800,000 / 400,000 / 0
Year (x-axis): 1976, 1986, 1996, 2006

Adelaide statistical division
Outer Adelaide statistical division

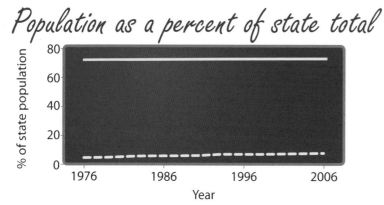

Population as a percent of state total

% of state population (y-axis): 80, 60, 40, 20, 0
Year (x-axis): 1976, 1986, 1996, 2006

Figure 16.6 Adelaide statistical division and outer Adelaide statistical division: population growth, 1976–2006. Source: Australian Bureau of Statistics, estimated resident population data.

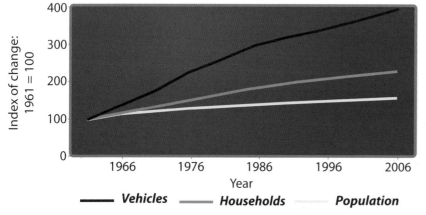

Growth of population, households and vehicles

Index of change: 1961 = 100 (y-axis): 400, 300, 200, 100, 0
Year (x-axis): 1966, 1976, 1986, 1996, 2006

—— **Vehicles** —— **Households** —— **Population**

Figure 16.7 South Australia: growth of population, households and vehicles, 1961–2006. Source: Australian Bureau of Statistics, census, 1961–2006.

per dwelling has fallen from 2.9 in 1976 to 2.4 in 2006 and is projected to decline even further to 2.2 in 2021.[8] The important point here is that in assessing the environmental implications of population increase, the growth rate of the population is not always the only appropriate measure to consider. Figure 16.7, for example, takes the year 1961 as a benchmark and compares the increase in population with the increase in the number of households. It also shows the growth in the number of vehicles, which also has implications for the environment, and growth here has been even more rapid than for households.

The decrease in the average number of people per dwelling has not been associated with a decrease in the size of dwellings, although allotment average size has decreased.[8] The increase in the land area under housing can potentially create environmental problems associated with stormwater and wastewater management. It is crucial that their management be designed into future housing development in Adelaide. Clever design can result not only in reducing the potential of damage but also lead to greater capture and storage of runoff.

Population and water demand in Adelaide

Writing 80 years ago, Fenner[9] stressed the significance of water conservation and supply in shaping population growth in Adelaide and South Australia generally:

> *This question of water supply, combined with and dependent on that of rainfall is the dominating geographic control of population distribution in the state.*

He shows how from the foundation of the state (1836) up to 1860, no systematic provision was made for water supply and settlers in Adelaide mostly obtained water from the River Torrens or from wells sunk in the irregular subartesian area underlying Adelaide. However, the growth of Adelaide's population rendered the construction of a reservoir, Thorndon Park, necessary (it was decommissioned in 1986). Figure 16.8 is a development and variation on a diagram drawn by Fenner to show the subsequent development of reservoirs to serve the growing Adelaide metropolitan population. There was a steady development of new reservoirs over the next century with Mount Bold (still the largest metropolitan reservoir) being commissioned in 1936. In the rapid population growth of the early post-World War II years there were four more reservoirs constructed in the Adelaide region but none built in the last three decades. The last reservoir (Little Para) was commissioned in 1976. Since then, population growth has been 24% but reservoirs have been commissioned to store additional water around Adelaide.

Fenner[9] identified four main clusters of reservoirs in South Australia, of which the group in the Mount Lofty Ranges (MLR) supplying Adelaide was the most important. It is interesting that he observed, in relation to the MLR cluster, 'the water supply from which most ultimately impose a limit

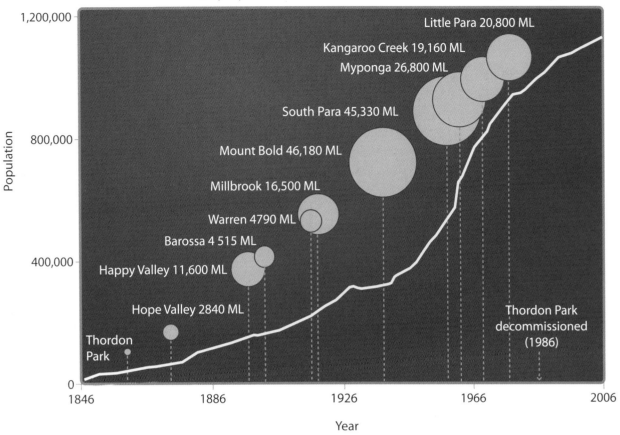

Growth of population and reservoir capacity

Little Para 20,800 ML
Kangaroo Creek 19,160 ML
Myponga 26,800 ML
South Para 45,330 ML
Mount Bold 46,180 ML
Millbrook 16,500 ML
Warren 4790 ML
Barossa 4 515 ML
Happy Valley 11,600 ML
Hope Valley 2840 ML
Thordon Park
Thordon Park decommissioned (1986)

Population: 1,200,000 / 800,000 / 400,000 / 0

Year: 1846 / 1886 / 1926 / 1966 / 2006

Figure 16.8 Growth of population and reservoir capacity, 1846–2006. Source: Australian Bureau of Statistics and SA Water; C. Fenner, *A geographical enquiry into the growth, distribution and movement of population in South Australia, 1836–1927, Adelaide,* 1929, p. 118.

on the growth of the metropolitan area'. Of course he was writing well before the advent of the Mannum–Adelaide Pipeline, which revolutionised Adelaide's water supply by allowing the city to draw on the Murray River and the runoff from the entire Murray-Darling Catchment rather than its own limited catchment area. Hence, increasingly Adelaide's reservoir system has not only had the function of capturing runoff in the Mount Lofty Catchment but as a storage area for water pumped from the River Murray.

It is apparent that water demand in Adelaide has increased because of the increase in population *and* a substantial increase in per capita consumption of water. Between the mid-19th century and the present there has been a trebling of average daily consumption in Australian cities from around 100 L/head to around 300 L, although peak consumption levels were recorded in the 1980s.[10] Changes in attitudes towards bathing and clothes washing, increased use outside the house, and the use of water for waste disposal contributed to this increase. The response to this increased demand has been to bring online new supplies of water, initially through construction of reservoirs and later through the construction of the Mannum–Adelaide Pipeline. In other words, the response has virtually all been on the left-hand 'supply' side of Equation 16.1, while very little has been done to respond to increased demand by developing policies to tackle that side of the equation.

In addition, most of the current discourse around how to tackle the 'water panic' that has gripped Australian cities in recent years has been directed at increasing the supplies of water. However, Troy[11] has argued strongly that the focus should be on tackling the 'shape of the demand' for water in Australian cities.

All (current) proposals also imply contradiction of the 19th-century solution to meet the demand for water by increasing supply. Before adopting any of these 'solutions', a different strategy is now required to significantly reduce consumption of potable water.

The current proposals for Adelaide and other Australian cities include construction of desalinisation plants. Troy[11] argues against them on the following points:

- they are expensive in environmental terms, having high levels of carbon emissions;
- they cannot easily be run efficiently in low-flow situations;
- in Sydney they will involve an extra cost to each household of $110/year; and
- in Adelaide we could add that the current proposed site (the abandoned oil refinery site at Port Stanvac) is in the south of the metropolitan area when the bulk of planned population growth is on the northern margins of Adelaide.

Table 16.3 Average water consumption

	NSW	VIC	QLD	SA	WA	ACT
Bathroom	26.3	26.5	26.0	18.5	22.4	18.7
Toilet	23.2	19.4	16.4	16.0	14.5	16.4
Laundry	16.2	15.3	13.7	16.0	18.5	11.7
Kitchen	10.0	5.1	12.3	12.3	10.6	5.8
Outdoor	25.3	35.7	69.0	62	66.0	64.4
Total	*101*	*102*	*137*	*123*	*132*	*117*

Table 16.3 States' average annual per capita water consumption (kilolitre) by location of use in 2001. Source: P. Troy, *Troubled waters: confronting the water crisis in Australia's cities*, Canberra, 2008, p. 197.

There are also proposals for water recycling and Troy[11] presents evidence that reverse osmosis systems do not eliminate pharmaceutical drugs from sewerage, and reports anxieties about the efficiency of such systems to eliminate the bacteria, protozoa and viruses commonly found in sewage.

There is clearly considerable capacity for greater capture and storage of rainwater and runoff at the individual household level (through rainwater tanks), and at the city level. However, Troy[11] argues for consideration of ways in which consumption can be modified, such as the development of dual-flow water systems, which recognise that only a small proportion of water now consumed in cities needs to be of potable quality. Installation of greywater regulating systems, dry composting toilets, and tanks would reduce the consumption of potable water by up to 70%. Thus far in Adelaide and other cities, efforts to reduce consumption have been focused outside of the house and have taken the form of watering bans. Table 16.3 shows that there is considerable difference between states in the proportion of household water use for different activities inside and outside the house.

Effectively tackling the consumption dimension of the water equation is an important priority in Australian cities like Adelaide. What is needed is a substantial conceptual shift:

> *A key motif that swirls around is that Australians need a 'culture change' in relation to water … we need to change attitudes of profligacy, developed in the well-watered ancestral lands of north-west Europe, and attune both attitude and practice to the realities of living on the driest inhabited continent on Earth.*[12]

Head[13] has demonstrated that such a cultural shift is already underway:

> *Water is now an important issue in both consciousness and practice of suburban householders … Commitments to reducing water consumption are manifest in a variety of changed practices, many of which are hidden in the rhythms of daily life.*

Clearly there is a need for policies and programs of government that build on and support this demonstrated tendency to reduce consumption.

Looking to the future

There is strong evidence that climate change is reducing rainfall across south-eastern Australia[14] so that the runoff feeding the MLR reservoirs and the River Murray is being significantly reduced. It is estimated that the annual mean flow in the MDB may drop between 10 and 25% by 2050, and between 16 and 48% by 2100.[14] Moreover, recent climate change research in South Australia suggests that a 10% reduction in rainfall produces a 20–30% reduction in runoff. What does this mean for Adelaide? A simplistic argument would run that population growth should cease, thereby curbing the increasing need for water. Yet such simplistic solutions have much wider repercussions for the city – not only economic and social but environmental. To stop Adelaide's population growth at present would have highly negative effects in a number of respects. The fact that baby boomers make up 37.6% of Adelaide's workforce, and that they will be leaving the workforce over the next two decades, combined with low fertility and net interstate migration loss, means that the city's workforce will decline and age significantly in the absence of net immigration gain. This would lead to a substantial economic and social burden to the city and its people as it strives to support over a quarter of its population. Moreover, Adelaide's capacity to adopt much-needed sustainable environmental strategies will be compromised by the lack of financial resources which would follow from a rapidly shrinking and ageing workforce.

The development of Adelaide (and South Australia) is constrained by environmental limits among which water is one of the most significant. However, short-term stopping of population growth is not a 'silver bullet' solution. Certainly, in the medium to longer term, the achievement of a demographically stable population for Adelaide (and Australia) should be a goal, however, the contemporary demographic imbalances necessitate growth of the working age population in the short to medium term.

During this period we must focus on the other elements influencing the need for water by making better and more sustainable use of the available water, and by developing an optimum portfolio of water sources for the city, with appropriate charging mechanisms for them. We must also reduce per capita levels of consumption of potable water. Providing water for an increasing and ageing population represents a significant challenge, but it is one which is achieveable through careful planning. The challenge cannot be met by the traditional solution of 'finding new water' but rather through changing the way water is consumed and by reducing levels of consumption per person.

The long-term goal of reaching sustainability requires Adelaide develop a more varied array of water sources and use them in more sustainable ways. It also requires Adelaide and Australia to work towards achieving a stable population with a balance between working and non-working age populations. To reach the latter will require population growth in the short to medium term because of the baby boom bulge. Planning for a sustainable future water supply for Adelaide is a complex process that needs to incorporate not only the environmental but also economic, social and demographic considerations in an integrated way.

References

1. T.G. Taylor, *Australia: a study of warm environments and their effect on British settlement*, Methuen, London, 1947.
2. H.A. Nix, 'Australia's renewable resources' in Day, L.H. and Rowland, D.T. (eds) *How many more Australians?: the resource and environmental conflicts*, Longman, Cheshire, 1988, pp. 65–76.
3. A.B. Pittock and H.A. Nix, 'The effect of changing climate on Australian biomass production: a preliminary study', *Climatic Change*, 8 (3), 1986, pp. 243–255.
4. Government of South Australia, 'Prosperity through people: a population policy for South Australia', 2004.
5. Government of South Australia, *South Australia's strategic plan, 2007*.
6. G.J. Hugo, 'Australia's state specific and regional migration scheme: an assessment of its impacts in South Australia', *Journal of International Migration and Integration*, forthcoming.
7. Australian Bureau of Statistics, *Australian social trends, 2001*, catalogue no. 4102.0, Canberra, 2001.
8. Government of South Australia, 'State of the environment report', 2008.
9. C. Fenner, *A geographical enquiry into the growth, distribution and movement of population in South Australia, 1836–1927*, Charles Fenner, Adelaide, 1929, p. 117.
10. G. Davison, (2008) 'Down the gurgler: historical inuences on Australian domestic water consumption' in Troy, P. (ed.) *Troubled waters: confronting the water crisis in Australia's cities*, ANU E Press, Canberra, 2008, pp. 37 66.
11. P. Troy, 'Conclusion' in Troy, pp. 185–208.
12. Troy, P., 'Introduction: the water services problem' in Troy, 2008, pp. 1 5.
13. L. Head, 'Nature, networks and desire: changing cultures of water in Australia' in Troy, 2008, pp. 67–80.
14. K. Hennessy, 'Climate change' in Newton, P.W. (ed.) *Transitions: pathways towards sustainable urban development in Australia*, CSIRO Publishing, Melbourne, 2008, pp. 23–33.
15. D. Bardsley, 'There's a change on the way: an initial integrated assessment of projected climate change impacts and adaptation options for natural resource management in the Adelaide Mount Lofty Ranges region', Department of Water, Land and Biodiversity Conservation report 2006/06, Adelaide, 2006.

Box references

Box 77: How much water does a person need?

1. United Nations, 'Agenda 21, chapter 18: protection of the quality and supply of freshwater resources: application of integrated approaches to the development, management and use of water resources', United Nations Department of Economic and Social Affairs, Division of Sustainable Development, 2004 (online).
2. World Health Organization, 'Guidelines for drinking-water quality, 3rd edition, first addendum to the third edition', 2006 (online).
3. G. Howard and J. Bartram, 'Domestic water quantity, service level and health', World Health Organization, Geneva, 2003.
4. P.H. Gleick, 'Basic water requirements for human activities: meeting basic needs', *Water International*, 21, 1996, pp. 83–92.
5. Government of South Australia, 'Reducing water use in the home' (online).
6. United Nations Statistics Division, 'Environmental indicators: water', 2007 (online).

CHAPTER 17

Water quality and public health

David Cunliffe
A. John Spencer
Andrew Humpage
Peng Bi
Nina Allen
Loc G. Do
Michael Burch
Ying Zhang

This chapter is a truncated version of the South Australian Department of Health (DoH)'s 'Public Health Bulletin: Water and Public Health', 2007.

The human right to water is indispensable for leading a life in human dignity ... It is a prerequisite to the realization of other human rights.[1]

Introduction

Safe water is essential for life. This chapter deals with a broad range of issues relating to the benefits of safe water, the harm of unsafe water, and the pressures imposed by climate variations. It is divided into five sections, discussing:

1. The public health benefits of safe drinking water.
2. Water fluoridation.
3. Blue-green algae: toxicity and public health significance.
4. Water recycling in South Australia: overview from a public health perspective.
5. Climate change and waterborne diseases.

The chapter also includes a case study of water treatment in Adelaide and number of boxes on the public health issues.

The public health benefits of safe drinking water

The public health benefits of safe drinking water are often underestimated in developed countries like Australia. The safety of drinking water supplies is usually taken for granted in most parts of Australia, with the exception of rural and remote areas. Even in times of drought the foremost concern for most people is water quantity rather than quality.

The benefits of safe drinking water are best illustrated by the contrasting conditions in the developed and developing world. The introduction of safe drinking water and sanitation in the developed world has had a greater positive impact on public health than any other single measure.[2, 3, 4] In the first 40 years of the 20th century, overall mortality rates in the United States fell by 40%, mortality from infectious diseases fell by 65%, and life expectancy increased by 26 years.[4, 5] Nearly half of the impact on mortality and life expectancy can be attributed to the provision of safe drinking water.[4] During this period, safe water was responsible for three-quarters of the decline in infant mortality, nearly two-thirds of the decline in child mortality, and the near eradication of typhoid.[4]

Control of typhoid was particularly important for Adelaide. Typhoid is a food and waterborne infection caused by the salmonella bacteria. Symptoms include fever, diarrhoea or constipation, abdominal tenderness, enlarged spleen, and a rose-coloured macular eruption (rash) confined to the trunk. In 1866 there was a significant typhoid outbreak in Adelaide with 209 deaths from typhoid fever; in 1882 more than 140 individuals succumbed to it with similar numbers each year for the rest of the decade; and again in 1898 there were more than 140 deaths from it.[6] With safe drinking water, typhoid was eliminated from the city.

In contrast, 1.8 million people, predominantly in developing countries, die every year from diarrhoeal diseases, about 90% of which are attributed to unsafe water, and inadequate sanitation and hygiene. Ninety per cent of the deaths are children aged under five years.[7]

Recent developments

There have been three significant developments in the last 10 years relating to provision of safe drinking water. The first occurred in September 2000 when the United Nations adopted the 'Millennium Declaration' including the goal that by 2015 the proportion of people without sustainable access to safe drinking water should be halved.[8] This measure was later extended to include improved sanitation. The second was the declaration of the International Decade for Action: 'Water for Life' 2005–2015, to promote and support achievement of these goals.[9] The third was the United Nations' adoption of the human right to water in 2002, which entitles everyone to sufficient and safe water for domestic use.[1, 10]

Advantages of safe drinking water

Analysis of interventions in undeveloped countries shows that improving water supplies improves water quality by 15–17%, hygiene by 33% and sanitation by 22–36%; and reduces diarrhoeal disease by 19–22%.[11] In addition to the direct impacts, the reduction in disease has associated benefits for both societies and individuals. Lower incidence of disease means less expenditure by the health sector on treatment of diarrhoeal disease including the cost of hospitalisation. In terms of individuals, lower disease means reduced medical expenses, income gains from fewer days lost from work, and reduced absenteeism from school.[12] Better quality water and improved sanitation also reduce risks associated with local food production.[9, 10]

Economic analysis of the advantages of safe drinking water and sanitation have shown that the returns on each $1 invested range from $5 to $28 in the developing world.[12] This is consistent with the benefits of safe water in the United States, where the return on each $1 invested in the early 20th century was estimated at between $7 and $40.[4]

In most parts of the developed world, including Australia, these advantages have been realised through measures that were pioneered in the 19th century, and implemented in the 20th century, to reduce endemic waterborne disease and the incidence of epidemics.

History of drinking water protection

Provision of safe water and sanitation was an important feature of the Roman Empire, as illustrated by the accounts of Sextus Julius Frontinus, the water commissioner of Rome at the end of the first century AD, including his text *De aquae ductu urbis Romae*. The importance of safe water and sanitation was then largely lost through periods such as the Dark Ages and not revived until more modern times.

In the 19th century the link between sewage, drinking water and the spread of infectious disease was established by work led by John Snow, William Budd, John Sutherland and Robert Koch.[13, 14, 15] Snow first published his theory that sewage contamination of drinking water supplies was responsible

Waterborne diseases in Adelaide

BOX 78

Considering the enormous global burden of disease produced by waterborne infection, the problem faced by Adelaide residents from these diseases is miniscule. Various parasites, protozoa, bacteria and viruses can contaminate water and lead to acute and chronic illness. Most prevalent are the diarrhoeal illnesses. Perhaps best known are cholera and typhoid. Cholera is endemic in many areas of the globe, particularly the tropics, and epidemics occur when sanitation systems break down, often in times of natural disasters and manmade strife. Cholera is spread through the ingestion of contaminated drinking water, when the organism gets into the lining of the small intestine where it prevents the normal absorptions of fluids. The bacterium produces a severe diarrhoea and the resulting dehydration can lead to death if measures cannot be taken to rehydrate the patient. The disease is usually self-limiting in terms of the gut infection with the organism, but the morbidity of cholera is due to its ability to produce a rapid, debilitating state of dehydration.

Typhoid is a serious health issue in regions where the disease is endemic. It is estimated the disease affects 16–33 million individuals annually with more than 500,000 deaths. The usual mode of transmission is faeco-oral with individuals often becoming infected as the result of drinking contaminated water. The organism is absorbed into the bloodstream where it multiplies and exerts a systemic effect – 'enteric fever'. Subsequent excretion into the gut may lead to profuse diarrhoea.

Whilst less dramatic, a much more common cause of diarrhoeal illness is *Escherichia coli*. Enterotoxigenic *E. coli* is probably the most common cause of traveller's diarrhoea. The usual mode of transmission is faeco-oral. The organism is not absorbed into the body and so there is no system infection, and the disease is usually self-limiting.

Whilst almost unheard of in Adelaide, cholera and typhoid do occur from time to time, usually in overseas travellers. Included in the notifiable diseases reported to the South Australian health authorities in 2007 was one case of cholera, five of *Salmonella typhi* (typhoid) and 459 of cryptosporidiosis. Other waterborne infections encountered in Adelaide include amoebiasis and giardiasis.

Peter Devitt

for the transmission of cholera in 1849. Over the next 15 years, Sutherland, Budd and Snow produced evidence that 'deficient and poisonous water' was involved in cholera epidemics in England. Sutherland presented the case that water represented a predisposing factor that contributed to outbreaks, while Snow was credited with arguing that water was a primary cause of illness. Sutherland, Budd and Snow undertook a number of investigations, culminating in the signature work of Snow that resulted in the removal of the handle from the Broad Street pump in London in September 1864, following investigations showing that 83% of cases in a cholera outbreak had used the pump as a source of drinking water. This eliminated a primary source of *Vibrio cholerae* by preventing access to water contaminated by a cesspool.

In Germany, Robert Koch also studied cholera, isolating the bacteria in 1884. In 1892 he investigated the incidence of cholera in two adjacent cities, Altona and Hamburg, both which pumped water from the Elbe River. Hamburg had a cholera outbreak in 1892 while Altona did not. The difference was that Altona filtered its water supply through a slow sand filter while Hamburg did not.[15]

These researchers established principles that remain in place today – that sewage contamination of drinking water transmits disease, that prevention can stop this transmission, and that treatment of water can remove contamination. Chlorination and other forms of treating drinking water were introduced in the early 1900s and increasingly became standard practice. Together with improved sanitation, they greatly reduced morbidity and mortality from diarrhoeal disease.

However, large challenges still remain. The advantage of reticulated water supplies is that large numbers of people can be supplied with safe drinking water; the disadvantage is that failures that allow entry of contamination can have very large impacts. Outbreaks of waterborne disease, while infrequent, are a reality. In 1993 over 400,000 people in Milwaukee contracted cryptosporidiosis from a highly treated drinking water supply.[16] In 2000 in Walkerton, Canada, over 2000 people fell ill and seven people died when a drinking water supply was contaminated by *Escherichia coli* O157:H7 and *Campylobacter jejuni* from animal manure.[16, 17] In Sydney, although there was no evidence of illness, three advisories to boil water were issued over a period of several weeks in 1998 following the reported presence of *Cryptosporidium* (which causes cryptosporidiosis) in the drinking water supply.[18] This event provided a wake-up call for Australia.

As well as the personal cost, outbreaks have very large economic impacts in terms of illness. The estimated cost of the Walkerton outbreak is over CAD$60 million, while the Milwaukee outbreak is thought to be US$96.2 million, including US$31.7 million of direct medical costs and US$64.6 million in productivity losses.[19] There were also costs associated with upgrading the drinking water system to minimise the likelihood of a repeat event. The Sydney incident did not cause any illness but it is estimated that associated costs exceeded A$60 million.[18]

Improved risk management

A movement towards improved risk management to provide much better protection and reliability of drinking water supplies started to gain momentum in the late 1990s. Before the Sydney incident there had been growing concern in Australia and internationally that too much emphasis was being placed on using water quality monitoring at customer taps as a management tool. The shortcomings of end-product testing include the following:

- test results only become available after drinking water is consumed;
- only a small proportion of water supplied can be tested; and
- tests do not exist for many parameters.

As illustrated by Snow's work in 1864, the risk management approach increases the focus on prevention of contamination, and on monitoring of treatment measures to provide evidence that systems are working. The aim is to assure the quality of water before supply to consumers and to minimise the likelihood of incidents that can jeopardise water quality.

Australia was in the vanguard in adopting risk management for drinking water supplies. Development of a framework commenced in 1999 and led to the incorporation of the framework for management of drinking water quality in the 2004 *Australian Drinking Water Guidelines*.[20] In the same year, the World Health Organization (WHO) published the third edition of its *Guidelines for Drinking-Water Quality* including water safety plans,[21] which are similar in theme to the Australian model.

In 2005 a software package, 'Community Water Planner', was published by Australia's National Health and Medical Research Council (NHMRC) to assist in implementation of the framework in small communities.[22] The framework has received a high level of acceptance and has been adopted almost universally in major urban drinking water supplies across Australia and, increasingly, in small communities. The framework has also been applied to very small supplies including domestic rainwater tanks.

One of the advantages of the framework is that it provides a quality assurance approach with reduced reliance on end-product testing. This is useful in rural and remote communities where distance can be a barrier to sample collection, transport and analysis. A risk management approach means that much can be done at a community and local level to assure water quality. Pilot projects with very small communities in remote parts of Western Australia, South Australia, the Northern Territory and Queensland have confirmed the practicality of its implementation.[23]

Conclusions – the public health benefits of safe drinking water

No other measure has had a larger impact on public health than the introduction of safe drinking water and sanitation. Safe drinking water reduces morbidity and mortality from diarrhoeal disease, reduces medical costs, and increases productivity and attendance at school. In the United States, the broadscale introduction of water treatment and

The number of rainwater tanks in urban areas is increasing, with concern for water conservation and new housing regulations that make installation mandatory. Traditionally, rainwater consumption has been most common in rural and remote areas where tapwater supplies are not available. Rainwater may also be a preferred drinking water source where tapwater supplies are of poor aesthetic quality. Although rainwater is subject to microbial contamination from birds and small animals that have access to rooftops, the absence of contamination from humans and farm animals means that the range of pathogens in collected rainwater is more restricted than that found in contaminated surfacewaters. The risk of illness from consuming undisinfected rainwater is therefore lower than for undisinfected surfacewater of apparently similar microbiological quality (as judged by the presence of faecal indicator organisms such as *E. coli*). The consensus among Australian health authorities is that a well-maintained rainwater collection system can provide an acceptable drinking water supply in situations where a disinfected tapwater supply is not available. However, where present, a disinfected tapwater supply is recommended for drinking due to its more reliable microbiological quality.

The published literature contains a small number of reports of disease outbreaks and individual cases of illness attributed to rainwater consumption.[1] In Australia, the bacterial pathogens campylobacter and salmonella have caused documented outbreaks of gastroenteritis linked to rainwater consumption.[2,3] These pathogens are commonly carried by birds and are likely to represent the greatest microbial risk from rainwater consumption. Given that individual rainwater tanks usually supply only a small number of people (typically a single household), the low number of outbreak reports does not necessarily rule out significant risks of illness from rainwater tanks. Routine passive surveillance systems are unlikely to detect small outbreaks, as only about 20% of people with gastrointestinal illness seek medical attention and only a small proportion of those who do have a faecal specimen examined for pathogens.[4]

Concern about the sustainability of water supplies across Australia has resulted in increased interest in the installation of rainwater tanks in urban areas, with most state governments now offering rebates for their installation. Given the potential for this to lead to increased consumption of rainwater, even in areas where a high-quality tapwater supply is available, health authorities believe the potential health risks should be better documented and understood. In addition, mandatory water-conservation measures such as rainwater tanks have been introduced into housing regulations in some jurisdictions, creating a different legal context from the traditional scenario of individual choice and responsibility. Concerns have been expressed about the supply of rainwater even for non-potable uses (e.g., in the bathroom), due to the likelihood that water will occasionally be ingested and illness may result. This has led some developers and water authorities to restrict the use of rainwater to activities with less potential for ingestion, or to install treatment mechanisms such as UV disinfection for rainwater supplies. Such measures impact on the capital and operating costs of alternative water supply systems. Thus, despite the long tradition of apparently safe rainwater consumption in Australia and elsewhere, there is still a need for a better quantitative estimate of the risk associated with ingestion of untreated rainwater. This will assist the public health community to make better-informed risk management decisions on appropriate uses for untreated rainwater in areas where a conventional disinfected tapwater supply already exists.

Martha Sinclair

disinfection in the first 40 years of the 20th century had a substantial impact on reducing mortality and extending life expectancy.[4] Similar results were expected in developed countries such as Australia. Comparisons with the impacts of unsafe water supplies in developing countries reinforce the benefits of safe drinking water. However, challenges still remain and outbreaks of waterborne disease continue to be a threat. Outbreaks in major urban centres have the potential to cause widespread and serious illness, with substantial economic and social impacts.

Water fluoridation

The United States Centers for Disease Control and Prevention in 1999 identified the 10 great public health achievements of the 20th century.[24] Only one, water fluoridation, was directly related to water (although one other achievement, the control of infectious diseases, is associated with access to clean water). Water fluoridation is the backbone of 'dental public health' in numerous countries including Australia. Internationally, water fluoridation has been joined by salt fluoridation in Europe and the central Americas, and milk fluoridation in South-East Asia and South America, as strategies to expose whole populations to a level of fluoride providing substantial protection against dental cavities.

In Australia, 70% of the population lives in an area with the water supply's fluoride concentration adjusted to between 0.6 and 1.1 mg/L of water.[25] In South Australia, water fluoridation was introduced in Adelaide in 1971 under Premier Don Dunstan. The implementation of fluoride in drinking water was first investigated by a select committee in South Australia in 1964 and fluoridisation was agreed to in 1968 under Premier Steele Hall. However, implementation of fluoridation was not without its opponents in South Australia. In the late 1960s there were calls for a referendum on this issue, and claims that the use of supplements were a better alternative. However, Dunstan and Steele Hall took a bipartisan approach, and established direct government control over water processing (so fluoridisation was not the subject of a bill before parliament), and the process was implemented. Since then, towns in the Iron Triangle, those close to Adelaide such as Mount Barker, and those along the River Murray have also introduced water fluoridation at 0.9 mg/L concentration. Now just over 89% of South Australians live in an area with water fluoridation.[25]

However, like several other public health strategies, such as immunisation, a vocal minority of people question the benefit of, and raise concerns about, the risk of water fluoridation. This has led to an almost continuous process of review.

Recent international reviews

The United Kingdom's National Health Service Centre for Reviews and Dissemination, University of York, was commissioned to provide a comprehensive systematic review of the safety and efficacy of fluoridation of public water supplies.[26, 27] The review focused on the scientific literature up to February 2000 on the benefit of water fluoridation for dental caries, and any negative effects. Analysis showed a beneficial reduction in dental caries but the degree of reduction varied. Prevalence of caries was reduced by a

mean of 14.6%, and the mean reduction in decayed, missing or filled teeth was 2.25 teeth. It was estimated that six people needed to be exposed to water fluoridation for one extra person to be free of caries.

Some 88 studies, most before and after designs, were included in the analysis of dental fluorosis. Dental fluorosis is a disturbance in the development of dental enamel, resulting in enamel porosities and an altered appearance of the tooth, ranging from small, whitish flecks to pitting and staining. Analysis showed a strong association between fluoride concentration in the water supply and any dental fluorosis. It was estimated that the number needed to be exposed at 1 mg/L fluoride in the water supply for any dental fluorosis was six, while 22 were required for fluorosis of aesthetic concern. There was no clear evidence of other adverse effects.

The United States' National Research Council (NRC) Committee on Fluoride in Drinking Water was charged to review the toxicologic, epidemiologic and clinical data on fluoride exposure (all fluoride exposures, not just water fluoridation), with particular reference to the United States' Environmental Protection Agency (EPA) basis for standards.[28] The focus was on the concentration at which no adverse health effects are expected to occur and where margins of safety are judged adequate. The review concluded that the maximum fluoride level goal should be lowered from 4 mg/L to prevent severe enamel fluorosis and the lifetime accumulation of fluoride in bone. It was concluded that the prevalence of severe enamel fluorosis is very low (near zero) at fluoride concentrations of 2 mg/L. This fluoride level is the same as the EPA's current guideline for managing drinking water for aesthetic, cosmetic or technical effects. At this level, less than 15% of children will experience moderate enamel fluorosis of aesthetic concern. However, the NRC concluded that the adverse effect on social functioning from moderate fluorosis was not known.

The NRC review clearly stated that their report does not address the lower exposures commonly experienced by the United States' population (and in Australia). This is important, as Griffin et al estimated that only 2% of United States children had fluorosis of aesthetic concern from water fluoridation.[29]

Australian reviews

Two reviews initiated by the NHMRC were published in the 1990s. The first was conducted by a working group established in 1989 in response to the claim that the effectiveness of water fluoridation has been greatly exaggerated.[30] The group examined historical studies, claims of scientific fraud or misuse of data, and time trends in dental caries rates. It also reviewed available evidence on any adverse health effects of fluoride exposure. The group concluded that:

- aggregate evidence established that water fluoridation at around 1 mg/L had, in the past, conferred a substantial protective effect against dental caries;
- water fluoridation remained the most effective and socially equitable means of achieving population-wide exposure to fluoride; and
- there was no evidence of adverse health effects attributable to fluoridated water (at 1 mg/L) and contemporary discretionary sources of fluoride.

Cyanobacterial blooms

BOX 80

Over-use of super-phosphate based fertilisers, land clearing, runoff from roads, drains, urbanisation, and the loss of aquatic species' diversity has caused significant problems for water management in the City of Adelaide. Cyanobacterial blooms are a common occurrence during the warmer months, most visibly in the River Torrens in the heart of the city. The main contributing species are *Microcystis*, *Nodularia* and *Anabaena*, which produce the toxins microcystin, nodularin and saxitoxin respectively. In recent history, dense cyanobacterial blooms have produced toxin concentrations high enough to kill domestic animals and wildlife and pose significant health risks to human consumers. Potential health problems include hepatoenteritis, liver damage, tumor growth, gastroenteritis, hepatitis, renal malfunctioning and hemorrhaging. However, examples of these health problems occurring are uncommon due to management through monitoring programs and restrictions on the usage of affected waters for domestic and municipal purposes.

Adelaide's water supply comes from the Mount Bold and Happy Valley reservoirs (supplying the southern and central regions of the city up to Grand Junction Road), and the Murray River Pipeline, and the South Para, Kangaroo Creek and Warren reservoirs (supplying the north of the city).

Treatment with algicides, such as copper sulphate, is the most common method for killing blooms in the reservoirs supporting Adelaide's water supply. However, prevention of blooms can be approached in a number of ways.

External control measures include:
- protection of catchment areas by changing land usage and sewage treatment;
- reducing nutrient inflows to river water; and
- nutrient retention using wetlands and pre-dams.

Internal control measures include:
- mechanical mixing;
- artificial aeration;
- selective water withdrawal; and
- nutrient precipitation and sediment dredging.

Biological manipulations suitable for Adelaide in the future include:
- promotion of growth/harvesting of water plants; and
- promotion of algal grazing by herbivorous zooplankton through changes in fish community structure (biomanipulation).

These principles must be approached under the auspices of ecologically sustainable development to produce a suitable result. This requires consideration of all aquatic

and associated terrestrial biota, water quality, commerce, and public amenity. Water quality managers are working to adhere to these principles to reverse some of the effects of agriculture, urbanisation and industrialisation.

Michael Sierp and Jian Qin

However, increases in the occurrence of dental fluorosis were linked to increased fluoride exposure, predominantly from 'discretionary' (toothpaste, drops/tablets etc) sources. The NHMRC called for effective control over discretionary sources of supplementary fluoride. This became the focus of action at a workshop on Appropriate Fluoride Exposure for Infants and Children, held in Perth in December 1993. The workshop discussed: the elimination of fluoride from infant formula powder; a delay in commencement and reduced age-specific dosage of fluoride supplementation; the provision of low-fluoride children's toothpastes; and the promotion of appropriate practices for children using toothpaste (e.g. use under parental supervision, pea-sized amounts, spit out, don't swallow).[31]

Many of these themes were continued in the second review, a systematic one conducted for the NHMRC in 1999, which gathered and evaluated the scientific evidence since 1990.[32] The evaluation concluded that water fluoridation continued to provide significant benefits in the prevention of dental caries. The benefit was strongest in children but could also be demonstrated in adults. There was further evidence of increased dental fluorosis. Reduction in excessive intake of fluoride from discretionary sources was again targeted.

Ongoing issues

Australian research supports a strong association between water fluoridation and a high population preventative fraction due to water fluoridation. This research was conducted using a cohort study with an emphasis on exposure measurement. There is a consistency in the four reviews that is noteworthy. Those that addressed benefits confirm the evidence for water fluoridation in protecting against dental caries, predominantly in children, but also among adults. A theme that emerged from these reviews is that the benefits of water fluoridation have diminished over time. Australian research, using a cohort study approach with an emphasis on exposure measurement, supports a strong association and a high population preventative fraction due to water fluoridation.[33] For instance, in South Australia exposure to water fluoridation was significantly associated with caries experience in the deciduous dentition at age eight years old. However, calculations showed that there was a truncation of exposure to water fluoridation due to residential mobility, the use of water filters, bottled or tank water consumption, and reduction in the use of tap water in overall fluid consumption. These issues contribute to a misclassification error in simple ecological comparisons.

All four reviews drew attention to the risk of fluorosis. The benefit–risk trade-off in exposure to fluoride is not a new issue. Henry Trendley Dean identified this by establishing 1 mg/L as the desirable fluoride concentration level in temperate climates to achieve near maximal reduction in dental caries without unacceptable levels of dental fluorosis.[34] It is important to recognise that Dean and other early fluoride researchers were not seeking to eliminate all risk of fluorosis. Instead, they sought an acceptable prevalence or severity of dental fluorosis, i.e. a level which would not cause community concern over the aesthetic appearance of teeth.

As additional discretionary sources of fluoride exposure were added in the 1970s and 1980s, the total exposure to fluoride increased among children aged 0–6 years old. While dental caries in Australia reached a low point in the early to mid 1990s, it was clear that the prevalence of any fluorosis and of potentially more severe fluorosis was high, noticed by parents who felt that altered the appearance of their children's teeth.[35]

As mentioned above, a loose network of dental academics, public dental services, the organised dental profession and the dental industry combined to reduce fluoride exposure from discretionary sources. The result has been a dramatic reduction in the prevalence and severity of dental fluorosis. Riordan in Western Australia and Do and Spencer in South Australia have documented the approximate halving of the prevalence of any fluorosis by the early 2000s.[36, 37]

Spencer and Do have identified a reduced risk of fluorosis from exposure to water fluoridation due to a decrease in certain risk factors. Some tooth brushing practices are no longer a risk factor because of the use of low fluoride children's toothpaste.[38] Fluoride supplements (drops and tablets), which are no longer used widely, have ceased to be a risk factor for fluorosis. Importantly, a history of bottle-feeding with infant formula has failed to be a risk factor in Australia, either in the early 1990s or 2000s. This is the evidence behind the current guideline that infant formula can be reconstituted using either fluoridated or non-fluoridated water. This specific issue was considered in the development of guidelines on the use of fluorides in Australia at a workshop in Adelaide in October 2005.[39] The reduced prevalence of fluorosis, together with changes in fluoride exposure due to reductions in certain risk factors, point to a reduced total exposure to fluoride among children.

These findings are also relevant to the issue of accumulation of fluoride in bone, and of any adverse consequences, issues that are still under investigation. The formal reviews concluded that there is insufficient evidence that water fluoridation is associated with hip fractures or bone cancers.[27, 28] A recent retrospective case-control study on osteosarcoma in the United States has attracted some attention.[40] This study found an increased risk for osteosarcoma associated with exposure to water fluoridation across the ages of 6 to 8 years in males. Methodological issues and the plausibility of the sex-specific finding need to be considered. However, more importantly, other researchers have not been able to replicate these findings with the same data, and a related prospective case-control study has failed to find any association.[41] No doubt the issue will continue to attract researcher and community interest.

Conclusion – water fluoridation

Water fluoridation has been subjected to great scrutiny with regard to risks. The only established potential adverse outcome is dental fluorosis. Yet little of the identified fluorosis is aesthetically of concern. Recent research has clouded judgement even about mild fluorosis being an adverse outcome.[42] A challenge for the water industry and public health agencies is to show an equal concern for those communities who have exposure to water with negligible fluoride and lack of protection against dental caries.

Table 17.1 Cyanobacteria and drinking water quality

Type	Compound	Organisms	Effects	Public health significance	Guideline status	Implications
Hepatotoxins	Microcystin	• Microcystis aeruginosa • Anabaena spp.	• Liver damage • Tumour promotion in animal studies	• Acute toxicity in large water supply systems unlikely • Chronic liver damage possible with chronic exposure • Possible human carcinogens	• WHO guideline (1988): 1 µg/L for microcystin-LR • Australian guideline (2001): 1.3 µg/L expressed as microcystin-LR toxicity equivalents	• Operators obliged to monitor for the compound • The guideline is equivalent to 6500 cells/mL for a highly toxic population of M. aeruginosa
Hepatotoxins	Nodularin	• Nodularia spumigena	• Liver damage • Tumour promotion in animal studies	• Acute toxicity in large water supply systems unlikely • Chronic liver damage possible with chronic exposure	• No Australian guideline • However, hazard assessment is often guided by that for microcystin	• Only likely to be a significant drinking water quality issue in freshwater (e.g., lakes of Lower River Murray)
Hepatotoxins	Cylindrospermopsin	• Cylindrospermopsis raciborskii • Aphanizomenon ovalisporum • Anabaena bergii	• Cytotoxic • Liver, kidney and other organ damage • Genotoxic	• Risk of acute toxicity via drinking water supplies dependent on circumstances • Animal studies to investigate sub-chronic exposure have now been undertaken	• No Australian guideline • However, the value proposed by Humpage and Falconer (2003) of 1 µg/L is used in some locations	• A significant water quality issue for tropical Australia • May be an issue in southern Australia if the projected climatic warming occurs
Neurotoxins	Saxitoxins (paralytic shellfish poison – PSPs)	• Anabaena circinalis	• Sodium channel blocking agent – acute poisoning results in death by paralysis and respiratory failure	• Acute toxicity in large water supply systems unlikely • Effects from chronic exposure not known • Public health significance unclear: no evidence of human illness from drinking water	• No Australian guideline • However, a 'health alert' of 3 µg/L for acute exposure is sometimes used for guidance	• Not likely to be as significant a drinking water quality issue as microcystins
Endotoxins	Lipopolysaccharides (LPS)	• Most cyanobacteria	• Implicated in gastrointestinal disorders; skin and eye irritation; respiratory symptoms	• Less toxic than hepato or neurotoxins • Recent studies indicate relatively low potency • Effects from chronic exposure not known	• No Australian guideline	• New research currently underway will help to clarify potential health significance

Table 17.1 Cyanobacteria and drinking water quality: major classes of toxins produced by cyanobacteria in Australia, their significance to water quality, comments on their guideline status and implications for management of water supply. Source: adapted from South Australian Department of Health, 'Public Health Bulletin: Water and Public Health', 2007.

Blue-green algae: toxicity and public health significance

Blue-green algae are not algae! These photosynthetic prokaryotes are more correctly called cyanobacteria. Many cyanobacteria can regulate their buoyancy to take best advantage of available light and nutrients as conditions change during the day. Because they tend to sink during the day and then rise again at night, scums are often found on the water surface in the morning. A light breeze can concentrate them on the edge of a waterbody, endangering thirsty animals or people swimming or wading. Cyanobacteria produce a range of chemicals or 'secondary metabolites'. Many of these are, probably accidentally, harmful to mammals and have therefore been labelled cyanotoxins.[43]

In the drought year of 1991, the Darling River was reduced to a trickle between weir pools, and a stretch of the river turned green. Many cattle and sheep died, a state of emergency was declared, and the army was called in to provide clean, filtered water for local towns. The culprit was identified as a toxic blue-green alga called *Anabaena circinalis*, and the headlines screamed 'Algae AIDS'.

The toxins responsible are known as *saxitoxins*, after the parent compound, or alternatively as paralytic shellfish poisons because similar compounds are responsible for paralysis that can occur when saxitoxin-contaminated marine shellfish are eaten. These are neurotoxins that specifically inhibit nerve signal transmission, leading to paralysis of various muscles including those responsible for inflating the lungs. Twenty-seven structural variants have so far been identified worldwide, although only about seven are commonly found in *A. circinalis*.

Background and Australian drinking water quality guidelines

Australia has adopted a provisional drinking water health alert level of 3 μg of saxitoxin per litre of water. This health alert level is designed to protect drinking water consumers from the acute neurotoxicity effects caused by saxitoxins rather than any chronic effects, for which there is little evidence. While saxitoxin produced by *A. circinalis* is arguably the most potent and widespread cyanotoxin in surfacewater in South Australia, there are a range of other cyanobacteria that also produce toxic compounds, which can have a deleterious effect on drinking water quality.[44] Table 17.1 presents a list of compounds found to occur in cyanobacteria in Australia and their public health and water quality significance.

The Darling River bloom was certainly not the first time these toxigenic organisms had caused problems for Australians. Indigenous populations of the Lake Alexandrina and River Murray regions of South Australia knew to avoid areas of green scum in the water. In 1878, horses, cattle and sheep began dying around this same lake. George Francis, a local assayer and chemist, described scum 'like green paint', detailing the effects when fed to a sheep.[45] He correctly identified the organism responsible as *Nodularia spumigena*.

The toxin produced by *N. spumigena*, nodularin, is structurally similar to the much larger group of cyanotoxins, called microcystins, which are the most commonly found and studied cyanotoxins worldwide. In Australia the microcystins are produced by *Microcystis aeruginosa*, an organism that is much more common in our drinking water sources than is *Nodularia*. These toxins are cyclic peptides, nodularin containing five amino acids whereas microcystins are composed of seven amino acids.[45] Due to specific cellular uptake requirements, they are primarily liver toxins. The WHO has derived a provisional safe drinking water guideline level of 1 μg/L for microcystin-LR.[43] This guideline level is the concentration in water that it is considered safe to drink over a lifetime without incurring adverse health effects.

However, worldwide there have been about 80 microcystin variants identified, each having the basic cyclic peptide structure but with differing amino acid constituents or side-chain substitutions. Their toxicity varies over about a four-fold range. The NHMRC has adapted the WHO guideline, proposing an Australian drinking water guideline of 1.3 μg/L for total microcystins. Recently, the International Agency for Research on Cancer classified the microcystins as 'possible human carcinogens' (Class 2B), based on consideration of the accumulated toxicological data. In particular, this decision was influenced by the finding that these toxins inhibit certain protein phosphatases, which are enzymes that are critically involved in cell-cycle regulation.[46] Some modification of the drinking water guideline values may now occur to take account of this.

The most serious Australian cyanobacterial poisoning event occurred in 1979 when many residents of Palm Island in northern Queensland became sick with a 'mystery disease'.[47] Symptoms included liver, intestinal and kidney dysfunction. These events occurred shortly after the water authority treated a bloom of cyanobacteria with copper sulphate, a means of dealing with blooms that is still used in many places. The effect is to lyse the cells, which releases toxins into the water. A retrospective study of the cyanobacteria in the reservoir identified *Cylindrospermopsis raciborskii* as being able to produce similar effects in mice to those observed in the human victims.[48]

In 1992 the structure of cylindrospermopsin was determined from material purified from a culture of *C. raciborskii* from the same reservoir on Palm Island. Related toxins have since been found in *Aphanizomenon ovalisporum* and *Lyngbya wollei*. Somewhat surprisingly, given the great structural variety of the other toxin types, only three cylindrospermopsins have so far been found, all with similar potency. They inhibit protein synthesis but also have a variety of other effects including genotoxicity. In animal experiments the acute effects occur predominantly but not exclusively in the liver. However, low-dose, longer-term exposure also induces significant alterations in kidney size and function.[49] Human cylindrospermopsin poisoning also induced both liver and kidney dysfunction. The WHO is preparing a summary document on these toxins but does not yet consider there to be adequate data on which to base a guideline value.[50]

The control of cyanobacteria in reservoirs

BOX 81

Blue-green algae are one of the oldest forms of life still thriving on earth. In fact, they are not algae at all but a type of photosynthetic bacteria (cyanobacteria), which is reliant on sunlight for energy. During the warmer summer months, with an increase in water temperature and perhaps a decrease in flow and/or depth of water sources, cyanobacteria can form 'blooms', which are of significant concern to water utilities.

Some blue-green algae have an advantage over other aquatic microorganisms, as the cells contain small pockets of air, or gas vacuoles, which allow them to travel up and down within the water column, encountering favourable nutrients and sunlight. As the sunlight hits the water surface, the cyanobacteria photosythesise to form carbohydrates, making the cells denser and slowly sink to the bottom of the waterbody. The lower parts of the water column are usually higher in nutrients, such as nitrogen and phosphorus, and the cells convert their carbohydrates to cell material as they assimilate the available nutrients. As they utilise the carbohydrates they become lighter once more, and ascend towards the surface. In this fashion they can utilise the energy and nutrient sources available more efficiently than other phytoplankton.

There are more than 150 different types (genera) of cyanobacteria and literally thousands of species. Although we don't really know, a large proportion of these could be capable of producing algal toxins. What we do know is that around 60–70% of the time these cyanobacteria will produce toxins. In other words, if you have cyanobacteria present in your water supply, there is a reasonable chance that you will also have algal toxins.

Algal blooms have traditionally been controlled by the addition of chemical ('algicides'). The most popular chemical is copper sulphate, which has been used widely to control blooms in drinking water sources for nearly 100 years, as it is effective, economical and generally safe for operators to use. The copper sulphate is usually spread or sprayed onto the surface of the water from a boat. Other popular algicides include a range of copper chelated chemicals. The effectiveness of copper-algicide treatment is controlled by factors such as the pH, alkalinity and dissolved organic carbon in the water.

Physical factors, particularly thermal stratification in the reservoir, affect the distribution of copper after application, contacting with the algae. Cyanobacteria are generally regarded as being relatively sensitive to copper. The important point to remember about copper is that it is a broad-spectrum aquatic biocide, and this can have significant adverse environmental effects. Local environmental regulations may determine the conditions under which algicides can be used.

As stratification of the water source can be an important factor in the uncontrolled growth of cyanobacteria, techniques to mix the stratified layers of a lake ('destratification') have been shown to reduce cyanobacteria in some cases. The thing about lake mixing to control the growth of cyanobacteria is that success will depend on a combination of the type of mixer or aerator (and how much energy it puts out), and the depth and clarity (i.e., light penetration) of the lake. Generally, mixing is not very helpful in reducing growth of buoyant cyanobacteria in shallow lakes (say <15–20 m for clear water), and works best in deeper lakes where cells are carried down out of the light for extended periods. Even then it might not eliminate them completely but only slow them down to more manageable numbers. The circumstances where success in controlling blooms has occurred are not well studied, but we know that destratification can work for control of a common nuisance species called microcystis, but is not as good for smaller, less buoyant cell types like cylindrospemopsis.

The maximum biomass of cyanobacterial blooms is ultimately determined by nutrients, so the reduction of nutrients such as nitrogen, and particularly phosphorus, coming into a waterbody remains the best sustainable long-term solution to reducing problems associated with cyanobacteria. Without a commitment to nutrient reduction from the catchment, other measures aimed at controlling cyanobacteria will be compromised.

Mike Burch and Justin Brookes

BOX 82

Copper in reservoirs:
the Warren Reservoir experience

In Australia, the eutrophication of waterbodies from the excessive input of nutrients is a significant and expensive environmental problem. The most common symptom of eutrophication is an increased abundance or 'blooms' of nuisance species such as blue-green algae (cyanobacteria).[1,2] While managing nutrient sources is extremely important, there is still a need for direct treatment of blooms when they occur.

For over a century the most extensively used algicide in lakes and reservoirs has been copper sulphate $(CuSO_4)_3$. The widespread popularity and use of $CuSO_4$ as an algicide relate to the chemical's toxicity to lower organisms (algae, etc.), and rapid precipitation. Nevertheless, copper (Cu) is a biologically toxic contaminant and there are specific guidelines for its concentration within the water and sediments.[4] Sediments deposited through time in reservoirs that have been dosed with $CuSO_4$ can provide valuable information on the fate of this contaminant after precipitation.

Warren Reservoir

Warren Reservoir is located approximately 60 km northeast of Adelaide in the Mount Lofty Ranges. The reservoir is commonly eutrophic, with frequent blooms of toxic cyanobacteria. In 2005, a 0.7 m long sediment core was raised from the reservoir as part of a paleolimnological study to reconstruct total phosphorus (TP) levels since the reservoir was filled in 1916. The research included creating a ^{210}Pb (a lead isotope) and ^{137}Cs (a caesium isotope) based chronologies of the sequence, and determining its geochemistry to infer changes in sediment sources using x-ray fluorescence (XRF).

Excess ^{210}Pb ($^{210}Pb_{ex}$) activities in the upper 50 cm of core were analysed, and showed a relatively constant decay with depth (depth vs $^{210}Pb_{ex}$ correlation coefficient r^2=0.0.9846, n=4). For this reason it was assumed the sedimentation rate at the top 50 cm of the core was constant. The first appearance of ^{137}Cs activity was found at 50 cm, which marks the nuclear testing period of 1950–1960 (peak at 1958).[5] The age of the sediments at the top 50 cm of the core were determined by interpolating the sediment age at 50 cm, from the Cs-137 data, to the top of the core at the sample collection date of 2005 .Furthermore, there is a strong correlation between known dates of $CuSO_4$ dosing and CuO sediment concentrations.

In terms of CuO concentrations in sediments, it is important to note that the <10 m fraction are likely to be overestimates relative to guidelines that prescribe concentrations for sediments overall (i.e., non-particle size specific). As a result, we studied a minimum total CuO concentration that assumes all the sediment CuO is found in the <10 m fraction. This conservative approach allows comparison of our data with sediment quality guidelines.

Copper concentrations in the core show strong correlations with copper dosing, which began in 1950. Copper concentrations in sediments deposited in the early 1960s follow the pattern of increased dosing during this period. Increased concentrations to a late-1970s peak correspond to increased dosing between 1978–1983, with a short-lived decline in dosing recorded in the sediments, followed by a peak in both dosing and concentration in the early 1990s.

Much of the sediment deposited in Warren Reservoir sediment since the late 1960s exceeds the 'low' Australian and New Zealand Environment Conservation Council and Agriculture and Resource Management Council of Australia and New Zealand guidelines (2000) for protecting ecosystem health. Concentrations have recently fallen below this threshold due to reductions in $CuSO_4$ dosing. However, as the guidelines note, it is difficult to discern whether such concentrations will do environmental harm. The strong correlation between historical $CuSO_4$ dosages and core geochemistry perhaps indicates that the bioavailability of $CuSO_4$ after precipitation is low. This conclusion warrants further investigation. Certainly, our study highlights the benefits of reducing $CuSO_4$ dosing, since recent reductions have resulted in commensurate declines in sediment CuO. Furthermore, it is clear that the pattern of copper deposition can be used as a 'chronometer' for sediments deposited in reservoirs with a copper dosing history. These sediments can further be used to address a variety of questions about the effect of changing catchment management.

Ashley Natt, John Tibby, Atun Zawadzki and Jennifer Harrison

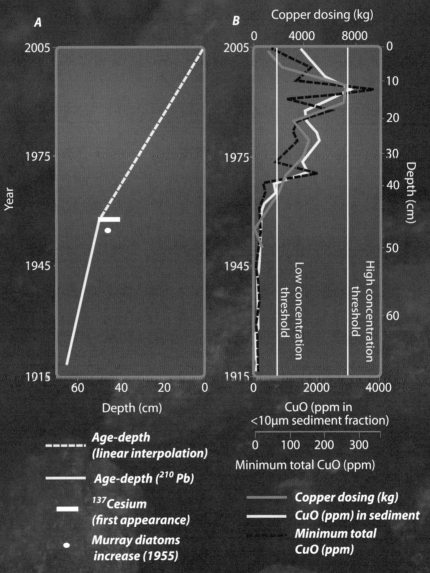

Warren Reservoir sediments

Figure A Age-depth relationship of Warren Reservoir sediments and location of chronological markers.

Figure B CuSO$_4$ dosing history of Warren reservoir and sediment CuO concentrations for the <10μm fraction and a (minimum) estimate of the total CuO concentration in the sediment. The white vertical lines represent the low and high interim sediment quality guidelines for protecting ecosystem health (Australian and New Zealand Environment and Conservation Council (ANZECC) and Agriculture and Resource Management Council of Australia and New Zealand (ARMCANZ), 2000).

Other routes of cyanobacterial poisoning

Oral consumption of cyanotoxin-contaminated drinking water is by no means the only route of exposure.[43] In Brazil in 1996 toxin-containing water was supplied to a dialysis clinic, and the poorly maintained onsite water purification system was overwhelmed. One hundred and sixteen people experienced visual disturbances, nausea and vomiting; 110 developed acute liver failure; and 60 deaths were attributed to acute intoxication by cyanotoxins. Both microcystins and cylindrospermopsin were found in the water filtration system.[51]

During recreational or occupational use of water bodies supporting blooms of toxic cyanobacteria, dermal and inhalation exposures may be more important. There is epidemiological evidence for an association between contact with water containing cyanobacteria and relatively mild and self-limiting dermal effects. Contact hypersensitivity in susceptible individuals has also been reported anecdotally.[52, 53] There has been very limited experimental work investigating the inhalation route, but the few results available suggest that microcystins can cause lesions in the nasal and respiratory lining, and that this may facilitate access of these and other toxins to the circulation.

Food may also be contaminated. Although contamination of meat and milk does not occur when cattle drink contaminated water, fish, prawns and shellfish have been shown to sometimes contain significant quantities of cyanotoxins, to the extent that populations relying on these sources of protein may receive potentially hazardous exposures. In Australia this route seems to be of most importance for Indigenous people leading a traditional lifestyle. A particularly First World exposure route can occur via 'blue-green algae supplements', which are sold as health foods. A number of these products have been found to contain microcystins in quantities high enough to be of concern, particularly for children.[54]

Current research

The scientific investigation of toxigenic cyanobacteria and their toxins, as well as our understanding of water treatment options for their removal, have taken great strides in the last 20 years. Australian researchers have been at the forefront of that endeavour. This has been greatly facilitated through bodies like the Cooperative Research Centre for Water Quality and Treatment, and by the Australian drinking water industry, which has a recognised duty of care to ensure drinking water is protected by comprehensive monitoring and alert response protocols that bring additional treatment options into use when needed. However, other potential routes of exposure have been less well researched; in particular, the potential risks due to inhalation of aerosols generated during water-skiing, washing of industrial equipment, spray irrigation and dust suppression remain largely undefined. Lastly, in the current climate of water shortage and drought, there may be an understandable temptation to concentrate on water quantity rather than quality, but this must be resisted if we are to avoid repeating the unfortunate events of the past.

Water recycling in South Australia: overview from a public health perspective

The provision of safe water and sanitation has been more effective than any other intervention in reducing infectious disease and increasing public health.[55]

Recycled water schemes have the potential to reduce demands on other water sources, particularly for purposes which do not require water of drinking quality. Recycled water schemes can also reduce discharges of nutrient-enriched wastewater into receiving waters, resulting in potential environmental benefits.

Use of recycled water for irrigation has occurred for many years. Advances in wastewater treatment technologies, current drought conditions and water restrictions have increased demand for alternative water sources and expanded the scope for recycled water use. The result has been a progression from applications such as restricted irrigation to toilet flushing, and, more recently, use of recycled water for drinking. The current plan for the addition of highly treated recycled water into south-east Queensland's drinking water supplies demonstrates the value of recycled water in today's climate. However, protection of public and environmental health is of paramount importance. Recycled water schemes must therefore be planned, designed, installed and managed to ensure that public and environmental health is not compromised.

This section provides an overview of recycled water schemes in South Australia, risk management approaches to recycled water use, and the South Australian DoH's role in managing recycled water from a public health perspective.

Recycled water in South Australia

South Australia is a national leader in recycled water use.[56] In 2006–2007, 29.7% of metropolitan and 19.1% of rural reated sewage was recycled by SA Water.[57] Recycled water is also used in areas outside SA Water's jurisdiction. For example, septic tank effluent disposal schemes (STEDS) service 10% of the population in South Australia, and more than 50% of the treated effluent produced from STEDS is reused.[58] Recycled water applications in South Australia are wide and varied, both in the type and volume of reuse. The main uses are irrigation of crops, such as woodlots, lucerne and grapevines, and of parks and gardens. This type of reuse can be achieved on a smaller scale than applications such as toilet flushing, which requires a more expensive treatment and management regime.

Some of the largest and most innovative schemes in South Australia are supplied with recycled water from SA Water's metropolitan sewage treatment plants at Bolivar, Glenelg and Christies Beach, and include:

1. The Virginia Pipeline Scheme (VPS) supplies treated sewage from Bolivar wastewater treatment plant (WWTP) for irrigation of commercial food crops.

BOX
83

Biomanipulation:
a potential tool to improve water quality

Biomanipulation is an approach that attempts to restore water quality in eutrophic lakes, including lake improvement procedures that alter the food web to favour grazing on algae by zooplankton. Biomanipulation is new to the lake management community, particularly to those without training in limnology. The theoretical basis underlying this concept is the Trophic Cascade Theory, which states that a rise in piscivore numbers decreases planktivore numbers, increases herbivores, and decreases the amount of phytoplankton, resulting in a high Secchi depth (water clarity). In essence, biomanipulation involves eliminating certain fish species or restructuring the fish community to favour the dominance of piscivorous fish instead of planktivorous fish. There seems to be little question that the biomanipulation approach will become increasingly popular not only because of its effectiveness but its lower cost and the absence of using machinery and toxic chemicals.

Strong top-down effects of fish on zooplankton have been demonstrated in many freshwater reservoirs. High fish predation pressure generally results in low zooplankton standing stocks. A reduction of predation by removal of fish will result in an increase in total zooplankton biomass and will also allow populations of large zooplankton to develop. A high-grazing efficiency should allow populations of large zooplankton to grow fast. This can result in a higher grazing pressure exerted on phytoplankton, and the shift towards a dominance of large zooplankton can be an important factor in successful biomanipulation.

In Australia, research on biomanipulation is limited but has yielded promising results. Studies conducted in some Australian reservoirs in New South Wales have shown that grazing by zooplankters has significant effects on phytoplankton, at least during some times of the year. In contrast to the situation found in lakes of the Northern Hemisphere, the planktivorous fish of Australian reservoirs are relatively small and there is a lack of large piscivorous species. It is possible to stock drinking water reservoirs with some large piscivorous fish such as Murray cod (*Maccullochella peelii peelii*) to control the abundance of planktivores. But, before applying biomanipulation as a management option, it is necessary to understand the fish community structure and to assess the fish stocks at each trophic level.

In South Australia, the quality of drinking water reservoirs is a great concern. For instance, Warren Reservoir in North Adelaide supplies drinking water to the metropolitan area. To reduce algal blooms in the reservoir, considerable money has been spent to place an artificial aeration pumping system extending along the dam wall. The underlying reasons for this aeration system are that aeration can: prevent anaerobic sediment conditions and reduce nutrient loading in water; and inhibit motile algae floating on the water surface and decrease algal productivity. Despite the artificial aeration system, algal blooms still occur in the reservoir and costly application of copper sulphate is still required to kill blooms. In addition, excess use of copper sulphate in drinking water could lead to accumulation of ionic copper and can be harmful to consumers.

In Warren Reservoir, a total of five fish species have been found with only one native fish, the flathead gudgeon (*Phylypnodon grandiceps*); the other four were exotic species – redfin (*Perca fluviatilis*), mosquitofish (*Gambusia holbrooki*), goldfish (*Carassius auratus*), and common carp (*Cyprinius carpio*). Redfin perch are the top predators, but the maximum size was less than 25 cm long. Most specimens of common carp exceeded 36 cm long, goldfish more than 20 cm long, and all other fish less than 3 cm long. These results clearly indicate the necessity for introducing large carnivorous fish species into the system for biomanipulation to control planktivorous fish.

Murray cod are Australia's largest freshwater fish, once abundant, but have declined in range and abundance, caused by overfishing and environmental changes. Artificial spawning has been successfully established in an attempt to increase the natural population. Murray cod could be a potential piscivorous fish candidate for biomanipulation in reservoirs. At Flinders University, the feeding biology of hatchery Murray cod as a top predator has been studied. The results show that hatchery-reared Murray cod is capable of preying on exotic species such as mosquitofish and goldfish, and has the potential to control exotic fish including mosquitofish, redfin fish and goldfish.

Jian Qin and Michael Sierp

2. The Mawson Lakes Dual Reticulation System supplies recycled water within the Mawson Lakes development for purposes such as toilet flushing and garden watering.
3. The Willunga Basin Pipeline distributes treated wastewater from the Christies Beach WWTP through the southern vales for irrigation of grapevines, fruit trees, nut crops and flowers.
4. Adelaide Airport is supplied with recycled water from Glenelg WWTP for toilet flushing and irrigation.

In areas not serviced by SA Water, wastewater systems are generally managed by local councils and, in some cases, private operators. These schemes service outer metropolitan, regional and rural townships as well as holiday shacks or other small communities. They are stand-alone systems consisting of collection, treatment and reuse infrastructure. Recycled water is typically used for applications such as the irrigation of crops (e.g. lucerne or grapevines), town ovals and parklands. Recycled water therefore provides economic and aesthetic benefits to local farming communities and townships, particularly in areas of low rainfall.

Onsite wastewater systems also serve approximately 150,000 South Australians living in areas not serviced by a community wastewater management system (i.e. sewer or STEDS). Typically, these are individual households and small commercial facilities. Many onsite wastewater systems provide treatment such as aeration and disinfection, after which the treated wastewater may then be used for restricted irrigation of household garden beds and lawns.

Human health risks associated with recycled water
The *Australian Guidelines for Water Recycling: Managing Health and Environmental Risks*[55] are currently being developed to support increased water recycling in Australia. Development of these guidelines is separated into two phases. Phase 1 deals with non-drinking water uses of treated sewage and greywater while Phase 2 addresses stormwater recycling, use of recycled water to augment drinking water supplies, and managed aquifer recharge. Phase 1 was published in November 2006 and Phase 2 is nearing completion.

The *Australian Guidelines for Water Recycling* are important to those organisations responsible for assessing health risks, identifying appropriate risk management requirements, and managing public concerns. The South Australian DoH contributed to the development of Phase 1 of the guidelines and is playing a similar role in Phase 2. This involvement forms part of the Water Proofing Adelaide strategy.[59]

The guidelines adopt a similar risk management framework to that used in the *Australian Drinking Water Guidelines*. They comprise 12 elements that fall into four main categories:[55]

- Commitment from all stakeholders to responsible use and management of recycled water.
- Recycled water system analysis and management.
- Supporting requirements such as employee training, community involvement, validation, and documentation and reporting systems.
- Evaluation and audit reviews.

Human health risks associated with use of recycled water are presented by microbial hazards (i.e. pathogenic bacteria, viruses, protozoa and helminths) and human exposure to these hazards.[55] The safe use of recycled water requires management of both aspects.

Risk assessment of a recycled water system therefore involves consideration of the type of treatment applied and the potential exposure to humans though irrigation or other uses.[60] For example, the level of treatment required for irrigation of a fenced woodlot with no public access is significantly less than that required for a system using recycled water for irrigation of leafy vegetables which are eaten raw, such as lettuces.

Other measures must also be built into recycled water schemes to control risks.[55] There must be control of trade waste discharges into the wastewater system because they may contain heavy metals or other contaminants. Treatment plant design must be robust and reliable enough to provide proven sufficient treatment levels for the type of recycled water end-use anticipated. Online monitoring of critical treatment processes which indicate treatment failure is also required.[55]

Department of Health's role in the use of recycled water
Recycled water schemes are regulated under the *Public and Environmental Health (Waste Control) Regulations 1995*. The DoH provides formal approval for all recycled water schemes involving treated sewage, under the regulations. In order to support safe recycling, the DoH works cooperatively with stakeholders while upholding public and environmental health as the primary objective. By providing advice for proponents throughout the project's planning, design, installation and operation stages, the DoH ensures that specific risk management goals are met for each system. The DoH also receives routine water quality monitoring reports and is notified of incidents associated with many of the recycled water schemes.

Each year the DoH approves approximately 100 wastewater system installations. Most of these approvals include some aspect of water recycling or have an impact on the recycled water system. Increasingly, proposals incorporate new technologies and innovative uses for recycled water, presenting an ongoing challenge to keep abreast of changing technologies and risk management methods. The DoH regularly reviews policies and procedures accordingly.

Conclusions – a public health perspective on recycled water
A key outcome of the Water Proofing Adelaide strategy is to increase the use of recycled water from about 14,000 ML/year in 2002 to in excess of 30,000 ML/year by 2025.[59] This projection reinforces the importance of recycled water in meeting South Australia's water resource needs over the next two decades. The DoH will play a vital part in ensuring that current and future recycled water schemes do not compromise public and environmental health.

Managing public and environmental health risks is paramount in the continuing success of recycled water schemes. The public expects to have access to safe water, and if this confidence is compromised it is difficult to regain.[55] Changing trends and technologies, increased confidence and, in some cases, complacency in recycled water use, along with other factors, contribute to varying risks in the use of recycled water.

Public health experts must apply risk management methods and keep in touch with changing technology and innovation when considering use of recycled water. Appropriate recycled water policies can then be determined in order to continue to protect public and environmental health.

Climate change and waterborne diseases

According to the most recently released report from the Intergovernmental Panel on Climate Change (IPCC) in 2007, the global average surface temperature has increased approximately 0.74 °C during the last century.[61] For the 21st century, based on the best estimate for a 'high scenario', it is estimated that there will be a 4.0 °C increase in temperature with a likely range of 2.4 to 6.4 °C, and the sea level may rise in the range of 26 cm to 59 cm. It is very likely that there will be an increase in frequency of warm spells, heatwaves and events of heavy rainfall. Human beings will suffer from more hot days, more extreme rainfall, and more floods and droughts, which may have effects, direct and indirect, on population health, including an increased prevalence of waterborne diseases.

Waterborne disease is a global public health problem. In the United States about 9 million cases occur every year.[62] Food-related diseases account for about 10% of morbidity and mortality in the United Kingdom, with costs of about £6 billion annually; and among these, many are waterborne infections.[63] In Australian communities, the cost of enteric infection, including both food- and waterborne infections, has been estimated at over A$2.6 billion a year.[64] According to the WHO, infectious diseases transmitted by contaminated water and food in developing countries are a constant and frequently fatal threat to population health.[65] For example, diarrhoeal diseases are estimated to cause an annual 2.7 billion cases and 1.9 million deaths, mostly in infants and young children.[65]

The control of waterborne diseases is still a big challenge; it is critical to understand the underlying causes, including the role of climate variability, in their transmission.[66] Although waterborne disease is multi-causal, the growth and spread of the responsible microorganisms are influenced by climate variations.[62, 67] Temperature and relative humidity directly influence the rate of replication of the pathogens and the survival of the agents in the environment. Rainfall, especially heavy rainfall events, may affect the frequency and level of contamination of drinking water. Moreover, climate change may influence water resources and sanitation so that water supply is effectively reduced. Such water scarcity may necessitate using poorer quality sources of freshwater such as water from rivers, which are often contaminated. All these factors may result in an increased incidence of food – and waterborne diseases.[62]

Relationship between climate variation/change and waterborne diseases

It is very difficult to identify where and how a change in climatic conditions might alter the hazards posed by waterborne diseases.[68] However, the relationship between climate variation and food- and waterborne diseases has been documented by limited studies conducted in Europe, Australia, the United States and Asia,[62, 69, 70, 71, 72, 73, 74] and the association between climate variation and bacillary dysentery (shigellosis) in China has also been reported recently.[75] Most of these studies suggested that climate change with increased temperature will bring about more cases of food- and waterborne diseases. One study indicated that each 1 °C increase in temperature is related to a 5% increase in the risk of severe diarrhoea.[71] Two studies suggested that other climatic variables such as rainfall also played an important role in enteric infection transmission,[76, 77] although previous studies reported inconsistent results.[70, 71]

The impacts of climatic variables on waterborne diseases vary across different climatic regions, including the thresholds of weather variables, lag times of effects, and seasonal patterns in a number of cases. Highly urbanised populations and well-managed water treatment systems in metropolitan areas may conceal the association between rainfall and waterborne diseases. More studies in different climatic areas are needed to further understand the relationship.

In Australia only a few studies have investigated the relationship between climate variations and food- and waterborne diseases. OzFoodNet, the Australian national food-borne diseases surveillance system, in 2002 recorded 14,716 cases of campylobacteriosis and 7917 cases of salmonellosis, the most notified enteric infections.[78] A study based in Queensland showed a positive association between salmonellosis and temperature.[72] In another it was reported that a 1°C increase in temperature might cause an approximately 10% increase in the number of salmonellosis cases in Australia.[75] Salmonella Mississippi, a major contributor to salmonella infections in Tasmania, has been reported to be associated with human contact with native animals and untreated drinking water, mostly that collected in rainwater tanks.[79] Climate change might have affected the growth and survival of S. Mississippi by changing the behaviour of both host animals and human beings, and increasing their exposure to contaminated drinking water.

The associations between campylobacteriosis and both temperature and rainfall have been studied at an international level, including in Melbourne and Brisbane, with varying results.[80] A weak association between time of peak of infection and high temperatures was reported by this study. However, a recent study in Adelaide found a reverse association between temperature and the number of cases, indicating a different response to climate variation in various regions.[81] Examination of the Southern Oscillation Index and the occurrence of hepatitis A in Australia has indicated that an El Niño event is statistically significantly associated with the transmission of the disease.[73] However, this correlation needs to be further examined, with consideration of other influencing factors such as human behaviours and food safety issues.

BOX 84

Influence of River Murray water on River Torrens' water quality

Transfers of River Murray water to the River Torrens commenced in 1955. This box focuses on the water quality impacts of this transfer on River Torrens storages.

River Murray water is delivered into the River Torrens at three entry points. Dissipators deliver water at Mount Pleasant and via Angas Creek, and scour valves deliver piped water direct to Millbrook Reservoir. Water is usually delivered simultaneously to all three distribution points to reduce pumping costs (and greenhouse gas emissions). Most commonly, water that has been delivered by dissapators is pumped from Gumeracha Weir (capacity 200 ML) to Millbrook Reservoir. This practice maximises flexibility in water distribution with water from Millbrook Reservoir able to be transferred to water treatment plants to both the north and south. Thus, Millbrook Reservoir and Gumeracha Weir represent key points in the Upper River Torrens distribution network.

Nutrient status has been identified as a crucial issue for natural resource management in the River Torrens system.[1] We examine the extent of influence that River Murray transfers have on phosphorus concentrations, since phosphorus is commonly the limiting nutrient in freshwater systems, with excess concentrations resulting in 'algal' blooms. Additionally, because phosphorus concentration in the River Murray tends to be correlated with suspended sediments (and thereby water clarity), it acts as a proxy for broader water quality impacts of River Murray water.

Analysis of water quality, sampled by an integrated sampler on the Torrens to Millbrook diversion, found that for the period 2000–2003, total phosphorus concentrations were considerably higher in water derived from the River Murray (282 µg l^{-1}) than the River Torrens (166 µg l^{-1}), a difference noted to a lesser extent in the concentration of suspended particulate matter (68 vs 43 mg l^{-1}). Since such data are only available for the very recent past,[1] we use a combination of historical data and information derived from the study of sediments to extend our understanding of the impact of inter-basin transfers of River Murray water back through time.

Figure 1 illustrates Total Phosphorus (TP) data for the decade 1990–2000 from the water column in Millbrook Reservoir, near the dam wall. Also shown by a dashed line

is the average TP concentration for this period. Prolonged periods of above average TP concentrations occurred particularly in late 1992/early 1993 and the latter parts of 1997 and 1998. Of these periods, only the period in late 1992/early 1993 is not associated with the transfer of substantial quantities of River Murray water into Millbrook Reservoir. These data, in combination with data from the samplers at the Millbrook off-take, strongly suggest that River Murray water plays an important role in nutrient enrichment of Millbrook Reservoir.

Evidence from diatoms (a type of algae that are sensitive to water and aquatic habitat change) deposited in the sediments of Gumeracha Weir also indicate that River Murray water has had important impacts on the biota of the system. As can be seen from Figure 2, there is a substantial change in the diatom community that occurs in the early 1980s. This change follows the highest volume of River Murray pumping on record, and occurs in the absence of major shifts in rainfall patterns. Hence, it appears likely that this change was brought about by the influence of River Murray water that at times has substantially higher nutrient and phosphorus concentrations. It is also notable that the Gumeracha Weir diatom record faithfully reflects the pattern of River Murray transfers with River Murray-derived diatoms.

Taken together, these data indicate that River Murray transfers have had and continue to have a notable impact on water quality and the biota of storages in the River Torrens. This knowledge should provide an incentive to reduce Adelaide's reliance on supplies from the River Murray; at a minimum, it provides impetus to examine ways of improving the regime of transfers from the River Murray. Such improvements could take account of the factors that influence the substantial intra- and inter-annual variation in TP concentrations in the Lower River Murray (including changes in flow source and velocity). This information, in combination with hydrological modelling, could be used to optimise inter-basin transfer of water to times when its influence on water quality is minimised.

John Tibby, Peter Gell and Asta Cox

Phosphorus in Millbrook Reservoir

Millbrook Reservoir
TP

Murray transfers to
Millbrook
Reservoir

Above average TP
associated with ...

■ **pumping**

■ **no/low pumping**

■ **ambiguous**

Mean TP

2000
1998
1996
1994
1992
1990

0   200   400   0   6000

μg l⁻¹   ML/month

Gumeracha Weir diatoms

Murray input to
Torrens
Catchment

Gumeracha Weir
Murray diatoms

Gumeracha Weir:
summary diatom
change

2000

1975

1950

0   100,000   0   25   50

ML/yr   %

Increasing presence
of River Murray
diatoms in relation to
total types of diatoms

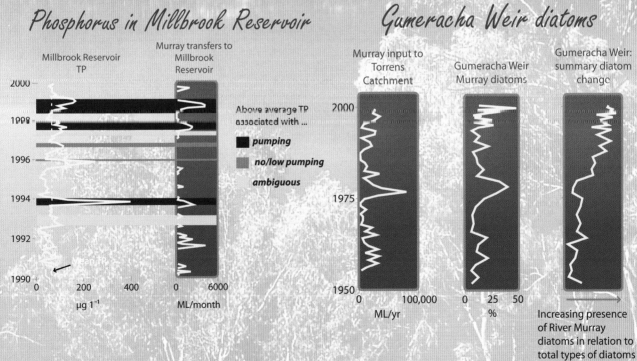

Figure 1. TP concentrations from Millbrook Reservoir in comparison to the record of pumping water from the River Murray. Highlighted are periods when above average TP is associated with, respectively, pumping, no pumping, or where a relationship is not clear. This graph shows that apart from the period in late 1992/early 1993 the high TP concentrations are associated with the pumping of high volumes of River Murray water into Millbrook Reservoir.

Figure 2. River Murray transfers to the River Torrens and summary diatom changes from the Gumeracha Weir diatom record. These graphs show substantial changes in the diatom population in response to the considerable River Murray transfers that occurred in the late 1970s.

Climate change will also increase sea level and river flow, which could have a significant impact on vector-borne diseases. The rise of sea level will favour the growth of saltwater mosquitoes, which may then bring extra cases of Ross River virus infection. An association between high tides and the number of Ross River virus infections has been reported in Queensland and Western Australia.[82, 83] A positive relationship has also been identified between flow in the River Murray and Ross River virus infections in South Australia,[75] where a change in local weather may have favoured the breeding of freshwater mosquitoes and altered their distribution in the region.[84]

Extra health burdens from waterborne diseases caused by climate change

Compared with chronic conditions, infectious diseases do not account for a large proportion of the health burden in developed countries. However, this proportion might increase substantially in the future due to climate-related infections, including waterborne diseases. In Australia, if there are no effective public health interventions, the morbidity burden for salmonellosis associated with future climate change is projected to increase markedly compared with 2000.[75] Given higher population density, poorer living conditions, and poorer environmental hygiene, developing countries may face more serious problems caused by future climate change than developed countries.

Children already bear a large proportion of the health burden due to food- and waterborne diseases. In the next few decades population growth may have an additional adverse impact on the burden of diarrhoeal diseases in children. The health burden of climate-related food- and waterborne diseases in the elderly may also increase due to ageing of the population.[75] Health policies that focus on preventing and reducing the risks that climate change poses to population health, in particular to the health of vulnerable sub-populations including children and the elderly, should be a consideration for governments at all levels.

Climate change may also affect freshwater ecosystems, facilitating the growth and spread of microorganisms, increasing contact opportunities with contaminated water and food, and posing extra risks on population health. For example, the potential impact of climate change on both the quality and quantity of fish production has been reported, although evidence is limited.[85] Extreme weather events, such as flooding, storms and El Niño events, may be more frequent due to climate change, and may cause additional health burdens.

Implications and recommendations

Further studies in Australia on climate change and waterborne diseases could take advantage of the high quality of current disease surveillance data, which includes the onset of disease and the likely location of origin of infection. In addition, integrating local climatic variables into disease surveillance systems could strengthen related research power and thus benefit disease prevention and control. Future studies should examine different serotypes of pathogens (e.g., the more than 2000 serotypes of

Salmonellae in the natural environment), which might have different sensitivities to climate variation.[66]

Adaptations to minimise the health burden of waterborne diseases should include improved access to safe water, improved food processing and storage, improved sanitation and sewage disposal facilities, and increased provision of and access to high-quality health care. At the present time, one of the most important long-term climate change adaptations is to reduce the burden of diarrhoeal diseases by improved public health and environmental management. A strongly integrated public health approach, involving microbiologists, epidemiologists, public health practitioners and experts from environmental science disciplines, is necessary to prevent and control the risk of further increases in the burden of such diseases attributable to climate change.

References

1. United Nations Committee on Economic, Social and Cultural Rights, 'General comment no.15: the right to water', parts 11 and 12 of the International Covenant on Economic, Social and Cultural Rights, 2002.
2. Centres for Disease Control and Prevention, 'Achievements in public health 1900–1999', *Morbidity and Mortality Weekly Report*, 48 (29), 1999, pp. 621–628.
3. R.B. Levin, P.R. Epstein, T.E. Ford, W. Harrington, E. Olson and E.G. Reichard, 'US drinking water challenges in the twenty-first century', *Environmental Health Perspectives*, 110, 2002, pp. 43–52.
4. D. Cutler and G. Miller, 'The role of public health improvements in health advances: the twentieth-century United States', *Demography*, 42, 2005, pp. 1–22.
5. G.L. Armstrong, L.A. Conn and R.W. Pinner, 'Trends in infectious disease mortality in the United States during the 20th century', *JAMA*, 281, 1999, pp. 61–66.
6. Family History SA, '19th-century South Australian history timeline' (online).
7. World Health Organization, 'Water, sanitation and hygiene links to health: facts and figures 2005', 2004 (online).
8. United Nations, 'United Nations Millennium Declaration', 2000 (online).
9. United Nations, '"Water for life" decade 2005–2015', United Nations Department of Public Information, 2005 (online).
10. World Health Organization, 'Right to water', 2003 (online).
11. L. Fewtrell, R.B. Kaufmann, D. Kay, W. Enanoria, L. Haller and J.M. Colford Jr, 'Water, sanitation, and hygiene interventions to reduce diarrhoea in less developed countries: a systematic review and meta-analysis', *Lancet Infect Dis*, 5, 2005, pp. 42–52.
12. G. Hutton and L. Haller, 'Evaluation of the costs and benefits of water and sanitation improvements at the global level', World Health Organization, Geneva, 2004.
13. S. Snow, 'Commentary: Sutherland, Snow and water: the transmission of cholera in the nineteenth century', *International Journal of Epidemiology*, 31, 2002, pp. 908–911.
14. G.D. Smith, 'Commentary: behind the Broad Street pump: aetiology, epidemiology and prevention of cholera in mid-19th century Britain', *International Journal of Epidemiology*, 31, 2002, pp. 920–932.
15. National Research Council, *Drinking water and health*, National Academies Press, Washington, 1977.
16. S.E. Hrudey and E.J. Hrudey, *Safe drinking water: lessons from recent outbreaks in affluent nations*, IWA Publishing, London, 2004.
17. D.R. O'Connor, 'Report of the Walkerton Inquiry: the events of May 2000 and related issues', Ontario Ministry of the Attorney General, 2002.
18. J.L. Clancy, 'Sydney's 1998 water quality crisis', *Water*, 27 (2), 2000, pp. 55–66.
19. P.S. Corso, M.H. Kramer, K.A. Blair, D.G. Addiss, J.P. Davis and A.C. Haddix, 'Cost of illness in the 1993 waterborne cryptosporidium outbreak, Milwaukee, Wisconsin', *Emerging Infectious Diseases*, 9, 2003, pp. 426–431.
20. National Health and Medical Research Council and Natural Resources Management Ministerial Council, *Australian drinking water guidelines*, 2004.
21. World Health Organization, *Guidelines for drinking-water quality*, 3rd edition, Geneva, 2004.

22. National Health and Medical Research Council, 'Community water planner', 2005 (online).

23. R. Grey-Gardner, 'Remote community water management project', National Aboriginal and Torres Strait Islander Environmental Health Conference, Cairns, Queensland, 2007.

24. United States Centers for Disease Control, 'Ten great public health achievements of the 20th century', *Morbidity and mortality weekly report*, 48, 1999, pp. 933–940.

25. J. Armfield, 'The extent of water fluoridation coverage in Australia (letter to the editor), *Australian and New Zealand Journal of Public Health*, 30, 2006, pp. 581–582.

26. United Kingdom National Health Service, Centre for Reviews and Dissemination, 'A systematic review of public water fluoridation', report no. 18, University of York, York, United Kingdom, 2000.

27. M.S. McDonagh, P.F. Whiting, P.M. Wilson, A.J. Sutton, I. Chestnutt, J. Cooper, K. Misso, M. Bradley, E. Treasure and J. Kleijnen, 'Systematic review of water fluoridation', *British Medical Journal*, 321, 2000, pp. 855–859.

28. United States National Research Council (of the National Academics), *Fluoride in drinking water: a scientific review of EPA's standards*, The National Academics Press, Washington, DC, 2006.

29. S.O. Griffin, E.D. Beltran, S.A. Lockwood and L.K. Barker, 'Esthetically objectionable fluorosis attributable to water fluoridation', *Community Dentistry and Oral Epidemiology*, 30, 2002, pp. 199–209.

30. National Health and Medical Research Council, 'The effectiveness of water fluoridation', Canberra, 1991.

31. Dental Services, Health Department of Western Australian and the University of Western Australia Dental School, 'Consensus conference on appropriate fluoride exposure for infants and children: programme and abstracts, Perth, 2–3 December 1993', Perth, 1993.

32. J.T. Ahokas, L. Demos, D.C. Donohue, S. Killalea, J. McNeil and C.R. Rig, 'Review of water fluoridation and fluoride intake from discretionary fluoride supplements', commissioned by the National Health and Medical Research Council, Key Centre for Applied and Nutritional Toxicology and Department of Applied Chemistry, RMIT University and Department of Epidemiology and Preventive Medicine, Monash University, Melbourne, 1999.

33. L.G. Do and A.J. Spencer, 'Risk-benefit balance in the use of fluoride among young children', *Journal of Dental Research*, 86, 2007, pp. 723–728.

34. H.T. Dean, *The investigation of the physiological effects by the epidemiological method*, American Association for the Advancement of Science, Washington, DC, 1942.

35. A.J. Spencer, 'Water fluoridation in Australia', *Community Dental Health*, 13, (suppl. 2), 1996, pp. 27–37.

36. P.J. Riordan, 'Dental fluorosis decline after changes to supplement and toothpaste regimens', *Community Dentistry and Oral Epidemiology*, 30, 2002, pp. 233–240.

37. L.G. Do and A.J. Spencer, 'Decline in the prevalence of dental fluorosis among South Australian children', *Community Dentistry and Oral Epidemiology*, 35, 2007, pp. 282–291

38. A.J. Spencer and L.G. Do, 'Changing risk factors for fluorosis among South Australian children', *Community Dentistry and Oral Epidemiology*, 36, 2008, pp. 210–218.

39. Australian Research Centre for Population Oral Health, 'The use of fluorides in Australia: guidelines', *Australian Dental Journal*, 51, 2006, pp. 195–199.

40. E.B. Bassin, D. Wypij, R.B. Davis and M.A. Mittleman, 'Age-specific fluoride exposure in drinking water and osteosarcoma (United States)', *Cancer Causes Control*, 17, 2006, pp. 421–428.

41. C.W. Douglass and K. Joshipura, 'Caution needed in fluoride and osteosarcoma study (letter to the editor)', *Cancer Causes Control*, 17, 2006, pp. 481–482.

42. L.G. Do and A.J. Spencer, 'Oral health-related quality of life of children by dental caries and fluorosis experience', *Journal of Public Health Dentistry*, 67, 2007, pp. 132–139.

43. I. Chorus and J. Bartram, *Toxic cyanobacteria in water: a guide to their public health consequences, monitoring and management*, World Health Organization, London, 1999.

44. I.R. Falconer, *Cyanobacterial toxins of drinking water supplies: cylindrospermopsins and microcystins*, CRC Press, Boca Raton, Florida, 2005.

45. G. Francis, 'Poisonous Australian lake', *Nature* (London), 18, 1878, pp. 11–12.

46. Y. Grosse, R. Baan, K. Straif, B. Secretan, F. El Ghissassi and V. Cogliano, 'Carcinogenicity of nitrate, nitrite, and cyanobacterial peptide toxins', *Lancet Oncology*, 7 (8), 2006, pp. 628–629.

47. S. Byth, 'Palm Island mystery disease', *Medical Journal of Australia*, 2, 1980, pp. 40–42.

48. P.R. Hawkins, M.T.C. Runnegar, A.R.B. Jackson and I.R. Falconer, 'Severe hepatotoxicity caused by the tropical cyanobacterium (blue-green alga) cylindrospermopsis raciborskii (Woloszynska). Seenya and Subba Raju isolated from a domestic water supply reservoir', *Applied and Environmental Microbiology*, 50, 1985, pp. 1292–1295.

49. A.R. Humpage and I.R. Falconer, 'Oral toxicity of the cyanobacterial toxin cylindrospermopsin in male Swiss albino mice: determination of no observed adverse effect level for deriving a drinking water guideline value', *Environmental Toxicology*, 18 (2), 2003, pp. 94–103.

50. I.R. Falconer and A.R. Humpage, 'Cyanobacterial (blue-green algal) toxins in water supplies: cylindrospermopsins', *Environmental Toxicology*, 21 (4), 2006, pp. 299–304.

51. W.W. Carmichael, S. Azevedo, J.S. An, R.J.R. Molica, E.M. Jochimsen, S. Lau, K.L. Rinehart, G.R. Shaw and G.K. Eaglesham, 'Human fatalities from cyanobacteria: chemical and biological evidence for cyanotoxins', *Environmental Health Perspectives*, 109 (7), 2001, pp. 663–668.

52. L. Pilotto, P. Hobson, M.D. Burch, G. Ranmuthugala, R. Attewell and W. Weightman, 'Acute skin irritant effects of cyanobacteria (blue-green algae) in healthy volunteers', *Australian and New Zealand Journal of Public Health*, 28 (3), 2004, pp. 220–224.

53. I. Stewart, P.M. Webb, P.J. Schluter and G.R. Shaw, 'Recreational and occupational field exposure to freshwater cyanobacteria: a review of anecdotal and case reports, epidemiological studies and the challenges for epidemiologic assessment', *Environmental Health*, 5 (6), 2006, pp. 1–13.

54. D. Dietrich and S. Hoeger, 'Guidance values for microcystins in water and cyanobacterial supplement products (blue-green algal supplements): a reasonable or misguided approach?', *Toxicology and Applied Pharmacology*, 203 (3), 2005, pp. 273–289.

55. Natural Resource Management Ministerial Council and Environment Protection and Heritage Council, *Australian guidelines for water recycling: managing health and environmental risks*, 2006.

56. South Australian Water Corporation, 'SA Water wins $2m grant for recycled water project', 2004 (online).

57. South Australian Water Corporation, 'SA Water annual report 2005/2006', 2006 (online).

58. Local Government Association of South Australia, 'Review of STEDS in South Australia', vol. 1, final report, Adelaide, 2003.

59. Government of South Australia, 'Water proofing Adelaide: a thirst for change 2005–2025', 2005 (online).

60. J. Ottoson and T.A. Stenstrom, 'Faecal contamination of greywater and associated microbial risks' *Water research*, 37, 2003, pp. 645–655.

61. Intergovernmental Panel on Climate Change, 'Climate change 2007: the physical science basis. Summary for policymakers. Contribution of Working Group I to the Fourth Assessment Report of the IPCC', Cambridge University Press, Cambridge UK, 2007 (online).

62. J.B. Rose, P.R. Epstein, E.K. Lipp, B.H. Sherman, S.M. Bernard and J.A. Patz, 'Climate variability and change in the United States: potential impacts on water- and food-borne diseases caused by microbiologic agents', *Environ Health Perspect*, 109, 2001, pp. S211–S221.

63. M. Rayner and P. Scarborough, 'The burden of food related ill health in the UK', *Journal of Epidemiology and Community Health*, 59, 2005, pp. 1054–1057.

64. L. Neville and J. McAnulty, 'Communicable enteric disease surveillance, NSW, 2000–2002', *NSW Public Health Bulletin*, 15, 2004, pp. 18–23.

65. World Health Organization, 'Global defence against the infectious disease threat' (online).

66. S. O'Brien and H. Valk, 'Salmonella: "old organism", continued challenges', *Eurosurveillance*, 8, 2003, pp. 29–31.

67. F. Curriero, J. Patz, J. Rose and S. Lele, 'The association between extreme precipitation and waterborne disease outbreaks in the United States, 1948–1994', *American Journal of Public Health*, 91, 2001, pp. 1194–1199.

68. Health Canada, *Climate change and health*, Research Report, Health Canada, 2004.

69. C. Tam, L. Rodrigues, S. O'Brien and S. Hajat, 'Temperature dependence of reported campylobacter infection in England, 1989–99', *Epidemiology and Infection*, 134, 2006, pp. 119–125.

70. R. Kovats, S. Edwards, S. Hajat, B.G. Armstrong, K.L. Ebi and B. Menne, 'The effect of temperature on food poisoning: a time-series analysis of salmonellosis in ten European countries', *Epidemiology and Infection*, 132, 2004, pp. 443–453.

71. W. Checkley, L.D. Epstein, R.H. Gilman, D. Figueroa, R.I. Cama, J.A. Patz and R.E. Black, 'Effects of El Niño and ambient temperature on hospital admissions for diarrhoeal diseases in Peruvian children', *Lancet*, 355, 2000, pp. 442–450.

72. R. D'Souza, N. Becker, G. Hall and K. Moodie, 'Does ambient temperature affect foodborne disease?', *Epidemiology*, 15, 2004, pp. 86–92.

73. W. Hu, A.J. McMichael and S. Tong, 'El Niño Southern Oscillation and the transmission of hepatitis A virus in Australia', *Medical Journal of Australia*, 180, 2004, pp. 487–488.

74. M. Patrick, L. Christiansen, M. Wain, S. Ethelberg, H. Madsen and H.C. Wegener, 'Effects of climate on incidence of campylobacter spp. in humans and prevalence in broiler flocks in Denmark', *Applied and Environmental Microbiology*, 70, 2004, pp. 7474–7480.

75. Y. Zhang, 'Climate variation and infectious diseases: Australian and Chinese perspectives', PhD thesis, University of Adelaide, Adelaide, 2007.

76. R. Singh, S. Hales, N. de Wet, R. Raj, M. Hearnden and P. Weinstein, 'The influence of climate variation and change on diarrheal disease in the Pacific Islands', *Environmental Health Perspectives*, 109, 2001, pp. 155–159.

77. Y. Zhang, P. Bi and J.E. Hiller, 'Climate variation and salmonellosis transmission in Adelaide, Australia: a comparison between regression models', *International Journal of Biometeorology*, 52, 2008, pp. 179–187.

78. The OzFoodNet Working Group, 'Foodborne disease in Australia: incidence, notifications and outbreaks', annual report of the OzFoodNet network, 2002.

79. R.A. Ashbolt and M. Kirk, 'Salmonella Mississippi infections in Tasmania: the role of native Australian animals and untreated drinking water', *Epidemiology and Infection*, 134, 2006, pp. 1257–1265.

80. R.S. Kovats, S. Edwards, D.F. Charron, J. Cowden, R.M. D'Souza, K.L. Ebi, C. Gauci, P. Gerner-Smidt, S. Hajat, S. Hales, G. Hernández Pezzi, B. Kriz, K. Kutsar, P. McKeown, K. Mellou, B. Menne, S. O'Brien, W. van Pelt and H. Schmid, 'Climate variability and campylobacter infection: an international study', *International Journal Biometeorology*, 49, 2005, pp. 207–214.

81. P. Bi, A.S. Cameron, Y. Zhang and K.A. Parton, 'Weather and notified *Campylobacter* infections in temperate and sub-tropical regions of Australia: an ecological study', *Journal of Infection*, 57 (4), 2008, pp. 317–323.

82. S. Tong, P. Bi, K. Donald and A.J. McMichael, 'Climate variability and Ross River virus infection', *Journal of Epidemiology and Community Health*, 56, 2002, pp. 617–621.

83. M. Lindsay, J. Mackenzie and R. Condon, 'Ross River virus outbreaks in Western Australia: epidemiological aspects and the role of environmental factors' in Ewan, C.E., Calvert, G., Bryant, E. and Garrick, J. (eds) *Health in the greenhouse: the medical and environmental health effects of global climate change*, AGPS, Canberra, 1993.

84. D. Harley, A. Sleigh and S. Ritchie, 'Ross River virus transmission, infection and disease: a cross-disciplinary review', *Clinical Microbiology Reviews*, 14, 2001, pp. 909–932.

85. F.J. Wrona, T.D. Prowse, J.D. Reist, J.E. Hobbie, L.M.J. Lévesque and W.F. Vincent, 'Climate impacts on arctic freshwater ecosystems and fisheries: background, rationale and approach of the Arctic Climate Impact Assessment (ACIA)', *Ambio*, 35, 2006, pp. 326–329.

Box references

Box 79: How safe is rainwater consumption?

1. M.I. Sinclair, K. Leder and C.H., (2005) 'Public health aspects of rainwater tanks in urban Australia', occasional paper 10, CRC for Water Quality and Treatment, 2005 (online).

2. A. Merritt, R. Miles and J. Bates, 'An outbreak of campylobacter enteritis on an island resort, north Queensland', *Communicable Diseases Intelligence*, 23 (8), 1999, pp. 215–219.

3. R. Taylor, D. Sloan, T. Cooper, B. Morton and I. Hunter, 'A waterborne outbreak of Salmonella Saintpaul', *Communicable Diseases Intelligence*, 24 (11), 2000, pp. 336–340.

4. G.V. Hall, M.D. Kirk, R. Ashbolt, R. Stafford, K. Lalor and O.W. Group, 'Frequency of infectious gastrointestinal illness in Australia, 2002: regional, seasonal and demographic variation', *Epidemiology and Infection*, 134, 2006, pp. 111–118.

Box 82: Copper in reservoirs

1. D.I. Smith, *Water in Australia: resources and management*, Oxford University Press, South Melbourne, 1998.

2. J.P. Smol, *Pollution of lakes and rivers: a palaeoenvironmental perspective*, Arnold, London, 2002.

3. G.T. Moore and K.F. Kellerman, 'Copper as an algicide and disinfectant in water supplies', *Bulletin of the Bureau of Plant Industries*, 76, 1905, pp. 19–55.

4. Australian and New Zealand Environment and Conservation Council and Agriculture and Resource Management Council of Australia and New Zealand, *Australian and New Zealand guidelines for fresh and marine water quality*, vol. 1., 2000 (online).

5. J. Olley, A.S. Murray, P.J. Wallbrink, G. Caitcheon and R. Stanton, 'The use of fallout radionuclides as chronometers' in Gillespie, R. (ed.) *Proceedings of the Quaternary dating workshop*, Department of Biogeography and Geomorphology, Australian National University, Canberra, 1990, pp. 51–55.

Box 84: Influence of River Murray water on Torrens Water quality

1. S. Roberts, B. Hart, M. Grace and I. Rutherfurd, *Relative importance of sources of nutrients and sediments in the Upper Torrens Catchment, South Australia*, Water Science Pty Ltd., Mulgrave, Victoria, 2005.

CHAPTER 18

Maintaining green infrastructure

Keith E. Smith

Introduction

The current drought is having a major impact on Adelaide. With limited water, the structural integrity of urban infrastructure and many older buildings and houses is deteriorating. However, public debate about water supply and conservation typically descends into a discussion about providing water 'sustainably' for 'critical human needs'. In some high-profile water cases, like the health of the Murray River, environmental concerns are now considered together with economic and human needs. Yet the urban ecosystem, our parks and gardens and the services they provide, is conspicuously absent from the current dialogue.

Urban vegetation is deteriorating and dying. The vital functions performed by this vegetation, our 'green-infrastructure', are not widely recognised and are under threat due to their omission from the water debate. While the effects of the drought are obvious on the built environment, the long-term impacts of a degraded green infrastructure are less well understood. A reduction in the quantity and quality of the vegetation in gardens, parks and along streets will radically change the landscape and ecology of Adelaide and the quality of life of its citizens for generations to come.

In modern cities, built infrastructure including roads, supply and drainage of water, communications and energy systems are essential to urban life. These systems are usually well managed; infrastructure is maintained and eventually replaced on consideration of financial and lifecycle parameters and constraints. Rarely are similar asset management approaches applied to the essential components of urban vegetation or green infrastructure.

Cities also metabolise. Water, energy, materials and wastes are constantly moving into and out of these systems, and can be measured as a balance, and managed according to needs. If our green infrastructure is to adapt and thrive into the future, we must manage water as a key input to this system.

This chapter provides a reference for communities considering the impacts of climate change, drought, and other pressures such as land-use change on the vegetation cover and ecology of their neighbourhoods and the broader Adelaide area. It highlights:

- benchmarks for monitoring changes to land cover types across metropolitan Adelaide;
- the effectiveness of revegetation schemes across the metropolitan Adelaide region;
- identified and recorded changes to Adelaide's urban vegetation cover over a 10 year period;
- the status of the vegetated landscape in metropolitan Adelaide
- the richness and complexity of the Adelaide landscape, both in its own right and as an important landscape that connects to surrounding ecosystems; and
- the use of vegetation type, structure and cover as a useful indicator of the health of the urban landscape.

Chapter photograph: Kensington Gardens.

The importance of Adelaide's urban forest

The government of South Australia noted in 2002 that the state faced key challenges in three areas: economic growth and change; sustainable use of natural resources; and management of urban issues to maintain and improve people's quality of life.[1] Each of these challenges remain in 2009, particularly as the community's awareness of environmental issues and focus on the quality of the urban environment continue to grow. Increasingly, the quality of vegetation in parks, gardens and streetscapes is cited by communities and interest groups, such as real-estate and development industries, as highly important in determining the desirability of areas in which people choose to live.

With the exception of some older heritage streetscapes, which consist of large numbers of similar species of mature trees, economic values can be difficult to ascribe to urban vegetation. This is especially true for vegetation cover that contains mixed species, types and plant ages. As a consequence, urban vegetation may be managed in a piecemeal way with little regard for the economics and practicalities of eventual replacement until significant numbers of trees fail or die.

Urban vegetation is becoming increasingly valued by communities around the world. Johnson[2] notes that street trees provide a range of benefits and services, including:

- improved visual amenity, human health and increase in real-estate values;
- provision of screening for increased privacy;
- reduced heating and cooling costs (through shading and windbreak effects);
- increased personal comfort through shade and amenity;
- increased safety through increased passive surveillance, community socialisation and networking;
- trees planted as personal or community memorials have significant historic or heritage value;
- provision of leaf mulch and compost materials that can be used for soil enrichment and maintenance;
- reduction or eradication of glare and reflection of heat and light, moderating climate both locally and globally;
- enhancing views, screening undesired views, and complementing or softening architecture;
- improving air quality by fixing carbon dioxide and carbon monoxide, producing oxygen, and filtering dust;
- provision of habitat for wildlife, and provision of associated educational opportunities;
- reduction of water runoff, helping to prevent erosion, dryland salinity and related groundwater issues; and
- provision of employment (equipment manufacturers, retailers, educators, researchers, arborists, mechanics and horticulturalists).

Economic analysis of urban forest benefits

Vegetation is an essential component of the urban environment, affecting the enjoyment and quality of life of people living in cities around the world. It provides important environmental and social services, such as noise and pollution abatement, carbon sequestration, stormwater retention, habitat enhancement, and certain psychological benefits. The American Urban Forests organisation[3] describes the important links between urban vegetation and water thus:

Building better communities is not just a matter of finding the right mix of land use, economics, and quality of life for its citizens. All of these essential components are built upon the existing natural systems that affect the water we drink and the air we breathe. The complex interactions of our waterways, vegetation, soils and air systems are inextricably linked to the built elements of our cities. How they function greatly affects our communities and its citizens.

The *economic benefits* of urban vegetation have been analysed by American Urban Forests to provide a financial perspective of the services provided by urban trees for cities in the United States. American Forests' software package, *CITYgreen©*,[4] calculates in dollar terms the benefits of urban trees for carbon sequestration, stormwater control, ultraviolet radiation control, heat reduction, and absorption of suspended particulate matter.

Analysis of urban forests in the United States credits urban vegetation with provision of substantial services of considerable community value, as shown in the following examples:

- The city of Detroit's vegetation cover provides 191 m³ feet of stormwater management benefits, valued at US$382 m; 2.1 m pounds of air pollution removal, valued at US$5.1 m annually; and stores 1.2 m t of carbon and sequesters 9334 pounds of carbon annually. (American Forests, 2006.)
- The city of San Diego's vegetation provides 82 m³ feet of stormwater management benefits, valued at US$164 m; and 4.3 m pounds of air pollution removal, valued at US$10.8 m annually.[5]

Hewett[6] notes that these values are calculated using specific local criteria including soil, plant, climatic conditions, and other information specific to the study localities – criteria that would require extensive conversion for application to urban vegetation in Australian cities. Table 18.1 lists vegetation assessment parameters that should be considered for Australian conditions.

The considerable value of urban vegetation in terms of stormwater management and associated pollution reduction is largely unrealised locally, with water demands of urban vegetation typically seen as a liability rather than an opportunity for bioremediation of polluted water from sources such as road surfaces.

The gross annual benefit of a typical Adelaide street tree (a four-year-old Jacaranda) in 2002 was calculated at $171. This benefit consisted of energy savings due to reduced air-conditioning costs ($64), air quality improvements (CO_2 reduced power output $1), air pollution treatment $34.50), stormwater treatment ($6.50), and aesthetics and other benefits ($65.00).[7]

Although urban trees have significant additional values, which are not considered in the economic assessment by Killicoat et. al., consideration of the figures suggests the following conservative (but nonetheless alarming) value of tree loss due to climate change and drought can be expected: for 10,000 trees lost in the urban landscape, the annual loss of environmental services (at 2002 values) will be approximately $1,700,000 a year, every year. The figure of 10,000 trees is conservative (0.5% of Adelaide's tree population) if we consider that the loss of trees associated with drought and climate change is not limited to just street trees but also includes trees in parks, reserves and home gardens. Although it is likely that councils will replace street tree losses over time, the same can not be assumed for private gardens.

A decade of change in Adelaide's urban forest

Following is an analysis of the change in Adelaide's urban vegetation cover over the decade from 1990 to 2000 by Smith.[8] This analysis provides a benchmark against which future change in vegetative cover might be measured; it is provided to encourage and assist with the management of green infrastructure as a recognised asset in parallel with civil infrastructure such as pipes, bridges and roads.

Before European settlement, 21 native vegetation associations occurred in and around the Adelaide area, ranging from open forests and woodlands along South Road at Black Forest, to shrublands at Merino.[9] Today, only 2.7% of the original vegetation cover of the Adelaide Plains remains, the remainder being planted vegetation in parks and gardens. Oke[10] noted that the dominant land use for the Adelaide area was cropping and pasture, closely followed by urban habitats.

Tables 1 and 2 summarise the extent of Adelaide's vegetation, in four distinct layers, based on the cover of the tallest plant stratum. The dominant land cover types for metropolitan Adelaide and a description of each are as follows:

- **Urban forests** are mapped where trees up to 30 m form the tallest stratum and foliage cover as projected on the ground is approximately 30% or greater. Urban forests are usually limited to stands of mature, dense vegetation (both native and exotic), and include mangrove areas around the Barker Inlet, trees in mature urban gardens, and suburban areas with older trees and street tree plantations, especially around the Adelaide foothills.
- **Urban woodland** areas occur where trees or large shrubs form the tallest stratum and foliage cover is between 20–30%. In urban areas, this land cover type is mainly confined to Metropolitan Open Space reserves, gully escarpments, and older suburbs around the Adelaide foothills. Urban woodlands also include areas of dense coastal shrublands and vegetation around reservoirs.
- **Urban open woodland** occurs over most residential areas in metropolitan Adelaide, where trees or large shrubs form the tallest stratum and foliage cover is up to 20%. This land cover type comprises predominantly urban vegetation, including tree and shrub plantings

BOX 85

Grange Golf Club
aquifer storage and recovery

With many of South Australia's water resources currently under stress, the Adelaide and Mount Lofty Ranges Natural Resources Management (AMLRNRM) Board is promoting and supporting projects that use urban stormwater as an alternative water source.

Large quantities of stormwater run off from the urban parts of Adelaide during winter and eventually drain out to Gulf St Vincent. While this stormwater contains pollutants, such as from road runoff, it does have the advantage of containing low levels of dissolved salts.

Unfortunately, stormwater flows mainly during winter and tends to come in large volumes over short periods. To effectively capture stormwater it is necessary to have large wetlands, where the stormwater can be collected and filtered during rainfall, and a suitable aquifer where the water can be stored over winter.

For these reasons, aquifer storage and recovery (ASR) schemes require not only a supply of stormwater but sufficient open space to construct artificial wetlands. Golf courses are therefore ideal for stormwater reuse schemes, particularly since they also have a requirement for large volumes of water for irrigation during the summer.

The Grange Golf Club
Located west of Adelaide near West Lakes, the Grange Golf Club is one of South Australia's premier sporting venues. Like most large recreational areas in urban Adelaide, the club has traditionally used groundwater for irrigation during summer. Having identified the potential problems associated with continuing to draw on the locally stressed aquifer, the club committed itself to investigating sustainable water use.

The club identified an opportunity to reduce its dependency on groundwater and utilise stormwater through a wetland and ASR scheme. It had two of the prime prerequisites for potential success: sufficient open space and an ongoing need for large volumes of water.

Long-term vision
In 2002, the (then) Torrens Catchment Water Management Board, now part of the AMLRNRM Board, began discussions with the Grange Golf Club about the potential for a wetland and stormwater reuse project.

During 2003–2005, they jointly funded studies into the feasibility of harvesting stormwater from local urban catchments.

These studies showed that it would be feasible to harvest stormwater from a 450–500 ha urban catchment upstream of the golf club by diverting the water into artificial wetlands and then using the treated water through an ASR scheme.

In a true example of government and industry partnerships working towards innovative water resource management, the $2.8 million infrastructure project received funding and support from:

- the Grange Golf Club (about $1 million or 30%);
- the state government (through the previous Catchment Management Subsidy Scheme, now the Stormwater Management Fund, administered by the Department for Transport, Energy and Infrastructure, the AMLRNRM Board);
- the federal government (through the Water Smart Australia program); and
- the City of Charles Sturt.

Construction commenced in 2006, and the project was launched on World Wetlands Day in February 2007.

Delivering on natural resources managment goals
This scheme will deliver on many of the key goals encapsulated in the board's initial Natural Resources Management plan, including:

- the construction of wetlands at strategic locations for water quality improvement;
- the facilitation of viable reuse opportunities and water conservation practices; and
- the fostering of an informed, committed and involved community.

The Grange Golf Club Wetland and Stormwater Reuse Scheme is an exceptional outcome that will provide significant benefits to the environment, the club, and the local community for many years to come.

Scheme components
The project harvests around 300 ML of urban stormwater each year by diverting water from the Trimmer Parade and West Lakes Boulevard stormwater catchments into the golf course wetland. That is an annual equivalent of around 150–300 Olympic-size swimming pools and about the same volume of groundwater that the golf club was previously extracting from the underlying aquifer to irrigate its two courses.

Pumps and rising mains have been constructed to divert stormwater from both the Trimmer Parade (to the south) and Brebner Drive (to the north). Stormwater flows by gravity to two pump sumps located on the club grounds, and is then pumped via about 1500 metres of underground pipework to an inlet zone, where coarse sediment is trapped and removed.

After passing through the inlet zone, stormwater enters three macrophyte cells with dense bands of wetland vegetation to improve water quality. These are located along Frederick Road, and cover an area of 24,000 m^2. Finer sediment particles are trapped in these cells, which provide a natural system for initial, secondary and final cleaning of the stormwater. Most of the pollutants in stormwater are attached to particles of soil and organic material which settle out in the wetlands and are captured by the dense reedbeds as the water flows through them. The final of the three macrophyte cells is a deep pool that holds the treated water prior to injection into the underlying aquifer.

The term 'aquifer storage' refers to the artificial recharge of an underground aquifer by allowing water to flow or be injected into it via a bore. Because water generally flows very slowly through aquifers, any water injected into it forms a 'bubble' around the bore, allowing it to be retrieved or 'extracted' when needed for irrigation. In the final component of this scheme, water is pumped from the deep pool to two ASR wells.

The 'recovery' phase occurs when this bubble of water is pumped out at a later date. In this way, the aquifer is being used like a water storage reservoir to hold water collected during the winter for retrieval and use during the drier Adelaide summer.

Biodiversity outcomes

The wetland project has resulted in an increase in local biodiversity and has recreated native aquatic habitats, replacing in some way the vast reedbeds that previously occupied this region.

The future survival of the local provenance native plant species tangled lignum (*Muehlenbeckia florulenta*) is being protected through this project. From the last known remaining plant on the Adelaide Plains, 350 have been propagated and planted.

In the original Reedbeds, about 90 different plant species were recorded. Approximately half of these have been replanted in the wetlands so as to recreate the original Reedbeds habitat, albeit on a reduced scale, at the same time providing water quality improvements. A full range of plant types from trees and shrubs, to reeds, rushes, bog plants and underwater plants have been planted. Also being reintroduced are locally extinct reedbed plant species such as river eelgrass (*Vallisneria spiralis*) and silky tea-tree (*Leptospermum lanigerum*).

The wetlands have been designed and constructed to produce aquatic habitats not favoured by mosquitoes, and have been naturally populated with fish and macroinvertebrates that are mosquito predators.

Project benefits

- A major reduction in water use from the locally stressed aquifer.
- Immediate groundwater pressure improvement.
- Contribution to long-term salinity reduction in the aquifer.
- A sustainable irrigation water supply for the club.
- A reduction in stormwater inflows to West Lakes Lake, the Port Adelaide River and the Barker Inlet, and from there to the Gulf St Vincent.
- An aesthetic asset for the club.
- An increase in biodiversity through the recreation of native aquatic habitats.
- Demonstration of best practice in environmental protection and water conservation.
- A facility that can be used by local schools and the community to increase awareness of water conservation and ecological issues.

The Grange Golf Club Wetlands and Stormwater Harvesting Project is the first in a series of three wetlands projects. The other two are at the Royal Adelaide Golf Club and the Glenelg Golf Club.

Keith Downard

Table 18.1 Vegetation assessment parameters

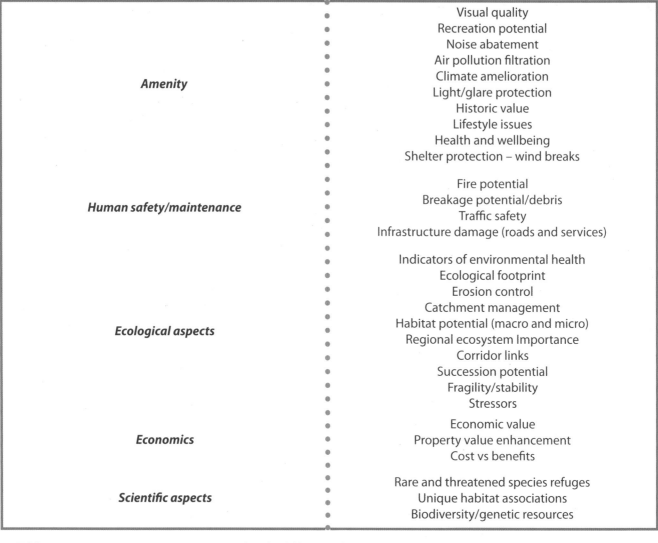

Amenity	• Visual quality • Recreation potential • Noise abatement • Air pollution filtration • Climate amelioration • Light/glare protection • Historic value • Lifestyle issues • Health and wellbeing • Shelter protection – wind breaks
Human safety/maintenance	• Fire potential • Breakage potential/debris • Traffic safety • Infrastructure damage (roads and services)
Ecological aspects	• Indicators of environmental health • Ecological footprint • Erosion control • Catchment management • Habitat potential (macro and micro) • Regional ecosystem Importance • Corridor links • Succession potential • Fragility/stability • Stressors
Economics	• Economic value • Property value enhancement • Cost vs benefits
Scientific aspects	• Rare and threatened species refuges • Unique habitat associations • Biodiversity/genetic resources

Table 18.1 Vegetation assessment parameters that should be considered in urban ecosystem management. Source: adapted from Goodwins and Noyce, 'A "greenness" rating for the suburbs of metropolitan Adelaide: an assessment of the vegetation cover of the suburbs of metropolitan Adelaide using a geographical information system', South Australian Office of Planning and Urban Development, Adelaide, 1993, p. 35.

dominated by mixed woodland formations and introduced exotic vegetation. Urban open woodlands often occur over grasslands, parklands or pasture. This category also includes sparse open woodlands over modified pasture in rural areas, as well as recent vineyard and orchard plantings. The extensive cover of woodlands across Adelaide is a composite of urban woodlands and open woodland areas in 2000. These woodlands are most dense along the Hills Face Zone and into the higher rainfall areas of the Mount Lofty Ranges; they also extend over large residential areas of the plains. Woodland areas become more 'patchy' in the more recently developed urban areas to the north and south of the city.

- **Urban grassland** includes irrigated and non-irrigated turf and open grassland areas such as reserves, golf courses, and institutional and industrial lands in urban areas, as well as irrigated and non-irrigated crop and pastures in horticulture and rural zones. It also includes

disturbed sites, such as new housing development sites, often characterised by weedy cover. Urban grassland areas vary from grass cover up to 100% of the ground (irrigated grasslands) to areas with less than 10% cover (weedy grasslands).

The extent of Adelaide's vegetative cover is summarised in Table 18.2; a breakdown of these figures by local government area is given in Table 18.3. From Table 18.3 it can be seen that the combined woodland coverage increased significantly in most council areas over the 10 year period but especially in the older, more heavily vegetated areas of Burnside, Mitcham and Walkerville. Conversely, many inner-metropolitan councils recorded a slight reduction in grassland cover during the same period, mainly due to the success of various revegetation and tree planting programs. This reduction in grassland areas should be monitored and managed in the future, as it has the potential to reduce the diversity of land cover types and associated habitats. The significant increases

Table 18.2 Metropolitan land cover

Land cover type	% cover metro area 1990	% cover metro area 2000	Total change
Urban forests	6.15	5.72	< 0.43
Urban woodlands	11.58	14.72	> 3.14
Open woodlands	13.65	15.99	> 2.34
Urban grasslands	15.62	14.83	< 0.79
Total	47.00	51.26[1]	4.26

1 Note: Figures do not add up to 100% as only vegetation cover is considered. Buildings, roads, water and other surfaces are excluded from the classification.

Table 18.2 Adelaide metropolitan land cover types by percentage cover. Source: Smith, 'An urban land cover analysis for the Adelaide metropolitan ecosystem', unpublished masters thesis, 2006.

in grasslands recorded in the outer metropolitan councils of Mallala, Light and Barossa are a result of changes to cropping and pasture regimes in these areas, rather than any net increase in grasslands.

While the changes to each of the land cover types appear quite small for forests, woodlands and grasslands in Table 18.2, they are significant when it is considered that these changes occurred over a relatively short period of 10 years. The combined urban woodlands increased by almost 5.5%, which represents a considerable increase in the 'greenness' of Adelaide.

Vegetation canopy cover targets

The benefits provided by urban vegetation are proportional to the extent of cover, so an important question is: what constitutes an appropriate target to set for vegetation cover in Adelaide? The urban environment is highly variable in microclimate, rainfall, soils and built forms, so a single figure for vegetation cover across the entire region is inappropriate and undesirable.

It would be unreasonable to prescribe for coastal areas, such as Port Adelaide, a mandated combined woodland coverage of approximately 34% such as that which exists in Burnside. Similarly, it would be unreasonable to set higher targets in high-density residential or industrial areas where local high demands were placed on public open space for active recreational pursuits.

Efforts to prescribe canopy cover targets must consider local variations in climate, rainfall, soils and topography. For example the soils found in coastal areas (between 0–10 m above sea level) vary considerably from those found further inland; and rainfall generally increases with elevation, especially towards the Mount Lofty Ranges.

Very little work has been done locally on this issue, though targets are regularly set for cities overseas. In a recent ecosystem analysis of San Diego in California, American Forests[5] set the following generalised canopy target goals for the city, taking into consideration the existing tree cover, climate, rainfall and geography:

- 25% tree canopy cover overall;
- 30% tree canopy cover in suburban residential areas;
- 20% tree canopy cover in urban residential areas; and
- 10% tree canopy cover in central business districts.

While San Diego has a climate somewhat similar to Adelaide, its ecology and plants are very different and it may be inappropriate to attempt to apply these targets under local conditions. It is important to note that the density of vegetation in much of Adelaide is already *artificially* high and that the city appears now more like a dense forest. Prior to European settlement it consisted mainly of open woodland.[11] Sustainable vegetation cover targets, which will not overuse scarce resources such as water, may be lower than existing levels of vegetation cover; they may approximate predevelopment figures in some areas.

Even over the relatively short period of the last decade, dramatic increases in the extent and density of urban vegetation can be clearly seen through comparison of the pre-European vegetation with the extent of urban woodlands. In the Northern Adelaide Plains, large areas of low woodlands and shrublands have been replaced by urban woodlands. This trend is especially evident for the tussock grassland systems (*Stipa* and *Danthonia spp.*) and the previously extensive *Melaleuca halmaturorum* low woodlands.

A combination of extensive land-filling activity, levelling and urban expansion in these areas over many years has altered the hydrology, and urban woodlands are expanding in these areas in association with increasing urban development.

Adelaide's urban forest

The term 'urban forest' is often used to refer to the dense cover of vegetation over metropolitan areas, which can be seen from vantage points such as Adelaide's Mount Lofty summit. Girardet[12] comments:

Adelaide is fortunate in having a significant cover of urban vegetation and it has been noted that it looks like a city in a forest from the air, having in excess of 20 million trees growing in the metropolitan area.

Table 18.3 Land cover changes

Council name	AREA (HA)	% FOREST 2000	% WOODLAND 2000	% OPEN WOODLAND 2000	% GRASSLAND 2000	FOREST DIFFERENCE 1990–2000	WOODLAND DIFFERENCE 1990–2000	OPEN WOODLAND DIFFERENCE 1990–2000	GRASSLAND DIFFERENCE 1990–2000
Adelaide Hills	47,193	18.39	34.15	14.34	15.93	- 0.81	- 7.33	- 1.32	- 0.98
Alexandrina	41,942	5.92	14.17	9.64	18.78	+0.96	+1.58	+1.41	-2.00
Adelaide	1527	1.33	21.28	17.56	14.40	-1.28	+8.10	+0.85	-2.42
Burnside	2748	5.51	34.42	37.25	7.09	-1.75	+9.83	-1.74	-4.44
Campbelltown	2423	1.42	16.41	41.84	8.88	-0.59	+4.28	-2.26	-5.48
Marion	5572	0.04	3.71	29.03	7.36	-0.03	+1.25	+2.26	-4.13
Mitcham	7569	12.16	38.31	26.63	6.86	-0.26	+10.31	-1.83	-3.32
Charles Sturt	5345	0.12	4.62	31.39	12.65	+0.02	+1.45	+0.94	-7.18
Holdfast Bay	1345	0.17	7.71	41.64	·8.06	-0.21	+1.14	-2.35	-6.44
Norwood, Payneham, St Peters	1515	0.25	14.13	41.42	6.18	-0.31	+4.35	-0.69	-4.41
Onkaparinga	51,541	3.12	13.27	17.64	13.97	-0.34	+3.65	+3.69	-3.16
Playford	32,242	1.42	8.01	14.76	15.13	-0.90	+1.78	+3.97	+0.95
West Torrens	3681	0.06	5.27	26.81	14.54	-0.03	+2.33	+1.38	-3.08
Port Adelaide Enfield	9081	0.75	2.82	20.28	8.08	-0.05	+0.68	-0.81	-4.32
Prospect	781	0.02	5.58	45.13	7.37	+0.01	+1.23	+0.15	-4.16
Salisbury	14,595	7.58	4.91	20.78	12.01	-0.70	+2.18	+3.89	+0.01
Tea Tree Gully	9397	2.70	14.93	28.69	10.71	-0.40	+4.20	+5.34	+0.59
Unley	1440	0.31	18.13	45.71	4.44	-0.52	+5.28	-1.73	-2.58
Gawler	4113	0.10	3.19	16.05	14.23	-0.04	+2.12	+6.94	+1.02
Walkerville	357	0.45	28.59	48.22	4.92	-0.40	+9.40	-3.15	-4.08
Mallala	18,844	1.30	1.56	4.77	16.24	-0.44	+0.66	+2.83	+11.50
Mount Barker	27,081	6.26	16.15	12.48	21.78	-1.61	+2.11	+2.10	-3.49
Yankallila	2807	1.33	5.42	5.00	13.14	-0.01	+1.63	+0.55	-0.05
Light	7971	0.00	1.64	4.42	12.41	-0.03	+0.95	+2.17	+6.25
Barossa	10,866	0.53	9.90	11.41	12.51	-1.16	-1.47	+3.49	+2.25

Table 18.3 Metropolitan Adelaide suburban land cover changes from 1990 to 2000. Source: Smith, 'An urban land cover analysis for the Adelaide metropolitan ecosystem', 2006.

Introduction

Most of us have played the game 'rock, paper, scissors', where, on a simultaneous showing of hands, the eventual fate of the participant is determined. Scissors cut paper, paper covers rock, rock blunts scissors. A similar game is being played out in the streets of Adelaide right now, in the midst of the worst drought in history, record heatwaves and tough new water restrictions. It is called 'tree, water, bluestone', and depending on how we as a community play the hand, the fate of many urban trees, scarce water resources and valuable old buildings will very soon be decided.

Why do the changing climatic and political circumstances affecting the availability of water in parks and gardens pose such a serious threat to our city? What opportunities, arising from these threats, present themselves to us, so we can provide an even better environment for Adelaide's residents in the years ahead?

In the game 'tree, water, bluestone', during drought and irrigation restrictions, urban trees can extract moisture from the earth to such a degree that the subsoil shrinks: 'tree beats water.' Changing moisture regimes around the footings of buildings, particularly those with bluestone or poorly engineered footings, can cause significant structural damage: 'water beats bluestone.' Buildings are considered to be irreplaceable and of great value, unlike most trees, so whenever there is an actual or perceived conflict, the tree goes: 'bluestone beats tree.'

The state government has recently imposed water restrictions over all of Adelaide in an effort to ensure adequate but basic supplies for drinking, health and industry, now and into the future. Our dependence on the flow down the Murray, from New South Wales and Victoria to the pumping station at Morgan, means South Australians have to be seen to be doing the 'right thing' by our interstate cousins, who are also doing it tough. Hence the imposition of restrictions on watering home gardens and public open spaces.

The budgeted saving in water of this policy is 20% of the average annual Adelaide consumption, which in turn is only 1% of total annual allocation from the Murray-Darling Basin. Of course, that seems an insignificant saving, and in the event of prolonged drought these restrictions will mean little more than the difference between running dry before lunchtime or straight after! There are many more learned and considered analyses of the politics of water, but here it is important to contrast the relatively minor effect of this policy on future water supply with the dramatic and costly impact, evident at the start of 2008 in the streets and houses of Adelaide.

How trees affect soil moisture (tree beats water)
This is dependent on:
- the suction potential in natural and built environments;
- Adelaide's reactive soils which can lead to soil shrinkage;
- exotic vs native plant water use and response to drought; and
- the extent of the urban forest in private and public hands, and increased dependence on irrigation.

How soil moisture affects footings (water beats bluestone)
This is dependent on:
- the effects of vegetation on the swelling and shrinking of soils;
- the history of footings design in Adelaide; and
- the effect of current policy on shrinkage and rebound of subsoils.

Influence of built environment on urban forest policy (bluestone beats tree)
This is affected by:
- response by architects and engineers.
- exclusion zones;
- litigation; and
- impermeable paving vs turf and gardens.

Emerging opportunities in water management
- Allocation of water for trees in open space.
- Stormwater harvesting.
- Development of appropriate irrigation practices.
- Identification of new tree species that survive without irrigation in normal years but at reduced suction pressures so that water can be applied before soil moisture reaches critical values.

David Lawry

Figure 18.1: Adelaide Plains, 1890. Artist: James Ashton. Source: Adelaide and Mount Lofty Ranges Natural Resources Management Board.

This thick cover of vegetation over Adelaide is a relatively recent phenomenon. Figure 18.1 shows what the Adelaide Plains looked like in 1890, near the location of the current Adelaide Airport. This image is in stark contrast to the dense areas of vegetation that currently exist along the River Torrens and the mature street and garden trees of the surrounding suburbs. The number and diversity of plant species has also increased dramatically since 1836, with the number of tree species increasing by approximately 160% to include 38 native species and 60 introduced species.[13]

Vegetation cover varies widely across the city. Part of the reason for this diversity of land cover types – and a concentration of woodland areas around the centre of Adelaide – is the historical sequence of Adelaide's residential development. The older residential suburbs found around the CBD have had time to develop mature gardens and vegetation cover, while the relatively newer suburbs to the north and south have had less.

While projects such as the Three Million Trees Program are important within the metropolitan open space system, it is useful to note that this equates to only approximately 9% of the total area of the Adelaide ecosystem. The remaining 91% consists of buildings and grey infrastructure, residential gardens, streetscapes, and industrial and institutional lands. How these areas are managed are key drivers in determining how the metropolitan ecosystem will function as a whole.

There are significant implications associated with these statistics in managing vegetation cover in the urban landscape. Adelaide has extensive plantings of exotic and deciduous trees in its streets and private gardens, especially in older suburbs, with an emphasis on native plantings in public lands and open space reserves. In metropolitan Adelaide a broad range of Australian species were planted as street trees throughout the 1960s to the 1980s including *Acacia, Eucalyptus, Melaleuca* and *Lophostemon*, especially in the newer and rapidly expanding suburbs to the north and south of the CBD.

Adelaide also has large numbers of exotic deciduous street trees including plane trees *(Platanus sp.)* and ornamental pears *(Pyrus sp.)*, particularly in older central suburbs and gardens. Recently there has been considerable discussion about the need to plant local indigenous species as it is assumed that these trees will perform better under local conditions. However, Young and Johnson[14] point out that this view fails to consider the radical changes urbanisation has had on soils, hydrology and the local microclimate conditions. Indeed, many Australian species, such as the larger eucalypts, struggle with these altered conditions and can have significantly increased maintenance costs associated with road and infrastructure damage as they mature.

Young and Johnson also note that the argument against deciduous species based on leaf litter is flawed, as all species, including Australian trees, create leaf litter. Further, unlike deciduous species, evergreen trees extract relatively little nutrient from their leaves before shedding. The leaves of evergreen trees fall most often during summer when water levels in creeks and watercourses are low. When creeks have

Growing trees from the gutter

BOX 87

In an average year, 160 GL of water flows down Adelaide's gutters and into Gulf St Vincent. This runoff carries a high load of nutrients and sediment which are destroying the marine environment. Further pressure on this delicate ecosystem will arise from the discharge of highly concentrated brine from the proposed desalination plant designed to replace 50 GL of this loss per annum. The recommendations from the Adelaide Coastal Waters Study final report (ACWS, November 2007) are for an urgent reduction in the volumes of stormwater discharge to bring about a 75% reduction in nitrates, a 50% load reduction in particulate matter, as well as reduced flows of organic and mineral toxicants to coastal waters. The state government, in its 2004 'Water Proofing Adelaide' blueprint, set a target of 20 GL of stormwater reuse by 2025, only a 12% reduction on current outputs. Urban sprawl and infill will only add to the torrent of stormwater, so there is little chance that any improvement in the marine environment as envisaged in the ACWS can be achieved. No doubt there will be increasing adoption of well-established 'wetland' technologies to clean some of this polluted rainwater before sending it on to the ocean, or preferably to the aquifer. This is a very capital-intensive strategy suited to a few locations where land is available, and the environmental benefits are limited to the creation of aquatic habitats. What Adelaide needs are new, at-source, low-cost, readily implemented strategies that deliver multiple benefits to the community and the environment. Taking water from the gutters and putting it into the subsoil adjacent to street trees is the option currently being investigated in a growing number of TREENET trials.

Trees can each easily transpire 100 kL of water annually, so that 10,000 trees will take up at least 1 GL. The modest 20 GL reuse envisaged by 2025 could be taken up by 20% of our estimated local government tree population, by diverting stormwater only a metre or so from the gutter into aggregate-filled trenches, swales or permeable pavements. This removes the road-generated pollutants and organic-based nutrients at the source and puts them into the subsoil where they are broken down and taken up by the tree, or bound tightly to the surfaces of clay particles. Many complex designs for achieving this have been proposed by engineers and landscape architects but their high installation and maintenance costs have until now discouraged widespread adoption by councils. The current drought and restrictions that have drastically reduced the irrigation of front gardens have created the need to trial systems that will replace the traditional source of supplementary water for street trees. TREENET has a nominal target of achieving this – $200 per tree – by retrofitting existing kerbs and verges, and negligible costs for new installations. A much needed regeneration of Adelaide's ageing urban forest will provide the opportunity to make these practices standard. There will be significant savings in expenditure on repairing uplifted kerbs and footpaths; the root systems will run parallel to this infrastructure in response to the relocation of water resources to the driest and thirstiest zone currently in the urban environment – the curiously named 'nature strip'.

Trees can provide additional significant benefits other than simply providing an alternative for stormwater and pollutant discharge to the marine environment; in fact, some could argue that trees leave all alternative uses of stormwater in the shade. A University of Adelaide study[1] conservatively estimated the value of Adelaide's street trees in 2002 at $171 each, based on the cost of replacing the services they provide. Apart from habitat benefits for humans and wildlife, and a myriad of other services, the direct influence of trees on climate and hydrology are standout advantages. In transpiring all that water, trees are like giant evaporative coolers. Combined with services such as shading and controlling air movement, trees reduce the temperature of the city and suburbs by at least 5°C. This results in a massive reduction in energy use for air-conditioning. They also sequester carbon dioxide, considered to be a major contributor to global warming.

Finally, trees have a direct benefit in the control of stormwater by intercepting rain in the canopy, thus reducing the amount and rate at which stormwater threatens our marine environment.

David Lawry

low water levels, pollution levels are most concentrated and can be increased by falling leaves.

Deciduous species also provide structural diversity in the landscape, and a range of services including reduced cooling costs in summer when they are in leaf, and reduced heating costs in winter when they have dropped their leaves.

1. Analysis of the figures in tables 18.1 and 18.2 reveals that the metropolitan Adelaide landscape changed considerably between 1990 and 2000. Urban woodland areas increased in most council areas, with the greatest change resulting from urban open woodlands being more heavily planted to become urban woodlands.
2. In 1990, urban woodlands and forests covered approximately 31.4% of metropolitan Adelaide, comprising 6.1% urban forests, 11.6% urban woodlands, and 13.6% urban open woodlands. Urban grasslands covered approximately 15.6% of the metropolitan area.
3. By 2000, urban woodlands and forests covered approximately 36.4% of metropolitan Adelaide. Urban forests covered approximately 5.7%; urban woodlands increased slightly to 14.7%. Urban open woodlands increased slightly to 16.0%, while urban grassland areas had reduced to cover 14.8% of the metropolitan area. The total forests and woodland area across metropolitan Adelaide increased from 31.4% (1990) to 36.4% (2000).
4. Urban forests declined slightly from 6.2% in 1990 to 5.7% in 2000, while urban woodlands and open woodlands increased largely due to urban expansion and maturation of plantings in new developments such as Golden Grove. Urban grasslands also declined slightly from 15.6% to 14.8% of the metropolitan landscape in 2000.
5. The total vegetation land cover for Adelaide increased approximately from 47.0% to 51.0% of the total area in 2000. These increases are significant, having occurred over a relatively short time span of only 10 years, pre drought and water restrictions. Trees and vegetation growing in gardens and on private lands (approximately 90%) comprise the majority of Adelaide's urban vegetation cover.
6. The effects of drought on Adelaide's green infrastructure are likely to be severe and ongoing. The lack of water has already resulted in the death of grasses, shrubs and many of older trees. As the drought continues and the soil dries out at deeper levels (around the tree roots), older and larger trees are likely to continue to show effects of these impacts and continue to deteriorate and die over many years. Serious social, economic and ecological impacts are likely to follow the deterioration of urban forest cover, in terms of stormwater management, temperature modification associated with the loss of shade, air quality, and community life in the region.

Conclusion

In their book *Resetting the Compass*, Yencken and Wilkinson[15] note that two-thirds of the Australian population already live in the five largest cities and that continued migration to urban areas is an increasing global trend. The quality of the urban environment already influences the health of the majority of Australia's population, as well as a significant and growing proportion of the world's people.

The World Bank[16] estimates that cities around the world will more than double in area over the next 20 years and could eventually cover more than 1 million km^2. They may possibly consume as much as 5–7% of the world's total arable land, depending on the future rate of expansion of arable land (currently 2% per annum).

An additional challenge for Adelaide will be the potential impacts of climate change on the city's vegetation. A recent report by the CSIRO[17] notes that for Adelaide and the Mount Lofty region the projected annual rainfall could decrease by between 1 to 10% by 2030 and by 3 to 30% by 2070; and that by 2030 the annual temperature could increase between 0.4 and 1.2°C (0.8 and 3.5°C by 2070).

As we move into the century of climate change, increased understanding about the importance of the urban ecosystem and its connection with surrounding ecosystems (and the human and other communities they support) will require more sophisticated forms of environmental management. Sustainable environmental management will include the management of a sustainable level of urban vegetation across metropolitan Adelaide.

Globally, many rapidly expanding urban areas are losing their trees and forests at an alarming rate. Governments do not yet fully account for the economic, social and environmental costs associated with the loss of associated ecological services. Vegetative cover is a convenient indicator that can be used to monitor changes to the level of such services. By expanding on Goodwin and Noyce's[18] 'greenness rating' work to classify vegetation cover into structural types, the values attributed to urban forests, woodlands, open woodlands and grasslands can be individually monitored and managed into the future. The classification is particularly useful for comparing and contrasting changes in urban land cover over time, both within and between areas and across the greater metropolitan region of Adelaide generally.

In the ongoing debate about water we must include consideration of Adelaide's green infrastructure. The services urban vegetation provides to the community must be considered equally alongside currently accepted 'essential services'. Sustaining urban vegetation requires the support of legislation, policy and planning. Legislation and water policy must provide both the impetus and resources to maintain adequate and appropriate urban vegetation.

References

1. Planning SA, 'Planning strategy for the development of metropolitan Adelaide (draft for consultation)', Adelaide, 2002.
2. T. Johnson, 'Why plant street trees?' in Daniels, C. and Tait, C. (eds) *Adelaide, nature of a city: the ecology of a dynamic city from 1836 to 2036*, Biocity: Centre for Urban Habitats, Adelaide, 2005.
3. American Forests, 'Urban ecosystem analysis: SE Michigan and City of Detroit', US Department of Agriculture, Washington DC, USA, 2006.
4. American Forests, CITYgreen, 5.0 [software], US Department of Agriculture, Washington DC, USA, 2005.
5. American Forests, 'Urban ecosystem analysis: San Diego, California', US Department of Agriculture, Washington DC, USA, 2003.
6. P. Hewett, 'The value of trees: the big picture' in *3rd National Street Tree Symposium*, TREENET, Adelaide University, 2002, pp. 9–16.
7. P. Killicoat, E. Puzio and R. Stringer, 'The economic value of trees in urban areas: estimating the benefits of Adelaide's street trees' in *3rd National Street Tree Symposium*, pp. 90–102.
8. K.E. Smith, 'An urban land cover analysis for the Adelaide metropolitan ecosystem: monitoring vegetation change in the urban environment 1990–2000', Masters thesis, University of Adelaide, Adelaide, 2006.
9. Urban Forest Biodiversity Program, 'Conserving Adelaide's biodiversity', information package, Adelaide, 2000.
10. R. Oke, 'Conserving Adelaide's biodiversity: a planned approach', Department for Environment and Heritage, Adelaide, 1997.
11. C.J. Tait, C.B. Daniels and R.S. Hill, 'Changes in species assemblages within the Adelaide metropolitan area, Australia, 1836–2002', *Ecological Applications*, 15, 2005, pp. 346–359.
12. H. Girardet, 'Creating a sustainable Adelaide: a green city', Government of South Australia, Adelaide, 2003.
13. C.J. Tait, 'Trees and urban green areas' in Daniels *et al Adelaide, nature of a city*.
14. K. Young and T. Johnson, 'The value of street trees' in Daniels *et al Adelaide, nature of a city*.
15. D. Yencken and D. Wilkinson, *Resetting the compass: Australia's journey towards sustainability*, CSIRO Publishing, Melbourne, 2000.
16. World Bank, 'The dynamics of global urban expansion', Transport and Urban Development Department, Washington DC, USA, 2005.
17. R. Suppiah, B. Preston, P.H. Whetton, K.L. McInnes, R.N. Jones, I. Macadam, J. Bathols and D. Kirono, 'Climate change under enhanced greenhouse conditions in South Australia: an updated report on: assessment of climate change, impacts and risk management strategies relevant to South Australia', CSIRO Marine and Atmospheric Research, Canberra, 2006.
18. D. Goodwins and T. Noyce, 'A "greenness" rating for the suburbs of metropolitan Adelaide: an assessment of the vegetation cover of the suburbs of metropolitan Adelaide using a GIS', Office of Planning and Urban Development, Adelaide, 1993.

Box reference

Box 87: Growing trees from the gutter

1. P. Killicoat, E. Puzio and R. Stringer, 'The economic value of trees in urban areas: estimating the benefits of Adelaide's street trees' in *3rd National Street Tree Symposium*, pp. 90–102.

CHAPTER 19

Perceptions of water in the urban landscape

Andrew Lothian
June Marks
Gertrude Szili
Matthew W. Rofe

Introduction

Water symbolises life. The three case studies presented in this chapter illustrate differing aspects of how we interact with water. There are a vast range of needs and values that water fulfils – the utilitarian requirement for drinking, cleansing, transport, recreation and fitness, health and restoration, image promotion, aesthetic and even spiritual values.

Section 19.1, Aesthetics of water in the urban environment, examines our historical association with water through springs, fountains, rivers, gardens and baptisms, and poses the question: why does water attract us as much as it does? It also examines the psychological and physiological benefits of viewing water in an urban context, together with aspects lacking in appeal. The value adding that water provides to real estate is significant. Finally, the study puts forward practical suggestions on how we can maximise the aesthetic benefits provided by water.

Section 19.2, Greenwashing the Port Adelaide Waterfront Redevelopment, focuses on the use and abuse of water as an agent in image creation, and the use of advertising to create favourable images of the development. Suffering from a history of neglect, pollution and a woeful image, the promotion of the redevelopment of the Port Adelaide waterfront posed substantial challenges. The paper examines the extent to which the image proffered differs from the reality. But the redevelopment is about the future, and perhaps, in time, the substance might merge with the image.

Section 19.3, Adelaide, the water recycling capital, examines the major projects that have transformed wastewater into a resource. The Virginia Pipeline takes, treats and distributes wastewater from the Bolivar wastewater treatment plant (WTTP) to Virginia for use on market gardens, while Mawson Lakes reuses much of wastewater for gardens and toilets. The study presents evidence from various surveys demonstrating strong community support for the use of recycled water. The study gives hope for the future – that in our water-scarce state, we can do much to make better use of our water.

Section 19.1: Aesthetics of water in the urban environment

Andrew Lothian

> *Water though not absolutely necessary to a beautiful composition, yet occurs so often, and is so excellent a feature, that it is always regretted when missing: and no large place, or indeed small place, can be imagined where water may not be agreeable; it adapts itself to every situation; it is the most interesting object in a landscape, captivates the eye at a distance, invites approach, and is delightful when near: it refreshes an open exposure; it animates a shade and cheers the most crowded view; in form, in style and in extent may be made equal to the greatest compositions or adapted to the least; it may spread in a calm expanse to soothe the tranquility of a peaceful scene.*

> – Thomas Whateley, English 18th-century writer.

The significance of water to people

Water has long held a place of deep significance to people. Historically, springs were important to the Greeks; the mythical 'fountain of youth' and 'water of life' were among the mythologies of the ancients; while the Garden of Eden is associated with an eternal spring. Fountains are symbols of purity and throwing coins into them for a wish or good luck probably originated in appeasing the gods of the waters. Eden's four rivers are often depicted in Islamic carpets; Persian gardens were rectangular and comprised four quarters separated by canals with a central pool – these walled gardens were called *pairi-daēza* (enclosed gardens), translated by the Greeks as *paradeisos*, the origin of our word *paradise*. For the ancient Egyptians, the Nile marked the division between life on the eastern bank and death (eternal life) in the tombs of the western bank. The three monotheistic religions, Judaism, Christianity and Islam, all use water literally and symbolically to mark significant human events including Christian baptism (symbolic of purification and rebirth). Both Jews and Moslems wash in water prior to marriage, and all wash the body before burial. The Villa d'Este at Tivoli, outside Rome, built in the 16th century by Cardinal Ippolito II d'Este with extensive gardens dominated by gravity-powered fountains, included waterfalls, cascades, a water organ, jets and sprays, and a terrace of the Hundred Fountains. The Villa d'Este follows a long tradition of the use of water in gardens, extending back to Mesopotamian, Egyptian, Greek and Roman gardens.

In the 18th century, Jean-Jacques Rousseau developed an almost pantheistic love of nature, with God and nature being one, which profoundly affected the human view of nature. In 1765, he lived for a few months on an island on Lake Bienne in Switzerland; the following passage describes the intensity of his experience:

> *I often sat down to dream at leisure in sunny, lonely nooks … to gaze at the superb ravishing panorama of the lake and its shores … When evening fell, I came down from the higher parts of the mountains and sat by the shore in some hidden spot, and there the sound of the waves and the movements of the water, making me oblivious of all other distraction, would plunge me into delicious reverie. The ebb and flow of the water, and the sound of it … came to the aid of those inner movements of the mind which reverie destroys and sufficed me pleasantly conscious of existence without the trouble of thinking …*

Rousseau's romanticism helped transform European appreciation of nature, and his influence continues today.

The English landscape garden designers, of which Capability Brown and Humphrey Repton were the most eminent, transformed English estates from formal avenues of trees to natural recreations of Italianate landscapes based on the paintings of Claude Lorraine. Brown excelled in the use of water in his gardens by establishing sinuous lakes, clumps of trees and sweeping lawns in tranquil pastoral scenes. It was his boast that 'Thames could never forgive him for the glories of Blenheim'.

This everything water

BOX 88

Gold lip pearl shell, *Pinctada maxima*, has been harvested by Aboriginal people from the seas off the Dampier Peninsula in north-western Australia since before white contact. Pearl shell was valued for its brilliance, collected, shaped, engraved and used in ceremony or for personal adornment by the Dampierland people. It was used for many purposes – in rain-making rituals and initiation ceremonies, and for purposes associated with tribal law, medicine and love magic. In all these uses the shining surface of pearl shell reflects the associations with physical and spiritual wellbeing flowing from the Ancestral Creation period,[1] thus ensuring that pearl shells are perceived by Aboriginal people as particularly potent objects. For the people of the Kimberley, the luminous fluidity of mother-of-pearl, shifting and shimmering under the play of light, is associated with water. Kim Akerman notes in his 1993 monograph on Kimberley pearl shell that, irrespective of use, for Aboriginal people pearl shell is water, an emblem of the very basis of life.[1]

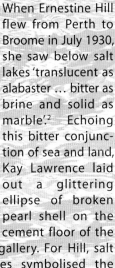

This belief was eloquently expressed by Walmajarri Elder Mumbadadi at Christmas Creek, south of Fitzroy Crossing in 1990: 'This is for everybody – man and woman. This is rain. This everything water.'[1]

In ceremony the gleaming, iridescent surface of the shell, held in the hand and flicked swiftly back and forth, sweeping an arc through space, flashes like the lightning that brings rain. Broken shell, glittering in the shifting sand of a dune or in the cracked mud of a tidal estuary, could reveal a midden or a site where digging would reveal a trickle of freshwater.

The exhibition *This everything water* explored the material qualities and metaphorical associations of pearl shell with water, through pearl shell ornaments engraved by Nyigina Law Man, Butcher Joe Nangan, and Bardi and Djawi Elder and Law Man, Aubrey Tigan, and artwork made from pearl shell by Kay Lawrence.

Across Australia and outside the cities, the country is dotted with dams to catch the rain and runoff to provide water for farms and pastoral leases. Glassy at noon, these sheets of water shimmer under a breeze at dusk. In dry years they are often empty, and in some parts of Australia these glittering ellipses may instead be saltpans, the result of excessive irrigation that has caused salt from ancient seabeds to rise and poison the land.

When Ernestine Hill flew from Perth to Broome in July 1930, she saw below salt lakes 'translucent as alabaster … bitter as brine and solid as marble'.[2] Echoing this bitter conjunction of sea and land, Kay Lawrence laid out a glittering ellipse of broken pearl shell on the cement floor of the gallery. For Hill, salt lakes symbolised the 'dream and bitter reality' of the changes made to land by the pastoral industry. This crisis, compounded by drought, is even more compelling in 2009, where, in a dry continent, all life depends on water.

Water is everything.

This is an abridged version of an essay in the catalogue that accompanied the 2008 Adelaide Bank Festival of Arts exhibition *This everything water* was held at the South Australian School of Art Gallery from 28 February to 28 March 2008.

Kay Lawrence

Why do we like water?

Delight in water does not necessarily result from its utility value – for transport, fishing, recreation and industry – which occurs unrelated to its aesthetic worth. Water is essential for life – some believe *this* explains human delight in viewing water. However, this is an insufficient explanation; there are many other such necessities for life (e.g., food, shelter) which do not engender delight as does water. We require water to live, but why do we enjoy seeing the sea even though it is undrinkable?

Some believe that human preferences for beautiful landscapes, which include water, can be explained by an evolutionary perspective: what we find beautiful is survival enhancing. Research has found that clear, flowing water and large bodies of water are preferred over swamps and small ponds, suggestive of survivability. While there is some supporting evidence for the evolutionary explanation, it is by no means conclusive; for example, why do we find appealing arid areas such as the Flinders Ranges, scarcely a conducive environment for life?

None of the conventional explanations of the appeal of water – cultural, religious, utilitarian, evolutionary – provide an adequate rationale. The appeal of water is deep-seated within the human psyche for reasons which remain unfathomable (no pun intended!) Recent research indicates that viewing natural scenes, including water, triggers physiological processes that influence the immune system and engender a sense of wellbeing.

What do we like about water?

Many studies of landscape aesthetics have examined the role water plays in enhancing scenic attractiveness. The presence of water in a scene has often been found to be the most important attribute, ahead of trees and vegetation, hills and valleys, and attractive land uses. Water adds interest, aesthetic pleasantness, and is enjoyed immensely. Water symbolises nature, and this always enhances the appeal of the scene.

In contrast to viewing carparks, sides of buildings, housing, shopping centres and the like, views out of the windows of offices, schools and particularly hospitals and prisons towards scenes containing trees, greenery and water are restorative and calming.

Studies of university students facing exams found that viewing scenes containing vegetation and water reduced their stress levels and enhanced their feelings of wellbeing. A study of patients in a hospital found that those with windows looking out at a natural scene had shorter post-operative hospital stays, received fewer negative evaluative comments in nurses' notes, and fewer analgesics than an equivalent number of patients in similar rooms with windows facing a brick wall. Another study using photographs of different scenes found patients viewing a natural scene with a water or forest experienced much less post-operative anxiety than if they viewed an abstract painting (which provoked high anxiety!), or a blank wall.

Similar findings of positive emotional states from viewing nature have been found in studies of dentists' waiting rooms and of people disabled by accidents or illness. Prisoners in cells with a view of nature have lower symptoms of digestive illness and headaches, and fewer sick calls. There is also anecdotal evidence that natural scenes and settings with water features induce positive emotional states conducive to creativity and problem-solving. How many Nobel Prizes had their origins in a walk in the park?

The provision of Central Park in New York City stemmed from the belief of its planner, Frederick Olmsted, that viewing nature was an essential counter to the stresses of cities, and this in 1865! Adelaide is fortunate indeed that Colonel Light established a wide parkland belt around the central city.

A study in Ann Arbor, Michigan, found that motorists preferred to use a parkway to drive to a large shopping centre than a state highway. The highway was lined with manmade structures and had little appeal, whereas the parkway ran past wooded, undeveloped scenes including along a river. Interestingly, though, the highway was shorter and quicker, but even then the majority of drivers chose the parkway.

Serenity is the word often used to describe places with water. 'There is simple serenity in the perfect expression of the horizontal', wrote an English author. Such places function as 'natural tranquilisers', providing people with relief from stress and pain. Studies have shown that whereas urban scenes often induce feelings of sadness, water has a stabilising effect on emotions and in particular in sharply reducing feelings of fear.

Water is particularly significant for outdoor recreation – estimates suggest that about quarter of all activities are water-dependent, including swimming, canoeing, kayaking, sailing, model boats, nature study and bird watching, even riding on Popeye!

Moving water, such as waterfalls and rapids, can produce feelings of awe. Turbulence focuses attention but reduces the sense of tranquillity associated with calm water. The scenic appeal of moving water increases up to a point, but then tends to diminish when rushing, turbulent water (such as floods) can engender fear. A study in Colorado using videos of moving water found that the rating of scenic attractiveness rose with waterflow up to around 1100–1500 cubic feet per second (31–42 m^3/s), and then diminished as flow increased (Figure 19.1). The peak viewing time thus corresponded with two short periods annually, occurring during the rising and falling waters of the late-spring peak runoff. These scenic quality estimates corresponded with the findings of economic studies using people's willingness to pay (WtP) for the maintenance of in-streamflow levels. The WtP rose to a point of flow and then decreased as flow increased.

The scenic attractiveness of water increases with the length of the water/land edge and the area and depth of the water. Many straightened streams reduce the water/land edge. Placid waterbodies such as ponds and lakes engender tranquillity and serenity. Water (along with trees and flowers) helps people cope with the stresses of urban living.

What don't we like about water?

While clean, sparkling, clear water is found to be attractive, dirty, poor-quality water, coloured water, stagnant water and algae growth on the water are all regarded as unattractive. Litter, erosion and channelisation also are negatives.

Water adds value

Given that water adds appeal to a scene and people enjoy water in their environment, what effect does it have on house prices? Houses near water or with a water view would be expected to attract higher prices than houses without.

Studies support this. A study of houses with a clear view of Lake Erie[1] in the United States found that the view added nearly 90% to the value of a house. A study in Auckland, New Zealand,[2] found that wide views of the harbour added 59% to waterfront property values, and that the effect diminished away from the water. A second study in New Zealand[3] found the value of the water view related to how prevalent water views were in the city; in Wellington, where nearly 20% of houses had a water view, it added less than 7% to its value, while in Auckland, where 12.5% had a water view, it added nearly 10%, and in Christchurch, where only 2.5% of houses viewed the water, it added nearly 11%.

In Adelaide, a simple comparison of average house prices in suburbs between Marino and Glenelg (near the coast) with those of the adjoining suburbs indicated a 22% premium for the coast (Figure 19.2). Similarly, a comparison of average house prices for three suburbs built around significant water features with those of nearby suburbs indicated a 29% increment (Figure 19.3). Adding water to housing development clearly has real-estate appeal. It is very fortunate that the provision of such features coincides with good environmental practice through the retention of stormwater.

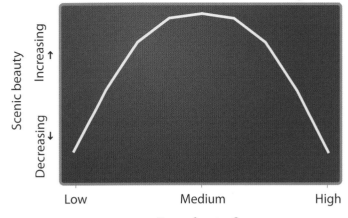

Figure 19.1 Influence of river flow on scenic beauty. Source: adapted from T.C. Brown and T.C. Daniel, 'Landscape aesthetics of riparian environments: relationship of flow quantity to scenic quality along a wild and scenic river', *Water Resources Research*, 27 (8), 1991, pp. 1787–1795.

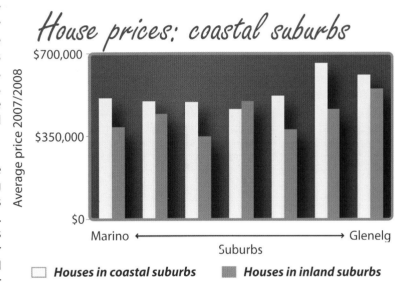

Figure 19.2 Comparison of average house prices for houses in coastal and adjacent inland suburbs from Marino to Glenelg. Source: Domain.com (website).

How can we maximise the aesthetic benefit of water?

It is striking that compared with cities in with more temperate climates, Adelaide does not have extensive waterbodies, such as lakes and rivers. The Torrens Lake and the River Torrens have significant water colour issues, diminishing their attractiveness. Efforts to reduce pollutant inflow into the Torrens would have a positive effect on the aesthetic quality of the water. The urban creeks across the Adelaide Plains flow only temporarily after rain and have been largely confined in concrete channels. Their potential role in enhancing urban amenity have been largely lost in the older suburbs of Adelaide by urban development. The creation of waterbodies at new urban developments, such as Mawson Lakes and Andrews Farm, demonstrate the appeal that water provides, as well as its real-estate value. The lessons learned from these developments are being applied in new urban areas where creeks have been retained as part of the open space and waterbodies have been formed.

Urban wetland initiatives are of great aesthetic benefit, such as the Urrbrae Wetland, the Warriparinga Wetland, the Paddocks, extensive wetlands at Salisbury, and many small wetlands established over recent years. The layout of wetlands should aim to maximise the visibility of the waterbody to passers-by and, wherever possible, lakes and wetlands should be created. For example, the Parklands Creek, which enters the southern end of Victoria Park, could be diverted into a large water-holding basin, which, as well as providing flood relief and aquifer recharge, would serve as an attractive waterbody in an otherwise open paddock.

In designing waterbodies, the water/land length should be maximised through indentations rather than a smooth, round edge, as this maximises their visual appeal. The waterbody should be extensive and islands established, which provide excellent visual contrast and further edges. Water depth is important, partly to ensure the retention of water for as long as possible, and also because deep water is more visually attractive. Measures to trap silt, debris and pollutants before the water enters the waterbody will provide cleaner, more attractive water. Fountains provide a very effective means of enabling many people to enjoy the psychological benefits of water in the city. The emphasis of fountains should be the water not the structure that conveys it. Pools of water should be clear and deep. The sound of moving water provides additional cues.

Conclusions

Adelaide has a Mediterranean climate with moist winters and hot, dry summers, conditions in which waterbodies can play a significant role in enhancing the liveability and attractiveness of the city.

In the past, water management has largely focused on maximising its utility as a resource, minimising the negative effects of stormwater, and protecting the quality of surfacewater. Technical, engineering-based standards and solutions are employed together with complex legal sanctions and requirements. Given the beneficial human values and wellbeing associated with water, including recreation and its restorative effect from illness and stress, the conventional provisions should be broadened to include psychological and physiological considerations.

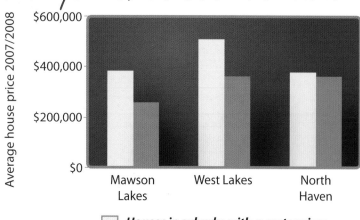

House prices: suburbs with water

Figure 19.3 Average house prices in Adelaide suburbs with a significant water feature, compared to houses in adjacent suburbs.
Source: Domain.com (website).

Section 19.2: Greenwashing the Port Adelaide Waterfront Redevelopment

Gertrude Szili and Matthew W. Rofe

This section is a truncated version of 'Greening Port Misery: marketing the green face of waterfront redevelopment in Port Adelaide, South Australia'.[4]

Introduction

Following the decline of the industrial economy, industrial regions have become synonymous with social, economic and environmental decline. As a result of global economic restructuring, recent trends in urban policy have revealed an explicit emphasis on the redevelopment and revitalisation of these under-utilised industrial landscapes. Indicative of these landscapes are ports and other neglected waterfront sites. The redevelopment of the Port Adelaide waterfront serves as a prime example of such a post-industrial transformation that purports an economically vital, socially empowering, and environmentally progressive future. Central to this revitalisation is the renegotiation of the city's meaning and relationship with the Port's inner harbour.

The site

Located approximately 14 km north-west of Adelaide, Port Adelaide is on the cusp of an urban 'renaissance',[5] spearheaded by the redevelopment of 52 ha of former waterfront industrial land for mixed residential and commercial use. The redevelopment is overseen by the state government's Land Management Corporation (LMC) in collaboration with the private sector development company Urban Construct. Trading as Newport Quays Consortium, the redevelopment will realise approximately A$1.2 billion in reinvestment capital (LMC, n.d.), and will be undertaken in six stages, with completion anticipated in 2014. It is envisaged that the realisation of the master-planned vision will drive the 'grime out' of Port Adelaide.[6] In the rhetoric of urban revitalisation, this claim is not unique. However, what is interesting about the Port Adelaide redevelopment is the centrality of the environment as a guiding premise of the project. Indeed, the language and imagery associated with this project are saturated with explicit and implicit environmental claims. Thus, although there will be some environmental improvement, the extent to how sound and achievable they are – or whether they represent a veneer of environmentalism to legitimise development and boost sales – is questionable.

A legacy of misery

Legally proclaimed a port in 1837, Port Adelaide has been stigmatised as a bleak place. From colonial settlement, it was described as 'a most unwholesome and unsavoury spot'.[7] Settlers who were anchored in the inner harbour described it as a 'narrow dirty ditch, fringed on both sides with odious mangrove

BOX
89

The modified environmental conditions created by land use (e.g., nutrient and sediment runoff from agriculture and urbanisation) cause massive change to marine ecosystems. Habitats such as coral reefs and kelp forests change, as do the plants and animals that rely on them for food and shelter. While the link between human expansion and coastal changes appears obvious, we continue to be surprised by the ecological outcomes.

Debate about Australia's future population growth[1] single out our coastal zone as shouldering much of our cultural 'sea change'. The 'State of the Marine Environment' report identifies 'declining marine and coastal water/sediment quality, particularly as a result of inappropriate catchment use practices' as a primary concern in managing Australia's marine environment.[2] As Australia's coastal populations continue to increase, coastal marine resources will change more and more as a function of declining water quality.

There is growing recognition of the ecological contexts (e.g., biogeography and geomorphology) that make coastal systems more vulnerable or resistant to increasing human dominion of the world's coastlines. The magnitude and onset of human-driven change to biogenic habitats is not always proportional to human population size. The impacts of human acitivies, such as the discharge of pollution and over-fishing, are relative to

local environmental conditions. In these cases, changes in nutrient levels and food chains do affect rates of biomass production and loss, but not independently of the environment.[3,4] For example, sudden increases in nutrient concentrations (e.g., runoff)[5] can cause disproportionately larger responses on nutrient-poor coasts.[6] Therefore, the distribution of research efforts on the ecological impacts of humans on marine environments should not necessarily be based on the concentration of human populations in the adjacent coastal areas (i.e., Sydney in Australia).

The ecological consequences of nutrient enhancement is different between Adelaide and Sydney. Both are located at the same latitude (~34°S) but in different biogeographic provinces, and are subject to contrasting nutrient conditions[6] and productivity.[7] The coasts adjacent to Adelaide have limited nutrient reserves[8] and low chlorophyll a concentrations.[9] There have been massive losses of habitat (kelps and seagrasses) adjacent to land with point-source and diffuse coastal runoff. In contrast, studies of urban discharge in Sydney highlight the lack of effect of outfalls around the city.[10] Similarly, there have been only vague or small effects identified on kelp and benthos elsewhere on the east coast.[11,12]

Sean Connell

trees'.[8] A lack of sewerage or waste disposal except for the outgoing tide, which 'turned the mangrove swamps into a smelly abomination'[9] added to this stigma. The alternate place name this engendered was simply 'Port Misery'.

By the end of the 19th century, the Port had grown from an 'eerie tangled mangroves and banks … of mud'[10] into a thriving industrial port. However, the stigma of Port Misery remained. Problems of delinquency, prostitution, unemployment, union militancy and violent protests were consistently discussed in social commentaries. The Great Depression and World War II brought the stigma of Port Misery once again to the fore. Shortages of employment and capital consequently restricted development.[11]

During the 1950s, the ailing waterfront was dealt a further blow when the introduction of new containerised cargoes initiated a shift of primary port activity to Outer Harbour. To stem this decline, public/private sector initiatives to establish the Port as the major retail centre emerged during the 1960s and 1970s. However, this was never realised as the proposals were initiated during periods of economic downturn.[12] At present, substantive concerns from the local community have arisen over the state of the Port's physical environment, particularly over water and soil quality and potential groundwater contamination from former and current industrial sites. Recent investigations by the Environment Protection Authority (EPA)[13] support these concerns detailing clear signs of environmental stress and ecosystem decline around the inner harbour. Water and sediment analyses have revealed contamination by a range of toxicants including heavy metals such as copper, lead and cadmium, and maritime anti-fouling agents such as tributyltin.[13] Thus, the adage of Port Misery is not simply echoed through the decaying factories and derelict buildings, but haunts the landscape today as the spectre of pollution.

In order to halt these processes of decline, the policy-lead revitalisation of the Port Adelaide waterfront aims to reposition the landscape as a region ripe for reinvestment. Through the physical remediation of the inner harbour, and discursive messages mobilised through advertising and promotional materials, the development consortium renegotiates the city's identity by asserting a positive image aligned with a post-industrial rhetoric, as opposed to the industrial discourses of decline. Central to this transition is the renegotiation of the city with its environment.

Greening Port Misery

According to marketing documents, the reversal of industrial environmental degradation rests in large-scale revitalisation projects. Such discourses represent highly strategic forms of environmental reimaging. Just as revitalisation is depicted as producing cosmopolitan landscapes of professionalism,[14] so too does revitalisation recast the brownfields of industrial decline with a green hue. This practice is commonly referred to as 'greenwashing'.[15, 16] A critical examination of the literature pertaining to greenwashing[4] offers considerable insights into the process of revitalisation and the way in which the environment is manipulated through advertising campaigns currently active in Port Adelaide.

Many corporations, despite their new green image, are not environmental saviours but remain the peddlers of polluting and unsustainable technologies.[17] Urban regeneration similarly professes to improve the environmental situation of decaying industrial landscapes. However, the extent to which remediation and environmentally sound practices are adopted, especially in highly entrepreneurial situations, is questionable. For Port Adelaide, the transition to a clean post-industrial city involves a direct engagement with, and repudiation of, the city's industrial heritage and legacy of pollution. Central to this transition is the renegotiation of the city with its environment, particularly the extensively polluted Port River.

Striving to renegotiate the city's identity, the meaning and use of the Inner Harbour is carefully being remediated and repackaged from an industrially polluted body of water to that of a pristine recreational amenity. Images commonly associated with a 'pristine' environment, such as blue water, vividly green vegetation and emblematic animals are strategically employed to engender a sense of environmental health (Figure 19.4). Specifically, images of a healthy waterway are consistently drawn upon in the marketing materials for the redevelopment. For example, emotive visions of a waterfront that reflect the lifestyle aspirations of potential investors saturate a recent advertising campaign promoting the redevelopment's second stage. Such images are in stark contrast to an industrially polluted waterfront, and signify the transformative qualities of revitalisation and the rebirth of once derelict and polluted industrial spaces. However, as Beder[18, 19] alerts us, green marketing is quintessentially a corporate concept, where city marketers and developers take advantage of a 'green' consciousness to increase profits without making any substantial and fundamental changes to the environment. Augmenting this, some of the marketing images used to promote Newport Quays do not illustrate actual imagery of Port Adelaide. Thus, such images may rather be powerful chimeras that distort public perception of corporate concerns.

Supplementing the mobilisation of environmental signifiers is a range of more explicitly apparent messages. The LMC's flagship marketing brochure, 'Developing Opportunities',[20] is itself deeply embedded within the signification of environmentalism. Three of the seven core objectives of the Port Adelaide redevelopment listed in the brochure are explicitly environmental and acknowledge the need to remediate the contaminated land prior to development.[4] However, local voices question the extent of this remediation. As one prominent local government representative observed:

> … they have said that they are going to remediate the land, but they are not actually too sure about how they are going to do it because it is mega, mega 'bucks'.

When confronted about the details of the remediation process, an LMC informant acknowledged that the 'cost of land remediation has been very high' and that this placed limitations upon the remediation undertaken. Thus, reflecting the local government representative's concerns, there is doubt that significant environmental improvement will occur as, ultimately, 'corners will be cut to save money'.

Coastal development

BOX 90

Many residents of Adelaide were concerned when the Holdfast Shores development was built adjacent to the beach at Glenelg. Not only did it block the view of the sea down Anzac Highway, it seemed to encroach upon public open space. There are also questions – within the context of global warming – about the risks that sea level rise will pose for the development.

As part of state-wide survey in 2005 of the visual impacts of coastal developments, Dr Andrew Lothian of Scenic Solutions included scenes of the Holdfast Shores development. Photographs were taken of the scene and two images prepared – one with the development and one without. In some scenes, existing developments were removed while in others, images of the development were inserted. Participants in the survey were asked to rate the scenic quality of the scene on a 1 (low) to 10 (high) scale.

The following scenes show the Holdfast Shores development from varying distances along the beach, together with the average rating of over 1600 participants.

In the distant scene the presence of Holdfast Shores reduced the beach's scenic quality by 28%; in the closer scene it reduced it by over 40% – a very large reduction.

In similar scenes, developments at various other localities along Adelaide's coast were examined and in most cases the presence of significant developments such as marinas and high-rise buildings significantly detracted from the ratings.

These results indicate the significant impact that development can have on coastal amenities. Further information on the coastal development study is available at Scenic Solutions (online).

Scenic quality rating: 4.46

Scenic quality rating: 6.18

Scenic quality rating 3.62

Scenic quality rating : 6.14

Andrew Lothian

Community suspicion has been further fuelled by a lack of information concerning the touted environmental remediation entering the public domain. While the brochures and press releases tout 'the extensive clean-up of the land',[21] no detailed information on how this will be achieved is contained therein.

Conclusion

Stigmatised for over 150 years as a morally corrupting and polluted landscape, the Port's current transformation aims to shed the Port Misery mantle in favour of a cleaner post-industrial identity. This transition involves both a physical recreation of the Port's landscape and a discursive reorientation of its identity. Central to this reimaging is a renegotiation with the Inner Harbour from a working industrial and polluted watercourse to a clean, post-industrial space of leisure and consumption. To achieve this, the Newport Quays Consortium has carefully lent a green hue to their activities, thereby legitimising and promoting the redevelopment of former industrial land. Metaphorically, this green hue washes away the dirty brown landscape of the Port's industrial heritage with a glossy, clean and green corporate image.

In an era where there is increasing public concern over environmental affairs, it is not surprising that development consortia have clambered to illustrate their environmental sensitivities. While it is not refuted that the Newport Quays project will bring some form of environmental remediation to the Port Adelaide waterfront, the extent of the environmental claims made by the consortium remains contentious. Through the analyses of various promotional materials and key stakeholder interviews, the extent to which these purported benefits are valid for Port Adelaide are problematised. Thus, it is contended that the veracity of such claims are problematic, distort environmental truth and, as such, constitute a form of greenwashing. However, as a largely discursive practice greenwashing is essentially rhetorical in nature seeking to head-off demands for any real environmental reform or sensitivity. Rather than revealing a meaningful engagement with, and concern for, the environmental condition of the Port, the environmental rhetoric mobilised in relation to the Newport Quays redevelopment is considered to be an astute and strategic marketing ploy.

Section 19.3: Adelaide, the water recycling capital
June Marks

Adelaide has a history of leading Australia in developing 'dual reticulation' recycled water schemes. Currently, 21% of all treated wastewater is recycled. Secondary-treated effluent from the Bolivar (WWTP) found its first customers on the Northern Adelaide Plains (NAP) in the 1970s, where the water sustained a restricted range of crops grown for the commercial market. While growers were lobbying for the now well-established Virginia Pipeline Scheme (VPS), a consortium of government agencies turned on the tap of a demonstration system for the New Haven Village. From 1995 through to 2001, New Haven was arguably the only cluster of private homes in any city in the world to recycle water

Figure 19.4: Environmental imagery used to promote the Port Adelaide waterfront redevelopment. Photograph by Szili, 2004.

sourced from highly treated effluent for toilet flushing. This was a forerunner for the then proposed Multi Function Polis (MFP), a state-of-the-art city whose design would be based on the principles of sustainable development. Although the MFP was abandoned by its Japanese developers, Mawson Lakes arose from its ashes. And while Mawson Lakes dual-delivery system was being laid in the north, growers in the McLaren Vale wine region to the south of the city had successfully negotiated a scheme recycling water from the Christies Beach WWTP.

What do the general public and water recyclers think of the idea of using alternative sources of water? This section will outline findings from social research that includes those involved with the Virginia scheme and householders who recycle water at Mawson Lakes.

The Virginia Pipeline Scheme

For many years, several growers in the Virginia maket garden area north of Adelaide gained access to effluent channeled from the Bolivar WWTP for discharging to Gulf St Vincent. They eventually lobbied for a higher-quality effluent, to enable them to grow an unrestricted range of crops. This, together with depleting groundwater resources on the NAP, the EPA ruling for SA Water to reduce nitrogen levels in the effluent to protect the gulf, coupled with the opportunity to expand horticulture production in the region, lead to the establishment of the VPS. Funding was obtained through a federal Better Cities program, the state government through SA Water, and the private sector, through WRSV. The scheme involves over 100 km of pipeline at a cost of $22m and a $30 m dissolved air flotation and filtration (DAFF) plant at the Bolivar WWTP to achieve class A water quality. There are over 280 connections to approximately 250 customers of the pipeline, which distributes around 15 GL of recycled water per year.

Social research conducted on the scheme five years after it came online late in 1999[22] and in 2006[23] confirm that growers appreciate the benefits as well as the responsibilities involved in using an alternative water source. The former research involved interviews with 75 stakeholders (growers and other key players). The 36 growers interviewed were from various farming practices: 13 broadacre, 12 glasshouse, six perennial and five perennial-glasshouse producers. Their cultural backgrounds include Italian, Greek, Vietnamese, Cambodian and Anglo-Saxon. The types of crops grown range from heavy vegetables (potatoes, carrots, broccoli) to salad (tomatoes, lettuce) through to permanent crops (olives, almonds, wine grapes).

Another main finding is that the successful establishment of the scheme can be attributed to ongoing communication between all those involved. The VPS operates on a build, own, operate, transfer (BOOT) arrangement, with SA Water assuming responsibility around 2020. There is a healthy working relationship between the current owners of the pipeline delivery system, WRSV, the Bolivar WWTP (managed by United Water), and the Virginia Irrigation Association. The Department of Health played a key role in establishing and now maintaining water quality and the EPA, the Virginia

Horticulture Centre and other agencies, promote practices that will safeguard environmental health. There is some concern about soil salinity and rising watertables, however, it is acknowledged that a range of factors, not simply recycled water – for example all water sources, fertilisation and irrigation regimes that potentially have an environmental impact – require good practice, monitoring and further research. A glasshouse grower suggests:

Some people still complain about this and that thing. But it is good, it depends how they grow. You've got to know what to do.

In general, growers are enthusiastic about the availability and quality of the recycled water. One glasshouse grower reports:

For me, personally, it has been a godsend. If I didn't have 'Bolivar water', I wouldn't be growing anything. It is hard to get a borewater quota.

A broadacre producer reflects:

It is world-class water. We can use it to produce vegetables and it hasn't killed anybody.

Initially, there was some anxiety about public perceptions and market acceptance. To allay concerns, scientists from the CSIRO worked to investigate and communicate research and information to wholesalers and growers. No incidents were reported, only claims were made and subsequently addressed. Some 13 wholesalers, quality assurance agents and key retailers were interviewed or contacted to gauge current acceptance of the produce. It was reported that, like all produce, vegetables and fruit grown with recycled water is subject to quality assurance protocols and testing to ensure compliance with health standards. The produce has been mainstreamed to the extent that some wholesalers had to really think about why class A recycled water would be any different in quality from dam, river or mains water. In addition, all produce is washed in drinking water before entering the market. Recent social research confirms that city consumers are accepting of crops grown with recycled water – two-thirds nationally and three-quarters of the Adelaide population.[24, 25] Equipped with knowledge of the quality assurance regimes and competent management involved in schemes like the VPS, it is likely that grower and wholesaler reassurance as well as public acceptance can be further consolidated.

In conclusion, a glasshouse grower forecasts the future of water recycling for crop irrigation:

The reclaimed water is the best thing that ever happened to Virginia ... It might be the best thing that ever happened to South Australia.

Certainly, the success of the scheme has positively influenced similar schemes within the state and around Australia, including the Werribee Scheme in Victoria. A recent government announcement has confirmed that a 20 km extension of the pipeline to Angle Vale will receive over $4 m of Commonwealth and state funding. The extension should be completed in time for the 2008–2009 seasonal demand for water, and will eventually make a further 3 GL of water available.

Water recycling in the home

As with crop irrigation, the public represents no barrier to the practice of using recycled water within their own properties. At the national level, over 90% of Australians are willing to use recycled water for garden irrigation, hand watering of gardens, toilet flushing and car washing, and some two-thirds would consider using it for their laundry.[25] They are also just as keen to use water from other alternative options, such as onsite greywater systems, rainwater tanks, stormwater and desalinated seawater. Research results for the Adelaide public echo the national results (Table 19.1).

Trust in recycled water

Social research in the United States and Australia with leaders of water recycling, a review of public perceptions, and research with water recyclers confirms that trust is fundamental for the acceptance and establishment of water recycling schemes. Factors that build and maintain trust will allow early public consideration of the problem being addressed and involve the community in planning, implementing and managing recycled water. Regulations or guidelines need to be clear and widely understood and adhered to by practitioners, plumbers and the public. Transparency in the whole process of delivering and managing recycled water is vital. Familiarity with the new water source should be fostered through education and experience. Finally, maintaining trust depends on the accountability of all those involved in the proposal, ownership and management of the system, as well as the responsible use of the water.

Mawson Lakes

The Mawson Lakes scheme was funded by developers Delfin Lend Lease, the LMC, SA Water, and the City of Salisbury. SA Water has assumed ownership and responsibility for the scheme which is operated by United Water. Class A recycled water from the Bolivar WWTP is mixed with stormwater collected from City of Salisbury-constructed wetlands, then disinfected and stored before being distributed to households at Mawson Lakes through a dual distribution system. According to SA Water,[26] the recycled water is suitable for watering lawns, gardens, flushing toilets, washing cars,

Table 19.1 Willingness to use different water sources

Water source	Water use (non-drinking)	Without hesitation (%)	Some qualification (%)	Total willing (%)
Recycled water (class A water from sewage effluent)	Toilet flushing	77	21	98
	Garden irrigation	82	15	97
	Hand watering	80	15	95
	Car washing	76	13	89
	Washing machine	38	37	75
	Hand washing clothes	33	37	69
	Retrofitted to existing home for garden watering only	69	27	96
	In a new home, garden watering, toilet flushing and rainwater for other uses	77	21	98
Greywater (used shower, laundry water)	onsite units for garden and toilet flushing	68	30	98
Stormwater	for garden and toilet	89	9	98
Water source	Water use	Without hesitation (%)	Some qualification (%)	Total willing (%)
Rainwater	Drinking water	77	21	98
Stormwater	All household uses including drinking	25	52	77
Desalinated water	All household uses including drinking	57	36	93
Recycled water	Mixed with current sources for all uses including drinking	25	53	78

(Number of respondents = 356)

Table 19.1 The people of Adelaide's willingness to use different sources of water within their homes.

The power of reticulated water

BOX 91

A strong visual representation of the power and influence of reticulated water in the 1870s is illustrated in two images of the 'Heywood' residence. The house is located near Heywood Park, Unley, on the banks of Brownhill Creek, and featured in the movie *Shine* (1996). This was a simple, symmetrical, single-storey stone residence with 2.8 m wide peripheral verandahs. It was erected in the dormitory village of Unley in around 1858 to house William Hawke and his family.

The upper picture was taken in around 1870, with a buggy enclosed on the left verandah. Note the semi-arid ornamental plant selection and the absence of lawns. Upon Hawke's death in 1872, timber merchant Simon Harvey purchased the residence and renamed it 'Dorset House'. Harvey renovated the residence, adding the cast-iron lace facia decorations, and extensively redeveloped the garden, celebrating reticulated water with the installation of a focal point fountain and extensive irrigated lawns, as evidenced in this photograph from around 1880. In 1897, William Haslam MLC acquired the residence and reverted its name to 'Heywood'.

David Jones

Images reproduced with the permission of Necia Gilbert (dec.).

filling ornamental ponds and water features (with no fish), and the watering of fruit, vegetables and flowers. However, it is not suitable for drinking, cooking, kitchen use, personal washing, washing clothes, household cleaning, swimming pools, washing pets, filling ponds (with fish in them), children to play with, or any other recreational use.

SA Water's approach to its responsibility echoes the research findings on trust. The five factors for trust building clearly shape the information provided to residents. Transparency of all facets of the scheme and building familiarity with recycled water frame the facts given. SA Water's website information for Mawson Lakes residents[26] references water laws and regulations, compliance to rules and regulations, and accountability (Mawson Lakes Recycled Water Supply Agreement; Plumbing Guide and Checklist).

Mawson Lakes householders trust SA Water at a level of great and moderate trust (85%), similar to that for the general Adelaide population (82%). There is also an association between communication and trust. Some 82% state they had received updates of information from SA Water, and there is greater trust in SA Water indicated by 53% who say they require no further information. A higher level of trust is also found with 63%, who know that the authority provides a phone number for advice on any problems, and the third who believe that SA Water is open to improvements in the service. The strongest correlation with trust in the authority is indicated by those who give a high rating to the quality of recycled water and satisfaction with the service provided. (Several indicators of satisfaction with the scheme are indicated below and in Table 19.2.)

The Mawson Lakes respondents of the mail-back research had just over one year's experience of recycling water when they considered their level of confidence in using it for various applications. Around three-quarters or more state they have great confidence in recycling water for toilet flushing, irrigation and hand watering the garden. Their total confidence is at a level similar to the willingness of the general Adelaide population shown in Table 19.1. However, they are much less supportive of using recycled water for the laundry. Only 36% are confident to use it in the washing machine, and just less than a third have confidence in handwashing clothes with recycled water.

Mawson Lakes residents also responded to questions about the benefits, perceived health risks, and their levels of satisfaction with recycled water. Like the rest of Adelaide, Mawson Lakes is subject to permanent water conservation measures, however, it is not affected by Level 3 water restrictions (three hours of watering per week with hand-held hose or drippers). This research was conducted before the introduction of Level 3 restrictions. Based on their experience, they were asked for the benefits of recycling water. Some of the 297 respondents described more than one benefit, while 9% were not sure or found no benefit. The results in Table 19.2 are calculated on a total of 371 items that have been placed into six categories. The 338 respondents were asked the extent to which the recycled water supplied at Mawson Lakes presented a risk to them and their families; 330 answered and 85% perceived little or no health risks associated with the use of recycled water (Table 19.2). In terms of satisfaction, most respondents were very satisfied with both the quality of recycled water and the level of customer service (Table 19.2).

Table 19.2 Mawson Lakes residents' opinions

Perceived benefits of recycled water	Percentage of respondents
Saves water	39%
Good for the environment	26%
Cost savings	20%
Value recycling	9%
Enables garden watering	3%
Feels good factor	3%
Total	**100%**
Perceived health risks of using recycled water	Percentage of respondents
No health risk	20%
Low health risk	65%
Moderate health risk	14%
High health risk	1%
Total	**100%**
Satisfaction with water and service	Level of satisfaction
Recycled water quality	61% of respondents rated 15 and above (scale 1–20)
Service satisfaction	67% of respondents rated 5 and above (scale 1–7)

Table 19.2 Selected results of a survey of the residents of Mawson Lakes.

Future outlook

Extending the use of recycled water for laundry applications is under investigation in the water industry because it offsets the demand on the drinking water supply at a steady rate throughout the year, in contrast to the seasonal demand for irrigation. Those with experience of recycling water for uses such as garden watering and toilet flushing are probably acting responsibly in their lack of support for what is currently an illegal use of the same water. It is worth looking at explanations put forward by respondents. In the national study, reasons given for some hesitation or concern for uses that include the laundry relate to a lack of trust (familiarity, needing reassurance of safety), health risk, water quality and the water source.

The categories are similar for Mawson Lakes recyclers, for example:

1. Trust factors:
   'Not sure about handwashing of clothes in recycled water based on current advice.'
2. Health risk:
   'Clothes worn by people – possible skin irritation problems.'
   'Just concerned re. irritants in chemicals used for the recycling process.'
   'Just the fact that if used for clothes could ruin clothes and any germs would be on skin etc.'
3. Water quality: *'Recycled water is not of good enough quality for us to use for laundering purposes.'*
   'The possibility of discolouration of clothing.'
   'May cause staining with light coloured clothing.'
4. Water source: *'I feel a bit uncomfortable putting my hands in the water. Similarly, I'm not sure clothes would be really "clean" if washed in recycled water especially if its source was sewage effluent.'*

Research on the controversial topic of recycling water for drinking, and specifically through indirect potable reuse systems (mixing recycled water of drinking standard with traditional sources), documents repeated cases of public disapproval and rejection (from Bruvold, 1972[27] through to current work). The national survey explores a range of alternatives to mains water. Rainwater for consumption and personal use, in combination with recycled water for toilet flushing and garden watering, is the preferred alternative source, as found for the Adelaide population (Table 19.1). Support for desalinated seawater follows where around half are willing without hesitation (57%, Adelaide). Stormwater and recycled water are the least preferred sources: around a quarter have no hesitation (national and Adelaide). Yet, these results for recycled water – indirect potable reuse – are high compared to previous published research.

When trust is explored in relation to confidence in using recycled water for showering, cooking and drinking, the response is more measured. Just over three-quarters of the samples (national and Adelaide) have confidence in showering. This diminishes somewhat for cooking (around 55%). Finally, 44% in the Adelaide population have confidence in drinking, 42% nationally. Are water recyclers more trusting of the concept? Results from the Mawson Lakes research suggest this is not the case; they are more reserved (61% have confidence in showering, 42% cooking, 36% drinking). This, again, may be moderated by the fact that they are already recycling water in a responsible manner that avoids human consumption.

So, while Adelaideans are willing to use alternative sources of water for a range of uses around the home, and support its use for crop irrigation, the same support is not found when bodily contact or consumption is suggested. There are sociocultural bases for this response. The introduction of recycled water or other alternatives includes change in infrastructure for treatment, distribution and end use, and behavioural change – of all parties involved. Acceptance will depend on the extent to which the proposal is in tune with already established routines or, if ambiguous, whether it is worth further consideration. Proposals that violate cultural norms are more likely to be rejected.

Recycling water for irrigation and toilet flushing appears to fit with current cultural values of recycling and conserving, and cultural norms of where wastewater should flow. After a year of Level 3 water restrictions, the impact of the drought on our streetscapes and parks is sadly apparent. Mawson Lakes is an oasis demonstrating what could be achieved. Using the water for laundering clothes, however, seems to fall within the realm of ambiguity. There is a certain amount of willingness but final acceptance may depend on working the trust factors through community engagement. In time, the same may apply to showering, cooking and drinking. For now, Adelaide is more willing to adopt desalinated seawater and is open to any other source to revive drought-stricken gardens, and to relieve the back-breaking task of bucketing drinking water.

Conclusion

Water, as these three studies demonstrates, is an incredibly diverse and essential resource, but one which demands creativity and commitment to use effectively. One lesson from these studies is not allowing one use of water to limit its use for other differing requirements. A statesman, Abba Eban, once said, 'Humans and nations don't behave rationally until they have exhausted all other options'. The same can be said of our use of water. For far too long we have used and abused it and failed to recognise and appreciate the full range of needs and values that it can fulfil. We need to protect and sustain it so that water can continue to sustain us into the future.

References

1. M.T. Bond, V.L. Seiler and M.J. Seiler, 'Residential real estate prices: a room with a view', *Journal of Real Estate Research*, 23 (1/2), 2001, pp. 129–137.
2. S.C. Bourassa, M. Hoesli and J. Sun, 'What's in a view' FAME research paper no. 79, International Center for Financial Asset Management and Engineering, Geneva, 2003.
3. S.C. Bourassa, M. Hoesli and J. Sun, 'The price of aesthetic externalities', FAME research paper no. 98, International Center for Financial Asset Management and Engineering, Geneva, 2003.
4. G. Szili and M.W. Rofe, 'Greening Port Misery: marketing the green face of waterfront redevelopment in Port Adelaide, South Australia', *Urban Policy and Research*, 25 (3), 2007, pp. 363–384.
5. R. Hoyle and P. Starick, 'The renaissance of a port', *Advertiser*, 4 July 2005, p. 10.
6. L. Craig, 'Grime out, people in for the Port's grand plan', *Advertiser*, 30 July 2002, p. 3.
7. W. Harcus, *South Australia: its history, resources and production*, Sampson Low, Marston, Searle & Rivington, London, 1876.
8. T.H. James, *Six months in South Australia: with some account of Port Philip and Portland Bay, with advice to emigrants*, J. Cross, London, 1838.
9. M. Page, *Port Adelaide and its Institute 1851–1979*, Rigby, Adelaide, 1981.
10. D. Whitelock, *Adelaide, 1836–1976: a history of difference*, University of Queensland Press, St Lucia, 1977.
11. B. Samuels, *Mudflats to metropolis: Port Adelaide 1836–1986*, B & T Publishers, Adelaide, 1986.
12. J. Pascoe, 'The redevelopment of the Port Adelaide centre: urban conservation and the planning of heritage', unpublished Honours thesis, Flinders University, Adelaide, 1990.
13. Environment Protection Authority, 'Port River waterways', unpublished report, 2004.
14. M.W. Rofe and S. Oakley, 'Constructing the Port: external perceptions and interventions in the creation of place in Port Adelaide, South Australia', *Geographical Research*, 44 (3), 2006, pp. 272–284.
15. A. Rowell, 'The spread of greenwash' in Lubbers, E. (ed.) *Battling big business: countering greenwash, infiltration and other forms of corporate bullying*, Scribe, Melbourne, 2002, pp. 19–25.
16. S. Beder, 'Greenwash', in J. Barry and E.G. Frankland (eds) *International encyclopaedia of environmental politics*, Routledge, London, 2002.
17. J. Greer and K. Bruno, *Greenwash: the reality behind corporate environmentalism*, Third World Network/Apex Press, New York, 1996.
18. S. Beder, *The nature of sustainable development*, Scribe, Melbourne, 1996.
19. S. Beder, *Global spin: the corporate assault on environmentalism*, Scribe, Melbourne, 2000.
20. Land Management Corporation, 'Developing opportunities', Adelaide, n.d.
21. E. Doherty and B. Heggen, 'New era on the waterfront', *Advertiser*, 12 September 2004, p. 35.
22. J.S. Marks and K.F. Boon, 'A social appraisal of the South Australian Virginia Pipeline Scheme: five years' experience, report to Land and Water and Horticulture Australia Ltd', Flinders University of South Australia, Adelaide, 2005.
23. G.B. Keremane and J. Mckay, 'Role of comunity participation and partnerships: the Virginia Pipeline Scheme', *Water*, 33 (7), 2006, pp. 29–33.
24. J.S. Marks, B. Martin and M. Zadoroznyj, 'Acceptance of water recycling in Australia: national baseline data', *Water*, 33 (2), 2006, pp. 151–157.
25. J. Marks, B. Martin and M. Zadoroznyj, 'How Australians order acceptance of recycled water', *Journal of Sociology*, 44 (1), 2008, pp. 83–99.
26. South Australian Water Corporation, 'Mawson Lakes', Government of South Australia, 2004 (online).
27. W.H. Bruvold, 'Public attitudes towards reuse of reclaimed water', contribution no. 137, School of Public Health, University of California, Berkeley, 1972.

Box references

Box 88: This everything water

1. K. Akerman and J. Stanton, 'Riji and Jakoli: Kimberley Pearlshell in Aboriginal Australia', monograph series 4, Northern Territory Museum of Arts and Sciences, Darwin, 1994.
2. E. Hill, *The great Australian loneliness*, Robertson and Mullens Ltd, Melbourne, 1948.

Box 89: Humans, water and marine ecosystems

1. B. Foran and F. Poldy, *Future dilemmas: options to 2050 for Australia's population, technology, resources and environment*, CSIRO Publishing, Canberra, 2002.
2. L.P. Zann, *Our sea, our future: major findings of the state of the marine environment report for Australia*, Department of Environment, Sport and Territories, Canberra, 1995.
3. S.D. Connell, 'Subtidal temperate rocky habitats: habitat heterogeneity at local to continental scales' in Connell, S.D. and Gillanders, B.M. (eds) *Marine ecology*, Oxford University Press, Melbourne, 2007, pp. 378–401.
4. S.D. Connell, 'Water quality and the loss of coral reefs and kelp forests: alternative states and the influence of fishing' in Connell, S.D. and Gillanders, B.M. (eds) *Marine ecology*, 2007, pp. 556–568.
5. B.D. Russell and S.D. Connell, 'Response of grazers to sudden nutrient pulses in oligotrophic v. eutrophic conditions' in *Marine ecology progress series*, 349, 2007, pp. 73–80.
6. B.D. Russell, T.S. Elsdon, B.M. Gillanders and S.D. Connell, 'Nutrients increase epiphyte loads: broad-scale observations and an experimental assessment', *Marine biology*, 147, 2005, pp. 551–558.
7. S.C. Jeffrey, D.J. Rochford and G.R. Cresswell, (1990) 'Oceanography of the Australasian region' in Clayton, M.N. and King, R.J. (eds) *Biology of marine plants*, Longman Cheshire, Melbourne, 1990, pp. 244–265.
8. D.J. Rochford, 'Nutrient status of the oceans around Australia, report 1977–1979', CSIRO Division of Fisheries and Oceanography, Hobart, 1979.
9. M.G. Chapman, A.J. Underwood and G.A. Skilleter, 'Variability at different spatial scales between a subtidal assemblage exposed to the discharge of sewage and 2 control assemblages', *Journal of Experimental Marine Biology and Ecology*, 189, 1995, pp. 103–122.
10. D.E. Roberts, A. Smith, P. Ajani and A.R. Davis, 'Rapid changes in encrusting marine assemblages exposed to anthropogenic point-source pollution: a 'Beyond BACI' approach', *Marine ecology progress series*, 163, 1998, pp. 213–224.
11. P.A. Ajani, D.E. Roberts, A.K. Smith and M. Krogh, 'The effect of sewage on two bioindicators at Port Stephens, New South Wales, Australia', *Ecotoxicology*, 8, 1999, pp. 253–267.
12. D.B.C. Gorman, Russell and S.D. Connell, 'Land-to-sea connectivity: linking human-derived terrestrial subsidies to subtidal habitat-change on open rocky coasts', *Ecological application*, in press (2009).

CHAPTER 20

Reporting the water debate

Cara Jenkin
Clare Peddie

'Water: there's nothing like a good drought to focus the attention on the issues.' – South Australia's water commissioner, John Radcliffe.

Introduction

Since the drought officially began in 2006, nothing has stimulated public debate on a single issue as much as water. While SA Water supplies have been guaranteed for urban use and have not yet dwindled to critically low volumes, water has become one of the top issues for the community. Water was debated from the seats of state parliament, where it was one of the most common issues raised during 2007, to seats at the family dinner table. As the drought continued, the popularity and prominence of water as an issue for discussion did not wane. The issue dominated letters to the *Advertiser's* 'Letters to the Editor' section and continually replaced other topical issues, including the Victoria Park grandstand development, law and order, terrorism and natural disasters, such as bushfires, as the main opinion topic of the day. Water was raised at least weekly, if not daily, on ABC Radio's morning program. Everyone in the community, at social group meetings, on the sports field, at family gatherings, among work colleagues and groups of friends, had a grumble about what was being done to secure water supply – and everyone had a solution.

This chapter outlines the response by the general community to South Australia's water woes. As the *Advertiser's* environment reporters, we were at the forefront of community debate. We received calls from members of the public and talked to the experts and people on the street. We researched and identified the reactions and issues for our articles and witnessed how South Australia was learning to live with, or without, its water.

The debate started slowly, much like the drought. Demand for water began to outstrip supply in the 1990s, and the environment was the first to suffer. The over-allocation of water meant that even in wetter years the river's environment was not receiving the water volumes to which it was accustomed. Calls were made to 'save the River Murray' by environmentalists, and residents became aware the river was suffering. Then, when the rain stopped coming and the drought was linked to climate change, the problem was dramatically exacerbated. More urgent calls were made to save the river. However, the need to supply water to towns came first and to farmers second; the environment was last on the urgency list.

It was labelled the 'one in a thousand year' drought by Premier Mike Rann, describing the serious extent to which people believed the crisis had reached. No one in living memory could recall such a severe drought and Murray-Darling Basin Commission records show inflows were at their lowest since the statistics were first kept one hundred years ago.

Chapter photograph: by David Radzikiewicz, A7Designs.

In the past five years, awareness of water issues has grown tremendously. Water restrictions, the drought, and, more recently, climate change, have become topics of widespread general interest. Everyone has an opinion about what the problems are and what should be done to solve them.

Wake-up call

South Australia relies heavily on the River Murray for water, so when it became clear the river was in trouble, people were forced to finally pay attention. In 1981 there was not enough water flowing to the ocean to keep the river mouth open, and it silted over – the 'Murray Mouth' was closed.

The government decided to bring in heavy machinery to dredge the mouth, physically moving sand and silt from one place to another so as to keep a small channel open. The problem returned in 2002 and again the diggers were brought in. It was supposed to be a temporary solution, but dredging continues to this day.

At the time, the public was outraged. The main question was how the River Murray could have been left to deteriorate to such a state that water could not flow through the mouth. But as there was a manmade solution to keep the water flowing through the mouth, which appeared to be working, the issue dropped off the agenda. However, this should have been the wake-up call to spur more long-term and comprehensive action.

At the political level, this situation caused some important changes to improve water management in the Murray-Darling Basin, although not far-reaching enough. In 1994, the Council of Australian Governments agreed to a national 'Water Reform Framework'. Water rights were separated from land rights, enabling trade. In 1997 the Murray-Darling Basin Commission brought in 'the cap' to limit extraction, in recognition that supply was limited and there would be times when water-users would not be able to access their full allocation. In 2003/2004 the National Water Initiative was established to drive reform further, with funds to support innovative approaches to dealing with the issues.

The Wentworth Group of Concerned Scientists was formed to counter some of the irrational arguments being put forward in the media, such as 'turning the rivers inland' and 'towing icebergs'. The group's 'Blueprint for a National Water Plan' and other documents about land management appeared to have some influence on government. The knowledge and standing held by the group also received widespread media coverage and their messages were debated by the public.

Meanwhile, irrigators were becoming increasingly concerned that their water would be taken away from them and resisted changes that would see water diverted to other uses, such as the environment. The Murray-Darling Basin Ministerial Council's 'Living Murray' initiative established in 2002 was not well received in many rural communities, as evidenced in newspaper articles in Adelaide and regional areas at the time. During 2007 and 2008, irrigators were quoted in news stories in the *Advertiser*, ABC and commercial television network news bulletins as being angry when the environment, including Ramsar-listed wetlands and regions, which have received no water for several years, was given a small allocation under the Living Murray, believing the water should be given to them. Farmers wondered why water was allowed to be 'wasted' to green-up environmental areas of the river. But in 2003 the ministerial council took a historic first step to improve the environment at six icon sites including the Chowilla Floodplain and the Murray Mouth, Coorong and Lower Lakes. An average of 500 GL of water a year was promised over five years (but so far little of that water has been delivered). In the 2007/2008 financial year, 133 GL was listed as ready to be released for the environment, but the total amount provided was subject to the drought and water allocations.

In October 2003, the state government initiated permanent water conservation measures to reduce water consumption in Adelaide; the 2002 drought had proved that Adelaide did not have an endless supply of water. The government also launched its Water Proofing Adelaide plan to explore water issues and develop a strategy to support Adelaide's growth and ensure water was used more wisely. Efficiency, better management of existing water resources, and finding new sources of water, such as stormwater and wastewater recycling, were new initiatives suggested to become more widespread by 2025. When the final strategy document was released in 2005, then environment minister, John Hill, and then water minister Michael Wright, said the plan was to ensure Adelaide had enough water to meet its needs up to 2025 and beyond, without the need for extreme water restrictions such as total bans on the use of water for gardening and external purposes except in very dry years. The plan did not account for those very dry years arriving now.

In mid 2007, the Howard federal government instigated its $10 billion National Water Plan to take control of the water resources, take back over-allocated water, and determine a future plan which would guarantee water security for town users, irrigators and the environment. By the end of 2007, it had not taken effect and only a small portion of funds had been earmarked to be spent by 2010.

The Rudd federal government began work on the plan in the first six months of its term and committed to buying water back from willing sellers. It deemed the first buyback a success, but again irrigation groups were concerned. They issued media releases and were interviewed by journalists across the nation about their fears that irrigators would be disadvantaged by government buybacks. Meanwhile, environmental groups and scientists urged the government to buy back water even more quickly. Many conservationists called for all licences to be bought back immediately and redistributed to cure the over-allocation once and for all.

All the political progress made since the mid 1990s was not fast enough for the River Murray, nor for the public. They are both feeling the effects now and the public have demanded why more progress was not made earlier and why the warning signs were not addressed more quickly.

Country vs city

The water issue has exposed and exacerbated the divide between city and country. There are always many exceptions among city and country people, but generally those who live in the metropolitan area tend to be accustomed to having services provided to them.

Those who do not live in the metropolitan area are used to drought. Farmers on the land are not connected to mains water systems and must rely on rainfall for household water. Residents of bigger cities, which do have mains water, sometimes come from farming families or backgrounds or have friends and relatives who do live on the land. They are used to cutting back, conserving or rationing water, as they are with many other goods and services. Lifestyle essentials and services many city folk take for granted are often stripped from country residents, be it the ability to duck down to the corner shop when the milk runs out, or to have a practicing GP within a two-hour radius. If there is a drought, country people know they will have to conserve the water in the rainwater tank until it fills up when it rains again, and so they use less in the shower, kitchen and garden. Adelaide residents have always had water 'on tap', so when their ready access to water is taken away, they expect the government to provide it.

There is also the city versus country divide about *how* the available water should be consumed. For example, the state government pumped millions of litres of River Murray water into Adelaide's reservoirs during winter 2007 to ensure there was enough water to meet urban water needs in this financial year and the next. At the same time, irrigators along the Murray were watching the water flow past their properties, only to sit in a storage facility in the Adelaide Hills for 12 months or more. Irrigators thought the water could have been used to keep their crops alive – and even produce fruit to generate an income for the year. Adelaide residents saw the water was available in storage and demanded access to it to keep lawns, flowers and trees alive through summer. Debate raged in the media that some water should be released to households to keep their gardens alive, even in the midst of winter when rainfall alone should be enough to keep most plants alive at least until spring.

The state government relaxed water restrictions, which kept Adelaide gardeners happy, but irrigators were incensed. They believed their livelihoods were drying up for the sake of keeping Adelaide residents in pretty surroundings.

At the time, scientists and water experts labeled it the 'wrong debate', warning that if the drought continued for another year, water that had been allowed for irrigator or gardens might be needed for 'critical' water supplies, such as drinking or washing.

Both issues were symptoms of 'affluenza' in the metropolitan community. It was a similar scenario to increasing levels of debt and credit card use. Water was wanted now, and the consequences worried about later.

Us or them?

The perception that water is being sucked unfairly upstream, and that more should be provided to South Australia, is still rife in the community. In particular, rice and cotton farmers who use more water per hectare than our fruit and vegetable growers, but less of an area, still receive the blame. However, rice and cotton farmers have also had their water allocations cut – to zero. Most farmers were not able to plant a crop in the 2007/2008 season because there was no water. Rice and cotton crops are annual crops, unlike citrus and grapes that need water during the year just to keep the trees and vines alive without producing a crop; yet the rice and cotton growers are still seen as the 'bad guys' and the public has demanded their allocations be bought by the government for South Australian irrigators or Adelaide residents to use. What is not thought out is that because of the drought none of that water exists for South Australia anyway. Even if the water in Cubbie Station in Queensland was released for the basin, none would reach South Australia.

The perception that rice and cotton growers use more water, and that it is an inefficient method of farming, is also false. While the latest farming figures show the amount of water per hectare is 12.1 ML/ha for rice and 6.7 ML/ha for cotton, improvements to agricultural practices in the past two decades have allowed greater rice yields from less water. Australia's cotton industry still produces higher yields per megalitre of water compared to other countries.

Meanwhile, irrigation of dairy pasture is the single biggest user of water in Australia. While it consumes only 4 ML/ha, there is more land to water and so more water is used. The year 2005 recorded the biggest grape glut experienced by the wine industry, and growers were forced to sell their grapes for low prices or even allow them to rot on the ground. All the water that had been used to irrigate grapevines during the year had returned a marginal profit, if any.

Again, these examples show the community's attitude to blame others, rather than ourselves, for water wastage and over-use. It also shows the community's belief that their own needs, such as gardens, should be put ahead of others livelihoods. The community was not taking responsibility.

Water restrictions

The dynamics of the water debate changed with the introduction of household water restrictions. Residential water use inside the home was not restricted as it was outside the home for washing cars, topping up pools and, particularly, garden watering. The state government encouraged people to use less water inside the home through rebates for water-efficient appliances; while media campaigns, such as the *Advertiser*'s shower timer giveaway, also encouraged water saving within the home.

People now understood that the environment was suffering, water supplies were dwindling, and everyone had to do their bit to save water. Initially, the restrictions were a bit of a novelty, and were not too difficult to follow.

But as time wore on and restrictions became more and more onerous, householders started to worry about the health of their gardens, and point the finger at other water-users who they believed should have to cut back first. This was especially proven when a total ban on outside watering was implemented. Despite it being winter, many people in the public felt suffocated that they could not water their gardens at all, turning on others who they believed were wasting or did not need the water as much as they did.

Water restrictions were imposed on everyone, regardless of whether restrictions occurred through lower irrigation allocations, SA Water garden rules, or country farmers who only had access to the rainwater collected in their tanks. City residents were angry, believing their gardens had been singled out under water restrictions. The first quandary with this anger was that many residents had not and did not act *themselves* to provide the mechanisms for their own provision of water for their garden's needs, such as installing rainwater tanks or greywater systems, which would allow gardeners to water whenever they pleased. They demanded the water be provided to them to be used however they choose, promising to minimise water use.

The fact is, gardens *do* use a lot of water: 65% of water provided by SA Water is to households compared to 9% to industry and 6% to commercial businesses. Studies by SA Water, as well as other water corporations and researchers across the country, have found that 40% of an average household's water is consumed outside the home through washing cars or paved surfaces, but in particular on gardens (far from 1%, the oft-quoted and never credited figure used by callers to talkback radio and even prominent politicians). By cutting back on water outdoors, public health would not be put at risk; and residents still had access to water to cook, shower and wash clothes, for any size household or family. Therefore, the biggest, easiest and quickest overall savings were figured to be made by restricting outside water use by households.

Yet many householders failed to see the reasons why those particular restrictions were implemented. Furthermore, there was little communication to the public of the facts

BOX

92

Changing water consumption behaviour

Changing water consumption behaviours is one important objective of a sustainable water management strategy. For example, consumers can use low-quality water such as recycled water or stormwater to flush the toilet and water the garden rather than use 'clean' water. However, the efficient use of this methodology requires sufficient statistic data, and an understanding of public preferences and consumer behavioural motivations, which are not well investigated or reported by studies.

The economic view believes that users react systematically to a change in cost of consumption. However, apart from the general statement that prices affect water use, it is difficult to assess the strength of this relationship. While some studies suggest a relationship between volumetric pricing of water and water use reduction, others report money is found not to be a predominant driving force of behaviours when it comes to environmental resources. This contradictory relationship is not well explained nor understood. Little is known about which factor leads to the largest reduction of water use among many driving forces including education, regulatory mechanisms and metering installments.

In 1999, Charlottetown, Canada, had a flat rate price but had one of the lowest per capita water use rates in Canada. In Toronto in 1999, only three quarters of the population was metered, but used less on average than citizens of Victoria, who were fully metered. While these contradictions may be artefacts of data collection, they still need to be explained. In England and Wales the residential water use increased with water price increases, which was explained by the fact that metering was not universal in that situation, and concluded that price not based on volume of water used is ineffective for water demand management.

In Denmark, a tax on water use was found to have been responsible for 40% decline in use, while education was responsible for 60%. In South Australia, consumption patterns have been linked to the receipt of water bills. When water usage was mapped in relation to the billing cycle, lower consumption followed 'excess' water bills. More generally, while coupled price Increases and reduced water consumption have been observed in a number of European countries, public education and other instruments have usually been applied simultaneously, such that the effect of a single instrument has not

been shown to be very significant, and the declines in usage cannot be satisfactorily explained.

These facts imply that while water consumption behaviour may be addressed through pricing tools, there is ample room for non-pricing measures (e.g., education). Economic instruments as a tool for water resource management have limitations and cannot be proposed to replace other management tools. A better understanding of consumers' perception of what would affect their consumption behaviour can help to predict consumer behavioural changes in response to different management instruments. With this question, a survey was implemented in the May 2007 to explore what would encourage the Adelaide community to change their water consumption behaviour.

What would encourage you to change your behaviour in order to assist with improving water quality?

	Responses	
	number	per cent
Information and education	637	62.1
Enforcement or fines	211	20.6
Levy	105	10.2
Don't know	36	3.5
Nothing	33	3.2
Don't care	4	.4
Total	**1026**	**100.0**

Information and education were found to be the most important tool. Enforcement and fines could work (28.3%), but levy only accounted for 14.1%. The results tell us that fines or levies were perceived as much less important than information and education in working towards water consumption behaviourial change. Under these circumstances, economic tools would have limited scope in changing water consumption except that they can provide opportunities for revenue increases. Therefore information and education based polices should be introduced at a large scale.

Zhifang Wu and Jennifer McKay

to make them understand why these restrictions were in place. The *Advertiser* was one media outlet that published several stories outlining where and how much water is used around the home to highlight why the restrictions stood and why they should be adhered to rather than criticised. However, that debate was largely overshadowed by the efficiency of dripper irrigation versus buckets.

This occurred when a ban on outside watering was in force in winter 2007. The debate quickly swung from what Adelaide should be doing to secure its water supply in the future and prevent water scarcity from occurring again, to if garden watering should be allowed using buckets or drippers. Drippers were hailed as the more efficient form of garden irrigation. However, when water restrictions were relaxed at the end of September 2007, and dripper use was allowed, water consumption rose substantially. SA Water figures show in the first weekend of the relaxed restrictions, 820 ML of water was consumed, compared to 572 ML from the weekend prior. South Australia was using more water. Even being allowed to use drippers wasn't enough for gardeners. In late 2008, there were calls to relax restrictions further to allow gardeners to water twice a week instead of once. The government agreed to do so and water use again soared, suggesting that some gardeners resumed heavy watering, regardless of whether their gardens required it or not.

Water consumption

Throughout the water debate there were signs that people were willing to reduce water use and do their bit. While daily water consumption increased when water restrictions were first implemented in October 2006, the implementation of tougher restrictions saw daily consumption fall below the daily average, and this was maintained for 95% of the time. Many people allowed their lawns to 'brown off', as they could not be watered with sprinklers, and that action was a major water saving. Other residents stored buckets in the shower to catch excess water and used it on pot plants instead of filling watering cans from the outside tap. Nurseries recorded a strong surge in sales for native and low water-use plants, as gardeners opted for gardens that did not need irrigation. There was also a rush for rainwater tanks. Major companies developed water wise products, such as water-saving washing machine powders, and water wise machines such as dishwashers increased in popularity. This has all led to a reduction in water consumption.

However, more and more water was being used each day as the weather warmed up. During winter, when outside watering was banned, Adelaide was consuming less than 300 ML a day; during December that figure averaged at 500 ML/day and spiked at 600 ML/day on dry weekends of warmer temperatures. All this despite months of public debate and concern over water security. So the question begs: how well has the community been engaged by the public debate?

The solution?

The community has recognised that they need to use less water. However, the savings being made were not enough to save South Australia and protect it from water scarcity.

Desalination was heralded as the solution by the public and politicians since the drought began. Everyone, except for a small minority of environmental conservationists, appeared to want a desalination plant to provide water to Adelaide. The public saw every other major capital city in Australia had either built, was building, or had approved a desalination plant to act as a water source for residents – South Australians demanded their government do the same. Yet desalination was not put forward as a solution to completely protect the city against water shortages. Water security minister Karlene Maywald said it would not reduce Adelaide's reliance on the River Murray, as SA Water would still take its water allocation from the river to provide the majority of Adelaide's water supply. The state government's desalination working group even highlighted that a desalination plant would not prevent water restrictions, producing only 45 GL/year (less than one quarter of the city's annual consumption so far this decade of 200 GL – and the population is increasing). Desalination is an energy-intensive process and unless it is powered by alternative energies, it would increase the state's greenhouse gas emissions and put further stress on the environment. Desalination is also a very expensive process – this message has been portrayed by scientists and politicians and the media since desalination was first mooted as a solution for Adelaide. But, despite its shortfalls, the public seem to demand a desalination plant.

Building a desalination plant could be compared to the pipelines built from the River Murray to Adelaide and other South Australian towns. Here is the same case of residents living in an area of limited water resources demanding a new and more reliable source of water, apparently not content to live on what is available locally. There is little information about the effects of desalination on South Australia's environment, such as emissions and the effect of brine discharge on marine environments, so how do we know we're not making the mistakes of our forebears? It is the same approach referred to earlier in this chapter – take water from somewhere else, deal with the consequences later.

The fact is that the choices being made today could have serious consequences for future generations. In a similar way, when our forebears cleared native vegetation for agricultural land and development they disturbed the salt/water balance, creating salinity problems. Saline land has since been abandoned; some has been restored through careful land management, but it is always far more difficult to fix the problem than to avoid causing it in the first place.

Meanwhile, land clearing has caused erosion and allowed for the release of more carbon dioxide into the atmosphere, exacerbating climate change. Further, by sending wastewater and stormwater out to sea for more than a hundred years, we have killed seagrasses and transformed the marine environment, reducing the habitats of fish and other important marine organisms. Pollution also affected our enjoyment of the coastline. Today it is astonishing to think that anyone would have been prepared to send untreated, raw sewage into the ocean, but it still happens in some parts of Australia. In Adelaide, we treat it to some extent first, but we still allow stormwater to wash various contaminants into the ocean. Instead, we could be catching the water, cleaning it and putting it to good use – at a fraction of the cost of taking water back out of the ocean for desalination and purification.

The mindset is to build a desalination plant as fast as you can, without considering the significant polluting effects of such a facility, because other cities have them. The mindset also supports the expansion of Mount Bold Reservoir, although conservationists have warned a threatened species would become extinct. But because the plant is not well known and its wellbeing seems insignificant to the need of humans, the public is not concerned about it. Both plans could have more serious repercussions that are not yet realised.

This mindset has also supported the closing of wetlands along the River Murray to prevent water loss from evaporation. At first, it was suggested that this would be good for the environment, mimicking a natural wetting and drying cycle, however, scientists are beginning to discover that the plants and animals these wetlands support may be lost forever.

In an attempt to find another source of water, residents across Adelaide are installing backyard bores at an unprecedented rate. Groundwater is believed to be an unlimited supply of water for the garden, with households proudly displaying signs saying 'Borewater in Use' to show they are not breaking water restrictions. However, scientists are warning of the effect on surfacewater. Any water scientist will explain that water above and below ground is connected. Underground springs feed into catchments through creeks. Underground water also supports deep-rooted vegetation, much of which has suffered in this drought. If billions of litres of water are taken out of the ground each year, there will be repercussions. Groundwater is replenished by rainwater and if rainwater is no longer flowing into catchments, it will not flow into underground reservoirs. This process can take hundreds, thousands or even millions of years. 'Water Proofing Adelaide' acknowledged most underground resources in the regions supporting metropolitan Adelaide were already over-used or at least at capacity. In late 2007, after it was revealed that groundwater was at its lowest level since records began, the state government finally accepted it could be a problem and banned the sinking of more backyard bores.

Water restrictions are unpopular in the community, as evidenced earlier in this chapter when an outdoor ban on watering was enforced. Yet many householders do not want the expense or hassle of installing rainwater tanks and greywater systems when water is already supplied to the house through the mains system. Another possibility, reusing wastewater and treating it to a standard suitable for drinking water and supplying it through Adelaide's mains system, has been resoundingly rejected by residents through media debate and some community polls. As yet, no referendum has been held to gauge community opinions officially. There is a mental 'ick' factor of drinking what was once sewage, whether or not it is scientifically justified. Using recycled stormwater for use on parks and gardens across Adelaide has gained immense popularity, spearheaded by more than 20 years of practice and research undertaken by Salisbury Council. Already the state government has ruled it out as a source of drinking water, and has continued to maintain its stance during the drought, but it is still supporting research into the idea. It is not known either officially or unofficially if the Adelaide public would be receptive to drinking recycled stormwater.

Surprisingly, both recycled sewage and stormwater are already in Adelaide's water supply. Treated sewage is dumped in the River Murray upstream by New South Wales towns, houseboats have dumped their loads in the past, and animal manure from farming properties bordering the river also makes its way into the water. In the Adelaide Hills, rivers running into reservoirs flow through agricultural land and pick up muck, although authorities have now demanded farmers fence off their stock from the rivers. People do not seem to believe it is a problem to be drinking from these water sources, but are sceptical of new schemes that would see us drinking recycled sewage or stormwater.

Greywater is being used as an optional source of water by householders to some extent, but is not yet widespread. Again, the expense and hassle of establishing greywater systems has prevented many people from using it as a major water source.

These options are much less expensive to develop and implement by governments, organisations and households than desalination, yet because they are perceived as socially unacceptable, or demanding on the individual, they have not had serious consideration as a major source of Adelaide's supply.

The response

Water has become much more important as an issue to governments since the water debate began. Water is now recognised by both state and federal governments as an issue worth its own portfolio, and ministers are responsible for developing plans on water security.

Some action is being taken. All state governments have agreed to a federal takeover of the Murray-Darling Basin and work is in the pipeline to buy back water allocations and upgrade infrastructure to provide more water for the environment. A desalination plant for Adelaide has been fast-tracked to be operational by late 2010, and water pipelines will be extended to reuse more recycled water for agriculture.

Water has also become a major issue for the media to highlight. The *Advertiser* reports daily on water issues, about possible solutions to be raised and debated in the community, as well as the current situation to inform and encourage the debate.

As the River Murray struggles and media try to communicate the severity of the drought, images of bare riverbanks – including the *Advertiser's* most controversial image of a dry Murray riverbed interstate in September 2007 – and falling water levels are used to urge people to cut back on their water use. These images have created a false perception among city folk that there is no water in the River Murray for recreation. Consequently, tourist numbers have reduced dramatically to regions such as the Riverland.

Each time the *Advertiser* printed a photo of low river levels, and the location captioned and attributed to in the article, residents from river regions upstream of Blanchetown were not happy. They believed the media was destroying tourism to their regions, which is a major contributor to their economies, especially in times of drought. They wanted the media to explain the River Murray is full of water, thanks to the locks which are keeping water upstream to ensure a supply, and that houseboating, water skiing and fishing can continue. So, images of a River Murray full of water are now flashed in the media and city residents see there is plenty of water and claim the drought cannot be that bad after all. Many city residents use this information to demand more water for household use, or even flout water restrictions, because they see the river has not yet reached a state where it is running dry.

How can the media convey the message that the River Murray is in trouble, that our water supply is fading, when images of a river full of water is splashed in papers and on television screens? How can both messages be conveyed at the same time?

In time, if the drought continues, community attitudes will change. With less and less water becoming available, and the continuation of the water debate, people will be forced to take responsibility for water, become more waterwise, and adopt new water resource options. If the drought breaks, the community will hopefully learn from the past, but will implement water saving measures and new water resources less urgently to help them in the future.

Hopefully, in the next drought, the same debate does not resurface.

CASE STUDY 7: Letters to the Editor

Cara Jenkin and Clare Peddie
Cartoonists: Michael Atchison and Ross Bateup

Cartoons from the *Advertiser*
1/1/07, 29/1/07, 23/2/07, 29/8/07, 12/9/07, 27/9/07

Overview

As part of its involvement with presenting and discussing the water debate in South Australia, the *Advertiser* regularly publishes Letters to the Editor. These are unsolicited, short comments by members of the public, often reflecting the mood of the community. The *Advertiser* also communicates the mood of the debate by publishing cartoons that satirically portray an aspect of the debate. As with a well-written letter, a great cartoon can capture the essence of a situation and present it humorously, yet be challenging and thought provoking, commenting on and judging the outcome of a public debate. Both the Letters to the Editor and published cartoons are well recognised to have a powerful ability to sway decision-makers and direct the debate. Here, we provide examples of material published in the *Advertiser* in 2006 and 2007, which played important roles in crystallising aspects of the water issue in the hearts and minds of the general public.

**The *Advertiser*, edition 1, state
Monday, 17 July 2006, p. 16**

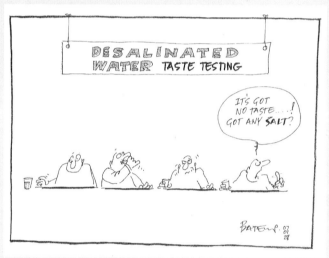

Tougher restrictions

WITH rainfall being very light for this time of the year and South Australia's largest reservoir, Mount Bold, being only 16% full, isn't it about time the Rann Government showed courage enough to impose far tougher water restrictions in the metropolitan area? It seems that up to the present, Rann and company have taken no notice of my many published warnings on this crucial matter; apparently content to hammer the River Murray's not-so-healthy water volume?

* W.B. Wreford, Morphett Vale
**The *Advertiser*, edition 1, state
Wednesday, 11 October 2006, p. 16**

Do the right thing

NOW that water restrictions are to be introduced, it will be interesting to see if residents who live along the banks of the Murray continue to water their lawns and gardens during the middle of the day, as they were on Tuesday last week when the temperature was 37°C and the wind was blowing from the north. This was happening along the Murray while farmers on the Eyre Peninsula were spiralling downwards into more financial crisis, but yet these farmers are still paying the Save the Murray River Levy.

* Troy Richardson, Cleve
**The *Advertiser*, edition 1, state
Friday, 9 May 2007, p. 18**

424

Conserving water

THANK YOU for printing my letter 'The Ice Age' (The Advertiser, 23/9/1999), in which I expressed confidence 'that one day our water resources will be managed responsibly'. Well, here I am $13,000 later, recycled water system using every last drop of wastewater. I have installed a 5600 gallon rainwater tank using rain and not interfering with catchment volume. I have mulched all of our 300 native plants trying to establish a better environment. I have spent a lot of time and money to conserve our precious resource. I am such a fool. I should have put in a swimming pool, like the two houses in my neighbourhood have in the past four weeks I guess they needed their pools more than I needed to water my plants. Please don't tell me we are short of water. I am doing my best.

* Richard Webb, Cockatoo Valley
The *Advertiser*, edition 1, state

Action on water

I AM confused by the recent State Budget. We have a proposal for a hospital to be built in a decade's time that no one wants, named after someone that everyone respects. We have state debt at levels comparable to the time of the State Bank collapse that can apparently be sustained by the economy as long as 'rainfall returns to more normal patterns'. What we don't seem to have are any plans to significantly address the present and ongoing water crisis, and the effect that it has on the society and economy. Reassuring, feel-good TV advertisements and hope for rain are no substitute for immediate and substantial action on water for the entire state and competent economic management.

* Richard Parkes, Adelaide
The *Advertiser*, edition 1, state
Tuesday, 4 September 2007, p. 17

Aquifer phenomenon

AN ever increasing number of aquifers are being sunk in Adelaide backyards. The government is now considering new regulations for this unregulated phenomenon. Why so late? This is our precious water, just like the surfacewater. Or do we wait until the aquifers have run dry as well?

* Dalia Gordon, Marion
The *Advertiser*, edition 1, state
Friday, 15 June 2007, p. 16

Hostile buckets

THE water 'buck' stops with Premier Mike Rann. While a great many in the community would be sympathetic to the Government's position because of the rapidity with which our water crisis has developed, few have sympathy for lugging buckets of water to keep their gardens alive. The buckets should go, drippers should come back, and, as other readers have suggested, some form of household rationing adopted because of the crisis-like nature of our lack of water. Also, perhaps Premier Rann could exhume Professor Peter Cullen's 2004 'Water Challenges'.

He gave the Government 18 recommendations in order to solve our water crisis. Few, if any, have been followed. To be fair to the Government though, even Professor Cullen did not foresee drought in both sources of our water; the Hills catchments and the River Murray.

* Richard Bennett, McLaren Vale.
The *Advertiser*, edition 1, state
Friday, 16 November 2007, p. 16

National water scheme

NO doubt many of the election pork barrelling items are important. However, all the pollies have missed the point – it's the water, stupid. If the flow of river water is not managed now, then what future do we have? It's now or never to negate the interstate rivalry about water and get a national scheme, or do we need internal warfare?

* Wayne Lodge, Athelstone
The *Advertiser*, edition 1, state
Wednesday, 12 December 2007, p. 16

Expensive options

THE proposed outrageous water charges, to pay for the new desalination plant, prove beyond all doubt this State Government's money-handling skills are wanting. Why does it continue to take the expensive options? Overpasses are much cheaper than underpasses. Cost blowouts and a stubborn refusal to face tax reform all point to the sooner this Government is consigned to the opposition benches for a lengthy period, the better off this state will be.

* Kim Bray, Elizabeth Vale
The *Advertiser*, edition 1, state
Monday, 10 Septmeber 2007, p. 17

Population pressure

HERE I am watching good water flowing down the local creek. My neighbours are enjoying their pool, the doctors and lawyers, who have plenty of money, water their gardens with a bore and the residents in Melbourne and Sydney can use their drippers. BHP can establish a desalination plant for uranium mining, but we are told this may be environmentally unsound for water supply for the people of Adelaide. The government tells us we need to increase our population so that we can prosper, yet as a poor lawn contractor, I will need to go to university to understand this logic.

* Steve Davies, Westbourne Park

Tank option

IF the government was serious about the water crisis, it would make it compulsory that when a new house is being built, it includes a water tank in the plans. The cost of the house should also include having the tank connected. Why not go one step farther and have an underground tank with pump and filter that collects water from the kitchen, shower and bath, so it can be used in gardens. It seems to me that the government may feel it will lose money if it did this, as we would not have lower water bills.

* Veronica Farrell, Hackham

Free market

WITH the drought projected to last another 10 years, along with restrictions, (the *Advertiser*, 7/9/07), a sustainable regime needs to be implemented that relies on willing participation rather than threats of punishment. The most durable and robust regime in history is the free market. The State Government should charge a flat fee for extraction of both ground and riparian water. Adelaide and other cities would be subject to the same pricing regime as irrigators and holiday shack owners. Cities could then oncharge for infrastructure and delivery costs if they so chose. The free market should be the determinant if farming is viable in the state, not political allocation of both water and tax dollars.

* Steve McClure, West Croydon

Wet advice

WITH the farcical water restrictions with a 'one size fits all' mentality, one wonders which school Minister Maywald's advisers went to. It could not have been the Pughole Public School, because at that school they at least taught you to add and subtract and how to work out averages. For the three months of June, July and August, Port Augusta has had a total of 19.6 mm of rain while Adelaide, over that period, has had 158.4 mm. Yet these very intelligent advisers of Ms Maywald seem to think that Port Augusta gardens should be no worse off than those in Adelaide. Let us hope Premier Rann's advisers are a little better at mathematics than Ms Maywald's advisers. You don't need to be Albert Einstein to find these figures either; they are on the Bureau of Meteorology website. Oops! I should not have said that. The Rann Government will probably have the site closed down so we cannot make comparisons.

*Malcolm Butler, Port Augusta

Dry witless

WE stand on the threshold of an environmental disaster, a combination of the drought and water mismanagement. Academics have dire predictions for us. Then we have the bleatings of Ms Maywald. She is out of her depth in more ways than one. People in her electorate are still using overhead sprinklers, totally impractical and inefficient. The government has spent millions on an unnecessary tram extension, while our water system continues to decay. The problem here is Mr Rann is interested in flashy short-term achievements, rather than subtle, long-term improvements. Nero fiddled while Rome burned. Rann fiddles while Adelaide dehydrates.

* Mal Kean, Glenelg North

No supply solution

TALK about bucketing water to gardens makes great emotive Talk Back and letters to the Editor. The State Government must be laughing? Better to fend off these questions and raise people's emotions because they can be dismissed. The crux of the matter is, this State Government has, for two years, told us how drastic things are and hoped that rain would come to ease the problem. It hasn't, and in the two years, there is not one definitive water supply solution set in concrete by the government.

* Johanna Sampson, Adelaide
The *Advertiser*, edition 1, state
Thursday, 13 September 2007, p. 17

Grower suffers

POPULIST Rann, the man with the tram, does it again. Gardeners in Adelaide win again, allowed to water for three hours once a week, while we, as fruit growers, should be so lucky. I get enough water to allow me to water for three hours a week for six weeks only. The water for your gardens will cost me five years' income and 3000 stone-fruit trees and my family home, a total loss to me of $500,000 because 16% allocation doesn't allow me to keep them alive. I hope all you whingeing do-gooders rot in your gardens while my family suffers at your expense. To Mr Rann, maybe a class action against your government for failing to deliver on the water allocation that was sold by the SA Government to us might work. See you in court.

* Kris Werner, Waikerie

CHAPTER 21

Water management networks

Richard D.S. Clark
Jerome J. Argue

Water management networks

Introduction

This chapter outlines the historical context of Adelaide's water management networks, including surfacewater drainage, potable (drinkable) water supply reticulation, wastewater (including sewage) collection and treatment, and groundwater reserves. Each of these has been developed since European settlement, exploiting to varying degrees the available resources. As the population has heightened the pressure on these resources, an increased bureaucracy has developed to manage these systems, straddling both state and local governments. However, a piecemeal approach has characterised the development of these management systems, resulting in fragmented responsibility, management and ownership.

Surface drainage systems

At the time of European settlement much of the coastline now bounding metropolitan Adelaide consisted of dune systems backed by extensive swamps and lagoons. Behind these lay the relatively narrow wedge of outwash plains, on which Adelaide is now located. Only the larger of the winter flow events, generated from rainfall on the escarpment and hills to the rear of the plains, had sufficient momentum to reach the coast, with the majority of smaller flows being lost in the gravels and fault lines along the foot of the escarpment. Larger flows, generated only after long periods of heavy rainfall, had sufficient volume to pass through and replenish the extensive reedbeds, eroding the seasonal build up of coastal sandbars and escaping into the sea.

The Adelaide Plains were crossed by a network of creeks and streams including the numbered creeks, First to Fifth, as well as more imaginatively named ones including Sturt Creek, Brownhill Creek, Keswick Creek, the Torrens River, and so on. Largely dry during much of the year, the creeks and rivers would flow after significant rainfall, often overtopping their channels to escape over poorly defined overbank pathways.

In pre-European times, most rain falling onto the plains area, where Adelaide would eventually be located, would not have runoff. Instead, the water would have been caught up in the vegetation, infiltrating into the soil or coalescing into shallow depressions from where it would later evaporate. Only during heavier or longer rainfall, when the ground had a chance to wet up, would runoff have gone into the creeks. As a consequence, only a small fraction of the area's total rainfall was discharged into the gulf as runoff.

As the township of Adelaide started to develop, the vegetation, soils and gravels that had previously soaked up and held the rainfall from running off were replaced with hard paved surfaces associated with houses, roads and industrial development. Instead of holding the rainfall where it fell, these surfaces collected and shed the rainfall in ever increasing proportions. Before drainage was effected to lead the excess water away, the actions of many feet, hooves

and wheels resulted in mud, puddles, bogs, inconvenience and disease. However, even then, more formalised albeit uncoordinated drainage did not help much because runoff from upstream developments only ended up on downstream land, thus starting off the cycle of disputes and litigation – a constant consequence of Adelaide's colonial surface drainage schemes.

The combination of horse-drawn traffic, unsealed roads, and primitive garbage disposal services meant that early urban stormwater runoff was highly polluted. Technologies were inadequate to treat this runoff, so that, with the exception of small amounts of roof runoff caught in rainwater tanks, it was not seen as suitable for water supplies. Since better-quality water could initially be found at no great distance from the town, stormwater rapidly acquired the reputation of having nuisance value only, to be disposed of as far away and as fast as possible. Drains were often used for refuse disposal, and thus became unsightly and smelly.

There were, however, examples of early understandings of water quality issues. During the 1880s the local council contructed a manure storage facility at the upper margins of Brownhill Creek, where market gardens had been operating over the previous 40 years. The facility was designed to prevent downstream pollution of Brownhill Creek, and can still be seen today.

Wherever possible, undergrounding of drains in closed pipes became the logical means of drainage, having the added advantages of freeing high-value land for alternative uses. As a consequence, development was allowed to extend ever closer to the original, ephemeral drainage paths, significantly reducing the flowpath area necessary to pass larger and rarer floods. The unique nature of South Australia's surface drainage paths was not well understood by its settlers. Frequent flooding of properties, leading to economic and human loss, have been regular features of Adelaide's history, a circumstance as true today as at any time during the past.

Whilst the conditions that lead to the adoption of these design practices and solutions have now changed, most stormwater is still channelled via underground drains to the most convenient point of disposal on a creek or river channel, and thence to the gulf.

Legacies of poor design abound, such as along the Port Road drainage system. A central trunk stormwater main, running from Hindmarsh to its outfall into West Lakes, was recently found to exhibit the following examples at different locations along its length: lateral drains entering at a level much lower than the main trunk line; decreases in pipe capacity at key road intersections; undersized pipes; and no water-quality improving devices. Such problems are not unique across the Adelaide Plains. Indeed, it has been estimated that approximately $160 million (2007) would be required to adequately fix all of the known problems in stormwater drainage systems and to raise them to an acceptable standard. This would, however, deal with nuisance flooding only. Major events, such as the 1 in 100 year flood, would be

Chapter photograph: Kangaroo Creek Reservoir.

largely unaffected by this work as most of the flow during these events is conveyed across the surface rather than underground in pipes.

The development of the metropolitan area has also involved the progressive in-filling of most of the coastal and riverine wetlands (swamps). In the periods between and after the world wars, much widening, deepening, straightening and lining of the previous natural drainage channels took place, in order to minimise the likelihood of flooding within the remnant rivers and creeks receiving the enormously increased stormwater runoff. The rich heritage of native species associated with the wetlands, lagoons and flow channels have therefore been early and continuing victims of urbanisation, through loss of habitat through pollution, erosion, replacement by concrete, or invasions by competing non-indigenous species.

While the original Adelaide city was located on a high, well-drained ridge well above flood levels, its outward growth has been increasingly over the unstable outwash fans and poorly defined floodplains of the creeks fanning from the Adelaide Hills face and the Mount Lofty Ranges catchments. Today, in the event of any heavy rainfall, larger and larger volumes of flood inflows more rapidly into a drainage system designed to handle inflows from a much smaller city.

Gradually, properties on the Adelaide Plains previously above flood levels experienced floods more frequently as upstream development has proceeded, and proceeds, in a largely unregulated manner. The policy of urban consolidation, commenced over the past decade or so, will further aggravate this situation, as will any potential increase in the frequency of extreme weather patterns brought about by future climate change.

The raised nature of many flowpaths (for example, parts of the Torrens River are up to 3 m above surrounding topography), and the likelihood of debris damming of intakes to culverts and underground drains, in association with angled street layouts, mean that the pathways for the escape of large flood flows becomes highly unpredictable. High velocity and deep flows are likely to escape along unpredictable paths and flood over much wider areas than are formally identified on past flood maps.

The past two decades have seen increasing concern about the deterioration of water quality derived from the highly urbanised catchments. A movement to improve the quality of discharge to sensitive marine environments, like the Port River, together with the observed loss of seagrass meadows in the gulf, has led to the construction of numerous artificial wetlands near the outfalls, particularly in the northern part of Adelaide but also across the metropolitan area. This has provided opportunity for storage and reuse of stormwater runoff, changing its perception from being a waste product to a useful resource to be exploited for the benefit of the community.

Groundwater in the Adelaide region

Adelaide is fortunate among Australia's cities, along with Perth and Melbourne, in the extent and quality of its underground water resources. A consequence of geological history, numerous levels of aquifers (underground water reservoirs) are present below virtually all of the metropolitan extent of Adelaide, north to south. These systems are fed by surfacewater recharge along the original surface streams, creeks and rivers, as well as infiltration into fracture zones associated with the extensive faults that trend north–south across the plains.

Two broad aquifer systems are defined. Quaternary aquifers, shallow and generally unconfined, radiate from the original surface drainage flowpaths; near the creeks and rivers their quality has generally been suitable for irrigation. Further from the recharge sites, the quality has typically deteriorated. Quaternary aquifers are generally used by 'backyard' users to supply useable water for irrigation and other uses.

Much deeper and more extensive is the system of confined Tertiary aquifers. Over millennia, the seawater filling these aquifers has been displaced by freshwater entering the ground along the fractures associated with, primarily, the Eden–Burnside Fault, which runs along the foothills, north to south. Between this fault system and the parallel Para Fault, which runs beneath the city, the Tertiary aquifers are relatively shallow (tens of metres). West of this, between the city and the coast, the system drops to a depth in excess of 100 m and a thickness of over 150 m.

Water quality of the Tertiary aquifer system varies with location and depth, but much of the recoverable water has salinity of between 800 and 1500 parts per million, compared with seawater of 35,000 parts per million.

This water is mainly used by industries, including Coca Cola and Coopers Brewery, and there are an estimated 140 operating bores into this aquifer system, extracting approximately 9 billion L of water per year. This compares with an estimated annual recharge of approximately 6 billion L/year.

On the Northern Adelaide Plains, over-exploitation of this resource has resulted in a decrease in the water surface level in the aquifer to the point where extraction is much more costly and difficult. Consequently, treated wastewater is now utilised as an additional water source for the market garden industry.

The western extent of the Tertiary aquifer system is presently unknown, but some local experts believe that it is likely to extend beneath the gulf, possibly to the Yorke Peninsula. However, even taking the extent between the hills and the coast, the amount of water contained within the Tertiary aquifer system has been estimated by hydrogeologists to be approximately 3000 GL, or enough to supply Adelaide for 15 years at present consumption rates. To date little use is made of this large underground reservoir. Apart from the industrial users mentioned previously, a number of aquifer storage and recovery (ASR) schemes have been constructed in Adelaide's west, taking collected stormwater runoff, treating it using constructed wetlands and sand filters, and injecting the water, under pressure, into the Tertiary aquifer. It is subsequently extracted and typically used for irrigation of golf courses and horseracing facilities, replacing potable water use.

Historically, however, groundwater resources have been turned to in times of severe drought. Since 1915, six recorded instances of the use of Tertiary aquifers to supply drinking water have occurred, including approximately 40 billion L extracted during the 1967 drought. Typically, this water has been injected directly into the water supply system, resulting in no known recorded health issues.

Potable (drinking) water supply systems

Adelaide's original water supplies were provided by shallow wells, rainwater tanks, and by carting water from the nearest permanent waterhole in a nearby creek or river. With the increasing pollution of local, shallow groundwater and waterholes, and the rising demands for larger, better-quality and more reliable supplies, the first diversion weirs and reservoirs were constructed on the local creeks and rivers emanating from the Mount Lofty Ranges. Reticulation of piped water to all properties soon followed, seen as a service to be provided freely by government as a means of reducing outbreaks of disease, improving health standards, fighting fires, and encouraging expansion into new areas.

Southern Australia suffers the world's strongest cycles of flood and drought. Compounding this, as far as water supply is concerned, a seasonal mismatch soon developed as demands for garden and crop irrigation increased in parallel with the free availability of 'unlimited' water. The provision of large volumes of reservoir storage therefore became a fundamental requirement in order to balance out the natural annual variability of inflows and to cater for the increasing mismatch between winter runoff and peak summer water demands.

Following the rapid post-World War II expansion of Adelaide, easily developed reservoir sites in the Mount Lofty Ranges had reached their maximum capacity. While some work was undertaken later, such as the construction of the Kangaroo Creek Dam on the Torrens River, Adelaide's total hills' reservoir storage now totals some 200 GL, or approximately one year's supply of water. This contrasts with eastern states capitals, with multiple years' storage in their near-city storages.

While supplementation of Adelaide Hills' storages was therefore necessary, alternative sources of groundwater, stormwater or recycled wastewater were deemed impractical or inadequate. Attention turned to the River Murray as an alternative supply; thus the multiple offtake pipelines to tap into its 'unlimited' flows were constructed, principally in the reach between Lock 1 at Blanchetown and the Lower Lakes, cut off from the ocean by a series of barrages near Goolwa.

Adequately favourable agreements were reached with the upstream states on the sharing of the waters of the River Murray, with South Australia participating financially and administratively with the joint construction and operation of several major storages and works in the upstream states. Thus South Australia, the driest state in Australia, has been able to secure one of the most reliable reticulated water supply systems of any Australian capital city. When its access to interstate storages is taken into account, Adelaide has a per capita storage capacity of about 550 kL, at least equal to that reported for Melbourne and Sydney,[1] albeit most of the capacity is located outside the state. More than 95% of South Australia's population is now reliant, at least in part, on River Murray water for its drinking water supply.

Throughout Adelaide's development, groundwater has continued to be a major source of water for peri-urban horticulture and agriculture. Where groundwater is of particularly good quality, or can be accessed cheaply, it has continued to be used by certain industries and households, even in those areas served by reticulated potable water. However, in general, there is little interaction between the South Australian government's reticulated water supply systems and the groundwater systems beneath the city.

Because only one pipe was provided to each customer of the water supply service, only one quality of water could be provided. With rising standards of living, the quality of water demanded for drinking and washing also rose. Water filtration was now standard, along with disinfection, despite the fact that well over half the water supplied was and continues to be used for toilet flushing and garden watering and does not need to be of such high quality.

Whilst public health was a main reason for establishing the network of water supply pipes delivering high-quality potable water to each house, the size, flow rate and costs of the reticulation system are actually dictated by the subsidiary objectives of irrigation and fire-fighting. These objectives dictate that pipe diameters and balancing storages must be sized many times larger than would be required for meeting only the normal, in-house requirements for potable water.

Prior to the considerations of 'Water Proofing Adelaide',[2] the last major planning for future water supply to Adelaide was undertaken in the late 1970s, confirming that the River Murray could continue to supply Adelaide into the foreseeable future. However, it was not long after this that concerns were raised about the rising salinity of the river and the need to construct salt-diversion works, particularly in South Australia.

Neither the spectre of rising salinity nor the later occurrences of algal blooms in the River Murray could cause a succession of state governments to waiver in their reliance on the river. Only in very recent years has the effect of sustained drought and the potential of permanent climate change caused questioning of the future reliability of this source, and a wholesale reassessment of the available alternatives.

As part of this recent review, the Water Proofing Adelaide project was undertaken by the state government. It promised much in terms of a detailed analysis of Adelaide's future water supply options but, to date, has delivered very little. More recent worsening of the drought gripping the Murray-Darling Basin, combined with a local (hills catchment) drought, has seen a rapid re-evaluation of supply options, with desalination coming into the water supply mix for the first time since being proposed (and being shelved) in the 1980s. At the time of writing, a desalination plant is proposed for Adelaide, expected to supply approximately 25% (potentially 50%) of the current average annual potable water needs of Adelaide.

Recent legislative changes have seen the introduction of the rainwater tank policy to South Australia. One kilolitre tanks are now mandated on all new dwellings, plumbed into the house to provide toilet flushing and laundry services. Adelaideans are already inveterate rainwater tank owners, with the highest percentage ownership of any capital city (approximately 48%). However, too often tanks are used only to provide water for the kitchen kettle, significantly reducing their potential impact on reducing potable water use. Further, the potential for flood reduction, combined with potable water replacement, has not been fully explored through the use of suitably sized rainwater tanks.

Sewage collection, treatment and disposal systems

The frequent outbreaks of contagious disease, especially during hot summers, in Adelaide in the mid-19th century were the catalyst for the adoption of reticulated sewerage to replace the previous pit toilets and night-carting systems.

A major public debate took place, during which it was resolved, in a prescient decision, to adopt a system of sewer drainage quite separate from that of stormwater drainage. Adelaide was fortunate in this decision, since cities that adopted the combined sewer system have experienced severe problems: continued growth has added increased volumes and peak levels of both sewer and stormwater flows to the combined systems, resulting in frequent overflows containing diluted raw sewage. Adelaide has largely avoided this problem, along with only a handful of other cities around the world outside Australia.

Initially, the sewage was transferred to sewage farms on the outskirts of urban areas, where the produced manure was used as fertiliser. These operated relatively successfully but were seen to reduce amenity as development continued to spread, surrounding the sewage farms. Ultimately, the farms were driven to their final locations adjacent to the sea. The farming of waste products as a parallel activity was abandoned and the disposal of effluent and manure was effected by offshore discharge into the sea.

As living standards rose and recreation on the beaches and coastal waters increased, the standard of wastewater treatment, and/or the distance of the outfall from the coast, was forced to increase. In the 1980s, the discharge of nutrient-rich and turbid effluent to the sea was linked to the die-back of seagrasses off the coast. While the contribution to this phenomenon is still being investigated, the wastewater discharge linkage resulted in a major upgrade in the standard of treatment required to be given to sewage, and the discharge of digested sludge to the sea ceased. Dried sludge is now increasingly sold as fertiliser and soil conditioner, with the remainder disposed in landfill.

The upgrading of the treatment processes has been accompanied by an acceleration in the reticulation of treated effluent for irrigation to the peri-urban areas to the north and south of the city. Many of these areas may be urbanised in the foreseeable future as the city continues its population growth, but for now the reuse of highly treated effluent, and the associated upgrade of existing treatment plants, provides a valuable source of water to agricultural pursuits such as market gardening, viticulture and horticulture. Effluent is even mixed with treated stormwater to provide water for external uses to urban development, such as at Mawson Lakes. South Australia leads the nation in terms of total effluent reused with approximately 20% of Adelaide's wastewater treated and reused in this manner. This is set to rise with the construction of the Glenelg to Adelaide Pipeline, recycling up to 50% of the effluent from the Glenelg wastewater treatment plant.

Salinisation of treated effluent, already an issue in reuse opportunities, looms as a long-term problem to be addressed in the recycling of treated effluent. Many residential in-house and industrial processes add salts to used water before it is disposed of via sewers. Leakage of saline groundwater into underground sewer pipes further adds to the salinity, as does evaporation from settling ponds at treatment plants. The final salinity of effluent available for recycling is then only suited for certain crops and uses, without 'shandying' the treated effluent with water from other sources.

For areas without access to the sewer system, the septic tank became the preferred method for sewage disposal. This system involves the settling out and anaerobic breakdown of wastewater in leaky underground tanks on individual properties. Where the absorption capacity of the soils on the site are inadequate to accept all of the discharge from the tank, septic tank based systems are subject to leakage and overflow.

A solution to this has been the establishment by local councils of septic tank effluent disposal systems, which collect the overflow effluent from individual houses for treatment in communal settlement and aeration lagoons, typically before release back to the natural drainage system.

This overall approach to wastewater management has resulted in a highly centralised collection, treatment and disposal system. While recent advances in reuse opportunities have removed a significant proportion of treated effluent from ocean disposal, the vast majority is still disposed of in this manner.

The myth of economy of scale when applied to total water systems

The astute reader will have discerned that water infrastructure, be it waste, storm or potable, has developed based on the notion of 'big is beautiful'. Adelaide is presently committed to a policy of centralised collection, treatment and disposal. This pervades our institutions, as well as our thinking, and only allows for solutions on a grand scale. But is this necessarily so, particularly when applied to water?

It is well understood and accepted that economies of scale exist in relation to many components of water systems. A simple example can be seen in relation to a standard length of pipe. As the pipe diameter increases, both the cost of the pipe per unit length, and the energy required to pump a unit amount of water through the pipe, reduce per unit of water passed through the pipe. Thus, other things being equal, the inhabitants of a large city would pay less for their water if it has to be imported over a certain distance via a large diameter pipe than would the inhabitants of a smaller city importing the same amount of water per inhabitant over the same distance via a smaller pipe.

This argument of 'economy of scale' has often been used blindly and incorrectly to support the continuation of past practices and to argue against any alternatives put forward that require the scale of operations to be reduced. The flaw in the 'big is best' argument can readily be seen when it is realised that the economies of scale of any individual components of a system do not necessarily apply to the system as a whole – particularly when it comes to pipelines.

Going back to the example above, if either the large or the small cities can locate water closer to it, then the length of pipelines can be reduced. By recycling all water onsite, no offsite pipes might be required at all. Since, in the existing systems, about 70% of all water supply infrastructure costs are associated with pipes and water pumping, up to 70% of all present costs could be foregone if all city inhabitants could recycle their own water without the involvement of a water authority. This argument, of course, is misleading if the additional costs of establishing the other components of treatment and storage, which would be required on each individual site, exceed the savings on the foregone external pipelines. The treatment and storage components required for the onsite systems will lose real economy of scale. Additionally, there may not be space on properties for the infrastructure, and the site-owner may not have the skill or desire to maintain and operate the system. Thus, while taking the argument to the limit of a single onsite system is presently unrealistic, the nature of the trade-off between different system costs can be seen. In fact, the relationship between scale of operation and total cost has been investigated by the CSIRO and others. All agree that many factors influence the relationship, but in general it is found that:

1. Using presently available technologies, the minimum cost of operations reaches a minimum at a medium scale, of the order of 500–5000 houses.

2. The cost differentials due to economies of scale are only small under most urban situations.
3. There is much more scope to reduce costs by focussing on cost reductions, which could be introduced via innovations in treatment and storage technologies, changes to systems' design, and changes in operations.

Focus on cost alone is, of course, unwise without also considering the benefits being provided by the expenditure. This is where stormwater and wastewater recycling have an enormous social and environmental advantage over the existing systems. The important point to be made, however, is that the cost of water supply services per se will generally *not* have to rise on account of reducing the scale of operations. There is no simple argument dictating that the cost of small systems will be higher than the existing large-scale systems, based on scale alone. The cost of the alternative smaller-scale systems will be about the same as the existing larger systems. Therefore, any future arguments regarding alternative water supply strategies should focus much more on the cost to benefit ratio, particularly the cost to sustainability ratio, which can undeniably be shown to favour the smaller-scale systems based on local stormwater capture and wastewater recycling.

Concluding remarks

Over the last 150 years there has been significant investment in infrastructure, predicated on a paradigm of collection and rapid disposal. Therefore, significant opportunity has been lost in creating alternative water supplies based on changing perceptions of what is waste.

Implicit in this is an acknowledgement that filtered, chlorinated water is not the appropriate resource to be using for toilet flushing or the range of external activities that (on average) consume approximately 40–50% of the water delivered to average urban allotments.

The steps that have been taken to collect, treat and reuse stormwater, together with improvements in wastewater treatment, are commendable, but much remains to be done if we are to change the dependence of Adelaide on River Murray water, and to meet our water needs more sustainably with confidence. Climate change, whatever it may yield in the future, must also feature in a reconfigured water supply system.

Whatever choices and decisions are made, it is essential that these be made in light of a coherent, strategic and rational plan for our water future. Past practices, perhaps appropriate for a growing city during the late-20th century, must be reconsidered, taking into account reduced rainfall, increasing energy costs, rising sea levels, and potentially increasing flood risks.

Integration of all aspects of water – from rainfall through to discharge of ultimate waste products – must be achieved, so that multiple use of water is the norm. Decentralised systems would seem to be a necessity of the future, already pointed to in programs such as Water Proofing Northern and Southern Adelaide.

Overarching all considerations, however, is the issue of sustainability – of water resources, ecology, energy, and the full range of human activities present in a modern city. To this end, a robust and abiding definition of what a sustainable future might look like must be argued and developed, shaping decision-making at all levels.

References

1. R.M. Argent and J. Langford, 'Elements of the urban watercycle: a review', Hydrology and Water Resources Symposium, 1996: Water and the Environment, preprints of papers, Institution of Engineers, Australia, Barton, ACT, 1996, pp. 67–72.
2. Government of South Australia, 'Water proofing Adelaide: a thirst for change 2005–2025', 2005 (online).

CHAPTER 22

A sustainable future for water

Richard D.S. Clark
Jerome J. Argue

Introduction

Adelaide water planning sits at a crossroads. We have determined to double our population while evidence of climate change is clearly apparent, which will increase our temperatures, reduce our rainfall, and increase the severity of storms – all exacerbating our already precarious water supply and flooding situations. Sea level rise threatens our low-lying coastal areas, along with the wastewater plants and stormwater outfalls there.

A very uncertain future for the Murray-Darling River system throws into question an underlying premise of water security policy in South Australia, and Adelaide in particular. Where will our potable water come from if we can no longer rely on the Murray? Does it make sense, anyway, to pump water from a source 70 km away while enough stormwater and wastewater runs into the gulf in an average year to supply Adelaide's needs?

As we move into an uncertain future, it is evident that our infrastructure systems, which deliver services we consider essential, must be put onto a more sustainable footing. This chapter outlines the vital questions of how we achieve sustainability of our water supply infrastructure. Many questions remain unanswered, and research is needed to provide a sound scientific basis for decision-making. But this should not be a reason for inactivity.

A sustainable future for water in Adelaide

Traditionally, the volumes of stormwater runoff and sewage generated as a consequence of the growth of the city have been regarded as too polluted to be used for water supplies, and have been discharged in ever increasing quantities to downstream rivers, estuaries or the sea. As a consequence, water supplies have had to be imported to the city from upstream rural catchments in continually larger amounts and from ever increasing distances, as demand has continued to grow.

While the established water supply systems have brought about great benefits in public health, prosperity and convenience, the environments from which the upstream water is diverted, and the downstream environments that receive the un-used and polluted waste flows, have all suffered.

In recognition of this situation, and in light of advances in technology, these practices are increasingly being recognised as unsustainable. With advances in water treatment it would be possible to pursue a totally different strategy, in which excess stormwater and wastewater, generated within the city, be treated to high qualities and made the primary sources of future water supplies. All the water services to which we have become accustomed can continue to be provided, except that environmental impacts would be greatly reduced. Once such change has taken place, the costs of establishment and

operation of this 'new generation' of water systems may be no greater, or may be even less, than present-day costs.

The adoption of any strategy for urban water management involving such a major paradigm shift could only proceed after its feasibility and implications have been fully investigated. Fortunately, while the adoption of stormwater and wastewater as primary sources of water would be new, many aspects of such a system have already been well investigated and the feasibility, costs and benefits can be estimated with reasonable certainty.

Unfortunately, the emergence of the present severe drought, with its revelations of extensive over-allocations from the River Murray, has brought problems with water security demanding a quick response. Worse still, some of the necessary in-depth investigations to pave the way for the introduction of more sustainable water systems have never taken place, despite having been proposed and part-demonstrated for more than 15 years. Instead, projects such as the Adelaide desalination plant, increased storage in the Adelaide Hills, and a north–south connector pipeline have been announced and are underway with little or no regard as to how they might fit into a more sustainable future.

Desalination in particular is costly and carries a need for high energy input and a range of significant environmental risks, which have the hallmarks of the old unsustainable thinking. Unfortunately, these projects now appear to be a political reality. Only time will tell whether they can be executed in a manner that is compatible with sustainability rather than being a short-term technological 'fix', required as a consequence of a long, stangnant period of inadequate long-term water systems investigation and planning.

Opposing the continuation of such projects into the future are the concepts contained within the new generation of decentralised and multipurpose water systems. These seek to reduce costs and increase social and environmental benefits through an examination of the working of the total watercycle as it affects individual households, communities, towns, cities and regions. A necessary starting point for Adelaide will be to recognise the particular hydrological regime in which the city is located.

Adelaide's unique hydrological situation

One of the main barriers to progress in water reform is the tendency for the established water industry to seek 'one size fits all' solutions to its many problems. Thus, the same solutions tend to be slavishly copied from area to area across the whole of Australia with little regard to local conditions (i.e., the industry acts globally without thinking locally first).

Adelaide has several features that set it apart, hydrologically speaking, from all the other capital cities in Australia. Some of these features, and their implications, are central to the investigation of sustainable urban water systems. Unfortunately, since sustainability itself has not been adequately defined or investigated by the state's water planners, it is not surprising that these features have

Chapter photograph: Mount Bold Reservoir.

The Paddocks

BOX 93

The Paddocks in Para Hills is the first integrated wetland and stormwater management system park designed in South Australia, and possibly Australia. Designed in 1971–1972 by City of Salisbury landscape architect Barrie Ormsby, in consultation with indigenous plant expert and landscape designer Ray Holliday, it comprises a series of ephemeral and permanent wetlands that control and filter stormwater pollutants and flood peaks amid a park-like setting of indigenous and salt-tolerant vegetation, with sporting ovals and picnic grounds.

Formal design works were undertaken in 1975, with construction commencing in 1977 on a flat to gently sloping landscape with a shallow watertable and evidence of salt scalding.

This cultural landscape is now host to the first major aquifer storage and recovery project in Adelaide.

David Jones

not been recognised, investigated and incorporated into a coherent water policy across the full water management system.

Australia rejoices in recognising itself as a land of droughts and floods. With respect to the former, it is well recognised that Adelaide and its hinterland share the lowest rainfall of all the Australian capital cities. However, it is less well recognised that, at the other end of the water spectrum, Adelaide has most of its urban area built over unstable flood debris fans and old floodplains.

This situation leads to a very high but uncertain flood risk level for much of the metropolitan area, particularly, but not confined to, the western suburbs. For years Adelaide's grave flood situation has been raised by hydrologists and others, only to see a lack of political will to tackle the problem. It is no exaggeration to claim that the past water management regimes have delivered Adelaide a situation where it cannot claim to be sustainable in the face of either 'present and perceived danger' levels in respect of either floods or droughts.

Moreover, as the city and its population expands, and the potential for intensified floods and droughts associated with climate change is predicted, these inherent problems will only worsen. While our focus during the early years of the 21st century has been much engaged with drought, if a storm such as occurred in February 1925 were to occur again, it is reasonable to suggest that many lives could be lost, property damage would be extensive, and questions of responsibility would be raised.

Other hydrological features of Adelaide, critical to any investigation on sustainability, are listed below. These will become central factors in any argument put forward for the adoption of a more sustainable water system for Adelaide.

1. Despite its low rainfall, the total of water flowing by gravity (sewer and stormwater) through Adelaide, sourced from within Adelaide or its immediate surroundings, is well in excess of its total demand for water; this situation will remain even if Adelaide were to double or triple in population.
2. Adelaide is largely unique in its access to extensive, unpolluted aquifers suitable for storage and recovery of bulk water.
3. In common with the other southern mainland capitals, Adelaide's most severe floods occur in summer, while the majority of its total rainfall and consequent runoff occur in winter.
4. Of all the capitals, Adelaide's winter rain has the lowest intensity, and large proportions of the consequent runoff can be harvested using smaller-capture storages than in other capital cities.
5. The total volume of water storages needed to capture sufficient of the water flowing by gravity through Adelaide, and thus to supply all of its water needs, is much less than the volume of storage that would be required to solve its flooding problems. Because the winter flows and summer floods occur in different

seasons, the necessary storage volumes to mitigate flooding could be used for capture of water supplies, with no additional cost except for providing long-term drought storage in the underlying aquifers.

Five key considerations for the adoption of sustainable water systems for Adelaide

1. Definition and analysis of sustainability for urban water systems

A simple definition of sustainability for urban water systems is provided by consideration of the outputs sought for the system itself, namely:

The system should be capable of sustaining all of the following services, equitably and at minimum lifecycle cost, under future scenarios of population growth, climate change, technological advance and social adaptations.

1. Reliable, affordable and healthy supplies of water for supporting urban activities and lifestyles.
2. The conveyance of excess water and wastewater away from living areas to avoid disease, inconvenience and long-term effects on soils, plants and animals.
3. Minimal damage from floods (or sea storms) up to a nominal recurrence interval of about one in 100 years, but also building in the ability to survive more infrequent but 'catastrophic' storm events occurring once in 500 years.
4. Avoidance of any damage to ecosystems caused either by the excessive diversion of water from them or to them. The damages may be caused by either the amount or quality of the water diverted.
5. Minimum contributions to greenhouse gases involved with the construction and operation of the systems.

To this primary list can be added secondary outcomes, brought about by the improvement of water supply systems, such as access to water for recreation, amenity and aesthetic satisfaction.

The key to achieving sustainability for urban water systems lies in the design of an innovative system of water infrastructures and operations that can provide *all* of these services at minimum cost.

Having opened the door to the golden future of urban water sustainability, several axioms can now be advanced about the nature of the 'sustainable' system.

1. Logical and simple analysis dictates that the minimum cost system will be one that manages the total watercycle, commencing with rainfall, and addressing all the many pathways that water follows on its passage through the urban area.
2. The design of the city itself becomes quite crucial, since it must cater for the storage, cleansing and transmission of water at all stages along these pathways, through the fabric of the city.
3. Since rainfall, as the starting point of the cycle, is spatially distributed, and all the services sought are also

spatially distributed, spatially distributed systems, in which infrastructure is decentralised and replicated and (wherever possible) serves multiple purposes, will be the cheapest and most sustainable option.

4. Flood control and management will become key to the identification of the 'best' system, since it is associated with the fewest choices of system design. Floods are inevitable; the only means for reducing their impacts in urban areas involve minimising the peak runoff rates and keeping flood risk materials, properties and lives out of flood pathways. These two fundamental requirements are key to the remainder of the jigsaw puzzle, since:

   a. as described above, the amount of storage required for flood control is usually greater than that required for any other of the sustainability objectives. The best place for storage is on floodplains, serving multiple purposes in pollution control, biodiversity and water harvesting. Thus the identification and recovery of floodplains and the provision of sufficient storage in, on, along and under the floodplains becomes the single greatest multipurpose, low cost/high benefit task for the architects of the new generation of sustainable cities;

   b. once the floodplains are recovered and storage is provided to reduce flood peaks, the rest of the sustainability jigsaw puzzle falls into place, since many of the other performance criteria can 'piggy back', at very little additional cost, on the achievement of the flood-proofing criteria.

5. Adelaide has two features, previously mentioned, that fit with this strategy:

   a. Adelaide's climate dictates that although the majority of rainfall occurs in winter, severe floods in South Australian towns and cities will inevitably occur in summer, more than likely in February. While it is not yet confirmed, it seems unlikely that this will change as the climate changes. The importance of this is that the storages needed to reduce the most severe flood peaks to acceptable levels, which must be dotted about at key locations along the drainage paths, will only be required for flood mitigation in summer. These storages can be used for harvesting stormwater for water supplies in the winter; and

   b. Adelaide is underlain by aquifers for low-cost, long-term bulk storage beneath the floodplains.

The design of the city must cater for the incorporation of water storage in all its possibilities and forms, including permeable paving, roof gardens, rainwater tanks, wetlands, floodplains, aquifers, and flood-storage dams. Similarly, water treatment must be carried out in a parallel fashion, by avoiding unnecessary pollutants (composting toilets, waterless urinals), source controls, wastewater treatment in basements of buildings, and the 'natural' treatment associated with most storage forms (permeable paving, wetlands, aquifers and floodplains).

Stormwater harvesting from impermeable surfaces and wastewater recycling are totally compatible with sustainability, since these waters are generally in excess to the natural flows and are associated with damaging effects, which can be reduced by the processes of storage and treatment, required if they were used as water supplies in the near vicinity of their production.

An appropriate slogan and yardstick for sustainable water management is 'manage your own water before coveting someone else's water'. Sustainability can be achieved by using imaginative and innovative ways to provide all the traditional water services in a single, replicated, decentralised system at reduced economic and environmental cost.

2. The present and likely future water balance and storage needs for Adelaide

To what extent would stormwater and wastewater be sufficient as primary sources of water supply to the city?

Many estimates have been made of the various components of the water budget for Adelaide. A comprehensive present-day budget has been given by Clark,[1] which also addresses the question of the location and amount of storage required under various strategies, in which stormwater and wastewater could be used to replace reliance on River Murray water.

Clark's estimates show that the present availability of flows through Adelaide are more than twice the amount presently used, and if sufficient storage were available to balance out the variability of these flows, Adelaide would not need to import any water from the River Murray nor contemplate desalination of seawater.

With respect to future water budgets, under the 'most likely' future climate predicted by the CSIRO, the runoff from the rural and hillsface catchments will reduce by about 32% by the year 2050. The runoff from the urban catchments, however, will only reduce by about 17%, since they are far less influenced by soil moisture and evaporation.

Using this information, it is possible to extrapolate Clark's present-day water budget to cater for a growth of Adelaide's population to up to 4 million persons, taking into account the effects of increased population density on stormwater generation and the reduced demand for irrigation of smaller gardens.

Table 22.1 shows Clark's water budget figures extrapolated to the year 2050 under an assumed population of 2 million persons, but modified to take climate change into account. The figures illustrate the increasing proportions of the total budget that stormwater and treated effluent in particular) will provide to the future water budget. These are the only sources that automatically 'grow' as the city grows, and therefore have inherent characteristics of sustainable water sources.

Figure 22.1 shows that under an assumed year 2050 climate, and even allowing for only 67% of all surface flows to be harvested, the availability of local water would remain greater than the demand for a population up to of 3 million persons (assuming the climate had not further deteriorated by that time).

Table 22.1 Water demand and availability in 2050

Water demand	
Supply of water used for internal purposes and returned via the sewage system	176
Supply of water used for external purposes and lost by evapotranspiration	117
Total	**293**
Water availability	
Mount Lofty Ranges flow to the Adelaide reservoirs (excluding Myponga Reservoir)	87
Runoff from the hillsface catchments immediately behind Adelaide	22
Runoff from the urban catchments on the Adelaide Plains	110
Flows of treated effluent discharged from the sewage treatment plants	166
Total	**385 GL/a**

Table 22.1 Water demand and availability (gigalitres/year) in Adelaide for the year 2050. Estimates have been made for Adelaide with a population of 2 million persons with an estimated year 2050 climate.

3. Aquifer storage and recovery as the key to sustainable urban water management

Clark has calculated the storage that would be required if the four sources of water listed above in Table 22.1 (only) were used for supply. Various assumptions were made, based mainly on the experience gained by the demonstration projects at Mawson Lakes and similar schemes elsewhere. It has been assumed that:

- Because of a shortage of open space, only a maximum of 70% of the stormwater could be harvested by wetlands situated throughout the urban areas. Water captured in these wetlands during the winter seasons would be stored by a network of distributed bores in the lower Tertiary (T2) aquifer. Recovery of this water during summer or drought periods, as required, would be via the same bores, or a subsidiary network of bores situated closer to the locations where the water was required. The space requirement for wetlands would be about 5% of the urban area. This could be reduced by the adoption of permeable paving, underground 'buffer' tanks and rainwater tanks.

- Only 95% of the treated effluent could be recovered and reticulated to meet non-potable purposes. For maximum efficiency, these uses would include not only the irrigation of gardens and open spaces but also toilet flushing (as for Mawson Lakes). Excess treated wastewater generated in winter would be stored in and recovered from the upper Tertiary (T1) aquifer.

- The existing reservoirs in the Mount Lofty Ranges would continue operations and Mount Lofty flows would continue to be supplied as the main source of potable water. However, without back-up from the River Murray, after a long sequence of dry years the reservoirs would fail, even when supplying potable water only, and thus the best-quality stormwater would be withdrawn from the aquifers and treated to potable standard to augment the potable supplies.

Clark's investigations showed that because of the far higher reliability of stormwater and wastewater flows, the amount of storage required to carry supplies from these sources through drought periods would be about 1/5th of that required for traditional sources, only a small proportion of the total storage of water held naturally in these aquifers.

Since the T1 and T2 aquifers beneath Adelaide are already 'full to capacity', only limited amounts of additional water can be pumped in. To date, no problems have arisen with the pumping in and recovery of the relatively small amounts involved in the present level of aquifer storage and recovery (ASR) activities. However, as the amount to be exchanged rises, a program of aquifer 'dewatering' (of the poorer quality in-situ water) may be required to make space for the better-quality stormwater and treated wastewater.

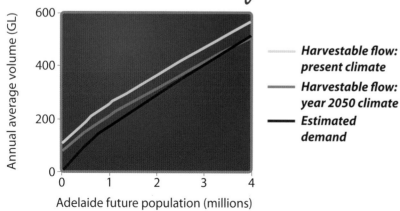

Adelaide – harvestable flow v demand

Legend:
- Harvestable flow: present climate
- Harvestable flow: year 2050 climate
- Estimated demand

x-axis: Adelaide future population (millions)
y-axis: Annual average volume (GL)

Figure 22.1 The volume of harvestable stormwater versus the estimated demand for water in Adelaide in 2050. The data has been extrapolated to 4 million persons under present and estimated year 2050 climates (with only 67% of surface flows captured).

Greenfields Wetlands

BOX
94

The Greenfields Wetlands, located adjacent to Port Wakefield Road near Mawson Lakes, was successfully developed in 1984–1985. The proposal sought to: develop an unique freshwater wetland to host local ecology and avifauna; address an engineering/drainage facility essential to service the future development of the adjoining areas on low-lying land; create a major passive recreational and educational venue; and to visually renovate a barren area into a vegetated ecological landscape.

The design concept sought to provide a wetland system with a range of habitats, combined with the capacity for ponding stormwater runoff.

As a wetland, habitat diversity was essential. The development included open water, mudflats, reed and sedge beds, fringing trees/shrubs to the water's edge, areas subject to short-term inundation and areas above inundation.

As a stormwater retention pond, the design assumed the wetlands were at full capacity, with a top water level of 2 m and additional scope for storm surges, resulting in an overall top water level of 2.5 m. Some low areas, not required for storm storage capacity, were also provided, requiring filling to a minimum level of 3 m to provide sufficient head for drainage flows into the storage pond. Outflow from the stormwater-retention areas will occur when water levels in Dry Creek Drain at Port Wakefield Road permit.

The Greenfields Wetlands builds from knowledge learnt from the Paddocks project.

David Jones

Clark investigated various combinations of storage for supplying the different quality demands from the different sources. Because stormwater and wastewater have a far lower temporal variability, storage requirements are minimised when the use of these sources is maximised. The modelling showed that even under the 2050 scenario it was likely that sufficient aquifer storage could be accessed, but that desalination of some of the brackish water involved might be required. However, because the salinity concentration would be only 1/10th of that of seawater, the energy and environmental impacts would be far less that that of seawater desalination.

Because the injection and recovery from the aquifers at any one bore can only be undertaken at a relatively slow rate, many hundreds of bores would be required, distributed across the western plains area. The spatial distribution of the bores would be compatible with the spatially distributed nature of the small-scale systems in providing their full range of services in water harvesting, flood mitigation, biodiversity enhancement and water supply.

The feasibility of storing large quantities of water in the aquifers beneath Adelaide is the key to the feasibility of the 'new generation' systems proposed herein. While there is little evidence to suggest that large-scale ASR could not be made feasible, successive state governments have not undertaken the necessary investigations to either prove or disprove the hypothesis. In view of the crucial nature of storage to successful water management, and the nature and location of the aquifers directly beneath the city, this ongoing lack of investigation must be addressed.

4. Single or multiple water supply pipe systems?

The present water supply system provides one quality of water – drinking quality – via a reticulation network that connects every customer to the major supply system. The single quality of water is used for all purposes, even though most purposes (e.g. irrigation of gardens) do not require such a high standard. The potable water that is reticulated is gravity-fed from the raw water source catchments to treatment plants located along the eastern side of Adelaide, where it is treated by disinfection, aeration, coagulation and filtration to drinking water quality. This water is then gravity and/or pump-fed by an extensive pipe network to each service point.

If stormwater and wastewater become primary sources of supply, a total redesign of the water systems is needed. Both the sources of these supplies and the demands for the services are spatially distributed, and the storage and treatment infrastructure can be provided most economically when sited, at a small scale, within the local urban fabric.

At present, at least 70% of all the capital costs of urban water systems lie in the pipes and channels that bring water into or take water back out of the urban area. If the stormwater generated locally can be captured, treated, stored, used for in-house 'first-class' purposes, disposed via the wastewater system, then recaptured and treated again, stored again and reused again, the offsite pipe networks would be very significantly reduced in size and cost.

Several researchers have shown that the savings in the size and length of the import and export pipes and channels can more than counter any increased costs associated with the greater complexity of the integrated systems. However, several options are possible in respect to the design of the reticulation systems:

1. Since the existing single pipe system exists, it may be cheapest to treat all additional water to drinking standard and inject it into the pipe network close to the location it is harvested.
2. Only 'raw' water is reticulated to all houses via the single pipe system, with treatment being provided by onsite house-scale treatment plants. Drinking water might be sold bottled.
3. Second-class water for garden watering and toilet flushing could be reticulated via a second pipe running in parallel with the existing pipe network. This is the solution provided at Mawson Lakes and Rouse Hill in Sydney.

It is probable that different approaches could be adopted for different locations and for different stages of the progressive restructuring of the system. Once again, these alternatives appear not to have received serious investigation by the current water planning authorities.

5. Planning for a sustainable water city

How could the changeover from the old system to the new system be planned and scheduled?

It is apparent that the time for research and investigation of a new water management paradigm is here. To date, insufficient resources have been applied to the problem of the definition and analysis of water sustainability, thus the basic data and systems design principles for future planning are still missing.

It is suggested that the South Australian Water Resources Council (or similar) could be re-established, but with much wider powers, as the prime coordinating body for water in South Australia under a rejuvenated state water planning regime. It must be adequately resourced and funded to undertake investigations and to set overall directions. Its earliest deliberations should be to identify definitions, principles and policies for water sustainability. It must draw on data and expertise in all areas related to water – hydrology, water systems, economics, urban and regional planning, and the like. It must feed its directions into the highest levels of decision-making in the state.

A citizen's water advisory group might be considered as an adjunct to the Water Resources Council. This group could be charged with identifying the pathways whereby citizens pay for water services through numerous levies, rates and service changes, paid to different agencies. Since sustainability will be measured, in part, by the reductions in costs and increase in benefits, and citizens are the final recipient of these (and are presently very poorly informed about them), such an advisory group should be able to add greatly to the public's participation in the water reform process.

Under directions from the Water Resources Council, the various responsible government departments must then cooperate within their areas of expertise and responsibility to undertake the directions set by the council. The urgently required work program should address:

- the capacity for storage in the aquifers;
- the amount of surface storage (and floodplain clearing) required to deal with flooding under climate change scenarios;
- establishment of improved water sensitive urban design features and opportunities compatible with future urban and regional water sustainability;
- analysis of the economics of alternative systems compatible with the agreed directions for sustainability;
- establishment of programs for progressive restructuring of the city and its water systems to bring about the desired level of sustainability;
- the removal of impediments and barriers to reform, cooperation and competition, as appropriate; and
- the establishment of priority directions for research into areas that restrict the uptake of water sustainability.

The main agencies involved should be local government, the Department of Water, Land and Biodiversity Conservation, the Environment Protection Authority, the Adelaide and Mount Lofty Ranges Natural Resources Management Board, Planning SA, Economic Development Board, Department of Treasury and Finance, and the universities. Casual investigation indicates that the present systems might require considerable restructuring if these changes were pursued to any extent. Finally, such a shift needs visionary and talented leadership. History has taught that fundamental and enduring change needs 'champions' to drive it forward. While a number of voices are presently being heard, no one has yet stepped forward to lead change in South Australia. Characters of the stature of the late Peter Cullen are few in number, but just such a person is needed to act as a rallying point for systemic change to take our city into a sustainable water future.

References

1. R. Clark, 'Water proofing Adelaide: modelling the dynamic water balances', AWA regional conference, Glenelg, South Australia, 2003.

CASE STUDY 8
An integrated model for managing Adelaide's water using interconnected stormwater collection and aquifer recharge systems

Colin Pitman

Introduction

The waterflow pattern over the floodplain on which much of metropolitan Adelaide sits has altered dramatically since 1836. Prior to European settlement, the floodplain consisted of a complex mixture of habitats including grasslands and open woodland, and a series of swamps and waterways leading to the shoreline of sandhills and mangrove forests. Waterways leading to those swamps ensured the sandhills and the highly permeable Quaternary (high-level) aquifers of Adelaide were full of water for most of the year. Many of the Quaternary aquifers extended to the surface in what are now the western suburbs of Adelaide.

European occupation has resulted in the development of a relatively impervious floodplain, which has dramatically increased the amount of runoff occurring during rainfall. Urbanisation also has resulted in a range of pollutants being discharged to the sea, the most destructive of which are the silts, suspended solids and nitrates, as well as the effect of low-salinity water entering the gulf in large volumes. The other consequence of urbanisation has been the extraction of water from the Quaternary and Tertiary aquifers along the Adelaide shoreline, increasing over the last 50 years with extractions by primary producers and more recently by industry. While the aquifers in Adelaide are generally of very ancient origin, their ability to recharge under decreasing rainfall and increasing sealing of the ground surface places them under considerable threat as a sustainable resource. Rainfall runoff is now directed into pipes and drains, no longer allowing infiltration to replenish the nearsurface aquifers.

It is highly desirable that sustainable management of water resources be introduced into the consumption of water throughout metropolitan Adelaide, taking into account both aquifer sources and water from supplies external to metropolitan Adelaide, in particular the Adelaide Hills Catchment and the Murray-Darling System. A water balance model has been outlined in the 'Water Proofing Adelaide' document published in 2004, describing the manner in which Adelaide's water resources from the River Murray, Adelaide Hills and, to a lesser extent, groundwater are currently sourced. It has become clear in the time since that document's release that water supplies to Adelaide for both drinking and industrial uses are under threat due to variability in rainfall affecting inflows to the River Murray, Adelaide Hills and urbanised catchments.

Prior to the release of the Water Proofing Adelaide targets, only a token amount of stormwater was recycled. However, since the adoption of the targets in 2004, the situation has not changed! Approximately 100 GL of stormwater per year discharges to the sea, and most of the capture and recharge to aquifers has been limited to the northern suburbs. Likewise, wastewater discharge to the sea is currently approximately 70 GL/year, of which approximately 18 GL is captured and recycled. The model described below outlines how stormwater can be recycled in conjunction with wastewater across metropolitan Adelaide. The variability of stormwater supply in Adelaide is very much dependent upon a number of key factors relating to the manner in which stormwater can be captured, stored and supplied. The following discussion makes reference to the manner in which contemporary methods of stormwater capture, storage and delivery can provide a seamless supply of water to customers across metropolitan Adelaide. This method of supply takes into account the manner in which fit-for-purpose water can be supplied, based on quality, needs and volumetric demands.

It has been demonstrated that across metropolitan Adelaide, and in particular the northern suburbs, stormwater can be captured and biologically treated through the multi-barrier process to supply high-quality water on a fit-for-purpose basis to customers. The manner in which this form of capture, storage and supply can occur in other locations in Adelaide requires innovative thinking in relation to the specific local nature of the supply systems and the availability of land to biologically treat the water at that locality. Therefore, alternative techniques, which are used across the world to capture and treat stormwater in wetlands to a level suitable for industry and potable use, must be considered. In addition, economic evaluation of the cost to produce this water has been undertaken at a range of locations, demonstrating that in all cases water can be produced at a lower price (with a considerable profit margin) than that which is currently offered to the market.

Recently there has been considerable debate surrounding the ability of inner-urban Adelaide to capture and treat stormwater in the same manner as in the outer-growth suburbs of the north. However, a system to enable maximisation of capture and distribution of water to treatment plants, either through biological treatment or through partial biological/partial mechanical treatment systems, can be established across metropolitan Adelaide. The methodology proposes interconnection of all systems across Adelaide, extending from the Gawler River in the north to the Onkaparinga Catchment in the south. It also proposes a capacity to supply treated and untreated water to an appropriate treatment site near the centre of Adelaide or other locations in the north or south of Adelaide. Equally, the supply side of the system includes a similar distribution grid enabling supply to key customers on a fit-for-purpose basis across the city. Such supply systems have a capacity to supply water into the potable water grid, under particular circumstances, by using treatment systems currently common in the market.

Capture and storage and supply of water using integrated catchment management

There are three methods whereby flows can be captured prior to treatment and then injected into the aquifer.

Category 1 – traditional capture treatment and injection

Stormwater can be captured, treated and injected into large streams where open space for the biological treatment may or may not be available.

Category 2 – splitting streams from one stream to three streams

These are systems where land can be purchased to enable the capture of large volumes of water, which is then subject to pretreatment and injection without the extensive purchase of existing developed properties. The streams for which this can be considered are Gawler River, Little Para, Smith Creek, Helps Road System, Dry Creek, Lower Torrens, Onkaparinga River, and Christies Creek.

The system involves expanding the existing channel to create three channels of flow. It involves the widening of the existing channel while retaining the existing riverbed. The widening provides for the diversion of water to holding and biological treatment systems and subsequent injection systems on a progressive basis along the length of the river or creek. This technique enables long lengths of rivers to capture up to 100% of the water flows. The technique does require clay lining to the receiving basin and the treatment basin to ensure the Quaternary aquifers are not a source of saline flows or receiving points for water from the system. The system will be shortly introduced to the Gawler River to demonstrate the manner in which this can occur along its length. The other rivers mentioned have equal opportunities to undertake the same form of storage, pretreatment and injection at localities which are generally where the aquifer has a large receiving volume capacity.

Category 3 – weir and flash pump systems

Weir and flash pump systems are currently used in the City of Salisbury, Holland, Singapore and Germany. The system employed by Salisbury Council at Dry Creek and Cobbler Creek involves the controlling of flow in the system using detention basins and controlled release weirs to batch-process water and move it to a receiving point, enabling flash pumping to nearby pretreatment systems. Where these systems have been used they can either include prefiltration systems, such as in Berlin, to remove silts and clays, followed by biological treatment systems. Salisbury Council has introduced a number of controlled release systems in their catchment, which are electronically operated and batch processed to supply water to the wetland at a lower level. This enables the maximisation of water capture when peak storm events occur. In Salisbury, existing flood-control dams have been used to batch process stormwater and maximise the capacity to recycle water in the existing wetlands. This system does not rely on the water being pretreated within the catchment under consideration. The streams in Adelaide where this could be considered include the following:

- North-west Arm Drain, City of Port Adelaide Enfield;
- Torrens/Port Road System in conjunction with the River Torrens;
- Cheltenham and Port Road System in conjunction with Torrens Drainage;
- River Torrens in conjunction with the Adelaide Airport and Port Road systems; and
- a number of systems in West Torrens in conjunction with the Adelaide Airport System (Sturt Creek, Field Creek, Onkaparinga River).

Storage and supply

Aquifer storage requires capture using one of the following forms of devices:

1. A progressive supply system from within the subcatchment to primary biological treatment systems and storage in localised, low-pressure portions of the aquifer.
2. Controlled release of stormwater from stream weirs with a pump-to-treatment system, similar to that in option 1 (above).
3. The diversion of waters from major stream systems to treatment systems adjacent to the stream using the technique in option 2, with the alternative to divert to parallel, adjacent catchments and treatment systems, with subsequent injection into low-pressure portions of the aquifer.

The distribution of stored water through the aquifer

In all storage options, a dynamic model must be developed to describe the injection, storage and transfer of water from the aquifer to the customer supply system. The design of the scheme must also recognise that in some cases pressure release will be required to remove highly saline water from the aquifer, and to control the storage capacity in localised areas in order to reduce local pressure in the aquifer.

Across Adelaide, aquifers, particularly the deeper Tertiary aquifers, are known to be able to accommodate elastic and artesian pressurisation. However, this capacity is not infinite and there will be occasions when it will not be possible to store all stormwater at a particular time, for example during peak flow periods. Dynamic models are used in the northern suburbs to describe the manner in which this form of injection and release can occur, so as to maximise injection without creating localised artesian effects.

The technique of distribution of water to localities more appropriate for treatment (option 3) but remote from the stormwater source has been practised successfully in Germany. Here, flash pumping has been introduced to supply water to treatment systems located within dispersed communities. The treatment systems are wholly engineered to be an acceptable part of the urban landscape. Opportunities for undertaking such a scheme in the Adelaide metropolitan region can be identified, commencing in the north. Examples include:

Gawler River

The Gawler River system is ideally suited to the option 3 treatment. The majority of the stormwater flow occurs at flood peak time and the technique of capture and treatment can utilise the diversion of these flows, allowing pretreatment using wetlands prior to aquifer injection. The Gawler River also provides an opportunity because significant open space is still available along the length of the river to provide for capture and treatment. There is therefore the opportunity to inject water into the aquifer in areas with relatively low pressure. The aquifer depletion in this area is very high currently, as is the demand for water. The opportunity exists to combine stormwater and wastewater in a large mixing facility to provide fit-for-purpose water. If it was interconnected with the existing Salisbury system, it could build the water supply capacity in the north to benefit industry, parks and reserves, schools, new residential growth areas, and primary producers. It offers the capacity to provide economic supplies to existing urban customers either by pretreatment and injection into the existing SA Water grid, or into an independent supply system (such as Salisbury). The Salisbury and Playford network grids, currently under construction, incorporate injection systems with cross-linked supplies to supply commercial, residential and local government consumption. The capacity within the Salisbury and Playford systems, with new cross-links in the north of Adelaide, can be expanded as new customers are added, and advantage is taken of the availability of open space to provide increased biological treatment capacity. Both systems can be linked to the proposed (option 3) Gawler River system.

Barker Inlet

The Barker Inlet system has a number of fragmented catchments, some of which have well-defined pretreatment systems already installed, while others await approval. The most notable of these small catchments are the Barker Inlet Wetlands, the Torrens Road drainage system, Range Wetlands, and the wetlands of the North Arm. This collection of small, interconnected systems with the proposed Port Road system and Cheltenham Wetlands, has the capacity to pump into the aquifer in the local vicinity, and therefore link with the northern suburbs system currently under construction.

Central Adelaide

The channels of the Central Adelaide Plains, including the River Torrens system and the Keswick Creek/Brownhill Creek systems, have a large capacity to provide water for capture, storage, pretreatment and biological treatment at Adelaide Airport. There are a number of other locations that could be considered, however it would be more appropriate if one, or at most two, focus points are provided for treatment and injection. The controlled capture and pumping from a primary treatment near the outlet of each flow system – into systems that have pretreatment process at Adelaide Airport – are the most economically viable.

Southern systems

The Field River, Onkaparinga River and Christies Creek are close enough to link with the central Adelaide system, to inject into localised aquifers. There is therefore a large capacity to supply the Southern Vales and the urban growth of the southern suburbs. Hence major opportunities exist for pretreatment and injection into local aquifers at selected locations in the Southern Vales.

Summary of production systems

The capacity to interlink connections from northern suburbs to central suburbs and the Southern Vales is therefore possible. Both existing and new pipe networks should be explored to achieve this interconnection. Of particular potential is the Birkenhead to Port Stanvac pipe, currently abandoned as a result of the closure of the petroleum refining plant at Port Stanvac. The rehabilitation of this pipe could provide a significant opportunity for establishing a cross-connection system between northern, central and southern systems.

Supplying aquifer-stored water to users

Supply systems can be in the following form:

1. **Direct supply** involves the extraction of water from the aquifer after injection, and its connection by either an existing or new distribution main to the appropriate customer.
2. **Traded supply systems**, where the injection credit at one location is traded, in the same aquifer, to another location, subject to appropriate hydrodynamic pressures and salinity in the aquifer.
3. **Trading systems**, where the injection regime is traded to another location, where there exists an aquifer with a higher salinity than the customer requires. The desalination of the groundwater at that location using mini desalination plants could provide high-quality water, potentially usable for drinking purposes. The brine concentration in groundwaters in metropolitan Adelaide are approximately 3000–4000 parts per million compared with seawater, with a salinity of 37,000 parts per million. A pilot system demonstrating this form of treatment is about to be installed in the City of Salisbury using wind generation to drive the desalination component of the project. The energy consumption of such a system is very low, with the cost of production as low as 40 cents per kilolitre.

The method of supply using direct supply systems can be undertaken utilising a ring main, similar to that undertaken in the cities of Salisbury, Tea Tree Gully and Playford. In each case it is necessary that the supply system will supply major consumers in order to meet its economic targets. Subsidiary consumers could be connected if the economics of this supply are found to be acceptable. The City of Salisbury is currently undertaking trials to examine the reticulation of existing residential areas.

Conclusion

This case study outlines an integrated methodology whereby different methods of stormwater collection, treatment and storage in the aquifer, followed by a range of distribution systems to end-users, can be developed. This integrated management system provides flexibility to account for the different collection storage and distribution conditions across the Adelaide metropolitan area while still allowing a high level of interconnectivity. Such a system makes use of the storage capacity of the Tertiary aquifer by using current and innovative technologies developed and employed in the City of Salisbury and elsewhere in the world, and will be economically viable.

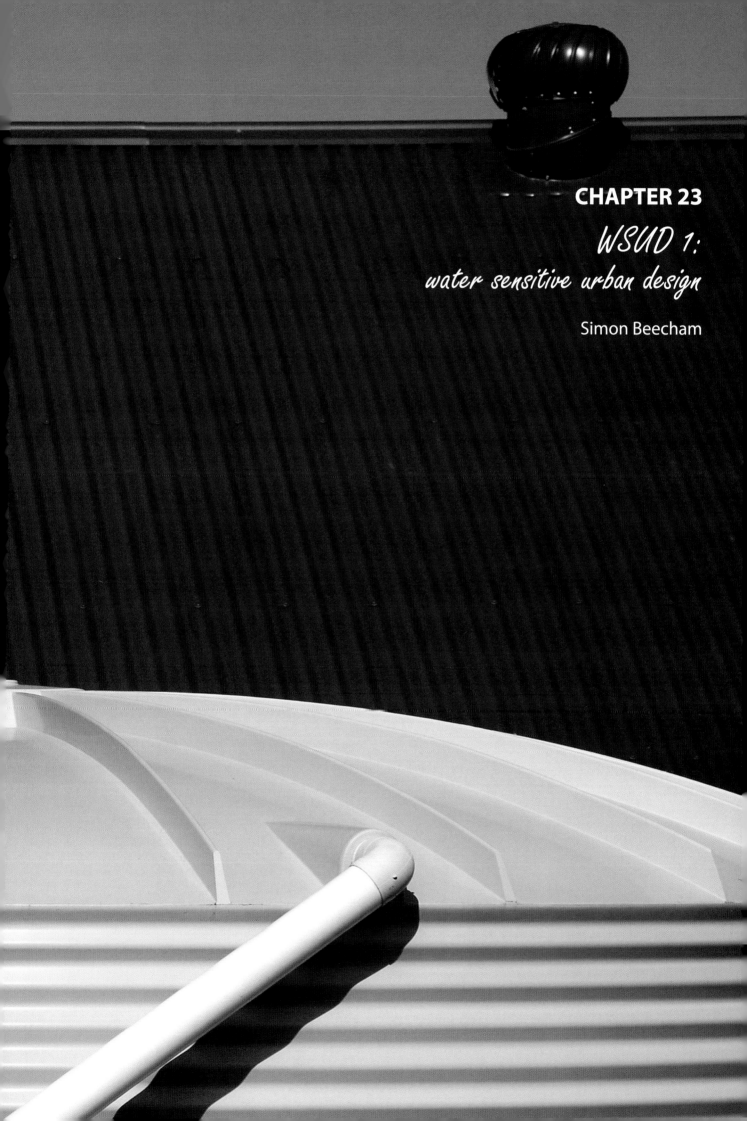

CHAPTER 23

WSUD 1:
water sensitive urban design

Simon Beecham

Introduction

Water sensitive urban design (WSUD) is a new way of managing urban water that has evolved from advances in engineering, biotechnology and environmental management. This chapter looks at the development of WSUD as a technology and some of the current impediments restricting its wider application. Available standards and guidelines, as well as lifecycle costs and the future directions of WSUD are also discussed.

Traditional potable water supply, wastewater and stormwater drainage infrastructure systems in Adelaide are based on European experiences. These engineered water supply, wastewater and stormwater systems are efficient at collecting and transporting water but they constitute a profound change from the natural watercycle.

Being located in the driest inhabited continent, most Australian towns and cities rely on importing and storing enormous volumes of freshwater from rivers, groundwater and dams. At the same time, large volumes of stormwater and wastewater are discharged, unused, from the same urban areas. Figure 23.1 compares Adelaide's water use with Australia's two largest cities, Sydney and Melbourne. In all of these cities, the combined volumes of discharged stormwater and wastewater greatly exceed the amount of water imported into the metropolitan area for water supply.

For both Melbourne and Sydney, about three-quarters of the water imported into the cities is discharged as wastewater effluent; in Adelaide the wastewater discharge is around half of the imported water volume. While wastewater is treated before being discharged to coastal waters or rivers, the majority of stormwater from these cities runs off into receiving waters without any quality improvement.

Now with Australia's rapidly growing urban population, there is an increased demand for new infrastructure, together with an increasing pressure on the current ageing infrastructure and receiving waters. Many of Australia's rivers, lakes and coastal waters are degraded due to the effects of urban stormwater infrastructure and runoff. Major changes are needed in the way urban landscapes in Australian cities are designed, and how water, in all its forms, is managed.

Problems associated with urban stormwater runoff

As urban areas develop in a catchment, the area of impervious surfaces, including roadways and roof areas, tend to increase. One of the many effects of this is an increased amount of stormwater runoff.

It is a difficult task to estimate the volume of stormwater that runs off the Adelaide metropolitan area every year because of the impact of rainwater tanks and other storages. The annual runoff changes from year to year but recent estimates range from 80 to 150 GL/year. This huge volume of runoff picks up a variety of pollutants from our roofs, roads and other urban surfaces, washing them into Adelaide's creeks and eventually into the ocean.

Conventional stormwater drainage systems, often referred to as 'pit and pipe' or 'conveyance systems', practise a 'catch and convey' approach to stormwater management. They are designed to achieve only the primary aim of stormwater management, which is flood prevention. They 'catch' or drain the rapidly developing overland runoff from the surface of an urban catchment and 'convey' it to natural streams and watercourses, where it flows away from the catchment area. This inevitably leads to pollution of receiving waters.

Although demonstrably successful in preventing flooding, such an approach to stormwater management has proved problematic in the broader sense. First, simply disposing of stormwater runoff into urban streams via engineered systems significantly alters the natural hydrology of a catchment. Second, urban stormwater is known to contain a variety of pollutants that are major contributors to the degradation of streams, wetlands and their accompanying habitats. Third, the current water shortage in Australia is increasingly permeating the consciousness of the public, and simply disposing of stormwater runoff is generally no longer acceptable to the community.

The need for stormwater reuse

The Adelaide community is now more aware of the importance of resource conservation and reuse, especially regarding water. This may be due to water restrictions and the proposal of more complex water supply solutions including desalination and new dam construction. It is evident that treating stormwater as a waste product to be disposed of quickly into urban streams is becoming a less acceptable stormwater management option.

In 2006, Australia's prime minister John Howard identified water use as 'our greatest conservation challenge'. He went further to say:

> … we need nothing short of a revolution in thinking about Australia's urban water challenges. Two assumptions have dominated water infrastructure in our cities. The first is that water should only be used once. The second

Water use, wastewater and stormwater

Gigalitres

600

400

200

0

Adelaide Sydney Melbourne

☐ **Water use** ▨ **Wastewater** ■ **Stormwater**

Figure 23.1 Water use, stormwater and wastewater flows (sewage) in Adelaide, Sydney and Melbourne.

Urban redevelopment projects

BOX 95

Recent years have seen the incorporation of integrated stormwater management systems into a number of high-profile urban redevelopment projects. Early projects such as Lynbrook Estate, a demonstration project by VicUrban in 1997, incorporated water sensitive urban design (WSUD) initiatives in outer-suburban housing estates, delivering both environmental benefits and marketing advantages through the creation of 'green streets' and water features. Since then a number of innovative projects have been implemented in more highly urbanised inner-city settings.

Docklands, Melbourne

The Docklands project comprises the redevelopment of the 200-ha Victoria Harbour precinct adjoining the Melbourne CBD. Commenced in 2002 by the former Docklands Authority, Docklands is a leading example of innovation, incorporating WSUD practices into a large, contemporary urban development. Key stormwater management initiatives include treating runoff to reduce pollutant loads entering the Yarra River and Port Phillip Bay, and the harvesting and reuse of stormwater for irrigation of public spaces. An innovative stormwater management system is integrated into Docklands' public spaces. Above-ground constructed wetlands, designed as contemporary landscape features, treat runoff and act as WSUD signature elements. Below ground, an innovative system of storage and 'green pipes' facilitate the movement of water around the site and store it for local irrigation reuse. Streetscape-scale bioretention systems are also part of the wider stormwater management system.

Street tree bioretention pits have been incorporated into the design of the extension of Bourke Street into Docklands, possibly the first street tree stormwater treatment application in Australia. Bioretention systems have also been integrated into the streetscape design of Batmans Hill Drive, intended as a high-profile southern gateway to Docklands, with high-quality and well-defined vehicular, pedestrian and cycle environments. The road is lined on one side with a linear bioretention swale and on the other with individual tree pits. The swale is at road level, protected from traffic by a row of bollards. The individual tree pits are defined by raised, semi-circular elements, which protect pedestrians from vehicular traffic, take up the grade change between footpath and soil level, act as seating, and provide a sculptural element in the streetscape. The project is an example of best practice in achieving multiple objectives through an integrated design.

Another innovative feature in Docklands are the bioretention basins integrated into the design of the Docklands Point carpark, which treats carpark runoff. The bold design, incorporating a gabion basket to take up the grade change from the surrounding surfaces, fits well into the surrounding formal environment.

Victoria Park, Sydney

In Sydney, the Victoria Park project comprises the redevelopment of 25 ha of former industrial land, three kilometres south the Sydney CBD. The low-lying, flood-prone land was originally a swamp and a natural wetland system. The initial masterplan was modified in 1999 to better integrate a water management system into the public domain of streets and open spaces. The scheme by Hassel, in association with the Government Architects Office and consultants Ecological Engineering, built upon the inherent physical constraints of the site. An innovative community water management system was developed for the whole site, which aimed to more closely mimic the original natural water processes by retaining, filtering, moving and recycling stormwater across the site in a controlled fashion. These processes have been successfully integrated into a high-quality public domain, in which the watercycle is made visible. Victoria Park has become a benchmark for inner-urban development. Its success can be attributed to the strong collaboration between the developer, designers, artists, engineers and ecologists.

The wide, east–west streets of Victoria Park were reconfigured to drain to a central bioretention swale, with permeable 'saw tooth' kerbing. The swales were designed to capture the most contaminated first-flush stormwater, which filters to an underground storage tank for reuse. Excess flows are captured by a system of inlets below the pedestrian bridges over the swales, then conveyed to dished detention basins in the local parks. Runoff is further treated by a system of constructed wetlands integrated into public spaces, and then recycled for irrigation use. The pumping system also draws irrigation water from the aquifer during low rainfall periods.

Conclusion

WSUD initiatives are now being integrated into the design of inner-city urban redevelopment projects in the form of integrated stormwater management systems. As well as achieving the desired environmental outcomes, such an approach also creates a high-quality public realm for the benefit of both residents and visitors to these cities.

Martin Ely

BOX 96

Retrofitting exsiting streets with water sensitive urban design

Water sensitive urban design (WSUD) can help us achieve 'green streets' in times of water restrictions. Stormwater runoff that usually flows down our drains and out to sea can instead be harvested to irrigate street trees and other landscaping. At the same time, runoff can be filtered and cleaned of pollutants before being returned to aquatic ecosystems.

Streetscape-scale applications

In recent years WSUD treatments have evolved from large-scale 'end-of-pipe' solutions, such as constructed wetlands, to smaller-scale applications which can treat runoff from local catchments 'at source' and be integrated into the design of urban streets and public spaces. These smaller-scale applications deliver a range of benefits; as well as protecting downstream waters through pollutant removal and slowing stormwater flows, they can harvest runoff for local landscape reuse, reducing the use of mains water for irrigation. They also help increase awareness of the connections between human activities and the watercycle by making the processes more visible.

Bioretention system applications can be integrated into the design of new streetscapes, or 'retrofitted' into existing streets. They offer opportunities for innovative and creative design solutions and considerable design flexibility, being adaptable to many scales and spatial configurations. Streetscape-scale WSUD applications can take a number of forms, however, the most popular are bioretention systems, also known as 'rain gardens', which can be scaled to confined spaces and adapted to a range of urban situations. Bioretention tree pits are a recent innovation, involving the redesign of tree root zone environments as stormwater treatment systems, allowing the incorporation of stormwater management into confined urban street spaces.

Metropolitan Melbourne

Melbourne is recognised as a leading city in urban stormwater management, in which WSUD best practices have become institutionalised. The process has benefited from a number of 'champions' including Melbourne Water, local councils and environmental engineering consultants. The process began in the late 1980s with recognition of the impacts of urban runoff on the Yarra River Catchment and Port Phillip Bay. An early initiative was the 1997 development of an outer-suburban subcatchment-scale demonstration project at Lynbrook Estate by VicUrban, which incorporated a stormwater 'treatment train' at streetscape level. Since then a number of innovative streetscape-scale projects have been implemented in more highly urbanised settings, in partnership between Melbourne Water and local councils.

Kingston City Council

Kingston City Council is recognised as a leading organisation in Victoria for advancing WSUD. The council's 1380 ha of road surfaces discharge large volumes of stormwater runoff and pollutants into Port Phillip Bay. Funding from the Victorian Stormwater Action Program was the catalyst for the construction of a trial infiltration trench in Riviera Street, Cheltenham, in 2001. The council has now installed some 84 rain gardens in 11 suburban streets across eight suburbs. These have been retrofitted into existing residential street verges or traffic calming devices as part of the council's innovative road reconstruction program.

In Stawell Street, Mentone, bioretention basins were installed in the street's wide verges during street reconstruction in 2004. The basins were designed to fit into the available space between existing mature street trees and driveways. In Brisbane Terrace, Mordialloc, rain gardens were incorporated into traffic calming outstands being installed in response to local safety issues. The project has been a success, with the heavily planted installations creating a green landscape, which is appreciated by local residents. The project is also a good example of a multifunctional project that achieves aesthetic, water quality and traffic calming outcomes.

Due to its pioneering role, Kingston City Council has been able to review its constructed projects and learn from early mistakes. Key findings have included the need for community acceptance through appropriate design and resident education, as well as the need for long-term maintenance.

Lower Yarra councils

Melbourne Water, in partnership with local councils, has implemented a number of other streetscape-scale demonstration projects aimed at improving the health of the Yarra River. The 2006 Yarra River Action Plan allocated

$10 million to implement WSUD projects in the four lower Yarra River councils.

In the City of Yarra, the 2003 reconstruction (incorporating rain gardens) of Cremorne Street, Richmond, was the first example of streetscape-scale applications in a highly built-up commercial area with heavy traffic. The project was jointly funded by council and a Victorian Stormwater Action Program grant. Rain gardens were installed in outstands in the parking lane. The change in level between the footpath and filter bed was addressed with post and wire fencing, however, deliberate and accidental damage has reduced the aesthetic value of the streetscape. The Yarra City Council also has a program of installing WSUD treatments during the periodic renewal of city assets, such as the incorporation of rain gardens in traffic calming devices at the intersection of Napier and Kerr streets.

The City of Stonnington has adopted a Sustainable Water Management Strategy aimed at incorporating WSUD elements into all its facilities. Bioretention tree pits were installed in the 2007 upgrade of Glenferrie Road, Malvern, an intensively used shopping street. The design comprises a tree pit integrated with a grated kerb inlet, grated tree pit cover, and tree guard.

In Orchard Street, Armadale, rain gardens have been incorporated into the reconstruction of a suburban street to help reduce flooding and treat road runoff. The materials and plantings chosen complement the surrounding contemporary streetscape.

City of Melbourne

A number of high-profile bioretention projects have been installed in the City of Melbourne, in partnership with Melbourne Water. In 2006, bioretention tree pits were installed in Little Bourke Street in the heart of the Melbourne CBD. The project is of significance as the first example of a WSUD application in such a highly urbanised area. The project faced the challenge of integrating the design into the infrastructure and functioning of the narrow, busy street, and also meeting the high standards of aesthetics and urban design

required in the Melbourne CBD. Particular challenges included the need to relocate underground services and also to achieve the required infrastructure to connect with the existing stormwater system. The tree pits, installed in the footpath, are covered with a heavy-duty bluestone cover, hinged to allow opening for the removal of accumulated litter, with the sunken tree pits also acting as litter traps.

Bioretention street tree pits were also installed in residential Acland Street, South Yarra, in 2007. The street trees are located in the parking lane rather than the footpath, with runoff flowing directly from the road surface into the grated tree pit cover. Each tree is protected from traffic by a pair of bollards. This installation is part of the council's approach to the greening of city streets, incorporating WSUD measures into street tree planting to achieve water quality objectives while at the same time reducing watering requirements.

In 2006 an innovative rain garden was installed in Davidsons Place in the Melbourne CBD. Davidsons Place is a narrow, high-walled laneway with a hard-surfaced, low-amenity environment but with a growing residential population. The opportunity was taken to install a rain garden that would improve the quality of urban runoff, while creating urban greenery and a gathering space for residents. The rain garden is comprised of a raised planter bed at the end of the lane, which collects stormwater from surrounding roofs via downpipes and spreaders. The water is then treated by being filtered through the raised garden bed before returning to the stormwater system, and at the same time irrigating the single tree and underplantings. The raised bluestone wall, built at seating height, also provides a gathering place for local residents.

This overview identifies the scope of emerging applications of WSUD at the streetscape level, through retrofitting of existing streets and thus creating incremental changes to the urban fabric. Lessons have been learnt from early projects, which were largely engineering driven. The most successful recent projects exhibit a highly collaborative and innovative approach, achieving a range of social, economic and environmental outcomes.

Martin Ely

is that stormwater should be carried away to rivers and oceans as quickly as possible. Neither assumption is suited to a world in which we should judge water by its quality and not by its history. Compared with desalination, for example, water recycling and capturing stormwater have much more to recommend them as strategies for solving Australia's urban water challenges.

It is clear that among the alternative water supply options under discussion, stormwater reuse is being considered. Both large- and small-scale systems for stormwater reuse are being pursued to alleviate the Australian 'water crisis'.

Water sensitive urban design

The term water sensitive urban design (WSUD) was first referred to in the early 1990s in various publications exploring concepts and possible structural and non-structural practices in relation to urban water resource management. Parallel movements, such as Sustainable Urban Drainage Systems (SUDS), were also developing in Europe and the United States. SUDS are now generally referred to as SuDS to reflect the wider application of *Su*stainable drainage systems. Both WSUD and SuDS embrace the concept of integrated land and water management and in particular integrated urban watercycle management. This includes the harvesting and/ or treatment of stormwater and wastewater to supplement (normally non-potable) water supplies.

The first objectives for WSUD in Australia were articulated by Whelans Pty Ltd in 1994 as being:

I. To manage a water balance by:
   • maintaining appropriate aquifer levels, recharge and streamflow characteristics in accordance with assigned beneficial uses;
   • preventing flood damage in developed areas; and
   • preventing excessive erosion of waterways, slopes and banks.
II. To maintain and, where possible, enhance water quality by:
   • minimising waterborne sediment loading;
   • protecting existing riparian or fringing vegetation;
   • minimising the export of pollutants to surface or groundwaters; and
   • minimising the export and impact of pollutants from sewage.

III. To encourage water conservation by:
   • minimising the import and use of potable water supply;
   • promoting the reuse of stormwater;
   • promoting the reuse and recycling of effluent;
   • reducing irrigation requirements; and
   • promoting regulated self-supply.
IV. To maintain water-related environmental values.
V. To maintain water-related recreational values.

WSUD represents a modern approach to the management of urban water. Broadly, it seeks to implement the principles of ecologically sustainable development with respect to water resource management. WSUD focuses on the interaction between the urban built form and the natural watercycle. It may be regarded as an alternative to the aforementioned 'catch and convey' approach to stormwater management.

Water sensitive urban design technologies

So what are typical WSUD components? For stormwater systems, Beecham[1] includes rainwater tanks, grassed swales, biofiltration swales, bioretention basins, sand filters, infiltration trenches and basins, vegetated filter strips, permeable pavements, wetlands and ponds.

WSUD embraces the concept of integrated land and water management, in particular integrated urban watercycle management. WSUD is a natural systems based approach comprised of controls in 'treatment trains', which use a combination of source, conveyance and discharge controls. Controls used in WSUD treatment trains employ the use of primary, secondary and often tertiary treatment devices (see Table 23.1).

Source controls are used to minimise the generation of excessive runoff and pollution of stormwater at or near its source. Source controls are located at the individual building lot and can apply to single dwellings, multi-storey apartment complexes, or commercial and industrial premises. Source controls are typically a combination of primary and secondary treatment devices. It is now well documented that physical trapping of gross pollutants, coarse sediments and sedimentation should occur before any biological or chemical process can be applied effectively.

Conveyance controls are used in transporting the flows from the allotments, roads and public open spaces. These controls

Table 23.1 Water treatment devices

Primary	Secondary	Tertiary
rainwater tanks	grass swales	wetlands
litter baskets	filter strips	ponds
racks and booms	infiltration trenches and basins	sand filters
sediment traps	porous pavements	
gross pollutant devices	extended detention basins	
catch basins		
oil and grit separators		

Table 23.1 Typical primary, secondary and tertiary treatment devices.

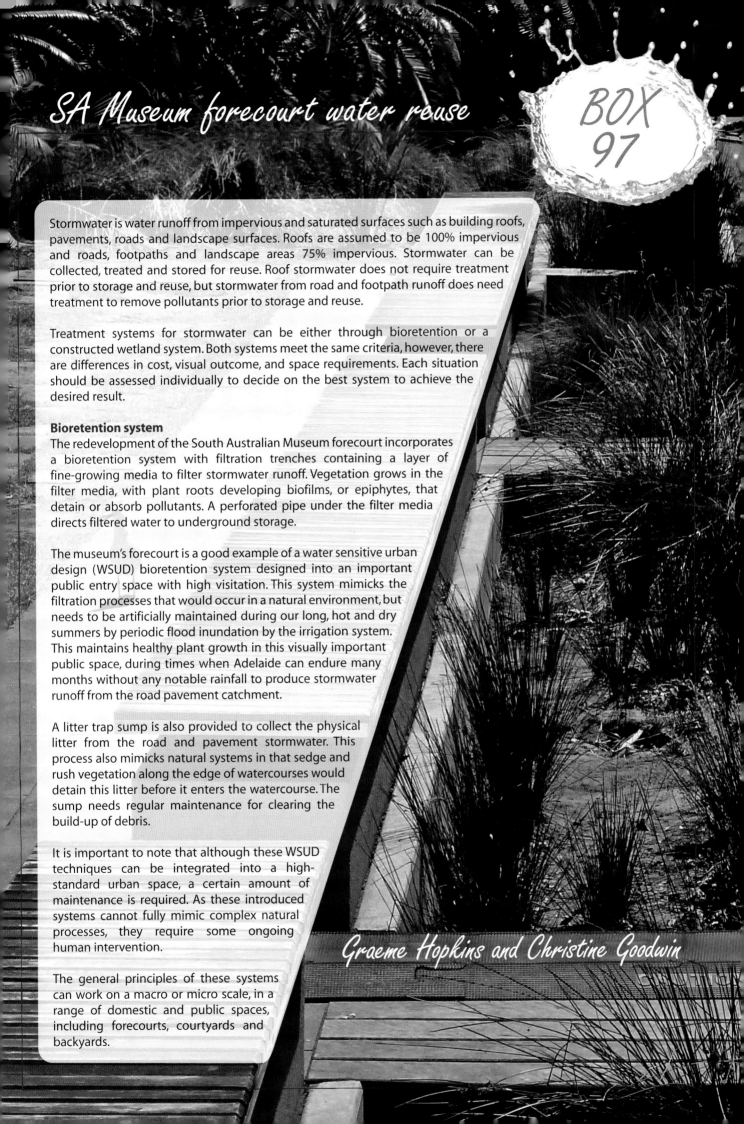

SA Museum forecourt water reuse

BOX 97

Stormwater is water runoff from impervious and saturated surfaces such as building roofs, pavements, roads and landscape surfaces. Roofs are assumed to be 100% impervious and roads, footpaths and landscape areas 75% impervious. Stormwater can be collected, treated and stored for reuse. Roof stormwater does not require treatment prior to storage and reuse, but stormwater from road and footpath runoff does need treatment to remove pollutants prior to storage and reuse.

Treatment systems for stormwater can be either through bioretention or a constructed wetland system. Both systems meet the same criteria, however, there are differences in cost, visual outcome, and space requirements. Each situation should be assessed individually to decide on the best system to achieve the desired result.

Bioretention system

The redevelopment of the South Australian Museum forecourt incorporates a bioretention system with filtration trenches containing a layer of fine-growing media to filter stormwater runoff. Vegetation grows in the filter media, with plant roots developing biofilms, or epiphytes, that detain or absorb pollutants. A perforated pipe under the filter media directs filtered water to underground storage.

The museum's forecourt is a good example of a water sensitive urban design (WSUD) bioretention system designed into an important public entry space with high visitation. This system mimicks the filtration processes that would occur in a natural environment, but needs to be artificially maintained during our long, hot and dry summers by periodic flood inundation by the irrigation system. This maintains healthy plant growth in this visually important public space, during times when Adelaide can endure many months without any notable rainfall to produce stormwater runoff from the road pavement catchment.

A litter trap sump is also provided to collect the physical litter from the road and pavement stormwater. This process also mimicks natural systems in that sedge and rush vegetation along the edge of watercourses would detain this litter before it enters the watercourse. The sump needs regular maintenance for clearing the build-up of debris.

It is important to note that although these WSUD techniques can be integrated into a high-standard urban space, a certain amount of maintenance is required. As these introduced systems cannot fully mimic complex natural processes, they require some ongoing human intervention.

The general principles of these systems can work on a macro or micro scale, in a range of domestic and public spaces, including forecourts, courtyards and backyards.

Graeme Hopkins and Christine Goodwin

use a combination of primary and secondary treatment devices. Discharge controls occur at the point where stormwater leaves the total site or catchment. These controls usually employ a combination of secondary and tertiary treatment devices.

Rainwater tanks

Rainwater is usually collected from roofs and stored in tanks for both indoor and outdoor uses. The use of the rainwater determines the water quality requirements and treatment train required. Subject to prefiltering of the first flush and appropriate treatment systems, water quality that meets the national health guidelines can be achieved for both potable and non-potable uses.

Bioretention and biofiltration systems

Bioretention basins are relatively small-scale vegetated WSUD systems with functionality to harvest, filter and purify stormwater through filter media and underground drainage, before either storing for reuse or discharging the stormwater into receiving waters. These installations are mostly located along streets or in carparks, or integrated with traffic islands in streetscape designs. As bioretention basins can be of different sizes and shapes, they have the flexibility to be easily included in locations within urban developments as relatively newly defined land uses. If stormwater conveyance is incorporated into the design, the system is often referred to as a 'biofiltration swale'.

The development of WSUD in the Australian context is rapidly changing the appearance of urban landscapes at small and large scales with the appearance of relatively new vegetated stormwater systems. This also changes the nature of traditional green spaces in our towns and cities, positively influencing biodiversity and habitat values. Due to the lack of guidelines to inform design and management of vegetated WSUD systems for biodiversity improvements, current practice lacks understanding of the ecological performance of these systems and the possible solutions for their future ecological sustainability.

At the University of South Australia, research is underway to investigate how to design and manage these systems to optimise their capacity as ecologically sustainable urban landscapes. Using terrestrial invertebrates as biodiversity indicators, this work aims to evaluate the influence of some habitat factors on the ecological and biodiversity values of bioretention basins and their corresponding urban green spaces. Distribution patterns of invertebrate assemblages within the systems are also being investigated.

Grass swales

Grass swales are grass-lined channels for conveying runoff from roads and impervious areas, which can be used as an alternative to or in combination with a conventional kerb and gutter system. The main advantage of grass swales is that they decrease stormwater flow velocities; they also settle out suspended particulates due to the lower velocities, and the grass acting as a filter.

Filter strips

Filter strips are grassed or vegetated areas that treat overland flow by removing sediments, particulate matter and associated pollutants like nutrients, metals and bacteria. The use of filter strips to reduce particulates and associated pollutants has shown varied results, despite similar parameters used during experimental testing. Parameters such as length of filter strip, slope angle, vegetative characteristics, catchment characteristics, and runoff velocity are all recognised as factors that affect the pollutant-removal efficiency of filter strips.

Litter baskets, racks and booms

A litter basket is a wire, plastic or geo-fabric collection device installed within a stormwater pit that collects litter, organic material and coarse sediments either from a paved surface or from the piped stormwater system. Litter baskets and pits require regular maintenance, particularly after storm events and in the autumn months. When full, these devices reduce pit inlet capacities and cause flooding upstream. This also remobilises any caught material.

Litter racks are a series of metal bars located perpendicular to flow across a channel or pipe. These devices capture large-sized litter, organic matter and debris and are usually used upstream of other secondary or tertiary controls. Sediment is often trapped upstream of the rack. Litter racks require regular maintenance, particularly after storm events, to eliminate potential flooding upstream due to blockages, pollutant breakdown, odours and remobilisation. These devices are often difficult to clean.

Litter booms are floating devices with mesh skirts attached and weighted below the float to capture floating litter and debris. These devices are usually placed downstream of outlets within channels or creeks. These devices require frequent maintenance particularly after storm events and are also difficult to maintain as they are usually only accessible by boat.

Porous and permeable pavements

Pavements are an intrinsic but seldom thought about part of life, particularly in urban areas. However, for developers, industrial facilities, and local authorities addressing stormwater and associated water quality guidelines and regulations, pavements stay very much at the forefront of planning issues. Pavements designed for use by vehicular traffic typically consist of a subgrade, one or more overlying courses of compacted pavement material, and a surface seal. An integral aspect of conventional pavement design involves preventing entry of water to the pavement via the seal to protect the integrity of the underlying base course, sub-base and subgrade.

Conversely, permeable pavement has quite different objectives and design requirements. The joints between the surface pavers are not sand-filled and are designed to infiltrate stormwater through to the underlying layers. Water passes to the open-graded single-sized gravel sub-structure, and is drained through the subgrade. The pavements therefore perform a dual function – supporting traffic loads

BOX
98

Permeable pavements

Two permeable pavement parking bays were constructed along Kirkcaldy Avenue, Grange, in 1999. The bays were designed to collect, treat and infiltrate runoff generated on the roadway and the parking bays themselves. The scheme reduces both storm runoff (peak flow and volume) and pollution conveyance to downstream waterways.

- The catchment consists of the two parking bays and approximately 90 m of Kirkcaldy Avenue carriageway (limited to half the road pavement); a total of 650 m^2.

- The pavement comprises:

  - permeable pavement blocks (80 mm deep);
  - no sand jointing between pavers;
  - 6 mm screenings (50 mm depth); and
  - 20 mm aggregate-base course, surrounded by geotextile fabric (150 mm depth).

- Runoff from the upstream road surface drains to the kerb and gutter and is directed to the permeable pavement surface. It then infiltrates through this surface and fills the storage voids in the pavement base course material before being slowly infiltrated into the underlying native soil.

- The pavement is designed to retain all runoff generated from the catchment area during storm events up to and including a five-year average recurrence interval (ARI) event. This translates to the capture of more than 95% of all stormwater runoff.

- In dry periods between storms, water stored in the pavement infiltrates into the underlying soil, making this storage available for the next event.

- In large storms (>5 year ARI) short-term ponding of up to 40 mm in depth occurs on the pavement. Excess runoff continues past the parking bays and enters the existing stormwater pipe network via side entry pits at the northern end of Kirkcaldy Avenue.

- Allowance is made in the design for a 90% reduction in infiltration capacity over the 20-year design life. To avoid exceeding this level of blockage, it is recommended that the permeable surface is cleaned twice a year with a mechanical suction brush.

- Pollutants carried in the stormwater are contained on site, the majority being trapped within the pavement layers. This serves to improve water quality in the nearby creek system.

- The system provides increased soil moisture, which is available to trees and grass in the vicinity of the parking bays.

David Pezzaniti, John Argue and Simon Beecham

and stormwater drainage. Pollutants within the stormwater infiltrate the pavement, with the majority being trapped within the pavement layers.

In Australia, permeable pavement systems are an emerging technology; there are few installations more than 10 years old. Urban stormwater runoff contains significant concentrations of suspended sediments and gross pollutants. As such, there is a perception that permeable pavements used as source control devices, and designed to infiltrate runoff, will tend to clog quickly, resulting in high maintenance and replacement costs. Permeable pavements are therefore most suited for use in developed areas, as runoff from fully developed catchments is likely to contain lower levels of sediment loading due to reduced construction activity.

Both porous and permeable pavements are onsite stormwater management technologies, applied in a similar manner to conventional pavement surfacing.

The key advantage of porous and permeable pavement is that they allow for the retention or detention of stormwater onsite without compromising the amenity of above-lying land. They also introduce a variety of potential water quality treatment processes to intercept the contaminants contained in urban stormwater runoff prior to infiltration.

In terms of terminology, 'permeable pavement' refers to the application of impervious materials (such as individual pavers), which are physically shaped and/or arranged to allow infiltration through gaps in pavement material blocks themselves, such as deliberate indentures into the paver ends. Individual gaps between pavers when they are laid are clearly visible.

'Porous pavements' are made using materials that will allow water to pass through the pavement, based on the properties of the material itself. They are generally made of concrete or asphalt products with no or minimal 'fines' (fine particles) resulting in a network of continuously linked voids to allow the passage of fluids through the pavement surface.

A typical porous or permeable pavement construction is depicted in Diagram 23.1. After excavating to the total desired depth, a suitable base course material is laid. This must be a free-draining material that will provide structural support. In some applications there may be multiple base course materials applied. Depending on the structural demands, this layer is usually in the order of 300 mm in thickness and consists of 20–40 mm aggregate.

On top of the base course, a geotextile is applied. Although not essential, it provides an additional element of filtration where water quality is of concern. It also provides a distinct layer separating the base course from the bedding course. A bedding course is placed on top of the geotextile. This layer, of 3–7 mm aggregate, is approximately 30 mm thick in most applications. Pavers may be placed directly on top of this layer. Sand is not recommended for most applications due to a susceptibility to clogging.

Water reuse

Rainwater tanks are widely used for harvesting and reusing stormwater. However, residents in Adelaide are used to the long dry periods that occur between successive storm events. This means that large tanks are required to secure a reliable supply of rainwater. As more and more houses are constructed in Adelaide, average lot size has decreased. Quite often two houses are built on lots that used to contain a single dwelling. This means that the available space for siting a large rainwater tank is becoming harder to find.

A solution to this dilemma is to place the storage underground. Porous and permeable pavements present a unique opportunity to harvest and store urban stormwater that would otherwise contribute to overland runoff into the conventional stormwater pipe and channel network. With minimal surface infrastructure, porous and permeable pavements provide a serviceable, hard-standing area that facilitates water harvesting, treatment and reuse.

There are several options for the design and construction of such a system. After infiltrating through the pavement surface, the stormwater can be stored in a submerged tank, or in proprietary plastic cell systems. It can also be stored in a matrix of base course aggregate contained within an impermeable membrane. This latter and arguably more cost-effective concept is shown in Diagram 23.2.

Researchers at the University of South Australia recently completed construction of a conceptual prototype facility featuring a pavement the size of a standard car space (in accordance

Porous or permeable pavement cross-section

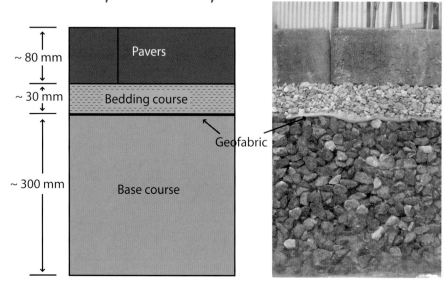

Diagram 23.1 Typical cross-section of a porous or permeable pavement construction.

with appropriate standards: AS/NZS 2890.1) Storage is provided within a limestone base course aggregate material. Excavations were undertaken according to plan, ensuring adequate water storage volume was available. Laboratory tests indicated that voids in the base course aggregate material would be around 40% of the overall sub-base volume. This initial testing suggested the overall storage of this facility is approximately 3600 L. The stormwater harvested from this system is used to irrigate plants in a nearby plot of land.

In constructing the pavement, a procedure similar to the construction of a standard pavement was followed. The excavation varied between 650 mm and 750 mm depth overall, sloping downward to a sump at one end, where a submersible pump was located. A 0.75 mm polypropylene material was used to line the excavation. To access the stored water, a stormwater pipe was installed prior to filling. The pipe was perforated at sufficient intervals to allow the ingress of water while preventing the ingress of aggregate stones to the submerged pump reservoir. Following the placement for the storage access pipe, 14 t of 20 mm dolomite aggregate was placed on top of the liner and compacted using a vibrating plate compactor. Permeable and porous pavers are laid in a similar fashion to conventional concrete block paving units: the upmost layer of aggregate (the 'laying course') is screeded to a desired level and pavers are placed according to the manufacturer's instructions. The pavement itself is then compacted with a vibrating plate compactor. The installation of appropriately designed edge restraints completes the construction.

A standard solution

There is no Australian standard that deals explicitly with WSUD; even the 'Australian Standard for Stormwater Drainage' (AS3500.3.2), which deals with property drainage, does not mention WSUD explicitly. As a precursor to standards, we are left with guidelines and policies. John Argue, from the University of South Australia, is coauthor and chief editor of one of the first Australian WSUD manuals, 'Water Sensitive Urban Design: Basic Procedures for "Source Control"'.[2] There are now also WSUD guidelines and manuals available for most cities in Australia. Various states and territories also have their own water recycling strategies.

However, Sharon Beder of Wollongong University[3, 4, 5] warns us that standards and legislative requirements, while necessary, are insufficient to ensure ecological sustainability. In particular, standards and criteria tend to shift responsibility for designing environmentally sound engineering works away from the engineers (who can feel satisfied they have discharged their responsibilities if they meet the standards, whether or not those standards are sustainable). Standards tend to be based on what has been economically achieved before, whereas engineers can push open the boundaries of achievement.

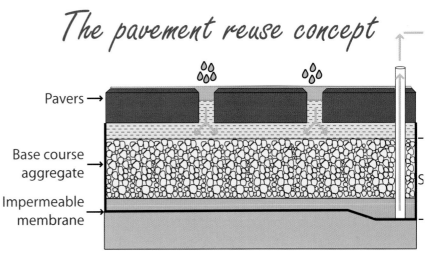

Diagram 23.2 A porous or permeable pavement cross-section with underground stormwater storage.

Costing the earth

Economic considerations have also traditionally been an integral part of engineering design and a force for refinement and sophistication of design methods. But many engineers find difficulty in viewing the environment as a set of resources for the economic system. Not so the Commonwealth government economist David James,[6] who argues:

> Any economist will tell you that the environment really ought to be classified as a resource; hence it should fall fairly and squarely into the policy arena concerning the allocation of resources. Management of the environment is essentially an economic problem.

Parallels can be drawn with society's underpricing of natural resources such as water and sand. Economists argue that environmental degradation has resulted from the failure of the market system to put any value on the environment, even though the environment does serve economic functions and provides economic and other benefits. It is argued that because environmental 'assets' are free or underpriced, they tend to be overused or abused, resulting in environmental damage. Because they are not owned and do not have price tags, there is no incentive to protect them. This is a view shared by business people. The Business Council of Australia[7] argues that it is not economic growth that is the real problem:

> Rather, it is that important environmental assets tend not to be priced in a market like other assets. These assets are common property – they belong to everybody, and to nobody. Without ownership rights there is not the incentive for any person or group to look after them properly ... if the environment has a zero price to users it will eventually be used up.

In order to fix this perceived problem, sustainable development involves putting a price on the environment and charging people to use it, privatising the commons, and creating artificial markets and price mechanisms through economic instruments and tradable rights to pollute. The idea is that 'the power of the market can be harnessed' to [achieve] environmental goals. However, Sharon Beder

Table 23.2 Water supply costs

Supply method	Average incremental cost[1] ($/kL)
Desalination	3.1 to 4.7
Rainwater	2.53 to 4.72
Greywater	2.65 to 3.47
Treated wastewater	1.38 to 2.34
(current excess water rates)[1]	(0.6 to 1.00)[2]

1. Indicative comparative only (Connell Wagner 1993 inflated to 2002 values).
2. The cities of Brisbane, Newcastle and Sydney all fall within this range.
Table 23.2 Indicative water supply replacement costs ($ per kilolitre). Source: L. Howells, N. Collins and M. Babister, Institution of Engineers Australia (Sydney Division) Water Engineering Panel submission to the NSW Parliamentary Inquiry on Urban Water Management, 2002.

argues that while such an approach means more account will be taken of the environment, at the same time there will be a devaluation because it brings the environment down to the same level as other commodities that can be bought and sold. The environment's value is reduced to an economic value and it is then treated as a substitutable part of the economic system. The imperative that environmental deterioration might once have had for social and political change has been dissipated by the cleverly constructed notion of sustainable development.

Despite Beder's opinion, engineers persist using rational economic assessment techniques such as lifecycle cost-analysis to evaluate WSUD and conventional drainage systems.

Lifecycle costs for water sensitive urban design
One of the challenges for WSUD, as mentioned earlier, is to develop clear costings. Table 23.2 shows indicative costs (in 2002 terms) for the various supply methods as an average incremental cost.

However, the current excess water rates do not generally include the full replacement cost for the infrastructure to produce the water supply. Looking only at separate storm-water drainage systems, Lloyd and Associates[8, 9] give the lifecycle costs for conventional downstream treatment, and distributed WSUD approaches (Table 23.3). The conventional approach represented underground concrete pipes discharging directly to an ornamental lake; the downstream treatment approach represented underground concrete pipes discharging to a stormwater treatment wetland system (includes litter trap, inlet zone and macrophyte zone), prior to entering an ornamental lake; and the distributed approach represented bioretention systems discharging to an urban wetland (only a macrophyte zone required), prior to entering an ornamental lake.

The lifecycle costs of rainwater tanks have been investigated in many studies; the figures vary from $1 to $14/kL with the majority being between $1 and $2/kL. Similarly, a number of studies have investigated the real cost of wastewater recycling at the household scale. Simon Fane from the Institute for Sustainable Futures in Sydney[10] investigated the theoretical relationship between the scale of an effluent

reuse system and the pathogen risks it presents at various effluent qualities. His results are shown in figures 23.2 and 23.3. Fane found that with the inclusion of pathogen risks as a costed externality, taking a decentralised approach to urban water reuse is advantageous in most cases. However, he recognised that exact evaluation of increased waterborne disease due to effluent reuse is problematic.

Challenges for the future

In Australia, the barriers hindering the wider adoption of WSUD include:

- perception of increased costs;
- lack of design guidance;
- lack of design experience;
- lack of available modelling tools;
- operation and management uncertainty; and
- adoption and ownership.

The past issue of a combined lack of design guidance, design experience and modelling tools is slowly being resolved. More guidelines are becoming available, as are modelling tools, which are discussed at length in this paper, and design experience is increasing with the number of WSUD projects undertaken. Effective WSUD is not easy to achieve and requires an innovative approach. Innovation, by definition, involves risk taking, and risk can occasionally lead to failure. However, it can also lead to success. Both success and failure are important components of experience. A perception of high costs also discourages designers from using WSUD. WSUD practitioners in Australia and their counterparts in the United Kingdom have described six key sustainability challenges for WSUD:

- Develop clear lifecycle costings of WSUD and traditional drainage systems, taking into account the full range of benefits, downstream implications and maintenance implications of each system, enabling holistic and comparative cost–benefit risk assessment.
- Increase awareness about multiple benefits of WSUD, which must include the multiple values they bring to society, ranging from flood storage, water quality treatment, as well as wildlife and amenity value.

BOX
99

Bioretention systems:
an unsealed carpark

A vegetated bioretention system was constructed for an unsealed carpark at the Mawson Lakes campus of the University of South Australia in 2005. The system was designed primarily to reduce the high levels of suspended solids generated from the surrounding unsealed surface, prior to discharge into the conventional drainage system. The vegetated strip also reduces the volume and peak flow of the runoff generated by the carpark, and promotes the reduction in vehicle associated pollutants and nutrients originating from the surrounding landscape.

- The catchment consists of approximately 1500 m² of carpark and surrounding footpath, located at the centre of the Mawson Lakes campus, and is used regularly.

- The unsealed surface is a major source of sediment during storm events, and water quality treatment was essential to minimise the quantity of suspended solid levels entering the local council stormwater drainage system.

- The parking areas are graded at a 2% slope towards a centrally located bioretention system.

- The bioretention system is gradually sloped towards a stormwater drainage pit, and was designed to remove the sediment using filtration and interception by the vegetation.

- The system consists of a 3 m wide, V-shaped channel filled with propagating sand and planted with a variety of shrubs and reeds. The channel is situated on top of a 600 mm wide by 500 mm deep infiltration trench.

- The trench is filled with 14 mm aggregate, separated from the surrounding soil by a woven geotextile. A 160 mm diameter slotted drainage pipe is located at the bottom of the trench, connected to the downstream drainage pit.

- In minor storms the runoff is retained by the system and provides a passive water source for the plants. Sediments and other pollutants are removed by infiltration through the substrata and the remaining clean water is collected by the subsurface drainage pipe.

- The retention function reduces the peak flow from the site, allowing for nutrient uptake by plants and microbial breakdown of any hydrocarbon pollutants.

- In large storms the system is designed for short-term ponding, which is controlled by the design of the drainage pit. Excess runoff exits the system through an overflow entry pit.

- Typical treatment efficiencies of bioretention systems are 90% removal for suspended solids, 80% removal for total phosphorus and 40% removal for total nitrogen.

David Pezzaniti and Simon Beecham

Table 23.3 Lifecycle costs for stormwater systems

Scenario	Capital cost	Annualised maintenance cost net present worth *
1. Conventional	$1,310,261	$410–$925 ($1–$2 per household)
2. Downstream approach	$1,824,823	$2815–$9430 ($6–$21 per household)
3. Distributed WSUD approach	$1,782,249	$2050–$7380 ($5–$16 per household)

* (60 yr LCC @ 10% & 4% discount rate)

Table 23.3 Comparison of lifecycle costs for conventional and WSUD stormwater systems. Source: S. Lloyd, T. Wong and C.J. Chesterfield, 'Opportunities and impediments to water sensitive urban design in Australia', proc. 2nd South Pacific stormwater conference, Auckland, 2001, pp. 302–309.

- Embody WSUD within appropriate legislation. It is no longer acceptable that the 'best practice' norm should be unsustainability, and so there is a need to revise relevant legislation (e.g. planning) such that WSUD measures are accepted as 'best practice'.
- Establish protocols for the adoption and maintenance of WSUD. This is essential to ensure their ready acceptance, and for all to be clear about maintenance requirements and responsibilities. Guidance needs to be clear, and to be appropriate for a range of situations. The wider benefits of WSUD and other incentives need to be established such that ownership and management of WSUD schemes rests with the organisations, both public and private, appropriate to the individual scheme.
- Increase funding to WSUD from other areas of public expenditure. WSUD addresses a range of problems and also delivers a range of benefits to different aspects of society – flooding and pollution control, wildlife, amenity and landscape. It will be important to create a mechanism for providing funding from existing streams of public expenditure that have not traditionally been associated with the single issue of urban water.
- Overcome the technical shortcomings of WSUD. It will be necessary to continuously improve WSUD design to deliver multiple benefits, and provide guidance and demonstration projects on their effectiveness in a range of situations.

Meeting the challenges

In South Australia, the Department of Planning and Local Government (DPLG) is currently engaged in the 'Institutionalising WSUD project', which aims to ensure water-friendly design will be incorporated into all of Greater Adelaide's suburbs, houses and commercial and industrial precincts and buildings. The project's objectives are to economically and safely:

- improve water quality;
- minimise the use of potable ('mains') water;
- maximise water reuse;
- reduce pollution;
- reduce peak flows and flood risk;
- restore and maintain water for the environment; and

- provide visual and potential recreational benefits in local communities.

The project aims to tailor WSUD to local conditions in the metropolitan Adelaide and near country (outer metropolitan Adelaide) areas, and to provide a consistent reference tool for planning practitioners and the development industry. The objectives will be achieved through three key initiatives:

1. The development and promotion of a WSUD framework document, which details legislative, policy and system requirements necessary to ensure the consistent implementation of WSUD to all forms of urban development in the Greater Adelaide region.
2. Building the capacity of organisations and individuals to implement WSUD. This will be done by preparing specific technical guidelines and accompanying documents suitable for application in Greater Adelaide, and developing a WSUD training package to improve the application of WSUD 'on the ground' in new developments and in redevelopments in the Greater Adelaide region.
3. The DPLG will prepare WSUD Better Development Plan (BDP) module for incorporation into local area development plans. The module will ensure that new development proposals will be consistently assessed against clear and unambiguous WSUD policies.

This WSUD 'package' includes a draft WSUD framework, draft technical documents, and a draft BDP planning policy library module.

Project completion (including the framework, technical guides and BDP module and capacity building activities) is scheduled for late 2008.

At the federal level in 2008, the Rudd Government established a $12.9 billion water investment program over 10 years. This includes $1.5 billion in new urban water investment to help secure water supplies for Australian households and businesses, comprising:

- $1 billion: National Urban Water and Desalination Plan.
- $250 million: National Water Security Plan for cities and towns.

• $250 million: National Rainwater and Greywater Initiative.

The private sector, water utilities companies and governments will be able to apply for funds, which will be delivered through grants and tax offsets.

Through the National Rainwater and Greywater Initiative, families will receive support to install rainwater tanks and greywater systems, which will reduce their use of drinking water. Households will be able to receive rebates of up to $500 for the purchase and installation of rainwater tanks and greywater products.

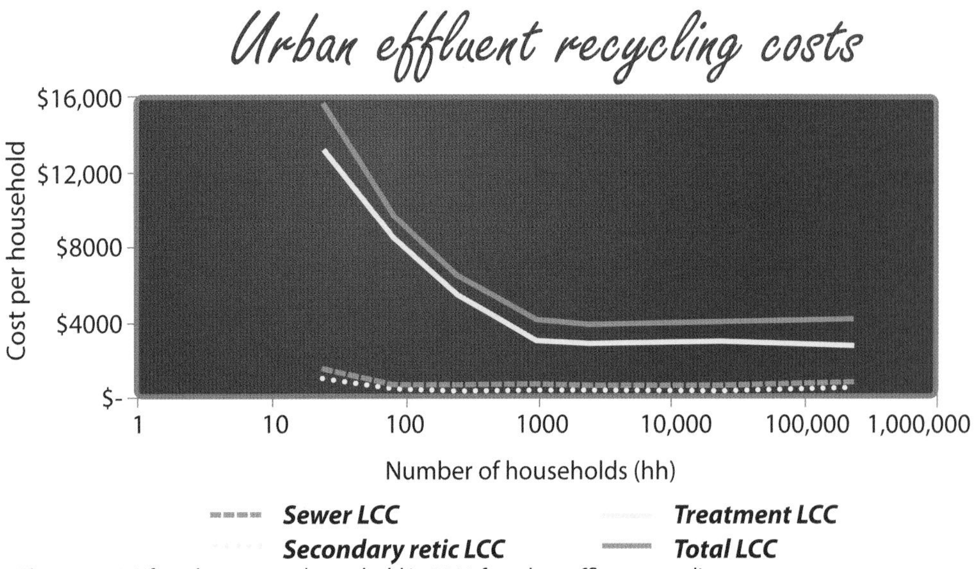

Conclusions

Conventional management of our urban drainage waters has lead to significant impacts on receiving aquatic ecosystems. Largely evolving from advances in engineering, biotechnology and environmental management, alternative approaches such as WSUD and SuDS appear attractive, but their adoption has been slow. Impediments include a combined lack of design guidance, design experience and modelling tools. Most of all, however, WSUD requires an innovative approach and therefore a degree of risk-taking.

More guidelines and modelling tools have recently become available, which have been examined in this paper. The major challenge with WSUD is now to assist designers to develop their cross-disciplinary skills and to encourage a widespread adoption of the technology.

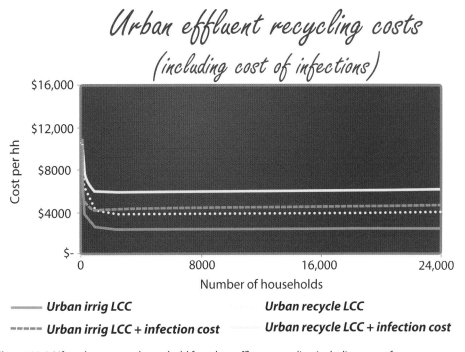

Figure 23.2 Lifecycle costs per household in 2001 for urban effluent recycling. Source: reprinted from S.A. Fane, N.J. Ashbolt and S.B. White, 'Decentralised urban water reuse: the implications of system scale for cost and pathogen risk', IWA, Berlin World Water Congress, Berlin, 2001.

Figure 23.3 Lifecycle costs per household for urban effluent recycling including cost of infections. Source: reprinted from Fane, Ashbolt and White, 'Decentralised urban water reuse', 2001.

References

1. S. Beecham, 'Water sensitive urban design: a technological assessment', *Journal of the Stormwater Industry Association: Waterfall*, 17, 2003.
2. J.R. Argue, 'Water sensitive urban design: basic procedures for 'source control' of stormwater: a handbook for Australian practice', Urban Water Resources Centre, University of South Australia, Adelaide, 2004, pp. 28–29.
3. S. Beder, 'Making engineering design sustainable', *Transactions of Multi-Disciplinary Engineering Australia*, GE17 (1), 1993, pp. 31–35.
4. S. Beder, 'The hidden messages within sustainable development', *Social alternatives*, 13 (2), 1994, pp. 8–12.
5. S. Beder, 'Costing the earth: equity, sustainable development and environmental economics', *New Zealand Journal of Environmental Law*, 4, 2000, pp. 227–243.
6. D. James, *Economics, environment and sustainable development*, Resource Assessment Commission, Canberra, 1991, p. 4.
7. Business Council of Australia, 'Achieving sustainable development: a practical framework', Melbourne, 1991, p. 9.
8. S. Lloyd, *Water sensitive urban design in the Australian context*, Cooperative Research Centre for Catchment Hydrology, Melbourne, 2001.
9. S. Lloyd, T. Wong and C.J. Chesterfield, 'Opportunities and impediments to water sensitive urban design in Australia', *2nd South Pacific Stormwater Conference*, Auckland, 2001, pp. 302–309.
10. S.A. Fane, N.J. Ashbolt and S.B. White, 'Decentralised urban water reuse: the implications of system scale for cost and pathogen risk', *Berlin World Water Congress*, International Water Association, Berlin, 2001.

CASE STUDY 9: Streetscape-scale applications of water sensitive urban design

Martin Ely

Streetscape scale bioretention systems (rain gardens)

Water sensitive urban design (WSUD) can help us achieve 'green streets' in times of water restrictions. Stormwater runoff that usually flows down our drains and out to sea can be harvested to irrigate street trees and other landscaping. At the same time, runoff can be filtered and cleaned of pollutants before being returning to aquatic ecosystems.

The traditional management of urban stormwater has been driven by the attitude that stormwater has no value as a resource, is environmentally benign, and does not contribute to the amenity of urban areas. In response to society's requirement for 'dry feet', conventional engineering practices have emphasised highly efficient drainage systems, which collect runoff as quickly as possible and dispose of it downstream, using a system of underground pipes and above-ground channels. Such systems, however, have had serious environmental consequences. To the originating area, the loss of local stormwater is a loss of a potential resource. To the downstream areas, environmental degradation of aquatic ecosystems can result due to increased runoff and pollution. There also has been a loss of visual connection between human activities and their impacts on aquatic environments, with the processes hidden below ground.

WSUD has emerged as a new paradigm for the sustainable management of the urban watercycle, based on a new attitude – that stormwater is a valuable resource. Key WSUD objectives include reducing the environmental impacts of stormwater runoff on aquatic environments, and better utilising stormwater as a resource. In recent years, WSUD treatments have evolved from large-scale 'end-of-pipe' solutions, such as constructed wetlands, to smaller-scale applications that can treat runoff from local catchments 'at source', and which can be integrated into the design of urban streets and public spaces.

These smaller-scale applications deliver a range of benefits. As well as protecting downstream waters through pollutant removal and retarding of stormwater flows, they can also harvest runoff for local landscape reuse. This can reduce the use of mains water for irrigation, creating 'self-watering' landscape features, enabling the greening of streets in times of water restrictions. They also help increase awareness of the connections between human activities and the watercycle, by making the processes more visible.

Streetscape-scale WSUD applications can take a number of forms; the most popular are bioretention systems, also known as 'rain gardens', which can be scaled to confined spaces and adapted to a range of urban situations. Bioretention systems filter stormwater runoff through a vegetated soil media layer. The filtered water is then collected via perforated pipes and discharged back into the stormwater system or stored for reuse. Temporary ponding above the soil media, in an 'extended detention zone', provides additional treatment by sedimentation. Bioretention systems, however, are not intended to function by infiltration. Treated water is returned to the stormwater system rather than into the surrounding soil and groundwater. Bioretention systems also typically include a high-flow bypass, to capture the most contaminated 'first flush' during rain events, while diverting excess flows to the main stormwater system.

Vegetation growing in the filter media enhances the function of the rain garden in a number of ways. Plant roots help remove pollutants through a combination of physical, chemical and biological processes, and also prevent erosion of the filter media, maintaining its porosity through continuous root growth. An appropriate filter media is therefore required, balancing the need for efficient flow through the soil profile with the need to retain sufficient water in the soil to sustain plant growth. Sandy loams are typically most suitable, however a specialised multilayered soil media can be designed to suit local conditions. A wide range of plants can be used in bioretention systems but species that tolerate periods of drought and inundation are preferred to the more aquatic species, as the former act as indicators of system failure due to clogging of the soil media.

Bioretention systems can take the form of bioretention basins or linear swales. Bioretention tree pits are a recent innovation, allowing the incorporation of stormwater management into confined urban street spaces. Bioretention systems can be integrated into the design of new streets, or retrofitted into existing streetscapes. Streetscape-scale applications, however, present a number of design challenges not faced in larger-scale applications. Successful design, therefore, requires an innovative approach, and close collaboration between designers, engineers and environmental specialists at all stages of the project.

The design must provide a sufficient footprint and filtration depth that meets functional water treatment criteria. It must then be adapted to the surrounding urban environment, including the constraints of confined space and interaction with existing services. As rain gardens are linked to the existing drainage infrastructure, the invert level of the drainage system will limit filtration media and under-drain depth. Rain gardens can be integrated with other streetscape elements, to reduce their footprint, for example in traffic-calming devices. Interactions with street users must also be addressed; of particular concern is public safety and liability. Rain gardens require an extended detention zone set some distance below footpath level, creating a potential tripping hazard. The required grade change can be addressed in a number of ways, depending on the availability of space, including sloped surfaces, fencing or permeable covers.

The final concern relates to aesthetics, a significant factor in gaining community support. Installations in highly urbanised areas may require a more formal, geometrical and hard-edged design than in suburban streets, where a more naturalistic approach may be appropriate. Rain gardens can be integrated with other design elements, such as seating, creating visual interest in the street. Self-irrigated landscape features enhance opportunities for urban greening. A high standard of detailing is also required. Some early examples of rain gardens used standard civil engineering details and failed to enhance the streetscape or gain community acceptance. Some early details also resulted in higher ongoing maintenance requirements. More recent examples have adopted a creative approach to materials, detailing and planting.

Successful design outcomes require collaboration between engineering, urban and landscape, design and environmental disciplines. Site-specific designs should respond to local context and involve a creative response to site constraints. Successful installations should value add to the streetscape in terms of function, aesthetics and ecological sustainability, delivering multiple benefits to the community.

Australian examples

Melbourne is recognised as a leading city in terms of urban stormwater management, in which WSUD best practices have become institutionalised. The process has benefited from a number of champions, including Melbourne Water, local councils and the environmental engineering firm Ecological Engineering. The process began in the late 1980s with recognition of the impacts of urban runoff on the Yarra River Catchment and Port Phillip Bay. An early initiative was the development of an outer-suburban subcatchment scale demonstration project at Lynbrook Estate by VicUrban in 1997, which incorporated a stormwater 'treatment train' at streetscape level. Since then, a number of innovative streetscape-scale projects have been implemented in more highly urbanised settings, in partnership between Melbourne Water and local councils.

with the heavily planted installations creating a green landscape appreciated by local residents – a good example of a multifunctional project that achieves aesthetic, water quality and traffic calming outcomes. Due to its pioneering role, Kingston City Council has been able to review its constructed projects and learn from early mistakes. Key findings have included the need for community acceptance through appropriate design and resident education, and the need for high standards of long-term maintenance.

Melbourne Water, in partnership with local councils, has implemented a number of streetscape-scale demonstration projects aimed at improving the health of the Yarra River. The 2006 Yarra River Action Plan allocated $10 million to implement WSUD projects in the four lower Yarra River councils.

Figure A Rain garden in Fitzroy, Napier/Kerr streets.

Kingston City Council in Victoria is also recognised as a leading organisation for advancing WSUD. The council's 1380 ha of road surface discharge large volumes of stormwater runoff and pollutants into Port Phillip Bay. Funding from the Victorian Stormwater Action Program was the catalyst for the construction of a trial infiltration trench in Riviera Street, Cheltenham, in 2001. The council has now installed over 84 rain gardens in 11 suburban streets across eight suburbs. These have been retrofitted into existing residential street verges or traffic calming devices as part of the council's innovative road reconstruction program.

In Stawell Street, Mentone, bioretention basins were installed in the street's wide verges during street reconstruction in 2004. The basins were designed to fit into the available space between existing mature street trees and driveways. In Brisbane Terrace, Mordialloc, rain gardens were incorporated into traffic calming outstands which were being installed in response to local safety issues. The project has been a success,

In the City of Yarra, the 2003 installation of rain gardens in the reconstruction of Cremorne Street, Richmond, was the first example of streetscape-scale applications in a highly built-up commercial area with heavy traffic volumes. The project was jointly funded by the council and a Victorian Stormwater Action Program grant. Rain gardens were installed in outstands in the parking lane. The change in level between the footpath and filter bed was addressed with post and wire fencing, however deliberate and accidental damage has detracted from the aesthetic value of the streetscape. The council is also installing rain gardens during the periodic renewal of city assets, such as the corner of Napier and Kerr streets, Fitzroy (Figure A).

The City of Stonnington has adopted a sustainable water management strategy that aims to incorporate WSUD elements into all of its facilities. Bioretention tree pits have been installed in the 2007 upgrade of Glenferrie Road, Malvern, an intensively used shopping street. The design comprises a tree pit integrated with a grated kerb inlet, grated tree pit cover, and tree guard. In Orchard Street, Malvern, rain gardens have been incorporated into the reconstruction of a suburban street, to help reduce flooding and treat road runoff. The materials and plantings chosen complement the surrounding contemporary streetscape.

A number of high-profile bioretention projects have been installed in the City of Melbourne, in partnership with Melbourne Water. In 2006, bioretention tree pits were installed in Little Bourke Street in the heart of the Melbourne CBD. The project is of significance as the first example of a WSUD application in such a highly urbanised area. The project faced the challenge of integrating the design into the infrastructure and functioning of the narrow but busy street, and also of meeting the high standards of aesthetics and urban design required in the Melbourne CBD. Particular challenges included the need to relocate underground services, and to achieve the required levels to connect with the existing stormwater system. The tree pits, installed in the footpath, are covered with a heavy-duty bluestone cover, hinged to allow opening for the removal of accumulated litter, with the sunken tree pits also acting as litter traps.

Bioretention street tree pits were also installed in residential Acland Street, South Yarra, in 2007. The street trees are located in the parking lane, rather than the footpath, with runoff flowing directly from the road surface into the grated tree pit cover. Each tree is protected from traffic by a pair of bollards. This installation is part of the council's approach to the greening of city streets, by incorporating WSUD measures into street tree planting, to achieve water quality objectives, while at the same time reducing watering requirements.

In 2006, an innovative rain garden was installed in Davidsons Place in the Melbourne CBD. Davidsons Place is a narrow, high-walled CBD laneway with a hard-surfaced, low-amenity environment, but with a growing residential population. The opportunity was taken to install a rain garden that would improve the quality of urban runoff while also creating urban greenery and a gathering space for residents. The rain garden comprises a raised planter bed at the end of the lane, which collects stormwater from surrounding roofs via downpipes and spreaders. The water is treated by filtering through the raised garden bed, before returning to the stormwater system, and at the same time irrigating the single tree and underplantings. The raised bluestone wall, built at seating height, provides a gathering place for local residents.

The Docklands project comprises the redevelopment of the 200 ha Victoria Harbour precinct adjoining the Melbourne CBD. Commenced in 2002 by the former Docklands Authority (now Vic Urban), this project is a leading example of innovation in incorporating WSUD practices into a large, contemporary urban development. Key stormwater management initiatives include treating runoff to reduce pollutant loads entering the Yarra River and Port Phillip Bay, and the harvesting and reuse of stormwater for irrigation of public spaces. An innovative stormwater management system is integrated into Docklands public spaces.

Above ground, constructed wetlands, designed as contemporary landscape features, treat runoff and act as WSUD signature elements; below ground, an innovative system of underground storage and 'green pipes' facilitate the movement of water around the site and its storage for local irrigation reuse. Streetscape-scale bioretention systems are also part of the wider stormwater management system.

Street tree bioretention pits have been incorporated into the design of the extension of Bourke Street into Docklands. This was possibly the first street tree stormwater treatment application in Australia. Bioretention systems have also been integrated into the streetscape design of Batmans Hill Drive, intended as a high-profile southern gateway to Docklands, with high-quality and well-defined vehicular, pedestrian and cycle environments .The road is lined on one side with a linear bioretention swale, and on the other with individual tree pits. The swale is at road level, protected from traffic by a row of bollards. The individual tree pits are defined by raised semi-circular elements that protect pedestrians from vehicular traffic, take up the grade change between footpath and soil level, act as seating, and provide a sculptural element in the streetscape. Another innovative feature in Docklands are the bioretention basins integrated into the design of the Docklands Point carpark, which treat carpark runoff. The bold design, incorporating a gabion basket to take up the grade change from the surrounding surfaces, fits well into the surrounding formal environment.

In Sydney, the Victoria Park project comprises the redevelopment of 25 ha of former industrial land three kilometres south of the Sydney CBD. The low-lying, flood-prone land was originally a swamp and a natural wetland system. The initial masterplan was modified in 1999 to better integrate a water management system into the public domain of streets and open spaces. The scheme by Hassel, in association with the Government Architects Office and consultants Ecological Engineering, built upon the inherent physical constraints of the site. An innovative community water management system was developed for the whole site, which aimed to more closely mimic the original, natural water processes by the retaining, filtering, moving and recycling of stormwater across the site in a controlled fashion. These processes have been successfully integrated into a high-quality public domain, in which the watercycle is made visible. Victoria Park has become a benchmark for inner-urban development. The success of the project can be attributed to the strong collaboration between the developer, designers, artists, engineers and ecologists.

The wide, east–west streets of Victoria Park were reconfigured to drain to a central bioretention swale (Figure B), with permeable saw tooth kerbing. The swales were designed to capture the most contaminated first flush stormwater, which filters to an underground storage tank for reuse. Excess flows are captured by a system of inlets below the pedestrian bridges over the swales, then conveyed to dished detention basins in the local parks. Runoff is further treated by a system of constructed wetlands integrated into public spaces, and recycled for irrigation use. The pumping system also draws irrigation water from the aquifer during low rainfall periods.

Also in Sydney, the Darlinghurst Road upgrade in Kings Cross included street tree bioretention pits, intended to help reduce stormwater pollution of Sydney Harbour. The design included the development of a new 'suspended slab' tree grate system to allow the trees to be set down below footpath level. The project was intended as a pilot program for review prior to its application elsewhere. Problems encountered included some difficulties connecting to the existing stormwater. The project is of significance in the development of an innovative technical system to better integrate WSUD measures into the streetscape, in both visual and functional terms, while sustaining tree growth.

below pavements, resulting in infrastructure damage. The trial is investigating options to harvest road runoff, clean it by filtration, while at the same time irrigating street trees and redirecting root growth to minimise infrastructure damage. Three different kerb inlet designs were installed. Collected runoff filters to a gravel-filled trench, sized to capture and store first flush runoff. The system operates by infiltration into the local soil to sustain tree growth rather than by returning filtered runoff to the stormwater system.

A stormwater harvesting system has also been incorporated in part of the Melbourne Street 'subtropical boulevard' entry to the Brisbane CBD. Stormwater runoff enters through a kerb inlet that removes litter, then into a filtration pit that removes sediments, then through filters through the planting bed to a collection pipe, and to an underground store for irrigating planting beds during dry periods.

In Adelaide, a bioretention swale was incorporated into the redesign of the South Australian Museum forecourt on North Terrace, the city's main cultural boulevard. The swale is part of a water management system that captures and treats runoff from surrounding roofs

Figure B Victoria Park bioretention swale showing permeable kerb and timber bridges.

and hard surfaces, then stores it underground for irrigation reuse. A key objective of the project was to showcase the museum's commitment to ecologically sustainable development, especially water conservation and reuse. The bioretention swale is designed as a concrete-lined trench, to prevent stormwater infiltration impacting on the adjacent heritage buildings. The concrete edging also provides the base for integrated public seating. The project was the result of collaboration between the designers (Taylor, Culitty, Lethlean), civil and structural engineers, and the consulting firm Ecological Engineering. Early integration of the water management system into the project, rather than as an add on, assisted in its retention in subsequent cost reviews.

Also in Adelaide, a stormwater harvesting system is being trialled in suburban Claremont Street, Urrbrae, undertaken through a partnership of Transport SA, the City of Mitcham, TREENET, and the Centre for Water Management and Reuse at UniSA. Street trees suffer during periods of drought while tree roots flourish in locally resource-rich areas such as kerb–footpath junctions and in the zone immediately

CHAPTER 24

WSUD 2:
community applications

Graeme Hopkins
Christine Goodwin

Introduction

The principles of water sensitive urban design (WSUD) are extremely important within the urban context, so we need to look at how these principles can be implemented in a practical and effective manner to create a sustainable urban environment. In this chapter we will consider a methodology that demonstrates the integration of natural systems, or green infrastructure, within the urban fabric. These solutions can clean, store and reuse water, filter out pollution and provide green corridors for biodiversity.

In cities like Adelaide, there is an increasing demand on open space and green areas due to infill development and moves towards higher-density living to cater for an increased population. This intense urban development increases the amount of stormwater runoff from buildings, and reduces opportunities for its infiltration into the ground due to the large amount of impervious paved surfaces. Greater pressure is placed on existing stormwater infrastructure, with associated problems such as flooding and polluted water from the urban catchment entering the River Torrens and other urban streams. Adelaide's urban ecosystem, like that of many other cities, is in an unbalanced state, and consumes considerable amounts of energy, resources and human endeavour to keep it functioning.

The process or methodology to achieve urban ecological sustainability is known as 'landscape urbanism', involving reintroducing natural systems rather than the existing artificial systems within the city, and integrating WSUD principles and practices with ecological systems for the benefit of the total environment, including humans and our complex cultural interactions. Landscape urbanism requires us firstly to understand how natural processes work in the landscape, and then to interpret these processes within an urban context – to create an urban ecosystem. The basic premise of this discussion of landscape urbanism is that the city is a living ecosystem.

Of particular interest here is the interrelationship between water and the urban ecosystem. In this chapter we will focus on water, but will also briefly look at the other major natural systems, or biogeochemical cycles, that interconnect with water – nitrogen, carbon and oxygen. We will also consider the effects of human interventions on these natural cycles, and how the adverse effects of some of these interventions can be ameliorated by establishing green infrastructure in the city. A current project in Adelaide is used as a case study to demonstrate the role of water in the interconnections between various ecosystems and with the major structural elements of the urban form. It will be demonstrated how the relationship between natural systems, including their reinterpreted urban forms, and the public infrastructure of the city gives rise to new urban strategies for green infrastructure.

The urban ecosystem

Urban areas, from cities to settlements, are constructed and managed by humans, but still function as ecosystems. These urban population centres, with the necessary supporting built infrastructure, have impacts not only within the urban environment, but also on the agricultural and natural landscapes surrounding them. The difference between natural landscapes and human landscapes (or urban ecosystems), is gradually being seen as less clearly defined as the impact of humans touches almost all parts of the earth. The complex task of creating functioning urban ecosystems that allow for human activity and natural processes requires a 'synthesis of social, political, and economic factors, as well as issues related to urban wildlife and water management'.[1] Water, although a critical element in any ecosystem, cannot be studied in isolation and instead must be seen *within* the context of the total environment.

To understand the basic concept of landscape urbanism – that cities are living ecosystems – we first must understand how these ecosystem processes work. A natural ecosystem is made up of natural systems consisting of biotic factors (plants, animals and microorganisms) functioning together with the abiotic factors (air, water, rock, energy), and the cultural components of their particular environment (or biotope). A fundamental ecosystem concept is that living organisms are constantly interacting with every element within their environment. Similarly, smaller ecosystems interact with others through cycles of energy and chemical elements within larger ecosystems that make up the earth's biosphere. Ecosystems are considered to be open systems as both energy and matter are transferred in and out, with resulting change and evolution over time. The transport and transformation of substances, known as 'biogeochemical cycles', includes the hydrologic (water), nitrogen, carbon, and oxygen cycles.

Urban ecosystems, like all other ecosystems, do not exist independently, but interact with other ecosystems. To demonstrate the impact of human intervention within the urban ecosystem we have selected the biotic factor of plants and the abiotic factor of soil as major elements, and will examine the relationships between them and the biogeochemical cycles of water, nitrogen, oxygen and carbon. We will then consider the impact of human intervention within an urban context, and also review the potential implications of climate change and its predicted acceleration.

Plants and biogeochemical cycles

The relationship between the biotic factor of plants and each of the major cycles is summarised as follows:

Watercycle

Plants have a complex interactive relationship with water. The surface of plant leaves intercepts rain on its way to the ground, with some of the water evaporating from the leaf surface into the atmosphere. Some water is taken up directly by the root system and water is stored in the plant tissue. Plants release water back into the atmosphere through transpiration, which in turn becomes available to form clouds and to fall as rain.

Nitrogen cycle

Although necessary to living organisms, nitrogen is inaccessible to most organisms without first being fixed by soil bacteria living on the roots of plants, particularly legumes. These bacteria chemically combine with nitrogen

Chapter photograph: by David Radzikiewicz, A7Designs.

to form nitrates and ammonia, which are accessible to plants, and in turn these are made available to those organisms that feed on the plants.

Oxygen cycle
Photosynthesis by plants is the most significant way oxygen is made available in the atmosphere. Some oxygen is bound to water molecules from plant transpiration and evaporation, and some is bound to carbon dioxide and released during animal respiration. To put this into a human scale, an 'adult requires approximately 150 kg of oxygen each year' while a 'healthy 10 m tall deciduous tree produces approximately 100 kg of oxygen each year'.[2]

Carbon cycle
Plants draw carbon dioxide from the atmosphere and photosynthesise it into carbohydrates. Some of this carbohydrate is consumed by plant respiration and the rest is used for plant growth. In turn, animals eating the plants consume the carbohydrates and return carbon dioxide to the atmosphere during their respiration.

In summary, water is an integral component of the earth's natural cycles of transport and transformation of substances, and is indispensable to the life of plants. Of course, this also applies to other biotic factors, such as animals and humans, so it is critical that we develop and maintain a balance between plants, water and the other biogeochemical cycles within our urban ecosystems.

Soil and biogeochemical cycles
The relationship between the abiotic factor of soil and each of the major cycles is summarised as follows:

Watercycle
Soil has an interactive relationship with water, through the processes of infiltration and evaporation. When it rains, water is stored in open pores in the soil, while some water evaporates back into the atmosphere. Once the soil is fully saturated, water flows over the surface, spreading its benefits, but also potentially causing erosion and flooding.

Nitrogen cycle
Soil bacteria are necessary to fix nitrogen into a form suitable for uptake by plants.

Oxygen cycle
In addition to the photosynthesising role of plants, oxygen is also released into the atmosphere by the weathering of carbonate rock (limestone).

Carbon cycle
Soil microorganisms oxidise carbohydrates when they decompose animal and plant remains. Carbon dioxide is released to the atmosphere through the process of soil respiration.

It can be seen that water and soil have an interactive relationship that provides the right conditions for other biogeochemical cycles to perform their functions. It is therefore necessary for us to ensure that water has access to, and also escape from, the soil within our built-up urban environment.

Human intervention
Typically in an urban ecosystem buildings cover much of the area and the groundsurface is largely impervious, so the area available for planting or biomass is reduced. Consequently, there is a reduced potential for photosynthesis by plants and carbon/oxygen exchange in the atmosphere. Plants draw about one quarter of the carbon dioxide from the earth's atmosphere through photosynthesis, therefore a reduction in biomass also reduces the amount of carbon dioxide absorbed.

Plants intercept and absorb rainwater on its way to the groundsurface. By reducing the biomass there is a greater amount of rainwater reaching the surface in any one spot; this increased stormwater may cause flooding and affect the balance of water runoff and infiltration. Covering the natural surface with hard surfaces and buildings also reduces air infiltration into the soil, affecting the amount of vegetation that will grow. Therefore, less nitrogen is fixed in the soil and less nitrogen is returned to the atmosphere through decomposing organic wastes.

Soils become potentially susceptible to soil compaction due to lack of vegetation and soil pores filled with water. Soil then resists further rainwater infiltration, which creates greater surface runoff, compounding the situation. This is particularly the case in short, intense downpours where reduced infiltration creates large volumes of runoff moving at a greater velocity, increasing the potential erosion effect downstream in the stormwater chain. If the infrastructure is not well designed, increased stormwater will create problems well beyond the physical boundaries of the urban ecosystem.

Increased impervious surfaces also create a greater build-up of heat, causing greater evaporation in nearby soils as well as on adjacent surfaces, due to the deflection of radiation. Reduced soil moisture content impacts on the growing conditions of plants. Build up of heat on surfaces radiates heat to the surrounding atmosphere – a process known as the 'heat island effect', resulting in increased ambient temperatures and altered microclimates. Increased pollutants and changes in atmospheric conditions affect plant growth and animal habitats.

Climate change
The predictions of rising temperature and more days of higher temperature, coupled with less total rainfall but more severe rain events, brings many challenges to the design of urban environments in South Australia. As mentioned above, intense downpours place great pressure on stormwater systems, and increased ambient temperatures impact on the growth of vegetation. More extreme conditions will have even greater impacts, with potential hazards for human inhabitants.

Climate change is, of course, impacting on the wider environment as well as our urban ecosystems. Tim Flannery says that 'global warming could not have come at a worse time for biodiversity' as 'in the past when abrupt shifts of climate occurred, trees, birds, insects – indeed entire biotas – would migrate the length of continents as they tracked conditions suitable for them'.[3] Today most of the landscape is modified by human activity, restricting the potential

BOX
100

Christie Walk

Christie Walk is a rare example of an inner-city ecological housing development. It consists of 27 dwellings of various configurations on a site of just 2000 m² (half an acre). The four stand-alone houses have strawbale walls, the four terraced townhouses and the three-storey apartment building of six units are constructed in load-bearing aerated concrete blockwork, and the five-storey apartment building fronting Sturt Street is a steel-framed structure with aerated concrete block cladding. All the buildings are double glazed and insulated above building code requirements, and avoid the use of toxic materials in their construction.

The buildings are designed to work with the climate and have recorded extremely low energy use compared with conventional South Australian housing, and they are water efficient. The development applies four linked strategies to deal with water:

1. Water use is kept to a minimum with low-flow showerheads and minimum appliance provision, i.e. most of the dwellings, typically two-bedroom units, have only one bathroom with a low-flush WC, washbasin and shower.

2. It captures as much rain and stormwater as the compact site will allow and uses the captured water to flush toilets and to irrigate garden areas, particularly the community produce garden and the intensive (deep soil) roof garden. The water is stored in two 20,000 L underground concrete tanks under the southern carpark/courtyard which are topped up from mains supply as needed.

3. The substantial planted areas use rainwater directly, with water falling onto the pedestrian pathways and running off into the adjacent garden beds.

4. The landscaping favours native and low-water use species.

The roof garden is South Australia's first 'true' roof garden, i.e., it is not a patio with pots, and its planting is predominantly natives with many indigenous plants. Its water supply is augmented from the stormwater capture because the roof garden environment is slightly dessicated (particularly because of wind exposure) compared with ground-level gardens.

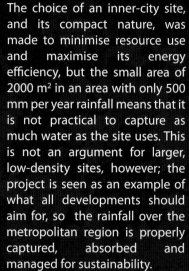

The choice of an inner-city site, and its compact nature, was made to minimise resource use and maximise its energy efficiency, but the small area of 2000 m² in an area with only 500 mm per year rainfall means that it is not practical to capture as much water as the site uses. This is not an argument for larger, low-density sites, however; the project is seen as an example of what all developments should aim for, so the rainfall over the metropolitan region is properly captured, absorbed and managed for sustainability.

The design concept for the site was to mimic natural systems, with the idea that the percentage of water that was captured for use on the site, the percentage that percolated into the ground, and the amount that evaporated, was equivalent to what would happen in its predeveloped, pre-European state.

A Dutch syphon system captures wastewater from the three-storey strawbale house and releases it in 20 L pulses to compensate for the extremely low gradient of the sewer pipe. The device was approved at a gradient below the normal limit as the first South Australian installation to be tested in 'normal' use.

It was always intended that the development would capture and treat wastewater, and in its early stages there

was a sewer mine on site thanks to a Coast and Clean Seas grant (the only government funds in the project apart from the solar rebates). Because this system was sub-optimal and proved inappropriate for the scale and type of wastewater generated, and consumed too much electrical power for its pumps and the electrolytic filter, it was decommissioned by the community corporation.

Stage 3 of the project is the five-storey apartment building, which contains 13 apartments, a community meeting room and kitchen, small library space, toilets and offices for Urban Ecology Australia's interpretive facility, the Centre for Urban Ecology. It also contains a laundry to assist in minimising water use by making the best collective use of resources.

During the development of stage 3, the not-for-profit, self-funded cooperative development entity, Wirranendi Inc., in association with Urban Ecology Australia, negotiated the sponsorship of a full onsite sewage treatment system with SA Water. This system was intended to be an in-ground Biolytix facility under the northern carpark/courtyard. It was supported by Adelaide City Council, who planned to use the treated wastewater to irrigate nearby Whitmore Square. All the necessary design work was undertaken, as was the legal work necessary to mix this unique blend of public infrastructure, private/community land use, and local government enterprise. Unfortunately a combination of circumstances including, perhaps, a loss of corporate memory by SA Water as their management underwent a series of changes, led to the withdrawal of sponsorship. This would have been a unique project; because it was prepared in sufficient detail to enable it to proceed to construction, it exists, at least on paper, as an example of the kind of innovative water infrastructure provision that our future cities are likely to require.

Paul Downton

abundance and movement of flora and fauna. It is therefore imperative that we provide opportunities for movement along corridors, especially those containing natural systems and water. This will contribute towards a more balanced and sustainable urban infrastructure, and reduce our broader adverse environmental impacts.

Green infrastructure for a sustainable urban ecosystem

As already mentioned, human intervention creates unbalanced ecosystems where the open systems of energy and matter transference break down, causing failure in natural systems, such as flooding, erosion and heat build-up. Usually, artificial systems are developed to replace the natural system without a thorough understanding of the natural system, which can create problems along the stormwater chain. For instance, the practice of piping stormwater away from an area concentrates water into large volumes and speeds up its flow, creating an erosion problem as well as volume issues elsewhere in the urban environment, such as the River Torrens Catchment.

Stormwater retention techniques that can be used instead of stormwater pipes can be grouped into four categories: those that prevent runoff from surfaces; those that retain and store runoff for infiltration or evaporation; those that detain or temporarily store runoff and then release it at a measured rate; and those that convey water from where it falls to a retention or detention area.

If cities and urban centres are seen as a water supply catchment area and a water storage facility, stormwater becomes a resource to be valued and utilised. By using an integrated system through mimicking the natural ecosystems and their interdependency and interaction with each other, a balanced urban ecosystem can be developed.

Green infrastructure is a process of designing new natural systems in the urban environment, or using natural systems to replace or augment existing supporting built infrastructure systems. These natural systems are smaller ecosystems – components of the urban ecosystem – that utilise the complex interactions of the earth's natural cycles (hydrologic, nitrogen, oxygen, carbon and others). These elements in the natural systems provide processes that repair, transform and transport matter, clean water and break down pollutants. Such processes are vital for a healthy sustainable urban ecosystem.

Landscape urbanism in practice

Our focus here is on water, so we will use the two premises of *water catchment* and *water storage* to see how the issue of stormwater can be the catalyst for a range of green infrastructure innovations in the public realm. A current demonstration project being developed and installed in the City of Adelaide involves the redevelopment of a laneway. The existing one-way paved road is to be removed and replaced with a low-speed shareway/plaza, to be used by cyclists, pedestrians and emergency vehicles. The project utilises a layering methodology, in which many natural systems interact and depend on each other to work as a living ecosystem.

Diagram 24.1 Adelaide demonstration project.

The urban design approach for this project, using natural systems or green infrastructure, is based on water being transferred and interconnected to numerous smaller ecosystems. This approach accommodates the usual human interventions that one would expect in a city environment, including the replacement of natural groundsurface with impervious hard paving, reduced vegetation or biomass, and stormwater and pollution issues. Other issues not as readily visible to passers-by include the lack of water infiltration into the soil, no nitrogen fixing in the soil, and no soil respiration. There is a compounding effect on other systems, local climatic conditions, and eventually on the whole biosphere.

This project demonstrates an integrated natural system or green infrastructure solution that cleans, stores and reuses water, as well as providing green corridors for biodiversity and pollution filtration. A number of social and cultural layers, including crime prevention and open space recreation opportunities, are overlaid with the biotic and abiotic layers, to create an integrated and interdependent urban ecosystem. This provides multilayered benefits to the urban environment.

Design layers in a laneway project

The laneway diagrams in this chapter identify 12 design layers that contribute to interconnected natural systems. Energy and matter are transferred between these systems to reduce pollutants and excess stormwater runoff. The reflected heat load, the heat island effect, is also reduced.

Each of the 12 layers is shown in Diagram 24.4, depicting how water flows through the laneway and is stored and regulated. By exploding the various layers into single elements we will see the complex influences that each layer has, and their interdependency with each other and with water.

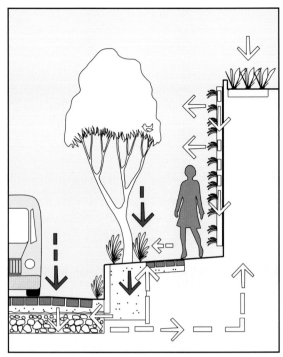

Diagram 24.2 Low-speed road, pedestrian and cycle linkage.

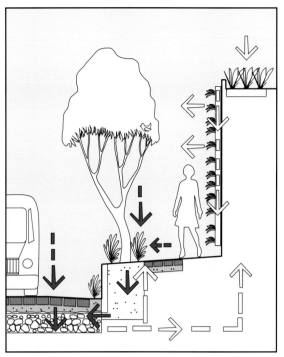

Diagram 24.3 Permeable paving.

Pedestrian and cycle linkage (Diagram 24.2)

As an integral component in the urban ecosystem, the complex needs of humans must be accommodated. In our demonstration project, easy access is provided for pedestrian and low-speed bicycle movement, to encourage people to use these forms of transport rather than motor vehicles. In Jan Gehl's report, 'Public Spaces and Public Life', it was identified that there 'are not enough north–south connections offered in the original plan of Colonel Light. Through time various routes have been created through the blocks, but the general state of these important links is poor and their status is unclear'.[4] The new shareway/plaza being developed provides a much needed north–south link between two major streets, enabling a more relaxed traverse across the city, with the potential for future linkages to extend the pedestrian and cycle network to incorporate other under-utilised access ways.

The Gehl report described many of the minor streets of Adelaide as having the 'character of dark, back alleys and are certainly not places one would like to walk through late at night'.[4] Increased pedestrian activity provides greater safety to those using the space and discourages anti-social behaviour, which, in turn, benefits the surrounding businesses and properties, not to mention the people using these streets.

The pleasing environment of a vegetated corridor will attract more people to use this shareway/plaza both as a commuting link and as a place to relax and spend time. Just as people are drawn to sites within the natural landscape that have water or water-associated systems, this revitalised laneway provides the attraction of water in a more sustainable way than the more traditional fountain or water feature.

Low-speed road (Diagram 24.2)

Like all pedestrian shareway/plazas, access for services and emergency vehicles must be included in the design. The pavement structure must be designed for the heavy load of vehicles, such as emergency fire engines, and this includes the permeable pavement over the water storage. Integrating all the potential access requirements is an essential part of the urban design, however, vehicle access should not appear to dominate the public space. This space is predominantly a pedestrian and low-speed cycle area, but at the same time occasional access by vehicles needs to be easily identifiable to the users. There are a number of ways to achieve this, such as using different types or colours of pavers to distinguish areas used by vehicles. The paving can still be permeable and graded towards the bioretention strip for water storage and harvesting. This natural water storage and cleansing system is the critical component of the project, as it supports the habitat corridor and provides opportunities for biodiversity.

Permeable paving (Diagram 24.3)

This paving system allows rainfall to soak through the paving and to be filtered by a coarse gravel substrate. Water may either infiltrate to the underlying soil or be stored for release or recycled for use as non-potable water, such as toilet flushing, by the adjacent buildings. Either concrete or clay modular pavers are installed with gaps between the pavers to allow the water to penetrate into the substrate below.

The permeable paving would be located in areas of mainly pedestrian usage, where the level of impaction is less constant than a main road regularly used by heavy vehicles. These areas are graded to a biofilter or bioretention strip, for cleaning and removing pollutants and sediments from the surface stormwater.

This system can also utilise heavier soil types to an advantage by providing an impervious surface close to buildings,

Diagram 24.4 Under-pavement water storage.

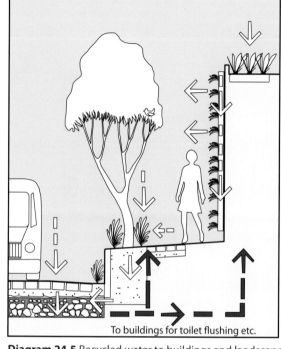

To buildings for toilet flushing etc.

Diagram 24.5 Recycled water to buildings and landscape.

allowing water to move through the substrate on a gradient away from the buildings. A thin, impervious sheet layer over the soil surface removes the water, and also removes possible damaging reactive soil movements away from any building footings. Wetting and drying causes reactive soils to shrink and expand dramatically, creating cracking problems for buildings. The extremes of drying and infiltration close to building footings can be reduced, and the water is directed to a bioretention strip where it can be treated and reused.

Under-pavement water storage (Diagram 24.4)

Water is stored under the paving in a shallow tank-like excavation, lined with a thin impervious membrane, at a depth of approximately 300–1000 mm. Stormwater flows between the gaps in the permeable paving system into a bedding layer of fine gravel, and then through a geotextile layer over the compacted, coarse gravel substrate that forms the storage tank. The fine gravel and geotextile layers remove the pollutants through a biological process, with microorganisms attached to the geotextile layer. The cleaned water is stored in the compacted gravel substrate by displacing the air in the gaps between the compacted gravel. Water can then be moved to other locations or to neighbouring buildings, as discussed earlier.

This water can also be used to irrigate the bioretention strip in the long, hot summer months. Given Adelaide's unique climate, and with climate change predictions of more extremes in the variation of rainfall, some human intervention in supplementing water supply in this urban environment is required to maintain suitable growth for sedges and rushes in the bioretention strip.

Recycled water to buildings and landscape (Diagram 24.5)

Water stored in the under-pavement gravel storage tank can also be used for any non-potable purpose, such as reuse in the nearby buildings, or for irrigation. This water will be important for the vegetation in the bioretention

strip during summer, as this vegetation needs to be kept alive and healthy so it can perform its functions efficiently. The water can also be used for the street trees when needed, which will create savings for public authorities in water usage, especially in times of drought and water restrictions. The water can be moved around the city via simple agricultural polypipe technology, using a solar-powered pump. Water can be transferred from one location to another, or could even be used to top up Lake Torrens if required.

Biofilter or bioretention strip (Diagram 24.6)

The biofilter strip in our demonstration laneway project is designed as a linear bioretention system to collect surfacewater, especially the first flush runoff which contains pollutants from vehicles and other urban grime and chemical substances. The strip is filled with an engineered soil medium for maximum filtration, and vegetated with sedges and rushes that combine with the soil medium to remove pollutants more efficiently. The strip also contains street trees and provides food and water for their growth; the trees in turn add to the filtration capacity of the biofilter strip with the formation of natural biogeochemical cycles.

Street tree corridor (Diagram 24.7)

Street trees provide a visual as well as functional green corridor throughout this project. These trees are part of the natural systems, especially the selected endemic species of the Adelaide Plains that integrate and support the bioretention strip, both functionally and by providing habitat space and opportunities.

Recent research in Australia and overseas on street tree performance has links with the green infrastructure methodology. The trees become an integral part of the natural system. The research says that street trees need more soil to fulfil their water needs, and not be impaired by hard urban design: 'By increasing the soil we will eliminate the need in

Why we need green roofs

In urban areas roofs represent 40–50% of impermeable surfaces, giving green roofs a potential to influence and control rainwater runoff from these surfaces. One of the major drivers for the green roof movement, especially in North America, has been stormwater management in urban areas. Green roofs can be an important element in an integrated water sensitive design and planning approach, as the roof is often the first point of contact for water in the stormwater chain.

What is a green roof?

Green roofs are made up of a series of layers – living vegetation growing on top of a substrate, over a layer of drainage, over a waterproofing membrane, on top of a building. Green roofs reduce runoff volume, delay runoff and reduce the rate of runoff.

There are various types of green roofs, classified by the thickness of the substrate. An extensive roof has 50–150 mm substrate, an intensive roof is 150–300 mm, and an elevated landscape is 600 mm or greater. In general, the deeper the substrate, the greater the rainwater absorption (retention), and the better its stormwater management potential.

How do green roofs manage rainwater runoff?

Rain falling on the roof can be intercepted in four ways:

- rainwater may land on the plant surface and then be evaporated away;
- rainwater that falls on the roof substrate can be absorbed by the substrate pores or taken up by absorbent material in the substrate, or evaporate;
- rainwater taken up by the plants is either stored in the plant or transpired back to the atmosphere; and
- rainwater can also be stored and retained by the roof's drainage system.

Green roofs are able to reduce the amount of water runoff by using their absorption and evaporative ability, and also by storing it before a delayed release of water to the existing stormwater drainage system. The storage/detention ability of green roofs has many variables that need to be taken into account to assess the retention or detention rate. These factors include:

- the season;
- number and type of layers used in the system;
- depth of substrate;
- angle of the roof's slope;
- physical properties of the growing media;
- type of plants incorporated in the roof;
- intensity of rainfall; and
- local climate.

Overseas studies generally agree that green roofs lead to yearly reductions in runoff of up to 60–80%. Seasonal variations for rainwater retention between summer and winter are considerable, due to the greater amount of water being returned to the atmosphere in summer through evaporation and transpiration. Stormwater monitoring of two ecoroofs in Portland, Oregon, showed that for small to moderate storms the green roofs

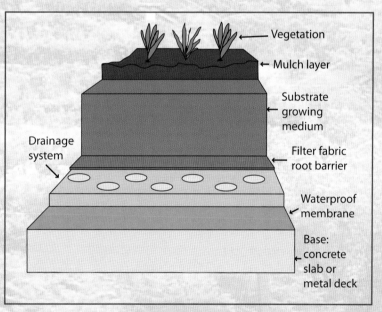

retained virtually all the rain that fell on them. In large rain events, once the substrate water-holding capacity was reached, any extra water would simply run off, therefore reducing the overall percentage of retention. The gradual release over a period of time evens out the peaks of heavy rainfall characteristic of large storms.

Green roofs could help reduce the pressure on the existing stormwater infrastructure in urban Adelaide.

Graeme Hopkins and Christine Goodwin

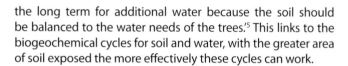

Diagram 24.6 Biofilter or bioretention strip.

Diagram 24.7 Street tree corridor.

the long term for additional water because the soil should be balanced to the water needs of the trees.[5] This links to the biogeochemical cycles for soil and water, with the greater area of soil exposed the more effectively these cycles can work.

An example is that a street tree with a 6 m diameter canopy requires approximately 9 m³ of root space/soil for healthy growth. This volume of soil can take any shape, therefore a long, rectangular prism would be ideal for the urban space. This can be achieved by integrating the bioretention strip and using new technologies to mimic natural systems, such as introducing linear trench tree pits where the tree planting occurs, and connecting trenches, allowing root space and drainage into the bioretention strip. This gives greater soil, air and water contact for many of the biogeochemical cycles to operate. This will require root barrier systems to stop roots filling up the bioretention engineered substrate system.

To achieve the required root soil volume for healthy trees, the urban pavements need to be supported by artificial structural systems instead of the common practice of compacting the soil for the paving. This structural system allows for air and water movement throughout the soil profile without compromising the paving structure. If the paving was permeable, the rainwater would infiltrate into the root space of the tree – an added bonus. The soil would balance the water needs of the trees, requiring less artificial watering. These street trees can also perform the function of deep soil infiltration, through their root systems penetrating the subgrade and allowing water into the existing groundwater table.

In addition, street trees perform many other functions of shading, reducing heat island effect, removing airborne pollutants, enhancing carbon dioxide/oxygen exchange, providing habitat, and providing a pleasant place to be in, as described in other layers of this green infrastructure system.

Biomass removal of airborne pollutants (Diagram 24.8)

It is internationally documented that street trees are beneficial in reducing the heat island effect by lowering ambient air temperatures and interrupting glare and reflection from surrounding hard surfaces. Trees also 'act as efficient filters of airborne particles because of their large size, high surface to volume ratio of foliage, frequently hairy or rough leaf and bark surfaces',[6] and when planted in rows or as loose plantations, they intercept and retain atmospheric particles as wind currents pass between them or are pushed up and over the trees.

Recent Indian research suggests that a thin screen of trees is more effective at removing and retaining airborne impurities and dust particles than a densely planted plantation that creates more air turbulence and subsequent relocation of the particles. It has also been found that 'smaller compound leaves are generally more efficient particle collectors than larger leaves'[6] and that different shaped leaves have different abilities to capture dust particles. Chakre cites forest research by Dochinger in the United States dating back to 1973, who found that 'the filtering effects of evergreen trees are better than the deciduous trees'.[6] The selection of tree species and the design of their siting is therefore very important in maximising potential benefits.

In our demonstration project the street trees and low vegetation in the bioretention strips provide the above benefits, as well as increasing the carbon dioxide/oxygen exchange through photosynthesis, and all the other benefits of a natural system within the urban ecosystem. Through natural air filtering processes, street trees reduce the concentration of 'toxic gases and particulate matter such as nitrogen dioxide, sulphur dioxide, ozone, carbon monoxide, dust and ash from the air we breathe'.[2]

Microwetlands

BOX
102

Microwetlands are small, constructed wetlands, designed to utilise the bioretention benefits of natural wetlands. They are different to rain gardens because they treat and detain water then allow it to move on in the water drainage system, without infiltration. Common uses for these wetlands are along road verges, downstream of large paved areas or plazas, and on green roofs. They assist in stormwater control by reducing the amount and rate of water runoff.

Stormwater should be intercepted by microwetlands as close as possible to its source. The wetlands then utilise physical, chemical and biological processes to treat the stormwater. Microwetlands are multipurpose and have many benefits including:

- flood protection;
- flow attenuation;
- water quality improvement;
- landscape amenity; and
- wildlife habitat.

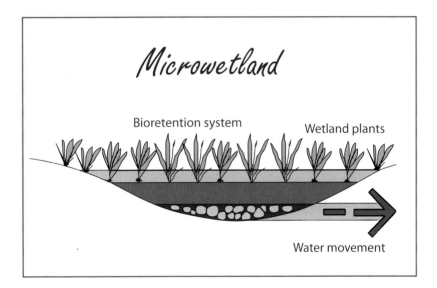

In terms of design, the inlet to microwetlands needs to be resistant to the velocity of water entering, and also capable of removing sediments from the water before it enters the wetland zone. In areas where public safety is an issue the wetland can be filled with rock and will still function normally. The wetland floor needs to be an impermeable layer. Above this layer, a filter medium of open sandy/loam organic soil will provide the best conditions for both wetland plants and essential organisms.

In Adelaide's climate, microwetlands will usually be ephemeral and their appearance will change with the seasons, bringing variety and ongoing interest to the landscape. Wetland biota needs to be able to tolerate periods of inundation followed by extended dry periods.

Microwetlands can take a natural wetland appearance or be designed as a contemporary urban element. They can be used in small-scale works, especially in residential areas in place of rain gardens where conditions require no infiltration.

Graeme Hopkins and Christine Goodwin

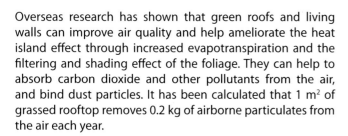

Diagram 24.8 Biomass removal of airborne pollutants.

Diagram 24.9 Habitat corridor and biodiversity.

Overseas research has shown that green roofs and living walls can improve air quality and help ameliorate the heat island effect through increased evapotranspiration and the filtering and shading effect of the foliage. They can help to absorb carbon dioxide and other pollutants from the air, and bind dust particles. It has been calculated that 1 m² of grassed rooftop removes 0.2 kg of airborne particulates from the air each year.

The air cleaning ability of vegetation is also being utilised inside buildings, particularly in Europe and North America, in the fight against 'sick building syndrome'. The US Environmental Protection Agency says that about one third of absenteeism due to illness is caused by poor air quality in buildings, and a European study showed that absenteeism can be reduced by 15% by greening indoor space.[7] An indoor four-storey high biofilter living wall at the Guelph-Humber Building in Toronto has been installed to filter the air inside the building and reduce the air-conditioning load. Research on the wall has shown that 'the system can remove half of the benzene and toluene in the air during a single pass and up to 90% of the formaldehyde'. This air filtering is a natural process that is 'carried out by the microbes living on and in the plant roots (to) break down these volatile organic compounds as air is drawn through the system'.[7] This natural air cleaning process applies to outdoor living walls, as well as to closed indoor systems. Living walls can be seen as large air cleaners for the city's air supply.

Habitat corridor and biodiversity (Diagram 24.9)
The urban ecosystem is already home to many flora and fauna species that have adapted to the harsh conditions of the built environment. Unfortunately many of these species are invasive or destructive, and in the case of introduced non-native species, often put extreme pressure on competing native species. For instance, the feral pigeon (*Columba livia*) has successfully made the move from its historical cliff-dwelling habitat to the rooftops and façades of city buildings.

The reasons why some species are so successful at colonising urban spaces is being reassessed through a 'habitat template' approach, so that buildings and urban spaces can be designed to attract the species we *want* into the city. This approach recognises that 'many urban habitats, while lacking historical continuity with the habitats they replaced, may be (as far as some species are concerned) functionally equivalent to other kinds of natural habitats'.[8] However it needs to be remembered that a basic requirement of all species is access to water.

Corridors of vegetation, especially native or endemic, provide opportunity for the integration of natural systems, plant communities, animals, birds and insects, so that they can interact with each other to survive and operate as a working ecosystem. Animals, birds and insects can breed and pass through the city, with connections between habitat corridors and the surrounding parklands recreating or reinterpreting some of the pre-settlement fauna movement patterns.

Green corridors (networks or greenways) that interconnect natural or vegetated open space in the midst of heavily developed or exploited areas can be utilised by both resident and migratory biota. Where it is not possible to create continuously connected corridors, they can be linked by stepping stone patches of vegetation. In fact, it has been argued that a close mosaic of stepping stone habitat patches may be as effective as a continuous strip in allowing many species to permeate the whole area. These stepping stones serve as refuges for metapopulations (a system of local populations linked by dispersal) and can provide a rescue effect, allowing small populations a higher probability of persistence than if they were isolated.

The metapopulation theory (Diagram 24.10), developed by Ehrlich in the 1970s, states that the 'species in metapopulation are usually thought of as having interdependent extinction and colonization processes, where individual populations may "wink out" (go extinct) and later "wink on" again (be

The Metapopulation Theory

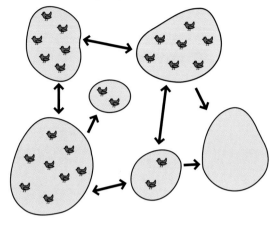

Diagram 24.10 The Metapopulation Theory. The green areas represent stepping stone patches of habitat.

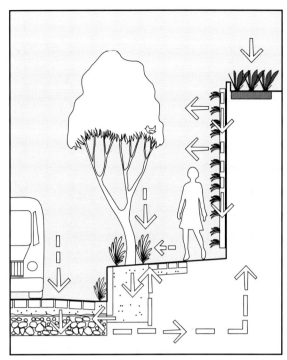

Diagram 24.11 Green roof connection.

recolonised from another population that is still extant)'.[9] For instance, research and projects by the Rana Creek group based in California, has shown that stepping stone vegetation patches, individually as small as 0.1 ha, and adding up to a total of just 1 ha, could provide enough host plants and nectar sources to ensure survival of the Bay Checkerspot butterfly.

It is important that habitat corridors and stepping stone patches be physically close enough to each other for target species to move between them. Visually oriented species such as birds and some insects need to be able to see the next vegetation corridor or patch, otherwise their movement may be inhibited and the patches made more isolated. It is optimal to have a cluster of patches rather than a linear row, to provide alternative routes between patches. This habitat interconnectivity is further increased with green roofs and living walls. The vegetated corridors and patches are all dependent on a sustainable supply of water, made possible by a practical implementation of the city as a water supply catchment area and storage facility.

Green roof connection (Diagram 24.11)

Green or vegetated roofs provide many environmental benefits, including stormwater and water quality management, increased biodiversity, improved air quality, insulation to buildings, and reduced air temperature. They form an integral layer in city-wide green infrastructure. In particular, bushtops – where a local ecosystem is created or reinterpreted to provide habitats for endemic species – provide further opportunities for biodiversity. A habitat template approach can be used to analyse the natural habitats of target species, and to create a new urban habitat template to mimic these habitats on the rooftop. Plants need to be selected so that they match the environmental conditions of the built environment, which in most situations is relatively harsh. The bushtops do not need to be connected, but if located on a series of isolated building roofs they form stepping stone habitats, which can be integrated with adjacent habitat corridors.

Researchers in Europe have found that the deeper the substrate of a green roof is, the greater the vegetation biodiversity possible – and this also refers to the animal biodiversity. The main target animal groups for green roofs to date have been insects (bees, butterflies, grasshoppers etc.), spiders, birds and lizards. It has been shown that a large number of mobile organisms such as spiders, beetles and bee species readily colonise green roofs and can establish themselves systematically. Green roofs on buildings adjacent to our demonstration laneway project will add to the stepping stone mosaic by providing more habitat options and increasing biodiversity in the urban environment.

Green roofs can play an important role in providing sufficient oxygen for humans, within an otherwise oxygen-depleted urban environment. For instance, it has been calculated that 1.5 m² of uncut grass produces enough oxygen per year to supply one human with their yearly oxygen intake requirement.[9] Green roofs can take a number of forms, ranging from 'extensive' (with a thin layer of grasses or sedums) to 'intensive' (with a wide range of plants growing in a deeper medium), through to bushtops (with integrated flora and fauna habitats).

Importantly in our focus on water, green roofs are able to reduce the amount of water available for runoff by using their absorption and evaporative ability. The water is used to sustain life on the green roof, and can be stored for a period of time before a delayed release to the existing stormwater drainage system.

Living wall connection (Diagram 24.12)

Living walls provide a physical link between the ground plane of vegetation and their habitats with the rooftops of buildings, where green roofs and bushtops provide another layer of green infrastructure. Living walls consist of suitable plants growing in the ground and trained up trellises or other wire structures, or modular panels with plants growing in a lightweight medium, held in a sturdy framework. Water is

Diagram 24.12 Living wall connection.

Diagram 24.13 Green roofs and living walls stormwater management.

efficiently supplied to these plants via low-volume drippers. The lush growth provides habitat to lizards, insects and birds, and allows their movement up to the rooftops. It has been shown in Europe that small lizards will climb at least four storeys to find a suitable habitat. The inclusion of living walls in our demonstration project creates stepping stone corridors vertically as well as horizontally.

Water collected in the under pavement water storage facility in the laneway is pumped into the living wall dripper system, or other sources of greywater can be used. For instance, microwetlands constructed on rooftops can provide recycled water for irrigating living walls. What is not absorbed by the planting medium or the plants is filtered back into the under pavement storage system rather than the stormwater system, and so is available for further recycling and reuse.

Green roofs and living walls stormwater management
(Diagram 24.13)
Green roofs can be an important element in an integrated water sensitive design and planning approach, as the roof is often the first point of contact in the stormwater chain. By intercepting the rain runoff at the source, the green roof eliminates the potential multiplying effect further down the stormwater chain. Vegetation assists the management of stormwater by reproducing many of the hydrological processes normally associated with the natural environment. Such elements that contribute to this process include:

- rainwater lands on plant surfaces and then evaporates;
- rainwater that falls on the roof substrate can be absorbed by the substrate pores or taken up by the absorbent material in the substrate, and some evaporates back into the atmosphere;
- rainwater taken up by the plants is either stored in the plant or transpired back into the atmosphere; and
- rainwater can also be stored and retained within the roof's drainage system.

Storage detention or retention rates for green roofs depend on many variable factors. As Adelaide, or indeed Australia, does not have any detailed research, we can only rely on current observation on existing green roofs and general trends in research in North America as a guide. This research has identified considerable seasonal variations for rainwater retention between summer and winter due to the greater amount of water being returned to the atmosphere in summer through evaporation and transpiration. In a study involving stormwater monitoring of two ecoroofs in Portland, Oregon, it was found that the summer retention rate can be up to 100%, and down to 69% at other times of the year for a 100–150 mm thick green roof.[10] This study also showed that for small to moderate storms, the green roofs virtually retained all the rain falling on them. In large rain events, once the substrate water holding capacity was reached, any extra water would simply run off, therefore reducing the overall percentage of retention.

A further development in technology is the installation of a layer of water-absorbing material within the soil profile. A green roof that has installed this technology in Adelaide has displayed evidence of 100% retention in summer and near that for the rest of the year. Given this development, various technologies can be used as a design tool for achieving whatever retention rate is desired. Green roofs retain rainwater, moderate the temperature of the water, and act as a natural filter for any runoff. This occurs through infiltration and the bioretention process in the substrate.

A recent innovative development in green roof design is to install constructed wetlands as an integral part of the green roof system. Taking the known benefits of constructed wetlands and adding this element to the stormwater management role of green roofs is an important WSUD technique. A demonstration case is the John Deere Works in Mannheim, Germany, where it was decided to begin treating

Porous and permeable paving

BOX
103

Paved, impervious surfaces have a significant impact on the water cycle by not allowing rainfall to soak through to the subsoil. Consequently, the volume and velocity of stormwater discharge is increased. This paving can be used to address these issues by allowing rainfall to soak through the pavement. Water is filtered by a coarse sub-base and may be allowed to infiltrate the underlying soil or be stored for later reuse. These systems also reduce the pavement's heat absorption and the reflection of heat into the surrounding atmosphere. Porous and permeable pavements can take numerous forms:

- porous asphalt or concrete (with no fines);
- concrete and clay modular pavers, installed with gaps between pavers;
- grid or lattice systems of concrete or plastic filled with soil or gravel (these systems may be vegetated); or
- gravel surfaces.

Porous and permeable pavements can be used for water reuse or storage. By introducing an impermeable membrane to the bottom and sides of the sub-base, a tank effect is created. This water can be gradually released or accessed for treatment and recycling.

Permeable paving is not suitable for use on land that slopes greater than 10% or 5 degrees, as water runoff may exceed infiltration and cause erosion. Additionally, it should not be used where there is a very high watertable, where there is a soil salinity hazard, where heavy clay soils are likely to collapse, or where the soil has a hydraulic conductivity of less than 0.36 mm/hour (unless using for a tanking storage system).

Porous and permeable pavements are well suited to public areas, including footpaths, carparks and plazas. In some instances, trafficable street surfaces are also suitable, particularly where there are low traffic volumes. Street parking lanes can use permeable paving and street trees will benefit from infiltrated water. Residential driveways are also suitable, and some of these systems can be visually softened with low vegetation growing within the pavement.

Porous

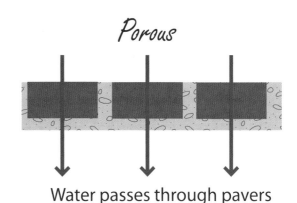

Water passes through pavers

Permeable

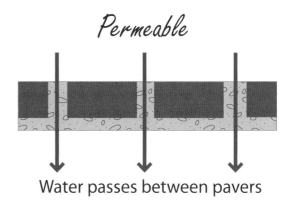

Water passes between pavers

Porous/permeable paving

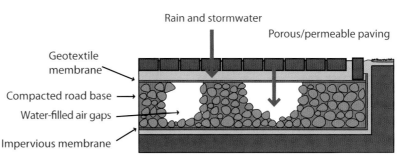

Rain and stormwater

Porous/permeable paving

Geotextile membrane

Compacted road base

Water-filled air gaps

Impervious membrane

Graeme Hopkins and Christine Goodwin

the wastewater from their manufacturing and assembly operations with a constructed wetland on the factory roof. The wetland includes a combination of sedges, rushes and irises grown hydroponically in 50 mm of water. It breaks down carbon and nitrogen compounds present in the wastewater while sequestering phosphates and heavy metals.[11]

Green infrastructure mimicking natural ecosystems

In the previous section we analysed the design layers that make up the green infrastructure of the demonstration laneway project; now we need to look at how this green infrastructure mimics natural ecosystems to create an integrated urban ecosystem. This will allow us to extend the landscape urbanism principles and processes to a city-wide scale, incorporating new projects and retrofitting of existing infrastructure to restore a sustainable balance to the urban ecosystem.

Many of the design layers described above mimic natural systems in principle, but new technologies are used to implement the principles. For example, water is stored in the gravel under the pavement and is retained onsite for reuse, including watering the plants. This mimics the natural process of rainwater infiltrating the soil and being stored within the pores of the soil, and then being reused by naturally occurring vegetation.

The bioretention strip is also mimicking what happens in a natural system of filtration. The substrate is modelled on natural draining soils, and the planting system is based on the wetland environment where sedges and rushes act together with soils to remove the pollutants. Microorganisms attach to the sedge and rush root systems and utilise oxygen release through the roots for their growth processes. This is a natural system for wetland vegetation, and the selected vegetation in the bioretention strip mimics this process.

The bioretention strip also allows for the soil to perform its natural biogeochemical cycles with water, nitrogen, oxygen and carbon. The natural biogeochemical cycles are also maximised through the design of the bioretention strip, where the soils are engineered for maximum infiltration. Therefore, a relatively small surface area of exposed soil will still perform at a high level, compared with a larger, degraded natural environment.

The street trees that grow in the bioretention corridors perform many roles, including providing shade, reducing the heat island effect, removing pollutants, and providing habitat. The trees' natural cycles of carbon dioxide/oxygen exchange are even more critical in the urban context than in the natural environment, helping to counteract the extreme atmospheric conditions created by highly industrialised population centres. The trees shade the ground, reducing heat build-up, and, in addition, their air circulation system acts as a heat pump within the tree canopy space. This is an action where the tree's natural system removes hot air and replaces it with cooler, more humid air. This natural technique is useful in urban courtyards and in this particular case, the laneway, by removing the hotter air associated with hard surfaces. Through transpiration the air is humidified – water vapour is added – making trees an important part of the watercycle.

The green roofs and living walls mimic natural systems in many ways, but are particularly important for the removal of pollutants through capturing pollutant particles on plant leaves and deposits on the roots, where microorganisms feed to convert these particles into water, carbon dioxide and nutrients. This process again provides water and water vapour as an end product. The green roofs and living walls have all the same natural systems as the ground level environment, but are set up artificially.

The stormwater chain in a natural system is the capture and retention of stormwater at its source, and its reuse or infiltration. In this demonstration project, this natural process is mimicked by using green roofs and living walls to capture and retain the rainwater that falls on them, thus eliminating a large volume of stormwater at its source. Paved areas are the next major source of potential stormwater build-up, so in this project the permeable paving and bioretention strip allows infiltration and under-pavement storage, mimicking the natural principle of infiltration instead of runoff that is implemented through a constructed urban landscape.

Water is central to all these natural systems, from the management of stormwater to using and producing water through natural processes. All green infrastructure requires the efficient use of water, and preferably at its source rather than transporting it elsewhere. Water is certainly too valuable to be wasted.

Maintenance of green infrastructure

Like all infrastructure, green infrastructure also needs a certain amount of maintenance, even though these systems mimic natural ecosystems and principles. The systems created are usually not complete or holistic solutions, so some artificial support and maintenance is necessary. The current single issue system of piping away stormwater requires a high level of maintenance, with burst pipes to be repaired, erosion at the outlet, pollutants entering waterways, and so on. Green infrastructure, through its layering of elements, has the potential to greatly reduce maintenance and associated costs.

A typical example that demonstrates this is the bioretention strip, where the engineered substrate is designed for an infiltration rate as required, and therefore does not retain moisture over a long period of time. Vegetation that is planted in these strips needs to be able to withstand inundation, followed by long periods without water. In Adelaide's long, hot summers, some supplementary watering is required to maintain plant vigour and amenity in the high-exposure urban realm. In a non-maintained natural system these plants would die off and the foliage would become ground mulch, with the plants regenerating after further rain. The periods of dry vegetation would be visually unacceptable in the high-amenity urban environment of a pedestrian plaza, so this supplementary irrigation can be classified as maintenance to maintain the standard required. As the water required for this irrigation is harvested onsite, the cost and energy expenditure is relatively low.

If artificial habitats are created, such as bat boxes, then a certain amount of maintenance will be required. Bats are part of a natural system and because they feed on mosquitoes and flying ants, they serve a useful purpose in reducing the numbers of these insects. These systems should be monitored regularly and any necessary action taken to assist with the healthy survival of relevant fauna and flora.

As shown, green infrastructure systems usually require less energy input and maintenance, but this maintenance may be needed at specific times or for components of the natural system that cannot be adequately mimicked.

Conclusion

This chapter has shown how the relationship between natural systems, including their reinterpreted urban forms, and the public water infrastructure of the city gives rise to new urban strategies for sustainable green infrastructure. Water, an essential element to any ecosystem, cannot be looked at in isolation, but rather as a critical part of a balanced, living urban ecosystem.

An important principle to understand and implement is that cities are water catchment and storage areas. By mimicking nature we can introduce natural systems into this urban catchment, using the hard, paved surfaces and vertical and elevated planes of the urban form to collect, store, cleanse, recycle and reuse this urban stormwater as a vital resource. Creating ground-level vegetated habitat corridors with bioretention strips provides multilayered strategies for water treatment, biodiversity and connectivity. Utilising the vertical and elevated planes of the built form to perform stormwater management, water recycling, and to change the microclimate adds further layers to the green infrastructure.

The concepts explored in this chapter are being successfully implemented, with ongoing monitoring, in the demonstration laneway project in the City of Adelaide. To take this project beyond an isolated scenario, a city-wide green infrastructure strategy, or set of strategies, must be developed. Around the world, green infrastructure planning is gaining momentum, with cities like New York, Chicago, Toronto, Singapore and Tokyo developing environmental plans to encourage and support integrated green infrastructure.

In South Australia this process requires the collaboration of local government, various state government authorities, and the private sector. Government departments and agencies need to take a lead role in planning and implementing this green infrastructure strategy. For instance, the Department of Planning and Local Government (DPLG) could lead and coordinate a city-wide program to strategically place green infrastructure in areas where stormwater management systems to protect the River Torrens can be provided. The majority of pollution in Lake Torrens is carried by urban runoff.

Given the complexity of this urban river system, its catchment, the range of existing landscape and urban design elements along its course, and the community's expectations and requirements, there is obviously no single solution. A broader look at the urban catchment is needed if the excessive level of nutrients and pollutants entering the river system is to be addressed. The implementation of WSUD in the urban catchment, especially retrofitting within existing infrastructure, is vital to control the inflow of pollutants.[12]

The DPLG is already taking a lead role in green infrastructure strategic planning through the introduction of its program Bushtops for Green Roofs and Walls, which includes a green roof and living wall incentive program for government buildings and projects. However, under this new incentive program, government projects can access design and development services from the DPLG. This program is linked to other city and urban developments, such as the new hospital featuring green roofs at various levels and proposed redevelopment infill residential projects with green roofs and walls as part of the open space recreation requirements. Together, these initiatives will form links for stepping stone corridors throughout the city.

The city is a living ecosystem. Carefully considered urban design from a landscape urbanism perspective will help to maintain a sustainable balance, with water being a central element to the green infrastructure.

References

1. E. Mossop, 'Landscapes of infrastructure' in Waldheim, C. (ed.) *The landscape urbanism reader*, Princeton Architectural Press, New York, 2006.
2. K. Young and T. Johnson, 'The value of street trees' in Daniels, C.B. and Tait, C.J. (eds) *Adelaide, nature of a city: the ecology of a dynamic city from 1836 to 2036*, BioCity: Centre for Urban Habitats, Adelaide, 2005.
3. T. Flannery, *The weather makers*, Text Publishing, Melbourne, 2005.
4. Gehl Architects, 'Public spaces and public life', South Australian Government, City of Adelaide and Capital City Committee, 2002.
5. J. Urban, 'Growing the urban forest', Healthy trees for a beautiful city Symposium, Toronto, 2004.
6. O. Chakre, 'Choice of eco-friendly trees in urban environment to mitigate airborne particulate pollution', *Journal of Human Ecology*, 20 (2), 2006, pp. 135–138.
7. A. Vowles, 'Guelph-Humber plant wall a breath of fresh air' at Guelph, 10 November 2004 (online).
8. J. Lundholm, 'Green roofs and facades: a habitat template approach', *Urban Habitats*, 4 (1), 2006, pp. 87–101.
9. Green Roofs for Healthy Cities, 'Green roof policy development workshop: participants' manual', 2006.
10. D. Hutchinson, P. Abrams, R. Retzlaff and T. Liptan, 'Stormwater monitoring two ecoroofs in Portland, Oregon, USA', Greening Rooftops for Sustainable Communities, Chicago, 2003.
11. Earth Pledge, *Green roofs: ecological design and construction*, Schiffer Publishing, USA, 2005.
12. C. Goodwin and G. Hopkins, 'Blooming summer: Australian Institute of Landscape Architects', 2007 (online).

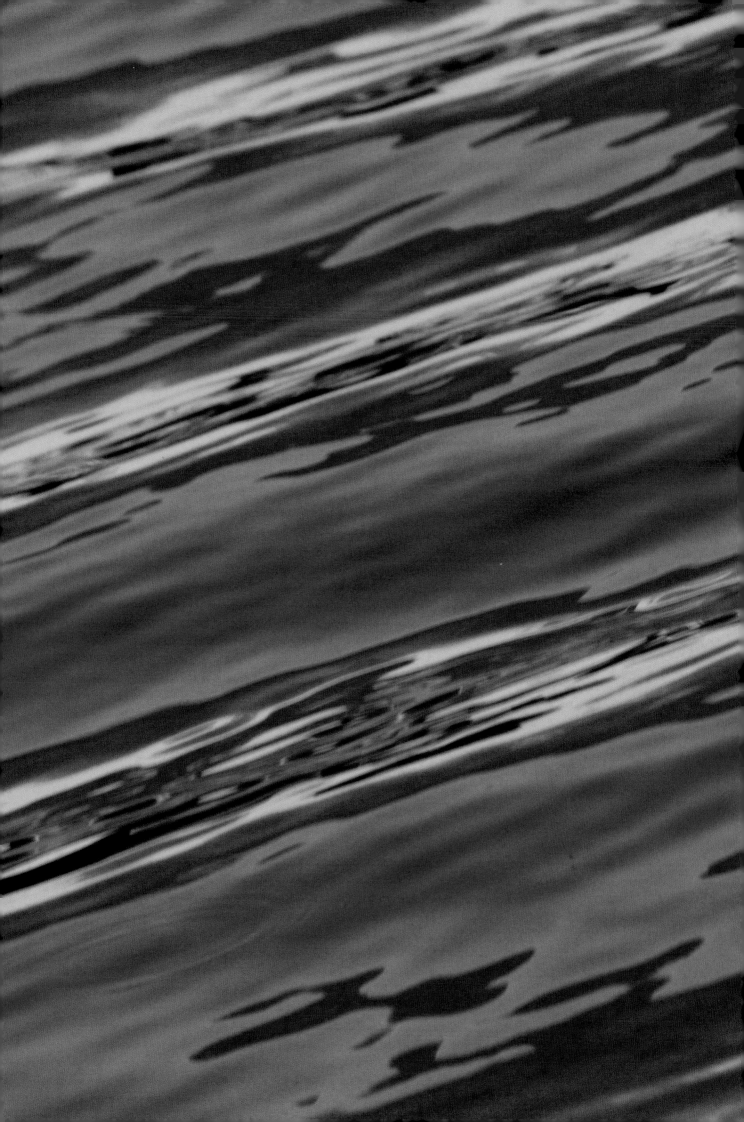

CHAPTER 25

Designing a sustainable water industry

Joe Flynn

Adelaide a model sustainable water city?

'If you want to see one of the world's great examples of how a city can manage water sustainably then you need to go to Adelaide. What they have done is amazing. Population growing, climate change worsening and yet the city gets by without water restrictions and is officially recognised as Australia's best example of a healthy, sustainable, biodiverse hot spot.'

Now, in 2009, it is difficult to imagine a day when someone could say that. The Murray, the source of 91% of Adelaide's water (in 2007), is below empty as we drain her wetlands. We are starving the Coorong, which cannot sustain birds like the fairy tern, whose population in 2007 numbered six and faces a grave risk of extinction.[1] Sea life habitat in the Gulf St Vincent is in decline partially because of the pollutants in our stormwater and sewage we dump there. And, in our desperation to reduce water demand to meet supply, we have made it a crime for a child to play under a sprinkler in their backyard.

The media constantly reinforce these issues, which can all seem quite overwhelming. Yet for all of these problems, the solutions *are* known. We know what needs to be done to become a sustainable water city.

While this chapter weaves together many industry views and experiences on how we can become a sustainable water city, it is not presented as an agreed industry view or the view of a particular organisation. The author takes responsibility for the opinions and ideas, while acknowledging that a lot of the ideas draw on the work undertaken by studies listed in the footnotes.

We need to optimise the way we use water rather than just conserve water. We need to reduce the amount of water we take from rivers and bores to sustainable levels. And, we need to supplement the increasingly unreliable, climate-dependent water we draw from the Murray River and from the Adelaide Hills catchments with new sources, including:

- purifying and reusing more of our wastewater;
- harvesting more urban stormwater;
- buying water from River Murray irrigators; and
- creating drinking water through desalination of seawater.

All of these sources of water and the required technologies to procure them are known, proven and operating to varying degrees within Adelaide. In some of these new technologies, Adelaide leads Australia. We recycle more water than any other major city in Australia (30% in 2007/2008).[2] Many of the new technologies are distributed, meaning they are being applied by households, businesses, local and state government – all of whom are working to care for the environment.

Technology is only part of the answer. Over the last decade Adelaide has led Australia in reforming our century-old water management system by enabling new players to develop new water supply solutions, often involving collaborative Commonwealth, state government, local government and private sector initiatives.

This decentralised approach has unleashed innovative, sustainable solutions, such as one of Australia's largest recycling projects, the Virginia Pipeline Scheme (VPS), which takes sewage and wastewater, separates the waste, then purifies and pumps it through a privately owned pipeline and distributes it to agricultural users and market gardeners. The nutrient-rich water can be used productively to grow produce instead of being dumped into the gulf as pollution that destroys fish habitat, seagrass, and fish life.

Another great example is the pioneering work Adelaide is doing to harness nature and to work with the natural watercycle through stormwater harvesting.

A sustainable water city will have many diverse, small sources of fit-for-purpose water. However, solving the bureaucratic challenges and finding the political will to make this happen on a large scale has proven very difficult.

What prevents this happening on a large scale? Tough political decisions like putting water prices up; higher prices that don't increase government dividends but pay for the sustainability; investing in water infrastructure and sources that create urban wetlands, keep more water in the Murray, and stop dumping pollutants into streams and the Gulf; water sources that create instead of destroy the habitats of birds, fish and plant life.

Some politicians believe it is more important to provide cheap water, even if it is unsustainable, rather than providing sustainable water that costs about an extra $1 for 1000 L or an extra $20 a month for the average Adelaide home. In January 2008, the South Australian government confirmed this:

The use of other sources would have economic consequences, as the river provided a relatively cheap water supply … infrastructure for recycling water, stormwater harvesting and desalination were expensive … changing the city's entire supply from the River Murray to these other sources would be expensive and would have economic consequences in terms of the price charged for water.[3]

Water governance or water supply problem?

If you relied on the media for understanding water you would probably think that Adelaide is running out of water. The story generally goes something like: increasing population, climate change and drought are combining and we have no option but to drastically reduce our water consumption. While some of this is true (we do have an increasing population, and climate change and drought are facts of life, and we definitely need to optimise water use), the conclusion that we are running out of water is wrong. We have a *water governance* problem not a *water supply* problem.

A water governance problem doesn't mean the people who work as managers in the myriad of current and previous government organisations involved with water are the

What is desalination?

BOX 104

Desalination is the process of reducing the concentration of salt and other minerals in water. Desalination technologies are used worldwide, with a capacity to produce around 40,000 GL of desalinated water every day,[1] mostly from seawater but also from brackish water and treated wastewater. Most desalination occurs in coastal regions, particularly in the Middle East, with other major plants in Europe and North Africa and major growth occurring in the United States, China and Australia.[2]

There are many ways to desalinate water. The most appropriate method to use is dependent on the site, the volume and quality of water required, the quality of the influent water, and the availability of energy to power the process.[3] The various methods of desalination fall into two main categories:

1. Desalination based on changing the physical state of water. For example, distillation, the most commonly used desalination technology, uses heat to vaporise freshwater and separate it from brine via one of various processes: Multi-Stage Flash; Multi-Effect Distillation; and Vapour Compression).[1,3] Alternatively, freezing water naturally excludes salts, but is difficult to put into practice and therefore has limited applications.[3]
2. Membrane-based desalination. Reverse Osmosis is a popular method of desalination for new instillations and involves pressurising water and forcing it through a membrane that prevents salts from passing through.[1,3] Another method is Electrodialysis, using electrical currents to direct salts through membranes, leaving freshwater behind.[1,3]

There are advantages and disadvantages in the use of desalination to supplement city water supplies. The primary advantage of desalination plants over other water supply options is that their production is not weather-dependent; they can produce water regardless of rainfall in accessible catchments. Thus, desalination can provide a high level of water security. This feature of desalination makes it a very attractive technology in southern Australia where water resources are under pressure from growing populations, rainfall is highly variable, and the changing climate is creating uncertainty over future rainfall.[4] Disadvantages that must be considered are:

1. Capital investment – it is expensive to build a desalination plant.[4]
2. Ongoing costs, particularly in areas with variable rainfall – when local rainfall is good, desalination becomes redundant but the plant must be maintained.[4]
3. Energy use – desalination requires large amounts of electricity.[2,4]
4. Degradation of the marine environment – including damage to organisms and changed environmental conditions at water intakes;[2] damage caused by the discharge of brine (desalination plants have a wastewater flow of increased salt concentrations);[2] damage where discharges contain chemicals from pre-treatment and cleaning processes for plant pipes and membranes;[2] and damage to the organisms when discharges are elevated in temperature (temperature is only an issue for thermal plants).[2]

Many of the environmental consequences can be mitigated through good design and operation.[2] For example, wastewater flows can be diluted and diffused. It must be stressed that the downsides listed above regarding the natural environment may be offset by environmental gains. That is, by using desalinated water a city may be able to reduce abstractions from surface or groundwater resources. Further, advances in desalination technology will improve its efficiency, and alternative energy sources, such as solar, wind and geothermal energy,[1,3] that can provide heat or electricity, reduce the problems associated with high-energy use.

It can be difficult to assess the viability of a desalination plant. The pros and cons are not directly comparable and the outcome of any analysis will depend heavily on the weightings given to each variable. Most importantly, the viability of a plant must be assessed alongside all other water supply options, and with a rigorous environmental impact assessment.

Philip Roetman

BOX
105

The pros and cons of desalination

Desalination is attractive since it is not dependent on the climate and it allows us to leave more water in the River Murray. However, energy consumption and brine disposal can contribute to environmental problems if not properly managed.

Adelaide's Port Stanvac desalination plant will produce 50 billion L of drinking water annually by taking about 350 million L of seawater out of Gulf of St Vincent everyday, filtering it to remove solid impurities and then squeezing the water through a very fine membrane. More than 99% of the dissolved salts cannot pass through the membrane, but water molecules can. The salinity is reduced from about 38,000 mg/L to less than 400 mg/L. About 45% of what comes through the membrane, or 150 million L/day, becomes drinking water, and the salt-concentrated liquid, often referred to as brine, will be returned to the gulf at approximately twice the salinity of the gulf.

Net impact: the gulf will have the same amount of salt dissolved in it but the concentration will be raised slightly until effectively dispersed. The impact of the brine could be a problem for the fish and seagrasses if the natural salt balance of the sea was altered significantly. The solution to this typically involves a pipe pumping the brine hundreds of meters out to sea where it then fans out into a number of sprays or diffusers. Typically, the discharge conditions set by the Environmental Protection Authority would require that the salinity return to background levels within a few hundred meters of the discharge zone.

On the energy front, desalination will require less than 4 MW/h of electricity per million L of drinking water produced. The alternative, pumping and treating water from the Murray, requires about 1.9 MW/h per million L. So the net impact is: desalination requires an extra 2.1 MW/h after deducting the avoided energy of not having to pump and treat from the Murray, and that could be sourced from a low-emission, renewable source like wind or solar. The net annual extra amount of electricity required to desalt 50 billion L of seawater instead of pumping it from the Murray would be equivalent to running a 12-watt globe for every person in Adelaide.

On average, 40% or 88 million tonnes of water is pumped to Adelaide homes from the River Murray every year, increasing to 179 million t in a dry year. Pumping from the River Murray consumes less electricity than desalination but it also reduces the available water for the natural river ecosystems and contributes to the decline of the health of the river.

Joe Flynn

Wind-powered desalination

BOX 106

There is enough power in the sky to desalinate seawater or to distil wastewater.

The $387 million, 24 MW Perth scheme produces 123 ML/day by reverse osmosis, powered by electricity from 37 wind turbines nearby. This meets 17% of Perth's needs. Other existing schemes are in the Torres Strait Islands, Hamilton Island and Dalby.

In South Australia today, 386 MW of electricity is being produced by wind turbines, with a further 341 MW planned. To provide the water for 17% of the 1,079,200 people in Adelaide would take 17 MW.

It is possible to use the power of the wind to distill water using a vertical axis turbine to stir and heat water directly, with no need to create electricity. The vapour given off is condensed by the incoming cool water, leaving impure water behind.

The background photograph shows a 6.5 kW prototype. Starting in winds of only 4 m/s and stalling itself in winds over 12 m/s, the rotor will gather 85% of the power in the winds near Adelaide, and protect itself from damage in high winds. The rotor fatigue life is over 20 years.

Thermal desalination, unlike reverse osmosis, needs no filters and minimal maintenance.

The unit is placed on flat ground without foundations, supported by the water ballast tanks. It can be moved easily if it is found to be in a local wind shadow.

The drinking of dirty water worldwide remains the biggest cause of death today after natural causes. These units are particularly suited to remote Australia and villages in the underdeveloped world, for example the arsenic-poisoned wells in Bangladesh.

Around the world, 80% of people live within 30 km of the sea. To reduce the problem of disposal of more saline water, units could also be mounted on a barge. Moored offshore, they would slowly fill themselves up with pure water, pumping the excess saline water back into the sea. When full they could be towed to dock or be emptied into another boat.

Another widely used existing technology that could be combined with wind power is vapour compression distillation, which uses the heat given off when vapour is compressed to heat the incoming water. These units could yield average outputs of 20,000 L a day of pure distilled water.

With adequate investment in technology, drinking water can be generated without the cost of electrical drives or the use of fossil fuels generating CO_2.

Charlie Madden

problem, it means the *system* is the problem. The system of governance includes policy, regulation, accountability, pricing, etc., and it is failing to manage our water sustainably.

Our water governance system wasn't designed. It is a bureaucracy that has slowly evolved over almost two centuries, bringing us up to now – where we have ended up with a mixed bag of good intentions, politics and self-interest that is trying, but failing, to resolve the competing needs of our community, our environment and our economy.

Recycling 100% of our wastewater and harvesting 75% of our stormwater[4] – a goal suggested by the Adelaide and Mount Lofty Ranges Natural Resources Management Board – would potentially provide 190 billion L of water, more than twice the 88 billion L we pump from the Murray in an average year.[5] Adelaide's water scarcity disappears if we are prepared to *pay* for the infrastructure to harvest, store and purify our stormwater and wastewater. We could give water back to our stressed Murray and our diminishing borewater reserves and revive the rivers and the animal and plant life that depend on them.

Desalination is another source – if we are prepared to pay for sustainable technology including brine spray diffuser networks and using electricity from wind or other renewable sources.

However, when it comes to new sources of water there is an elephant in the room or, more precisely, in every home's kitchen – agriculture.

CSIRO research indicates a typical 230 g home serving of steak requires 4640 L to produce.[6] If you eat a steak once a week, that will require 242,000 L of water a year, almost the same amount consumed by the average Adelaide home in 2007 (246,000 L).[2]

Producing beef, dairy or wool takes a lot of water to grow the plants the animals need to consume, however, producing less will not necessarily get more water into the rivers or storage. If it is irrigation water or catchment water – which stops inflows into rivers – being used to grow pastures, there is a genuine alternative use. However, if it is rainfall that is enabling pasture growth (for the animals to feed on), in many cases there is no genuine alternative use for that rainwater and the animal products may be a very good use of the vegetation and hence water. Knowing that diverting any water from an ecosystem has an effect, what level of trade off should we be making to obtain our food and clothing?

The vast majority of water taken from the Murray is used for agriculture, in fact 65% of all the water used in Australia is used for agriculture.[9] In 2005/2006 Adelaide accounted for less than 1% of the total water taken out of the Murray-Darling (Adelaide and associated country areas diverted 74 GL out of 9243 total diversions from the Murray-Darling Basin).[5] So, the amount Adelaide households are trying to save (state government targets) through permanent reductions in outdoor water use, 10 billion L, is 0.1% or 1/1000th of the total water taken out of the Murray-Darling.[8]

In South Australia, agricultural water use accounts for 75% of our water use, and more than half of that (54%) is used by dairy, livestock and pasture.[9]

Food is a necessity and the water to grow it is a necessity, we just need that food to be produced as efficiently as possible.

Nationally, a 17.3% improvement in the efficiency of our agriculture sector would be equivalent to all the water used by all the households in Australia![7] In South Australia, a 2% improvement in the efficiency of the water used by our dairy, livestock and pasture sectors would save 11 billion L of water, which is more water than the 10 billion L Adelaide is trying to save through permanent reductions in outdoor water consumption.[8]

There is plenty of knowledge about how to make agriculture more efficient. Irrigated land represents 0.5% of all agricultural land yet it produces 23% or 9.1 billion of Australia's gross value agricultural production.[9] While some farmers have invested in drip irrigation, more than 90% of irrigated land uses inefficient flood or spray, in which typically 44% of water is lost in getting the water delivered to the plant roots.[8]

South Australian pasture farms consume more than twice the water per hectare than equivalent farms in NSW.[10] The difference could partially be explained by South Australian soil types and the drier climate, but it also suggests there is room to improve the way we deliver and apply water. We also must confront the fact that some areas of South Australia may never be efficient pasture farms, which leads to the issue of food miles (how far food has travelled before you buy it).

Buying local has a lot going for it – freshness, local economic development and community cohesion – however, the much-touted claim that eating local reduces fossil fuel consumption has been turned on its ear by researchers at Lincoln University in New Zealand. In 2007, they published a peer-reviewed study challenging the food miles assumption. Instead of measuring a product's carbon footprint through food miles alone, they took a holistic, whole of life view to include all aspects of production including water use, harvesting techniques, fertiliser outlays, renewable energy applications, means of transportation and the kind of fuel used, the amount of carbon dioxide absorbed during photosynthesis, storage procedures, and dozens of other cultivation inputs.

Their conclusion: lamb raised in New Zealand and shipped 18,000 km by ship to the United Kingdom produced 625 kg of carbon dioxide equivalent per tonne of lamb, while British lamb produced 2580 kg.[7] So, when viewed holistically, it is four times more energy-efficient for someone in London to buy lamb imported from the other side of the world than to buy it from a farm just up the road. Similar figures were found for dairy products and fruit.

The key point: what food we eat, as well as how and where we grow it, makes a far, far bigger impact on our water usage than the water we use at home.

BOX 107

Wastewater management is a complex set of organised efforts and actions requiring proactive participation and contribution of various governmental and non-governmental stakeholders. Public Private Partnership (PPP) structures help in dealing with the complexities of implementing a wastewater reuse scheme, and the Virginia Pipeline Scheme (VPS), operating in the Northern Adelaide Plains (NAP), is a perfect example of this approach.

The VPS is built on the build, own, operate, transfer (BOOT) model for private sector participation. A tri-sector partnership between the SA Water (public sector), Water Reticulation Services Virginia (WRSV) (a private company), and the Virginia Irrigators Association (VIA) (representing the irrigators in the region) was instrumental in the successful implementation of the scheme.

As part of its environment improvement program, SA Water constructed a dissolved air flotation and filtration (DAFF) disinfection plant to treat lagoon effluent from the Bolivar wastewater treatment plant (WWTP), which produces class A reclaimed water that can be used for irrigation instead of being disposed of into the receiving waters.

The private water company won a contract from SA Water to access the output from the WWTP for 20 years, signed up clients for the reclaimed water, and built the water distribution system. As per the initial agreement, the project will be returned to the ownership of SA Water by WRSV at the end of the contract term.

As of 2005, the scheme supplied class A treated water to around 250 growers in the NAP. Most of these growers are market gardeners and have different cultural backgrounds. But the diversity of irrigators has not affected the functioning of the scheme. The scheme has been operating successfully since its inception, demonstrating the existence of a high level of community social capital.

Ganesh Keremane

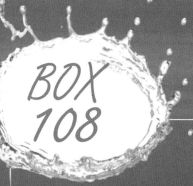

BOX 108

Adelaide, Australia's largest water recycler

Commissioned in 1998, the Virginia Pipeline Scheme (VPS) is one of the world's largest recycling schemes. A network of pipelines of more than 100 km supplies over 15 billion L per year of high-quality class A recycled water, suitable for use on food crops (consumed raw), from Adelaide's Bolivar wastewater treatment plant (WWTP) to market gardens and farms 35 km away on the Northern Adelaide Plains. More than 200 growers covering an area of 200 km^2 use the recycled water for horticulture irrigation. Irrigated crops include potatoes, various vegetables, olives, and grapes.

Importantly, it reduces the effluent Adelaide dumps in the gulf and the damage this causes to fish and their habitat.

A private consortium owns and operates the pipeline; at the end of the contract term, ownership transfers to the people of South Australia via the government. The project relied upon SA Water, using the expertise of their internationally connected operations and maintenance contractor United Water, upgrading the Bolivar WWTP to include a dissolved air flotation filtration process, resulting in high-quality treated effluent suitable for agricultural irrigation.

Recycled water typically has high levels of salinity that can build up in the soil so this risk is managed by an irrigation management plan, supported by an information and education program for growers and Environment Protection Authority monitoring. The scheme also reduces the pressure on groundwater resources in the Virginia region, the previous source of irrigation water.

The Willunga Basin Water Company operates a similar scheme taking recycled water from SA Water's Christies Beach WWTP, 10 km north of the Willunga Basin, and pumping it via 120 km of pipelines to more than 80 users, whose properties cover more than 1500 ha, to irrigate grapes, nuts, flowers, and community parks and gardens.

The pipeline, which began operating in 1999, continues to expand to provide others in the region with access to recycled water.

Joe Flynn

Harnessing nature to purify and store water

BOX 109

The City of Salisbury is implementing Australia's first totally integrated water management plan to efficiently harvest and manage systems for rainwater, stormwater, groundwater, recycled wastewater and potable water.

The city area covers some 161 km², extending from the Para Escarpment and the foothills of the Mount Lofty Ranges in the east to the shores of Gulf St Vincent in the west. Water flow from the Mount Lofty Ranges is harnessed and regulated in a series of flood-control dams constructed in the upper reaches of the catchment. These dams, constructed to handle a 1 in 100 year flood, divert the water into a series of pipes and open channels, planted with reedbeds to filter and cleanse the water of nutrients and heavy metals.

These pipes and channels are the arteries for the flow of stormwater into a network of some 20 harvesting sites, or wetlands, developed by the City of Salisbury. Stormwater is further cleansed in the wetlands and grassed swales, which act as self-sustaining filtration and water treatment systems.

Treated stormwater is pumped via distribution mains directly to industrial and commercial users with high-water dependency, while some is used for irrigating council parks and reserves, along with schools and sports grounds.

Cleansed of pollutants, much of the stormwater is 'banked' in underground limestone aquifers to be accessed for industry and the community in drier months through aquifer storage and recovery.

Experiments are being conducted with the CSIRO to determine the practicality of injecting cleansed stormwater into the aquifer where, through natural processes, it may be converted into drinking water.

Already, the City of Salisbury's integrated management plan is yielding 7500 million L a year and the city will add another 25,000 million L as part of the Water Proofing Northern Adelaide plan. This will make the City of Salisbury almost self-sufficient and will reduce demand on the River Murray in a non-drought year by 75%.

Joe Flynn

Adelaide needs to understand what are the foods we have a comparative advantage in producing, and become a sustainable food producing hub for this produce, while withdrawing from that produce we are not suited to growing. We must understand that buying local is not always beneficial for the environment and that while we encourage sustainable local food systems we must also allow them to develop in tandem with what could be more sustainable national or global alternatives. Renowned water expert Peter Cullen described the situation as one of inevitable adjustment.

So what can inefficient farms do to improve? They could leave their water allocation in the river and sell the rights to that water to someone else, perhaps to a more efficient farmer who can afford to buy the water, as they grow a crop that uses less water or the same crop but are more profitable as they have invested in precision irrigation technology. In 2008, 1.68 billion L worth of water changed hands in this manner. If SA Water's pipe network was open to third party use they could sell the water to communities in Adelaide who may want to maintain school grounds or parks with no restrictions. Perhaps they could sell the water to what is the most pressing need – the environment – and just leave the water in the river for nature to use. In 2004 the federal government allocated $3 billion to start buying back water entitlements for environmental flows. Unfortunately, at time of writing, not $1 has been spent.

The farmer selling the water could use the money to invest in precision irrigation technology, or they may come to the conclusion that their farm is unsustainable and choose to use the money to invest in another livelihood.

The point of trading is that water gets used where it is most valuable, including the critical value of maintaining river systems we depend on, often referred to as 'environmental flows'. It moves from someone who gets little value from it to someone who uses it in a smarter way – and is prepared to pay for it.

So, remembering the Water Proofing Adelaide strategy is trying to save 10 billion L of water through permanent reductions in outdoor water consumption, how much would it cost Adelaide to buy 10 billion L of water from irrigators wanting to sell their water? At the average 2008 price of $660 per megalitre, $6.6 million would buy 10 billion L – the equivalent of every person in Adelaide paying 12 cents per week.

Unleashing innovation – making things happen

Climate change has brought into sharp focus the need to live sustainably with the ecosystems that support life. As Nicholas Stern said:

> The scientific evidence is now overwhelming: climate change presents very serious global risks, and it demands an urgent global response. This response will require … price signals and markets for carbon, spurring technology research, development and deployment, and promoting adaptation.

The same tools of price signals and markets need to be applied to water.

What do those terms mean for Adelaide? They mean competing suppliers should be able to supply water and compete for the business of retail suppliers, who in turn would compete for individual customers of households and industry, selling the water to reflect the cost of production and scarcity. Suppliers could be households, schools, councils or business. Sources could be recycled water projects, aquifer storage, desalination, dams, or farmers who want to sell their River Murray entitlements.

What difference would water trading, competition and market pricing make to a typical Adelaide householder? Imagine this, you are at home and the phone rings … (see Box 144: Nature calling).

The fictitious example is based on a fictitious company called Nature's Water selling new water from aquifer storage reuse, a technology that Adelaide has pioneered and leads Australia in. A similar example could be made up based on selling recycled water or desalted water. The company could be a community cooperative, local government, a school, or anyone who wanted to start a water business and meet the health and license access regulations. Unfortunately, the legislation and regulations that govern how water is managed would make a company like Nature's Water unviable. The government-owned pipes are effectively a monopoly that can only be used by the government. This roadblock needs to be solved if we are to encourage diversity.

What is my role? I'm confused

In many of life's essential services, governments around the world have figured out that their role is best served by being the regulator who sets the standards, monitors compliance and enforces the law. That's how it works for food, electricity and housing. Anyone can produce and sell food, electricity or housing as long as it is safe and complies with government regulations. If you don't comply, the government may fine you, remove your license to supply, or even send you to prison.

Infrastructure industries like telephones, rail, gas and electricity have also enabled competitors to share one delivery system. You can buy your home telephone services from any number of telephone retailers who all share the same wire that runs to your house. You can buy your electricity or gas from any number of energy retailers who share the gas pipes or electricity lines that run to your house. Electricity suppliers include anyone from homes who sell their surplus solar-generated electricity back into the grid to privately owned wind farms, local governments who generate electricity from methane gas harvested from rubbish dumps, to government-owned generators.

This is called 'disaggregation'; if water was disaggregated it would mean the monopoly pipes part of the business would be separated from the competitive part, supply and retail. The experience from disaggregating other infrastructure industries shows that this process:

Solving urban water storage

BOX 110

An Adelaide company has developed a low-cost method to enable the capture and subsequent use of stormwater in urban areas. Until now, the only means of detention and storage have been large, purpose-built detention basins or wetlands, and most heavily developed urban landscapes lack the space for these structures. It has been difficult to use stormwater because of the problems in intercepting, storing and transporting it, however, Suburban Water has worked out a novel solution to these problems with its virtual detention structure. Existing stormwater drainage systems are used to distribute the captured water to areas of potential demand.

Rainwater harvesting systems developed in Adelaide are enabling businesses and homes to harvest runoff from their property or carpark, and then treat, store and maintain the rainwater to provide a high-quality alternative to mains water for outdoor use. Homeowners could benefit by being paid for the water provided by their tanks through a rebate on their water bills.

The product is a completely integrated water storage and reuse system, which incorporates a range of smart integrated technologies that function to maintain water quality within the storage. If the average Adelaide quarter-acre block harvested 50% of their 455 mm annual rainfall, they would harvest 230,000 L of water, considerably more than the 2007/2008 average South Australian household consumption of 194,000 L.

Joe Flynn

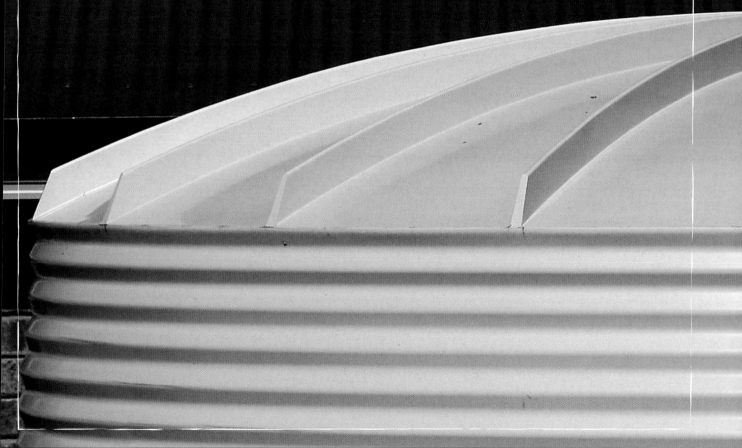

- separates policy and regulation from ownership;
- encourages competition and innovation; and
- encourages efficiency through clear focus versus the many, often conflicting objectives, that government-owned entities must pursue.

These lessons are very relevant to the problems we face with water supply and management.

We know what happens when Australia's state governments can't supply enough water – they blame the drought, use legislated authority to impose restrictions, and fine customers if they breach them.

How are supply shortages caused by hot weather dealt with in an independently regulated industry such as electricity? In January 2006, Adelaide experienced a four-day heatwave of consecutive 40 degree days, which led to widespread interruptions and 560 customers losing power for more than 24 hours. The result: the government regulator forced the privately owned electricity company to pay customers more than $1,000,000 in guaranteed service level payments and within three months the company announced they were investing $42 million to improve reliability.[11]

The National Water Commission's chairman, Ken Matthews, has highlighted the lessons we can learn from other reform.

For too long, water has been the forgotten cousin of infrastructure. Water reform has lagged behind the reforms undertaken over the last twenty years in energy, road, rail and Australia's ports. There is much to be learned from the reforms in the energy market. Both water and electricity are essential services. Both have been subject to public ownership, a lack of competition and no interstate trading. In the 1980s it was believed fundamental reform could not occur in either sector for a range of political and technical reasons. Yet one sector – the electricity sector – did embark on a fundamental reform journey. Whilst the reforms are still being progressed there has been major change – increasing competition, opening up trading between states and new ownership arrangements. The water sector now needs to actively embrace a similar reform agenda.

Australian state government-owned water utilities work under very difficult and conflicting economic, environmental and political objectives, and when faced with water scarcity during a drought have responded by imposing restrictions and ensuring their revenue is protected by having a large gap between the price charged to customers and the cost of supply. This allows the reduced supply to be balanced with reduced demand, gets the city through a drought, and largely preserves the dividend paid to the owner – the state governments.

In 2006/2007, while Adelaide was living with Level 2 and Level 3 water restrictions, SA Water collected a total of $678 m from customers and gave 30% of this ($208 m), to the government in dividends.[2] While a lot in cash terms, this represents a 6% economic rate of return[12] on the assets, and is money the government can spend on health and education.

Unfortunately, the dividend comes at an enormous cost. New jobs that could come from a more dynamic water industry are constrained, the community suffers with parks, sports grounds and gardens turning brown, businesses relying on water, like nurseries, suffer under restrictions, the Murray continues to suffer as alternative sources of water aren't encouraged, and we continue to treat the vast majority of our stormwater and used water as waste instead of recycling it. Innovation and sustainability is stifled.

Importantly, it needs to be emphasised that this isn't about the people in the current or previous government doing anything wrong. Quite the opposite; the public service is filled with talented people doing exactly what the governance system they operate in requires them to do. It is the governance *system* that needs fixing!

If our government water supply wasn't a monopoly and instead operated in a competitive market, new sources of supply would be created, like stormwater harvesting, desalination, recycling or purchasing water from irrigators who want to sell their River Murray entitlements. New water would continue to be made available to the residents of Adelaide until the incremental cost of new sources of supply reaches the 'ouch' point (when the cost matches the price consumers are prepared to pay). This is what happens with life's essentials like food and housing – why not do the same with water?

A common reason cited is that social justice demands keeping water as cheap as possible. An alternative would be to let water bills go up to pay for sustainable water and at the same time introduce a levy, which would help large families and low-income people with their water bill.

Research indicates that Adelaide households are a long way from the 'ouch' point, and may be willing to pay an additional $270 per person a year to avoid restrictions.[13] Are the politicians right in judging what the community is prepared to pay for? What is the strength of community concern about water and the willingness of citizens to adopt new water systems? Do people buy rainwater tanks in response to government subsidies or by calculations of 'payback' in savings from reduced water charges, or is it just the desire of people to act, to 'do something' as a personal contribution to a perceived, collective problem?

Future sources of water for Adelaide

If we diversified our sources of water for Adelaide, how much would we pay? In 2008 the typical residential water bill was $299.94.[17]

One consequence of Australia's low water prices is that alternative solutions look uneconomic. An alternative would be to bill customers the true economic and environmental cost while also providing targeted support to those in society who can't afford it or are in remote locations. Alternative local water supply solutions could be accurately assessed against the true cost and the most economic solution that meets the environmental and social needs of the community.

Precision irrigation: an urban and rural tool

BOX 111

Adelaide is home to some of the world's leading precision irrigation technology providers, which take the guesswork out of watering.

Adelaide companies like WISA, Micromet, Sentek and AquaSpy are global leaders of monitoring, control, communication and delivery systems, enabling rural irrigators such as farms and market gardens, as well as urban irrigators like golf courses, parks, sports fields and home lawns and gardens, to reduce water consumption while improving turf or crop health.

The technologies can monitor soil moisture content, communicate with weather satellites, monitor salinity, and then use software to calculate and direct when water is applied, at what rate it is applied, and precisely how much is needed. Water is delivered exactly where and when it is needed, in precisely the right volume. Benefits typically include water savings of 20–70%, crop yield increased by up to 50%, deeper and stronger root growth, reduced pumping costs, less mowing time for turf, and greater resistance to disease.

Joe Flynn

As an example, look at how rental assistance functions. The price of rental accommodation is set by the market when supply meets demand, and the government then provides targeted rental assistance to those who need it. Why not apply the same principle to water?

Figure 25.1 lays out the water options – including purchasing more water from the Murray River, desalination, recycling, new reservoirs and reducing losses – that were considered in the Water Proofing Adelaide strategy. For each of these it shows the cost for each tonne (equal to 1000 L) of drinking water delivered to an Adelaide home, and estimates how many billion litres could be provided annually.

The assumptions in the graph prepared in 2004 have been overtaken by the speed of the dramatic downturn in the inflows into the Murray (impacting how much we can reliably draw on) and the decision to invest in a 50 GL desalination plant for Adelaide. The 'current sustainable supply' of 218 GL referred to in Figure 25.1 needs to be revised down by about 50% to reflect best-practice climate change modelling and costs for the various supply sources updated. The cost of existing supply for 250 kL has increased from $1.25 to $1.68/L, reflecting the 2008 water charges. However, the trend the graph presents is still valuable information: there are many alternative sources of water available if we are prepared to invest in them.

What the graph shows is that if water prices are increased by 50–100%, a range of new supply options become economically viable. Once we start paying $3 (or less with the decreasing costs of technology) for every 1000 L of drinking water, desalination, recycling, reuse of stormwater all become feasible. This would be about an extra $22 a month for the average Adelaide home. It also shows how individual home-based solutions like greywater reuse and rainwater tanks cost around five times as much per litre of water than shared community solutions.

Designing sustainability into an industry

The 1992 Rio World Environment Summit and the 2003 Johannesburg World Summit on Sustainable Development endorsed a number of objectives to improve water management, including the transfer of operational responsibility as far as possible to the private sector. This objective is to separate the specifier and regulator of water services (government) from the provider or operator (the private sector). This is a reform trend that became popular through the 1980s and 1990s as governments in Australia and around the world improved the delivery of government and monopoly services by adhering to market mechanisms and addressing regulatory failure.

To the credit of the current and previous South Australian governments, the benefits of private sector involvement in the water industry have been recognised, and South Australia has led the nation in creating alliance partnerships where the private sector and government can work together. This has opened the door to the VPS, the Rural Water Filtration Project, Mawson Lakes, the Willunga Basin Water Company,

and the outsourcing of metropolitan water operations and maintenance. Collectively, partnerships have helped reduce operating costs and improve water quality while stimulating local jobs. Over the same period, national and international exports of water-related technologies and services from South Australia grew 1600% from $25 m in 1996 to $400 m in 2008.

One of the attractions of alliancing is that the contractor's revenue is typically tied to performance. If performance of the water network declines, payments to the contractor decline, providing an enormous motivation to improve their profits by improving the performance of the water system. This motivation is difficult if not impossible to replicate inside public sector monopoly organisations where people are generally more risk averse and following policy can take precedent over pursuing continuous improvement.

Competition and involvement of the private sector in water is about unleashing the incentives that will encourage and reward a broader range of innovation. It is about bringing discipline to the provision of services, measurement of performance, and consequences when perform standards aren't reached.

Charles Landy, an Adelaide thinker in residence, highlighted the importance of Adelaide encouraging innovation:

> If incentives were given for people to be creative it would lead people and organisations to tread new pathways and establish new preferred routes … Being creative is a mindset … Creative people work at the edge of their competency, not at the centre of it. This sits with difficulty in large organisational structures, and in particular public organisations, whose attitudes to risk are tempered by accountability issues, although these can sometimes be a cover for avoiding action. Risk, failure and bureaucracy are uneasy bedfellows as people rarely acknowledge failure as a learning device. Failure is seldom total and elements of a failure may be useful for future projects. Importantly, we need to distinguish between competent and incompetent failure. Aversion to risk taking can mean that an institution has no internal R&D mechanism to weed out future failure. For Adelaide, which has a particularly strong administrative culture simply by dint of its large public sector, these issues linked to the creativity agenda pose strong challenges.[15]

The principles of this change in governance can be thought of in biodiversity terms. The variety, distribution and abundance of water suppliers and water sources can be treated as a proxy for living organisms in an ecosystem. We know that maintaining biodiversity of living organisms promotes sustainability and resilience of ecosystems. The same principle would suggest that a diversity of water suppliers and water sources promotes sustainability and resilience of our water. If the opposite of biodiversity is a monoculture, then that is what a monopoly approach represents – and we know monocultures leave an ecosystem vulnerable to plagues and changes in the environment.

Prices of alternative sources of water

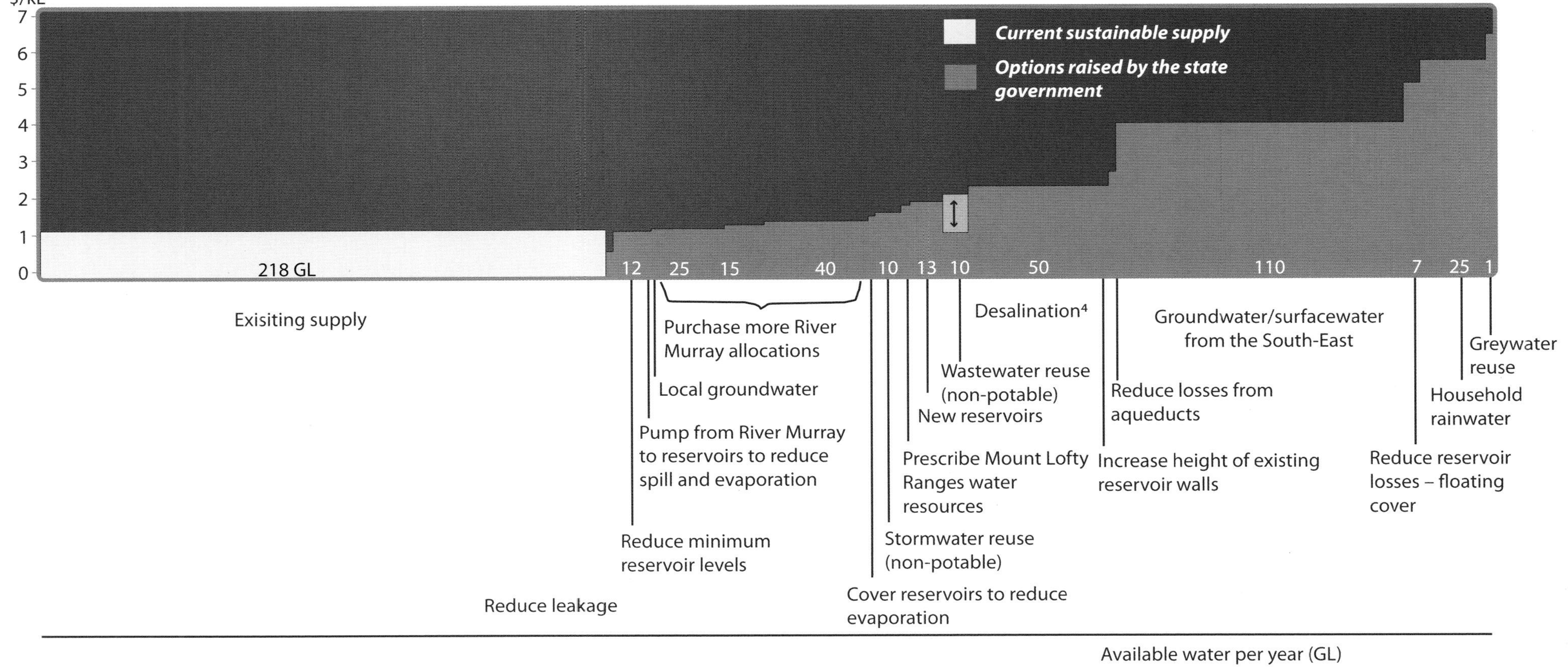

Cost to Adelaide consumers
$/kL

Current sustainable supply

Options raised by the state government

218 GL

Existing supply

Available water per year (GL)

Purchase more River Murray allocations

Local groundwater

Pump from River Murray to reservoirs to reduce spill and evaporation

Reduce minimum reservoir levels

Reduce leakage

Desalination⁴

Wastewater reuse (non-potable)

New reservoirs

Prescribe Mount Lofty Ranges water resources

Stormwater reuse (non-potable)

Cover reservoirs to reduce evaporation

Groundwater/surfacewater from the South-East

Reduce losses from aqueducts

Increase height of existing reservoir walls

Greywater reuse

Household rainwater

Reduce reservoir losses – floating cover

1. Based on usage of 250 kL pa and the 2003–2004 price of $0.42/kL up to 125 kL, $1kL >125 kL and a fixed fee of $135 pa.
2. A preliminary estimate suggested 3 GL of leakage reduction is possible at ~$0.7/kL, further reductions may be possbile but at a higher cost.
3. These options would decrease the security of supply from reservoirs.
4. Costing of desalination based on 50 GL, potential available water may be unlimited.

Figure 25.1 Current price of Adelaide water compared to alternative sources of supply. The numbers at the bottom of this graph indicate the available water (GL) per year that each different option could provide. For example, using groundwater/surfacewater from the South-East of South Australia adds an extra 110 GL to the existing supply of 218 GL.
Source: Business Council of Australia, 'Water under pressure: Australia's manmade water scarcity and how to fix it'. An adaptation from BCG research using additional information from 'Waterproofing Adelaide: exploring the issues – a discussion paper', South Australian government, 2004, 2006; WSAA facts, Water Services Association of Australia, 2003.

503

Table 25.1 Water reforms

Step	Explanation
Give the environment what it needs	We need to care for systems that sustain our existence and immediately reduce the stress on our rivers, lakes and groundwater. We do that by buying water entitlements from willing sellers, reducing over-allocated groundwater entitlements, and returning the water to the environment. When recycled water is applied to agriculture it should be directly linked to a reduction in groundwater entitlement.
Encourage diverse suppliers and diverse supply sources	Replace physical water restrictions with properly functioning, competitive water markets where water can be traded and suppliers can compete to supply water from sources like recycled water projects, aquifer storage, desalination, dams, and trading entitlements. These suppliers could be individuals, schools, community groups, councils, and businesses. They would share appropriately controlled safe access to the publicly owned pipeline network and would compete for the business of retail suppliers, who in turn would compete for individual customers of households and industry.
Independent regulation	Environmental standards, price, access, quality, and other important controls and rules of supply and competition should be governed by an independent regulator with financial penalties for non compliance. Water pricing will result from the scarcity of supply and the costs of the new sources of supply. Targeted support would transparently assist regional locations and low income people to ensure water for basic household needs remains affordable for all.
Remove all barriers to water going to where it is needed	To ensure water is used where it is most valued, we need a shared water registry and transfer system to distribute the 154,023 entitlements for 21,642,769 million L of water entitlements in New South Wales, Victoria and South Australia. This is critical to returning over-allocated systems to environmentally sustainable levels of extraction and forms part of the 2004 Intergovernmental Agreement on a National Water Initiative.
Simplify water recycling and aquifer storage reuse	There are many impediments that make water recycling, stormwater harvesting and aquifer storage reuse complicated and slow, like complicated regulatory processes and low urban water pricing. Removal of such impediments is needed to promote these sustainable sources.
Expand the National Water Initiative	The National Water Initiative – Australia's overarching water policy that drives the agenda for state governments – needs to be expanded to cover urban water issues. An output of this will be to improve our urban water planning to world's best practice.
Make water reform happen	Ministerial-level responsibility for water policy is spread across a mix of water, environment, infrastructure and planning at both the state and federal levels. This leads to complexity in policy development and implementation. The Council of Australian Governments should increase its focus on water to ensure progress is being made and any impediments are being removed. Action milestones are needed to measure the progress towards removing trading impediments and the introduction of effective urban water access regimes. Outcomes need to be defined like reducing stress on the Murray and achieving urban water supplies that can meet future needs. Incentive payments from the commonwealth to the states should be based on the achievement of the agreed milestones and outcomes.

Table 25.1 The water reforms necessary to create a sustainable water management system.

We now face a horrible readjustment in rural Australia, similar to that faced in Goyder's time, when three to four wet years led to unrealistic expectations. We have had 40 years of unusually wet conditions in the Murray-Darling Basin, from 1950–1990, but the last 15 years have seen the basin drying. Our refusal to recognise this meant we have allowed the major storages to empty, and they will not refill again without a run of wet years. The lack of water is having a devasting impact on irrigators and communities, as we have already seen the death of some permanent plantings in the basin, and more can be expected due to lack of water and salinisation of water.

Irrigation landscapes will change. We expect a reduction in the area of permanent plantings and perhaps more emphasis on annual crops that can be planted once water availability for the season is known. Irrigation properties may become larger, to cope with a mix of perennial and annual plants and more opportunistic irrigation. There will be an overall contraction in the area irrigated, leading to issues of stranded assets and increasing operating costs to those remaining.

There will be ongoing downward pressure on prices, especially in export markets, with the value of water increasing due to scarcity and pressures for cost-reflective pricing. There will be opportunities to leave farming, and strong pressures to improve water-use efficiency – in some enterprises such as dairy there is room for this, whereas many efficiency improvements have already been made in rice and cotton enterprises. We may see dairy farms leave irrigation areas and become purchasers of fodder from irrigators and other sources. New irrigation technologies like subsurface tape may see crops move from heavier to lighter soils.

A critical issue for irrigation is whether to spend funds refurbishing old irrigation districts, where properties may be too small and irrigation layouts inappropriate for current irrigation technologies, or to allow water to trade out and develop new irrigation enterprises on Greenfield sites. Refurbishing last century's infrastructure may be silly when whole regions must be reconfigured with bigger farms and different irrigation layouts.

We have a tremendous opportunity to build an irrigation sector that can double the wealth obtained from around half the water.

Joe Flynn and Peter Cullen

BOX 113

Waterfind Environment Fund

Healthy Rivers Australia is an Adelaide-based, not-for-profit organisation dedicated to improving the health of Australian river systems by building Australia's largest environmental water bank. The water bank acts as an insurance against drought and ensures there will be some environmental water available for environmental projects well into the future.

In 2008, Ian Denny, a shack owner on the River Murray, was the first donor of permanent water, assisting Healthy Rivers Australia to make their first delivery of environmental water to a drying wetland called Little Duck Lagoon, near Berri in South Australia.

In 2008, Healthy Rivers Australia recruited over 30 volunteers prepared to offer their time and their farm dams as temporary refuges for endangered native fish suffering through low river levels.

Haigh's Chocolate have found a great way to donate. Everytime you purchase a Haigh's Chocolate Murray Cod, a percentage goes to Healthy Rivers Australia.

Healthy Rivers Australia chair, Dr Mark Siebentritt, said anyone can contribute to the Healthy Rivers Fund and all donations over $2 are tax deductible.

Dr Mark Siebentriatt said the web-based system provided a way for urban and rural communities to become more involved in helping to improve the health of Australia's river systems.

'We can use donations of money or water and all our environmental projects are fully accredited,'
Dr Siebentriatt said.

Joe Flynn

Nature calling

BOX 114

'Hi, this is Nature's Water. Would you like to say goodbye to water restrictions and do your part to help the Murray River by buying all your water from a sustainable source?'

'Maybe. Tell me more.'

'How it works is we make our own drinking water by harvesting all that rainwater that used to run out to sea. We let nature purify it in wetlands, then we store it in underground dams, then when we need it we pump it up, purify it to World Health Organization standards and put it into the Adelaide water pipes.'

'Sounds pretty good.'

'It is good. By buying your water from Nature's Water you do your part to beautify Adelaide and create more wetlands around to city, which means more frogs and birds. We're actually really proud of our wetlands and invite our customers to visit them and see what you have helped create. Also, by switching to Nature's Water you help the Murray River get healthy again. In 2007, 91% of your water came from the Murray. Another groovy sustainable thing we do is all the electricity we use comes from renewable sources like wind and solar so you do your bit to cut greenhouse gases. You help Adelaide become a sustainable city.'

'How does that water get to my home?'

'Well if you think of Adelaide's water pipes as a river, we add water to that river and you can then get that same amount of water out somewhere downstream.'

'That makes sense but what's the catch? Do I still have to have water restrictions?'

'No more water restrictions is one of the great advantages with Nature's Water. Of course, we encourage our customers to follow the government's water conservation principles, like not watering your garden in the middle of the day and using five-star water appliances and so on, but if you become a Nature's Water customer you can use water whenever you want and say goodbye to compulsory water restrictions.'

'Mmmm, my lawn and garden will love that, not to mention being able to let the kids play under the hose. How much will it cost?'

'You will still pay whatever you pay the government to be connected to their pipes. On top of that our drinking water prices start at 0.2 cents a litre compared to the 0.1 cents a litre you pay now. For an average Adelaide home it will be less than $20 extra a month to have no restrictions and know you're doing your bit to help the Murray.'

'What do I have to do to switch to Nature's Water?'

'Its really easy. I just take a few details from you and leave the rest up to us. The only difference you'll notice is you will stop receiving your usual water bills and just get one bill from us. Oh, and the other difference you will notice is a few more birds and wetlands around the city, and no more water restrictions.'

Joe Flynn

A plan to make sustainability happen

If 'sustainable development' describes progress that meets the needs of the present without compromising the ability of future generations to meet their needs, then sustainable water management needs to become much more adaptive and holistic and adaptive to a growing population, climate change, community attitude, politics, and our economy. Holistic in counting environmental needs and costs, in treating groundwater, rivers and wetlands as interconnected, and in connecting the separate pieces of government that collectively form our water governance system.

At a practical level, we need to lift water restrictions while maintaining the education and promotion of sensible, permanent water conservation measures such as avoiding sprinklers in the middle of the day. We need to continue reducing our reliance on climate-dependent water we draw from the River Murray and from the Adelaide Hills catchments by developing a diverse supply mix including:

- buying water from Murray River irrigators;
- creating drinking water through desalination of seawater;
- purifying and reusing more of our wastewater; and
- harvesting more urban stormwater.

Table 25.1 shows the reforms required to embed this adaptive governance system, which will cope with the competing water needs of the environment, society and the economy.

Finding the will to make it happen

None of the steps outlined above are fresh ideas. Industry and various water experts have been promoting them, in part or in full, for some years. We could probably make it all happen in less then five years if we really wanted to. To date we have been prepared to make the environment pay, to make future generations pay, and to make our economy pay, initially through ignorance of the consequences of our actions but that excuse ran out long ago. In recent years we simply haven't been prepared to find the political will to make the necessary reforms and start paying for sustainable water solutions.

We can blame our politicians for procrastinating but in a democracy like Australia, governments largely reflect the will of the people and that means we have failed to give politicians the clear message that we, the community, demand and are prepared to pay for sustainable water management.

We won't solve our water problems by using the same thinking that created the problem. A fresh approach is needed. Once we may have done that out of goodwill towards the environment. We now have the knowledge that we *need* to do this to not only sustain our environment but also to sustain our economy.

It is up to each one of us to make a difference in our sphere of influence – at home, at school or at work. To be the voice for change and be the change we want to see.

References

1. D. Paton, 'Lessons learned from the Coorong' (presentation), 2007 (online).
2. South Australian Water Corporation, 'SA Water annual report 2006/2007', 2008 (online).
3. C. Jenkin, 'Restrictions will stay', *Advertiser*, 16 January 2008.
4. Adelaide and Mount Lofty Ranges Natural Resources Management Board, 'Regional NRM plan (draft for consultation)', 2007.
5. Murray-Darling Basin Commission, 'Water audit monitoring report 2005/2006', Canberra, 2007 (online).
6. W. Meyer, 'Water for food: the continuing debate', CSIRO Land and Water, Adelaide, 1997 (online).
7. C.M. Saunders and A.J. Barber, 'Comparative energy and greenhouse gas emissions of New Zealand's and the United Kingdom's dairy industry', AERU research report no. 297, Agribusiness and Economics Research Unit, Lincoln University, Christchurch, 2007.
8. Government of South Australia, 'Water proofing Adelaide: a thirst for change 2005–2025', 2005 (online).
9. Australian Bureau of Statistics, 'Water account, Australia, 2004–2005', 4610.0, Canberra, 2006 (online).
10. P. Cullen, 'Water challenges for South Australia in the 21st century', Department of the Premier and Cabinet, Adelaide, 2004 (online).
11. Essential Services Commission of South Australia, 'Inquiry into ETSA Utilities' network performance and customer response, January 2006', Adelaide, 2006.
12. Water Services Association of Australia and National Water Commission, 'National performance report 2005–2006', Melbourne, 2007.
13. Allen Consulting Group, 'Saying goodbye to water restrictions in Australia's cities: key priorities for achieving water security', Infrastructure Partnerships Australia, 2007.
14. The South Australian Centre for Economic Studies.
15. C. Landy, 'Rethinking Adelaide: capturing imagination', Department of the Premier and Cabinet, Adelaide, 2003 (online).
16. National Water Commission, 'Australian water markets report 2006–2007', Canberra, 2008.
17. Water Services Association of Australia, 'National performance report 2007–2008 urban water utilities', 2008.

Box references

Box 104: What is desalination?

1. World Health Organization, 'Desalination for safe water supply: guidance for the health and environmental aspects applicable to desalination', 2007 (online).
2. S. Lattemann and T. Höpner, 'Environmental impact and impact assessment of seawater desalination', *Desalination*, 220, 2008, pp. 1–15.
3. URS Australia, 'Economic and technical assessment of desalination technologies in Australia: with particular reference to national action plan priority regions', prepared for Agriculture, Fisheries and Forestry Australia, Canberra, 2002 (online).
4. Productivity Commission, 'Towards urban water reform: a discussion paper', Melbourne, 2008.

Introduction

This brief discussion is about local government and water use, including technologies and attitudes, in relation to the urban landscape. An historical overview is combined with a look at the current situation and emerging issues for the future.

Material has been gathered through a series of interviews with eight of the 16 Adelaide-based metropolitan councils. Interviewees included local government staff such as engineers, horticulturalists, planners, landscapers, irrigation specialists and environment officers working with the cities of Playford, Tea Tree Gully, Salisbury, Adelaide, Burnside, Unley, Mitcham and Onkaparinga. Councils were selected to reflect a diversity of geographic and demographic circumstances across the metropolitan area. Many thanks to the staff members of these councils for their willing and cooperative contributions.

The information regarding metropolitan councils' water consumption at a consolidated level has been sourced from SA Water.

Looking back

During the early years of the 20th century, Adelaide was watered by impact sprinklers, hoses with bayonet fittings, travelling tractors, fluming and coupling pipes, rows of sprinklers shifted regularly by caretakers or gardeners dedicated to particular ovals or reserves, with equipment often stored in site huts.

Some areas of Adelaide were very green and others not irrigated at all. Prior to the 1960s for example, Hazelwood Park in Burnside received no irrigation. The park only became green and ornamental with the installation of the swimming pool. Verges, median strips and many reserves around Adelaide were not watered at all. As development progressed and the metropolitan area grew, so did the spread of green open spaces.

During the 1950s, 1960s and 1970s many of the councils with larger geographical areas employed large numbers of field staff to look after the many new ovals and reserves. This required continuous watering regimes, which still used manual systems and dedicated staff.

The aim of the Adelaide City Council in caring for the Adelaide parklands, the 'gardens of the people', was to keep them green. The desire for green is perceived to be a result of the influence of European landscapes; in Adelaide there were lawns and ovals, and many planted trees and European-style garden beds. As all irrigation systems were manual, thousands of tap covers still dot the parklands.

Similarly, from early days the Salisbury community fought the 'dust bowl' of northern Adelaide by establishing large areas of green grass that were extensively irrigated.

Mitcham Council's experience was a little different in that until the 1980s they managed minimal green space areas; approximately 15 areas, consisting of ovals and community spaces, were irrigated. During the 1980s there was a change in community perceptions about green space; expectations became greater and people wanted more. Prior to the 1980s having open space was sufficient, but suddenly it all had to be green. Leading up to the year 2000, Mitcham Council installed more than 80 automatic irrigation systems as the demand for green spaces increased. This demand is understood to be the consequence of developments like West Lakes, which were based on the leafy, green 'California' marketing model. Unley Council's experience was similar to that of Mitcham, with the area of community green space increasing significantly in the 1980s.

Likewise, Tea Tree Gully found that the 1980s Golden Grove development changed community expectations of open space in their council area, leading to the irrigation of places that historically had never been watered, such as verges. In the Golden Grove development flood irrigation was the norm, swales were watered, water regularly flowed down the road and tractors became bogged in summer. Prior to the development open space tended to be dry in summer and green in winter, with exceptions for some irrigated reserves. The legacy of this development lives on, with Golden Grove consuming more water than the rest of the council area even today.

In general, from the 1960s the emphasis was on green, green and even greener. Should a patch of grass turn brown, the answer was to turn on the sprinklers for longer. There was little science involved with irrigation, although some councils chose not to water if it was raining. During most of the 20th century, water was simply considered an abundant resource to be used to maintain expanding green acreages of grass and gardens. The community increasingly expected that ovals and reserves, lawns and garden beds should be well irrigated and lush.

Adelaide was watered from various sources. Water for irrigation was either from the mains system (River Murray and local catchments), pumped from bores or pumped from rivers. The Adelaide Cricket Ground has been irrigated for over 100 years from the Torrens Lake, and this is still the case today although blue-green algal blooms in the lake affect its reliability. Historically Tea Tree Gully drew significantly from the Torrens and essentially all areas close to the river were watered from it. Small rivers and creeks were also important water sources. Prior to reticulated water becoming available in the 1870s, Unley was supplied largely from Brownhill Creek and its tributaries. Bores were sunk all over Adelaide, some proving more successful than others, and many are still used by councils for irrigation today.

From the 1980s automatic pop-up sprinkler systems replaced manual systems, with several councils reporting that while making life easier these systems tended to result in over-watering. During the 1980s and 1990s several councils connected their irrigated spaces to the externally controlled Micromet pop-up sprinkler irrigation system, designed to match irrigation patterns with weather patterns. Sites irrigated under Micromet control have included sports ovals, reserves, parks, median strips and verges. For Playford Council, installing Micromet was an opportunity to rationalise water use, as well as taking account of public perception that it was

Water Proofing the North

BOX 115

A new major project to the north of the city in the council areas of Tea Tree Gully, Playford and Salisbury will see the development of Adelaide's largest integrated water management system, which has the potential to significantly change the landscape over the next 10 years. The Water Proofing the North project will build on the already significant number of stormwater recycling projects in the region and integrate these into a system of 28 wetlands and aquifer storage and recovery (ASR) projects on the Torrens River, Dry Creek, Helps Road and Smith Creek catchments. Local aquifers will be used to store this water once it is treated and harvested, and a series of bores will extract the water during summer for a range of 'fit-for-purpose' uses including irrigation of parks, private gardens, industry and horticulture. The storage of water in aquifers will provide a 'water bank' where water can be injected underground for use in future dry years. In Tea Tree Gully, a recycled water distribution main will be built to distribute recycled stormwater and treated effluent from a septic tank and effluent disposal system (STEDS) treatment plant to irrigate local parks and schools. The use of stormwater and recycled effluent will provide security of water supply.

It is expected that the project will:

- Reduce the region's dependence on potable mains water by up to 12 GL per annum.
- Provide an annual stormwater resource of approximately 18 GL/year that will be fit for use in the irrigation of local community open space, industrial applications, residential non-potable uses, and commercial irrigation.
- Treat septic tank effluent to provide 0.6 GL of irrigation water in Tea Tree Gully.
- Provide a new water resource of more than 4 GL to be used in over 115 recreation reserves and parks, 54 school grounds, seven lake systems, and extensive public reserves, including linear parks.

- Secure water for environmental flow requirements in over 40 km of major urban watercourses.
- Subsitute approximately 2 GL/year of groundwater use in the region, and provide an additional 5 GL/year stormwater recharge to assist with a sustainable groundwater yield. In addition to these volumes, a one-off 25 GL of water will be injected by ASR to provide a water bank, where water can be stored for future dry periods.
- Reduce the outfall of untreated stormwater from the urban environment into the Barker Inlet by over 20 GL/year and remove over 40 t/year of associated nutrients and sediment loads.
- Reduce costs to the community and industry by providing a cheaper alternative water supply to mains water while providing environmental services, including environmental flows.

Currently, this region has a population of 285,000 people; it is expected that this project will provide sufficient water for the projected population of 350,000 by 2030, while reducing demand on external supplies. It is likely that with the additional water security offered by this supply there will be a rapid development of high-quality, irrigated urban parks and reserves in the region.

The project will transform large areas of the urban landscape by providing opportunities for wetlands, water features, irrigated parks and recreation areas. Water Proofing the North will be an important catalyst for urban renewal projects in areas such as Playford and Salisbury. Coupled with similar proposals for projects in the south of the city, they have the potential to redefine the urban landscape in these areas over the next 20 years.

Keith Smith and Chris Kaufman

unnecessary to water when it was raining. Another factor was the cost of water. The Micromet system allowed councils to select 'levels of green', depending upon the requirement of the site. For example, sporting fields and high-profile sites were kept at a high level of green. Some councils continue to use the Micromet system while others have found that it does not suit their particular needs.

Also from the late 1980s, subsurface drip irrigation systems were installed in most council areas. New technologies developed quickly throughout the 1990s and automatic irrigation systems became more efficient and less wasteful. During the 1990s changes were increasingly implemented, variously due to the cost of water, the cost of staff leading to staff reductions and to the fact that staffers were better educated and knowledgeable about the water needs of their open spaces.

Salisbury Council was particularly proactive in the area of water conservation from the mid 1980s. From 1983, with their first water cost study, there was talk of rationalisation of turf. Policy changed in Salisbury and, increasingly, areas were watered only for functional requirements. From then on much turf was replaced by trees which were hand-watered for establishment, and non-irrigated grassy dryland areas, with a turf care officer employed in the late 1980s. Throughout the 1990s inefficient irrigation systems were replaced and although open space area increased and annual water costs doubled, between 1983 and 1991 water usage in Salisbury reduced overall by 58%. While the first wetlands in Salisbury were constructed in the 1970s, water harvesting began in the mid 1990s and many more wetlands have been built since then.

Since 2000

Since around the year 2000, change has occurred quickly. All councils have introduced changes and many have undergone rapid developments in areas such as water harvesting and recycling, reductions in water use, improved intelligent irrigation systems, system audits, smart scheduling, moisture sensors and probes, rain sensors, subsurface irrigation, better maintenance practices, monitoring, analysis of soils and site conditions before irrigation systems are installed, and introduction of the irrigated public open space (IPOS) code of practice. An initiative of the state government's Water Proofing Adelaide strategy, the IPOS code encourages irrigation efficiency by addressing turf and irrigation management through sound planning, programming and monitoring. This code describes the specialist science of irrigation of parks and reserves and has become important in local governments achieving irrigation efficiencies.

Councils are growing more creative and opportunistic in their searches for alternative water supplies. Tea Tree Gully uses recycled water collected in and taken from an old clay and mineral quarry for irrigation, water trucks and street sweepers. The River Torrens is still used to water the Linear Park, and two district sports fields direct irrigation water into subsurface drainage tanks, which are mixed with mains water, treated, pumped back up and reused.

Aquifer storage and recharge (ASR) is increasingly being investigated and utilised, although Tea Tree Gully notes a concern that the aquifer is already experiencing an increase in algae, which is blocking pumps. Staff suggest that these aquifers are very old, and because we are using them but still lack sufficient knowledge about them, contamination appears to be emerging as a problem.

While some councils have been able to make good use of borewater, Tea Tree Gully has not had much luck. The bores began to fail in the early 1990s due to declining water levels and increasing salinity. In 2005 several bores were reopened and new bores sunk, but by the end of 2007 the water quality had declined to the point that they became unusable.

For Salisbury Council the focus since 2000 has been on developing alternative water supplies, an example being the Mawson Lakes Recycled Water System. By 2006, about 30% of Salisbury's total water consumption came from 18 major bores. The extensive construction and use of wetlands for harvesting has enabled the council to keep water costs under control while effectively managing their significant flooding and stormwater issues. In January 2008 this water independence allowed the planting of petunias in the Mawson Lakes Boulevard for the Tour Down Under. While still adhering in the most part to water restrictions, Salisbury Council does maintain the flexibility to use alternative water supplies as they wish. In fact, the council currently recycles more than 1500 ML of water per annum, using recycled water for irrigation of open spaces including council reserves and schools. The council also sells recycled water to a range of industries and to residential subdevelopments for non-potable use.

Salisbury's 2005 Integrated Water Cycle Management Plan examined water in a more holistic manner than previously attempted by any local government, integrating demand for water with stormwater, flooding, groundwater and bores. This plan was a precursor to the Water Proofing Northern Adelaide project, outlined later.

Councils with large geographical areas have some advantages over smaller councils, often having access to land on which to develop wetland or ASR projects. The wetland area required to clean stormwater is 1–3% of the area of contribution of the catchment, so that a catchment of 100 ha requires 1–3 ha of wetland. Councils such as Burnside, Unley and Mitcham cannot supply sufficient areas of land.

In Mitcham, therefore, recent history is a little different. Without access to recycled water or much opportunity for harvesting, Mitcham's changes since 2000 involve system upgrades, analysis of soil structures and better soil management, better plant selections and horticultural practices, use of the IPOS system and water sensitive urban design (WSUD), such as harvesting stormwater direct from gutters to water street trees and reserves. As with other council areas, the amount of green space in Mitcham continues to increase as development increases. This appears to be linked to the current assumption that development must have green areas such as lawns and garden beds.

Rainwater reuse in Clarence Gardens

BOX 116

Flooding of Bailey Reserve used to cause frequent game cancellations for Cumberland United Soccer Club, but not anymore … an innovative rainwater reuse system has been installed to:

- prevent rainwater from the roofs of the soccer and bowling clubs from flooding the soccer pitch and training grounds;
- supply indoor water requirements for water-efficient showers, toilets and handbasins; and
- recharge the aquifer to supplement oval irrigation requirements.

The system captures all of the rainfall from the soccer and bowling clubrooms in separate 20,000 L above-ground rainwater tanks. The tanks are interlinked and a pressure pump supplies the majority of fixtures in the soccer club. The club is well suited to rainwater supply as it is utilised in the winter season – when the rain is available. The system is sized to supply the clubrooms for the whole season, with automatic back-up mains supply in dry years. Water efficiency at the clubrooms has been improved through the installation of auto-off handbasin taps, low-flow showerheads, and dual-flush cisterns. Overall, the soccer clubrooms are expected to run on rainwater for the whole season, with mains water savings of 275,000 L.

The rainwater tanks have an unusual feature – the overflow pipes are two-thirds up rather than at the top. Rather than wasting space, this provides a buffer during heavy rainstorms. Outflows are gravity fed into the aquifer. This recharge is expected to save 540,000 L of borewater currently used for irrigation. As an additional benefit, since rainwater is very low in salinity, the overall quality of the borewater may improve.

Aquifer recharge with clean rainwater is much simpler than many aquifer storage and recovery schemes. As the rainwater is already potable, there is no general requirement for expensive treatment and ongoing system monitoring. This makes aquifer recharge accessible for small sites.

The system was installed prior to the onset of the heavy winter rains, and the soccer club has reported excellent results with no games cancelled.

The project was funded through the City of Mitcham and a Commonwealth government community water grant.

Jake Bugden, Sarah Gilmour and David Deer

BOX 117

Urban watercourse vegetation

Urban watercourses bear little resemblance to their condition prior to European settlement. In the early years of settlement, grazing of stock destroyed much of the native riparian vegetation, and weed species were introduced. The subsequent urbanisation of the catchment has:

- increased the variability of the flow;
- increased the load of sediment and litter;
- increased the load of dissolved and suspended pollutants;
- introduced a range of exotic riparian species; and
- led to the concrete lining or piping of watercourses.

All of these factors make the task of establishing riparian native vegetation difficult.

Within the existing system of degraded watercourses, opportunities still exist for re-establishing native riparian vegetation. These opportunities exist wherever watercourses are open channels and not lined with concrete.

Urban watercourse sites have some advantages for vegetation establishment in that they are smaller and have a higher profile than rural watercourses. For these reasons, more resources can often be made available for a given area and can be applied for a longer period of time. Supplementary watering is often possible over the first summer. It is possible to develop vegetation of higher quality at urban sites than can usually be achieved at rural sites.

The City of Burnside is gradually improving the vegetation quality at various urban watercourse sites. These sites provide examples of the approaches taken to watercourse stabilisation and revegetation in creeks of different flow regimes.

Example – watercourse at Beaumont Common

In its original state, the watercourse at Beaumont Common flowed very infrequently as it was relatively high in the catchment and surrounded by woodland vegetation. It was not wet enough to cause a change in the grey box canopy and its only noticeable influence on the vegetation was to encourage the growth of scattered rushes along its alignment. With stock grazing and then urbanisation, the watercourse became a small, eroded gully through a mown park, which still retained some of the grey box canopy.

In 1998, Burnside Council decided to undertake an engineering project to stabilise the creek. To mitigate the erosive forces of the water, a structure was built to direct high flows into a pipe buried beneath the creek and to direct low flows over the surface. The profile of a gentle swale was created where the erosion gully had developed. The ability to control overland flow enabled the development of vegetation without the risk of erosion.

Initially, only trees were planted – as trees do not interfere with weed control – and for the first 12 months the main focus was on weeds. Weeds were thoroughly sprayed out and frequently followed up. Rye-corn was sown in late winter to provide a non-persistent groundcover and weeds were sprayed amongst it. The only indigenous species to regenerate naturally during this phase of the project was the native coloniser (*Lythrum hyssopifolia*). Plants of this species were marked and protected during weed control.

As the water regime in the watercourse is variable and subject to prolonged periods without flow, we propagated local, drought-tolerant aquatic species. The species selected were *Cyperus vaginatus* and *Juncus subsecundus*, with some *Carex tereticaulis*. A range of local dryland species were planted in the adjoining higher areas. These were planted in spring 1999 and watered for the first summer. Native grasses were sown in subsequent years as areas became free of weeds.

The key to success is to frequently revisit the site and remove every weed as soon as it appears. Recognising natural regeneration is critical, too, and short pieces of cane were used to mark any naturally generating local species.

Management now involves patrols for weeds and the cutting back of vegetation after the lifecycle is completed, and responding to requests from residents to crown lift trees to create a more open landscape. The patrolling for weeds is the most important because weeds can come down in the water and become established without being noticed. A single weed soon becomes an infestation if not removed immediately.

Conclusion

The key lessons from our work at watercourse sites in Burnside are:

- erosion: build enough hard engineering to hold the ground.
- erosion: build porous rather than concrete structures.
- erosion: use rye-corn to prevent minor erosion.
- weedy trees: remove ash trees immediately, phase out other species.
- weeds: don't fight them, eliminate them.
- weeds: remove them early and often.
- regeneration: learn to recognise and protect regenerating native plants.
- planting: propagate local species from local seed sources.
- planting: chose species to match existing water regime not the pre-European water regime.
- management: prune and cut back to maintain aesthetics.

Andrew Crompton

In Unley the story is similar. Irrigation has been and is being upgraded, efficiencies improved and remote-control watering has resulted in less wastage. Unley has moved from an over-watering situation prior to 2000 to significant under-watering. This has occurred fundamentally through the introduction of water restrictions. Fifty per cent of open spaces are no longer irrigated and the remaining 50% receive considerably reduced irrigation.

Unley has little open space – about 3% of the council area. The two ovals, Goodwood and Unley, are still considered 'fit for use' but are under increased stress. Not only are the ovals receiving less water, but as other previously well-irrigated areas now receive less water and are losing their green, more schools, clubs and the general public utilise the only two green ovals.

Unley is known for its trees, most of which used to be watered indirectly through the irrigation of parks or gardens. Park trees no longer receive this water as the council deals with severe water restrictions, and to keep the trees alive through the drought, the council has instigated a tree program whereby 45 park trees of importance and in need of special attention have been selected. These trees receive mulch, decompaction, grass removal, drip irrigation, and soil conditioner. The number of trees selected for this attention is only restricted by the availability of resources. Through these tough times the emphasis is changing from maintaining turf to ensuring the survival of tree stock.

Rainfall has always been somewhat unpredictable and since European settlement flooding has historically been an important issue in many parts of Adelaide. The city and greater metropolitan area is constructed on a floodplain. A massive construction of drains took place, largely in the 1930s, to get rid of the water as quickly as possible. Now we are faced with the challenge of how to slow and capture this water in a largely built-up urban environment covered in hard surfaces. In Unley, as in Mitcham and Burnside, more WSUD is now being incorporated into streetscapes in order to make the best use of rain and stormwater when it is available. One example in Burnside is the Linden Gardens, where permeable paving allows water to be collected into underground wetlands, which then provide water to the landscape.

Tea Tree Gully claims to be the first council in Australia to use subsurface drip irrigation on a sports oval. In their experience this works well, toughens the plants, reduces maintenance and vandalism, and results in water saving of 30–40%. In this council area, four sports fields are currently irrigated by subsurface drip irrigation with more in development.

Playford Council has one large oval and several reserves on subsurface irrigation. The challenge in relation to ovals has been the timing of topdressing so it does not blow away. The installation of a subsurface system costs approximately four times that of an overhead irrigation system, however water savings contribute to the environmental benefits of installing such systems and are likely, in most cases, to outweigh the financial costs. Reserves in Playford are now well served by subsurface systems; new reserves, as in many council areas including Tea Tree Gully, are mostly irrigated with subsurface systems.

It should be remembered that the funding capabilities of local government to install water-saving initiatives are often limited. Such programs compete for funding with other community projects and services such as playgrounds, libraries, events and infrastructure repairs. There has long been a challenge in selling environmental initiatives and attracting community support for their benefits.

Plant selections

Another area of change for many councils is in plant selection. Landscaping is now carried out predominantly with drought-tolerant plant species, including more use of local indigenous species.

In Mitcham Council, until the late 1990s tree selections consisted mainly of West Australian species, many of which did not perform so well in Adelaide. In recent years an emphasis on the whole ecology of landscapes has gained momentum, and, whereas not so long ago the focus was on trees, understorey is now considered important wherever its inclusion is possible. Most reserves now use native and local plant selections, with a decline in the use of exotic species. Such new plantings often receive irrigation only for the first two years for establishment purposes.

In Unley since the late 1990s there has been a big shift towards the use of local indigenous plants, led more by individual enterprise than by policy. Approximately 10% of open space is under indigenous plantings (including understorey for habitat), which is approximately double that of most other councils.

The Windsor Street Linear Trail in Unley has been particularly successful. The re-landscaping of this street and verge involved the removal of mature oaks and elms to make way for local plants. However, while the project has resulted in water savings, it is constructed on a concrete culvert base with poor drainage and, combined with the need to keep the plants looking good so that the community enjoys them, this has meant that the project does consume more water than other local indigenous plantings may require. The biodiversity and education outcomes of the Windsor Street landscape are considered by Unley Council to be valuable, while the site is also an important seed bank used for propagation of threatened plants.

In Onkaparinga Council area, plant selections have also changed in the last few years. While the 1990s saw a shift to greater use of native plants including some indigenous plant species, more recently there has been a new focus on suitable indigenous plants, with research and trials identifying species that are adaptable and appropriate for the area.

Landscaping in the Tea Tree Gully Council area predominantly uses drought-tolerant plantings, including local indigenous species, in all new areas. Approximately 5% of council land demonstrates this relatively new landscaping style.

Playford Council has incorporated local indigenous species into recent horticultural plantings and landscaping projects, using these to help demonstrate sustainable design. Rosewood Park and the Little Para River Seed Orchard

Breakout Creek Wetlands

BOX 118

The first stage of the project has delivered many benefits for the environment and the community. Significant riparian improvements allow for increased use by the comunity, and a noticeable increase in the variety of native wildlife species has been reported.

Breakout Creek forms the last few kilometres of the River Torrens before it meets the sea. Prior to the development of Adelaide, the river terminated in a series of coastal wetlands and swamps associated with the coastline and dune system. During early settlement the area was cleared and drained for agriculture and flooding became a major problem. The existing channel, with high levee banks to provide flood protection, was constructed in 1937.

In 1996 the Adelaide and Mount Lofty Ranges Natural Resources Management (AMLRNRM) Board identified Breakout Creek as a potential site for riparian and water quality improvement works. A concept plan, which involved works for the complete length of Breakout Creek from Henley Beach Road to the River Torrens outlet, was released for community comment in April 1997. In May 1998 the board approved a shorter 'demonstration project' as a precursor to the full-scale project.

The first stage of the Breakout Creek Wetlands project, a 500 m stretch of the River Torrens upstream of Henley Beach Road at Lockleys, was completed by the board in 1999. This demonstration site involved the widening of the existing waterway and deepening of sections to form wetlands with public access paths, viewing areas and riparian vegetation. During construction, the area was drained and a temporary weir was constructed upstream to prevent flows through the site.

The project was designed to recreate a viable ecosystem within the riparian zone, which could be be accessed and enjoyed by the local community. To ensure flood risks were managed, detailed computerised flood modelling was undertaken to simulate the behaviour of the wetland during a flood.

Following the success of stage one of the project, the second stage of the wetland has been designed for a 700 m section between Henley Beach and Tapleys Hill roads. Before detailed planning commenced in 2005, the board commissioned a social impact assessment to ensure that proposed improvements were of the greatest possible benefit to the community, which involved a range of community consultation activities to identify, explore and document issues and ideas, in particular the impact of ceasing horse agistment in the area.

Stage two includes a number of habitat islands, reedbeds and deeper water pools, maximising opportunities for a variety of habitats to be formed in the river channel while maintaining flood protection.

Benefits of the project include:

- improved wetland and aquatic habitats in an urban environment;
- improved recreation opportunities for the local community;
- improved water treatment and quality, especially during low-flow periods;
- increased plantings of local native plants, including wetland species, following the removal of more than 30 weed species;
- improved public access to the site through the addition of viewing platforms and a network of paths within the wetland;
- improved native fish migration along the Torrens; and
- improved flood management.

Keith Smith

are examples of blending improved plant selections and horticultural practices with biodiversity benefits. Community interest and participation has been essential to the success of these projects and the council is keen to encourage residents to adopt more sustainable landscaping and gardening practices.

Adelaide City Council reports a strong move towards more sustainable plant selections, especially the use of native species in the parklands. With three main landscape styles – recreational, natural and cultural – and the 'cultural' landscapes still tending to be European, the council and community now favour landscapes with a more Australian character. Adelaide City Council leads the state with its water conservation and sustainable landscaping incentive schemes for residents and businesses. Financial incentives are offered for the installation of water conservation measures both inside dwellings and in gardens. Sustainable practices such as low water-use plant selections are encouraged and rewarded, with the campaigns delivering community education as well as reimbursements.

The only council consulted that is not evolving their plant selections is Salisbury, as staff consider that they already use the most appropriate plants.

The exceptions to the shift towards using more native and indigenous plants are often private developers over whom councils have little or no control. While local government may attempt to influence them, the perception of several councils is that private developers are reticent about changing to more appropriate and low-water-use plant selections.

Turning it off

Irrigation to non-functional areas is increasingly being turned off by local government and considerable amounts of water are being saved each year. This is due largely to water restrictions, but also to the introduction of better practices and systems such as the IPOS code of practice, and to changes in attitudes to landscapes and water.

As discussed earlier, until the 21st century, the community expected everything to be lush and green, especially in areas such as Golden Grove. Now, in most places, those expectations have changed. Adelaide City Council reports that a significant change in community expectations occurred during 2007.

Water use was reduced during 2006/2007 and by 2007/2008 water to 82 reserves in the Mitcham area had been turned off, leaving approximately 54 sites with irrigation. Of these, the emphasis was on sport and key recreation facilities. The water use target for 2007/2008 was just over 50% of the average usage for the previous four years.

Since 2005 Playford Council has significantly reduced the number of reserves and non-functional sites being watered, turning off approximately 60% of previously irrigated areas. In the City of Onkaparinga, where a vast amount of development is in progress (with approximately 15 subdivisions at the time of writing), and with 1800 ha of reserves, 400 parks and 25 sports sites, many changes have been made in recent years. All turfed roundabouts, verges and median strips, and all

other non-active use areas are no longer irrigated. Numerous new technologies have been or are being trialled, including a combination of climate-controlled site methods using the internet. In early 2008 water was reported to be sourced from approximately 87% mains, 0.5% stormwater, 10% borewater, and 2% reclaimed. While during 2007 water was turned off to reserves, increased development resulted in a slight increase in water use.

Similarly, Burnside Council is progressively reducing water use, minimising turf and installing subsurface irrigation when areas are upgraded. Adelaide City Council reduced water use by over 30% through instigation of a rating system in 2006/2007 to better determine the amount of irrigation allowed to individual landscapes, and by turning off some areas altogether. Most water features such as fountains have been turned off.

In 2004, the Adelaide City Council began using the Maxicom system, a weather-based computerised watering system for parklands, which measures soil moisture and rain and irrigates to optimum levels; this is still being rolled out. Another significant change for the Adelaide City Council is reduced reliance on the River Torrens, due to increased algal blooms. This has meant increased reliance on mains water for irrigation. Management of the River Torrens is an important issue for the Adelaide City Council; appropriate management that balances the amenity value of the resource with ecological health is an ongoing necessity.

The Adelaide parklands provide the venues for major state events as well as recreational uses. These events require the landscape be in excellent condition, maintained, and that rehabilitation is carried out afterwards. Such events include the Tour Down Under, Clipsal 500, WOMAD, the Fringe and Adelaide festivals, circus events, cultural festivals, and Elder Park events. In addition to these major events, the parklands support many local sporting events every week. Given the complexity of users and the fact that the council provides services to Greater Adelaide and South Australia, the council's water footprint is high. The council is proactive, however, and boasts the greatest area of 'green' (environmentally sound) office space in Australia (as of early 2008).

The community response to water use and landscape change has been positive in most council areas, with residents reporting broken pipes immediately and complaining quickly if too much water is being applied to a site. If a council applies water at will – when the community in general is restricted – this is perceived as an unacceptable double standard, regardless of a council's water savings elsewhere. It is also suggested that as local government is the closest tier of government to the community, it is expected to provide leadership and guidance for the community on such an important issue.

The elements of the community most resistant to change include those who bought into the lush, green developments such as Golden Grove and West Lakes with the expectation that they were to be maintained, as well as those living in more affluent areas who can afford to pay extra water costs and wish to maintain their high property values. It would seem that the group proving to be least compromising remains

Recycled water to benefit Adelaide's parklands

A major new environmental project for Adelaide has commenced through close cooperation of the Adelaide City Council and SA Water. The Glenelg to Adelaide Parklands Recycled Water Project (GAPRWP) will treat and transport high-quality recycled water from the Glenelg wastewater treatment plant (WTTP) to the Adelaide parklands and the CBD. Piped underground from Glenelg, the system will reduce the parklands' dependence on water sourced from the River Murray, Adelaide Hills' catchments, aquifers, and the River Torrens. In addition to reducing the amount of potable water used for irrigating the parklands, the project will also make water available for non-potable use in new developments in the city.

The parklands are a unique feature of Adelaide; they not only act a demonstration site for sustainable best practice but hold a special place in the hearts of those who visit, study, work and live in Adelaide. Parks are becoming more valued as spaces for respite, recreational activities, education and social gatherings as the population of Adelaide increases.

The parklands contain significant stands of trees, which demonstrate key landscape features of South Australia and have significant heritage value. The trees rely on a regular water supply, and new plantings require watering to establish during dry periods.

Adelaide's parklands are the site of national and international events, attracting many visitors to Adelaide. Thus, an assured water supply will help support the ongoing economic prosperity of South Australia.

The parklands are also heavily used and relied upon by many sporting clubs, schools and businesses, and are a highly valued location for family gatherings. In the 2005–2006 financial year, there were almost 800 events held in the parklands, including 130 significant or major events and over 200 weddings – providing a total of over 2.7 million users. There are also 33 sporting licence holders, eight lease holders, and many subleases and casual users. The sporting clubs encourage physical activity and can help reduce obesity levels in our community, which in turn have positive physical and mental health benefits for the state.

Through delivering 3800 ML of class A recycled water per year to the parklands and city, this project will be a major catalyst for South Australia to demonstrate national leadership in water recycling, and is the first large-scale water recycling scheme in an Australian capital city.

It is planned that 1000 ML will substantially offset potable supplies used for parklands irrigation, currently drawn from local catchments and the River Murray, thereby releasing water supplies that will enable savings for community use. A further 1000 ML will replace River Torrens and groundwater drawings, improving environmental flows and reducing algal blooms in the Torrens Lake. In addition to providing this sustainable and long-term solution for irrigating the parklands, the project has the capacity to be expanded to enable additional water conservation projects, such as providing water for cooling towers, fire sprinkler systems, and toilet flushing in new buildings in and around the CBD.

Outfalls to the Gulf St Vincent from the Glenelg WTTP will be reduced as well, reducing pollution of local seagrass beds and marine ecosystems.

This project is both timely and necessary, given the ongoing national drought conditions and the increasing realisation of the impact of climate change on our cities. Major construction works began in 2008 and recycled water will be available for use in the parklands by 2010.

The main infrastructure components of the project include the development of a new filtration treatment plant on the site of the existing Glenelg WTTP, construction of an underground pipeline from Glenelg to the Adelaide parklands, construction of an underground storage tank, pump station, and the underground network incorporating the whole of the CBD and North Adelaide.

This exciting project follows other major recycled water projects in South Australia, including the Virginia Pipeline Scheme (recycled water for irrigation of crops), Mawson Lakes (irrigation and dual reticulation), and the Willunga Basin (irrigation of vineyards).

Adrian Marshall

the latter. Burnside Council staff perceive, for example, that while their community is becoming more tolerant of less green, it is not as accommodating as the rest of metropolitan Adelaide. This phenomenon is similar for at least two of the local government groups interviewed.

The view of staff from several councils is that the drought and water restrictions have allowed them to turn off the taps and not cop the flack. For a long time councils maintained significant areas under irrigation that they considered should not have been irrigated. As a result of the 1980s and 1990s development push for green spaces, a great deal of water was poured unnecessarily onto reserves, road verges and other non-functional sites. State government-imposed water restrictions have meant that councils have been able to change this and not be criticised by the community. The following sentiment expressed by one council was echoed by others:

> The drought is the best thing that's happened to South Australia; it is the drought we had to have to make us realise how vulnerable we are as a state and that we can't continue to waste water.

The end result of water restrictions, along with changing practices and technologies, is a significant water saving over the past five years and in particular the past year. Total usage of mains-supplied water by the 16 Adelaide-based metropolitan councils has decreased from 7.2 GL (7182 ML) in 2003–2004 to 4.1 GL (4072 ML) in 2007–2008. This represents a 43% decrease in consumption.

During 2007–2008, mains water consumption by these 16 councils decreased by 36% in comparison with 2006–2007, a considerable reduction over just one year. While some water savings have resulted from improved practices in the built environment, such as harvesting stormwater for use within buildings, the majority of savings are attributed to increasingly efficient practices in managing public open space in response to water restrictions.

Looking into the future

The future of water supply in Adelaide looks vastly different to that of the past and present, and will continue to become much more sophisticated. Large-scale water projects will transform the way water is supplied to our cities, while technologies and better practice will increase efficiencies in the way water is recycled and applied on all scales.

Adelaide City Council look forward to irrigating the parklands using recycled water piped from the Glenelg wastewater treatment plant (WWTP). This will replace 80–90% of council irrigation by the end of 2009, preventing up to 50% of wastewater being pumped from the treatment plant out to sea every year. This water will also be piped to large developments, including the new SA Water building, and possibly to other councils, such as Unley, for irrigation purposes.

Another initiative of the Adelaide City Council is to use the backwash water from the Aquatic Centre. With 1 ML per week lost in backwash, projects will be developed that

reuse this water either for irrigation or for returning to the pool after treatment. Currently, a small percentage of this water is used for manual landscape irrigation. Similarly, Unley Council is looking at the reuse of backwash water from the Unley Swimming Centre for use either back into the pool, in watering new street trees, or possibly for street sweeping.

The Water Proofing Northern Adelaide plan is a large-scale joint project with multiple funding partners including federal, state and local governments and the Adelaide and Mount Lofty Ranges Natural Resources Management (AMLRNRM) Board. It involves Playford, Salisbury and Tea Tree Gully councils in developing wetlands, bores and replacing mains water with harvested and recycled stormwater by mid 2010. Twenty-four stormwater recycling projects, two wastewater recycling projects, and an experimental project to inject, store and cleanse stormwater in underground aquifers for recovery as potable water suitable for human consumption will save up to 30 GL of water annually.

In Salisbury approximately 60% of irrigation is currently (early 2008) from non-potable water, and the aim is to reach 95%, mainly from recycled water in wetlands. Salisbury is also looking forward to a centrally controlled irrigation network, incorporating fully controlled treatment and supply of recycled water. The future holds at least eight major ASR systems, more wetlands, recycled water to more residential homes, more monitoring and more sports fields, given the pressure of development and population growth.

The aquifer storage transfer and recovery (ASTR) research project is looking at injecting water into the aquifer for approximately nine months and then extracting it as potable water. This water may potentially be produced at less cost than desalination, which would open up the water supply to other significant water-users, such as the food industry.

Salisbury Council is considered a leader in collection and harvesting of stormwater. Flood-control dams are being automated and management plans activated to ensure optimal interception, treatment and release of water. Salisbury is cognisant of the potential for change and development, such as with the relocation of mining and defence technology companies to the area. Innovation, risk-taking and persistence are key to Salisbury Council's success in managing water.

The Water Proofing Playford project plans to substitute 80% of mains water with recycled stormwater, with which all reserves and major sports fields will be irrigated by 2010. Wetlands are in construction to capture and treat stormwater to a standard equivalent of water stored in underground aquifers.

It is interesting that just as rainwater tank installation regains popularity after many decades, new subdivisions in northern Adelaide will be exempted from having rainwater tanks as, instead, they will be connected to recycled piped water. The recycled water, a mix of treated sewage and harvested stormwater, will be a community resource.

Water yields in Playford are being modelled on reduced runoff, relative to a changing climate. The current model is

Re-establishing submerged aquatic vegetation in the Torrens Lake

BOX 120

Aquatic plants play an important role in the ecology of waterbodies, providing a food source, refuge and breeding habitat for native animals. They also have an essential function in maintaining healthy nutrient balance and water quality. The Torrens Lake is a central urban parkland feature of the River Torrens. Since 1999, blue-green algal blooms have been a recurring phenomenon in the lake. Blooms are prominent due to large nutrient loads from stormwater, warm summers with minimal rainfall, and limited shade. Calm conditions exacerbate the problem as cool water settles to the bottom of the lake (thermal stratification) and becomes low in oxygen, triggering the release of further nutrients from sediments. This, along with the warm water at the surface, promotes algal growth.

Immediate approaches to manage algal blooms include oxygenating the water and preventing thermal stratification by operating aerators and releasing water from reservoirs upstream (not necessarily feasible during a drought). Water 'treatments' can also be applied to remove nutrients from the water. However, to address the recurrence of blooms, long-term solutions are required that reduce the nutrient loads of water entering the lake, or improve natural nutrient regulation in the lake itself. One option is the re-establishment of submerged aquatic plants.

Submerged aquatic plants compete with blue-green algae for nutrients, and stabilise and oxygenate sediments, preventing sediment nutrient release. The submerged aquatic plant, curly pondweed (*Potamogeton crispus*) was once abundant in the lake. However, as an unfortunate result of sediment dredging in 1997 and 1998 to maintain access to the lake for recreational purposes, the species was eradicated. As natural recruitment has not occurred, active establishment methods are required to repopulate the lake.

In 2005, Adelaide City Council commenced a re-establishment trial that followed on the success of a marine eelgrass restoration technique from the United States. The transplanting eelgrass remotely with frame systems (TERFS) approach involves eelgrass shoots being tied to the bottom of a weighted frame, which is placed on the water body floor. The eelgrass root systems then integrate with floor sediment and plants become established.

Due to previous difficulties in establishing curly pondweed, locally extinct native eelgrass (*Vallisneria gigantea*) shoots were tied to wire frames placed at shallow locations on the Torrens Lake floor. The wire frames were covered with cages to prevent access to European carp and Australian black swans, which posed a perceived risk to plant establishment through sediment/root disturbance and grazing.

The plants established well and displayed healthy growth. The removal of the cages led to loss of eelgrass; however, partially lifting the cages to permit access to carp but excluding swans caused little damage – highlighting that swan grazing, not carp feeding, is likely the major limiting factor in eel grass re-establishment in the lake.

Reinstating a self-sustaining eelgrass population in the lake will require establishment across a large enough area to create a 'critical mass' to mitigate the impact of swan grazing in the absence of protective cages. Trials are being conducted using more flexible establishment methods that could be applied over larger areas. One involves placing eelgrass frames in deeper water, out of the reach of grazing swans. Another involves the use of a floating net barrier to exclude swans from the establishment area. Success of these trials is yet to be determined.

In 2006, the Torrens Taskforce was commissioned to seek solutions to improving the ecological health of the River Torrens and Lake Torrens. The Taskforce recommended the re-establishment of aquatic plants as a component in the long-term management of nutrient loads and algal blooms in the lake. Adelaide City Council will continue to work with other agencies to return aquatic plant communities to the lake, and to improve the ecological condition of the River Torrens.

Zoe Drechsler

based on a 20% reduced future rainfall. Playford Council staff note the challenges associated with the new systems and technologies, and emerging uses of water bodies such as wetlands. Implications for operations and maintenance, liabilities and insurance, and the impact of climate change on stormwater harvesting are largely unknown.

The Water Proofing the South project is similar to that for northern Adelaide. It involves extending the Willunga Pipeline from the Christies Beach WWTP (to be upgraded) into the southern suburbs to irrigate parks, gardens, school grounds, and for use in public toilets, saving 1 GL of mains water annually. Federal government funding will be supplemented by the City of Onkaparinga, SA Water, AMLRNRM Board, Willunga Basin Water Company and Flinders University.

The anticipated benefits of the project include increasing the reuse of recycled water in the Onkaparinga region from 4.4 GL/year to 8.8 GL/year by 2011, with provision for further expansion with an expected population increase in the region, reducing demand on potable water supply for urban use. This will enable expansion of the viticulture industry and conservation of ocean ecosystems in Gulf St Vincent by reducing in the volume of stormwater and effluent discharged to the gulf. In addition to greater access to reclaimed water through this scheme, Onkaparinga Council plans to harvest more stormwater and will also look at the potential for use of artificial aquifers.

Mitcham anticipates increasing its use of the IPOS code of practice with retrofitting of irrigation systems, better soil management, better plant selections, better horticultural practices, more stormwater capture and reuse, sewer mining, partnerships with shopping centres and other entities with large roof areas, changing green spaces to treed spaces, clearing waterways of exotic vegetation, further investigation of alternatives to mains water supply and continued monitoring and auditing. As an old council in the process of renewal, one particular challenge for Mitcham will be the management of recreation areas and the increased activity taking place on fewer sites.

Conclusion

Ahead of us in Greater Adelaide lies a brave new world. It is a world of rapid development to the north and south, urban renewal, changing water technologies, supplies and restrictions, and a changing climate. The pressures exerted by each of these are extraordinary.

The geographic and demographic circumstances of each local government area result in varied experiences, pressures, responses and opinions. The only certainty is that significant change is occurring and will continue to radically impact the water story of every local community and environment. Changes to landscape as well as to building design require innovative and flexible systems and thinking. Community education will be a critical element of future policy, programs and activity. Urban landscapes, human attitudes and behaviours are firmly locked into an evolutionary process, and, whereas not so long ago it was often the vision and passion of individual staff that resulted in change, the future appears to be far more connected to a collective sense of urgency in achieving greater sustainability. Many questions have arisen in this discussion, such as:

- To what extent should sustainability in water use be about reducing reliance and changing the way we use water rather than always looking for new sources?
- Given consideration of a changing climate with higher temperatures and reduced rainfall, should the widespread move towards using native and local indigenous plant species in landscaping instead become a move towards using more arid-land plants that will cope even better with less water in the landscape?
- In the event that water restrictions coalesce with a warming and drying climate, to what extent will demand for irrigated neighbourhood parks, reserves and sports ovals increase, while 'green' backyards decrease? And to what extent will this impact on the health of the urban ecology?
- Should the increasing availability of recycled water mean that councils can use more water on their landscapes, and make them 'pretty' again? The community has become used to a dryer standard and new developments are being encouraged to deliver more sustainable designs and plant selections, including no non-functional lawns or entry statements. Apart from needing regular water, irrigated lawn also requires use of machinery such as lawnmowers, involving fuel consumption and leading to greenhouse gas emissions. In addition lawns usually require fertilisers, leading to nutrient-rich runoff entering soils and waterways.
- If water becomes abundant again in the future through recycling, stormwater harvesting, desalination or even rainfall, will communities demand back their green spaces? Given that water restrictions have driven and are still driving many of the recent and current changes, it is suspected that if restrictions were lifted many in the community would want to return to the level of green they have enjoyed for so long.
- Given that current calculations reveal that around 160 GL of stormwater flow from Adelaide's urban environment into Gulf St Vincent, should we consider a shift in focus from dealing with water shortage to dealing with water surplus? If, for example, the majority of stormwater that flows along urban streets was captured and directed underground to quench the growing thirst of our street trees, a significant area of the urban environment could become a quite well-watered landscape.

These and many other questions are raised for discussion and debate as we plunge into this brave new world in pursuit of the 'holy grail', freshwater.

The Playford Greening and Landcare Group

BOX 121

A dedicated group of volunteers have been greening our parks and revegetating along our creeks, reserves and rural roadsides with the stated aim of making the City of Playford the greenest in the state. It all began in 1985 when our organisation was initiated as the Greening of Munno Para Group, with the support of the (then) mayor, a councillor, and a gifted community development officer.

The Greening of Munno Para Program was officially launched at Virginia by the (then) minister for the environment, Dr Don Hopgood. We were greatly assisted in the formative years by the Department for Environment and Heritage's Greening of Adelaide Unit, headed by Ian Hobbs, and the dedicated council staff who operated the council's plant nursery.

Subsequent years saw many successful revegetation projects completed. In February 1998, the Greening of Munno Para Group became Playford Greening and Landcare Group as a direct result of the amalgamation of the cities of Elizabeth and Munno Para.

In order to enhance biodiversity in the city we have established three seed orchards, on the central plains, the foothills, and alongside the Little Para River in Hillbank. The latter is used to showcase the various species suitable for the whole of the City of Playford.

Another aspect of our activities is to educate and inform the public on many environmental issues. To this end we conduct annual environmental forums – we are currently planning our 13th!

Networking with other like-minded groups in the region is another aspect of our work. We assisted with the establishment of the Shore Birds Trail at Thompsons Beach (with three other groups). We also have assisted the Two Wells Landcare Group with their wetland plantings between Dawkins Road and Hayman Road, near Two Wells, on three successive World Environment days – and what a sight it is after three years.

One of our members has been honoured by having a valley reserve named after him – Hillbank has the 'Ken Patterson, OAM Reserve' to be proud of.

As we collect seeds, sow the seeds, transplant the multiple germinations, weed the plants, then plant into our various projects, the group grows in knowledge and friendship, with laughter and goodwill strengthening an already unique group of dedicated individuals.

Pauline Frost

CASE STUDY 10: Using solar thermal technology to upgrade water quality or for small-scale desalination

Richard Thomson

Background

Limited water resources in many locations around the world are driving the development of techniques to increase water quality to agricultural grade or potable grade. There are a range of processes being investigated, such as linking large-scale reverse osmosis with wind farms to provide potential agricultural volumes of water, or solar thermal distillation, focusing on smaller volumes of high-purity water suitable for remote townships.

Desalination of seawater and upgrading of poor-quality water supplies based on evaporation and condensation has been a process applied effectively over many years. This technique can provide very pure water from essentially any contaminated source. However, the traditional process is marked by high energy consumption so more recently the thrust of investigations has been focused on the use of solar stills.

Simple but sophisticated solar stills are capable of producing up to several hundred litres of water per day. However, improved concepts are leading to solar thermal plants with substantially greater capacity, offering opportunities for significantly supporting water requirements utilising renewable energy.

The challenges for producing moderate volumes of purified water from the solar still at reasonable cost include improving efficiencies to reduce the specific energy consumption, and reducing the solar aperture per cubic m of produced water. Much of this work has been undertaken in Europe, with pilot plants being established for over 10 years. These units were designed as desalination plants, with the largest pilot plant being built in Oman.

Technologies

There are two basic ways to separate salt and other impurities from water – distillation and reverse osmosis (filtration). Reverse osmosis has become the accepted filtration method whereby the water is pressed through a membrane at a pressure of 50–60 bar, while the salts remain on the other side of the membrane. In comparison with distillation this method saves energy, but the membranes generally do not tolerate temperatures above 40°C and so is less suited to hot climates or sources of warmed water. This method also does not provide the high purity achievable with distillation.

For distillation, there are a number of techniques utilised to reduce the energy intensity of the process. Of these processes, multi-stage flash (MSF) evaporation is common for large-scale plants, overtaking the older multi-effect distillation (MED) and vapour compression technologies.

Multi-effect humidification (MEH) is designed to have the water evaporate at lower temperatures and ambient pressure, and is better suited to smaller-scale operations.

Projects and products

For large-scale plants, each has to be planned as an individual project and tailored to the demand profile. Smaller-scale operations are seeking to produce standardised units that can be connected in series to provide growth in capacity. The overall cost of the water produced is, as expected, a function of scale and the degree of renewable energy incorporated into the project. The range of projects being developed and the approximate cost of produced water are shown in Table A.

As described, the distillation processes of the small-scale systems make them a preferred choice to provide water for drinking or for use in hospitals, etc. Domestic water systems are best suited to the series designed units where multiples can be used together to support the demand. Importantly, also, these units need to have low maintenance requirements and high levels of reliability. The unit currently best meeting these requirements and also providing the most competitively priced water is the TiNOX MEH system, developed in Germany. These units consist of a modified standard 20 ft container in which the major components of condenser and evaporator are housed. This unit is heavily insulated to ensure optimum thermodynamic performance. An additional small unit is required to transport the pumps, valves and controllers. Thus the unit is ideally suited for transportation and minimum engineering onsite after delivery.

Material selection has been carefully considered to achieve the high levels of reliability and low corrosion impacts, with all parts in contact with brine being made from polypropylene or highly alloyed stainless steels. The complete container is lined with stainless steel and vapour tight. The unit is expected to provide 20 years service life.

Multiple-effect humidification

The diagram on p. 528 shows the basic elements of a system designed to use only renewable energy, based on the concept of MEH, whereby a closed casing is used for the heat and mass transfer required by the process, and the solar thermal collectors provide energy in a separate fluid circuit.

Evaporation and condensation surfaces are oriented in an optimised configuration with surfaces parallel to each other. This enables continuous temperature stratification along the heat and mass transfer surfaces, importantly allowing quite small temperature differences to be sufficient for the process to operate. This is possible because most of the energy supplied to the evaporator is recovered in the condenser, and this recovery can reduce net energy requirements for the system to below 90 kWh/m³. *TiNOX GmbH* introduced such systems as commercial products in 2005, with demonstration systems installed and commissioned in Jedda, Saudi Arabia. The objective is to develop such systems with a capacity from 1 to 50 m³ per day.

Table A Projects and products

Large-scale projects

Company	Project location	Technology	Stage of development	Water production	Production costs
Kernenergien, Germany	Aqaba (Jordan)	MED, hybrid system powered by gas and sun	Commercial project, construction commenced 2008, operation 2009	5000 to 7000 m³/day	$1.85–$2.50 per m³
Enercon, Germany	Aurich (Germany)	RO	Demonstration plant – 2005, commercial plant negotiations in progress	1200 m³/day with standard plant, smaller units available	n/a
WME mbH, Germany	Rugen (Germany)	Purely wind driven distillation under vacuum	Pilot plan in operation; first commercial plants expected this year	270 m³/day	$2.00 per m³
Synlift Systems GmbH, Germany	Tan-Tan Plange (Morocco)	RO, hybrid system powered by wind and grid electricity	Commence construction this year, first stage completed 2010	6000 m³/day by 2010 and 12,000 m³/day by 2020	$2.10 per m³

Small-scale projects

Company	Product name	Technology	Stage of development	Water production	Production costs
RSD Rosendahl, Germany	F8 and F10	Distillation directly in the collector	Commercial sales since 1999, so far 500 collectors sold	Typically 36L/day	$10–$25 per m³
Fraunhofer ISE, Germany	Oryx 150	Membrane distillation	Several systems in testing stage	100–500L/day for compact unit; up to 20 m³/day for multiple units	$25–$35 per m³
TiNOX GmbH, Germany	Mini Sal, Midi Sal, Mega Sal	Distillation using MEH	Commercial sales commencing	1–10 m³/day	$5–$15 per m³

Table A Large- and small-scale projects and products for solar desalination. For large-scale plants each one has to be planned as an individual project and tailored to the demand profile. Smaller scale operations are seeking to produce standardised units that can be connected in series to provide growth in capacity. The overall cost of the water produced is, as expected, a function of scale and the degree of renewable energy incorporated into the project. The range of projects being developed and the approximate cost of produced water are shown in this table.

MEH desalination

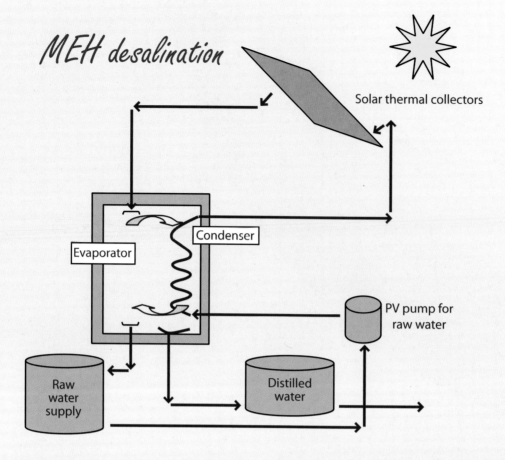

Solar thermal collectors

Condenser

Evaporator

PV pump for raw water

Raw water supply

Distilled water

Diagram A Multiple-effect humidification desalination using solar thermal and PV energy sources.

Commercial installations

The first commercial installations of MEH desalination using solar thermal and PV energy source systems were in Middle Eastern countries, with two units in Jeddah in 2005 and 2006. This followed about 10 years of operating demonstration units in the same environment. The commercial systems comprise MidiSal™ units, with design daily freshwater capacity of 5 m³ and is supplied by 140 m² of solar thermal collector and a 10 m³ heat storage tank, ensuring 24 hours/day water production using solar energy. It is estimated the true lifecycle cost of supplying townships and holiday resorts with water is of the order of $1 and $3.50/m³, which is within the cost range of such systems, while providing the environmental benefits of water production using renewable energy.

Potential applications in South Australia

There would appear to be a broad range of attractive opportunities to utilise MEH desalination using solar thermal and PV energy source technology. Examples include:

- Relatively remote, small coastal townships could use the process for desalination; inland towns could use the process for upgrading groundwater or recycled water.
- Small townships currently on River Murray water or having a water supply piped over long distances. Providing a local potable water source reduces the strain on the River Murray and also reduces pumping (non-renewable energy) and pumping costs.

- State parks and National Parks and Wildlife areas where water is not supplied locally.
- Holiday resorts, particularly those that have peak occupancy during summer; solar themal can provide the hot water load as well as potable water.
- Agricultural endeavours where the quantity of water consumed is not high but the regular supply of water is paramount for productivity.

CHAPTER 27

In-home solutions

Jon Kellett
Timothy McBeath

Introduction

Households consume more water in Adelaide than agriculture, industry, and public uses of water. Of the 300,000 ML of water consumed in Adelaide each year, around 135,000 ML or 45% is used in our homes.[1] In South Australia we use less water per person in households than the Australian national average; consumption is 94 kL per person while the national average is 103 kL per person per year.[2] However, by world standards, we are profligate users of water, consuming vastly more than most other countries in the world. The United Nations classifies countries according to their intensity of water use, providing statistics that rank countries by the volume of water delivered by water suppliers per capita, including that delivered to industry. Australia ranks as one of the highest at 598 m^3 per capita per year, in comparison to Singapore at 217, Norway 201, the UK 117, and Italy at 99 m^3 per person.[3]

In 2004, on average, each home in Adelaide used 280 kL of water. It is important to note that more water is used in dry years (up to 300 kL per household) than in wet years (as little as 265 kL per household).[4] Most of this water, around 86%, is taken from the mains supply.[1] Alternative sources of water include borewater, rainwater and recycled water. An increasing number of homes have rainwater tanks to supplement their supplies, but it is estimated that less than 1000 ML of rainwater is used each year.[5] This volume of water is difficult to calculate but is certainly increasing as more houses install and use tanks. Certainly, very few houses are completely self-sufficient for their water needs.

Figure 27.1 demonstrates how water is used in Adelaide households. It is noticeable that most of the water used in the home is not consumed by humans or animals but for purposes such as watering our gardens and flushing toilets (outdoor use is predominantly garden watering, along with car washing and the filling of swimming pools).[6] The quality of most of the water provided to our homes far surpasses what is required for these uses. In the case of toilet flushing, we effectively transform expensively treated potable water into blackwater at the touch of a button! In fact, over 90% of the water used in the home does not actually require expensive filtration and cleaning. With this fact in mind, we can begin to look for sensible ways to save water. These savings will reduce pressure on the environment (from where we divert water) and also potentially reduce our personal financial outlay on water.

In recent years public concern at decreasing rainfall, the plight of the River Murray, and the effects of government-imposed water restrictions have reduced demand on mains water supplies. Total mains water use in 2003/2004 was 14% down on 2002/2003, and average household use in 2005/2006 was 16% lower than the previous year. Being careful with water and changing our habits are important aspects of water conservation. Small-scale conservation measures such as turning off the tap while we clean our teeth, or saving the water used for washing the salad to water house plants are all evidence of personal concern for water. By practising such measures and focusing our attention on water conservation, we tend to find ways to achieve more significant savings.

Household water use

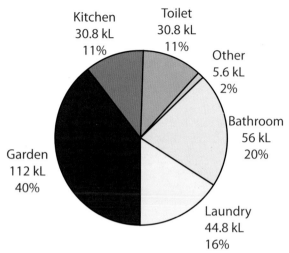

Figure 27.1 A comparison of average water use in different areas of a typical Adelaide household using 280 kL per year with no greywater used.
Source: Government of South Australia, 'Water proofing Adelaide: a thirst for change 2005–2025', 2005.

For example, one of our neighbours recently began diverting the wastewater from his evaporative air-conditioning system into his swimming pool.

The Australian Bureau of Statistics[7] has recorded changing attitudes to water conservation in the home, noting the most popular and widely used approaches. Nearly half of all households (46%) reported using one or more water conservation practices in 2004. The most popular measures adopted included using full loads when washing dishes and clothes, and taking shorter showers (18% of all households reported doing each of these). Recycling or reusing water was reported by 16% of all households, up from 11% in 2001 (Figure 27.2). These measures were particularly popular in Victoria, where more than a quarter of households undertook these activities.

Changing our everyday behaviour is only one way to conserve water. Clearly there are limits to how far we can take behaviour change. We expect a certain level of service and utility from water in our homes. While it may be reasonable to ask our teenage children to shower for five minutes rather than 15, we would not necessarily see it as reasonable to ask them not to shower at all! So we need to think about innovative ways to use water without reducing the quality of our lifestyle.

We can change the way we use water as, in recent years, we have changed the way we deal with waste. Since domestic waste separation and recycling were introduced, most people have quickly become accustomed to the waste hierarchy; that is, rather than simply dispose of our unwanted items in the bin and eventually into landfill, we should look to first reuse, then recycle, and only then think about disposal. Exactly the same logic can be applied to water use. Rather than purchase large quantities of expensively treated water and use it only once, we should be thinking about how we can maximise the use we get

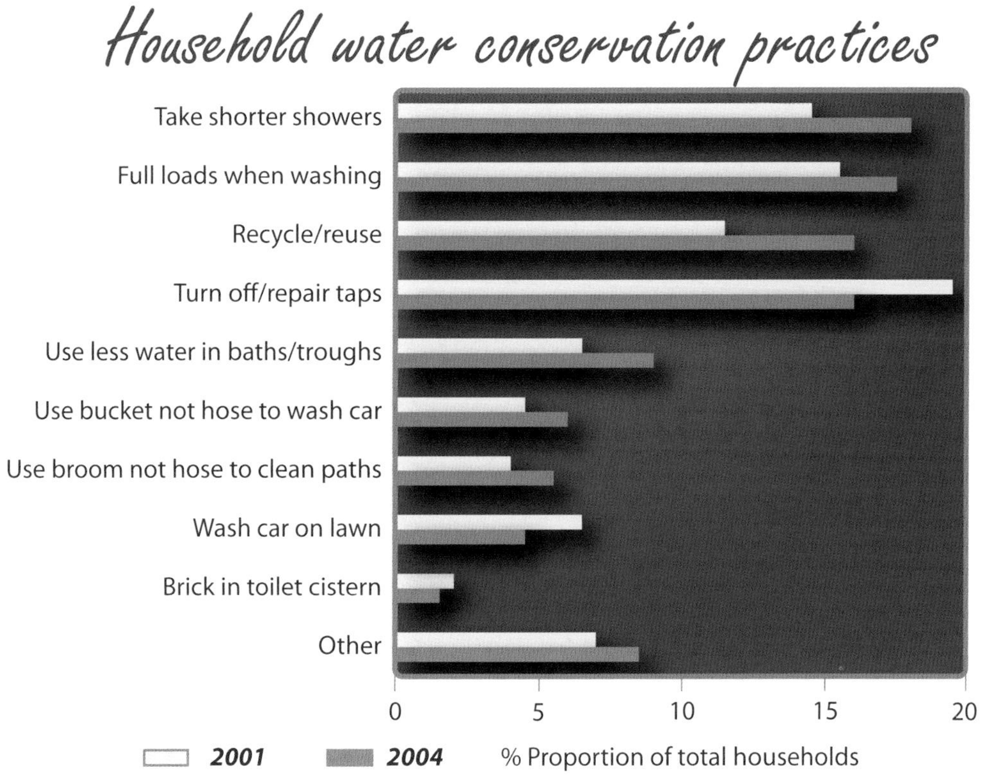

Household water conservation practices

Figure 27.2 The proportion of households practising various water conservation measures in 2001 compared to 2004. Source: Australian Bureau of Statistics.

aerating, which mix air with the water; and non-aerating, which simply reduce flow. Both types maintain water pressure so the quality of shower is not affected. Flow rates of 6–7 L/minute can be achieved with water-saving showerheads. Installing these devices, which cost as little as $20 each, will have significant effects on the bath/shower comparison. The five-minute shower is now likely to use less than half as much water and energy as a bath. A low-flow showerhead costing $30 can save $10 in water per year, so it pays for itself in three years. If every home in Adelaide were to install low-flow showerheads some 3000 ML of water could be saved each year, which represents 1% of total water demand for the city or over 2% of residential water demand.

out of each litre. Only when it is beyond any kind of safe use in the home should we let it go down the drain. At the same time, we should be seriously considering whether there are alternative sources of water for uses that do not require it to be treated to a high quality (e.g., watering plants or flushing the toilet). Let us consider each of the major household uses of water and look at ways in which we might apply the logic of the waste hierarchy to water. The key aims are efficiency, reuse and recycling.

Bathroom water use

Baths and showers account for 20% of average domestic water use each year. Over many years there has been a shift away from bathing and a growing preference for showering, which may be as much to do with time saving as water saving. Nevertheless, this is beneficial in water terms. An average-sized bath will normally require between 50 and 100 L of hot water plus 25 to 50 L of cold water. The comparison with a shower depends on the flow rate of the shower and the length of time spent under the shower. Old-style showerheads typically have flow rates of around 20 L/minute, so a five-minute shower will use 100 L of water. This is comparable with the typical bath, though experiments suggest that the energy demand is around one third less for the shower than the bath. Of course, the longer we stay under the shower the less favourable is the comparison with bathing, particularly if we share the water by bathing with a friend/family member or using the same water after someone has bathed. So time spent under the shower is critical. But major savings can be achieved with no loss of amenity if we install low-flow, water-saving showerheads. These come in two main types:

Laundry water use

Laundry represents 16% of home water use. Washing machines represent a hugely popular labour-saving device that have freed up time and liberated householders from hours of drudgery washing clothes by hand. However, up until recently, washing machines have been produced with their main design parameters being their effectiveness at cleaning clothes, convenience in use and purchase price rather than water and energy efficiency. A washing machine designed and built 10 years ago, which would be a standard feature of many Adelaide homes, typically uses 130 L of water per wash. Many of these machines are top loaders, bought for convenience-of-use in the typical Australian domestic laundry. In Europe, by contrast, most households do not have separate laundries and washing machines are usually located under a kitchen bench. Therefore, front-loading machines are the norm, which tend to be more water efficient than top loaders because of their configuration. The trend is now towards front loaders in Australia, and this is being helped by incentives and pressures on consumers to purchase more water-efficient machines.

The New South Wales government provides a $150 rebate to households who purchase efficient machines. Such schemes have been made possible by the Water Efficient Labelling Scheme (WELS), introduced to encourage manufacturers and consumers to think seriously about water-saving appliances in the home. All new washing machines offered for sale must now carry a mandatory water-efficiency label as well as an energy-efficiency rating. The WELS water-efficiency label is similar in appearance to the energy-efficiency label that washing machines and dishwashers must also carry. The more stars, the better. As well as a star rating, the labels also show a water-consumption or water-flow figure.[8]

It is important when choosing a new machine to divide the water consumption figure by the full-load capacity figure (usually given in kilograms) to obtain a comparative efficiency rating with other machines. Typically, a highly rated (4 star) machine would use around 7–8 L of water per kilogram of load, making it around twice as efficient as an old-style top loader. It is also important to ensure that such machines are used efficiently. Always wash a full load; if small loads are necessary, choose a machine with a half- or partial-load setting. Many machines now incorporate fuzzy logic to add a degree of 'intelligence'. Typically, the machine will add enough water to help match the size and absorbency of the load, and some will check for suds to see if an extra rinse cycle is required. Future innovations may include ultrasonic agitation and reuse of rinse water. If every household in Adelaide were to install a water-efficient washing machine then savings of around 6000 ML of water could be made each year, representing 2% of current total water consumption in Adelaide.

Kitchen water use

Water use in the kitchen represents around 11% of total domestic consumption. Here we are looking at multiple uses with differing characteristics. Food preparation and drinking water clearly demand high standards of cleanliness, which may seem difficult to economise on. Family health considerations dictate that cutting corners on water on these aspects of daily life would be inadvisable. Nevertheless, reductions resulting from more efficient technology are possible. Water-efficient taps can be a wise investment. Old-style taps, such as are fitted in many Adelaide kitchens, may have flow rates in excess of 12 L/minute. Many kitchen taps currently on the market, which have failed to attract a single WELS star, have flow rates of this order, while 6 star rated taps are typically around 4.5 L/minute and can be as low as 2.7 L/minute. They operate using the same principles of aeration or reduced flow as showerheads, and should show no reduction in utility in comparison to older models. Savings of around half of the water used for food preparation may be possible by using water-efficient taps.

More than half of kitchen water is typically used for washing dishes, either by hand or in dishwashing machines. In the latter case, similar technical advances to those outlined in respect to washing machines are of importance. Older-model dishwashers were often extremely water inefficient, while the imperative of the WELS scheme now demands that consumers are able to choose water-efficient machines with confidence that the claims about their efficiency are backed by rigorous testing. In assessing performance it is important to compare the load against water consumption, just as with washing machines. A modern 4 star or above rated dishwasher uses less than 1 L of water per place setting compared to an old-style or poorly rated modern machine, which is likely to use twice as much (between 1.75 and 2 L per place setting). Prospective purchasers are therefore advised to carefully assess the water use to place setting ratio as well as the overall water use per load factor.

Also, it should be normal practice always to run the machine when it is full. Never use a half-loaded dishwasher since this is an extremely inefficient use of water. Some new models are divided into two sections so that small loads can be run efficiently.

Toilet flushing

Toilet flushing accounts for 11% of our domestic water use. If we consider toilet flushing as a proportion of indoor domestic water use then the proportion rises to around 25%! Not only is this a significant proportion of domestic water demand but it represents a use for which there is very clearly no real need to use expensively treated water. As is often the case, two potential solutions present themselves. First, we can examine technological approaches that can reduce the amount of water we use in toilet flushing. Second, we can look at alternative sources of water for this use, thus avoiding the need to take water from the mains supply. This latter approach is examined below in the sections on greywater reuse. Here we examine the technology that exists to help reduce our water demand for this vital household function.

Toilet flushing technology has changed over the last 30 years. Until 1982, toilet cisterns in Australia were typically single-flush models that used around 11 L of water for each flush. Dual-flush toilets were introduced at that time, but by present-day standards were still fairly inefficient, using 11 L for a full flush and 5 L for a half flush. Since then, more efficient designs have been developed, tested and marketed. Toilets are rated using the WELS scheme but, unlike other appliances, a minimum efficiency has been specified for toilet flushes, so it is no longer legal to sell a new toilet with an average flush exceeding 5.5 L. State-of-the-art 4-star toilets now use an average of no more than 3.5 L per flush (4.5 full/3 L reduced). Research is being undertaken to reduce these water volumes even further using a variety of techniques such as integrating a handwash basin into the design and using its wastewater for flushing, urine-separating toilets, and air-assisted flushes. These latter two, which are not yet available in Australia, can use as little as 1.5 L per average flush. However, a number of technical and behavioural issues remain in their development and implementation, such as males being required to sit on the toilet to urinate, or paper being disposed of separately. It may be that the technology has reached its limits in terms of water saving for now, but it is clear that very significant savings have been achieved and that future technological advances could reduce demand further.

The rate of household conversion to water-saving toilet systems in Adelaide dictates how much water can be saved by these technologies. It has been estimated that a gradual introduction of toilet flushing technologies until 2020 could save between 164 and 587 ML/year.[9] At best, this represents 0.48% of current residential water demand, which does not appear to be much, however, we must recognise that the rate of change is slow. Average toilet life in the scenarios modelled was taken to be 15 years, and replacing a toilet is a significant investment for many households. If the projections are carried forward, by 2050 we see more

What is greywater?

BOX 122

Greywater is normally defined as any household wastewater that has not come into contact with toilet wastewater. Greywater normally comprises bathroom and laundry wastewater. Toilet wastewater is referred to as blackwater.

Neither of these are potable: that is, they should not be consumed by humans or animals. Greywater can contain pollutants including detergents, bleaches and disinfectants. Blackwater can be highly dangerous since it may contain faecal matter and dangerous microbial organisms such as *E. coli*. Water from kitchens is sometimes referred to as dark-grey water because it may contain a proportion of vegetable matter, which could stimulate the development of pathogens if stored.

The presence and number of coliform organisms in water are commonly used as indicators of unsafe water. Coliform counting procedures to ascertain the safety of water for human use were developed in 1905. Coliforms are the most widely used organisms as indicators of water contamination because they inhabit the intestinal tract of humans and other animals in large numbers. Consequently, the presence of coliforms in large enough numbers in a water sample indicates that it is unsafe for human consumption.

Jon Kellett and Tim McBeath

significant improvements of the order of 3% of domestic water use in Adelaide being saved.

Reusing water

Moving away from technological improvements that can be installed in the home to reduce water demand, we now need to consider the potential reuse of water from the various uses discussed previously. The focus of this discussion is on the reuse of greywater in the home, usually by retrofitting various devices within the property. However, we should not forget that new properties should be built and fitted out with reducing water demand in mind, so all of the technologies discussed already, plus a range of the ideas set out below, should become standard expectations for new home buyers, who should insist on their builders supplying them.

Retrofitting is the process of installing greywater systems to existing housing. Every household produces a large quantity of greywater every day. Typically this water comes from our daily routines of washing, showering and laundry. As a proportion of the total water supplied to the typical Adelaide household it accounts for about 36%. Greywater normally goes straight down the drain after use. Increasingly, this is seen as a waste of a resource which could be recycled within the domestic system for a variety of non-potable uses such as toilet flushing and garden irrigation. Policies are now in place to encourage such recycling systems in newly built dwellings but the majority of existing dwellings do not have such systems and retrofitting them can be technically difficult and expensive. In addition, most local councils have strict rules regarding retrofitted systems, which can deter people from installing them. Some commentators argue that the rules are unnecessarily stringent in that they regard greywater as a health risk when in fact it is not if it is reused in an appropriate way.

There is currently a debate regarding the health risks attached to greywater. It is argued that the coliform counting procedures used for over a hundred years significantly overestimate the risks associated with reusing greywater. The measurement of biomarkers in wastewater (such as fecal sterols) is a more appropriate measurement of the level of contamination in water.[10] Normally this approach demonstrates that greywater is not a health risk to humans and animals. While this does not imply that greywater is safe to drink, it does suggest that it may be safely used for a range of household uses such as toilet flushing and irrigation without the complex and often expensive treatments the current regulations often demand.

Greywater systems can vary, in utilisation of water source, size, complexity and cost. Basic systems include the simple connection of a greywater tube to a washing machine outlet. Complex systems include the permanent modification of household plumbing, and diverting and treating wastewater to a standard that is considered safe for toilet, laundry and other non-potable uses. The key problem for most people who want to use greywater for irrigation relates to the effects of washing powders, bleaches and detergents contained in the water. These problems can be addressed by the use of detergents that do not contain boron, phosphorus or salts, which are widely available. Bleaches also need to be avoided. Filtration through soil or specially prepared filter beds can also prove effective. There are five main categories of greywater systems, listed here beginning with the least expensive:

1. $20: garden watering via temporary laundry diversion.
2. $500: garden watering via direct diversion.
3. $1500: garden watering via storage.
4. $4000: house and garden use via storage (partial reuse).
5. $5000: house and garden use via storage (maximum reuse).

There is great variation in how long it takes to recoup the financial investment of installing a greywater reuse system. While the more expensive systems allow greater flexibility in how and when greywater is used, the cheaper systems often save the same amount of water. Only the less expensive systems repay the initial cost within a short time, however, all the systems discussed have significant water savings.

There are other, non-financial considerations involved in the installation of a greywater system. One important but non-financial spinoff of greywater use is the ability to water gardens during the most stringent water restrictions, to create green islands and to keep our garden city green. This benefit may, in fact, translate to a financial benefit through increased property value. Secondly, with a longer-term view, if more greywater recycling took place in homes and gardens, broad-scale savings could be made in the cost of delivering water to houses and removing wastewater from houses. The capacity of water supply and wastewater removal systems could be reduced to account for lower demand for mains water and the creation of wastewater.

Garden watering via temporary laundry diversions

These inexpensive tube systems expel untreated greywater onto the garden. Water is ejected by the washing machine pump at laundering times. These systems place the highest onus of responsibility onto the user to ensure that humans and animals do not come into contact with untreated greywater in compliance with human and environmental health regulations. Nevertheless, due to ease of installation and low cost, it is probable that they will become the most commonly used form of greywater reuse system. It is notable that even where plumbing is installed in a foundation slab, such systems can be installed in most houses including rental properties. Most householders can install a system themselves at no cost. All that is required is a hose pipe at a cost of around $20. Results of system installation are:

- water savings of around 45 kL/year (16% of household water use); and
- cost saving of $50/year (break even in around six months).

Garden watering via direct diversion

These systems directly divert greywater from bathrooms and laundries to gardens via subsurface irrigation. Filters can be installed to remove larger matter such as hair and lint. Diagram 27.1 depicts an installed water diverter. The

Water diverter

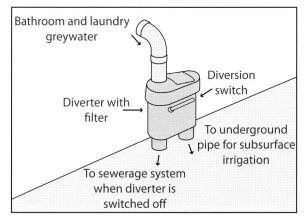

Diagram 27.1 A device for garden watering via direct diversion. Source: adapted from Nylex Water (online).

main advantages of this system are that it is cheap and safe and has South Australian Department of Health product approval. Systems cost around $200 plus installation (around $300, not including installation of an irrigation system). Results of system installation are:

- water savings of around 101 kL/year (36% of household water use); and
- cost savings of $111/year (break even in around 4.5 years).

Garden watering via storage

These systems save and store in tanks bathroom and laundry greywater for up to 24 hours prior to garden watering (in compliance with public and environmental health regulations). Greywater is not treated and must only be used for garden watering through subsurface irrigation. The advantage of such systems is that the storage enables garden watering at times other than when the greywater is generated, for instance at night during hot weather. Typical system cost is around $1500 including installation. Results of system installation are:

- water savings of around 101 kL/year (36% of household water use); and
- cost savings of $111/year (break even in around 13.5 years).

House and garden use via storage (partial reuse)

These systems treat bathroom and laundry greywater only (Diagram 27.2). While they save the same amount of greywater as other, less complex systems, they treat it to be used in toilets and laundries. They have a number of safety features including biologically treated media, which detain and remove impurities from greywater. A further advantage is that these systems can store greywater for garden watering or internal household use at any time. Typical system cost is around $4000 including installation. Results of system installation are:

- water savings of around 101 kL/year (36% of household water use); and
- cost savings of $111/year (break even in around 36 years).

The repayment period of this system precludes it as an investment through water savings alone. Purchasers may have other reasons, such as those outlined for the maximum reuse system below.

House and garden use via storage (maximum reuse)

A maximum greywater reuse system works in much the same way as is depicted in diagram 27.2. It can also treat dark-grey (kitchen) water to produce high-quality water suitable for reuse internally in toilets and laundries as well as for outdoor uses such as car washing and garden watering.

Additional filtration is required to achieve this result. A typical system costs around $5000 and would provide:

- water savings of around 131 kL/year (47% of household water use); and
- cost savings of $145/year (break even in around 34.5 years).

The system is targeted at people who have sufficient income to fund a low-return investment. System choice could be based on considerations such as a commitment to a verdant garden, an ethos of environmental sustainability, or as part of a strategy to ensure a cost-effective retirement.

Rainwater tanks

One of the more obvious solutions to reducing domestic demand on the public supply is to collect our own water. Australians have a long history of using rainwater tanks. In many country areas this was, and sometimes still is, the only source of freshwater. After falling out of fashion in the postwar period, rainwater tanks are undergoing a revival, partly as a result of government financial incentive schemes recognising their potential contribution to reducing water demand. South Australia has the highest proportion of households with rainwater tanks in the nation (45%).

While many people who have long used rainwater tanks swear by the quality of the water and claim it tastes better than mains water, the key uses for rainwater are seen as non potable. SA Water, for example, does not advise drinking rainwater, but it does recognise the potential of rainwater for garden irrigation, toilet flushing and domestic hot-water supplies. It is important that, whatever the potential use of the water, the tank is covered and fitted with a mesh screen, which captures dirt and moss from roofs and gutters and inhibits the passage of particulate matter, insects and vermin into the tank.

Since mid 2006, all new housing developments in South Australia are required by state building regulations to have a rainwater tank installed, which must be plumbed in to at least one eligible fixture including toilet cisterns, laundry cold water supplies and hot-water services. However, the bulk of the housing stock is pre-existing and likely to remain unchanged for the next few decades. Retrofitting of rainwater tanks to existing homes could therefore save more water in the short to medium term than would be achieved through installing tanks at new properties; though

House and garden greywater use

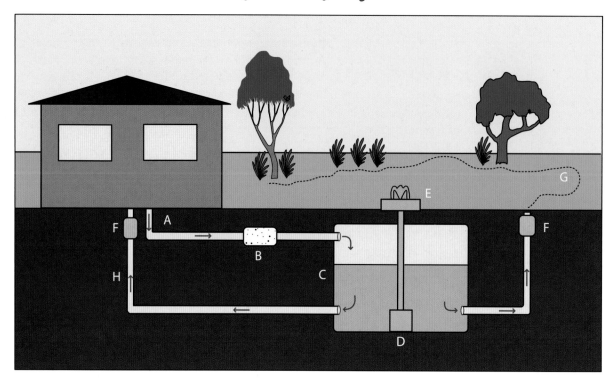

A: greywater from laundry and bathroom
B: filter to remove large particles
C: storage tank fitted with biomineral filters to remove contaminants
D: submersible pump ensures that grey water circulates

E: water feature keeps water moving and is aesthetically pleasing
F: disinfection chamber
G: subsurface garden irrigation
H: water for toilet flushing and laundry use

Diagram 27.2 A system for reusing greywater in the house and garden. Source: adapted from Enviro Water (online).

it remains vital that the new housing stock be constructed to a higher standard of water efficiency than in the past.

One inhibiting factor is space for the tank. Block sizes have become smaller as land prices have risen and planning policy has sought urban consolidation. The traditional large backyard, once typical of Adelaide suburbs, has shrunk. Tank manufacturers have responded to this problem by producing innovative designs for rainwater tanks that variously incorporate them into boundary walls, fit them under decks and pergolas, and make them design features within the garden. The huge cylindrical tanks partly responsible for the drop in popularity of urban rainwater harvesting in past years are now no longer obligatory. Nevertheless, storage capacity is important. Larger is better, though nowadays the total storage volume may be distributed between a number of tanks discreetly placed around the property rather than in one large one.

The amount of water we can collect depends on roof area and rainfall patterns. One square meter of roof area and 1 mm of rainfall allows collection of 1 L of water (Equation 27.1). The calculation for rainwater supply for a self-sufficient property at Clinton Shores on the Yorke Peninsula shows how the optimum tank size was determined.[11] The property is located on a 900 m² allotment, with a roof area of 315 m²; the rainwater tank itself has a roof area of 120 m², giving a total collection area of 435 m.² The average rainfall in the area

(Price Weather station data) is 333 mm per annum. So, in an average year, around 145 kL (435 x 333 = 144,855) of water can be obtained, providing an average household water use capability of 396 L per day. In this case, the property owner has deliberately set out to be entirely water self-sufficient, and by careful planning and management has achieved this aim. Total water use on the property is 262 kL per annum, slightly more than the current Adelaide household average but supplied entirely from rainwater harvested on site. It relies on extensive use of water-saving appliances as well as 90% recycling of greywater for garden irrigation. The tank, at 200 kL, is very large, and would be impractical in a constrained urban situation; however, it only needs to be so large in order to ensure security of supply.

A typical Adelaide household wishing to significantly reduce its reliance on mains water, but not wanting to be entirely self-sufficient, would require less than half of the storage capacity in the Clinton Shores example. The potential for undergrounding of water tanks, especially locating them underneath the house floor, is a key consideration for new

$$\frac{1 \text{ m}^2 \text{ roof area}}{1 \text{ mm rain}} \times = 1 \text{ L water}$$

Equation 27.1 Roof area and rainfall collection.

Environmental advantages

- Reduction of wastewater flowing out to the sea where it kills seagrass, which is the nursery for our fish stocks.
- Returning water to aquifers.
- Reduction of domestic potable water demand, allowing more environmental river flows.
- More lush gardens reduce dust, which is of great benefit to asthmatics.
- A reduction of the city's heat island effect may help to reduce the impact of heatwaves.
- Psychological boost of greener gardens and the more appealing aesthetic.

Environmental disadvantages

- Salt and other contaminants in washing detergents can build up in gardens.

Financial advantages

- Financial savings on water bills.
- Consistent watering during dry periods prevents building foundations from cracking and protects against subsidence.
- It is possible to grow flowers, fruits and vegetables which would not be possible with intermittent watering.
- Greener gardens encourage people to spend more time in them instead of spending money on alternative recreational activities.

Financial disadvantages

- Lack of rebates means that water-saving alternatives like rainwater tanks may be more attractive.
- Costs of coliform testing and application fees.
- Excessive payback period for top-of-the-range systems.

Social advantages

- Greener backyards are more inviting to socialise in.
- Healthier lifestyles: less dust and more greenery provide more opportunities for outdoor activities.
- Maintains Adelaide's historical image as a beautiful garden city.

Social disadvantages

- Greywater can become smelly if stored without treatment for more than 24 hours.

Jon Kellett and Tim McBeath

residential developments. It is also worth noting that many existing home owners have chosen to decommission backyard swimming pools. Rather than infill these, they might be better viewed as a water storage opportunity since they represent an already constructed water storage facility, only requiring covering and plumbing to be capable of providing the property with around 50 kL of water storage. This amount of storage capacity represents a quarter of that calculated for the Clinton Shores house and would allow a property with mains connection to be water self-sufficient for substantial periods during the year, only requiring mains water during extended dry periods, typically the summer months in Adelaide.

Installation of rainwater collection tanks is crucial if appreciable water savings are to be achieved in Adelaide households. The actual savings achievable will be partly dependant on the capacity of the rainwater tank and the nature of future rainfall patterns. If we assume an average of 10 toilet flushes per day at 3.5 L per flush, and that the longest dry period without rain is 100 days, then we would need at least a 3500 L capacity tank. This assumes no greywater is used for flushing and that rainwater is not used for any purpose other than toilet flush. The advice is always to purchase the largest capacity rainwater tank you can afford and accommodate in your space.

Since late 2007, the South Australian government has encouraged householders to invest in rainwater tanks, and many of the other water-saving domestic devices discussed, by offering financial rebates for purchase and installation. While the details of technological features, eligibility and cost change from time to time, details can be found on SA Water's website.

Conclusion

It is clear from this discussion that there are many ways in which we can reduce our household demand on the public water supply. It is impractical to expect that existing housing stock can become entirely self-sufficient in terms of water use, but it has been shown in a number of demonstration schemes that new homes can be designed to be independent of the public supply. These homes still provide the same level of service to their residents as we expect from a mains-connected property. It is also clear that there is probably something every household can do to reduce its water consumption. How far people are prepared to go will depend primarily on their commitment, their disposable income, and the current water efficiency of their home.

So, how much water could Adelaide households save if the measures above were applied? There are many variables to take into account, such as the varying age of the current housing stock, the range of plumbing fittings and domestic appliances within the homes of Adelaide, the pace of technological change, and changing government policies and subsidy regimes. Therefore there is no authoritative answer to this question. However, it is clear that greywater recycling has the potential to reduce average household use by 36% if all garden irrigation is obtained using recycled greywater. If water-efficient taps, showerheads, washing machines and dishwashers are installed then further savings can be made. Add to these savings a plumbed rainwater tank that provides toilet flushing for water-efficient cisterns and provides the water supply to the laundry, and we see further improvements.

Overall, it is feasible at the present time to reduce average household reliance on mains water supply by 50–65%. Consumers in Adelaide who take advantage of the range of government rebates available should be able to achieve this in a manner that is cost effective over the next few years, provided they intend to stay in their current home. Further, with possible savings of over half of the current domestic potable water use, more must be done to assist the retrofitting of greywater devices to existing and new housing. The overarching strategy should be to increase the conservation of a limited resource. Actions should include:

1. Legislative change and policy shift at all levels of government, based on the latest scientific understandings. Legislation should be formulated in clear, succinct and unambiguous language, which promotes the safe use of greywater without being overly restrictive.
2. A public awareness campaign should promote water conservation and the responsible use of greywater. It should dispel the myths and misunderstandings that currently prevent water recycling and the use of greywater.

While potential water savings will inevitably vary according to individual physical and financial circumstances, the information presented in this chapter demonstrates that a great deal can be achieved, often without major expense. It suggests that if the people of Adelaide take the potential for home water savings seriously then they can massively reduce their dependence on reticulated water supplies, save money, and, in the process, reduce Adelaide's dependence on water from the River Murray. All of this can be achieved without having to compromise comfort or basic health standards and with a minimum of behavioural change. In the longer term, builders should incorporate all of the ideas set out above in newly constructed homes. New homes built with sub-floor water storage reservoirs and integrated greywater recycling systems should become standard features in housing developments. Adelaide can become a model for the rest of the world's arid and climate-affected cities, demonstrating that even in a dry climate with acute water supply issues, life can be lived comfortably and at reasonable financial cost, while imposing minimum pressure on the environment.

References

1. Government of South Australia, 'Water proofing Adelaide: exploring the issues', discussion paper, 2004 (online).
2. Australian Bureau of Statistics, 'Water account, Australia, 2004–2005', 4610.0 (online).
3. United Nations Statistics Division, 'Environmental indicators: water', 2007 (online).
4. Government of South Australia, 'Reducing water use in the home', (online).
5. Government of South Australia, 'Rainwater use for households and industries', (online).
6. Government of South Australia, 'Saving water outside the home' (online).
7. Australian Bureau of Statistics, 'Year book Australia', 2008.
8. Department of the Environment, Water, Heritage and the Arts, 'Water efficiency labelling and standards (WELS) scheme', 2008 (online).
9. A. Schlunke, L.J. and S. Fane, 'Analysis of Australian opportunities for more efficient toilets', prepared for the Australian Government Department of the Environment, Water, Heritage and the Arts, Institute for Sustainable Futures, University of Technology, Sydney, 2008.
10. P. Ridderstolpe, 'Introduction to greywater management', Stockholm Environment Institute, Stockholm, 2004 (online).
11. D. Whitford, 'Clinton Shores living in the future residential estate: rainwater self sustainable water supply', D. Whitford, Adelaide, 2007 (online).

CHAPTER 28

In-garden solutions

John Zwar
Pamela Gurner-Hall

Introduction

Adelaide is a garden city. We might think of ourselves as a parks city because of the impressive parklands surrounding the inner heart of Adelaide, however, less than 3% of greater metropolitan Adelaide is remnant natural bushland and less the 6% of Adelaide is open parks. Why don't we have many parks? Because, until now, we haven't needed them. We had our backyards – our green spaces where we tend the plants, mow the lawns and socialise, vital to our wellbeing as individuals and communities. Now, the climate change and water restrictions we currently face, and the possibility of greater flooding events in winter, have forced a re-evaluation of garden design and construction. This chapter contains two parts. Firstly, John Zwar presents practical advice concerning how we can maintain gardens while responding to further changes in water availability. Secondly, Pamela Gurner-Hall reviews garden design past, present and future.

What's possible and what works: new directions in garden design

John Zwar

With an increasing awareness that water is a precious and scarce resource, water conservation has become a normal, everyday part of life. Water has historically been perceived by the urban public as being plentiful and cheap. We now realise this is not so. Various innovative measures appropriate for adoption in cities such as Adelaide, to minimise water usage without lowering the 'quality of life', have been successfully used in many Outback communities, including some in South Australia. These measures can be applied in our parks, reserves and home gardens, and are described below.

Plant species selection

It is important to base plant selection on local endemic species and on other plants, both native and exotic, from similar climatic regions. These will be hardy and should survive without irrigation in dry seasons if necessary, once they are well established. The inclusion of local endemic species will attract and benefit native bird and animal populations. Public plantings should include local species which will fit comfortably into the surrounding landscape, helping retain a 'sense of place'. Other native plants, tolerant of local conditions and selected for their low water requirements may also be included, along with exotic species selected for a specific purpose, such as a deciduous vine to cover a pergola. However, it is important that potentially invasive species are avoided when selecting plants for your garden. Local councils often keep lists of suitable and unsuitable plants. The 'Grow Me Instead' guide for gardeners in South Australia provides useful information on weed alternatives. Ideally, plants should be raised locally, as they will be hardier than those raised in a milder climate. Most of the garden should be planted with these hardy species.

Favourite plants, both native and exotic from wetter areas, may be included in a water-efficient garden, but these should be grouped in small areas where they will provide impact; for example, some tropical feature plants near a pool or outdoor living area, or some ferns and shade-loving plants in a shadehouse. These small 'oasis' plantings needing extra water can be irrigated separately from the larger plantings of low water-use plants. A small area of lawn, located where it will be of maximum benefit, is appropriate, and a small area of fruit trees and vegetables may be included and also watered separately. Incorporate plenty of organic matter, compost, manure and possibly water-storing crystals into the soil in vegetable gardens to retain soil moisture and reduce the need for irrigation. Light shade will reduce water needs and improve the quality of vegetables grown in hot summer conditions in Adelaide. Shade should ideally be provided by deciduous trees or shade cloth (around 50%; removed after summer), as these both allow maximum sun during winter. It is important to group plants with higher water needs together and to water them accordingly, rather than have them scattered throughout the garden where it would be difficult to water them adequately without over-watering hardier plants. These small 'oasis' plantings, if well placed can give the effect of lush surroundings without needing excessive water and may benefit from the protection of hardier surrounding plants.

Any existing native vegetation should be retained and incorporated into the landscape whenever development occurs. It is essential to carefully protect vegetation that is to be retained from root damage, soil compaction and damage to trunk stems and foliage, and to ensure that contractors are aware of these requirements. Retention and adequate protection of existing vegetation can be included in building and development contracts. Penalties for damaging vegetation may be appropriate. The retention of vegetation, including understorey plants, will protect soil and reduce dust during construction, helping to create an established feel before new plantings grow. Existing vegetation is obviously well adapted to the local soil, climate and rainfall, and provides valuable habitat for wildlife.

Although local endemic plants will survive on natural rainfall, it is important to realise that with denser planting than may occur naturally, it may be necessary to irrigate in order to maintain plants in good condition. During extended dry periods, one or two deep soakings with drip irrigation may be sufficient. Otherwise, with severe competition, the result could be unattractive, drought-affected plantings with die back and some losses. This may be unavoidable in severe drought, but with removal of dead plants and branches after the drought, the remaining plantings will recover.

Irrigation

Drip irrigation is the most effective method of watering trees and shrubs, and is ideal for establishing new plantings. It is much more water efficient than flood or sprinkler irrigation, estimated to use between 10 and 33% of water used by these systems respectively. It is cost effective, greatly reduces evaporative losses, saves time when compared to manual systems, and applies water only where needed. There is also the advantage of automatic irrigation control by using timer taps or automatic control units.

Drip irrigation can be installed as a permanent system below the mulch layer. It can be used regularly during the establishment of trees and shrubs and then less often as they mature. In times of drought, extra irrigations can be applied to keep landscaping in good condition, if sufficient water is available. Extra drippers should be installed as trees grow to encourage a spreading root system, and trees that are wind firm.

Subsurface irrigation of lawns and other plantings is effective and very water efficient, and is becoming widely used. In some northern towns extensive grassed areas and other plantings are successfully subsurface irrigated using treated effluent.

Saline water can be safely applied with drip irrigation without salt-burn to foliage or salt build-up in soils if drainage is good and plants are selected for salt tolerance. Treated and filtered wastewater can be similarly applied. Fertiliser can also be applied via the irrigation system. There is a wide range of drippers and fittings available and expert advice should be sought to ensure that an appropriate system is installed.

Irrigation guidelines vary depending on soil types, plant species, the density of planting, and the time of year. The general recommendation is to irrigate infrequently but thoroughly. Water in the top 5–10 cm of soil will evaporate within a few days during summer, so it is important to allow moisture to penetrate deeply. Mid autumn to mid winter planting is appropriate for most plants to ensure good root development by summer, though sub-tropical and tropical species are best planted in spring. Young plants may need a weekly (drip) irrigation of approximately four to six hours through the first summer. As plants grow, the intervals between irrigations and the duration of each irrigation can be decreased gradually. From the fourth year onwards, hardy shrubs and trees may receive a supplementary irrigation only if drought conditions prevail. This is intended to avoid stress to plantings, keep them growing actively, and maintain a fresh appearance. Irrigation frequency and duration will be determined largely by soil type. Some routine maintenance of drip irrigation systems and occasional cleaning of blocked drippers is needed to ensure trouble-free operation. Filters are necessary unless the water supply is filtered, and these need regular checks and cleaning.

Sprinkler irrigation systems for lawns should effectively cover the grassed area with sprinkler types that produce large droplets rather than a fine mist (much of which will drift away or evaporate without reaching the lawn). If possible, irrigation in windy weather should be avoided. Night irrigation is recommended.

Public education and community involvement

The promotion of water conservation and water-efficient gardening techniques in local communities is worthwhile and has positive benefits. A regular public commentary on local water consumption will keep residents informed and increase participation in water-saving measures. Print media opportunities include the production of local garden guide booklets and regular newspaper items offering water-saving suggestions. The distribution of water conservation brochures with water bills and in mail boxes helps maintain public awareness of the value of water and the need to use it wisely. Also useful are water conservation posters displayed in public places, and water conservation messages in the various media telling people what they can do to save water. Freely available, practical waterwise garden advice from appropriate agencies and businesses is also helpful in promoting water-efficient gardening. Appropriate public events should feature displays and promotion of water conservation. Opportunities can be developed to allow interested community volunteers to become involved in the promotion of water conservation in the wider community, in schools, garden clubs and societies, and at events.

In schools, water conservation can be incorporated into general teaching and promoted through talks and poster or essay competitions, and through visits to water facilities such as reservoirs and treatment plants. At Roxby Downs, students designed water conservation murals and painted them on concrete rainwater tanks at the school, produced a video, wrote water poems, and visited the desalination plant. Children take the water conservation message home with them! This work at schools is especially valuable, and should be actively encouraged. There is excellent detailed curriculum material available for students at all levels, including 'Watercare: a curriculum resource for schools' by Angela Colliver of the Department of Environment and Natural Resources, Adelaide, and 'Waterwise: water resource management and conservation' produced by Waterwise Queensland.

Reuse of treated sewage effluent

It is common practice to reuse treated sewage effluent for the irrigation of amenity plantings and sports fields in many towns in South Australia. This important source of water provides facilities which may not otherwise be available to the community, or would be much more expensive if potable water was used. The Department of Health requires that treated effluent is chlorinated to regulation standard; a separate reticulation system is necessary to convey effluent waters to the point of use.

Treated sewer effluent has been successfully utilised for oval and parkland irrigation at Woomera and Leigh Creek for more than 40 years. At Port Augusta, a fine-grassed 18 hole golf course (established over the past 30 years) utilises treated sewer effluent for irrigation at a fraction of the cost of potable water pumped from the Murray River. In Roxby Downs, treated sewer effluent is utilised for irrigating the oval and school sports field and part of the golf course. At Coober Pedy, where reticulated desalinated water and a sewerage system are relatively recent innovations, treated sewer effluent has been used to provide a one-hectare grassed playing field at the school, and a grassed town oval, surrounded by amenity plantings. At Port Augusta, sewer mining is practised and extracted effluent is treated in a small treatment plant located in a central reserve, and piped to various parks and reserves where it is applied efficiently in subsurface irrigation systems.

Most of these towns would not have these irrigated parks and sports facilities if treated effluent was not used, due to the high cost and limited supply of potable water. There is much more scope for the use of treated effluents in reserves in Adelaide, possibly based on the successful Port Augusta sewer mining model, which would reduce the pressure on existing sewage treatment plants and provide irrigation water for local reserves.

Water harvesting

Rainfall runoff from hard surfaces can be harvested by good design and landscaping. The water can be either allowed to run directly onto lower-lying planted areas or collected in storages for later reuse. Valuable water savings can be achieved by making use of rainfall and urban runoff in this way. This ancient and simple technique provides high-quality irrigation water, is

BOX 124

Water restrictions and sustainable gardens

Water restrictions are a part of life now; however much we dislike or are inconvenienced by them, we have to live with them. The plight of our rivers and the below-average rainfalls around much of the country mean that, at least in the short term, water restrictions are inevitable.

Much of South Australia is characterised by semi-arid or arid environments. Water is rarely plentiful for very long and substantial freshwater use by a growing human population is something the environment can no longer afford. To live in harmony with our environment, there are changes we simply have to make. But they need not be difficult and our gardens can still thrive.

By using basic principles we can design, create and maintain gardens that will thrive and function in a reduced-water environment and still be beautiful. These principles involve appropriate plant selections and water conservation measures, all of which are easy to achieve.

The first principle concerns plant selections. We need to select plants that grow well on the natural rainfall, with minimal supplementary water in dry periods. There is a vast selection of plants that do just this, including groundcovers, grasses, low and medium shrubs and trees. These plants usually originate from places similar in climate and soils to our own. The important proviso here is that we select plants that are not invasive and will not become weedy, as we can easily create other, more difficult problems if we choose invasive plants.

To select plants that are water efficient, it is useful to look at the foliage. Pale-coloured foliage will reflect heat and light and therefore require less water. Small, waxy, hairy leaves will lose less water than broad, flat leaves. Needle-like leaves are also water efficient, and succulent leaves store water for dry times.

The second principle involves water conservation. This means using practical measures to ensure that plants and gardens make the best use of available water. There are a variety of ways to achieve this. First, plant in late autumn or early winter rather than during spring or summer. The ideal time is after the opening rains of the season. Most plants will benefit from getting established in their new home over winter before the long periods of dry and hot weather that come with summer. This greatly reduces the amount of water they will need through spring and summer. For the first or second summer, some plants will need supplementary water but early planting reduces this requirement.

Good placement of plants can make a difference. Group them according to their water and sunlight or shade needs, and make the best use of the naturally dry and damp areas in your garden. Adding compost to your soil

improves water retention, and it is important to mulch well and maintain the mulch cover. Coarse, organic mulch is best as it allows rain to penetrate the soil and also provides nutrients to the plant. It is wise to apply mulch when the soil is already damp. Efficient under-mulch drip irrigation is usually the most effective way to water plants should they need a drink, however, if restrictions preclude using dripper systems, use a bucket or watering can. Plants that have been planted at the right time of year and are well mulched should only need occasional watering, even in dry periods. Water must reach the root zones of plants, so direct any supplementary water carefully. It is usually most efficient to apply water in the early morning rather than in the heat of day.

While some lawns can be quite drought tolerant, it is sensible to minimise their area and to have them only where they will be used. Drought-tolerant groundcovers are an excellent alternative. Lawns use less water if they are allowed to grow long, as the grass shades the soil and evaporation is reduced. Also be prepared to allow lawns to brown off a little in summer.

Another way to conserve water in the garden is to reduce the area of hard surfaces and use permeable pavers or soft surfaces for pathways. This allows more rain to soak into the ground rather than run off into stormwater drains. Using rainwater and recycled water wherever possible is also recommended.

There are many plants that will grow well in South Australian gardens and not use much more water than rainfall provides. Some of the plants that have been popular during the past century have required significant amounts of supplementary water, and much of this has been reticulated from our reservoirs and the Murray River. Such plants are not adapted to survive in our environment, and now it is time to select plants naturally adapted to need less water. In this period of freshwater uncertainty, there is no reason why we cannot still achieve beautiful gardens. It is a matter of working in harmony with the landscape rather than against it.

Sheryn Pitman

cheap, very effective, and can reduce the need for large, costly stormwater drainage schemes. A light rainfall of perhaps a few millimetres is of little use to plants, but by concentrating runoff from larger areas onto smaller planted areas, the volume of water supplied is equivalent to a larger rainfall event. Water that soaks in deeply is of real value to deep-rooted trees and excess may even recharge the aquifer. This freshwater has the added advantage of leaching accumulated salts out of the root zone of plants.

Water harvesting reduces the need for supplementary watering by making more effective use of the natural rainfall and runoff. Water harvesting schemes will vary in scale and nature depending on factors such as the size and topography of the catchment, the duration and intensity of rainfall, and the capacity of the site to accept onsite recharge. They can be small, like a suburban front garden or much larger like The Paddocks scheme at Salisbury. By careful design and good management it may be possible to retain all water falling on a site. There may even be the opportunity to direct runoff from adjoining properties onto the site as well.

In home gardens, all roof runoff should be directed onto gardens rather than into drainage systems. Through careful design, any risk of flooding and waterlogging can be minimised. If water harvesting was widely practiced in home gardens and parklands in Adelaide, aquifers under the city would be replenished rather than the stormwater running out to sea and creating environmental problems there. We must see our runoff water as an asset and, wherever possible, utilise all of it onsite for irrigation and aquifer recharge.

Various trials by the group TREENET and others aimed at directing some stormwater in streets into the soil to irrigate street trees are to be encouraged, and must be widely adopted to benefit trees, reduce stormwater runoff, and to utilise more of this resource. Rain gardens developed in shallow depressions on street verges and irrigated with stormwater utilise large volumes of runoff, irrigating attractive plantings with excess water soaking in deeply, helping recharge aquifers. In high rainfall events there may be some runoff but, as much water is utilised onsite, drainage systems to dispose of excess can be smaller and are cheaper to install.

Since European settlement, many amenity plantings in streets have been kept high and dry by kerbing. In our increasingly dry climate we must consider slotted kerbing and planting in depressions, or having no kerbing, allowing stormwater to run onto areas of tree planting and landscaping installed at intervals along streets. The rain garden concept should become standard practice in all streets in new subdivisions, and opportunistically installed in older streetscapes.

Carparks should be designed so that runoff is directed onto trees and landscaping, strategically planted in depressions, so shade is provided within the carpark and around the perimeter. Various types of pervious paving are available, allowing large volumes of water to penetrate, benefiting trees with underlying root systems. This water can soak into soil, providing aquifer recharge rather than running to waste in the sea. Successful trials of these paving materials have been conducted in streets and reserves in Burnside and Marion. They should be mandatory in public areas, and councils should require their use for paving around houses also.

On a larger scale there are many wetland developments and water harvesting schemes in Salisbury and other parts of Adelaide. These retain water onsite, creating attractive features in parks, creating habitat and often incorporating aquifer recharge. These systems work well in parts of Adelaide and must be adopted more widely throughout the metropolitan area and in other cities and towns, thus reducing stormwater runoff and the associated pollution of coastal water, while enhancing aquifer recharge.

The installation of rainwater tanks should be encouraged to provide water for garden and household use. Developers and local government should promote the installation of large underground tanks, purpose-built as part of house construction. These could be located under garages or driveways and covered with reinforced concrete, and store a much larger volume of water than smaller galvanised-iron or plastic, above-ground tanks. A pressure pump would enable use of this water in the house or garden. Tank overflow should be directed onto the garden, as should all roof runoff not caught in tanks.

It can be seen, then, that water harvesting schemes are often relatively simple and relatively cheap to develop. Large-scale wetlands (water harvesting schemes) are costly to develop, but can be cheaper to establish than the large stormwater drainage systems which they replace, with the environmental benefit of reducing stormwater outflow to the sea.

Mulching

Surface mulches are an essential component of all plantings as they decrease evaporation of water from the soil by as much as 70%, and greatly reduce the build-up of salts in the surface soil. Mulches reduce soil-surface damage due to raindrop impact and water erosion caused by uncontrolled runoff. They greatly reduce weed growth, help insulate soil and plant roots against temperature extremes, and can protect soil from wind erosion. Mulches should be spread to a depth of about 100 mm immediately after planting.

Many materials can be used for mulch. Organic mulches provide the greatest benefit as they eventually break down and improve soil, though will need occasional topping up to maintain an effective soil covering. Pine bark and woodchip are commonly used and do not break down quickly. Inorganic materials, notably gravel, pebbles and coarse sands are also commonly used in public landscaping. The use of continuous plastic sheeting can be harmful rather than helpful and is not recommended.

Ground-cover plants, especially dense-growing species, act as living mulches and provide similar benefits to other organic mulches. They should be planted in conjunction with supplementary mulching and have the added advantage of providing additional colour, texture and interest in planting schemes. In home gardens or small areas, a much wider range of organic mulches can be used, which may not be practical or available in sufficiently large quantities for use in large public areas. For example, light materials such as straw, animal manure and leaf litter tend to blow around in large spaces and break down quickly.

In urban centres, street verges and prominent areas where plantings may be dense and ornamental, suitable mulches should provide a continuous cover over the planted area. Where plantings are more utilitarian, such as buffer zones around urban developments or shelter belts, the aim should at least be to mulch around individual plantings (if it is not possible to mulch the entire area because of size, less dense plantings, and the cost involved). In buffer zone plantings it may be possible to utilise a range of mulching materials that may not be aesthetically acceptable in town centres or in more prominent areas.

Materials for use as mulch may be available locally, such as screened creek pebbles or gravels, or woodchip, which can be produced from tree branches and prunings. It may be necessary to purchase woodchip or pine bark or other mulches, but the cost can be justified because of the long-term benefits, particularly in reducing irrigation water usage.

Lawns and reduced grass areas

Irrigated grassed areas are costly to maintain and are high water-users. Lawns in both public areas and home gardens should be reduced to the minimum practical size and located where they will optimise usage and provide the most benefit aesthetically. Always aim to locate lawns or grassed areas with water harvesting in mind; that is, in locations which are slightly lower than the surrounds so they can receive runoff water from surrounding hard surfaces. In home gardens, a small, well-maintained lawn can provide a playing surface for children and complement outdoor living areas without needing large volumes of water. The maximum size lawn allowed in some towns in northern South Australia is 100 m^2.

Lawn substitutes. Some hardy native ground-cover plants make very effective and attractive lawn substitutes, especially in locations where usage and wear is minimal. They have the advantages of needing little water and no mowing or regular fertiliser applications or pest control treatments. Some produce attractive flowers. Various attractive paved surfaces (especially those allowing water to penetrate) and a range of ornamental mulches of different colour and texture can also provide effective lawn substitutes.

In high-use areas, lawns can be reduced in size or substituted with attractive paving, perhaps combined with overhead shade from pergolas or large trees. These protected locations can be further softened by the inclusion of suitable plantings in gaps left in paving or in raised planter boxes or beds.

Lawns on traditional kerbed street verges can be very wasteful of water, much of which runs off into gutters. Appropriate ground-covers may provide a suitable and attractive substitute. The rain garden concept in street verges, allowing water that runs off the road to irrigate slight depressions adjacent to the road, can include lawns as well as other plantings. Each time there is rain the plantings are irrigated; even in dry summers there is usually some rain, which will freshen up landscaping.

In both public and home gardens ornamental tree and shrub species requiring irrigation can be planted near or even in lawns and benefit from the extra water available. Limited numbers of showy ornamental trees with higher water requirements can be included in the landscape in this way. Hardier, dry land trees, shrubs and ground-covers may be planted in unirrigated areas surrounding lawns. Trees in lawns compete with grass and dense shade may prevent grass from growing under the canopy. In this case the area could be mulched and perhaps planted with low, shade-loving plants. Other trees with a higher canopy or producing only light shade would not affect grass growth. The lawn needs sufficient irrigation for grass and tree growth.

Lawns are one of the most expensive elements in landscaping, and by reducing the area of grass as described, considerable savings will be made not only in water usage but also in maintenance costs. Large, under-utilised lawns can be reduced in size or even eliminated and substituted with attractive dry land gardens, and the end result can be very pleasing. Artificial turf provides an excellent surface for sports facilities such as bowling greens and tennis courts, with the initial high cost being offset by the saving in water and intensive maintenance.

The drought- and moderately salt-tolerant kikuyu grass (*Pennisetum clandestinum*) is the hardiest and best-suited lawn grass for harsh summer conditions in Adelaide. Although it will brown off during prolonged periods without water, it will quickly green up with rain or irrigation. There are other hardy lawn grasses including sea shore grass (*Paspalum vaginatum* 'Velveteen'), which is both salt and drought tolerant, and soft foliaged cultivars of buffalo grass (*Stenotaphrum secundatum*) such as 'Sir Walter' and 'Palmetto'. Santa Anna couch, a *Cynodon* cultivar, is a hardy, finer grass that will survive dry conditions. The native ground-cover lippia (*Phyla nodiflora*) can be grown as a lawn or very compact, dense ground-cover. These turf species are more likely to survive drought and neglect than some of the softer, fine grasses, which need regular summer irrigation.

Before establishing lawn, incorporate organic matter and possibly water-storing crystals into the soil to help retain moisture in the root zone. Cut lawns longer in summer to shade and protect the root system. Regular, light fertiliser applications will keep a lawn healthy and better able to withstand dry conditions.

Irrigation timing

Grassed areas, vegetable gardens, shadehouses and other areas that are sprinkler irrigated should ideally be irrigated in the evening or early morning to minimise evaporative losses, which may be as high as 50% during the day. If the water supply is saline, overhead irrigation at night can reduce salt burn on foliage. This is relatively simple with automatically controlled irrigation systems. Sprinklers used should produce larger drops rather than fine mist, much of which drifts away and is wasted. Where possible, drip and other irrigation systems should also be operated at night. Overhead irrigation should always be carried out in calm weather, if possible, to minimise wind drift and wastage. Effective windbreak plantings or other shelter will protect lawns and gardens and reduce their demand for water.

Soil amendments

An understanding of soils is of value in implementing water conservation strategies. Soil compaction, either natural or compounded by construction equipment and traffic, greatly

BOX 125

Creating backyards for wildlife

Gardening is one of our favourite pastimes and is a great way to stay active and healthy. What we do in our gardens has the potential to benefit or harm the natural environment. By choosing to develop and maintain a garden in a wildlife-friendly way it is possible to:

- reduce your garden maintenance costs and time commitment;
- conserve local native plants and animals;
- make urban areas more ecologically sustainable; and
- save water.

If we can learn how to lessen our ecological footprint in our own backyards then we can apply those same skills and knowledge to improve the natural environment in our suburb or local district.

The need to conserve water is making us more aware that what we do in our gardens and homes has the potential to have an impact on the wider natural environment. The plants we choose to grow, how we structure our garden and the pest control methods we embrace will determine whether or not our gardens are friendly to visiting wildlife.

Establishing a 'backyard for wildlife' means that a garden can encourage a host of native animals – birds, butterflies, lizards, frogs and insects. Enjoying biodiversity in a garden is a real pleasure, and providing habitat for our unique and precious wildlife as part of the urban landscape is a generous gift to the future.

Backyards for Wildlife offers a vision that both looks back to our past as well as forward to a more environmentally sustainable future. No matter how big or small a backyard, an environmentally friendly garden can be created that will provide a home for our native animals and plants.

The early European settlers to Adelaide encountered an array of different environments: red gum forests, mallee and grey box eucalypt woodlands, shrublands, swamps, wetlands, coastal sand dunes, and mangrove forests. These different environments provided niches that supported a wide variety of plants and native wildlife species. The Adelaide Plains were one of the most biologically diverse regions in South Australia. Since 1836, however, around 97% of the region has been cleared for agriculture and urban development. Of the original 850 native plant species, 20% are now locally extinct and over 50% are rarely seen.

Increasingly, gardens are becoming a key element in the preservation of Adelaide's biodiversity. For example, gardeners are planting some of the lilies and groundcovers that are threatened in the wild. Planting local native species is a key factor in creating a wildlife-friendly garden. These species help to recreate the relationships that originally existed between plants and the local native wildlife.

Local native species are naturally adapted to the soils, rainfall and climate of a location. Using local native plants can drought-proof a garden, save water and improve biodiversity.

As Adelaide's freshwater supply is under increasing pressure due to climate change and drought, we have to change the way we use water in our homes and gardens. Before water restrictions came into effect, some people were using up to 70% of household water on their gardens.

As home gardeners have been affected by the water restrictions, so too have our native wildlife. Providing a source of water in a garden is important for wildlife, particularly over summer, and can be an attractive garden feature.

Over 270 bird species have been recorded in the Adelaide region, of which 16 are introduced and 76 have conservation significance. Birdbaths are a simple way to attract birds to a garden, especially during hot weather. Birds will become familiar with a regular source of water, and keeping a birdbath full and clean will enhance its attractiveness to them.

Butterflies enhance any landscape with movement and colour, and play an important role in the local ecosystem. About 20 butterfly species are commonly encountered in suburban gardens. Specific local plant species play a key role throughout the stages of a butterfly's lifecycle and are essential for attracting butterflies to a garden.

Ponds can support a range of local native plants, both aquatic and semi aquatic, that will remain green all year round as well as providing habitat for frogs and invertebrates. Adding native fish can control breeding mosquitoes and add colour and movement to a pond. Most of our native fish species are small and are ideally suited to aquariums and backyard ponds. Native fish are an excellent alterative to ornamental and exotic species, some of which have established in our watercourses with devastating impacts on native aquatic species.

Habitat loss poses the greatest threat to Adelaide's fauna. Protecting remnant habitat and planting local plants will help keep these animals as a part of our suburban environment. We have opportunity to implement water and biodiversity conservation at home by creating a backyard for wildlife, planting local plants, practicing responsible pet ownership, and taking an active role in protecting remnant native vegetation on private land, local reserves and along roadsides.

reduces the chances of successful plant establishment and growth. Retention and protection of existing vegetation during construction avoids soil compaction, retains plants for incorporation into the landscape, and reduces water and wind erosion. Compacted sites should be deep ripped, 0.5 m deep if possible, to break up compaction in areas to be planted. Additional topsoil stockpiled from areas being developed may be available for spreading to increase soil depth when construction is complete.

Gypsum applied at the rate of up to 1 kg per m² on poorly structured and poorly drained sodic soils will improve irrigation efficiency by increasing infiltration rates, especially if combined with cultivation, but results can be slow and variable.

Soil wetting agents and products such as water-saving crystals or gels incorporated into soil may considerably improve water infiltration and the water-holding capacity of most soils, and may be worth using in home gardens.

Various organic materials can be incorporated into soil which will improve water- and nutrient-holding capacity and improve drainage of heavy soils. Other organic amendments include the addition of compost as well as animal manures. If these materials are used, enough must be added to change the structure of the soil, and thoroughly mixed with existing soil to make a difference. Clay can be added to sandy soils and sand to clay soils, again in sufficient quantities to make a difference, and well mixed with existing soil. Obviously these soil amendments involve large quantities of expensive materials, and although they may be beneficial, an advantage of basing landscaping on local endemic and other suitable species known to be reliable in the area is that they will generally thrive without expensive soil amendments.

By using organic mulches in the landscape, and retaining organic waste onsite as mulch or compost, the soil will improve as organic matter breaks down, releasing nutrients. The addition of fertilisers, especially over a long period, can also considerably amend soils.

Soils are a valuable resource and should be protected during construction projects. Remove topsoil from construction sites, stockpile and respread as soon as possible after construction on areas to be landscaped. It may be necessary to rip compacted subsoil around new developments before respreading topsoil. Avoid mixing subsoil with topsoil. Ensure builders' waste and debris is removed from construction sites and not churned into soil or buried under a thin layer of imported soil at the end of construction.

Where extensive grading and re-contouring are undertaken, along with the installation of underground services, irrigation systems and possibly extensive agricultural drainage, topsoil should be stripped beforehand and respread on the re-contoured site afterwards, otherwise topsoil may be buried in some locations and exposed subsoil left in others, making plant establishment very difficult or completely unsuccessful.

Soil amendments are expensive and involve the use of energy. While there are situations where they may be necessary, such as during the development of sports facilities, it is often possible to achieve excellent results with minimal soil amendment if carefully chosen species form the basis of plantings,

following natural contours. This may be much more pleasing aesthetically than a scheme involving massive earthworks and grading, and will fit more comfortably into the surrounding landscape, especially if some pre-existing vegetation remains.

Greywater recycling

In drought times especially, many householders use greywater from the bathroom or laundry to irrigate their gardens. If using greywater, it is preferable to use biodegradable, phosphate-free laundry detergents. Householders may bucket the water out to their gardens, and use greywater hoses connected to the washing machine outlet. There are devices that connect to drainpipes and intercept greywater, diverting it for use on gardens, and much more elaborate systems intercepting greywater from drainpipes, treating and storing it in a tank and disposing it in subsurface irrigation systems so that there is no contact with the water. Plumbing and irrigation firms can advise on the various options. While health authorities do not normally condone the handling of untreated greywater, in dry times with restrictions on garden watering it is widely used by home gardeners to supplement garden irrigation. As greywater systems for home gardens become more refined they may become more widely used, and would certainly reduce the dependence on potable water in gardens.

Pruning and thinning

The water requirements of established trees and shrubs in gardens can be reduced by judicious thinning out of foliage and pruning. This reduction of foliage reduces transpiration and thus the water needs of pruned plants. Summer thinning and pruning is not appropriate for all plants, but many species will tolerate this treatment, which may allow them to survive with minimal irrigation in times of severe water restrictions.

Conclusions

These water conservation strategies are all reasonably simple, basic measures, which if applied widely would significantly reduce the volume of water needed for garden irrigation. They can all be applied without lowering our 'quality of life'. Indeed, if widely applied, the standard of our parks, reserves and home gardens may improve considerably, certainly in drought times, thus *increasing* our quality of life.

Gardens designed and other popular myths: legacies and future directions in urban residential garden creation

Pamela Gurner-Hall

The recent past, 2007–2008, was the driest year on record in Adelaide. Many of the water features originally taking pride of place in backyards and public spaces are now dry and desolate. This situation is not likely to change in the near future unless we have a record rainfall in the coming winter months or years. The many empty water features around Adelaide are clear examples of the folly of times past, when we either ignored our limited water resources or lived in hope for a wetter climate. They were designed with idealistic hope. Symbols of wishful thinking, they belong somewhere else.

Trees and plants of an inappropriate nature are still planted in many gardens. Silver birch (*Betula alba*) is a tree of the Northern Hemisphere; it loves cold, wet climates but does survive in more difficult climate zones. However, there is a difference between survive and thrive. When grown on the

Water for wildlife

BOX 126

Recent Adelaide summers have been blazingly hot, and water restrictions have limited garden watering to a few hours a week. The city is now marked by sunburnt bushes, barren stretches of clay where lawns used to flourish, and the wilted remains of daisies and groundcovers. We have all lost plants in our gardens and even large trees are dying of thirst. However, as we bucket water onto our favourite plants, we must also spare a thought for the other garden residents that suffer from thirst during our hot, drought-dominated summers.

Urban wildlife needs water, too. In the past, that dripping tap and the garden sprinklers, sprayers and drippers all provided enough water to support our garden populations of vertebrates and invertebrates. Bees would come to a dripping tap from considerable distances to drink, while common brown butterflies loved to dance in the spray of a hose or sprinkler. Small and large birds would drink from water pooling in rocks, the wheelbarrow, or little Johny's Tonka tip-truck. Most importantly, the small, dense bushes thrived despite the hot sun because of the regular watering, which provided a cool home for small skinks, blue-tongues, and many little birds. Because plants lose water through their leaves, the centre of a dense bush is both cooler and more humid than the outside air. The modifying, cooling and humidifying effects of plants on the local environment can be considerable. You can feel this for yourself on your next trip to the Adelaide Zoo. Feel the temperature and humidity in the shady rainforest aviary then walk quickly into the open seabird aviary next door. On a hot day, the temperature and humidity difference can be profound.

During recent summers, water restrictions and the absence of rain have resulted in the lack of pooling water for drinking. Large animals are forced to travel longer distances to find water. Koalas and blue-tongues have been rescued from fishponds and swimming pools in record numbers, thirsty brown snakes are now common garden visitors, and butterflies and birds struggle to survive. Empty frog ponds and dry lakes in our parks make it difficult for frogs to breed and survive. Many of those wonderful small or medium-sized, dense, shady bushes that protect our wildlife have died of thirst. Dead bushes do not provide protective habitat for small animals.

We need to remember that in a drought, wildlife die along with our plants. So top up that frog pond from your rainwater tank. We need to leave out water dishes for animals – elevated birdbaths for butterflies and birds, flat, shallow dishes on the ground for reptiles, mammals and insects. When we choose which plants we will save by bucketing water, consider whether it is also habitat for any animals. Save a dense, shady bush rather than a thin, straggly one. The wildlife enjoy a drink in your garden, and there is no better sight than a flurry of birds splashing in a bath, or the delicate flight of butterflies as they flit away after catching a quick sip!

Chris Daniels

Adelaide Plains, these trees require a lot of water. There is a presumption that if it is growing reasonably well with limited reticulated water it must be a dry-climate plant, whereas it is more often than not well established with its 'feet' in the aquifers, drawing huge amounts of water from our subterranean storage supply.

We persist with garden planting styles that are completely out of context with our climate. It is obvious that most of our residential gardens have been designed from an aesthetic approach with not much consideration for site or place. There is no doubt that European-style gardens appeal very much to many of us and fit very nicely with our inherited definition of a 'good garden'. Maybe it's time we redefined what a 'good garden' is.

To lawn or not to lawn, that is the question

There are many sound arguments for keeping a patch of lawn: for the kids to play on; to assist in cooling a house (if placed on a north-western aspect in Adelaide) and shaded in turn with a broad canopy tree, or, even better, if watered by greywater through a below-ground drip irrigation system or recycled grey/black water from an Envirocycle system or similar. Lawn can lower reflected heat and allow for water to penetrate and limit runoff into the stormwater system. These are all good reasons to consider keeping a small useable amount of lawn, however, lawns in front gardens, particularly with a southern aspect, have limited value other than aesthetic. The lawn in a front garden can be considered to have been grown for 'style' in most cases. There is an allegiance to some planting 'fashions', which are mindlessly repeated because they have become 'the norm'.

If planting lawn, the type of lawn selected is very important. Turf varieties are often more reliant on water than 'running'-type lawns such as couch or kikuyu. These lawn varieties will often brown off in the summer and return to a lush green once the rains come again.

While being able to water a lawn with recycled water may be good justification to establish one, consideration should be given as to what other 'useful' plants may be grown in its place (for habitat or production).

Streetscapes

Many gardens are designed from the street looking into the garden space. This is because most people still believe the old adage that the 'first impression of the house is the one that either sells it or not'. The idea has become to 'frame' the house with the garden. This results in endless gardens in a street, all looking similar and juxtaposed to the roadway.

This idea is not new. It has been around since the early 1800s (Gardenesque period) when J.C Loudon (1783–1843) decided that the working class of the industrial revolution deserved to have individuated spaces now that they had some income and a desire to emulate the gentry. Kitchen gardens were desirable but, like in the upper classes, were considered altogether too vulgar to be seen and hence were hidden behind walls 'out the back'. The front garden was a canvas to display wealth, high breeding or 'connections', and 'good taste' through a collection of beautiful, sometimes obscure plantings.

This excerpt from the *Suburban Gardener and Villa Companion*[1], published in London in 1836, sounds very familiar:

Limited plots of ground, whatever is their shape, grater variety of view will be produced by placing the house nearer one end, or nearer one side, than in the centre, If the garden were larger, or even of its present size, if circumstances were favourable, a small piece of water, supplied from a dripping rock, would have a good effect; and there might be a statue on a pedestal, surrounded with tazza vases o flowers, in the centre of the flower-garden The rest of the garden, with the exception of the surrounding border between the walk and the boundary wall, is entirely of turf, varied by choice ornamental trees and shrubs, including some rut trees and fruity shrubs. The standard roses and their fruity shrubs such as gooseberries, currants, raspberries, vacciniums, etc, of which there cannot be more than two orr three plants of each kind, stand in small circles, kept dug and manured …

We haven't moved very far from this view. Might it be time we created spaces more tempered by our experience and understanding; adopting an approach to streetscape and garden creation that considers the linear nature of the street, shared resources (including water), habitat, aspect, as well as beauty? A garden designed and predicated on performance will take on an aesthetic value by default. In turn, this garden can be improved on and more clearly defined, then layered with the detail that lends it its personality and sense of identity.

Public spaces and other influences: where do we get ideas from?

The Botanic Gardens of Adelaide has always played a very important part in educating the public about plants and gardens. Not only due to the history of structured education through apprenticeships, but through the imparting of experimental education, and creating spaces people can enjoy and participate in, viewing the beautiful (and the sublime). Many home garden features were direct or close copies of those seen in the Botanic Gardens, including the Grotto – a pond embraced by a bower of stone often displaying various mosses, ferns and water-edge plants.

There is still great opportunity to impart knowledge and experience through the Botanic Gardens. Most recently the redevelopment of the 'Mediterranean garden', complete with cobbled rill and plantings for drier gardens, illustrates the importance of its influence. Many public spaces around Adelaide are exhibiting very similar planting plans and styles (regardless of appropriateness).

Lush, green, highly decorative gardens are illustrated in many glossy magazines. In most recent years there has been a greater focus on appropriate plantings, sound design ideas and saving water, but there are others who do not make the most of the opportunity to positively contribute to garden awareness. Garden magazines are high on the list of influencing what we plant, when, where, and with what else. They are also high on the list of perpetuating the myths of what defines 'a good garden', by the traditional use of the language of aesthetics.

It remains rare that an article espousing the delights of garden creation centres upon its usefulness or its importance

Rain gardens

BOX 127

What are rain gardens?

A rain garden is a shallow, planted depression designed to take the excess rainwater runoff from a house or other building, assisting this stormwater to infiltrate the underlying soil and recharge the groundwater. The rain garden concept can be expanded to incorporate an entire garden or a city streetscape, but of particular interest is its small-scale application in the domestic garden, where there is potential for a very significant impact on stormwater management at the source.

Rain gardens are one element within a 'stormwater chain', a simple way to consider the way water moves through a property. The start of the chain might be a house, garden or shed, where water is introduced into the system; the end of the chain is the lowest point in the garden or stormwater system, where the water will flow to. Correctly designed rain gardens are very effective at capturing water, with minimal water overflowing, indicating that a rain garden reduces the peak flow rate and increases the lag time for water flowing into the stormwater system.

Rain gardens use the technique of retaining runoff for infiltration back into the soil. Through the chemical, biological and physical properties of plants, microbes and soil, the water is filtered before it enters the groundwater, so aiding pollutant removal.

Benefits for domestic gardens

In addition to retaining and filtering water onsite, rain gardens have a number of other attractive benefits for the domestic garden. By promoting more planting, rather than paved surfaces or high-maintenance mown grass, biodiversity is increased because habitats are provided for small animals, birds and insects. Rain gardens provide sensory pleasure and visual interest through the introduction of ephemeral water features into the garden; the cooling effect of this water can improve the microclimate of the whole garden. Rain gardens also provide creative opportunity for designers and home gardeners to experiment with different ways to feature water in their garden.

A wide range of plants is suitable for rain gardens, in particular many native species. General characteristics required for rain garden plants include: tolerance of short periods of inundation followed by longer dry periods; perennial rather than annual species; and deep fibrous root systems.

A demonstration rain garden is featured at the Royal Tasmanian Botanical Gardens, where it collects and filters runoff water from the training centre roof before its eventual percolation through the soil and back into the Derwent River. This rain garden is suitable for a domestic garden, and uses predominantly sedges and rushes with suitable growing characteristics. The soil is an open, sandy mix to provide good water filtration.

Design issues

Rain gardens are relatively easy to construct and maintain, with just a few issues to be considered. The rain garden should be sited at a relatively low point in the landscape. Any soil type can be used but the addition of sharp sand and organic compost will assist with infiltration in heavier soils. Plants may need to be watered to establish until it rains and mulch will help to maintain a constant moisture level. The surface of the inlet area should be of material that resists erosion, such as pebbles or rock lining, and compaction of the surface should be avoided – do not drive over your rain garden!

Conclusion

Rain gardens can be created within most domestic gardens and courtyards, and even kerbside planting strips – so they have an important part to play in a sustainable watercycle. Their evocative name also helps us to focus on the true value of our precious rainfall.

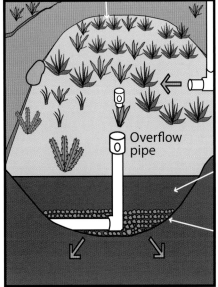

Wetland and native plants for inundation

Downpipe from roof

Overflow pipe

Engineered soil for bioretention process

Drainage rubble

Graeme Hopkins and Christine Goodwin

BOX
128

Gardening tips in the era of climate change

There are three things to consider when designing your garden in Adelaide in the 21st century: design for the climate; choose plants wisely; and understand the planting seasons. Below is an easy guide around these three points to help you construct your garden.

Garden design for climate change

1) Make a rain garden – a garden that grows, flowers, fruits and ripens in the rainy season.
2) Design and plan on the basis of a garden existing only on local rainfall, then a small amount of supplementary water can be applied to a restricted area selected for a particular planned purpose.
3) Adopt a general planning rule of equal thirds for consolidated zones that get: no water, some water and moderate water supply in summer.
4) Try to plan for onsite management of all garden waste as much as possible.
5) Plan for total onsite water management: rainfall harvest and storage, greywater recycling and safe septic discharge.
6) Make shade trees a high priority to shelter the house from western and northern exposure.
7) Use fruiting trees as garden trees: pomegranates, olives, almonds, apricots, peaches, nectarines, jujube, walnut, and citrus.
8) Make use of hard and soft paving to reduce dust and restrict grassed areas for essential play areas and pets.
9) Green is the most significant colour. Aim for a quiet, green background to outdoor living areas. Use urns and sculpture to create visual interest and focal points instead of floral displays.
10) Use mixed plantings rather than monocultures to discourage the spread of pests and disease.
11) Avoid the use of chemical controls.

Buy plants that will survive in a changed climate

Essential plant characteristics:

1) Waxy leaves: coprosma, olive, cotoneaster, oleander, rosemary, sedum, correa, hardenbergia, dianella.
2) Hairy leaves: geraniums and pelargoniums, santolina, senecio, oregano, salvia, eremophylla.
3) Silver leaves: artemisia, lavender, euphorbia, yucca, culinary sage, acacia, eucalyptus.
4) Succulent leaves: aloe, agave, echeveria.
5) Succulent stems: crassula, aeonium.
6) Succulent roots: day-lily, yucca, bearded iris, Algerian iris, spuria iris.
7) Bulbs: autumn, winter and spring flowering kinds such as amaryllis, nerine, sternbergia, jonquil, tazzetta narcissus, daffodil, babiana, sparaxis, ixia, wild tulip.
8) Plants with far-ranging, thongy root systems.
9) Plants that spread wide over the ground from a central growing point.

Disadvantageous characteristics:

1) Close, matted, fibrous root systems: azalea, rhododendron.
2) Thin-textured leaves: Japanese maple.

A gardener's calendar for climate change

January: rest in the shade with a cool drink and a good book.

February: order bulbs, hardy perennials, trees, shrubs, climbers, hedges, fruit trees, and nut trees for delivery in late April. Keep a watch for stressed plants and deal with accordingly – will it live or die? Keep resting. Get rainwater tanks cleaned out by an experienced contractor. Keep birdbaths topped up and clean.

March: commence gentle activity by preparing the ground for planting next month. Dig in compost to

improve the moisture-holding capacity of the soil. Check greywater recycling systems are operational and make repairs as needed. Nesting boxes may require cleaning and repositioning.

April: the most important season for planting all kinds of plants, but wait until the rains arrive. Plant shelter belts, regenerate native plant conservation areas and habitats for birds and animals. Construct simple habitat for small animals with flat rocks and hollow branch sections.

May: enjoy autumn flowers, plant winter greens, vegetables and herbs. Check new plants for wind and rock stability, stake where necessary. Patrol for snails and pests.

June: enjoy winter flowers, new growth and greenery. Control weeds and feral plants. Mow grassy parts. Compost weeds and garden trimmings.

July: enjoy early spring flowers, make notes to plant more of the plants that are performing well and plan consolidation of plant groupings for best impact and easy care. Plant a further rotation of winter greens, vegetables and herbs. Complete planting of new plants, considering how they will be cared for in the hottest months.

August: enjoy late spring flowers, prune plants where necessary immediately after flowering. Bucket-water new plants if the weather is dry and hot. Re-pot succulents and geraniums. Prune back any frost damage.

September: prepare for the hot, dry months ahead: mulch, mulch, mulch. It is too late to plant anything. Visit open gardens to gather design and planting ideas for gardens watered by local rainfall only. Deal with dry grass, branch and leaf litter, and clean up for summer fire season. Test run fire-fighting gear and emergency sprinkler systems.

October: dead head seed pods and seed heads, trim plants to send them into a summer rest period as the rains taper off. Decide whether or not any summer irrigation will be possible. What will be grown? Choose between vegetables, fruit and herbs, or flowers. Plant annuals as early as possible.

November: renew mulch, using deep litter coarse mulches to keep the soil cool and retain moisture. Condition new annual plantings for the hot, dry months by watering sparingly. Avoid promoting lush growth. Install new rainwater tanks.

December: prepare for the holiday season. Clean up garden furniture. Buy a yard broom to sweep paving and patios. Buy potted colour – geraniums, hippeastrums, succulents and architectural plants. Concentrate displays for maximum impact and ease of care. Wherever possible use large tubs and containers to hold 'dropped-in' herbs and potted colour. Keep out of the sun. Rest and enjoy the holiday.

Trevor Nottle

as habitat. 'Goodness' is rarely defined by a sound ethos of sustainability, which would include: the sensible use of resources; a contribution to the heat regulation of associated dwellings; the provision of habitat; a contribution to wildlife corridors; livability; and aesthetic value. It is not a matter of 'instead of', where performance is measured versus aesthetic value, it is a matter of *as well as*.

The movement towards Mediterranean climate zone planting combinations has had a great influence on plant material produced and sold to the public. It has been easy to promote, having been readily accepted by the popular media channels of television and print (magazines). These gardens in the main are pretty, look good on the page and screen, and provide a good story angle (new).

However, for a time (mid 1980s through to mid 1990s), the 'Tuscan' invasion took over; a classic example of 'too much of a good thing'. The sound thinking that grounded the movement towards Mediterranean climate zone plantings was extruded into the architecture (no eaves, small porticos, render), which is when the concept of 'shade' provided by trees was lost to the built pergola (later the Balinese Hut). Concurrently, there was a movement away from education of plants in the nursery industry, and a consequent reduction in training.

Popular media stopped talking about the plants themselves and started talking about 'architectural plants' (referring to what they looked liked). The combination of the architectural movement towards 'minimalisation' (starting with Tuscan and evolving) and the adaptation of formality (minimal) from traditional English to Mediterranean planting schemes – resulted in a huge loss of plant variety on the market (and what remains is in much greater volume).

Diminishing rain and strong indications that this is not a drought but a movement towards a drier climate or a different climate has meant that many of the plants we once relied on are no longer able to survive or thrive. The parkland giants, Eucalyptuses *maculatas* and *camaldulensis*, are now in steep decline, along with many other species. Couple this with less plant knowledge in the industry (and the public arena), and a movement supporting more numbers but less variety of plant material amongst 'designers', landscapers and developers, and we are building a very grave picture. As Milton Vadoulis of Vadoulis Garden Centre stated, 'Plants are part of the solution to climate change, not the problem.' (ABC radio, 2007.)

New directions: habitat

Due to dwindling water availability, the city and surrounds are being encroached on by bird and animal species which previously had been supported by surrounding regions where there was sufficient food and water supply. This has had an enormous impact on our residential and parkland gardens.

Rather than designer gardens driven from an aesthetic approach only, designing to provide habitat for a particular species, or number of species of birds and animals, allows for a substantial change in the choice of planting. Also, consideration to allocated space to accommodate our new city residents translates to different approaches to spatial design.

Plants need to be grouped together to accommodate protection and cover. For example, small birds need refuges where they are protected from predation. Flowering plants are selected to perform throughout various seasons to both attract and support the chosen species. Consideration of available resources is a natural follow on from this as the garden supports the habitat (water use being the main resource consideration).

For a garden to provide habitat for people, as well as wildlife, it must perform. A mindful selection of plants can provide shade, shelter, privacy and protection. The resulting habitat promotes biodiversity – an important role and a new direction for garden creation in our dry state. This garden will 'look good' but is defined by more sustainable criteria, rather than the strict measure of what currently defines 'a good garden'.

New directions: urban food production

Garden culture is demonstrated in many different ways depending on cultural traditions; for example, fruiting trees and shrubs under-planted with vegetables and herbs in front gardens abound around Stepney, Campbelltown and Prospect, where there are large Italian and Greek communities. These gardens are testament to the attractiveness, productiveness and wise use of water, differing substantially from the aesthetic approach of our predominately British heritage.

The creation of food gardens around our houses is regaining popularity. Up until recently the home vegetable garden was a typical feature of most residential gardens. The 1980s, the perceived era of prosperity, put a temporary end to that. People thought that the hassle of planting and maintaining a garden was an unnecessary drain on their constantly dwindling time resources.

Times have changed, and with the ever-growing threat of a peak oil crisis, coupled with rising prices for fruits and vegetables, we are seeing an advance in home-grown food production. This has come at a time of increased pressure on our shrinking water supply. It is a situation that demands an immediate shift in our approach to water harvest, storage, and water recycling within the boundaries of our own home gardens.

In the late 1960s and 1970s, Bill Mollison and David Holmgren coined the term 'permaculture', describing a sustainable settlement system developed to provide an alternative to broad-based reliance on agriculture. Permaculture focuses on the local production of food, drawing on the collective community knowledge, and recycling local materials and services, while minimising the impact on available natural resources. For some time permaculture has been more widely recognised and practiced internationally in Third World countries than in Australia and other Western countries. However, it is now in a growth phase here and will continue to influence the shift in urban food production and, hence, garden design.

I think as permaculture becomes accepted by the mainstream, permaculture designers, teachers and commentators need to head for the margins and edges (as they have always done) to keep breaking new ground. Most crucial will be promoting more "radical" ideas that will address climate change and encourage more

significant changes than the public are at present ready to accept.' – Steve Payne (editor of ABC Organic Gardener Magazine), March 2008.

What's possible in new directions for garden design?

Currently, despite the challenges or, rather, 'in spite of' the challenges, there are immense opportunities for garden design here in South Australia. New approaches must be adaptable (the true measure of sustainability), and authentic. So it follows that to base 'design' purely from the perspective of aesthetics won't take us very far. However, gardens based on their own potential to provide, *as well as* look good, will look good.

The Adelaide gardens of the future will certainly have a mixture of endemic Australian and exotic species of plants. The nursery industry will need to respond to a shift in demand created by a better-informed public with more plant selection. Localisation of expertise, service and supply will be a welcome factor, driven by the rising cost of transportation. Cooperatives for locally grown food exchange are emerging and will continue to emerge, as will more community gardens.

It is folly to think that this shift in the methodology of garden creation will generate a 'chaotic' aesthetic. Formality can still be a choice, but the parameters in which the formality is expressed will be less arbitrary and based on a site's macro- and micro-climates, the availability of resources, potential productivity, and the interface with homes and neighbourhoods (and therefore the region). In turn, these parameters will give rise to a more integrated and highly creative approach to garden design.

The successful creation of a sustainable city requires widespread acceptance and implementation of new gardening approaches, promoting participation from people with a wide range of economic capacities. If only those who can afford garden designs and works are able to contribute, the cumulative improvements to the city's resources and livability will be quite limited. Activities that will encourage extensive participation include workshops, public talks, one-on-one consultations, advisory services, and the sharing of access to new knowledge, techniques and technologies.

The challenge is immense, but as in all things, the opportunities are also immense. Garden designers, right now, are shifting in outlook and approach. Creativity is not dead or dying as a result of these changes; we can and will have wonderful gardens in the future, but they won't look like gardens of the past and it's about time!

References

1. J.C. Loudon, *The suburban gardener and villa companion*, London, 1838.

Figure 28.1: A mixed planting. This garden displays the traditional aesthetic approach to rhythm, tone, and texture with consideration of site conditions. A combination of endemic, Australian native and exotic plants has been used. The gravel driveway allows for water infiltration.

CHAPTER 29
Conclusion

The Editorial Board:
Christopher B. Daniels
Jerome J. Argue
Simon Beecham
Richard D.S. Clark
John R. Howard
David S. Jones
Richard Marks
Jennifer M. McKay
Philip E.J. Roetman
Keith E. Smith

Without doubt, water is the most important, integrated and complex environmental element in any city. We can attempt to understand the role of water from a utilitarian viewpoint, but as complex as that is, it is only a superficial analysis. Water is vital for drinking and waste removal but also for creating the nature of a city, providing aesthetic and social harmony. Water is also central to agriculture and industry. We now accept that we live in and are part of a natural landscape, and we must provide water for that landscape for its own sake. Furthermore, we now accept that water is not a limitless resource and its supply will become increasingly variable as a result of climate change. Today, we do not view water simply in terms of supply, treatment and management, but as a vital resource which must be managed sustainably. In this book we have examined water from a variety of perspectives with a view to generating an overall sustainable approach. In the Introduction we adopted the general definition of 'sustainability' as a system that 'meets the needs of the present without compromising the ability of future generations to meet their own needs (Brundtland 1987).

In this book we have adopted the following principles of 'water sustainability':

Under future scenarios of most likely population growth, climate change, technological advances and social adaptations, the urban water system should continue to provide, or be readily adapted to provide, all of the following outcomes into the foreseeable future:

1. A reliable, affordable, equitable and healthy supply of water that supports the social and physical environment.
2. Efficient management of excess water and wastewater to avoid disease, inconvenience and harm to people, biota, and the built environment.
3. Minimisation of damages from floods (or sea storms) to a risk level acceptable to the community, while building into the system the ability to survive more infrequent but 'catastrophic' storm events.
4. Avoidance of damage to ecosystems caused either by the excessive diversion of water from them or to them, and by either the amount or quality of the water diverted.
5. Minimum contributions to greenhouse gases involved with the construction and operation of water supply and management systems.
6. Water and wastewater systems that provide a maximum net benefit and consider full lifecycle costs.
7. Management of water and wastewater systems that encourage innovative responses to local conditions.
8. A community that embraces and contributes to water management in practice and in decision-making processes.
9. Sufficient access to water for recreation, amenity and aesthetic satisfaction.
10. To have learned from our historical mistakes and policies to provide a more secure and sustainable urban and rural habitats.

Clearly, the present water systems serving Adelaide are not sustainable because they do not address all of the sustainability criteria. At best they may address some of them, but in so doing they fail to meet or engage with others. Here we summarise the perspectives of each chapter to provide the reader with a synopsis of the views of the broad authorship. As you read the summaries, consider how they relate to the 10 sustainability criteria above. In this manner, the conclusion presents the core material necessary to inform both the community and its policy makers as they formulate, modify and enact the plans that shape Adelaide's watercycle for the foreseeable future.

The views of the authors

Chapter 2: The variable climate

Adelaide's climate is seasonal and highly variable. Recurrent droughts and floods are unpredictable and driven by atmospheric and oceanic circulations under the influence of sea surface temperatures in the Indian, Southern and Pacific oceans, the Southern Annular Mode, El Niño-Southern Oscillation, and the position of the subtropical ridge. Summer temperatures are warm to hot with heatwaves occasionally exceeding 40°C. Generally, winds are light and summer rainfall is limited to infrequent, short-lived thunderstorm events (occasional longer thunderstorm events can produce flash flooding). In autumn, days are balmy and winds are light. By late autumn Adelaide has usually received its first good rains for the year, but these are sometimes 'false starts' and winter rains are delayed or diminished. During winter, daily temperatures are typically in the teens and overnight temperatures rarely fall below zero. Frontal systems from the west deliver generally reliable winter rainfall, but heavy rains and flooding occur sporadically when cloud bands and cut-off low-pressure systems are drawn from the tropical north-west of Australia. Spring is the season of maximum windiness and rainfall decreases as daily temperatures increase.

Rainfall in Adelaide is highly variable over both time and space. Records show temporal variation from around 258 mm in 1967 to 883 mm in 1992. Importantly, rainfall is enhanced by the orographic lift created by the Mount Lofty Ranges. Accordingly, mean annual rainfall is lowest in the west and highest in the ranges to the east: Lefevre Peninsula receives around 400 mm, Adelaide CBD around 550 mm and Mount Lofty around 1100 mm. Generally, annual panevaporation exceeds rainfall around Adelaide, with only winter and late autumn months having greater rainfall than evaporation. With climate change, rainfall is predicted to decrease while there will be increases in temperatures, potential evaporation, and the number of extremely hot days.

Chapter 3: Catchments and waterways

Adelaide occupies alluvial lowlands between Gulf St Vincent and the Mount Lofty Ranges. Relatively high precipitation in the ranges feeds the streams of the lower country. There are four major catchments occurring in the metropolitan region – Northern Adelaide and Barossa, Torrens, Patawalonga, and Onkaparinga. The vast majority of streams are intermittent or ephemeral and many display a diminution in channel capacity downstream. Thus, there is a tendency for floods when heavy rainfall occurs. Before European settlement, many of the streams of the Adelaide region did not reach the sea; the streams either ran out of water as they flowed across the plain, losing water by evaporation or seepage into the underlying aquifers, or ended in wetlands. For example, the Torrens used to discharge into a low-lying marsh area called the Reedbeds until an artificial sea outlet was constructed.

European settlers have dramatically changed the landscape; urban development has greatly increased impervious surfaces and runoff, erosion in stream channels has accelerated sedimentation in downstream sections of river valleys, and streams have been channelised. These changes have vastly increased waterflow to the sea. Freshwater flows now generally occur as major pulses of turbid water and contain pollutants and nutrients. These changes have had detrimental impacts on local marine habitats.

Chapter 4:   Aquifers and groundwater

Around 20% of the total water supply for Adelaide and its surrounds is groundwater. It is used in a variety of commercial, industrial, residential and public activities. In past droughts, additional groundwater (up to 10,000 ML/year) has been directly injected into Adelaide's water supply.

There are three main types of aquifer beneath Adelaide and the Mount Lofty Ranges. On the plains and in the foothills, shallow Quaternary aquifers are between 3 and 50 m below ground. Domestic water-users take water from 'backyard bores', which tap into these aquifers and provide relatively low yields. Confined Tertiary aquifers lie below the Quaternary aquifers and are the largest and most important groundwater resource in Adelaide. These aquifers vary in thickness from about 20 to over 400 m and receive natural recharge from the aquifers of the Mount Lofty Ranges. In the ranges, fractured rock aquifers provide water for agricultural industries, local communities and natural ecosystems.

In light of its obvious importance, groundwater must be managed carefully, and be seen as a critical and connected part of the watercycle. There are opportunities for surfacewater to be stored in aquifers – as an alternative to dams and reservoirs – in order to minimise evaporative losses (Adelaide is a national leader in aquifer storage and recovery schemes), and, by exploring ways of better managing groundwater resources in Adelaide and its surrounds, ensuring they are not at risk of becoming overexploited. Water management plans and policies must recognise that groundwater, like surfacewater, is limited and requires attention and protection – from both water quantity and quality points of view. Importantly, groundwater resources were prescribed under the *Natural Resources Management Act 2004* in 2007, and future extractions for industrial and commercial purposes will be licensed and limited.

Chapter 5:   Biodiversity of the waterways

The waterways of Adelaide are home to a great many native and introduced plants and animals and provide significant corridors of habitat. The unusual geology, geography

and high winter rainfall of Adelaide have contributed to unique ecological communities. Typical riparian vegetation consists of an understorey including grasses, herbs and shrubs, with an overstorey of larger trees such as river red gums and introduced species such as willows. Birds are the most conspicuous fauna around water in Adelaide and there is a wide range of species. Mammalian fauna is less spectacular. Platypuses once occurred in the Mount Lofty Ranges but the only native mammal that still commonly occurs around and within Adelaide's waterways is the water rat. Introduced rodents are highly abundant and attract introduced predators such as cats and foxes. Reptilian fauna includes water skinks, tortoises and snakes. Several species of native fish no longer occur here, and many new species have been introduced. There are seven species of frogs found in the Adelaide metropolitan area, and a diverse array of invertebrates can be found within and around waterways.

Most of Adelaide's wetland habitats have been severely modified since European settlement, destroying their original ecological integrity. The most significant threats to the extant biodiversity include the clearance of vegetation, the intrusion of woody weeds, unrestricted grazing, polluted water, poor habitat management, unsuitable urban development, increased water extraction and stormwater runoff, and poor management of water temperature. Generally, watercourse management works should involve treating any erosion problems, protecting all remaining remnant riparian vegetation and avoiding activities that may initiate further erosion or decline in water quality. Appropriate revegetation will enhance the habitat for use by native birds, mammals, reptiles, amphibians, freshwater fish and invertebrates

Chapter 6:   Kaurna cultural heritage

The Kaurna people have resided on *Mikawomma* – the Adelaide Plains – for over 40,000 years. They have accumulated substantial knowledge of the natural systems and witnessed many changes to the landscape. To them, *Mikawomma* is their 'country', which they care for as custodians on behalf of their Ancestors and Dreaming characters. Therefore, because of this long-standing relationship, a complex set of codes and moral narratives have evolved that explain the role of the Kaurna people and their responsibilities to the care, conservation and maintenance of the landscape. The landscape possessed a diversity of water sources and forms, of varying fresh and saline levels, which were linked to food, ritual, death, and more often explained in myth, Dreaming story, and/or nomenclature. Water was carried and enabled life for human, animal and spirit alike. The protection and care of water was essential, irrespective of the harshness of its absence during drought or its abundance during the rains or floods, or its presence in narratives. Kaurna people often settled along the waterways of *Mikawomma* to utilise the many available food resources, and moved with seasonal changes.

Early European settlers were unfamiliar with the landscape and struggled to appreciate and understand it. Furthermore, the European four-seasoned calendar fitted uneasily on seasons of *Mikawomma*. Kaurna people have been apprehensive to convey knowledge to European settlers, feeling it could result in loss of 'knowledge' and create difficulty in living with the surrounding resources. Yet, this long-term understanding of the interactive changes in climate and landscape could help us to better manage our responses to the ongoing challenges of climate change.

Chapter 7: A history of water in the city

The availability of freshwater and drainage were both considered very important in the choice of the site for South Australia's capital, and these factors have continued to influence the development of Adelaide. Adelaide's history has been punctuated by times of water shortage, damaging floods and water quality issues. With the completion of the Thorndon Park Reservoir, reticulated water flowed into North Adelaide in 1859. Further reservoirs have been constructed at Happy Valley, Milbrook, Mount Bold, South Para and Myponga. Water shortages have led to the construction of two pipelines that divert water from the Murray River into Adelaide's water supply. Groundwater resources have also augmented supply, used mainly to irrigate crops and gardens, but sometimes used in mains supply. Water quality has been improved by the construction of six major filtration plants, and the *Waterworks Act* has served to preserve water catchment quality.

Adelaide's drainage systems have been developed to allow urban expansion and improve public health. Urban expansion in low-lying and flood-prone areas was allowed by the re-engineering of watercourses, the construction of stormwater drains and the diversion of runoff into the sea. The installation of sewer mains and a sewerage farm at Islington in the early 1880s meant Adelaide was the first Australian city with a complete and modern sewer system. Four wastewater treatment plants (WWPTs) now handle Adelaide's wastewater.

Since the 1980s, increasing attention has been given to the local environment and a more integrated approach to water management. Wetlands and 'soft' watercourses have been developed to manage stormwater, providing filtration and storage of stormwater, space for biodiversity and recreation, and natural infiltration into aquifers as well as managed aquifer storage reuse. Some recycled stormwater and treated wastewater is now used by local vegetable and grape growers, and new urban developments are using this resource to supply water for toilets, gardens, car washing, public parks and reserves. Recent strategies have emphasised the importance of public education and participation in water management alongside engineered solutions to water shortages. The importance of constructing better systems to take advantage of the watercycle is now recognised.

Chapter 8: The River Torrens 1: an unnatural history

The development of agricultural and urban settlement within the River Torrens Catchment since 1837 has progressively altered the river's hydrology, channel morphology and water quality. Flows in the River Torrens system could not have been predicted at settlement, and are still difficult to predict due to rainfall variability in the catchment, changing land use, and climate change. During the early years of settlement, the stream supplied freshwater to Adelaide while also being used as an open sewer and waste dump.

Unfortunately, flows were inconsistent and often insufficient to sweep all of the material away, making waste removal problematic until sewerage and drainage systems were operational. However, pollution from domestic, agricultural and industrial sources continued in many parts of the system. The removal of trees for construction materials and firewood increased channel bank erosion, and the river was used as a source of sandstones, gravels, sands and clays for construction.

From 1860, the hydrology and channel morphology of the Torrens river system was significantly modified by engineering works intended to provide Adelaide with reticulated water, mitigate flood risk and improve the aesthetic character of the river. Settlement would have destabilised the river system and initiated a period of significantly increased sediment yield and channel aggradation, leading to more frequent and severe winter floods. Urbanisation of the catchment has now greatly increased inflows to the system due to the runoff from impervious urban surfaces. By the 1980s, close to 20 weirs had been constructed on the river, and flows through the system were largely limited to this urban stormwater. The ecological integrity of the system was at a severe low. Through the 1980s and 1990s, the River Torrens Linear Park Project has had some success at addressing flood mitigation and the degradation of the river channel and its riparian landscapes.

Chapter 9: The River Torrens 2: contaminated sediments in the river

Sediments are the major repository of contaminants in river systems. Studies of the pollution of sediments provide baseline data for the assessment of the impact of land uses on rivers, and for the development of future environmental strategies. The sediments of the Torrens have been found to be contaminated by cadmium, lead, phosphorus and zinc; levels of chromium and copper are also of concern. In general, rural study sites present little cause for concern, and high values of copper, lead and zinc are probably a consequence of the mineralisation of the underlying rocks. By contrast, all downstream residential and industrial study sites exceed either the trigger value or the high guideline for one or more of the measured toxicants. It is probable that much of this contamination is a response to past pollution practices. Importantly, not every component of each toxicant is necessarily available to organisms. However, contaminants may be mobilised from the sediment body to the water column under anoxic conditions, most likely during the warmer half of the year. Thermal stratification is thought to contribute to the existence of anoxic conditions, and, in the Torrens Lake, a series of aerators has been installed to reduce this condition. Long-term monitoring is required to understand conditions in the remainder of the river system.

Unfortunately, these problems are likely to persist. There is little evidence that bed sediments are being moved downstream and flushed out of the system, or that they are being diluted by mixing with relatively uncontaminated sediments. The chains of weirs, sluices and reservoirs along the river must have reduced the magnitude of the flood flows that would have entrained the sediments under natural conditions while simultaneously acting as sediment traps to prevent pollutants being transported downstream.

Chapter 10: The River Torrens 3: creating a new riverscape

The River Torrens Linear Park is now a greenway and a core recreational venue for the people of Adelaide, although this has not always been the case. Before European settlement, the river was a major ecological corridor, and, although it was ephemeral, floods were regular. Development of the city saw vegetation removed, watercourse alignments changed, water extractions for irrigation and water supply, and a loss of biodiversity. Intensive agricultural, industrial and residential growth saw the watercourse become a dumping venue for stormwater, pollutants and hard wastes. Increased runoff from urban areas and the engineering of the watercourse only increased the velocity of flows, and both the frequency and intensity of flood damage.

Over the last 30 years the River Torrens Linear Park has been the largest integrated stormwater management project undertaken in Australia. Careful planning, land acquisition, flood mitigation works, landscaping and the construction of trails and recreation facilities have been undertaken. The river has been reinvigorated as an ecological corridor in the metropolitan area and, at the same time, the Linear Park has become an integral recreational venue and transport corridor. It is far from the pre-settlement river environment but is still valuable in contributing to our aesthetic enjoyment and psychological health, and to the ecological health of our city.

Chapter 11: Wastewater treatment and recycling

Adelaide, like all cities, must collect, convey, treat and dispose of wastewaters. These include sewage, which poses a risk to human health and the environment unless it is managed properly. However, with increasing water scarcity, treated wastewater is now being seen as a resource rather than a waste product, and the community has become increasingly supportive. The Adelaide metropolitan area sewerage system contains more than 6900 km of buried sewer pipes with more than 455,000 house and business connections, serving about one million people. It is serviced by four wastewater treatment plants (WWTPs) that employ various forms of the activated sludge process to purify wastewaters. The WTTPs discharge treated wastewater into coastal marine waters. These discharges have been reduced following the introduction of *Environment Protection Act* in 1992, and the establishment of the Environment Protection Authority (EPA). Consequent environment improvement programs have focused on nutrient reduction through plant upgrades, and increased reuse of wastewater has considerably reduced discharges.

Reuse schemes service agricultural, municipal, industrial and household users, under the control of the Department of Health and the EPA. In 2006–2007, 30% of the treated wastewater produced by the metropolitan WWTPs was recycled, making South Australia a national leader in this field. Important initiatives that contribute to Adelaide's wastewater reuse include the Virginia Pipeline Scheme, the Mawson Lakes Recycled Water Scheme, the Willunga Basin Pipeline Scheme, Water Proofing the South, and reuse from the Glenelg WWTP including the Glenelg to Adelaide Parklands Recycled Water Scheme. The major impediment to substantial expansion of wastewater reuse is the lack of demand from horticultural users during the winter months. Potential solutions to this barrier are storage of winter water in reservoirs or aquifers for reuse the next summer, or an increase in year-round industrial use.

Chapter 12: Who is responsible for water management?

Defining who is responsible for water management in Adelaide is complex. Many laws, policies and organisations are responsible for how water is managed, and these are influenced by laws and policies dating from earlier times when the demands for water were different.

National water management is undergoing significant change. As a downstream user of water from the Murray-Darling Basin, South Australian water security is reliant on national agreements and the management of water in other states. Commonwealth government responsibility over freshwater has been limited by the constitution, which provided that the power over water was to remain with the states. However, the *Water Act 2007* requires some state powers be referred to the federal government where the Murray-Darling Basin is affected. In fact, since 1994 there has been increasing federal influence, pushing the states to adopt certain principles such as ecologically sustainable development (ESD), full cost recovery, separation of water and land titles, and private sector provision of some aspects of water supply. The political compromises to achieve the Water Act are undergoing some strain, and these will play out over the next years.

South Australian management of water is also complex. In terms of water supply, SA Water, owned by the state government, is responsible for the delivery of water and wastewater services. SA Water has contracted United Water to manage the water supply pipes and to supply water to customers, while retaining ownership of the pipes on behalf of the state. A series of cascading plans impact on water resources, from a state strategic plan to the state Natural Resources Management (NRM) Plan and regional plans such as the Adelaide and Mount Lofty Ranges Regional NRM Plan. Regional plans are developed by regional NRM boards, which report to the minister for water security, the minister for the Murray, and the minister for environment and conservation. The content and operations of the NRM plans are directed by the Natural Resources Management Act 2004. Local government is obliged to refer to the NRM plans in its assessment of development applications. Additionally, the Water Proofing Adelaide strategy, managed by the Department of Water Land and Biodiversity and Conservation (DWLBC) has been developed to help secure Adelaide's future water supply.

Chapter 13: Climate change and water management

The earth's climate is naturally variable and changes over time, and this will continue. However, anthropogenic influences are now augmenting it. Warmer and drier conditions are predicted for Adelaide, with an increase in extreme weather events such as droughts and floods. These changes will present many challenges for water management including water quantity, water quality, recharge of groundwater resources, and problematic bushfires.

Planning for climate change is difficult considering our limited ability to predict future circumstances. Potential scenarios include a severe water deficit, so planners must consider infrastructure such as reservoirs and pipelines, built to last in excess of 100 years. Considerable work has already been done. Water Proofing Adelaide, released in 2005, is the South Australian government's blueprint for the management of Adelaide's water resources until 2025. It is based on better management of existing water resources, increasing water use efficiency and investigating alternative water sources. In 2007 the state government announced a further 4-Way Strategy, based on desalination, recycling, managing water use, and improving catchment management. Additionally, SA Water is developing a long-term Climate Change Strategy to 2100. As the state's major water provider, SA Water has the primary responsibility for greenhouse gas emissions associated with providing water and wastewater services for Adelaide. SA Water is working towards achieving the climate change targets and objectives of South Australia's Tackling Climate Strategy and the *South Australian Climate Change and Greenhouse Emissions Reduction Act 2007*.

To provide water infrastructure robust enough to deal with climate variability, climate change and population growth, significant investment is required. Even higher investment will be required if climate change is not properly addressed. An effective response to tackle climate change will require all government agencies, businesses and individuals to contribute. Each can play a part by supporting water conservation and the implementation of sustainable water technologies.

Chapter 14: What price water?
For many years water has remained at a low price for Adelaide's residents and businesses, regardless of its scarcity. This situation cannot continue as Adelaide continues to grow; low-cost water sources are limited and are predicted to diminish. An important mechanism to respond to a situation of decreasing supply and increasing demand is the market process. The price mechanism, if used appropriately, can complement other devices to help balance the demand for and supply of water. Accurate pricing of water provides an incentive to all users and potential suppliers to discover 'new' markets or supply possibilities, as water is traded, recycled, or reclaimed in ways suited to local conditions that have been previously overlooked or considered uneconomic. It is clear that the price of water in Adelaide needs to increase in real terms. Ideally this should continue over a period of years, and in a predictable and transparent manner, to allow consumers and suppliers time to respond. A balance must be struck between the need to signal scarcity and still allow access to affordable, clean drinking water.

The price mechanism is a powerful device for influencing individuals' behaviour. Market processes are important tools to create incentives. However, markets are interdependent and prices are relative. Market structures require complementary regulatory frameworks to deliver the major benefit of the market process – efficiency. Information, property rights and well-developed complementary markets are all necessary for market signals to work properly. Importantly, water infrastructure is costly. Suppliers will hesitate to expand supply or trial new water markets unless they are sure that such relative price rises will be maintained. A sustained regime of water prices that consistently signals its relative scarcity is therefore necessary before a widespread and systemic change occurs that rebalances the disequilibrium between supply and demand.

More work is needed to estimate the true cost of supplying water to Adelaide. In particular, the environmental costs of our reliance on the Murray need to be valued, as do the true costs of other supply options. This information needs to be gathered with some urgency; without it, pricing options cannot accurately reflect the true cost of supply. An integrated approach to managing water – including the full watercycle and what differentiates between water markets and different levels of water quality – also needs to be undertaken.

Chapter 15: Food glorious food
Agriculture is an important feature in and around Adelaide, with 40% of the Greater Adelaide area zoned as agricultural land, and much of the produce supplying the local market. This primary production includes market gardens, orchards, vineyards, pastures and livestock production including pigs, poultry, dairy, beef and lamb. The gross value of food production in the regions around Adelaide is around $2 billion per annum. On average it accounts for 28% of Adelaide's water consumption. This water is obtained from groundwater (56%), recycled wastewater (18%), local surfacewater (17%), and water from the River Murray (9%).

It will be challenging to maintain productivity as rainfall decreases with climate change and urban growth encroaches on productive lands, exacerbated by Adelaide's growing population. Improved water resource management is vital. Water extractions must be controlled and water allocation plans set in place. To maximise water savings, water-efficient crops suited to our soils and climate need to be planted, and a range of sustainable water management practices adopted including irrigation scheduling, mulching, leakage minimisation and the monitoring of rainfall, soil or plant moisture. In addition, greater utilisation of stormwater and wastewater will help to reduce the pressure placed on traditional water sources.

Many of the required changes are already starting to occur as the industry capitalises on a 'green' marketing image. Government could assist by developing or expanding initiatives such as water efficiency rebates, water audits, educational materials, changeover financial packages, or setting an enforceable industry water standard. A variety of research and technology transfer activities aimed at improving water use in agriculture are underway, including improved means to monitor and predict water requirements, differential watering techniques, new crop species, improved irrigation techniques, and better use of groundwater, wastewater and stormwater resources. These initiatives will help ensure a sustainable agricultural industry for Adelaide.

Chapter 16: Water and population

Water availability places an environmental constraint on human settlements. The amount of water needed to support a community is a function of the number of people, their per capita consumption, and the technology they employ to capture and maintain water resources. Land use, the size of allotments and the number of people in households influence the consumption of water. Trends towards smaller allotments reduce water usage, while reduced household sizes are increasing water consumption. Per capita consumption of water increased dramatically in Australia from around 100 L in the 19th century to 300 L in the 1980s. A cultural shift is underway with community commitments to reducing community consumption, however, a dramatic reduction has not yet been achieved.

During the 1990s Adelaide experienced low levels of population growth with an increase in older age groups and a corresponding decrease in younger age groups. South Australia has implemented a population policy to address the low population growth, with a target to increase the state's population to 2 million by 2051 (largely in Adelaide). Recently, population growth has increased; if it continues at the present rates population targets will be reached in the 2030s.

Population planning is an important part of integrated water management, along with consumption behaviours and technological changes. Water management policies have historically addressed water supply with little attention paid to reducing demand. However, a simple reduction in population would not immediately solve Adelaide's water problems as the ageing population would place a substantial burden on the economy of the state. The lack of financial resources would reduce potential for progress towards sustainable use of water. In the long-term, Adelaide should seek a demographically stable population, which will require growth in the short-term because of the baby boom bulge.

Chapter 17: Water quality and public health

Safe drinking water: No other measure has had a larger beneficial impact on public health than the introduction of safe drinking water and sanitation. Safe drinking water reduces morbidity and mortality from diarrhoeal disease, reduces medical costs and increases productivity. With safe drinking water available in Adelaide, disease outbreaks have been virtually eliminated. However, challenges still remain and ongoing vigilance is required to prevent outbreaks of waterborne disease.

Water fluoridation: Water fluoridation, introduced in Adelaide in 1971, is the backbone of 'dental public health'. A vocal minority of people raise concerns about the risk of water fluoridation, however, the only established potential adverse outcome is dental fluorosis, little of which is aesthetically of concern.

Blue-green algae: In South Australia there are a range of blue-green algae (cyanobacteria) that have the potential to produce toxic compounds, which, without appropriate treatment, could have a deleterious effect on drinking water quality. Cyanobacterial poisoning has caused serious human illness in Australia and led to deaths overseas. In the current climate of water shortage and drought, there may be a temptation to concentrate on water quantity rather than quality, but this must be resisted if we are to avoid problems caused by cyanobacteria.

Recycled water use: An increased demand for alternative water sources has expanded the scope for recycled water use. The result has been a progression from applications such as restricted irrigation to toilet flushing, and, more recently, use of recycled water for drinking purposes. However, protection of public and environmental health is of paramount importance. Recycled water schemes must therefore be planned, designed, installed and managed to ensure that public and environmental health is not compromised.

Climate change and public health: Climatic variations will influence the growth and spread of the microorganisms responsible for waterborne diseases. Further studies in Australia on climate change and waterborne diseases will benefit disease prevention and control.

Chapter 18: Maintaining green infrastructure

Urban vegetation, the 'green infrastructure', is a valuable resource that plays an important role in urban water management. Yet the urban ecosystem – our parks and gardens – and the services they provide, are conspicuously absent from the current dialogue on water in Adelaide. Urban vegetation is deteriorating and dying; the vital functions performed by this vegetation and the benefits it delivers are not widely recognised.

Urban vegetation is valued for its visual amenity, its shading and windbreak effects that increase the life of built infrastructure and reduce cooling and heating costs, its moderation of climate through transpiration and light reflection, its ecological roles, and its contribution to air and soil quality. The maintenance of urban vegetation provides employment and increases community socialisation and networking. In terms of water management, green infrastructure helps slow runoff through the urban landscape and promotes infiltration; it interacts with groundwater and can stabilise soils, reduce salinity, harvest excess water, and reduce erosion.

In modern cities, built infrastructure is usually well managed; infrastructure is maintained and eventually replaced on consideration of financial and lifecycle parameters and constraints. Rarely are similar asset management approaches applied to the essential components of urban vegetation or green infrastructure. Due to the complexity of its interactions with both natural and built systems, economic values can be difficult to ascribe to urban vegetation. However, the current drought is depleting Adelaide's green infrastructure, and the annual loss of environmental services has been estimated in the millions of dollars, even without considering replacement costs. The replacement of green infrastructure will take time and money, and care must be taken to ensure that vegetation is appropriate for local conditions now, and the drier conditions expected in the future.

Chapter 19: Perceptions of water in the urban landscape

People interact with water in many ways. There are a vast range of needs and values that water fulfils – the utilitarian requirement for drinking, water for cleansing, transport, recreation and fitness, health and restoration, image promotion – and water has aesthetic and spiritual value. We have an historical association with water through springs, fountains, rivers, gardens and baptisms. There are psychological and physiological benefits of viewing water in an urban context, as well as aspects lacking appeal. The value-adding that water provides to real estate is significant.

Suffering from a history of neglect, pollution and a woeful image, the promotion of the redevelopment of the Port Adelaide waterfront posed substantial challenges. The development used water as an agent in image creation, advertising to create favourable images, which differ from the reality. But the redevelopment is about the future, and perhaps, in time, the substance might merge with the image.

There is strong community support for the use of recycled water in Adelaide. Several major projects have transformed wastewater into a resource; in particular, the Virginia Pipeline that distributes wastewater from the Bolivar WTTP to Virginia for use on market gardens, and the development at Mawson Lakes that reuses much of its wastewater for gardens and toilets. Importantly, community perceptions of water and wastewater will help shape how we address water management in Adelaide.

Chapter 20: Reporting the water debate

Adelaide's water issues are a great source of community debate, especially considering the current drought conditions. Water restrictions and climate change are of widespread concern, as is Adelaide's reliance on the Murray River for water, at the expense of farmers and environmental requirements. There is some misunderstanding in the community about the environmental needs of the Murray River system, leading to disagreement over decisions to release environmental flows.

The water issue has exposed differences of opinions between the city and country residents. First, city residents generally demand the provision of water whether or not it has rained, whereas country residents are accustomed to dealing with times of limited water and are more supportive of water conservation measures. Second, country irrigators are frustrated by the water demands of Adelaide residents. Water restrictions in the city, first implemented in 2006, angered many city residents, but have been effective at reducing water consumption, although their effectiveness has been decreasing.

Many solutions to Adelaide's water shortages have been proposed, although some are problematic. For example, a desalination plant will have adverse impacts on the environment and will be costly to run, and increasing the capacity of Mount Bold Reservoir will impact local biota. However, even with the city's water security under pressure, residents of Adelaide have shown only a limited concern over these issues. Other solutions, such as the reuse of stormwater and wastewater, must be given the consideration they deserve. Community attitudes towards the environment and water reuse are changing, however it is unfortunate that only a prolonged drought seems enough to stimulate change (albeit small and slow change).

Chapter 21: Water management networks

The potential for reduced rainfall, along with increased energy costs and flood risks, dictate that an integrated approach to water systems needs to be taken, with multiple uses of limited water resources. Furthermore, economies of scale do not dictate that centralised water systems are more efficient than systems of smaller scales for the harvesting, reuse and recycling of water. However, Adelaide's water management networks are not designed to easily become integrated. First, the development of the stormwater drainage system has been piecemeal, leading to increased flood potential, large volumes of discharge and decreased water quality. Second, the sewerage system is highly centralised, separate from the stormwater drainage network, with most wastewater discharged off the metropolitan coast.

The River Murray is relied upon to augment water supply, although it is now realised that this resource is limited. Water storage is also problematic. Adelaide Hills reservoirs hold approximately one year's supply of water for Adelaide (much less than the storages of Australia's other capital cities). While Adelaide's Tertiary aquifers are underutilised, currently being used by only a few industrial users; these aquifers could potentially store large volumes of water and assist in an integrated management system.

Chapter 22: A sustainable future for water

Despite its low rainfall, the total of water flowing through Adelaide is well in excess of its total demand for water, but this water is not properly utilised. The present urban water systems are wasteful and cannot be sustained, as they are increasingly uneconomic and environmentally damaging. With advances in water treatment it is possible to pursue a strategy in which the excess stormwater and wastewater generated within the city be treated to high qualities and made the primary sources of future water supplies.

Large-scale stormwater harvesting and the reuse of treated wastewater will require changes to water infrastructure. Additional water storage can be provided through aquifer recharge and extraction. There are a number of options for piping networks to deliver recycled water to Adelaide residents, including installing a second pipe system to deliver non-potable water, or injecting treated water into the existing pipe system (either at potable quality or with further in-house treatment). Sewer mining and localised treatment could also be a feature of such a system, allowing for effluent to be recycled close to the location of its generation. Importantly, different solutions will be suited to different locations.

New systems must be capable of sustaining services, equitably and at a minimum lifecycle cost. Development must account for projected population growth, climate change, technological advances, and changes in community attitudes and the structure of society. The planning and implementation of new water systems necessitates more coordinated planning and has the advantage of progressive roll-out, with capital expenditure over a more extended period. We may need to establish a new government authority with increased public participation. Various government departments and local councils must cooperate to research and develop these strategies. Making Adelaide's water supply sustainable will require visionary and talented leadership.

Chapter 23: WSUD 1: Water sensitive urban design

Adelaide's water systems have been based on European models that profoundly modify the natural watercycle. The development of urban areas has increased the amount of impervious surfaces, which has increased stormwater runoff. Additionally, the stormwater drainage systems have been designed primarily for flood prevention and most runoff is discharged into the ocean via creeks, significantly altering the natural hydrology and carrying an environmentally damaging load of pollutants. The urban landscape must change in order to improve water management.

Water sensitive urban design (WSUD) has evolved from advances in engineering, biotechnology and environmental management. It embraces integrated urban watercycle management, including the harvesting and treatment of stormwater and wastewater to supplement water supplies. Objectives include managing the water balance, enhancing water quality, encouraging water conservation, and maintaining environmental and recreational values. WSUD components include rainwater tanks, grassed swales, biofiltration swales, bioretention basins, sand filters, infiltration trenches and basins, vegetated filter strips, permeable pavements, wetlands and ponds. These components are utilised in unison throughout the landscape (where combinations of components are described as 'treatment trains'). WSUD can rapidly change the appearance of urban landscapes, at small and large scales, with vegetated stormwater systems; however, implementation has been slow due to lack of design guidance, design experience and modelling tools.

Increased implementation of WSUD requires improved lifecycle costings (including externalities), increased awareness, appropriate legislation, established protocols for adoption and maintenance, increased funding, and technological advancement of WSUD. A positive development is that Planning SA is currently engaged in the 'Institutionalising Water sensitive urban design' project to ensure that WSUD will be incorporated into Adelaide's suburbs.

Chapter 24: Water sensitive urban design 2: community applications

In cities such as Adelaide there is an increasing demand on open space and green areas due to infill development and moves towards higher-density living catering for an increased population. This intense urban development increases the amount of stormwater runoff from buildings, and reduces opportunities for its infiltration into the ground due to the large amount of impervious paved surfaces. Greater pressure is placed on existing stormwater infrastructure, with associated problems such as flooding and polluted water entering urban streams.

Urban areas are constructed and managed by humans but still function as ecosystems. Adelaide's urban ecosystem is in an unbalanced state, consuming considerable amounts of energy, resources and human endeavour to keep it functioning. Following the methodology of 'landscape urbanism', we should reintroduce natural systems within cities, integrating the built environment with ecological systems for the benefit of the total environment, including humans and our complex cultural interactions. Water, an essential element to any ecosystem, cannot be looked at in isolation, but rather as a critical part of a balanced, living urban ecosystem.

WSUD principles can be implemented in a practical and effective manner to help create a sustainable urban environment. By mimicking nature we can introduce natural systems into the urban catchment, using green infrastructure. Hard, paved surfaces and vertical and elevated planes of the urban form can be used to collect, store, cleanse, recycle and reuse urban stormwater. Creating ground-level vegetated habitat corridors with bioretention strips provides multilayered strategies for water treatment, biodiversity and connectivity. As an additional benefit, green infrastructure systems usually require less energy input and maintenance than conventional city systems.

Chapter 25: Industrial and commercial use

In Adelaide we need to reduce the volume of water we use and use it more efficiently. We need to nurture diversity, collaboration and innovation. South Australia has benefited greatly from private sector involvement in the water industry, and this participation needs to be further encouraged.

There are many ways in which we can improve water supply and use. First, recycling wastewater and harvesting stormwater will reduce our reliance on the River Murray. Second, we need food to be produced as efficiently as possible, improving the way we deliver and apply water to crops and pastures; the food we eat and how we grow it has a huge impact on water usage. Third, we need to remove barriers to water going to where it is needed; water trading encourages water to be used where it is most valuable, including the critical value of maintaining river systems. Fourth, we need to diversify water suppliers and water supply sources; competing suppliers should be able to compete for the business of retail suppliers, who in turn would compete for individual customers of households and industry, selling the water to reflect the cost of production and scarcity. Water sources could be recycled water projects, aquifer storage, desalination, dams, or farmers with water entitlements. Fifth, water pricing needs to reflect water scarcity and the cost of supply (including the environmental costs); a comprehensive pricing structure will encourage the search for and adoption of alternative sources of water. Sixth, the community must demand and be prepared to pay for sustainable water management. Last, the government must respond with appropriate policy, regulation – and accountability.

Chapter 26: Council solutions

During the early years of the 20th century, all irrigation systems were manual. Some areas of Adelaide were very green and others not irrigated at all. As development progressed and the metropolitan area grew, so did the spread of green open spaces. Water for irrigation was either from the mains system, pumped from bores or from rivers.

From the 1980s the installation of automatic pop-up sprinkler systems tended to result in over-watering. Several councils connected their irrigated spaces to the externally controlled irrigation systems, designed to match irrigation patterns with weather patterns, and installed subsurface irrigation systems. Automatic irrigation systems soon became more efficient and less wasteful. During the 1990s, changes were increasingly implemented due to the cost of water, council staff reductions, and better education of staff.

Since around the year 2000, change has occurred quickly. All councils have introduced changes and many have undergone rapid developments in areas such as water harvesting and recycling, reductions in water use, improved irrigation systems, system audits, smart scheduling, moisture sensors and probes, rain sensors, subsurface irrigation, better maintenance practices, monitoring, analysis of site conditions, plant selection, and introduction of the irrigated public open space (IPOS) code of practice. Aquifer storage and recharge (ASR) and WSUD options are increasingly being investigated and utilised.

During 2007–2008, mains water consumption by councils decreased considerably. While some water savings have resulted from improved practices in the built environment, the majority of savings are attributed to increasingly efficient practices in managing public open space in response to water restrictions. Irrigation to non-functional areas is increasingly being turned off by local government, and considerable amounts of water are being saved each year. However, some sections of the community are resistant to change.

Chapter 27: In-home solutions

Domestic water use is the main consumer of water resources in Adelaide. Households use approximately 135,000 ML annually. Each home uses around 280 kL of water annually, most of which is taken from the mains supply, although some borewater, rainwater, and recycled water is used. Importantly, most of the water (over 90%) used by households is of higher quality than is required (it not being consumed by humans or animals). With this in mind, we can begin to look for sensible ways to save water without compromising water quality for drinking and food preparation.

In recent years, households have reduced water consumption by increasing water conservation practices and by adhering to government water restrictions. Household water fixtures and appliances play an important role and water-efficient models are becoming more popular. Rainwater can also be used to reduce household reliance on mains water. The South Australian government encourages householders to invest in rainwater tanks and water-saving devices by offering financial rebates for purchase and installation.

Greywater is increasingly being seen as a resource and reuse of this water is becoming more common. Some households reuse all of their greywater, whereas others may only reuse part of it. Minimal reuse typically entails a pipe from the washing machine outlet leading directly onto a garden. A maximum-use system has a storage capacity and reuses greywater outdoors and for some indoor uses. While complex systems allow greater reuse of greywater, they are more expensive to install.

There is potential for Adelaide households to reduce their reliance on mains water by as much as 50%. However, such change will take time. Retrofitting of water-saving devices, reuse systems and rainwater tanks into existing homes is costly and must be done equitably. Continuing government rebates and increased promotion and education will increase uptake by the community.

Chapter 28: In-garden solutions

Adelaide is a garden city. Although impressive parklands surround the city, less than 3% of metropolitan Adelaide is remnant natural bushland, and less than 6% is open parks. Backyards are our green spaces and they are vital to our wellbeing as individuals and communities. However, climate change and water restrictions have forced a re-evaluation of garden management, design and construction. We can and will have wonderful gardens in the future, but they won't look like gardens of the past!

Gardens have long been admired for their aesthetic beauty, but for gardens to be sustainable they must also be valued for climate amelioration of the built environment, provision of habitat and wildlife corridors, and their ability to produce food. Garden designs must be based on local conditions, the availability of resources, their potential productivity, and their interface with homes and neighbourhoods. Consideration of these parameters will give rise to a more integrated and highly creative approach to garden design.

Various innovative measures will reduce garden water usage without lowering the quality of Adelaide gardens. It is of foremost importance that we use plants suited to the local climate. Native and exotic plants can both be used, as long as they are not invasive. Often, thirsty lawns should be reduced in size. Sensible garden management should include soil improvement and timely mulching and pruning.

Important contributions to changing gardens can be made by television gardening and lifestyle shows, gardening magazines, the nursery industry, and the Adelaide Botanic Gardens. To engage the community and improve the water efficiency of Adelaide's gardens, consistent promotion and freely available advice are necessary.

In conclusion

The Editorial Board has identified a set of sustainability criteria that pertain to Adelaide's water systems. Addressing these criteria will assist us should the situation arise where we can no long rely on water from the Murray River. To be sustainable in our water use we need to provide more alternative supplies and use less water, and, most importantly, use water more wisely by ensuring we use it for the purpose for which it is fit. If we do not undertake these tasks it is likely that our only alternative will be to raise the price of water excessively to reduce demand, or simply have fewer users. Given that most – if not all – solutions to the water situation will impact on our urban lifestyle, sustainable water management plans will undoubtedly be highly complex to implement.

However, change is happening. We think about water differently now than we did even five years ago; and it is likely we will view water differently five years from now. We are beginning to better understand the natural systems and the importance of working with them so they can be used to our best advantage while ensuring their function remains intact. Natural systems, above and below ground, require careful management for their own sake, as well as ours.

Above all, we are beginning to realise that there is no single solution to our water problems and the road to an integrated water management system involves a major shift in community values, attitudes and behaviours. Some of our options are large scale, others are not. Some can be undertaken almost immediately while others will take a long time to be developed and/or will involve long-term incremental change. We must be receptive to and willing to adopt innovative options, technologies, economic imperatives and best management practices. We need to undertake more research and education to ensure that we all learn from our past mistakes. We must accept that water should be an integral component to all planning decisions. Moreover, sustainable water management will require the cooperation and active participation of the community and all levels of government. Everyone has a role to play.

Glossary

Word/phrase	Meaning
Activated sludge	A treatment process in which bacteria, sewage and oxygen are mixed in an aeration tank to enable the bacteria to biochemically oxidise the carbonaceous solid and dissolved material in wastewater. A portion of the active bacteria (the biomass) is separated from the wastewater stream in a settling tank (or clarifier), and recycled to the head of the process. The remainder are wasted. The process can be designed for removal of nitrogen and phosphorous, as well as carbonaceous contaminants.
Aquifers	Underground sediments or fractured rock that hold water and allow water to flow through them.
Alluvial sediments	Clay, silt or sand deposited by streams and rivers as they slow down.
ASR	Aquifer storage and recovery – a water storage alternative that allows for the injection of treated water into an aquifer for withdrawal when needed at a later stage.
Basin	An area drained by a given stream and its tributaries.
Blackwater	Wastewater from the toilet.
Bore	A hole drilled to extract groundwater.
Brownfield	Land that was previously used for industrial or commercial purposes and which may be contaminated.
Bulk water entitlements	A legal right under the *Water Act 1989* to harvest and use water.
Catchment	An area of land draining rainfall into a river or reservoir.
Catchment yield	The annual average volume of runoff from a catchment.
CO_2e	Carbon dioxide equivalent is as measure of the global warming potential (GWP) of a variety of greenhouse gases, and allows for a comparison of the greenhouse impact.
Demand management	An approach that is used to reduce the consumption of water.
Desalination	The process of removing dissolved salts from seawater (or brackish water) so that it becomes suitable for drinking or other uses.
Disinfection	The use of chlorine or ultraviolet light to control harmful microorganisms.
Diversion weir	A small weir across a river or stream diverting water into a tunnel or pipeline.
Ecology	The study of the inter-relationships between living organisms and their environment.
EIP	Environment improvement program. This is a document that sets out a list of actions and a timeframe for their implementation, aimed at improving the environmental performance of a facility or plant, as agreed with the Environment Protection Authority (EPA).
Environmental flow release	Release from a water storage intended to maintain appropriate environmental conditions in a waterway.
Eutrophication	Increase in the level of nutrients (mainly nitrogen and phosphorous) of a body of water, which results in a decrease of its quality.
EPA	Environment Protection Authority.

Word/phrase	Meaning
E. coli	*Escherichia coli* is a species of faecal coliform that is exclusively found in the human gut. Coliforms are numerous in sewage and easy to test for, so they are used as an indicator organism for the presence of faecal pollution. The presence of coliform organisms is taken as an indication that pathogenic organisms may also be present.
Faecal coliforms	A generic term for a number of different types of rod-shaped bacteria found in the human gut. Most coliforms are harmless to humans. Some faecal coliforms may be found in soil and may not always be associated with human wastes.
Filtered water	Water that has been passed through sand or membrane filters to remove impurities. Filtration is normally followed by disinfection.
Flow rate	Volume of water per unit of time (e.g., kilolitres or megalitres per day).
Gigalitre (GL)	A measure of water volume. 1 GL = 1000 million litres.
Greenfield	Undeveloped land usually outside the urban centres and perceived to be free of contamination.
Greenwashing	The unfounded promotion of a product, service or policy as environmentally sensitive to the end of creating a positive public or consumer image.
Greywater	Wastewater from the laundry, bathroom and kitchen.
Groundwater	Sub-surface water, particularly that which is in aquifers.
Hectopascal	A measure of atmospheric pressure where one millibar equals one hectopascal.
Intergenerational equity	Concept outlined at the Earth Summit of Rio in 1992 and predicated on the principle of equity between present and future generations.
Irrigation	The application of water to cultivated land or open space to promote the growth of vegetation.
Kilolitre (kL)	A measure of water volume. 1 kL = 1000 litres. There are 4.546 litres in 1 imperial gallon of water; so 1 kL = 1 cubic metre = 219.3 imperial gallons.
mg/L	The concentration of milligrams of substance per litre of water. For example, salinity can be measured as milligrams of dissolved solids per litre of water.
Megalitre (ML)	Megalitre, a measure of water volume. 1 ML = one million litres = 219,974 imperial gallons.
Millibar	A measure of atmospheric pressure where one hectopascal equals one millibar.
NTU	Nephelometric turbidity units is a measure of the turbidity of water. The term 'turbidity' refers to the lack of clarity of water due to the presence of fine suspended particulate matter in the liquid, which interferes with the passage of light through the water.
Non-potable	Non-potable water is not suitable for people to drink.
Pathogenic	Pathogenic organisms are microbes (bacteria, protozoa and viruses) that can cause disease in humans. Pathogenic organisms found in wastewater may be discharged by human beings who are infected with disease or who are carriers of a particular disease.

Word/phrase	Meaning
Polyelectrolyte	A synthetic, polymeric (long-chain molecule) chemical used as a flocculation aid to assist removing fine suspended solids and colloidal material in wastewater and water treatment plants.
Potable water	Water fit for human consumption.
Primary sewage treatment	The first major stage of treatment of sewage, usually involving the removal of settleable solids.
Precautionary principle	Where there are threats of serious or irreversible environmental damage, lack of full scientific certainty should not be used as a reason for postponing measures to prevent environmental degradation.
Retrofitting	Installation of fittings or appliances on existing buildings (e.g., dual-flush toilets).
Riparian zone	The land and vegetation along the banks of a stream or river.
Runoff	That part of precipitation which flows from a catchment area into streams, lakes, rivers or reservoirs.
Secondary sewage treatment	After the removal of suspended solids in primary treatment, secondary treatment involves oxidising organic matter in the sewage using bacteria, and the resultant growth is then removed by settlement or filtration.
Security of supply	Reliability or surety of meeting water supply demand. Storages provide the capability to ensure a certain level of supply is available despite seasonal variations in stream flow.
Sewer mining	Process of tapping into the underground sewerage network. This involves a compact treatment plant that collects a small proportion of the effluent. It has the benefit of reducing the overall load on the network and provides non-potable quality water where needed, however this water may also be treated for potable supplies in the future.
Stream flow	The flow in a stream or river.
System yield	The annual level of total demand that can be supplied by a water supply system subject to an adopted set of operational rules and to a typical demand pattern while maintaining a given level of security of supply.
Tariff	A system of charges for the provision of water supply.
Transfer/distribution system	A system of conduits (e.g., pipes, channels and aqueducts) used to supply water to customers. A distribution system is typically made up of large supply 'mains', which convey the water from the major storage points – perhaps to smaller service reservoirs; these can then feed into smaller 'service' pipes which deliver the water to the customers.
Tertiary treatment	Tertiary treatment is the advanced treatment process, following secondary treatment of wastewater, and is used to produce high quality non-potable recycled water. Tertiary treatment includes removal of nutrients such as phosphorus and nitrogen and practically all suspended and organic matter from wastewater. It typically involves the use of filtration, increasingly by membranes.
Treated effluent	The treated water discharged from a sewage treatment plant.
Treatment trains	A series of water sensitive urban design (WSUD) measures designed to manage to movement of water through the landscape. Measures may include components that allow the storgage, use and reuse of water.
Unfiltered water	Water harvested from uninhabited catchments and supplied without filtration but always disinfected.

Word/phrase	Meaning
UV	Ultraviolet light. This radiation can be used to disinfect both treated wastewater and drinking water. A pathogenic protozoa, *Cryptosporidium*, which is of concern in recycled water applications, is vulnerable to UV but resistant to chlorine. A disadvantage with UV is that it does not leave any disinfectant residual, unlike chlorine.
Volumes	Kilolitres (kL) = 1000 litres (or 1 cubic metre); megalitres (ML) = 1,000,000 litres (or 1000 cubic metres); gigalitres (GL) = 1000 million litres (or 1 million cubic metres)
Water harvesting	The process of collecting water runoff resulting from rainfall. This is undertaken primarily in protected catchments but is also used to describe the extraction of water from rivers or groundwater.
Wastewater	Contaminated water before it undergoes any form of treatment. The water may be contaminated with solids, chemicals or changes in temperature.
WSUD	Water sensitive urban design – a system of integrated urban water cycle management making use of various components of the urban landscape to direct, store, use and reuse water, particularly stormwater and wastewater.
WWTP	Wastewater treatment plant. Other similar terms include sewage treatment plant and water pollution control plant.
Water right	The property right that individuals and organisations have in the water resource under law.
Brine	Water containing salts.

Appendix 1

The various authors that contributed to this book provided water statistics from a range of sources and dates. Therefore certain statistics (e.g., total water use in Adelaide) vary in different chapters. The table below lists the statistics that differ throughout this book.

Note: The large variation for the *total Adelaide water use per year*, may be because some figures include rural consumption as well as urban consumption, while other figures only represent urban consumption.

Chapter	Water statistic
Total Adelaide water use/year	
25	218 GL
22	~200 GL
23	~170–180 GL
1	216,000 ML for urban consumption
27	300,000 ML
13	140 GL in 2007, 194 GL in 2002 and 173 GL for the 10 year average
14	300 GL
Amount of Adelaide recycled/reclaimed water from wastewater	
21	approx. 20% of wastewater
17	29.7 % of metropolitan wastewater, 19.1% of rural wastewater
11	25,000 ML
22	18 GL
Adelaide Hills reservoirs	
1	Use – 137 GL (average year)
15	Capture about 120,000 ML/year
Adelaide urban consumption as a percentage of total water use within the Greater Adelaide area	
15	66% in a dry year, 70% in an average year
15	72%
Adelaide water sourced from the Murray River	
15	28–55% (This is a lower range because it includes water sources for rural activities as well as urban consumption)
14	up to 80%
7	up to 90% on occasion
25	91% in 2007
11	up to 90% in dry years
25	40% or 88 million t/year
Mains water (Murray River and Hills catchments)	
15	68.1% in a dry year, 72.6% in an average year
4	68%
Water consumption by the Virginia Pipeline Scheme	
11	16,300 ML in 2006/2007
25	over 10 billion L/year
19	around 15 GL/year
Water discharged to sea	
22	approx. 100 GL/year including approx. 70 GL wastewater
3	177 GL/year (average for the period 1995–2005)

Index